Linear
Algebra with
Differential
Equations

Linear
Algebra with
Differential
Equations

DONALD L. BENTLEY
KENNETH L. COOKE

Pomona College

HOLT, RINEHART AND WINSTON, INC.
New York · Chicago · San Francisco · Atlanta · Dallas
Montreal · Toronto · London · Sydney

To my wife, Penny,
and my parents,
Viola L. and Byron R. Bentley

Donald L. Bentley

To my wife, Margaret,
and the memory of my parents,
Mildred B. and Sidney K. Cooke

Kenneth L. Cooke

Preface

This text was motivated by a desire to provide our students with an integrated approach to the material covered in introductory courses in linear algebra and differential equations. We felt that by breaking away from the tradition of two separate courses and exhibiting the interrelationship between the two subjects, we could provide the student with better motivation and more insight into the material. In addition, we hoped to achieve more efficiency in the presentation by calling upon the differential equations to illustrate concepts in linear algebra and using the linear algebra in the development of the theory of linear differential equations.

We feel we have had reasonable success in our objectives. Our colleagues at Pomona College have been using preliminary versions of the text for the last four years. Typically, we find we can finish in one year more material than we usually covered in one semester of a traditional differential equations course and a separate one semester course in linear algebra. Further, we believe that the students have achieved a deeper understanding of the material because of the integrated approach. Finally, the percentage of students finishing the full year sequence is greater than that of students who took both of the traditional courses.

Certain characteristics of the text have been dictated by the needs of colleagues in other departments. The first three chapters are devoted to a treatment of elementary first order differential equations. This provides a necessary prerequisite for material encountered in physical science courses taken concurrently by many of our students. If an instructor is not faced with this constraint, he can begin with Chapter 4, although we feel there is much to be said for a gradual immersion of the student into abstraction.

Although the text was designed to present linear algebra in combination with differential equations, it can be used for a traditional course in linear algebra by including Chapters 4–7, 10–11, and 14. Similarly, a course in differential equations with a linear algebra prerequisite could be constructed from the remaining chapters.

We feel the text includes more material than can be covered in a year's course at the sophomore-junior level. The instructor will have to make some choices for omission. We have indicated some of our choices for optional sections by an (*) in the table of contents. Although Chapters 4–6 serve as prerequisite to each of Chapters 7–14, the latter can be presented in any order with the exception of Chapter 9, which must follow Chapter 8, and Chapter 12, which must follow Chapter 11. Chapters 7 and 10 present independent proofs of equality of row and column rank. Finally, while Chapter 16 does not tie closely with the linear algebra emphasis of the text, it is included as a necessary tool for students expecting to use differential equations in later study.

We are deeply indebted to the faculty and students of the Pomona Mathematics Department. Each of our faculty members has taught from the manuscript and provided many valuable suggestions and exercises. The students of Pomona College have been most patient and encouraging in their support of our project. In particular, we thank Karen DeLay Archer and Katie Frohmberg for their typing of the manuscript, and Ruth Daniel, Gay Howland, and Robert Wesley, who devoted several summers to working on the text and exercises. Finally, we are most indebted to Mrs. Pat Kelly who, in addition to typing parts of the manuscript, provided organization and moral support throughout the total project.

Donald L. Bentley
Kenneth L. Cooke

April 1973

Contents

Chapter 1

Introduction to differential equations

1.1 Introduction

Before embarking on a course in linear algebra with differential equations it seems worthwhile to discuss the reasons for such an expedition. If one considers the material covered in a linear algebra or differential equations course separately, the value of the subject matter becomes quite apparent. Historically the development of the physical and biological sciences has been closely tied to the development of methods and theory in differential equations. In more recent years social scientists have found techniques in differential equations valuable in establishing and analysing models within their disciplines. Chapter 2 of this text illustrates how differential equations arise in model building.

Although the tools of matrix manipulation have found application in statistical inference since the late nineteenth century, it has been only recently that the subject matter in linear algebra has begun to blossom as a model building tool, primarily in the social sciences. Linear algebraic concepts play important roles in such areas as management science, statistical inference, and the biological and physical sciences.

In addition to their role in science, the theories of differential equations and linear algebra are closely related to many other areas within mathematics itself. For without going into detail here, there is a strong interconnection between differential equations and topology. The concepts of linear algebra provide numerous examples and ideas in the modern study of abstract algebra, are useful in algebraic geometry and in "functional analysis," and can be used in the study of circuits in graphs, and so on. In this text, we will point out only a fraction of the great wealth in the treasure chest of linear algebra. But the reader can be assured that as he proceeds in mathematics linear algebra will play an important role.

Traditionally the introductions to the subjects of linear algebra and differential equations have been presented in two independent courses. Sometimes, in a hasty effort to provide usable techniques, an instructor in the differential equations course will emphasize methods almost to the exclusion of general theory, whereas the material in linear algebra is frequently abstract and perhaps overpowering. Seldom does the student realize the relationships that exist between these two topics.

In this text many of the topics found in a first course in differential equations are presented as applications of the theory developed in a study of linear algebra. By combining the two topics it is hoped that the reader will find more mathematical motivation for the techniques in differential equations, and at the same time find relief from too much abstraction through some applications in the study of linear algebra.

While the main flavor of this text is that of linear algebra, the first three chapters are devoted to an introduction of topics in differential equations in order to develop a vocabulary to be used when treating differential equations throughout the remainder of the text. The material covered will also be useful in science courses which the student may be taking concurrently.

1.2   Basic terminology

The analysis of mathematical models for physical, biological, and social phenomena often involves the determination of an unknown function from an equation containing derivatives of the unknown function. Examples of such equations, called *differential equations*, are

$$\frac{dC(t)}{dt} = kC(t) \tag{1}$$

where $C(t)$ is the concentration of a drug in a biological system at time t and k is a constant characteristic of the drug; and

$$m \frac{d^2x(t)}{dt^2} + kx(t) = 0 \tag{2}$$

which describes the motion of a vibrating spring where m is the mass attached to the spring, k is a constant characterizing the resistance of the spring to stretching, and $x(t)$ is the distance the mass is displaced from an equilibrium position at time t.

One purpose of this text is to present methods for solving equations such as (1) and (2). Most of this presentation will take place within the framework of *linear algebra* since the techniques of solution will make use of some of the linear properties of the operation of differentiation; that is, if $x_1(t)$ and $x_2(t)$ are two functions of t and a and b are two real constants, then

$$\frac{d[ax_1(t) + bx_2(t)]}{dt} = a\frac{dx_1(t)}{dt} + b\frac{dx_2(t)}{dt} \tag{3}$$

DEFINITION 1.1   An equation involving a function and derivatives of that function is called a *differential equation*.

Equations (1) and (2) are differential equations. Other examples are

$$\frac{\partial^3 f(x, y)}{\partial x^2 \partial y} + \frac{\partial^2 f(x, y)}{\partial y^2} = 0 \tag{4}$$

and

$$\frac{d^2 y}{dx^2}\left(\frac{dy}{dx}\right)^2 + 3x^3\frac{dy}{dx} = \cos x \tag{5}$$

In Eq. (1) the function of interest is C and the derivative is taken with respect to the independent variable t. In Eq. (2) x is the function of interest. In Eq. (4) f is a function of two independent variables x and y. The derivatives in this case are partial derivatives and hence we are led to the following definition.

DEFINITION 1.2   A differential equation involving derivatives of a function of a single independent variable is called an *ordinary differential equation*. If the derivatives in the equation are of a function of two or more independent variables (that is, if the derivatives in the equation are partial derivatives), the equation is said to be a *partial differential equation*.

We have already seen examples of ordinary differential equations in Eqs. (1), (2), and (5). Equation (4) is an example of a partial differential equation. However, care must be taken in certain cases such as

$$\frac{d^2 f(x, t)}{dt^2} + 3x^2 f(x, t) = 0 \tag{6}$$

Equation (6) is an ordinary differential equation if x is a function of t. However,

$$\frac{\partial^2 f(x, t)}{\partial t^2} + 3x^2 f(x, t) = 0 \tag{7}$$

is a partial differential equation if x and t are independent variables. Throughout the rest of this text we shall deal primarily with ordinary (rather than partial) differential equations.

A second method of categorizing differential equations is by their order.

DEFINITION 1.3   The *order of a differential equation* is the order of the high-est order derivative appearing in the equation.[1]

If f is a function of x, then $d^n f(x)/dx^n$ is an nth order derivative. The differential equation

$$\frac{d^3 y}{dx^3} \cdot \frac{dy}{dx} + 3 \frac{d^2 y}{dx^2} = 0 \tag{8}$$

is a third order differential equation. Also

$$\frac{\partial^3 f(x, t)}{\partial x^3} + \frac{\partial^4 f(x, t)}{\partial x^2 \, \partial t^2} \cdot \frac{\partial f(x, t)}{\partial x} = 0 \tag{9}$$

is a fourth order partial differential equation.

Any first order ordinary differential equation is of the form

$$g\left(x, y, \frac{dy}{dx}\right) = 0$$

where g is some given function. In many cases, it will be possible to solve for dy/dx, and then a first order equation has the form

$$\frac{dy}{dx} = f(x, y)$$

for some function f. Most of the rest of this chapter will be devoted to a study of such equations.

In elementary algebra one of the primary problems is that of solving algebraic equations, that is, finding numbers that satisfy the equations. The study of differential equations is similarly concerned with determining functions that satisfy the differential equations. In many cases more than one function will satisfy a given differential equation. For example, consider the second order differential equation

$$\frac{d^2 y}{dx^2} - 2 \frac{dy}{dx} = 3y \tag{10}$$

It is easy to verify, by substituting into the equation, that $y_1(x) = e^{3x}$ and $y_2(x) = e^{-x}$ both satisfy Eq. (10). Further, if c is any given constant, then $cy_1(x)$ and $cy_2(x)$ both satisfy the equation. Finally, if c and d are any two constants, $cy_1(x) + dy_2(x)$ satisfies the equation. This additive property will be exploited in much greater depth after our development of linear algebra.

---

[1] In some cases we may wish to allow zero as the order of an equation, but in most cases the order is a positive integer.

DEFINITION 1.4   A *solution* of a differential equation is an explicit function, say $y(x)$, which satisfies the differential equation over a specified interval $a < x < b$. The function $y(x)$ is said to be a solution on the interval $a < x < b$.

Note that e^{3x} is a solution of $dy/dx = 3y$ as is $47e^{3x}$. However, ce^{3x} should not be called a solution if c is not specified, but rather is an expression denoting a set of solutions.

As a second example consider the differential equation

$$2x + 2y \frac{dy}{dx} = 0 \tag{11}$$

Let us consider the relation

$$x^2 + y^2 = c^2 \tag{12}$$

where c is a fixed positive constant. This relation can be regarded as implicitly defining y as a function of x. In fact, $y = \sqrt{c^2 - x^2}$ and $y = -\sqrt{c^2 - x^2}$ are two real functions defined for $-c \le x \le c$. Each of these functions satisfies Eq. (11) on $-c < x < c$; for example, if $y = \sqrt{c^2 - x^2}$, then

$$\frac{dy}{dx} = \frac{-x}{\sqrt{c^2 - x^2}}$$

$$2x + 2y \frac{dy}{dx} = 2x + 2\sqrt{c^2 - x^2} \cdot \frac{-x}{\sqrt{c^2 - x^2}} = 0, \qquad x \ne \pm c$$

It is even simpler to obtain Eq. (11) by *implicit differentiation* of Eq. (12), as discussed in elementary calculus. That is, we regard Eq. (12) as implicitly defining y as a function of x and then, differentiating both sides of Eq. (12), we have

$$\frac{d}{dx}(x^2) + \frac{d}{dx}(y^2) = 0$$

which yields Eq. (11).

If we assign different values to c, we obtain different solutions of Eq. (11). For example, $y = \sqrt{4 - x^2}$, $y = \sqrt{7 - x^2}$, $y = -\sqrt{\ln \pi - x^2}$ are solutions. The equation $y = \sqrt{c^2 - x^2}$ or the equation $y = -\sqrt{c^2 - x^2}$, with c not specified, generates a family or set of solutions. The graphs of these functions form two families of semicircles (see Figure 1.1) which fill the upper half-plane and lower half-plane, respectively. Note that these graphs join to form a family of circles $x^2 + y^2 = c^2$. However, each circle is composed of two solutions. The circle as a whole is not the graph of a single solution. In fact, where a circle crosses the x-axis, its slope is infinite, and therefore Eq. (11) cannot be satisfied at such a point.

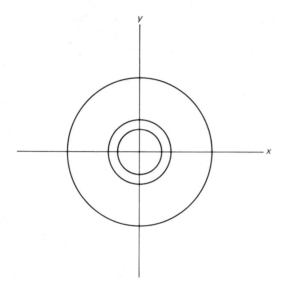

FIGURE 1.1

DEFINITION 1.5 An equation $y = f(x, c)$, where c is an unspecified constant, is said to define a *one parameter family* of solutions of a differential equation provided that for each value of c, $f(x, c)$ is a function which satisfies the differential equation on some interval. The unspecified c is called a *parameter*.

Example 1 Students of calculus are already familiar with a class of differential equations which consist of the first order equations of the form

$$\frac{dy}{dx} = g(x) \tag{13}$$

where g is a given function. Let us assume that $g(x)$ is continuous for $a \leq x \leq b$. According to the Fundamental Theorem of Calculus,[2] any solution y must be an integral of $g(x)$, and any integral of $g(x)$ is a solution. If we pick a fixed number x_0 in the interval $[a, b]$, we can write

$$y = c + \int_{x_0}^{x} g(u) \, du \tag{14}$$

where c is a parameter. This formula defines a one parameter family of solutions

---

[2] For example, see George B. Thomas, Jr., *Calculus and Analytic Geometry*, Reading, Mass., Addison-Wesley, 1968, p. 183.

of Eq. (13), and every solution is a function in this family. That is, Eq. (14) is a formula for all solutions of Eq. (13).

Exercises

1. Give the order of the following differential equations:

   (a) $\dfrac{d^2y}{dx^2}\dfrac{dy}{dx} + y^4x^2 - 3\left(\dfrac{dy}{dx}\right)^2 = 0$

   (b) $\dfrac{\partial^3 f(x, y, z)}{\partial x\, \partial y\, \partial z} - 3\dfrac{\partial^2 f(x, y, z)}{\partial z^2} = 3y^3$

   (c) $\left(\dfrac{d^3y}{dx^3}\right)x^2 + 3\dfrac{dy}{dx} + 13y^3 = 5$

2. Classify the following as ordinary or partial differential equations:

   (a) $\dfrac{\partial^2 z}{\partial x\, \partial y} + 3\left(\dfrac{\partial z}{\partial y}\right)^2 + \dfrac{\partial z}{\partial x} = 0$

   (b) $3\dfrac{d^2y}{dx^2} + 14\dfrac{dy}{dx} + 3y^3 = 0$

3. Verify that the given function is a solution of the given differential equation on the specified interval.

   (a) $y = e^{5x}$, $y'' - 7y' + 10y = 0$ on $-\infty < x < \infty$
   (b) $y = 0.2$, $y'' - 7y' + 10y = 2$ on $-\infty < x < \infty$
   (c) $y = \sqrt{9 - x^2}$, $x + yy' = 0$ on $-3 < x < 3$
   (d) $y = \dfrac{e^x}{x}$, $y' - y = -\dfrac{e^x}{x^2}$ on $0 < x < \infty$
   (e) $z = x^2 + e^y$, $\dfrac{\partial^2 z}{\partial x\, \partial y} = 0$ on $-\infty < x$, $y < \infty$.

4. Determine whether the given equation defines a one parameter family of solutions of the given differential equation.

   (a) $y = ce^{5x}$, $y'' - 7y' + 10y = 0$
   (b) $y = c\sqrt{9 - x^2}$, $x + yy' = 0$
   (c) $y = \dfrac{ce^x}{x}$, $y' - y = -\dfrac{e^x}{x^2}$

5. Find all solutions of $dy/dx = 1/x$ on $0 < x < \infty$ and on $-\infty < x < 0$. Find solutions which satisfy:

   (a) $y = 3$ when $x = 1$
   (b) $y = 3$ when $x = -1$

6. After each of the following differential equations a relation in x and y is given. Verify that if the relation defines y as a function of x, the function is a solution of the differential equation. Determine a one parameter family of solutions.

    (a) $yy' + \sin x \cos x = 0$; $y^2 + \sin^2 x = c$
    (b) $y'y = b$; $y^2 = 2b(x - a)$
    (c) $10x + 20 - 8(y - 1)y' = 0$; $5x^2 - 4y^2 + 20x + 8y = c$

7. Plot, on a single graph, solution curves for Exercise 6a for $c = 2, 1, \frac{1}{2}, 0$.

1.3 Solution curves

In the latter sections of this chapter we will show how certain classes of first order ordinary differential equations can be solved by various devices or tricks. This might lead the reader to suspect that mathematicians do not know any simple, general technique for producing all the solutions of an arbitrary first order equation (let alone higher order equations). This suspicion is correct. But actually it is not so much ignorance on the part of mathematicians as it is the inherent nature of the problem. Consider the example $y' = e^{-x^2}$ subject to the condition $y(0) = 0$. The solution is (see Example 1 in Section 1.2)

$$y(x) = \int_0^x e^{-t^2} \, dt$$

The reader will perhaps recall his amazement when he first discovered that this integral cannot be evaluated in "closed form" in terms of the "elementary" functions of analysis; that is, the polynomials, exponentials, trigonometric functions, and their inverses. In view of this it is not too surprising if the mathematician cannot produce an all-encompassing method for the vastly more difficult problem of solving the arbitrary first order equation $dy/dx = f(x, y)$.

What, then, should one do when confronted with a differential equation for which a solution is sought? Probably he should first see whether it falls into one of the classifications of equations for which the solution is known in terms of elementary or other known or tabulated functions. In this connection, he can use one of the methods presented below or consult tables of equations with known solutions,[3] analogous to the table of integrals familiar from the calculus. But if this fails there still is one recourse: he can attempt to construct an

---

[3] A most extensive compilation of differential equations with their solutions can be found in E. Kamke, *Differentialgleichungen, Lösungsmethoden und Lösungen, I.* Akademische Verlagsgesellschaft, Leipzig, 1943. Reprint Dover Publishing Company.

approximate solution. One way to do this is with numerical techniques. Another method is with the aid of the direction field, as will now be explained.

Our discussion of solving differential equations by consideration of the direction field is intended to give insight into the problem through taking a geometrical point of view. Let us consider a first order differential equation of the form

$$y' = f(x, y) \tag{1}$$

where f is a given function of two real variables. The *domain* of the function is the set of number pairs or points (x, y) at which the function is defined. For example, if $f(x, y) = (x^2 - 1)^{-1}(y^2 - 1)^{-1}$, the domain could be taken to be the interior of the rectangle bounded by the lines $x = \pm 1$, $y = \pm 1$, or even the whole x, y plane excluding these lines.

We now want to examine solutions of Eq. (1). For the sake of clarity, let us again define *solution* but in slightly different terms.

DEFINITION 1.6 A function $\phi(x)$, defined on an interval $a < x < b$, is called a *solution* of $y' = f(x, y)$ if $\phi'(x)$ exists for $a < x < b$, if each point $(x, \phi(x))$ lies in the domain, D, of f, and if

$$\phi'(x) = f(x, \phi(x)), \qquad a < x < b \tag{2}$$

Consider a graph of a solution $y = \phi(x)$. This is a curve in the x, y plane which exists for $a < x < b$ and lies in the domain D of f. We are thus led to the following definition.

DEFINITION 1.7 A *solution curve* or *integral curve* of $y' = f(x, y)$ is a curve which is the graph of a solution $\phi(x)$.

Consider a solution curve corresponding to a solution $y = \phi(x)$. At any point (x, y) on this curve, the slope of the curve is $\phi'(x)$ or by Eq. (2) it is $f(x, \phi(x)) = f(x, y)$. Therefore for each point (x, y) in D, the number $f(x, y)$ is the slope that a solution curve through (x, y) must have at the point (x, y), if indeed there is a solution curve through (x, y). In other words, the function $f(x, y)$ in Eq. (1) defines a direction or slope at every point (x, y) in D. We say briefly that the differential equation prescribes a *direction field* in D. A solution curve must therefore have the property that at every one of its points its slope is the same as that prescribed by the direction field. More briefly, a solution curve must at every point be tangent to the direction field.

By drawing a short line segment with slope $f(x, y)$ through each point (x, y) in D, it is possible to obtain a pictorial representation of a direction field. Moreover, it may then be possible to guess the appropriate shape of solution curves. One can say that a solution curve must wind its way through the plane in such a way as to be tangent to the direction field at every point.

Example 1   Investigate the differential equation $y' = x + y$. The direction field is shown in Figure 1.2. This figure is made up of a set of short segments

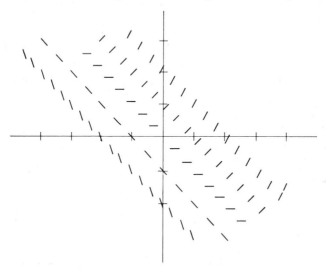

FIGURE 1.2

which are tangent to solution curves at various points in the x, y plane. For example, at $x = 1$, $y = -1$, the slope of a solution curve is $y' = x + y = 0$. At the point $x = 1$, $y = 1$ the slope of the solution curve is $y' = x + y = 2$. Figure 1.3 shows several curves which appear to be tangent to the direction field and therefore seem to be solution curves. We will see later that this equation has a one parameter family of solutions given by $y = \phi(x) = ce^x - x - 1$ where c is an arbitrary constant. For each value of c there is a corresponding solution curve, and therefore there are infinitely many solution curves. The curves in Figure 1.3 are a few of these.

We now see that for each given equation $y' = f(x, y)$ there is a corresponding picture similar to Figures 1.3 and 1.4. Each solution defines a solution curve. Moreover, for most of the differential equations we shall encounter we can find a whole *family* or set of solutions, depending on a parameter c. Consequently there will be a family or set of solution curves.

We shall now suggest several techniques useful in constructing the direction field and solution curves of a differential equation. The first is based on the following notion.

DEFINITION 1.8   An *isocline* of a differential equation $y' = f(x, y)$ is a curve defined by the relation $f(x, y) = $ constant.

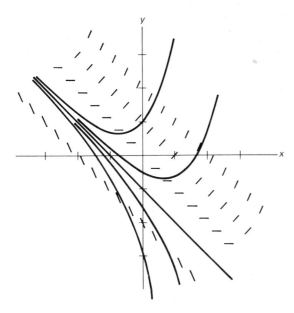

FIGURE 1.3

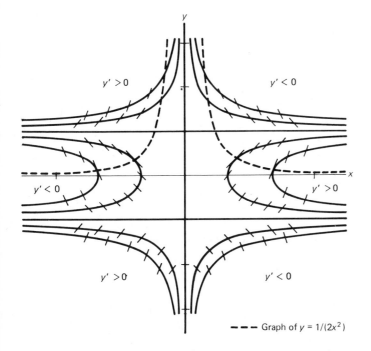

$y' > 0$

$y' < 0$

$y' < 0$

$y' > 0$

$y' > 0$

$y' < 0$

– – – Graph of $y = 1/(2x^2)$

FIGURE 1.4

Along an isocline, y' is constant and therefore all integral curves have the same slope wherever they intersect a given isocline. In fact, the word *isocline* means *of equal inclination*. Returning to the example $y' = x + y$, we see that the isoclines are the straight lines $x + y = c$ where c is constant. For example, taking $c = 1$ we find that $x + y = 1$ is the isocline corresponding to slope 1; taking $c = -1$, $x + y = -1$ is the isocline corresponding to slope -1, and so on. In drawing the direction field, it is helpful to sketch a few isoclines, and then mark short segments with the correct slopes at various points on these isoclines. It is especially helpful to draw the isocline corresponding to zero slope, since regions where the integral curves are rising (have positive slope) and where they are falling (have negative slope) may be separated by this isocline. Any maximum or minimum point on an integral curve must lie on the zero-slope isocline.

It may also be helpful to find the curve or curves in the plane on which $y'' = 0$. Such a curve is a locus of points where the concavity of solution curves may change.

Example 2   Consider the equation $y' = x(1 - y^2)$. It is clear that $y = 1$ and $y = -1$ are solutions. The zero-slope isoclines are $x = 0$, $y = \pm 1$. These lines divide the plane into six regions. In Figure 1.4 we indicate the regions where the integral curves have positive or negative slope. Also shown are the isoclines $x = \pm(1 - y^2)^{-1}$ and $x = \pm 2(1 - y^2)^{-1}$ corresponding to slopes ± 1 and ± 2, respectively. Furthermore, we have

$$y'' = 1 - y^2 - 2xyy'$$

Replacing y' by $x(1 - y^2)$ we get

$$y'' = 1 - y^2 - 2x^2y(1 - y^2) = (1 - 2x^2y)(1 - y^2)$$

Thus $y'' = 0$ on the curves $y = \pm 1$ and $y = 1/(2x^2)$. In Figure 1.4, one can observe a great deal about the qualitative nature of the solution curves. For example, a curve lying above $y = 1$ crosses the y-axis with zero slope, then as x increases the slope decreases until the curve $y = 1/(2x^2)$ is crossed. Since the integral curve cannot cross[4] the zero-slope isocline $y = 1$, and its slope is negative as long as $y > 1$, it seems clear that it must approach $y = 1$ asymptotically as $x \to +\infty$. Similarly, an integral curve which lies between -1 and $+1$ crosses the y-axis with zero slope and upward concavity. As x increases, the slope increases until $y = 1/(2x^2)$ is crossed, then decreases, and the curve approaches $y = 1$ asymptotically from below. Finally, for curves below $y = -1$, $y(x) \to -\infty$ along vertical asymptotes.

---

[4] In Chapter 3 we shall justify this assertion with the aid of a so-called uniqueness theorem.

In Example 2, the differential equation can be solved by the method to be presented in Section 1.7. The result is

$$y = -\frac{1 - ke^{x^2}}{1 + ke^{x^2}}$$

where k is an arbitrary constant. For $k = 0$ we get the solution $y = -1$. The solution $y = 1$ cannot be obtained from this formula by any finite choice of k (but letting $k \to +\infty$ we obtain $y = 1$ as limit). It is important to note that even if a first order equation cannot be explicitly solved by a known method, it can always be attacked by the graphical procedure discussed here, and frequently much useful information regarding the solutions can be deduced.

Finally, let us remark that it may be useful to find the curves on which y' becomes infinite (is undefined) as well as those on which $y' = 0$. The graph of a solution $y = \phi(x)$ cannot cross a curve on which $y' = \infty$, because at the crossing $\phi'(x)$ is undefined and therefore the differential equation is not satisfied. However, it may be possible that solutions join at such points to form continuous curves; see Exercise 4.

Exercises

1. For the equation $y' = x + y$ in Example 1, draw the isoclines corresponding to slopes 0, ± 1, and ± 2. Also, find the locus of points where $y'' = 0$. Explain carefully how one can show that

   (a) No solution curve can cross the line $x + y = -1$.

   (b) A solution curve lying above $x + y = -1$ has a single minimum and $\lim_{x \to \pm \infty} y(x) = +\infty$.

   (c) A solution curve lying below $x + y = -1$ has no maximum or minimum, but has an inclined asymptote as $x \to -\infty$. Also

   $$\lim_{x \to +\infty} y(x) = -\infty$$

2. For each of the following equations, find and graph several isoclines. Discuss the location of maxima, minima, and inflection points on the solution curves. Sketch the direction field and several integral curves. What information can you deduce about the presence of asymptotes or limiting behavior of the solutions as $x \to \pm\infty$? Are there any points through which no solution curve passes, or more than one solution curve?

   (a) $y' = 4x - y$ (b) $y' = \dfrac{4x}{y}$

   (c) $y' = 4x^2 - y^2$ (d) $y' = x^2 + y^2$

(e) $y' = \dfrac{4 - y}{y}$ (f) $y' = \sqrt{|y|}$

(g) $y + (x - 3)^2 y' = 0$ (h) $y' = \min(x, y)$

3. Discuss the set of solution curves for the following:

    (a) $y' = \sqrt{-y^2}$ (b) $y' = \sqrt{-y^2 - 1}$ (c) $y' = \sqrt{y^2 - 1}$

4. Verify that $y = (x - c)^{1/3}$, where c is a constant, is a solution of $3y^2 y' = 1$, for $x > c$ and for $x < c$. Show that each solution curve approaches the x-axis with infinite slope and that solutions can be joined at the axis to form curves which are continuous. Do these curves have a tangent line at every point?

1.4 First order linear differential equations

This section is concerned with the class of differential equations of the form

$$a(x) \frac{dy}{dx} + b(x)y = c(x) \tag{1}$$

These equations are called first order *linear* differential equations. They are important not only because they occur in various applications but also because their study introduces important concepts which will reappear later in the study of more general types of linear differential equations. At the moment we can explain the choice of the term *linear* by the fact that y and y' appear in Eq. (1) to the first degree only, and there is no term involving the product yy', so that $a(x)y' + b(x)y$ is a linear expression in the variables y and y'. Later we shall give a more general interpretation using the concept of a linear operator.

Over any interval in which $a(x)$ is nowhere zero, we can set $b(x)/a(x) = p(x)$ and $c(x)/a(x) = q(x)$ and replace Eq. (1) by

$$y' + p(x)y = q(x) \tag{2}$$

If $a(x) = 0$ at a certain number of isolated points, these points divide the interval of interest into subintervals. In each of these, $a(x)$ is not zero and the equation can be reduced to the form of Eq. (2) and solved by the methods we are going to explain. The only question remaining is whether these solutions can be pieced together to form solutions valid over the whole interval including points where $a(x) = 0$.

The term *homogeneous* stems from the Greek words *homos*, the same, and *genos*, a race, family, or kind. Unfortunately the word homogeneous is applied in many dissimilar situations in mathematics, which may lead to confusion unless the reader is careful to make the proper interpretation of the term. We now introduce one usage of the word.

DEFINITION 1.9 An equation of the form of Eqs. (1) or (2) is called a *first order linear* differential equation. If $c(x)$ or $q(x)$ is identically zero, the equation is called *homogeneous*.

Thus the homogeneous first order linear equations are of the form

$$a(x)y' + b(x)y = 0 \qquad (3)$$

or

$$y' + p(x)y = 0 \qquad (4)$$

We now show how to solve such equations. Let us suppose that we have already divided through by $a(x)$ so that the equation is in the form of Eq. (4). One solution is $y(x) \equiv 0$, the so-called *trivial solution*. To search for other solutions we rewrite the equation as

$$\frac{y'}{y} + p(x) = 0 \qquad (5)$$

which is valid except for points where a solution $y(x)$ is zero. To solve Eq. (5) we rewrite it in its differential form, namely,

$$\frac{1}{y} \, dy + p(x) \, dx = 0$$

By integrating we obtain

$$\int \frac{dy}{y} + \int p(x) \, dx = \text{constant}$$

$$\ln |y| = c - \int p(x) \, dx$$

$$|y| = e^c e^{-\int p(x)\, dx}$$

or

$$y = k e^{-\int p(x)\, dx} \qquad (6)$$

where k is a real arbitrary constant.

The fact that Eq. (6) does give a one parameter family of solutions of the homogeneous first order linear differential equation is easily checked by differentiating and substituting into the equation. Note that $k = 0$ yields the trivial solution. The above argument shows that if y is a solution which is never zero, y has the form of Eq. (6). Are all solutions of this form? It will be proved in the next section that they are.

Example 1 Solve the equation $y' + 3x^2 y = 0$. The solution can be obtained directly from Eq. (6) by substituting $p(x) = 3x^2$. For pedagogical reasons, it is perhaps better to go through the steps in the solution procedure which led to Eq. (6). Thus one obtains

$$\frac{y'}{y} + 3x^2 = 0$$

$$\ln |y| + x^3 = c$$

$$|y| = e^{c-x^3} = e^c e^{-x^3}$$

$$y = ke^{-x^3}$$

Example 2 $xy' + y = 0$. This equation is of the form (3) rather than (4). Rewriting the equation yields

$$\frac{y'}{y} + \frac{1}{x} = 0 \qquad (x \neq 0, \ y \neq 0)$$

$$\ln |y| + \ln |x| = c$$

$$\ln |xy| = c$$

$$y = \frac{k}{x} \qquad (x \neq 0)$$

Later we shall discuss the question of whether there are solutions defined at $x = 0$.

Exercises

1. Show that the following differential equations are linear and homogeneous. Then solve by using Eq. (6).

   (a) $y' + 3y = 0$

   (b) $\dfrac{dy}{dx} + xy = 0$

   (c) $\dfrac{dx}{dt} + (1 - t)x = 0$

   (d) $y \, dx + \tan x \, dy = 0$

   (e) $(1 - u^2) \, dv + v \, du = 0$

1.5 Nonhomogeneous first order linear equations

Next, consider the *nonhomogeneous* linear first order differential equation

$$y' + p(x)y = q(x) \tag{1}$$

We shall give two methods for solving this equation, both of which will be useful in other connections later. The first uses what is called an *integrating factor*. The idea is to multiply the equation by a function so chosen as to make both members of the equation readily integrable. A precise definition will be given in Section 1.10. Although for most differential equations it is extremely difficult to guess such a factor (see the exercises in Section 1.10) it is easy for an equation of the form in Eq. (1). In fact, the function[5] $\exp\left[\int p(x)\,dx\right]$ will do well, for if we multiply by this function we get

$$[y' + p(x)y]e^{\int p(x)\,dx} = q(x)e^{\int p(x)\,dx} \tag{2}$$

This is the same as

$$\frac{d}{dx}\left[ye^{\int p(x)\,dx}\right] = q(x)e^{\int p(x)\,dx} \tag{3}$$

Therefore if the function on the right of Eq. (3) has an indefinite integral which can be found, the solution y can be determined. In general, we can write

$$ye^{\int p(x)\,dx} = \int q(x)e^{\int p(x)\,dx}\,dx \tag{4}$$

or

$$y = e^{-\int p(x)\,dx}\int q(x)\,e^{\int p(x)\,dx}\,dx \tag{5}$$

In this formula, each integral sign represents an indefinite integral.

Examining the above argument carefully, we see that we have proved that if y is a solution of Eq. (1), then y must be of the form of Eq. (5). Conversely, if y is defined by Eq. (5), we can verify by differentiation that y satisfies Eq. (1). Thus Eq. (5) gives the family of all solutions of Eq. (1).

Example 1   Solve the differential equation $y' + y/x = x$, for $x > 0$. Here an integrating factor is

$$e^{\int(1/x)\,dx} = e^{\ln x} = x$$

Multiplying the differential equation by this factor we get

$$x\left(y' + \frac{1}{x}y\right) = x^2$$

$$\frac{d}{dx}(xy) = x^2$$

$$xy = \int x^2\,dx = \frac{1}{3}x^3 + c$$

$$y = \frac{1}{3}x^2 + cx^{-1}, \qquad x > 0.$$

---

[5] Here $\exp\left[\int p(x)\,dx\right]$ is used for typographical convenience as an expression for $e^{\int p(x)\,dx}$.

A second method for solving a nonhomogeneous linear equation is called the method of *variation of parameters*. To begin we recall that $k \exp\left[-\int p(x)\, dx\right]$ gives a one parameter family of solutions of the homogeneous equation $y' + p(x)y = 0$, the parameter being the constant k. In order to obtain solutions for the nonhomogeneous case, we search for a function $v(x)$ with which we can replace k so that the function

$$y = v(x)e^{-\int p(x)\, dx} \tag{6}$$

is a solution to the nonhomogeneous equation. The method gets its name from the fact that a function $v(x)$ replaces the parameter k.

If y is defined by Eq. (6),

$$y' = v'(x)e^{-\int p(x)\, dx} - v(x)p(x)e^{-\int p(x)\, dx}$$

Substituting into Eq. (1) we get

$$\left[v'(x)e^{-\int p(x)\, dx} - v(x)p(x)e^{-\int p(x)\, dx}\right] + p(x)v(x)e^{-\int p(x)\, dx} = q(x)$$

which reduces to

$$v'(x)e^{-\int p(x)\, dx} = q(x)$$

or

$$v'(x) = q(x)e^{\int p(x)\, dx} \tag{7}$$

Integrating both sides of Eq. (7) yields

$$v(x) = \int q(x)e^{\int p(x)\, dx}\, dx + c \tag{8}$$

Substituting into Eq. (6) gives a one parameter family of solutions

$$y = e^{-\int p(x)\, dx}\int q(x)e^{\int p(x)\, dx}\, dx + ce^{-\int p(x)\, dx} \tag{9}$$

The reader should observe that Eq. (9) is equivalent to Eq. (5), since the integral in Eq. (5) is an indefinite integral.

Example 2   Solve the differential equation

$$x\frac{dy}{dx} + (x + 1)y = x \tag{10}$$

First rewrite the equation as $dy/dx + (1 + 1/x)y = 1$ and compare this with Eq. (1). Then $p(x) = (1 + 1/x)$ and $q(x) = 1$. A one parameter family of solutions for the homogeneous equation is

$$ke^{-\int(1 + 1/x)\, dx} = k\left[e^{-[x + \ln|x|]}\right] = \frac{k}{|x|e^x}$$

Solving for $v(x)$ as given by Eq. (8) we obtain

$$v(x) = \int 1 \cdot e^{\int(1+1/x)\,dx}\, dx + c = \int |x|e^x\, dx + c$$

For simplicity let us restrict attention to positive values of x. Then

$$v(x) = (x - 1)e^x + c$$

The one parameter family of solutions given by Eq. (9) then becomes

$$y(x) = \frac{(x - 1)}{x} + \frac{c}{xe^x}, \qquad x > 0$$

It is not necessary to remember Eq. (9), since in each example the steps leading to Eq. (9) can be imitated. For example, the solution of Eq. (10) can be found by the following steps:

1. Divide by the coefficient of dy/dx and identify $p(x)$:

$$\frac{dy}{dx} + \left(1 + \frac{1}{x}\right)y = 1$$

$$p(x) = 1 + \frac{1}{x}, \qquad \int p(x)\, dx = x + \ln x$$

2. Write Eq. (6)

$$y = vx^{-1}e^{-x}$$

3. Differentiate and substitute into the differential equation:

$$y' = v'x^{-1}e^{-x} + v(-x^{-1}e^{-x} - x^{-2}e^{-x})$$

$$v'x^{-1}e^{-x} + v(-x^{-1}e^{-x} - x^{-2}e^{-x}) + vx^{-1}e^{-x}\left(1 + \frac{1}{x}\right) = 1$$

$$v' = xe^x$$

4. Solve for v by integration and so obtain y:

$$v = e^x(x - 1) + c$$
$$y = 1 - x^{-1} + cx^{-1}e^{-x}$$

Exercises

1. Solve each equation by identifying p and q and using Eq. (5).

(a) $y' + 3y = x$

(b) $\dfrac{dy}{dx} + xy = e^{-x^2/2}$

(c) $\dfrac{dy}{dt} + y = e^{2t}$

(d) $(\sin^2 u - v)\, du - (\tan u)\, dv = 0$

(e) $y' + \dfrac{y}{\sqrt{4x + x^2}} = \dfrac{2 + x}{\sqrt{4x + x^2}}$

(f) $y' + y \cot x = x^2$

(g) $x(\ln x)\, dy + (y - \ln x)\, dx = 0$

(h) $(x^4 + 2y)\, dx - x\, dy = 0$

2.   Solve each equation in Exercise 1 without using Eq. (5) by finding an integrating factor.

3.   Solve each equation in Exercise 1 without using Eq. (5) by using the method of variation of parameters.

4.   Consider the linear equation $y' + p(x)y = q(x)$. Suppose that on a vertical line $x = x_0$ we indicate the direction field at each point (x_0, y). Show that all lines drawn with these directions through the points (x_0, y) intersect in a common point. Explain how this fact can aid in the construction of the direction field.

1.6   Initial conditions

In many situations leading to the consideration of differential equations interest is centered not only on finding a one parameter family of solutions, but on locating that member (if any) of the family which satisfies certain specified conditions. For example, in Section 1.2 we mentioned the differential equation

$$\frac{dC(t)}{dt} = kC(t) \tag{1}$$

which describes a first order drug reaction. This linear homogeneous equation has a one parameter family of solutions given by

$$C(t) = ae^{kt} \tag{2}$$

where a is an arbitrary constant. The experimenter might be interested in determining that value of a such that the concentration of drug at a certain time t_0 be a given concentration C_0.

DEFINITION 1.10   The problem of finding a solution of a differential equation $y' = f(x, y)$ which also satisfies a given condition $y = y_0$ when $x = x_0$ is called an *initial value problem*. The condition $y = y_0$ when $x = x_0$ is called the *initial condition* and the point (x_0, y_0) is the *initial point*.

Example 1   Consider the equation for a first order drug reaction, Eq. (1). Suppose that the concentration at time $t_0 = 0$ is $C_0 = 5$. The initial value problem is to find $C(t)$ for which $C_0 = 5$ when $t = 0$. From Eq. (2) follows $C(0) = a$. Hence when $a = 5$ we obtain that member of the one parameter family in Eq. (2), namely, $C(t) = 5e^{kt}$, which satisfies our initial condition.

Example 2   Consider the initial value problem, $y' = x + y$, $y = -1$ at $x = 1$. We first obtain a one parameter family of solutions to the nonhomogeneous first order differential equation $y' = x + y$. Solutions obtained by the methods of Section 1.5 have the form

$$y = ce^x - x - 1$$

From the initial point $(1, -1)$ we obtain

$$-1 = ce^1 - 1 - 1 = ce - 2 \quad \text{or} \quad c = e^{-1}.$$

Hence the solution of the initial value problem is

$$y = e^{x-1} - x - 1$$

Further examples of initial value problems as they arise in relation to specific physical phenomena can be found in Chapter 2.

Exercises

1. Solve each of the following initial value problems.
   (a) $xy' - 2y = 2x^4$; $y = 8$ at $x = 2$
   (b) $(y + 2) dx + (x + 4) dy = 0$; $y = -1$ at $x = -1$

2. Suppose that in a first order drug reaction governed by Eq. (1), an experimenter measures $C = 5$ at $t = 0$ and $C = 3$ at $t = 1$. What are the values of a and k in Eq. (2)?

3. The following values are measured in an experiment on a certain drug. Graph $\ln C$ versus t and estimate a and k in Eq. (2).

t	0.40	0.60	1.00	1.50
C	1.22	1.36	1.66	2.11

4. Find the solution that satisfies the indicated conditions:
   (a) $(3xy + 2) dx + x^2 dy = 0$, $y = 1$ when $x = 1$
   (b) $xy' + 2y = 2x \cos 2x + 2 \sin 2x$, $y = 1$ when $x = \pi$

1.7   Separable equations

Many differential equations, especially those encountered in connection with practical problems, are very difficult to solve. Thus the subject of differential equations consists of a combination of theory and special techniques or devices ("tricks"). This text will, in general, emphasize linear differential equations which can be handled in a fairly systematic way. However, the knowledge of certain elementary and traditional techniques which might be applied in the nonlinear case is of value to the practitioner and can serve us by illustrating certain basic theoretical concepts. The next few sections are devoted to such techniques.

In many first order differential equations we can solve for dy/dx and then rewrite the equation in the form

$$M(x, y) + N(x, y) \frac{dy}{dx} = 0 \tag{1}$$

where M and N are, as indicated, given functions of x and y. In the special case of Eq. (1) where $M(x, y) = M(x)$ and $N(x, y) = N(y)$ the equation becomes

$$M(x) + N(y) \frac{dy}{dx} = 0 \tag{2}$$

DEFINITION 1.11   An equation which can be expressed in the form of Eq. (2) is said to be *separable*.

The reason for the term *separable* is evident if Eq. (2) is rewritten in its differential form

$$M(x) \, dx + N(y) \, dy = 0 \tag{3}$$

An equation of this form is called a differential equation. Its symmetric form suggests that we may regard either x or y as the independent variable. That is, we may ask for a function $y(x)$ which satisfies Eq. (2), or a function $x(y)$ which satisfies

$$M(x) \frac{dx}{dy} + N(y) = 0$$

Integration of Eq. (3) yields an implicit expression

$$\int M(x) \, dx + \int N(y) \, dy = \text{constant} \tag{4}$$

A function $y(x)$ which satisfies Eq. (4) will be a solution of Eq. (2), as we shall prove rigorously later in this section. First, we illustrate the method by several examples.

Example 1   Solve the equation

$$\frac{dC(t)}{dt} = kC(t) \tag{5}$$

Rewriting Eq. (5) in differential form yields

$$\frac{dC(t)}{C(t)} - k\,dt = 0 \qquad (6)$$

Note that care must be taken that $C(t) \neq 0$ for any t in order for Eq. (6) to be valid. On the other hand we see immediately that $C(t) \equiv 0$ is a solution of Eq. (5).

Since $dC(t)/C(t) = d\ln|C(t)|$, integration of Eq. (6) yields

$$\ln|C(t)| - kt = \text{constant} \qquad (7)$$

This yields $|C(t)| = be^{kt}$ for arbitrary positive constant b, or

$$C(t) = ae^{kt} \qquad (8)$$

for arbitrary real constant a. Equation (8) gives a one parameter family of solutions for $C(t)$. In this case we are fortunate in that our final result includes a solution $C(t) = 0$, which might have been lost because of our method of finding solutions. We shall prove shortly that this family contains all solutions of Eq. (5). One is not always this lucky and care is necessary.

Example 2 Solve the differential equation

$$\frac{dy}{dx} + y^2 \sin x = 0 \qquad (9)$$

By rewriting the equation in the form of Eq. (3) we get $y^{-2}\,dy + \sin x\,dx = 0$, provided we consider those x for which $y(x) \neq 0$. Integration yields

$$\int y^{-2}\,dy + \int \sin x\,dx = -y^{-1} - \cos x = \text{constant}$$

Therefore a one parameter family of solutions is $y = -(\cos x + a)^{-1}$ where a is an arbitrary constant. In addition $y(x) \equiv 0$ is a solution. It will follow from Theorem 1.1 that any solution which is nowhere zero must have the form $y = -(\cos x + a)^{-1}$.

Example 3 Obtain an implicit relationship for the solution of the differential equation

$$\frac{dy}{dx} = \frac{xy}{y-1}\ln x \qquad x > 0 \qquad (10)$$

Rewriting Eq. (10) in the differential form of Eq. (3) gives

$$\frac{y-1}{y}\,dy - x\ln x\,dx = 0$$

Integrating gives the desired relationship for $y(x)$,

$$\frac{x^2}{2} \ln x - \frac{x^2}{4} - y + \ln |y| = \text{constant} \tag{11}$$

It is not possible to obtain from this equation an explicit formula for y as a function of x using only the elementary functions used in calculus.[6]

We now introduce a theorem which guarantees the form of the solution of a separable equation.

THEOREM 1.1 Consider the separable differential equation

$$M(x) + N(y)y' = 0.$$

Assume that there exist functions $F(x)$ and $G(y)$ such that $dF(x)/dx = M(x)$ and $dG(y)/dy = N(y)$. If $y(x)$ is any differentiable function of x which satisfies the equation $F(x) + G(y) = $ constant on an interval $a < x < b$, then $y(x)$ is a solution of Eq. (2) on $a < x < b$. Conversely, if $y(x)$ is a solution of Eq. (2) on an interval $a < x < b$, then $y(x)$ must satisfy the equation $F(x) + G(y) = $ constant (for some constant).

PROOF: First we show that any function $y(x)$ satisfying the relationship $F(x) + G(y) = $ constant on an interval $a < x < b$ satisfies the differential equation on that interval. Since $F(x) + G(y(x)) = $ constant we obtain by implicit differentiation with respect to x the equation

$$\frac{dF(x)}{dx} + \frac{dG(y)}{dy}\frac{dy}{dx} = 0$$

Since $dF(x)/dx = M(x)$ and $dG(y)/dy = N(y)$, by hypothesis, this equation is the same as $M(x) + N(y)(dy/dx) = 0$, which is Eq. (2).

Next we shall prove that every solution $y(x)$ of Eq. (2) on an interval $a < x < b$ must satisfy the implicit relation $F(x) + G(y) = $ constant for some choice of the constant. Indeed, if $y(x)$ is a solution of Eq. (2) we must have

$$0 = M(x) + N(y(x))\frac{dy(x)}{dx}$$

$$= \frac{dF(x)}{dx} + \frac{dG(y(x))}{dy}\frac{dy(x)}{dx}$$

$$= \frac{d}{dx}[F(x) + G(y(x))], \qquad a < x < b$$

---

[6] We will not, as some authors do, refer to Eq. (11) as the solution of Eq. (10). Instead, we will call Eq. (11) a relation satisfied by solutions. The point is we are reserving the word *solution* for functions, not relations. (Of course, solution also refers to the process of deriving such functions or relations.)

Hence $F(x) + G(y(x))$ must equal a constant for $a < x < b$. This completes the proof of the theorem.   ∎

In this theorem it has been assumed that $dF(x)/dx = M(x)$ for *all* x and $dG(y)/dy = N(y)$ for *all* y. In examples this may be true only for some values of x and y, and then care must be taken to be sure that all solutions have been found. Let us reconsider Example 2 from this point of view. We have $M(x) = \sin x$, $F(x) = -\cos x$, so that $dF/dx = M(x)$ for all x. On the other hand, $N(y) = y^{-2}$, $G(y) = -y^{-1}$ and $dG/dy = N$ for $y \neq 0$. It follows from a reconsideration of Theorem 1.1 that if $y(x)$ is a solution which is nonvanishing (that is, nowhere zero) on an interval $b < x < c$, then $y(x)$ must satisfy $-y^{-1} - \cos x = $ constant, or $y = -(\cos x + a)^{-1}$, on $b < x < c$. These solutions are seen to be nonzero for $-\infty < x < \infty$. However, care must be used in the choice of the constant, for if $|a| \leq 1$, the solutions have vertical asymptotes at values of x for which $\cos x + a = 0$.

Conceivably there might be other solutions which vanish at one or more values of x. In fact, in Chapter 3 we shall establish general theorems which show that in this example the only solution which ever vanishes is the identically zero solution $y(x) \equiv 0$. Thus, $y = 0$ and $y = -(\cos x + a)^{-1}$ comprise the totality of solutions.

Theorem 1.1 is so formulated as to put emphasis on finding y as a function of x. In some cases, however, it may be convenient to invert the point of view. In Theorem 1.1 we could equally well have asserted that if $x(y)$ is any differentiable function of y which satisfies the equation $F(x) + G(y) = $ constant on $\alpha < y < \beta$, then $x(y)$ is a solution of the equation

$$ M(x) \frac{dx}{dy} + N(y) = 0 $$

on $\alpha < y < \beta$. This is illustrated in the next example.

Example 4   The equation $(3y^2 + 1)\, dy = 2x\, dx$ yields   the   relation $y^3 + y = x^2 + c$. This cannot easily be solved explicitly for y as a function of x, but can be solved for

$$ x = \pm(y^3 + y - c)^{1/2} $$

This function x satisfies

$$ 3y^2 + 1 = 2x \frac{dx}{dy} $$

Of course, the relation $y^3 + y = x^2 + c$ does define y as a function of x, as one can show from the *implicit function theorem* of calculus, even though we cannot easily explicitly exhibit this function. (However, see Exercise 7.)

Exercises

1. Solve each equation by the method of separation of variables. That is, find a relation $F(x) + G(y) =$ constant which must be satisfied by the solutions. Wherever possible, find y as a function of x or x as a function of y. It is not required (and it is not forbidden) to give the interval for which each solution is defined or to prove that all solutions have been obtained.

(a) $2x \, dx + y^2 \, dy = 0$

(b) $3xy + y' = 4y$

(c) $(\sin 3y)(\cos x) \dfrac{dy}{dx} = 4$

(d) $x\sqrt{x^2 - 9} \dfrac{dy}{dx} = \dfrac{1}{y} \sqrt{9 - y^2}$

(e) $\dfrac{y(2 + 3y)}{9 + x^2} + \dfrac{dy}{dx} = 0$

(f) $xy(\ln x)y' + 1 = 0 \qquad x > 0$

(g) $(\ln x) \, dx + \dfrac{dy}{y^2} = 0$

(h) $(x + 1) \dfrac{dy}{dx} = x(y^2 + 1)$

(i) $y' = \dfrac{y}{x(y - 3)}$

(j) $\sin 3x \, dx + 3 \cos 3x \, dy + y^2 \cos 3x \, dy = 0$

(k) $x^2(dx - x \sec^2 y \, dy) = 7 \sec^2 y \, dy$

(l) $x \ln y = y \dfrac{dx}{dy}$

2. Plot, on a single graph, the solution curves $y = -(\cos x + a)^{-1}$ of Example 2 for $a = 0, \frac{1}{2}, 1, 2$.

3. Solve the differential equation

$$xy' + y + xe^{-xy} = 0$$

[*Hint*: Set $z = xy$.]

4. Explain why the general first order equation

$$M(x, y) \, dx + N(x, y) \, dy = 0$$

cannot be solved by separation of variables. Give an example of such an equation.

5. The linear nonhomogeneous equation can be written

$$dy + [p(x)y - q(x)]\, dx = 0$$

Under what conditions is this equation separable?

6. Solve the initial value problem

$$4 \cos^2 y\, dx + \csc^2 x\, dy = 0; \qquad y = \frac{\pi}{4} \text{ at } x = 0$$

7. Look up the formulas for solving a cubic equation and use these to solve the equation

$$y^3 + y - (x^2 + c) = 0$$

for y.

1.8 Methods of substitution

Many problems in mathematics can be greatly simplified by a change from one coordinate system to another. A change commonly presented in basic calculus courses is from rectangular to polar coordinates. Changing coordinates can sometimes be used as a tool for converting a seemingly complicated differential equation to one in which the variables are separable (or to one of some other simple type). The method is called substitution and is illustrated by the following examples.

Example 1   Solve the differential equation

$$\frac{dx}{dt} = (x + t)^2 e^t - 1 \tag{1}$$

A cursory glance indicates the equation is not separable. We therefore "guess" a substitution $y(t) = x(t) + t$. Applying the chain rule we obtain

$$\frac{\overset{\cdot}{dy}}{dt} = \frac{dx}{dt} + 1$$

and substituting for x and dx/dt in Eq. (1) we obtain

$$\frac{dy}{dt} = y^2 e^t \tag{2}$$

which is separable. Solving by the method of Section 1.7 gives

$$y(t) = (a - e^t)^{-1} \tag{3}$$

as a one parameter family of solutions for y. However the original problem is to solve for $x(t)$. Hence replacing $y(t)$ by $x(t) + t$ in Eq. (3) we obtain the one parameter family

$$x(t) = (a - e^t)^{-1} - t \tag{4}$$

Also $x(t) = -t$ is a solution. This corresponds to $y = 0$, lost when separating variables in Eq. (2).

Example 2   As a second example of the use of substitution, consider the differential equation

$$y' = f(y/x), \qquad x \neq 0 \tag{5}$$

By making the substitution $y = vx$ and applying the chain rule we obtain $dy/dx = v + x(dv/dx)$. Substituting this result into Eq. (5) yields the equivalent relationship

$$v + x\frac{dv}{dx} = f(v) \tag{6}$$

which can be expressed (provided $f(v) \neq v$) as

$$\frac{dv}{f(v) - v} - \frac{dx}{x} = 0 \tag{7}$$

For any given function f, Eq. (7) is a separable equation which can be solved by the method of Section 1.7. Once a solution v is obtained the corresponding function $y = vx$ can be obtained; this will be a solution to Eq. (5).

Equation (5) of Example 2 is a special case of a class of differential equations, called homogeneous, which can be solved by the method of substitution. A treatment of this type of equation is developed in the exercises at the end of this section.

We close this section by applying the method of substitution to a second order differential equation which appears in problems in mechanics. A model in which this equation is developed is given in Section 2.3.

Example 3   Reduce the differential equation

$$\frac{d^2x}{dt^2} = \frac{k}{x^2} \tag{8}$$

to a first order differential equation. In attacking the problem, we make the substitution $v = dx/dt$. Applying the chain rule gives

$$\frac{d^2x}{dt^2} = \frac{dv}{dt} = \frac{dv}{dx} \cdot \frac{dx}{dt} = v\frac{dv}{dx} \tag{9}$$

Substituting this result into Eq. (8) we obtain

$$v\frac{dv}{dx} = \frac{k}{x^2} \tag{10}$$

This is a first order equation for the new unknown v, and moreover is separable. An implicit representation for v is given by

$$\frac{v^2}{2} + \frac{k}{x} = a \tag{11}$$

where a is an arbitrary constant. Once a solution for v is obtained from Eq. (11) the problem of solving for x still remains. But since $v = dx/dt$ we have reduced the second order differential equation to a first order differential equation which can then be attacked by other methods.

Exercises

1. Show that each of the following equations can be put in the form of Eq. (5). Hence reduce the equation to one that is separable, and solve.

   (a) $(xe^{y/x} + 2y)\, dx - 2x\, dy = 0$

   (b) $(2x + y)\, dx - (4x - y)\, dy = 0$

2. Solve each equation by making a change of variable which reduces the equation to one that is separable.

   (a) $(x - y)^2 \dfrac{dy}{dx} = a^2 \qquad (a = \text{real constant})$

   (b) $x\, dy + (x \cos x - 2 \sin x - 2y)\, dx = 0$

   (c) $\dfrac{dv}{dx} = \dfrac{x - e^v}{xe^v}$

   (d) $y^2(x\, dx + y\, dy) + 2x(y\, dx - x\, dy) = 0$

       [*Hint:* Introduce polar coordinates.]

   (e) $\dfrac{dy}{dx} + \dfrac{1}{2}x = \sqrt{x^2 + 4y}$

3. Prove: Let $y(x)$ be a solution of the differential equation $d^2y/dx^2 = F(y, dy/dx)$, in which the independent variable x does not appear explicitly. Assume that to each y there corresponds a unique value of dy/dx, and let $p(y) = dy/dx$. Then the given equation reduces to a first order differential equation for p. Also show that after p has been found, the first order equation for y is separable.

4. A function of x and y, $f(x, y)$, is said to be a *homogeneous function* of *degree* n if $f(tx, ty) = t^n f(x, y)$ for all specified t, x, and y. Show that each of the following functions is homogeneous and find its degree.

   (a) $xe^{y/x} + 3y \qquad x \neq 0$

   (b) $\ln x - \ln y \qquad x > 0, y > 0$

   (c) $x - 6y \cos (x/y) \qquad y \neq 0$

5. Show that if f is any function, then $f(y/x)$ is a function of x and y which is homogeneous of degree zero. ($x = 0$ and $t = 0$ must be excluded.)

6. Let $M(x, y)$ be a homogeneous function of degree zero. Prove that there exists a function f such that $M(x, y) = f(y/x)$ for all x, y with $x \neq 0$.
   [*Hint*: In $M(x, y) = M(tx, ty)$, let $t = 1/x$.]

7. A first order differential equation

$$M(x, y) + N(x, y)y' = 0$$

   is said to be a *homogeneous differential equation*, or to have homogeneous coefficients, if $M(x, y)$ and $N(x, y)$ are homogeneous functions of the same degree. Show that such an equation can be rewritten in the form of Eq. (5). It follows from Example 2 that every homogeneous equation can be reduced to a separable equation.

8. Show that a homogeneous equation $M(x, y) + N(x, y)y' = 0$ reduces under the substitution $y = xv$ to the separable equation

$$\frac{1}{x} + \frac{N(1, v)}{M(1, v) + vN(1, v)} \frac{dv}{dx} = 0$$

9. Verify that each of the following is a homogeneous differential equation. Then solve by making the substitution $y = xv$.

   (a) $\left(x + y \cos \dfrac{y}{x} \right) dx - x \cos \dfrac{y}{x} \, dy = 0$

   (b) $(x^2 + y^2) \, dx - 4xy \, dy = 0$

   (c) $(x + y) \, dy + (x - y) \, dx = 0$

   (d) $\dfrac{dy}{dx} = \dfrac{y^3 + 2x^2 y}{x^3}$

10. Solve the initial value problem

$$(x + y) \, dy + (y + 2x) \, dx = 0; \quad y = 1 \text{ at } x = 1$$

11. Using the technique of reduction of order solve the following:

   (a) $y'' - 3y' = x^2$
   (b) $y'' = y^2 y'$

1.9   Exact equations

An expression such as

$$M(x, y) \, dx + N(x, y) \, dy$$

where M and N are real-valued functions defined for (x, y) lying in some region of the plane, is called a *first order differential form in two variables.*[7] If such a form is set equal to zero, the resulting equation is a first order differential equation.

Recall that if $F(x, y)$ is a differentiable function,[8] its differential is given by

$$dF = \frac{\partial F}{\partial x}\, dx + \frac{\partial F}{\partial y}\, dy \tag{1}$$

Sometimes a differential form is the differential of a function $F(x, y)$. For example

$$x\, dy + y\, dx = dF, \quad \text{where } F(x, y) = xy$$

But there is no function F such that $dF = x\, dy - y\, dx$, as we shall presently be able to prove. This distinction suggests the usefulness of giving special attention to differential forms which are of the form dF for some F. Let $M(x, y)$ and $N(x, y)$ be defined for (x, y) in a given rectangle R in the x, y plane (that is, $a < x < b$ and $c < y < d$). We shall call the differential form $M\, dx + N\, dy$ *exact* in R if there exists a function $F(x, y)$ defined and having continuous first partial derivatives in R such that $dF = M\, dx + N\, dy$ in R. Since the differential dF of a function F with continuous first partial derivatives is given by Eq. (1), we can state this definition as follows.

DEFINITION 1.12   Let $M(x, y)$ and $N(x, y)$ be defined for (x, y) in a rectangle R. The *differential form $M\, dx + N\, dy$* is said to be *exact* in R if there exists a function $F(x, y)$ defined and having continuous first partial derivatives in R such that

$$dF = M\, dx + N\, dy$$

in R, or equivalently,

$$\frac{\partial F}{\partial x} = M, \quad \frac{\partial F}{\partial y} = N, \quad (x, y) \in R \tag{2}$$

DEFINITION 1.13   The first order *differential equation*

$$M(x, y)\, dx + N(x, y)\, dy = 0 \tag{3}$$

is said to be *exact* in R if the differential form $M\, dx + N\, dy$ is exact in R.

Exact differential equations are of importance because they arise in certain applications and because there is a simple procedure for solving them. This procedure is suggested by the fact that if Eq. (3) is exact, it may be written in

---

[7] Thomas, *op. cit.*, p. 518.

[8] See Thomas, *loc. cit.*, for the definition of differential.

the form $dF = 0$, from which it seems reasonable to conclude that $F(x, y) =$ constant. This last equation may be regarded as an implicit relation that determines a function $y(x)$ which is a solution. To be more precise, suppose $y(x)$ is a solution of Eq. (3) defined on an interval $a < x < b$, and is such that the graph of $y(x)$ lies in the rectangle R. Since y is a solution,

$$M(x, y(x)) + N(x, y(x))y'(x) = 0, \qquad a < x < b \qquad (4)$$

By virtue of Eq. (2), this can be written

$$\frac{\partial F\,(x,\, y(x))}{\partial x} + \frac{\partial F\,(x,\, y(x))}{\partial y}\, y'(x) = 0 \qquad (5)$$

or simply

$$\frac{d}{dx}\, F(x,\, y(x)) = 0 \qquad a < x < b$$

That is, the function $F(x, y(x))$ is a function with zero derivative on the interval $a < x < b$, and therefore by the Fundamental Theorem of Calculus $F(x, y(x)) = c$ where c is some constant. Thus the solution $y(x)$ must be a function which is given implicitly by the relation $F(x, y) = c$, for some value of c.

Looking at this in reverse, we see that if the implicit relation $F(x, y) = c$ defines a function $y = y(x)$ on some interval $a < x < b$, then $F(x, y(x)) = c$ for $a < x < b$. Differentiation of the function $F(x, y(x))$ with respect to x then yields Eq. (5), or equivalently Eq. (4), so that $y(x)$ is a solution of Eq. (3). Let us summarize these conclusions in the following theorem.

THEOREM 1.2   Suppose that Eq. (3) is exact in a rectangle R and let $F(x, y)$ be a function such that

$$\frac{\partial F}{\partial x} = M, \qquad \frac{\partial F}{\partial y} = N$$

in R. If on some interval $a < x < b$, $y(x)$ is a differentiable function defined implicitly by the relation $F(x, y) = c$, then $y(x)$ is a solution of Eq. (3) for $a < x < b$. Conversely, every solution $y(x)$ of Eq. (3) whose graph lies in R satisfies the relation $F(x, y) = c$ for some constant c.

Example 1   For the equation $y\, dx + x\, dy = 0$ we take $F(x, y) = xy$. Every solution therefore satisfies $xy = c$, the implicit relation. In this case we can solve explicitly for $y = c/x$.

Example 2   As a second example, consider any separable equation $M(x)\, dx + N(y)\, dy = 0$. If $F(x)$ is any indefinite integral of $M(x)$ and $G(y)$ is any indefinite integral of $N(y)$, the relations

$$\frac{\partial}{\partial x}\, [F(x) + G(y)] = M, \qquad \frac{\partial}{\partial y}\, [F(x) + G(y)] = N$$

are satisfied. Consequently all solutions must satisfy $F(x) + G(y) = c$ for a constant c. This is, of course, just the method previously described in Section 1.7.

Example 3 Consider the equation

$$x^{-1} \, dx + \sqrt{1 - y^2} \, dy = 0$$

This is an equation of the type in Example 2 (separable). Let us see how to choose M, N, and the rectangle R in this case. Let us take $M(x) = x^{-1}$, $N(y) = \sqrt{1 - y^2}$, and

$$F(x, y) = \ln |x| + \tfrac{1}{2}[y\sqrt{1 - y^2} + \sin^{-1} y]$$

The function F has partial derivatives $\partial F/\partial x = M$ and $\partial F/\partial y = N$ which are continuous in any rectangle R which excludes the line $x = 0$ and is contained in $|y| \leq 1$. Since Theorem 1.2 is therefore applicable, it follows that any solution $y(x)$ with graph in R must satisfy the relation $F(x, y) = c$, or

$$x = k \exp \{-\tfrac{1}{2}[y\sqrt{1 - y^2} + \sin^{-1} y]\}$$

Note that by taking $k = 0$ we get $x \equiv 0$; this is a solution of the original equation in the sense that

$$\frac{dx}{dy} + x\sqrt{1 - y^2} = 0$$

Example 4 Consider the equation

$$\frac{x \, dx}{\sqrt{x^2 - y^2}} + \left(1 - \frac{y}{\sqrt{x^2 - y^2}}\right) dy = 0$$

The reader can verify that

$$F(x, y) = y + \sqrt{x^2 - y^2}$$

has the given expression as differential. The function F is real-valued only for $|y| \leq |x|$. The equation is exact in any rectangle R interior to the set of points where $|y| \leq |x|$. The reader may wonder why we have framed our definitions and theorems in terms of rectangles rather than more general regions. In fact, we could allow arbitrary simply connected regions R, but have preferred at this stage to sidestep topological ideas such as connectedness.

In concluding this section, it is emphasized that a differential equation must be written in differential form before we may ask whether it is exact in the sense defined here. Also, if an exact equation is written in another "equivalent" form by multiplying by a factor, the resulting equation need not be exact also.

For example, the equation $y \, dx + x \, dy = 0$ is exact, but the "equivalent" equation

$$dx + \frac{y}{x} \, dy = 0$$

is not exact. In fact, if there were a function $F(x,y)$ such that

$$dF = dx + \frac{y}{x} \, dy$$

then by Eq. (2) we would have

$$\frac{\partial F}{\partial x} = 1, \qquad \frac{\partial F}{\partial y} = \frac{y}{x}$$

The first relation implies that $F(x, y) = x + g(y)$ for some function g, but then $\partial F/\partial y = g'(y)$ cannot be equal to y/x.

1.10   Construction of solutions of an exact equation

In all the preceding examples of exact equations it has been easy to find a suitable function F by inspection, but often this is not the case. We now seek to answer these two questions: (1) How can a given equation be tested for exactness? (2) If it is established that an equation is exact, how can a suitable function F be found? The first question is answered by the following theorem; the proof of the theorem provides a construction for F, thus answering the second question.

THEOREM 1.3   Let M and N be real-valued functions which have continuous first partial derivatives for (x, y) in a rectangle R. Then the equation

$$M(x, y) \, dx + N(x, y) \, dy = 0$$

is exact in R if and only if

$$\frac{\partial M}{\partial y} = \frac{\partial N}{\partial x} \tag{1}$$

for (x, y) in R.

PROOF: If $M \, dx + N \, dy$ is exact, there is a function F such that

$$\frac{\partial F}{\partial x} = M, \qquad \frac{\partial F}{\partial y} = N, \qquad (x, y) \text{ in } R \tag{2}$$

By differentiating the first equation with respect to y and the second with respect to x, we obtain

$$\frac{\partial M}{\partial y} = \frac{\partial^2 F}{\partial y \partial x}, \qquad \frac{\partial N}{\partial x} = \frac{\partial^2 F}{\partial x \partial y}$$

However, a theorem from advanced calculus[9] states that if F is a function with continuous second partial derivatives then

$$\frac{\partial^2 F}{\partial y \, \partial x} = \frac{\partial^2 F}{\partial x \, \partial y}$$

It follows that Eq. (1) is satisfied for every (x, y) in R.

To prove the converse statement, we assume that Eq. (1) is satisfied and seek to find a function F for which $dF = M \, dx + N \, dy$, or in other words for which Eq. (2) holds. The construction of such a function is most easily carried out by line integration. In order to make the basic idea clear to readers not familiar with line integrals, we have arranged this section in such a way that we do not have to rely on the theory of line integrals.

The intuitive idea used in finding F is that since $dF = M \, dx + N \, dy$, it should be true that $F = \int dF = \int (M \, dx + N \, dy)$. More accurately, if we fix one point (x_0, y_0) in R, then for any other point (x, y) in R,

$$F(x, y) - F(x_0, y_0) = \int_{(x_0, y_0)}^{(x, y)} (M \, dx + N \, dy).$$

In this equation, a clear meaning must be given to the integral. Intuitively, the equation states that the change in value of F between (x_0, y_0) and (x, y) equals the integral of dF in moving between these points. There are, of course, many paths on which a point can move from (x_0, y_0) to (x, y). We shall consider two of these. The first path, P_1, is along the vertical segment from (x_0, y_0) to (x_0, y) and then along the horizontal segment to (x, y). See Figure 1.5. On the horizontal segment, y is constant, and it appears that $dy = 0$ and $dF = M \, dx$. On the vertical segment, $dx = 0$ and $dF = N \, dy$.

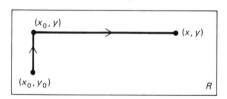

FIGURE 1.5

The preceding remarks, although not examples of the best mathematical rigor, suggest that we define $F(x, y)$ by the formula

$$F(x, y) = \int_{x_0}^{x} M(r, y) \, dr + \int_{y_0}^{y} N(x_0, s) \, ds \qquad (3)$$

[9] Tom Apostol, *Mathematical Analysis*, Reading, Mass., Addison-Wesley, 1957, p. 121.

Here we have arbitrarily selected $F(x_0, y_0) = 0$, since we can change F by an additive constant without changing the property $dF = 0$.

Now, however much one may approve or disapprove of the arguments given above, one must agree that Eq. (3) defines a function F for (x, y) in R. Can we now verify that F satisfies Eq. (2) and hence that $M \, dx + N \, dy$ is an exact differential? As a matter of fact, it follows at once by differentiating[10] Eq. (3) that $F_x(x, y) = M(x, y)$, where F_x denotes the partial derivative with respect to x, but

$$F_y(x, y) = N(x_0, y) + \int_{x_0}^{x} M_y(r, y) \, dr$$

That is, we get only part of what we want.

However,

$$M_y(r, y) = N_r(r, y)$$

from Eq. (1). Hence

$$F_y(x, y) = N(x_0, y) + \int_{x_0}^{x} N_r(r, y) \, dr$$

$$= N(x_0, y) + N(x, y) - N(x_0, y) = N(x, y)$$

proving that the equation is exact.   ∎

Example 1   Consider the equation

$$y' = \frac{x^2 - y}{x + y^2} \tag{4}$$

which can be written as

$$(y - x^2) \, dx + (x + y^2) \, dy = 0$$

Here $M = y - x^2$, $N = x + y^2$, and it is easy to verify that $\partial M / \partial y = \partial N / \partial x = 1$ for all x, y. In this case, the rectangle R of Theorem 1.3 can be

---

[10] By the Fundamental Theorem of Calculus,

$$\frac{d}{dx} \int_{x_0}^{x} f(r) \, dr = f(x)$$

Therefore

$$\frac{d}{dx} \int_{x_0}^{x} M(r, y) \, dr = M(x, y)$$

In general, let

$$I(z) = \int_{a(z)}^{b(z)} f(x, z) \, dx$$

Then

$$\frac{dI(z)}{dz} = \int_{a(z)}^{b(z)} f_z(x, z) \, dx + \frac{db(z)}{dz} f(b(z), z) - \frac{da(z)}{dz} f(a(z), z)$$

See D. Widder, *Advanced Calculus*, 2nd ed., Englewood, New Jersey, Prentice Hall, 1961.

replaced by the entire (x, y) plane. Since the equation is exact, a suitable F can be found from Eq. (3). Use of Eq. (3) with $x_0 = y_0 = 0$ yields

$$F(x, y) = \int_0^x (y - r^2)\, dr + \int_0^y s^2\, ds$$

$$= \left[yr - \frac{1}{3} r^3 \right]_0^x + \left[\frac{1}{3} s^3 \right]_0^y = xy - \frac{1}{3} x^3 + \frac{1}{3} y^3$$

Therefore any differentiable function $y = \phi(x)$ defined implicitly by the relation

$$3xy - x^3 + y^3 = c \tag{5}$$

will be a solution of the differential equation, and all solutions arise this way.

Although it is not easy to obtain explicit formulas for functions $y = \phi(x)$ which satisfy Eq. (5), such functions do exist and can be computed numerically. For example, suppose we seek a solution such that $\phi(1) = 1$, that is, $y = 1$ at $x = 1$. From Eq. (5) it is clear that c must be 3. We leave it as an exercise to compute this function for values of x near $x = 1$.

Example 2 An alternate method for solving Eq. (4) is as follows. We know that

$$\frac{\partial F}{\partial x} = y - x^2, \qquad \frac{\partial F}{\partial y} = x + y^2$$

Consider y fixed and integrate the first of these equations with respect to x. Then

$$F(x, y) = xy - \tfrac{1}{3} x^3 + f(y)$$

where f is independent of x. In this formula the integration constant must be allowed to depend on y, since in the integration y has been treated as a parameter; that is, a variable that does not enter into the integration process. But now

$$\frac{\partial F}{\partial y} = x + f'(y)$$

and since this must equal $x + y^2$ we deduce that $f'(y) = y^2$. A choice for $f(y)$ is therefore $y^3/3$, which gives $F(x, y) = xy - \tfrac{1}{3} x^3 + \tfrac{1}{3} y^3$ as before.

Example 3 Even though an equation $M\, dx + N\, dy = 0$ may not be exact, it is sometimes possible to guess a function $U(x, y)$ such that $UM\, dx + UN\, dy = 0$ is exact. Such a function is called an *integrating factor*. If U is nonzero in the rectangle R, the two differential equations have the same solutions. For example, the equation

$$y\, dx - x\, dy = 0$$

is not exact. However, multiplying by $U = x^{-2}$ gives

$$\frac{y}{x^2} \, dx - \frac{dy}{x} = 0, \qquad x \neq 0$$

which is exact. Solving by the method of Example 2, we deduce

$$F(x, y) = -\int \frac{1}{x} \, dy = -\frac{y}{x} + f(x)$$

$$M(x, y) = \frac{\partial F}{\partial x} = \frac{y}{x^2} + f'(x) = \frac{y}{x^2}$$

Therefore $f'(x) = 0$ or f is constant. Hence $y = cx$ defines a one parameter family of solutions.

Let us return briefly to Theorem 1.3. The difficult part of the proof was to show that if Eq. (1) holds, then there exists a function F such that $dF = M \, dx + N \, dy$. This was done by exhibiting such a function. For readers familiar with line integrals (others may skip this discussion), we shall try to make clear what is at stake here. It is a known theorem from advanced calculus[11] that if Eq. (1) holds (in any open simply connected region of the plane) then there is a real function F such that $F_x = M$ and $F_y = N$, that is, $dF = M \, dx + N \, dy$. This function F can be defined by a line integral

$$F(x, y) = \int_{(x_0, y_0)}^{(x, y)} (M \, dx + N \, dy) \tag{6}$$

Equation (1) is enough to guarantee that the value of the line integral is independent of the choice of path and therefore that F is unambiguously defined by Eq. (6).

Exercises

1. Test the following for exactness and find the implicit relation $F = c$.

(a) $e^x \cos y \, dx = e^x \sin y \, dy$

(b) $(ax + by) \, dx + (bx + ay) \, dy = 0$

(c) $(e^y + \cos x) \, dx + xe^y \, dy = 0$

(d) $y^3 \, dx + (3y^2x + y^3) \, dy = 0$

(e) $3x^2 \ln y \, dx + \frac{x^3}{y} \, dy = 0, \qquad y > 0$

---

[11] See Apostol, *op. cit.*, p. 294.

(f) $(\sec x \tan x + 2xy)\, dx + \left(x^2 + \dfrac{1}{y}\right) dy = 0, \qquad y > 0$

(g) $\dfrac{4y^2 - 2x^2}{4xy^2 - x^3}\, dx + \dfrac{8y^2 - x^2}{4y^3 - x^2 y}\, dy = 0$

2. Assume the equation $y^2 \cos x\, dx + yf(x)\, dy = 0$ is exact. Find all possibilities for $f(x)$.

3. Using Eq. (3), obtain the implicit relation $F = c$ for solutions of

$$e^{-(x+y)}\, dx + \left[ye^{-y} - \int_1^x e^{-(r+y)}\, dr\right] dy = 0$$

4. (Computational problem) Compute the solution of the equation in Example 1 such that $\phi(1) = 1$, at various points x_i near 1. Graph $\phi(x)$.

5. Find an integrating factor for each of the following equations, and solve.

(a) $y\, dx + (y^2 - x)\, dy = 0$

(b) $xy\, dx - y^2\, dy = 0$

(c) $(3x + 4y)\, dx + \left(2x + 2\dfrac{y}{x}\right) dy = 0$

6. Suppose $u(x, y)$ is an integrating factor and that $F(x, y)$ is a function such that $dF = uM\, dx + uN\, dy$. Prove that $u(x, y)g(F(x, y))$ is also an integrating factor, for any continuous function g.

7. Consider the equation $M\, dx + N\, dy = 0$ where M, N have continuous first partial derivatives on some rectangle R. Prove that a function u, having continuous first partial derivatives on R, is an integrating factor if and only if it satisfies the partial differential equation

$$N\frac{\partial u}{\partial x} - M\frac{\partial u}{\partial y} = u\left(\frac{\partial M}{\partial y} - \frac{\partial N}{\partial x}\right)$$

on R.

8. Prove that if

$$\frac{1}{N}\left(\frac{\partial M}{\partial y} - \frac{\partial N}{\partial x}\right)$$

depends on x only, there is an integrating factor u which depends on x only. Give a formula for u.

9. Apply Exercise 8 to solve

$$(2x^2 + y)\, dx + (x^2 y - x)\, dy = 0$$

Miscellaneous Exercises

Solve each of the following equations by one of the techniques of this chapter.

1. $x^2 \dfrac{dy}{dx} + y = e^{1/x}$

2. $y' - (\cot x)y = \sec x$

3. $(1 + ye^{-x}) \, dx - e^{-x} \, dy = 0$

4. $xy^2 + y^3 + 3x^2yy' = 0$

5. $3x^2(\tan y) \, dx + (x^3 \sec^2 y - y \sin y) \, dy = 0$

6. $\dfrac{dy}{dx} = \dfrac{x + y}{y - x}$

7. $(x^2 + 2x)e^y \, dx + x^2 e^y \, dy = 0$

8. $\ln (x^2) - 2 \ln y + y' \ln \dfrac{y}{x} = 0$

9. $e^u \dfrac{du}{dx} = -(e^u + x)^2 - 1$

10. $y' + \dfrac{1}{x\sqrt{9 - x^2}} \, y = \left(\dfrac{3 + \sqrt{9 - x^2}}{x} \right)^{1/3}$

11. $(2x + 2y)(dx + dy) = 0$

12. $\dfrac{y'}{x} + \dfrac{y}{x^2 + a^2} = \dfrac{1}{a}$

13. $(2x + y) \, dy + x \, dx = 0$

14. $(x^3 + y^3) \, dx - 3xy^2 \, dy = 0$

15. $\left(x^2 + \dfrac{y^4}{x^2} \right) dx + 2yx \, dy = 0$

16. $3xy + x^2 - 4y^2 \dfrac{dy}{dx} = 0$

17. $(\sin x)y' + y = 2$

18. $[y^2 \sin x + (x + y^2) \cos x] \, dx + 2y \sin x \, dy = 0$

19. $xe^v \cos (e^v) \dfrac{dv}{dx} + x^2 = \sin (e^v)$

20. $(xy^2 - 1) \, dx + x^2y \, dy = 0$

21. $2xye^{x^2y} + \cos x + x^2 e^{x^2y}y' = 0$

22. $\dfrac{dy}{dx} + \dfrac{y}{x^2 - 9} = (x + 3)^{1/6}$

23. $x^2(7 + 3y^2) - \dfrac{4x}{\sqrt{3 - 2x^2}} \dfrac{dy}{dx} = 0$

24. $\dfrac{3x^2 - 5xy - 2y^2}{4x^3} + \dfrac{1}{y} \dfrac{dy}{dx} = 0$

25. $x\,dy - y\,dx + x^3\,dx = 0$

26. $\left(\dfrac{\tan^{-1} x}{\cos y}\right) + y^2 \dfrac{dy}{dx} = 0$

27. $y' - \csc x - y \cot x = 0$

28. $\dfrac{(3x^2/2)\,y^2 + \cos x}{yx^3 + \sin y} = -y'$

29. $x\,dy - y\,dx = xy^2\,dx$

30. $e^x(y \sin x - 2 \cos x) \dfrac{dy}{dx} = 0$

31. $y' = \dfrac{x + y}{x}$

32. $(x^2 + 1)\,dy + 2xy\,dx = 0$

33. $\left(\dfrac{y^3}{x} + xy\right) \dfrac{dy}{dx} = 0$

34. $y' + (x^3 + \ln x)y = 0$

35. $y' = \dfrac{y}{1 + e^x}$

36. $\dfrac{dy}{dx} = \dfrac{x^2 \ln x}{y \ln y}$

37. $x\,dy - y\,dx - \sqrt{x^2 - y^2}\,dx = 0$

38. Show that a nonlinear equation of the form

$$y' + p(x)y = q(x)y^k$$

where k is any constant not equal to one (called a *Bernoulli* equation), can be transformed into a linear equation by the change of variable $v = y^{1-k}$.

39. Solve each of these Bernoulli equations:

(a) $xy' - y = x^2 y^2$

(b) $yy' + xy^2 = x^3$

40. An equation of the form $y' + ay^2 = bx^k$ where a, b, k are constants, $a \neq 0$, $b \neq 0$, is called a (special) *Riccati* equation.

(a) If $k = 0$, show that there is a solution curve through each point in the plane, and find the explicit form of the solution by separation of variables. Discuss $\lim_{x \to \infty} y(x)$ and $\lim_{x \to -\infty} y(x)$.

(b) If $k = -2$, show that the transformation $y = 1/z$ results in a homogeneous equation for z. Find the explicit solution when $a = 1, b = 2$.

41. An equation of the form

$$y' = f(x)y^2 + g(x)y + h(x)$$

where f, g, h are continuous functions on an interval $a \leq x \leq b$, is called a (general) *Riccati* equation. Show that if $f(x)$ is differentiable, and if we define

$$u(x) = \exp\left(- \int^x f(s)y(s) \, ds\right)$$

then $u(x)$ must satisfy the linear equation of second order

$$f(x)u'' - [f'(x) + f(x)g(x)]u' + [f(x)]^2 h(x)u = 0$$

Equations of this kind are studied in Chapter 9.

42. Let $y(x)$ and $z(x)$ be distinct solutions of a Riccati equation on an interval J.

(a) Show that $w(x) = [y(x) - z(x)]^{-1}$ is a solution of the linear first order equation

$$w' + [2f(x)z(x) + g(x)]w = -f(x)$$

(b) Show conversely that if w is a solution of this equation, and z is a solution of the Riccati equation, then $y = z + w^{-1}$ is a solution of the Riccati equation.

43. The Riccati equation

$$y' = y^2 - (2x + 1)y + (1 + x + x^2)$$

is seen by inspection to have a solution $z(x) = x$. Using the result in Exercise 42, find a one parameter family of solutions.

44. Show that any equation of the form

$$\frac{d^n y}{dx^n} = F\left(x, \frac{dy}{dx}, \ldots, \frac{d^{n-1} y}{dx^{n-1}}\right)$$

in which the dependent variable does not explicitly appear can be reduced to an equation of order $n - 1$ by the substitution $v = dy/dx$.

45. Given that the most general solution of $v'' + v = 0$ has the form $v = c_1 \sin x + c_2 \cos x$, where c_1 and c_2 are arbitrary constants, use the method of Exercise 44 to find the most general solution of $y''' + y' = 0$.

46. Consider an equation of the form

$$\frac{d^n y}{dx^n} = F\left(y, \frac{dy}{dx}, \ldots, \frac{d^{n-1}y}{dx^{n-1}}\right)$$

in which the independent variable x does not explicitly appear. Assume that to each y there corresponds a unique value of dy/dx, and let $p(y) = dy/dx$. Show that this substitution leads to the relations

$$\frac{d^2 y}{dx^2} = p \frac{dp}{dy}$$

$$\frac{d^3 y}{dx^3} = p \frac{d}{dy}\left(p \frac{dp}{dy}\right)$$

$$\ldots$$

$$\frac{d^n y}{dx^n} = p \frac{d}{dy}\left(\cdots \frac{d}{dy}\left(p \frac{dp}{dy}\right)\right)$$

and therefore that the equation can be reduced to an equation of order $n - 1$.

47. Using the method of Exercise 46, reduce the equation

$$\frac{d^3 y}{dx^3} = \frac{(d^2 y/dx^2)^2}{dy/dx}$$

and thus find the solutions $y = ae^{bx} + c$ where a, b, c are arbitrary constants. Verify the validity of the procedure in this case by showing that the relation between y and p is one-to-one.

Chapter 2

Applications of first order differential equations

2.1 Introduction

In Chapter 1 we developed methods for solving various types of first order differential equations. Chapter 2 is devoted to illustrating the use of these techniques in applications encountered in the study of the biological, social, and physical sciences.

Although this chapter may be omitted without disrupting the development of the theory contained in the text, the examples are included for the interest of the applications-oriented reader. In each example, we have assumed a previous exposure to the basic concepts involved.

2.2 Elementary mechanics

In the next few sections we shall apply our new-found skill in solving differential equations to certain scientific problems. First, we discuss some problems of the motion of particles. The basic equation of motion for a particle moving along a straight line (rectilinear motion) is

$$m\frac{d^2x}{dt^2} = F \tag{1}$$

where m is the mass of the particle, $x(t)$ is the position of the particle at time t, and F is the applied force. This is known as *Newton's second law*.

Example 1 Fall in a resisting medium Consider a particle of mass m falling under the influence of gravity in air, water, or some other medium. Such a medium offers resistance to the fall. There are then two external forces, the force of magnitude mg due to gravity, and a resistive force. We shall assume that this resistive force depends in some way on the velocity, that is, that it is some function of velocity v. If we denote this force by $h(v)$, we can write Newton's law in the form

$$m \frac{d^2x}{dt^2} = mg + h(v)$$

For example, suppose that the resistance is simply proportional to the velocity. Taking the positive direction downward (so that the gravitational force is positive), we then have $h(v) = -kv \equiv -k(dx/dt)$ since the resistive force will be upward (negative) if $v > 0$ and downward if $v < 0$. The differential equation of motion is therefore

$$m \frac{d^2x}{dt^2} = mg - k \frac{dx}{dt}$$

where k is some positive constant depending on the medium. Although this is a second order equation for x, it is a first order equation for $v = dx/dt$. This can be written

$$\frac{dv}{dt} + \frac{k}{m} v = g \tag{2}$$

This equation is linear and has integrating factor $e^{kt/m}$. Using the method of Section 1.5 we obtain

$$\frac{d}{dt}(ve^{kt/m}) = ge^{kt/m}$$

$$ve^{kt/m} = c_1 + \frac{mg}{k} e^{kt/m}$$

$$v = \frac{mg}{k} + c_1 e^{-kt/m} \tag{3}$$

where c_1 is an arbitrary constant. Since $v = dx/dt$, another integration gives

$$x = c_2 + \frac{mg}{k} t - c_1 \frac{m}{k} e^{-kt/m} \tag{4}$$

The constants c_1 and c_2 can be evaluated if appropriate initial conditions are given. For example, suppose the particle starts at time $t = 0$ at position $x = 0$ (a zero reference level) with velocity $v = v_0$. Then from Eq. (3) we obtain $c_1 = v_0 - mg/k$ and from Eq. (4) we obtain $c_2 = c_1 m/k = m(v_0 - mg/k)/k$. Hence

$$v = \frac{mg}{k} + \left(v_0 - \frac{mg}{k} \right) e^{-kt/m}$$

$$x = \frac{m}{k} \left(v_0 - \frac{mg}{k} \right) + \frac{mg}{k} t - \frac{m}{k} \left(v_0 - \frac{mg}{k} \right) e^{-kt/m}$$

From these formulas the position and velocity at any time $t > 0$ can be found. Moreover, certain broad characteristics of the motion are apparent. For example,

$$\lim_{t \to +\infty} v = \frac{mg}{k}$$

Hence there is a finite limiting speed which the falling particle approaches but does not exceed, and which is independent of the initial velocity.

Other examples of motions of particles in resisting media are given in the exercises.

Example 2   Escape velocity of a rocket   Consider the problem of determining at what velocity a rocket must be fired in order to leave the earth and never return, the *escape velocity*. To simplify this problem we take into account only the force of gravitational attraction of the earth on the rocket, which varies with altitude, and neglect such factors as air resistance. According to Newton's law of gravitation, this force is

$$F = -\frac{kmM}{x^2} \tag{5}$$

where m is the mass of the rocket, M is the mass of the earth, x is the distance from the center of the earth to the rocket (measured positively upward), and k is a constant. If this is the only force on the rocket, Newton's second law of motion becomes

$$m\frac{d^2x}{dt^2} = -\frac{kmM}{x^2}$$

or

$$\frac{d^2x}{dt^2} = -\frac{kM}{x^2} \tag{6}$$

If distance x is measured from the center of the earth, the initial conditions are

$$x = R, \quad \frac{dx}{dt} = v_0 \quad \text{at } t = 0 \tag{7}$$

where R is the radius of the earth. In this formulation we are assuming that the rocket is given an initial upward velocity v_0 at time zero and that there is no further burn (accelerative thrust).

Since $d^2x/dt^2 = -g$ at $x = R$, it follows from Eq. (6) that $k = R^2g/M$. Substituting for k in Eq. (6) yields

$$\frac{d^2x}{dt^2} = -\frac{R^2g}{x^2} \tag{8}$$

as the basic differential equation. To solve this second order equation, we use the substitution explained in Example 3 of Section 1.8. That is, we let

$$v = \frac{dx}{dt}, \qquad \frac{d^2x}{dt^2} = v\frac{dv}{dx} \tag{9}$$

Equation (8) reduces to

$$v\frac{dv}{dx} = -\frac{R^2 g}{x^2} \tag{10}$$

which is a first order separable equation. This substitution means that we are considering v to be a function of x rather than of t. As long as the rocket continues upward, that is $v > 0$, there will be a unique v corresponding to each altitude x. In other words, v will be a (single-valued) function of x, and the substitution in Eq. (9) will be legitimate.

Solving Eq. (10) we get

$$\frac{v^2}{2} = c + \frac{R^2 g}{x}$$

$$v = \frac{dx}{dt} = \sqrt{2}\left(c + \frac{R^2 g}{x}\right)^{1/2} \tag{11}$$

The positive square root has been chosen since we want $v > 0$. Let H be the maximum altitude attained by the rocket. Then $v = 0$ when $x = H$. Substituting these values in Eq. (11) we get $c = -R^2 g/H$, hence

$$v(x) = R\sqrt{2g}\left(\frac{1}{x} - \frac{1}{H}\right)^{1/2}$$

The velocity at the earth's surface is then

$$v(R) = R\sqrt{2g}\left(\frac{1}{R} - \frac{1}{H}\right)^{1/2} \tag{12}$$

Since Eq. (12) gives the initial velocity required to reach height H, the escape velocity is

$$\lim_{H \to \infty} v(R) = \sqrt{2gR}$$

This velocity is approximately 6.9 miles per second.

Example 3 Motion along a curve Suppose that a particle of mass m is constrained to move along a fixed plane curve (or space curve) under the action of an external force. The position of the particle at time t can be indicated by $s = s(t)$, the length of the arc measured from a fixed point 0 on the curve (see

Figure 2.1). Newton's laws of motion are still valid in this case, if appropriately interpreted. The appropriate interpretation of the second law is that the mass times the component of acceleration in any direction must equal the component of the external force in that direction. In particular, this must be true for the direction tangent to the curve on which the motion takes place. Thus

$$ma_T = F_T \tag{13}$$

where a_T is the tangential component of acceleration and F_T is the tangential component of force. Since $a_T = d^2s/dt^2$, as shown in calculus,[1] this equation becomes

$$m\frac{d^2s}{dt^2} = F_T \tag{14}$$

As an illustration of this general situation, consider the motion of a simple pendulum of length L (see Figure 2.2). The pendulum bob is considered to be a

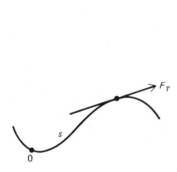

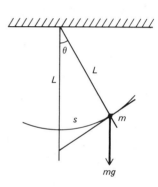

FIGURE 2.1   **FIGURE 2.2**

particle of mass m which moves along a circle of radius L. The arc length s is measured from the lowest point on the circle. The only external force is the gravitational force downward which has tangential component $F_T = -mg \sin \theta$ (negative since the acceleration must be negative when $s > 0$). Since $s = L\theta$, we have $d^2s/dt^2 = L\, d^2\theta/dt^2$ and the equation of motion is

$$mL\frac{d^2\theta}{dt^2} = -mg \sin \theta$$

or

$$\frac{d^2\theta}{dt^2} + \frac{g}{L}\sin \theta = 0 \tag{15}$$

The solution of this equation is carried out in the exercises.

---

[1] George B. Thomas, *Calculus and Analytic Geometry*, 4th edition, Reading, Mass., Addison-Wesley, 1968, p. 479.

Exercises

1. Find the velocity of a particle falling in a resisting medium in which the resistance is proportional to the square of the velocity. Show that the limiting velocity is $\sqrt{mg/k}$.

2. A body weighing 16 lb is dropped from rest through a liquid which offers resistance proportional to the velocity. If the limiting velocity is 1 ft per second, how long does it take the body to reach half its limiting speed?

3. Repeat Exercise 2 if the resistance is proportional to the square of the velocity.

4. A particle of mass m starts with initial velocity v_0 through a medium offering resistance proportional to v^n. No other force acts on the particle. Derive equations for v and x as functions of t.

5. A particle of mass m is projected vertically upward with initial velocity v_0. If it is acted on by the gravitational force mg and resistance kv, find the maximum height reached and the time required to reach that height.

6. A rocket is projected vertically upward from the earth's surface. The initial velocity is zero, but there is a thrust (force) of constant magnitude pm which is maintained until the rocket is a distance R from the earth's surface, where R is the radius of the earth. What is the smallest value of p for which the rocket will escape the earth's gravitation?

7. In the equation for the pendulum let $\omega = d\theta/dt$ and $d^2\theta/dt^2 = \omega\, d\omega/d\theta$. Show that the equation separates and solve for ω as a function of θ.

8. The result of Exercise 7 can be expressed as a first order differential equation in θ and t. Show that this equation separates and thus show that

$$t - t_0 = \int_{\theta_0}^{\theta} \frac{d\alpha}{\sqrt{2}[c + (g/L)\cos\alpha]^{1/2}}$$

The evaluation of this integral involves elliptic integrals.[2]

9. Assume that Lieutenant Van Kuran jumps from an airplane at an altitude of 2 km and wants to open his parachute at an altitude of 0.5 km. If the air resistance is proportional to the square of velocity, and if his terminal velocity would be $\sqrt{mg/k} = 100$ meters per second, then in how many seconds should he pull his rip cord?

---

[2] See *Handbook of Mathematical Functions*, M. Abramovitz and I. A. Stegun (eds.). U.S. Dept. of Commerce, National Bureau of Standards, Applied Mathematics Series Vol. 55. chapter on elliptic integrals.

2.3   Economic growth model

Theories of the growth of the national economy are fruitful areas for the application of differential and difference equations.[3] To illustrate how these equations arise from models of the economy we treat a model proposed by Domar[4] concerned with the growth required for the maintenance of full employment.

In order to attack the problem we make certain simplifying assumptions. While these assumptions will lead us to a model which is only an approximation of reality, the model will still be of value in lending insight into the behavior of the economy.

First, consider the economy from the point of view of the business sector where it is assumed the national income $Y = C + I$, where C is consumption of goods during a unit period of time, and I is the investment during a unit period of time by business in capital equipment, plants, inventories, and so on, for future consumption.

Next, assume there is a measurable (in dollars) quantity F which is the productive potential of the economy in the period of interest. Productive potential refers to the total amount of goods and services which could be produced if all the capital equipment and the total labor force available during the period were utilized—that is, if full employment were obtained. Hence at full employment the relationship $Y = F$ must hold.

Consider the effect of an increase ΔY in national income. A portion $a \, \Delta Y = \Delta I$ is invested by business and the remaining $(1 - a) \, \Delta Y$ is consumed. We assume the parameter a, the marginal propensity to invest by business, is constant. Hence we arrive at the equation

$$\frac{dY}{dt} = \frac{1}{a} \frac{dI}{dt} \tag{1}$$

Next, consider the relationship between the amount invested and the productive capacity. This relationship can be expressed as $\Delta F = \sigma I \, \Delta t$, where σ is a positive scalar indicating that the increase in productivity is proportional to the amount invested. From the above expression we obtain

$$\frac{dF}{dt} = \sigma I \tag{2}$$

Let us now assume an equilibrium condition exists at time $t = 0$ with $Y(0) = F(0)$, with investments prior to time zero of $I(0)$. We wish to obtain an

---
[3] Difference equations will be discussed in Chapter 13.

[4] E. Domar, "Capital Expansion, rate of growth and employment," *Econometrica* **14**, 137–147 (1946).

investment rate which will maintain this equilibrium. In other words, we wish to keep $dY/dt = dF/dt$. Substituting from Eqs. (1) and (2) we arrive at

$$\frac{1}{a}\frac{dI}{dt} = \sigma I$$

The solution to this differential equation is

$$I(t) = I(0)e^{a\sigma t}$$

In other words, investment must grow at an exponential rate in order to maintain full employment.

Exercises

1.   Suppose we generalize the Domar model by assuming $dF/dt = f(I)$. Determine the investment rate required to maintain full employment for $f(I) = e^{cI}$.

2.   The Domar model ignores the effect on F of population growth. If we assume an exponential population explosion in the work force of the form $L_0 e^{\lambda t}$, then $dF/dt = \sigma I + \lambda L_0 e^{\lambda t}$. Assuming equilibrium as in the Domar model, determine the required rate of investment to maintain full employment.

2.4   The oxygen debt

Differential equations often arise in models in the biological sciences. In this text we cannot begin to indicate the diversity of applications or the scope (and limitations) of mathematical methods as applied in biology and medicine, but we will present here one example in the hope that this will convey some of the spirit of what can be done. For further applications to biology, see the references listed at the end of the chapter.

During exercise, the human body (or other animal body) is able to incur an "oxygen debt." That is, the muscles can, for a time, use more oxygen than is being supplied to them. When exercise stops, this debt must be made up in order to replenish the stores of energy in the body that have been depleted during exercise. We propose to find an equation relating the amount of oxygen debt, the amount of work done, and the oxygen supply. The problem is an important one in practice, because the variation of oxygen debt when a patient exercises is sometimes used as an indication of the condition of his heart. We now define the following functions:

$x(t)$ = amount of oxygen debt existing at time t
$w(t)$ = amount of work done up to time t
$s(t)$ = amount of oxygen supplied to the body up to time t

Here "work" is intended in the precise sense defined in physics, but readers unfamiliar with the definition can think of work simply as a measure of the amount of exercise performed.

In a time interval of length Δt, say from t to $t + \Delta t$, let us suppose that a patient does an amount of work Δw, and that an amount of oxygen Δs is supplied. Then the oxygen debt will be decreased by Δs, and increased by an amount depending in some way on Δw. Let us make the simple assumption that this increase in oxygen debt is directly proportional to Δw and let k denote the proportionality constant. Then the net change Δx in oxygen debt in time Δt is

$$\Delta x = k \, \Delta w - \Delta s \tag{1}$$

If we divide this equation by Δt and pass to the limit as $\Delta t \to 0$, we obtain

$$\frac{dx}{dt} = k \frac{dw}{dt} - \frac{ds}{dt} \tag{2}$$

In physics the time rate of change of work is called *power* and denoted by P. Thus we can write

$$P(t) = \frac{dw}{dt} \tag{3}$$

and Eq. (2) becomes

$$\frac{dx}{dt} = kP(t) - \frac{ds}{dt} \tag{4}$$

In an experiment, the power $P(t)$ is measurable. For example, a patient may be asked to do exercise on an exerciser which is arranged so that the amount of power he is expending is directly measured. However, Eq. (4) still seems to contain two unknowns, $x(t)$ and $s(t)$. Let us now make a simple assumption regarding the amount $s(t)$ of "oxygen uptake," namely, that the amount Δs supplied in a short time Δt is proportional to the oxygen debt. As a first approximation this seems reasonable since the greater the depletion of energy the harder or more frequently the subject will breathe. Thus $\Delta s = cx(t) \, \Delta t$. Passing to the limit, we obtain

$$\frac{ds}{dt} = cx(t) \tag{5}$$

Equation (5) states that the rate at which oxygen is taken up is proportional to the deficiency of oxygen.

When ds/dt in Eq. (4) is replaced by cx, we get the equation

$$\frac{dx}{dt} + cx = kP \tag{6}$$

This is a first order linear differential equation. Given a knowledge of the power being expended, we can from Eq. (6) determine the amount of the oxygen debt at any time. Indeed, let us use the method of Section 1.5. The function e^{ct} is an integrating factor, and from Eq. (6) we have

$$\frac{d}{dt}(xe^{ct}) = kPe^{ct}$$

Integration yields

$$xe^{ct} = x_0 + k \int_0^t P(u)e^{cu}\, du$$

where $x_0 = x(0)$ is the amount of oxygen debt existing at zero time. Finally,

$$x(t) = x_0 e^{-ct} + ke^{-ct} \int_0^t P(u)e^{cu}\, du \qquad (7)$$

Equation (7) can now be used to determine $x(t)$ for various functions $P(t)$. The simplest case occurs when $P(t) \equiv 0$, yielding $x(t) = x_0 e^{-ct}$. This shows that after exercise is ended, the oxygen debt is replenished exponentially, if the foregoing theory is correct. It is beyond the scope of this text to discuss experimental procedures for testing the validity of the theory, but seemingly there is some confirmatory evidence. Of course, the reader must bear in mind that the theory we have given may well be oversimplified. For example, the hypothesis of proportionality in Eqs. (1) and (5) will not be strictly correct at all times, and there are various other factors which have not been taken into account (for example, time lags in carrying oxygen in the bloodstream). In more complicated mathematical models, many of these factors could be taken into account. Nevertheless, it is important for the student to realize that even very simple models may sometimes give insight into the processes under investigation, or may help the researcher to clarify his thinking about those processes.

The system described above is a simple example of a self-regulating or self-controlled process. That is, when an oxygen debt occurs, the body increases its rate of uptake to try to pay this debt. When the debt nears zero, the rate of uptake is lowered in turn. In this way, the body maintains a reasonably stable state. Figure 2.3 is a so-called *block diagram* to illustrate the process. The boxes or

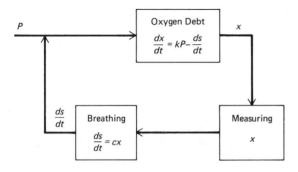

FIGURE 2.3

blocks are supposed to represent functions rather than physical locations or organs in the body. The diagram shows that when power P is expended, a debt x is created. This is somehow measured internally by the system, and a compensating uptake ds/dt occurs by breathing. P and ds/dt then interact to change the debt according to the law $dx/dt = kP - ds/dt$. Note that as one follows around the arrows in the figure, there is a closed loop.

The theory of regulated or self-regulating systems is called *control theory*. This is an important and growing subject of study in mathematics, engineering, biology, and economics.

Exercises

1. Suppose that a patient exerts constant power P for $0 \leq t \leq t_0$, and then rests. Find $x(t)$, and sketch a graph.

2. In reality, the oxygen debt $x(t)$ cannot be directly measured, but the uptake ds/dt can be. What kind of graph of ds/dt versus t should be observed in the resting phase?

3. Suppose that a patient is engaged in cyclical exercise so that $P(t) = B + A \sin \omega t$, say, where $B > A$. If $x_0 = 0$, find x and ds/dt.

4. Suppose it is assumed that ds/dt is proportional to x^β for some constant β not equal to 1. What equation replaces Eq. (6)? Solve this when $P(t) \equiv 0$.

5. Suppose the original model is modified by assuming that there is a time lag of T seconds (T being a constant) in communicating the fact of an oxygen debt and so increasing the breathing rate. Then one might assume that $ds/dt = cx(t - T)$. Find the equation which replaces Eq. (6) under this assumption. This equation is called a differential-difference equation. See Chapter 13 for a brief discussion of these equations.

2.5 Electric circuits

The study of the fluctuations of voltage and current in an electric circuit provides an interesting and useful application of differential equations. The basic components of an electric circuit with which we will work are the following.

1. A voltage source, represented diagrammatically by $\bigodot$, which generates a voltage rise $E(t)$ at time t.

2. A resistor, represented as $-\!\!\wedge\!\!\wedge\!\!\wedge\!-$, which creates a voltage drop $v_R(t) = Ri(t)$ where $i(t)$ is the current flowing into the resistor and R is a

parameter (constant) characteristic of the particular resistor and called the *resistance*.

3. An inductor, represented by $\overset{L}{\curvearrowright\!\!\!\!000\!\!\!\curvearrowleft}$, which creates a voltage drop $v_L(t) = L\, di(t)/dt$, where L is a parameter characteristic of the particular inductor, called the *inductance*.

4. A capacitor, represented by $\overset{C}{-\!|\,|\!-}$, which creates a voltage drop $v_C(t) = C^{-1} \int i(t)\, dt$ where C is a parameter characteristic of the particular capacitor called the *capacity* or *capacitance*.

The last relation can be written conveniently in terms of the *charge $q(t)$* on the capacitor. If q_0 represents the charge at time t_0, then

$$q(t) = q_0 + \int_{t_0}^{t} i(s)\, ds \tag{1}$$

We can now write

$$v_C(t) = \frac{1}{C} q(t), \qquad v_R(t) = R\dot{q}(t), \qquad v_L(t) = L\ddot{q}(t) \tag{2}$$

where the dots represent differentiation with respect to t. We leave to physics texts a discussion of appropriate units in which to measure these quantities.

The main problem we shall consider is that of determining the current $i(t)$ in a circuit when R, L, C and the values of q_0 and i_0 are known. The fundamental physical laws which enable us to write equations for the determination of $i(t)$ are called Kirchhoff's laws. At the moment we need only *Kirchhoff's second law: For any closed electric circuit, the sum of the voltage rises and drops must equal zero.* For example, consider the circuit represented in Figure 2.4. The

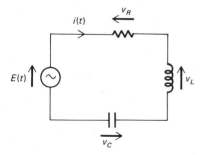

FIGURE 2.4

figure indicates that a generator, resistor, inductor, and capacitor are to be connected *in series* to one another, forming a closed loop. The arrows indicate the direction of rise of voltage across the components. The connecting wires are assumed to have negligible resistance, inductance, and capacitance.[5] For this circuit, by Kirchhoff's law

$$E(t) = v_R(t) + v_L(t) + v_C(t)$$

or

$$E(t) = L\ddot{q} + R\dot{q} + \frac{1}{C} q \tag{3}$$

If $E(t)$ is given, Eq. (3) is a second order differential equation for the determination of q. If q is found, then i can also be found since $i = dq/dt$ by Eq. (1).

An analogy exists between these equations and those describing the motion of a spring. Consider a spring to which is attached a mass m moving in a resisting medium (see Figure 2.5). The displacement x of the mass from its equilibrium position is governed by the differential equation

$$m\ddot{x} + b\dot{x} + kx = F(t) \tag{4}$$

where b is a constant associated with the resistive force, which we assume is proportional to $\dot{x}$, k is a constant associated with the elastic properties of the spring, and $F(t)$ is an externally applied force. Comparing Eqs. (3) and (4) we see that the voltage source creates an electrical "force" equivalent to the force applied on the mass m, the inductor plays the role of the mass, the resistor the role of the frictional material, and the capacitor the role of the spring.

To illustrate the application of differential equations to circuit theory consider the circuit of Figure 2.6.

FIGURE 2.5

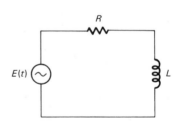

FIGURE 2.6

By Kirchhoff's second law, we obtain

$$E(t) = L\ddot{q} + R\dot{q}$$

---

[5] These are called *lumped parameter circuits* in electric circuit theory.

Although this is a second order equation for q, it is a first order equation for $i = \dot{q}$, namely

$$\frac{di(t)}{dt} + \frac{R}{L}\, i(t) = \frac{E(t)}{L}$$

This is a nonhomogeneous first order linear differential equation. From Section 1.5 we obtain the family of solutions

$$i(t) = ke^{-(R/L)t} + \frac{1}{L}\, e^{-(R/L)t} \int e^{(R/L)t}\, E(t)\, dt$$

for k an arbitrary constant. It is more convenient here to write this using definite integrals, for example,

$$i(t) = ke^{-(R/L)t} + \frac{1}{L}\, e^{-(R/L)t} \int_0^t e^{(R/L)s}\, E(s)\, ds$$

Then, setting $t = 0$, we see that k is the value of i at time zero. Denoting this by I_0, we finally obtain

$$i(t) = I_0 e^{-(R/L)t} + \frac{1}{L}\, e^{-(R/L)t} \int_0^t e^{(R/L)s}\, E(s)\, ds$$

For example, if the applied voltage $E(t)$ is a constant E_0, as is approximately the case for a battery, we obtain

$$i(t) = I_0 e^{-(R/L)t} + \frac{E_0}{R}\, (1 - e^{-(R/L)t})$$

Since R and L are positive, it follows from this equation that

$$\lim_{t \to +\infty} i(t) = \frac{E_0}{R}$$

Therefore the current flow becomes nearly constant after a lapse of time.

Exercises

1. An *RL* circuit such as in Figure 2.6 has no voltage source but has an initial current $i(0) = I_0$. Find the current at any time and draw a graph of $i(t)$.

2. In Exercise 1, how long does it take the current to decrease to half its initial value?

3. For the *RL* circuit of Figure 2.6, suppose that $I_0 = 0$ and $E(t) = E_0 \cos \omega t$ where E_0 and ω are constants. Show that

$$i(t) = \frac{E_0 L}{R^2 + \omega^2 L^2} \left(\frac{R}{L} \cos \omega t + \omega \sin \omega t - \frac{R}{L}\, e^{-(R/L)t} \right)$$

Observe that

$$\lim_{t \to \infty} [i(t) - i_s(t)] = 0$$

where

$$i_s(t) = \frac{E_0 L}{R^2 + \omega^2 L^2} \left(\frac{R}{L} \cos \omega t + \omega \sin \omega t \right)$$

The function $i_s(t)$ is often called the *steady-state current*.

4. Show that the result in Exercise 3 can be written

$$i_s(t) = A \cos (\omega t - \phi)$$

so that A is the *amplitude* and ϕ the *phase shift* of the periodic oscillation. Find A and ϕ in terms of E_0, R, L, and ω.

5. An *RL* circuit contains a battery and a switch in series as shown in Figure 2.7. The switch is in position A for $0 \le t < t_1$ and in position B for $t \ge t_1$.

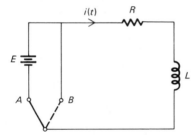

FIGURE 2.7

Assume that $i(0) = 0$ and that $i(t)$ is continuous for $t \ge 0$ even though the voltage is discontinuous at $t = t_1$.

(a) Without solving the differential equation, sketch a graph of $i(t)$.

(b) Solve analytically for $i(t)$ for the two intervals $0 \le t \le t_1$ and $t \ge t_1$.

6. An *RC* circuit containing a resistor, a capacitor, and a source of voltage in series is described by the differential equation

$$R \frac{dq}{dt} + \frac{1}{C} q = E$$

(a) Find q if $E(t) = E_0$ and $q = q_0$ at $t = 0$.

(b) Find q if $E(t) = E_0 \cos \omega t$ and $q = q_0$ at $t = 0$.

COMMENTARY

Books which emphasize the use of differential equations and linear algebra in the social sciences include:

GALE, D., *The Theory of Linear Economic Models*, New York, McGraw-Hill, 1960.

HORST, P., *Matrix Algebra for Social Scientists*, New York, Holt, Rinehart and Winston, 1963.

JOHNSTON, J., G. PRICE, and F. VAN VLECK, *Linear Equations and Matrices*, Reading, Mass., Addison-Wesley, 1966.

KEMENY, J. G., A. Schleifer, J. L. Snell, and G. L. Thompson, *Finite Mathematics with Business Applications*, 2d ed., Englewood Cliffs, New Jersey, Prentice-Hall, 1972.

KEMENY, J. G., and J. L. SNELL, *Mathematical Models in the Social Sciences*, Boston, Ginn, 1962.

Books which treat the applications of differential equations and linear algebra to the physical sciences include:

ARFKEN, G., *Mathematical Methods for Physicists*, 2d ed., New York, Academic Press, 1971.

MARGENAU, H., *The Mathematics of Physics and Chemistry*, New York, Van Nostrand–Reinhold, 1943.

MATTHEWS, J., and R. L. WALKER, *Mathematical Methods of Physics*, 2d ed., New York, Benjamin, 1970.

For a number of examples of the use of mathematics in biology and medicine, see:

DEFARES, J. G., and I. N. SNEDDON, *The Mathematics of Medicine and Biology*, Amsterdam, North-Holland, 1961.

THRALL, R. M., J. A. MORTIMER, K. R. REBMAN, R. F. BAUM (eds.), *Some Mathematical Models in Biology*, rev. ed., Ann Arbor, University of Michigan, 1967.

Among many books on control theory, we mention the following:

BELLMAN, R., *Some Vistas of Modern Mathematics*, Lexington, University of Kentucky Press, 1968.

BELLMAN, R., *Introduction to the Mathematical Theory of Control Processes*, New York, Academic Press, 1967.

MILSUM, J. H., *Biological Control Systems*, New York, McGraw-Hill, 1966.

Chapter 3

Existence and uniqueness theorems

3.1   Introduction

Chapter 1 was devoted to a discussion of several types of first order ordinary differential equations. The concepts of *solution*, *one parameter family of solutions*, and *initial condition* were introduced. For each type of equation, a technique was sought for generating a one parameter family of solutions. In this chapter a deeper and more inclusive study of the existence and number of solutions of differential equations will be undertaken. The objective is not to present new methods for finding solutions of particular types of equations, and indeed no such methods are given. Rather it is to see what kind of answers can be given to such questions as these: Given any differential equation, is there a way to tell whether it possesses a solution? How many solutions can it have? If an initial condition is given, how many solutions are there? If all solutions of an equation are graphed, what general properties can be discerned?

The answers to the above questions have important implications for the applicability of differential equations to science. One of the principal cornerstones of science is the principle of *determinism*, which we may take to be the principle that a system evolves from its present state in a regular and predictable way, given that we know the influences operating on it. For example, in the circuit theory problem of Section 2.5, the current at any time is believed to have a unique predictable value, depending on the types of the resistor, inductor, capacitor, and voltage source.

How is this principle expressed in the equations of the *mathematical model* which we employ as a description of the physical reality? The answer is that if we are given the differential or other types of equations which we take as descriptive of the system, and if we are given the condition or state of the system at some initial time, then the solution will be uniquely determined at future times. That is to say, if our model is to be of any value, there must be a solution, and just one solution, for any prediction to be possible. In addition, for the model to be considered satisfactory, the theoretical predictions must be in agreement with empirical observations of the physical system.

Thus we are led to establish these criteria which must normally be expected from a satisfactory model.

1. There must be solutions of the equations.
2. There must be exactly one solution corresponding to the specification of a state or initial condition.
3. The theoretical predictions based on the model must be reasonably in accord with empirical evidence.

In this chapter we shall present theorems that will provide us with suitable results in terms of (1) and (2). For example, for a model in which the differential equation has the form $y' = f(x, y)$, where f is continuous, (1) is always satisfied, and if in addition f has a continuous partial derivative, (2) is always satisfied. In these cases, then, the mathematical structure is in precise agreement with the physical notion of determinism. For any given model, (3) must be individually checked; no universally valid result can be given.

3.2 Review of previous results

For the types of equations discussed in Chapter 1, the questions of existence and characterization of solutions have been partly answered. For an equation $M(x)\, dx + N(y)\, dy = 0$ it was proven that if $F(x)$ is an indefinite integral of $M(x)$, and $G(y)$ is an indefinite integral of $N(y)$, then any solution $y(x)$ of the differential equation must satisfy the equation $F(x) + G(y) = c$ for some constant c. Conversely, if for a given c the equation $F(x) + G(y) = c$ determines a differentiable function $y(x)$ on $a < x < b$, then $y(x)$ is a solution of the differential equation for $a < x < b$. However, this result is not satisfactory unless it is known that $F(x)$ and $G(y)$ exist. Fortunately, one of the important theorems from calculus can be used here.

THEOREM 3.1 If $f(x)$ is continuous on an interval $a < x < b$, then there exists a function $\phi(x)$ on $a < x < b$ such that $\phi'(x) = f(x)$ on $a < x < b$.

In other words, every continuous function has an indefinite integral. It should be carefully noted that this theorem does not claim that the indefinite integral can be handily written down in an explicit formula. For example, $\phi(x) = \int (e^{-x}/x)\, dx$ cannot be expressed by elementary functions. Nevertheless, this function $\phi(x)$ exists, is well-defined, and its values can be computed with arbitrary accuracy (and are in fact tabulated in some books of tables) on intervals not containing $x = 0$.

Combining Theorem 3.1 with our previous theorem on the separable differential equation we have this corollary.

COROLLARY . Suppose that $M(x)$ is continuous for $a < x < b$ and $N(y)$ is continuous for $c < y < d$. Then there exist functions $F(x)$ for $a < x < b$ and $G(y)$ for $c < y < d$ such that any solution $y(x)$ of $M\,dx + N\,dy = 0$ must satisfy $F(x) + G(y) = c$ for some constant c.

Let us now turn to the consideration of first order linear differential equations.

The general form of such an equation is $u(x)y' + v(x)y = w(x)$. However, in discussing this equation we considered solutions only on intervals $a < x < b$ for which $u(x) \neq 0$, where the equation was rewritten as $y' + p(x)y = q(x)$.

An examination of Section 1.5 shows that if p and q are integrable functions, there is a one parameter family of solutions given by Eq. (9) of that section. Moreover, the method of deriving the formula shows that every solution is of this form. See Theorem 3.5. In those cases where $u(x) = 0$ at isolated values of x, we can ask whether solutions on the intervals between these points can be pieced together to form a solution over the real line.

Finally, consider the discussion of exact differential equations of the form $M(x, y)\,dx + N(x, y)\,dy = 0$. Since the equation is exact in $R = \{a < x < b, c < y < d\}$ there exists an $F(x, y)$ such that $\partial F/\partial x = M$ and $\partial F/\partial y = N$ on R. Theorem 1.2 states that any differentiable function $y(x)$ whose graph is contained in R and which satisfies $F(x, y(x)) = c$ is a solution of the exact equation. Further, every solution $y(x)$ whose graph lies in R must satisfy $F(x, y) = c$ for some c. Under what conditions will the equation $F(x, y) = c$ determine a differentiable function $y(x)$? Also, for a given c, can there be more than one solution?

Going beyond the forms of differential equations so far considered, can one ask similar questions for the broadest possible class of equations? As far as first order equations are concerned, we mean by this class those equations of the form $y' = f(x, y)$, where f is a member of the set of all functions of two variables, or more generally still the class of equations $g(x, y, y') = 0$, where g is a member of the set of all functions of three variables. However, it is easy to see that we cannot expect all such equations to possess a family of solutions, or even one solution. For instance, the equation $y' = \sqrt{-y^2}$, corresponding to $f(x, y) = \sqrt{-y^2}$, has only one real solution, the function $y(x) \equiv 0$, and the equation $y' = (-y^2 - 1)^{1/2}$, corresponding to $f(x, y) = (-y^2 - 1)^{1/2}$, has no real solution.

It would clearly be highly desirable to establish theorems, for as wide a class of equations as possible, which would answer the questions of existence and number of solutions. The purpose of this chapter is to state and illustrate two theorems of this kind.

3.3   Existence and uniqueness theorems

We shall now state two fundamental theorems which provide answers to the questions of existence and number of solutions of many first order differential equations. The proofs of these theorems depend on rather difficult mathematical reasoning and are postponed to the end of the chapter. The first theorem is as follows.

THEOREM 3.2   Suppose that the function $f(x, y)$ is defined, real, and continuous for (x, y) in the rectangle R defined by the inequalities $a \leq x \leq b$, $c \leq y \leq d$, and that (x_0, y_0) is a point in the interior of R. Then the initial value problem

$$y' = f(x, y), \qquad y(x_0) = y_0 \qquad (1)$$

has at least one solution $y = y(x)$. Further, the graph of the solution exists up to the boundary of R.

There are several noteworthy features of this theorem. First, it applies to every first order equation which can be expressed as in Eq. (1), provided only that f is a continuous real function in some rectangle containing the initial point (x_0, y_0). Second, it asserts the existence of at least one solution of the initial value problem but says nothing about the number of solutions. Third, it asserts that the graph of a solution, the so-called *solution curve*, which necessarily passes through the initial point (x_0, y_0), cannot stop suddenly in the interior of R. However, note that the solution $y(x)$ need not be defined on the whole interval $a \leq x \leq b$ since the graph may pass out of R over a horizontal boundary. (See Figure 3.1.) Of course, the theorem can make no assertion about the graph of $y(x)$ after it leaves R, not even to assert the existence of the graph, since the hypothesis contains no information whatsoever regarding $f(x, y)$ outside R.

Theorem 3.3 gives conditions sufficient to guarantee that the initial value problem has a unique solution, that is, exactly one solution.

THEOREM 3.3   If $f(x, y)$ and $\partial f(x, y)/\partial y$ are continuous in the rectangle R defined by $a \leq x \leq b, c \leq y \leq d$, and if (x_0, y_0) is an interior point of R, then the initial value problem $y' = f(x, y), y(x_0) = y_0$, has a unique solution. The graph of the solution exists up to the boundary of R.

The hypothesis of this theorem requires more of the differential equation, namely, that the partial derivative $\partial f/\partial y$ exist and be continuous in R. This

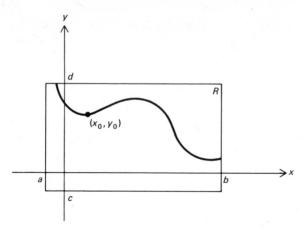

FIGURE 3.1

strengthening of the hypothesis makes it possible to give the stronger conclusion—that through a given initial point in R there passes exactly one solution curve.

Exercises

1. Consider the differential equation $y' = 1 + y^2$. Show that the hypotheses of Theorem 3.3 are satisfied in any rectangle and consequently there is one and only one solution curve through each point (x_0, y_0). Show, however, that no solution exists over an interval of x values of length greater than π.

2. In each of the following find and describe the set of all points (x_0, y_0) such that (x_0, y_0) can be placed in a rectangle R within which the hypotheses of Theorem 3.3 are satisfied. This set is the set of initial points for which Theorem 3.3 guarantees the existence of a unique solution of the initial value problem.

(a) $y' = \dfrac{x}{y}$

(b) $y' = \sqrt{\ln (y/x)}$

(c) $y' = \dfrac{1}{(y - x^2)(x^2 - 1)}$

(d) $y' = \max (x, y)$

(e) $y' = \dfrac{xy}{x^2 + y^2}$

3.4   Nonexistence and nonuniqueness

Although Theorems 3.2 and 3.3 assure the existence and uniqueness of solutions of a very large class of differential equations, it is by no means true that this is assured for every equation. We have already seen the example $y' = (-y^2 - 1)^{1/2}$, which has no solution at all. Here of course, the function $f(x, y) = (-y^2 - 1)^{1/2}$ has imaginary values for all real values of y, and therefore the theorems are not applicable at all.

It is also easy to give an example of a differential equation for which $f(x, y)$ is real-valued and yet there is no solution. For example, consider the equation $y' = f(x)$. All solutions are necessarily integrals of f, $y = \int f(x)\, dx$. Therefore, if f is a nonintegrable function[1] no solution can exist.

It is important to point out that by nonexistence of a solution we do not simply mean that we cannot find a convenient formula for the solution. For example, consider the initial value problem

$$(5y^4 - x)\, dy = (y - 3x^2)\, dx, \qquad y = 1 \quad \text{at} \quad x = 1 \tag{1}$$

This equation is exact, and the method of Chapter 1 yields the relation

$$y^5 - xy + x^3 = c$$

for the solutions. The initial condition implies $c = 1$, so that

$$y^5 - xy + x^3 = 1 \tag{2}$$

Now there is no algebraic procedure known by which this equation can be solved for $y = \phi(x)$, where $\phi(x)$ is given by an algebraic combination of elementary functions. However, the function

$$f(x, y) = \frac{y - 3x^2}{5y^4 - x}$$

and its first partial derivatives are continuous in some neighborhood of the point $(1, 1)$, and in fact in any domain which excludes the curve $5y^4 = x$. Therefore, Theorem 3.3 guarantees the existence of a unique solution of Eq. (1) through the point $(1, 1)$. The point of this example is that existence of a solution is not the same as existence of an explicit algebraic formula for the solution in the form $y = \phi(x)$. Worse yet, there are cases (in fact, it is usually the case) in which Theorem 3.3 guarantees the existence of a unique solution and yet there is not even an implicit relation such as Eq. (2) which we can find. In these cases, various approximation techniques must be used, as we shall later describe.

To illustrate the difference between Theorems 3.2 and 3.3, consider the initial value problem

$$y' = 3y^{2/3}, \qquad y(x_0) = 0 \tag{3}$$

---

[1] It is well known that there are nonintegrable functions. For example, the function defined by

$$f(x) = \begin{cases} 1 & \text{if } x \text{ is a rational number} \\ 0 & \text{if } x \text{ is an irrational number} \end{cases}$$

is not integrable in the sense usually defined in calculus.

It is easily verified that $\phi(x) = (x - x_0)^3$ is a solution. But so is $\phi(x) \equiv 0$. Hence, there are at least two different solutions of the initial value problem. The property of uniqueness of solution fails in this example because the function $f(y) = 3y^{2/3}$ is continuous but its derivative $f'(y) = 2y^{-1/3}$ is discontinuous at $(x_0, 0)$. Thus Theorem 3.2 can be applied, but not Theorem 3.3. Note, however, that in any rectangle $R = \{(x, y): a \le x \le b, 0 < c \le y \le d\}$ or $R = \{(x, y): a \le x \le b, c \le y \le d < 0\}$ both f and f' are continuous. Consequently there must, by Theorem 3.3, be a unique solution through any point (x_0, y_0) in R. This solution is

$$y = \phi(x) = (x - x_0 + y_0^{1/3})^3 \tag{4}$$

A few solutions are shown in Figure 3.2. Note that through each point $(x_0, 0)$ there is a solution curve $y = (x - x_0)^3$, as well as the solution curve $y \equiv 0$.

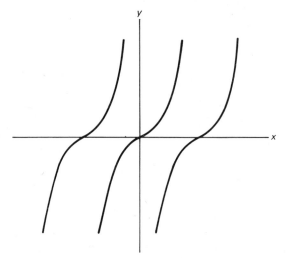

FIGURE 3.2

Exercises

1.   Show that the problem

$$\frac{dy}{dx} = y^{2/5}, \qquad y(0) = 0$$

does not have a unique solution.

2.   Show that the function

$$
\begin{aligned}
y(x) &= x + \tfrac{1}{27}(x + 3)^3, & x &< -3 \\
&= x & -3 &\le x \le 3 \\
&= x + \tfrac{1}{27}(x - 3)^3, & 3 &< x
\end{aligned}
$$

is a solution of the equation

$$\frac{dy}{dx} = 1 + (y - x)^{2/3}$$

Graph this solution.

3.5   Reprise: separable, linear, and exact equations

Let us show that the general existence theorems can be used to answer the questions raised in Section 3.1 concerning existence and uniqueness of solutions of separable, linear, or exact equations. First consider a separable equation

$$M(x)\, dx + N(y)\, dy = 0 \quad \text{or} \quad y' = -\frac{M(x)}{N(y)} \tag{1}$$

The function $f(x, y) = -M(x)/N(y)$ and its first partial derivative $\partial f/\partial y$ are continuous in any region excluding points where $N(y) = 0$, provided $M(x)$, $N(y)$, and $N'(y)$ are continuous. Therefore, given any (x_0, y_0) such that $N(y_0) \neq 0$ we can find a rectangle R containing (x_0, y_0) such that Theorem 3.3 can be applied. We can therefore strengthen Theorem 1.1 to the following form.

THEOREM 3.4   Consider the separable equation $M(x)\, dx + N(y)\, dy = 0$ and assume that $M(x)$, $N(y)$, and $N'(y)$ are continuous. Let (x_0, y_0) be any point such that $N(y_0) \neq 0$, and let R be a rectangle containing (x_0, y_0) such that $N(y) \neq 0$ for (x, y) in R. Then there is a unique solution $y = \phi(x)$ of Eq. (1) such that $\phi(x_0) = y_0$, and the solution curve exists at least up to the boundary of R.

If we combine Theorem 3.4 with the Corollary to Theorem 3.1, we see that we have the following information concerning Eq. (1). Assume that (x_0, y_0) lies in a rectangle R in which $M(x)$, $N(y)$, and $N'(y)$ are continuous and $N(y) \neq 0$. Then there is a unique solution $y(x)$ for which $y(x_0) = y_0$ and $y(x)$ must satisfy the implicit relation $F(x) + G(y) = c$ where $c = F(x_0) + G(y_0)$. In practice this means that we can separate variables and integrate to derive the implicit relation and be sure that the solution of the initial value problem is a function satisfying the relation.

For the linear first order equation

$$y' + p(x)y = q(x) \tag{2}$$

an even simpler result is possible. Using the formula in Section 1.5, we see that

$$y = \phi(x) = \exp\left[-\int_{x_0}^{x} p(s)\, ds\right]\left\{\int_{x_0}^{x} q(r) \exp\left[\int_{x_0}^{r} p(s)\, ds\right] dr + y_0\right\} \tag{3}$$

is a solution of Eq. (2) such that $\phi(x_0) = y_0$. Let us assume that $p(x)$ and $q(x)$ are continuous for $a < x < b$. Then $f(x, y) = q(x) - p(x)y$ and $\partial f/\partial y = -p(x)$ are continuous for $a < x < b$, $-\infty < y < \infty$. Then by Theorem 3.3 there is a unique solution such that the initial condition is satisfied. Thus Eq. (3) gives the one and only solution of the initial value problem. Let us state this as a theorem.

THEOREM 3.5   Consider the linear differential equation $y' + p(x)y = q(x)$ and assume that $p(x)$ and $q(x)$ are continuous for $a < x < b$. Given (x_0, y_0) such that $a < x_0 < b$, there is a unique solution $y = \phi(x)$ satisfying $\phi(x_0) = y_0$, and Eq. (3) gives this solution.

In other words, the method of treating the linear differential equation, as described in Section 1.5, always yields the unique solution of the initial value problem.

Similar considerations can be given for exact equations

$$M(x, y)\,dx + N(x, y)\,dy = 0 \tag{4}$$

However, instead of going into this matter we prefer to point out an interesting geometrical property of exact equations.

Suppose that the function $\phi(x)$ is a solution of an exact Eq. (4) on an interval $a < x < b$. Then as shown by Theorem 1.2, $\phi(x)$ satisfies $F(x, \phi(x)) = c$ for $a < x < b$, where c is a constant. That is, the function $F(x, y)$ takes the same value for every point (x, y) on the graph of $\phi(x)$, $a < x < b$. Therefore if we construct the three-dimensional graph of the surface $z = F(x, y)$, all points (x, y) on the graph of $\phi(x)$ are such that (x, y, c) is on the surface. A curve is called a *level curve* of F if all points above the curve on the surface are at the same height $z = c$. Thus a *solution curve of an exact equation $dF = 0$* is a *level curve of the function F*, or *part of a level curve*.

Example 1   The equation $x\,dx + 2y\,dy = 0$ is exact. We can take $F(x, y) = x^2 + 2y^2$, and all solutions satisfy the relation $x^2 + 2y^2 = c$, so that

$$\phi(x) = \pm\left(\frac{c - x^2}{2}\right)^{1/2}$$

The curves $y = \phi(x)$, or $x^2 + 2y^2 = c$, the level curves, are projections onto the x, y plane of the curves of intersection of the elliptic paraboloid $z = x^2 + 2y^2$ and the plane $z = c$. (See Figures 3.3 and 3.4.) Note that the level curve corresponding to $c = 0$ is in fact merely a point, and that the level curve corresponding to each positive c is an ellipse which has to be described by two solutions $\phi_1(x) = [(c - x^2)/2]^{1/2}$ and $\phi_2(x) = -[(c - x^2)/2]^{1/2}$.

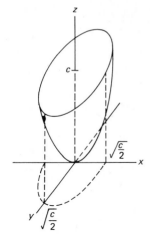

FIGURE 3.3   FIGURE 3.4

Exercises

1.   If $p(x)$ and $q(x)$ are continuous for $-\infty < x < \infty$, show that it follows from Theorem 3.5 that every solution of the linear equation $y' + py = q$ exists for $-\infty < x < \infty$.

2.   Can a nonlinear differential equation have a solution which exists for $-\infty < x < \infty$? Verify your answer.

3.   For each of the following functions $F(x, y)$, plot the family of level curves and find a first order exact differential equation for which the level curves define solutions.

   (a) $x^2 + 4y^2$ (b) $y - x^2$

   (c) $\dfrac{1}{x} \ln \dfrac{4}{y}$

4.   Over what x interval would you expect each of the following initial value problems to have a solution, and why?

   (a) $y' + e^x y = \cos x$ (b) $(x^2 - 1)y' + xy = 1$
   $\qquad\quad y(0) = 3$ $\qquad\qquad\quad y(0) = 3$
   (c) $y' + |x|y = \sqrt{x}$
   $\qquad\quad y(2) = 3$

3.6   Other existence and uniqueness problems

Theorems 3.2 and 3.3 do not always answer the questions of existence or uniqueness which arise. For example, consider the initial value problem

$$xy' = y, \qquad y = y_0 \quad \text{at} \quad x = 0 \tag{1}$$

Write the equation in the form $y' = f(x, y)$ where $f(x, y) = y/x$. Then clearly
in any rectangle R surrounding the initial point $(0, y_0)$ the function f is discontinuous, and the theorems cannot be applied. Does this mean that either existence or uniqueness fails? For this example, it does. There is a family of solutions $y = cx$. Given any point (x_0, y_0) with $x_0 \neq 0$ there is a unique solution
of this family. Clearly there are infinitely many solutions such that $y = 0$ at
$x = 0$. But for any initial point $(0, y_0)$ with $y_0 \neq 0$ there can be no solution,
since if there were its slope at $(0, y_0)$ would be $y_0/0$. Thus the initial value
problem in (1) has no solution if $y_0 \neq 0$ and infinitely many solutions if $y_0 = 0$.
We shall study problems of this kind more thoroughly in Chapter 16.

   In this chapter we have emphasized initial value problems, which are of
fundamental importance for theory and applications. However, various other
problems might be posed. For example, consider the equation $y' + y = 0$
with the auxiliary condition

$$\int_0^1 y(x)\, dx = 2$$

That is, we seek a solution of the equation such that the area under the graph
between $x = 0$ and $x = 1$ is 2. Since every solution of this linear equation has
the form $y = ce^{-x}$, the required value of c is seen to be $2(1 - e^{-1})^{-1}$. Hence
$y = 2(1 - e^{-1})^{-1}e^{-x}$.

   The rest of this chapter is devoted to giving a complete proof of Theorem
3.3.

3.7   The method of successive approximations

In this section we shall introduce one of the fundamental techniques of mathematical analysis, the so-called *method of successive approximations*. This method
will then be used to prove Theorem 3.3. The basic idea of the method is suggested by its name. One defines a sequence of approximations to a desired
quantity by some well-defined procedure. In favorable circumstances, the
sequence will converge to the desired quantity or come sufficiently close to it for
the purpose at hand. Clearly, this principle is so broad and so vague that there
is a possibility of applying it in a very wide range of problems, and this con-
tributes to its great importance.

   Let us now consider the initial value problem

$$y' = f(x, y), \qquad y(x_0) = y_0 \tag{1}$$

We assume $f(x, y)$ is continuous in a rectangle R (with sides parallel to the x and y axes) and that (x_0, y_0) is an interior point of R. We define a sequence of functions $\{y_0, y_1, \ldots\}$ by the equations

$$y_k'(x) = f(x, y_{k-1}(x)) \qquad k = 1, 2, \ldots$$
$$y_k(x_0) = y_0$$
$$y_0(x) = y_0 \qquad\qquad\qquad\qquad (2)$$

Let us explain Eq. (2) more fully. First, the function $y_0(x)$ is chosen to be the constant y_0. Then, $y_1(x)$ is determined by

$$y_1'(x) = f(x, y_0(x)) = f(x, y_0), \qquad y_1(x_0) = y_0$$

This equation determines a function $y_1(x)$ since it specifies its derivative everywhere and its value at one point. In fact,

$$y_1(x) = y_0 + \int_{x_0}^{x} f(s, y_0(s))\, ds$$

In the same way, once $y_1(x)$ has been found, $y_2(x)$ is determined, and so on. If $y_{k-1}(x)$ has been found, then $y_k(x)$ is given by the equation

$$y_k(x) = y_0 + \int_{x_0}^{x} f(s, y_{k-1}(s))\, ds \qquad k = 1, 2, \ldots \qquad (3)$$

Example 1 Consider the initial value problem $y' = y$, $y(0) = 1$. As we know from Chapter 1, the solution is $y = e^x$. Let us see what we get from the method of successive approximations. First, we have $y_0(x) \equiv 1$. We then have

$$y_1'(x) = y_0 = 1, \qquad y_1(x) = y_1(0) + \int_0^x y_0(s)\, ds$$

$$= 1 + \int_0^x ds = 1 + x$$

$$y_2'(x) = y_1(x) = 1 + x, \qquad y_2(x) = y_2(0) + \int_0^x y_1(s)\, ds$$

$$= 1 + \int_0^x (1 + s)\, ds = 1 + x + \frac{x^2}{2}$$

By induction one can easily show that

$$y_k(x) = 1 + x + \frac{x^2}{2!} + \cdots + \frac{x^k}{k!}$$

Clearly,

$$\lim_{k \to \infty} y_k(x) = e^x$$

In this example, the sequence of successive approximations converges to the solution of the initial value problem.

We shall now investigate the convergence of the sequence $\{y_k(x)\}$, in general. Besides assuming that $f(x, y)$ is continuous in the rectangle R, we also assume that the partial derivative $\partial f(x, y)/\partial y$ exists and is continuous in R (as in the hypothesis of Theorem 3.3). It is known from advanced calculus[2] that a continuous function on a closed rectangle has an absolute minimum and maximum thereon. Therefore, there are positive constants M and L such that

$$|f(x, y)| \leq M, \qquad \left|\frac{\partial f(x, y)}{\partial y}\right| \leq L \quad \text{for} \quad (x, y) \quad \text{in} \quad R \qquad (4)$$

Let (x, y) and (x, z) be two points in R. Then

$$f(x, z) - f(x, y) = \int_y^z \frac{\partial f(x, s)}{\partial s}\, ds$$

and so by (4)

$$|f(x, z) - f(x, y)| \leq \left|\int_y^z \frac{\partial f(x, s)}{\partial s}\, ds\right|$$

$$\leq \left|\int_y^z \left|\frac{\partial f(x, s)}{\partial s}\right| ds\right| \leq L|z - y|$$

The last inequality says that f cannot change by more than a fixed constant L times the change in its second variable (when x is held fixed), which is an obvious consequence of the fact that the rate of change $\partial f/\partial y$ is at most L in absolute value.

DEFINITION 3.1   A function f that satisfies an inequality of the form

$$|f(x, z) - f(x, y)| \leq L\,|z - y| \qquad (5)$$

for all (x, y) and (x, z) in a region is said to satisfy a *Lipschitz condition* in the region. The number L is called a *Lipschitz constant* for f in the region.

The argument above shows that if f has a bounded derivative $\partial f/\partial y$ in a rectangle R, then f satisfies a Lipschitz condition in R. However, some functions which do not have a bounded derivative may satisfy a Lipschitz condition. (See the exercises.)

In Example 1, all the approximating functions $y_1(x)$, $y_2(x)$, and so on, were defined for all x. However, in the general case it is conceivable that they would exist over shorter and shorter intervals, so that their limit (if any) could exist only at a single point. Lemma 3.1 shows that this distressing possibility cannot occur under the stated hypotheses. In this lemma, we will consider a rectangle R_0 of dimensions $2a$ and $2b$ placed symmetrically about a point (x_0, y_0).

---

[2] See, for example, Watson Fulks, *Advanced Calculus*, 2d ed., New York, Wiley, 1969, p. 229.

LEMMA 3.1   Let $f(x, y)$ and $\partial f(x, y)/\partial y$ be continuous in a rectangle

$$R_0 = \{(x, y): x_0 - a \le x \le x_0 + a, \qquad y_0 - b \le y \le y_0 + b\}$$

Let M and L be positive numbers for which the inequalities in (4) hold for (x, y) in R_0. Let α be the smaller of the numbers a and b/M. Then the successive approximations defined by Eq. (2) all exist and are continuous on the interval $J = \{x: |x - x_0| \le \alpha\}$ and satisfy

$$|y_k(x) - y_0| \le M\,|x - x_0| \le b \qquad (k = 0, 1, 2, \dots) \tag{6}$$

PROOF: We will first explain the reason for the choice of α in the lemma. We observe that the condition $|f(x, y)| \le M$ in R_0 implies that the slope of $y_k(x)$ is at most M in absolute value. Therefore the graph of $y_k(x)$ must stay between the lines of slope M and $-M$ through (x_0, y_0); see Figure 3.5. This is, in fact,

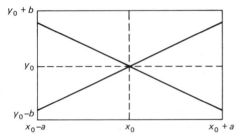

FIGURE 3.5

just the meaning of the inequality in (6). The length of the interval J depends on where these lines meet the boundary of R_0. If they meet the vertical sides as in Figure 3.5, we take $\alpha = a$, and the functions $y_k(x)$ can be defined over the whole width of the rectangle. On the other hand, if they meet the horizontal sides of R_0 as in Figure 3.6, we take $\alpha = b/M$. In this case we can define all $y_k(x)$ at least over $x_0 - \alpha \le x \le x_0 + \alpha$, but possibly no further since the graph of $y_k(x)$ might pass outside R_0 over a horizontal boundary.

Having explained the choice of α, we will now proceed with the proof. The proof will be by induction. For $k = 0$, the result is clear. Hence we suppose that $y_k(x)$ exists and is continuous on the stated interval and satisfies Eq. (6).

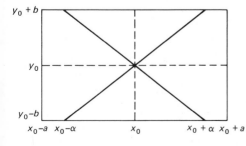

FIGURE 3.6

Then since $|y_k(x) - y_0| \le b$ for x in J, the point $(x, y_k(x))$ remains in the rectangle R_0 for all x in J. Consequently, $f(x, y_k(x))$ is well-defined for x in J. Since f is continuous, $f(x, y_k(x))$ is continuous, and therefore integrable. Hence

$$y_{k+1}(x) = y_0 + \int_{x_0}^{x} f(s, y_k(s)) \, ds$$

defines a function $y_{k+1}(x)$ for x in J. Moreover, $y_{k+1}(x)$ is continuous, being the integral of a continuous function. Finally,

$$|y_{k+1}(x) - y_0| = \left| \int_{x_0}^{x} f(s, y_k(s)) \, ds \right| \le \left| \int_{x_0}^{x} |f(s, y_k(s))| \, ds \right|$$

Since $(s, y_k(s))$ is in R_0, $|f(s, y_k(s))| \le M$. Therefore

$$|y_{k+1}(x) - y_0| \le M \, |x - x_0|$$

for x in J. This completes the induction.   ∎

LEMMA 3.2   Under the same conditions as in the preceding lemma, the sequence $\{y_k(x)\}$ converges uniformly on the interval J to a continuous function $y(x)$.

PROOF: From Eq. (3), written for $y_k(x)$ and $y_{k+1}(x)$, we obtain, for x in J,

$$y_{k+1}(x) - y_k(x) = \int_{x_0}^{x} [f(s, y_k(s)) - f(s, y_{k-1}(s))] \, ds, \qquad k = 1, 2, \ldots$$

Since the points $(s, y_k(s))$ and $(s, y_{k-1}(s))$ are in R_0 for s between x_0 and x, and $x \in J$, the Lipschitz condition in (5) yields

$$|f(s, y_k(s)) - f(s, y_{k-1}(s))| \le L \, |y_k(s) - y_{k-1}(s)|$$

Therefore, for $x \in J$,

$$|y_{k+1}(x) - y_k(x)| \le L \left| \int_{x_0}^{x} |y_k(s) - y_{k-1}(s)| \, ds \right|, \qquad k = 1, 2, \ldots \qquad (7)$$

Also, from Eq. (3) with $k = 1$ we get

$$|y_1(x) - y_0| = \left| \int_{x_0}^{x} f(s, y_0) \, ds \right| \le \left| \int_{x_0}^{x} |f(s, y_0)| \, ds \right|$$

Since (s, y_0) lies in R_0, $|f(s, y_0)| \le M$. Therefore

$$|y_1(x) - y_0| \le M \, |x - x_0|, \qquad x \in J \qquad (8)$$

Using (7) with $k = 1$, and applying (8), we get

$$|y_2(x) - y_1(x)| \leq LM \left| \int_{x_0}^{x} |s - x_0| \, ds \right| = \frac{LM}{2} |x - x_0|^2$$

By induction we can show that

$$|y_{k+1}(x) - y_k(x)| \leq \frac{ML^k}{(k+1)!} |x - x_0|^{k+1}, \qquad k = 0, 1, 2, \ldots, \quad x \in J \quad (9)$$

The inequality in (9) gives an estimate of how large the difference between two successive approximating functions can be. From this we also have

$$|y_{k+1}(x) - y_k(x)| \leq \frac{ML^k \alpha^{k+1}}{(k+1)!}, \qquad k = 0, 1, 2, \ldots, \quad x \in J \quad (10)$$

The infinite series

$$\sum_{k=0}^{\infty} \frac{ML^k \alpha^{k+1}}{(k+1)!} = \frac{M}{L} \sum_{k=1}^{\infty} \frac{(L\alpha)^k}{k!}$$

is a convergent series of positive constants [in fact, its sum is $M(e^{L\alpha} - 1)/L$]. By the comparison test, it follows that the series

$$\sum_{k=0}^{\infty} [y_{k+1}(x) - y_k(x)]$$

is absolutely convergent for all $x \in J$. In fact, it is uniformly convergent.[3] But the Nth partial sum of the series is

$$S_N(x) = \sum_{k=0}^{N-1} [y_{k+1}(x) - y_k(x)] = y_N(x) - y_0(x) = y_N(x) - y_0$$

Consequently from the convergence of $S_N(x)$ we deduce that $y_N(x)$ converges for every x in J, and in fact converges uniformly on J. We let $y(x)$ denote the limit

$$y(x) = \lim_{N \to \infty} y_N(x), \qquad x \in J \quad (11)$$

Finally, to show that y is a continuous function, we borrow another result from advanced calculus which states that if a sequence of continuous functions converges uniformly on an interval then the limit function is continuous on the interval. Since each approximation $y_k(x)$ is continuous for x in J, as shown in Lemma 3.1, this completes the proof. ∎

---

[3] We shall not attempt to explain the meaning of *uniform convergence* here. Readers unfamiliar with the concept may refer to advanced calculus texts, for example,

R. Creighton Buck, *Advanced Calculus*, New York, McGraw-Hill, 1965.
Watson Fulks, *Advanced Calculus*, 2d ed., New York, Wiley, 1969.
Joseph R. Lee, *Advanced Calculus with Linear Analysis*, New York, Academic Press, 1972.
Walter Rudin, *Principles of Mathematical Analysis*, New York, McGraw-Hill, 1953.

Next we show that the function y in Eq. (11) is a solution of the initial value problem.

LEMMA 3.3   The limit function y of the sequence of successive approximations satisfies

$$y'(x) = f(x, y(x)), \qquad x \in J$$
$$y(x_0) = y_0$$

PROOF: Consider

$$\left| \int_{x_0}^x [f(s, y(s)) - f(s, y_{k-1}(s))] \, ds \right| \le L \left| \int_{x_0}^x |y(s) - y_{k-1}(s)| \, ds \right|$$

The uniform convergence of y_k to y implies that for any given $\varepsilon > 0$, there is a $K_\varepsilon > 0$ such that

$$|y(s) - y_{k-1}(s)| < \varepsilon \quad s \in J, \qquad k \ge K_\varepsilon$$

Therefore for $x \in J$

$$\left| \int_{x_0}^x |y(s) - y_{k-1}(s)| \, ds \right| < \varepsilon |x - x_0| \le \varepsilon \alpha$$

Since α is fixed and ε is arbitrarily small, we conclude that

$$\lim_{k \to \infty} \int_{x_0}^x f(s, y_{k-1}(s)) \, ds = \int_{x_0}^x f(s, y(s)) \, ds, \qquad x \in J$$

Taking limits in Eq. (3), we obtain

$$y(x) = y_0 + \int_{x_0}^x f(s, y(s)) \, ds, \qquad x \in J \tag{12}$$

Thus, y satisfies the *integral equation* (12). Moreover, since $f(s, y(s))$ is continuous, differentiation of Eq. (12) yields

$$y'(x) = f(x, y(x)), \qquad x \in J$$

It is clear from Eq. (12) that $y(x_0) = y_0$.   ∎

Putting together the several lemmas above, we see that we have proved the following.

THEOREM 3.6   Assume that $f(x, y)$ and $\partial f(x, y)/\partial y$ are continuous in a rectangle $R_0 = \{(x, y): |x - x_0| \le a, |y - y_0| \le b\}$. Let $|f(x, y)| \le M$, $|\partial f(x, y)/\partial y| \le L$ for (x, y) in R_0, and let α be the smaller of the numbers a and b/M. Then the successive approximations defined by Eq. (2) all exist and are continuous on the interval $J = \{x: |x - x_0| \le \alpha\}$, and the sequence of approximations converges uniformly on J to a solution of the initial value problem $y' = f(x, y)$ for $x \in J$ and $y(x_0) = y_0$.

Consider the initial value problem in Eq. (1) and assume that $f(x, y)$ and $\partial f(x, y)/\partial y$ are continuous in R. Since (x_0, y_0) is an interior point of R, we can surround it by a rectangle R_0 lying entirely within R. From Theorem 3.6, we therefore deduce the existence of a solution of (1) over some interval $J = \{x: |x - x_0| \leq \alpha\}$. In order to complete the proof of Theorem 3.3, we must discuss the questions of uniqueness of solution and continuation of a solution up to the boundary of R. This is done in the next section.

Exercises

1. Show that $f(x, y) = x\,|y|$ satisfies a Lipschitz condition in any rectangle containing $(0, 0)$, even though $\partial f/\partial y$ does not exist. What happens to the Lipschitz constant as the length and width of the rectangle are increased?

2. Show that $f(x, y) = y^{1/3}$ does not satisfy a Lipschitz condition in the rectangle $-1 \leq x \leq 1, -1 \leq y \leq 1$.

3. Find a suitable Lipschitz constant L for each function on the given rectangle.

   (a) $f(x, y) = x^2 + y^2$ $\qquad 0 \leq x \leq 2, -1 \leq y \leq 3$
   (b) $f(x, y) = xe^{-y^2}$ $\qquad 0 \leq x \leq 10, -1 \leq y \leq 1$

4. Construct the first three successive approximations for each initial value problem. If possible, find an expression for the kth approximation and deduce convergence directly.

   (a) $y' = -y,$ $\qquad\qquad y(0) = 3$
   (b) $y' = x^2 + y^2,$ $\qquad y(0) = 2$
   (c) $y' = y \sin x,$ $\qquad y(0) = 1$
   (d) $y' = y^3,$ $\qquad\qquad y(0) = -1$
   (e) $y' = x\,|y|,$ $\qquad\quad y(0) = 1$

5. Construct successive approximations for $y' = y, y(0) = 1$, taking $y_0(x) = x$ instead of $y_0(x) = 1$. Do these converge to the solution?

6. Construct successive approximations for $y' = 3y^{2/3}, y(0) = 0$. Do they converge to a solution?

3.8   Uniqueness and continuation

In the previous section, the existence of a solution of the initial value problem

$$y' = f(x, y), \qquad y(x_0) = y_0 \tag{1}$$

was proved on an interval $J = \{x: |x - x_0| \leq \alpha\}$. We now want to show that this solution is unique. Our proof is based on the following important lemma.

LEMMA 3.4   Let u and v be continuous nonnegative functions on an interval $a \le x \le b$ and let c be a nonnegative constant. Suppose that

$$u(x) \le c + \int_a^x u(s)v(s) \, ds, \qquad a \le x \le b$$

Then

$$u(x) \le c \exp\left[\int_a^x v(s) \, ds\right], \qquad a \le x \le b$$

PROOF: Let

$$U(x) = c + \int_a^x u(s)v(s) \, ds$$

Then, by the Fundamental Theorem of Calculus, $U'(x) = u(x)v(x)$. Since $u(x) \le U(x)$ by the hypothesis, and $v(x)$ is nonnegative, it follows that $U'(x) \le U(x)v(x)$. Therefore

$$\exp\left[-\int_a^x v(s) \, ds\right] [U'(x) - v(x)U(x)] \le 0$$

or

$$\frac{d}{dx}\left[\exp\left[-\int_a^x v(s) \, ds\right] U(x)\right] \le 0$$

for $a \le x \le b$. Integrating from a to x we get

$$\exp\left[-\int_a^x v(s) \, ds\right] U(x) - U(a) \le 0$$

Since $U(a) = c$, we have

$$u(x) \le U(x) \le c \exp\left[\int_a^x v(s) \, ds\right] \qquad\blacksquare$$

THEOREM 3.7   Under the same hypotheses as in Theorem 3.6, the solution of the initial value problem in Eq. (1) is unique on the interval J.

PROOF: Suppose that y and z are two solutions of (1) existing on interval J. By implication, their graphs remain in the rectangle R_0 when $x \in J$. We have

$$y'(x) - z'(x) = f(x, y(x)) - f(x, z(x)), \qquad x \in J$$
$$y(x_0) - z(x_0) = y_0 - y_0 = 0$$

Therefore

$$y(x) - z(x) = \int_{x_0}^x [f(s, y(s)) - f(s, z(s))] \, ds, \qquad x \in J$$

For $x_0 \leq x \leq x_0 + \alpha$,

$$|y(x) - z(x)| \leq \int_{x_0}^{x} |f(s, y(s)) - f(s, z(s))| \, ds$$

Hence

$$|y(x) - z(x)| \leq L \int_{x_0}^{x} |y(s) - z(s)| \, ds$$

by use of the Lipschitz condition. Now we apply Lemma 3.4 with $u(x) = |y(x) - z(x)|$, $v(x) = L$, and $c = 0$. We obtain

$$|y(x) - z(x)| \leq 0 \exp [L(x - x_0)]$$

and therefore $y(x) = z(x)$ for $x_0 \leq x \leq x_0 + \alpha$. Similarly, $y(x) = z(x)$ for $x_0 - \alpha \leq x \leq x_0$. Thus, there cannot be two solutions of (1) differing anywhere on J.   ∎

In order to complete the proof of Theorem 3.3, it only remains to show that the graph of the solution exists until it reaches the boundary of the rectangle R. The point of concern is this. In our construction we proved that the solution $y(x)$ exists for $x_0 - \alpha \leq x \leq x_0 + \alpha$, where α is the smaller of the numbers a and b/M. In general, the points $(x_0 \pm \alpha, y(x_0 \pm \alpha))$ will be interior points of R. If they are, we want to show that the solution really exists beyond the points $x_0 \pm \alpha$.

LEMMA 3.5 Assume that $f(x, y)$ and $\partial f/\partial y$ are continuous in a rectangle R containing the interior point (x_0, y_0). Let $y(x)$ be a solution of the initial value problem in Eq. (1) which exists on an interval $\beta < x < \gamma$ (with $(x, y(x))$ in R). Then $\lim_{x \to \beta+} y(x)$ and $\lim_{x \to \gamma-} y(x)$ exist. The function $y(x)$ extended to the interval $\beta \leq x \leq \gamma$ by these limits is continuous and is differentiable on the right at β and on the left at γ, and satisfies Eq. (1) on $\beta \leq x \leq \gamma$.

PROOF: Let $\beta < x_1 < x_2 < \gamma$. Then, by integration of the differential equation, y satisfies

$$y(x_1) = y_0 + \int_{x_0}^{x_1} f(s, y(s)) \, ds$$

$$y(x_2) = y_0 + \int_{x_0}^{x_2} f(s, y(s)) \, ds$$

$$y(x_2) - y(x_1) = \int_{x_1}^{x_2} f(s, y(s)) \, ds$$

Since $|f(x, y)| \leq M$ for (x, y) in R,

$$|y(x_2) - y(x_1)| \leq M \, |x_2 - x_1|$$

Given $\varepsilon > 0$, choose $\delta = \varepsilon/M$. Then if $\beta < x_1 < \beta + \delta$, $\beta < x_2 < \beta + \delta$, we have $|y(x_2) - y(x_1)| \le M \delta \le \varepsilon$. That is,

$$\lim_{\substack{x_1 \to \beta + \\ x_2 \to \beta +}} |y(x_2) - y(x_1)| = 0$$

By the Cauchy convergence criterion,[4] it follows that $\lim_{x \to \beta+} y(x)$ exists. The other limit is proved in a similar way.

Now consider the function $\hat{y}(x)$ defined by

$$\begin{aligned}
\hat{y}(x) &= y(x), & \beta < x < \gamma \\
&= y(\beta^+), & x = \beta \\
&= y(\gamma^-), & x = \gamma
\end{aligned}$$

Clearly $\hat{y}$ is continuous. Moreover, for $\beta < x < \gamma$,

$$\hat{y}(x) - \hat{y}(\gamma) = \int_\gamma^x f(s, y(s))\, ds$$

Hence

$$\frac{\hat{y}(x) - \hat{y}(\gamma)}{x - \gamma} = (x - \gamma)^{-1} \int_\gamma^x f(s, y(s))\, ds.$$

As x tends to γ from the left, this tends to $f(\gamma, y(\gamma))$, by the integral mean value theorem. This shows that the left-hand derivative $\hat{y}'(\gamma)$ exists and equals $f(\gamma, y(\gamma))$. The right-hand derivative $\hat{y}'(\beta)$ can be discussed similarly.   ∎

Now consider the initial value problem $y' = f(x, y)$, $y(x_0) = y_0$, and assume that f and $\partial f/\partial y$ are continuous in a rectangle $R = \{(x, y): a \le x \le b, c \le y \le d\}$ containing (x_0, y_0) in its interior. We know that there is a unique solution y existing for $x_0 - \alpha \le x \le x_0 + \alpha$. Consider the point $(x_0 + \alpha, y(x_0 + \alpha))$. If this is an interior point of R, we consider a new initial value problem consisting of the equation $y' = f(x, y)$ and the initial condition that the solution pass through $(x_0 + \alpha, y(x_0 + \alpha))$. This point is the center of a rectangle R_1 lying entirely within R. (See Figure 3.7.) Moreover, the inequalities $|f(x, y)| \le M$, $|\partial f(x, y)/\partial y| \le L$, are still valid in R_1. Therefore, by applying the existence and uniqueness Theorems 3.6 and 3.7 to R_1 we can deduce that there is a solution y_1 which exists on some interval $|x - (x_0 + \alpha)| \le \alpha_1$ around the point $x_0 + \alpha$. By uniqueness, $y_1(x) = y(x)$ for $x_0 + \alpha - \alpha_1 \le x \le x_0 + \alpha$, and so we can regard $y_1(x)$ as a *continuation* of $y(x)$ to the larger interval $x_0 \le x \le x_0 + \alpha + \alpha_1$. This process can be repeated if $(x_0 + \alpha + \alpha_1, y(x_0 + \alpha + \alpha_1))$ is an interior point of R.

---

[4] See

Walter Rudin, *Principles of Mathematical Analysis*, New York, McGraw-Hill, 1953.

T. M. Apostol, *Mathematical Analysis*, Reading, Mass., Addison-Wesley, 1957, p. 66.

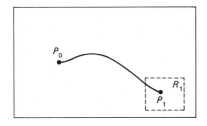

FIGURE 3.7

As long as the graph of the continuation $y(x)$ remains in R, this process of further continuation can be repeated. At each repetition the length of the interval of definition of $y(x)$ is enlarged by a positive amount. The question now is whether the graph $(x, y(x))$ must approach the boundary of R. Suppose it does not. Then there exists a number γ such that γ is the supremum (or least upper bound) of the values of x for which $y(x)$ exists. Clearly $\gamma \leq b$. By Lemma 3.5, $\lim y(x)$ exists as $x \to \gamma$ and, adjoining this value, $y(x)$ is defined and a solution of the differential equation on $x_0 \leq x \leq \gamma$. If $\gamma = b$ or if $y(\gamma) = c$ or d, the point $(\gamma, y(\gamma))$ lies on the boundary of R, as we wish to show. On the other hand, if $\gamma < b$ and $y(\gamma)$ does not equal either c or d, then $(\gamma, y(\gamma))$ is an interior point of R. But in this case, we could extend the solution beyond γ, contrary to the definition of γ. Thus, $(\gamma, y(\gamma))$ must lie on the boundary of R. This completes the proof of Theorem 3.3.

The proof of Theorem 3.2 involves still more difficult concepts, and is omitted.

Exercises

1. Find an upper bound for $u(x)$ on $x > 0$ if

(a) $u(x) \leq C_1 + C_2 \displaystyle\int_0^x u(s)\, ds \qquad (C_1, C_2 > 0)$

(b) $u(x) \leq C_1 + \displaystyle\int_0^x (C_2 + C_3 s) u(s)\, ds \qquad (C_1, C_2, C_3 > 0)$

2. Suppose that $v(x)$ is nonnegative and

$$u(x) \geq C + \int_a^x u(s)v(s)\, ds, \qquad a \leq x \leq b$$

What bound on $u(x)$ can be deduced?

3.   Suppose that $u(x)$ is nonnegative and

$$u(x) \le C_1 + C_2(x - a) + C_3 \int_a^x u(s) \, ds, \qquad a \le x \le b$$

where $C_1, C_2, C_3 > 0$. Show that

$$u(x) \le C_1 e^{C_3(x-a)} + \frac{C_2}{C_3} \left[e^{C_3(x-a)} - 1 \right]$$

4.   Suppose that $u(x)$ and $v(x)$ are nonnegative, $C > 0$, and

$$u(x) \le C + \int_a^x [u(s)]^2 v(s) \, ds, \qquad x \ge a$$

Assume that

$$\int_a^x v(s) \, ds < C^{-1} \quad \text{for} \quad x \ge a$$

Show that

$$u(x) \le \left[C^{-1} - \int_a^x v(s) \, ds \right]^{-1}, \qquad x \ge a$$

[*Hint:* Let $U(x) = C + \int_a^x [u(s)]^2 v(s) \, ds$.]

5.   Show that the solution of $y' = y^2$, $y(0) = 1$, exists on $-\infty < x < 1$ but cannot be continued to $x = 1$ or beyond.

6.   Find a maximal interval of existence of the solution of
(a) $y' = y^4$, $y(0) = 1$
(b) $y' = -e^y$, $y(0) = -1$

7.   Let y and z be solutions of

$$y' = f(x, y), \qquad y(x_0) = y_0$$
$$z' = f(x, z), \qquad z(x_0) = z_0$$

respectively on $x \in J$. Prove that (under the hypotheses of this section)

$$|y(x) - z(x)| \le |y_0 - z_0| \exp [L|x - x_0|]$$

Estimates of this kind are important because they show that the solution of a differential equation does not change much if the initial value changes little. In such a case, one customarily says that the solution depends continuously on the initial condition.

8.   Let y and z be solutions of

$$y' = f(x, y), \qquad y(x_0) = y_0$$
$$z' = f(x, z), \qquad z(\tilde{x}_0) = z_0$$

where, say, $\tilde{x}_0 < x_0$. Using the bound $|f(s, z(s))| \le M$, show that

$$|y(x) - z(x)| \le \{|y_0 - z_0| + M(x_0 - \tilde{x}_0)\} \exp [L \, |x - x_0|]$$

COMMENTARY

The science of differential equations originated with Isaac Newton, who about 1671 classified these equations into three types, which in our notation are

1. $\dfrac{dy}{dx} = f(x)$ or $\dfrac{dy}{dx} = f(y)$

2. $\dfrac{dy}{dx} = f(x, y)$

3. partial differential equations

Newton's general method was what we would call solution by infinite series (see Chapter 16). The special solution methods in Chapter 1 were discovered by early workers in mathematical analysis, such as G. W. Leibniz, various members of the Bernoulli family, and Jacopo Riccati, and were known by about the middle of the eighteenth century. Leonhard Euler introduced the notion of an integrating factor in 1734, and in 1739 began the study of linear differential equations with constant coefficients. In the 1760s, J. L. Lagrange proved that a general solution of a homogeneous linear equation of order n is of the form $y = c_1 y_1 + \cdots + c_n y_n$ where $y_1, \ldots, y_n$ is a set of linearly independent solutions and $c_1, \ldots, c_n$ are arbitrary constants (see Chapter 9). In 1774 Lagrange published the method of variation of parameters for the nth order nonhomogeneous equation, previously used in special cases by Euler and Daniel Bernoulli. For additional information and references on the early history of differential equations, see:

INCE, E. L., *Ordinary Differential Equations*, London, Longmans, Green, 1927, Appendix A. (Reprint by Dover, New York.)

EVES, H., *An Introduction to the History of Mathematics*, 3rd ed., New York, Holt, Rinehart and Winston, 1969.

The general study of existence-uniqueness theory for differential equations was begun in the nineteenth century by A. L. Cauchy, apparently the first mathematician to formulate the problem in so general a way and to prove an existence theorem. The method of successive approximations was advanced as a general principle by E. Picard and applied to obtain an existence theorem about 1890. The method was improved by E. Lindelöf in 1894, and so our Theorem 3.3 is sometimes called the Picard-Lindelöf Existence Theorem. Theorem 3.2, called the Peano Existence Theorem, was proved in 1885 by Giuseppe Peano using another method. For additional information see:

KAMKE, E., *Differentialgleichungen Reeller Funktionen*, Leipzig, Akademische Verlagsgesellschaft, 1930. (Reprint by Chelsea, New York, 1947.)

Among recent advanced books which discuss existence theory we mention:

CODDINGTON, E., and N. LEVINSON, *Theory of Ordinary Differential Equations*, New York, McGraw-Hill, 1955.

REID, W. T., *Ordinary Differential Equations*, New York, Wiley, 1971.

For surveys of work on differential equations, especially in the nineteenth century, see:

BOCHER, M., "Randwertaufgaben bei gewöhnlichen Differentialgleichungen," *Enzyklopädie der Mathematischen Wissenschaften*, II A 7a, 1900.

PAINLEVÉ, P., "Gewöhnliche Differentialgleichungen: Existenz der Lösungen," *ibid.*, II A 4a, 1900.

PAINLEVÉ, P., "Existence de l'intégrale générale. Determination d'une intégrale particulière par ses valeurs initiales," *Encyclopédie des Sciences Mathématiques*, II.16, 1910.

The lemma in Section 3.8 is known as the Gronwall inequality or sometimes as the Gronwall-Bellman inequality.

Vector spaces

4.1 Introduction

The material presented thus far has been from an area of mathematics closely related to the calculus, namely, differential equations. We now turn to a more abstract branch called linear algebra. The nature and significance of linear algebra will only become clear to the reader over a period of time. At the moment, we shall be content to say that it is an abstract, theoretical structure (like other forms of algebra), and that its abstractness makes it suitable for obtaining a unifying overview of many aspects of calculus, differential equations, and geometry. Initially, the reader may find no relationship between the topics of linear algebra and differential equations, and the change in direction may seem quite abrupt. However, we encourage the reader to try to search out relationships, some of which will become more evident as the material is developed.

This first section is devoted to a review of the concepts of planes, geometrical vectors, and ordered pairs of real numbers. Thereafter, we will explain far-reaching generalizations of these ideas. Consider first the concept of a plane in the sense explained in elementary plane geometry. That is, a plane is a collection of so-called points, lines, and so on, subject to certain axioms. (We assume that the reader has a good intuitive picture of what a plane is.) We let P^2 denote the plane and refer to this as *geometrical 2-space* (2-space means two-dimensional space). Note in particular that no coordinate system is associated with P^2. Also there is no associated algebra; that is, points or lines cannot be "added" or "multiplied." In the same way, we let P^3 denote *geometrical 3-space*.

We now recall the notion of geometrical vector. Let a fixed point O, the *origin*, be selected in a plane P^2. The notions of *line segment* and *directed line segment* are meaningful in P^2. A *geometrical vector* is a directed line segment beginning at the origin and terminating at some endpoint p. (See Figure 4.1.)

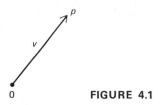

FIGURE 4.1

We can consider the collection of all such vectors. Each vector v has a length, which we denote by $|v|$, and a direction. If a fixed reference direction is given, the direction of v can be specified by giving the angle v makes with the reference. Two vectors are considered *equal* if and only if they are identical; that is, they have the same length and the same direction. For later purposes it is also convenient to define the *zero vector*, **0**, which is a *line segment of zero length*. We can think of this vector as the point O itself. No direction will be assigned to this vector.

Let v_1 and v_2 be geometrical vectors. It is customary to define the *sum* of v_1 and v_2 as the directed line segment from the origin to the point found by placing the initial point of a segment with length and direction of v_2 at the terminal point of the vector v_1. (See Figure 4.2.) This sum vector is denoted by $v_1 + v_2$, and the operation which produced the vector will be called *vector addition*, although the plus sign here does not have the same sense as in ordinary arithmetic. Note that if v_1 and v_2 have the same length but are oppositely directed, $v_1 + v_2$ is the zero vector, showing that if vector addition is to be defined for all v_1 and v_2, the zero vector is a necessity.

Another operation on geometrical vectors is *scalar multiplication*. For c any positive number and v any vector, the product cv is defined to be the vector in the direction of v with length $c|v|$. If c is negative, cv has length $|c||v|$ but the direction of cv is opposite to that of v. If $c = 0$, cv is defined as the zero vector. (See Figure 4.3.)

In this way, the set of all vectors from a fixed origin in a fixed plane has

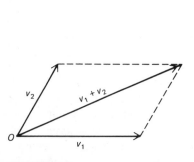

FIGURE 4.2

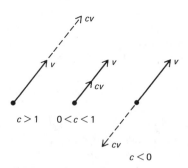

FIGURE 4.3

been given a pair of *operations*, namely, vector addition and scalar multiplication. We shall use the symbol $\mathbf{G}^2$ to refer to these geometrical 2-vectors when given these particular operations, or as we may say, this algebraic *structure*. Once a fixed origin has been given in a plane, there is a natural correspondence between vectors and points in the plane which associates each vector with its endpoint. In this sense, the points which make up P^2 and the vectors which make up $\mathbf{G}^2$ are more or less interchangeable. However, we regard P^2 and $\mathbf{G}^2$ as fundamentally different, because $\mathbf{G}^2$ has been given an algebraic structure but P^2 has not. Throughout much of this course we shall be interested in generalizations of these elementary concepts of vector, operation, and structure.

Once again let us consider a plane P^2 with origin O, but now let us place a pair of *coordinate axes* on the plane. As we know from analytic geometry, we can describe, or represent, each point p in P^2 by an *ordered pair* of real numbers, written (x, y). These numbers are called the coordinates of p with respect to the chosen coordinate axes. (See Figure 4.4.) If we were to take a second distinct

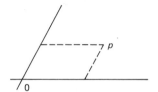

FIGURE 4.4

pair of coordinate axes, we would get a different coordinate pair representing p. Given an origin and specified axes, an ordered pair of real numbers is assumed to determine a unique point in the plane, and each point in the plane is assumed to be described by a unique ordered pair of real numbers. That is, it is assumed that there is a correspondence or relationship between the geometrical plane P^2 and the set of all ordered pairs of real numbers, which we shall hereafter denote by R^2. This is the fundamental assumption underlying the subject of analytic geometry. It should be noted that if the points of the geometrical plane are supposed to have some real or concrete existence, then it is not obvious that every ordered pair of real numbers is in correspondence with a point. We believe the best way to think of the matter is that P^2 and R^2 are both mathematical abstractions and (as shown in analytic geometry) we are free to assume the indicated correspondence.

In the same way, by choosing an origin and three coordinate axes we can establish a correspondence between geometrical 3-space and the set of all ordered *triples* (x_1, x_2, x_3). Furthermore, we can obviously consider also ordered 4-tuples (x_1, x_2, x_3, x_4) or, generally, ordered m-tuples $(x_1, x_2, \ldots, x_m)$ for any positive integer m. Although our ordinary geometrical intuition may balk at being carried into 4-space or m-space, the methods of this text will, as we shall see, suggest the possibility and desirability of using such geometrical language and concepts. Throughout this text, R^m will denote the set of all ordered m-tuples of real numbers.

The idea of setting up a correspondence between points in a plane and ordered pairs is familiar from analytic geometry. Now we want to show that a correspondence can be set up between the geometrical 2-vectors from the origin of the coordinate system and ordered pairs of real numbers. In fact, all that needs to be done is to associate with each vector v the coordinates of its endpoint (x_1, x_2). Clearly, once the coordinate axes have been specified, each vector corresponds uniquely to an ordered pair. (See Figure 4.5.) Moreover, suppose

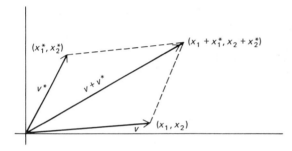

FIGURE 4.5

that the vectors v and v^* of $\mathbf{G}^2$ correspond to the ordered pairs (x_1, x_2) and (x_1^*, x_2^*), respectively. From Figure 4.5 we easily see that the vector $v + v^*$ corresponds to the ordered pair $(x_1 + x_1^*, x_2 + x_2^*)$. Thus, addition of geometrical 2-vectors corresponds to arithmetic addition of the coordinates of the endpoints of the vectors. It is also easy to see that if v is represented by (x_1, x_2) and c is any real number then cv is represented by (cx_1, cx_2). These ideas can be directly extended to geometrical 3-vectors, the set of which we denote by $\mathbf{G}^3$.

This discussion suggests that we consider the set of all ordered pairs of real numbers, and define two operations for members of this set. The *sum* of two ordered pairs is defined by the equation

$$(x_1, x_2) + (x_1^*, x_2^*) = (x_1 + x_1^*, x_2 + x_2^*)$$

The product of a scalar and an ordered pair is defined by

$$c(x_1, x_2) = (cx_1, cx_2)$$

We shall use the symbol $\mathbf{R}^2$ to refer to the set of ordered pairs of real numbers with the structure given by these two operations.

Clearly, we can extend this idea by defining $\mathbf{R}^3$ as the set of ordered triples of real numbers with operations defined by

$$(x_1, x_2, x_3) + (x_1^*, x_2^*, x_3^*) = (x_1 + x_1^*, x_2 + x_2^*, x_3 + x_3^*)$$
$$c(x_1, x_2, x_3) = (cx_1, cx_2, cx_3)$$

DEFINITION 4.1   If m is a positive integer, $\mathbf{R}^m$ is the set of ordered m-tuples of real numbers with addition and scalar multiplication defined by:

$$(x_1, x_2, \ldots, x_m) + (x_1^*, x_2^*, \ldots, x_m^*)$$
$$= (x_1 + x_1^*, x_2 + x_2^*, \ldots, x_m + x_m^*) \qquad (1)$$
$$c(x_1, x_2, \ldots, x_m) = (cx_1, cx_2, \ldots, cx_m) \qquad (2)$$

Observe that we have not attempted to define P^m or $\mathbf{G}^m$ for $m > 3$. It is easy to generalize the concept of ordered pair or triple to ordered m-tuple, but not easy to visualize an extension of the purely geometrical conception of space to higher dimensions. However, in working with the set $\mathbf{R}^m$ in this text, we shall find that we can formulate many geometrical concepts, for example length and angle, in a strictly algebraic way by reference to coordinates. This means that we can clothe $\mathbf{R}^m$ both in algebraic and geometric garb. In fact, we will presently be calling $\mathbf{R}^m$ *m-dimensional space* and referring to its m-tuples as *points*. In this way, the concept of higher dimensional space loses its mystery and becomes a useful tool in mathematics.

We can summarize our discussion by saying that we have dealt with various mathematical constructions, P^2, P^3, $\mathbf{G}^2$, $\mathbf{G}^3$, R^m, and $\mathbf{R}^m$ ($m \geq 1$). There is a natural correspondence between the points in P^2, the vectors in $\mathbf{G}^2$, and the ordered pairs in R^2 (likewise between elements of P^3, $\mathbf{G}^3$, and R^3), once an origin and coordinate axes are fixed. The objects in the spaces $\mathbf{G}^2$, $\mathbf{G}^3$, $\mathbf{R}^m$, are subject to certain operations called addition and multiplication by a scalar. In contrast, P^2, P^3, and R^3 are devoid of such operations.

Exercises

1.   Draw a diagram that illustrates the statement "It is also easy to see that if v is represented by (x_1, x_2) and c is any real number then cv is represented by (cx_1, cx_2)." Upon what theorem from elementary geometry does this result depend?

2.   Elaborate on the last sentence of this section.

3.   Perform the indicated operations on ordered m-tuples.

(a) $(2, -1) + 3(-1, 6)$

(b) $2(1, 2, 3) + (-4)(2, 1, 0)$

(c) $3(1, 0, -2, 3) + 4(-1, 1, -1, -2)$

4.   Let an origin and coordinate axes be fixed. Find the ordered pair and graph the vector which corresponds to each of the following:

   (a) The vector of length 2 which makes an angle of 30 degrees with the positive x-axis.

   (b) The vector which is the sum of the vector in (a) and the vector of length 2 which makes an angle of 135 degrees with the positive x-axis.

   (c) The vector of length 1 which when extended bisects the side AB of the triangle OAB with vertices at the origin and the points $A(3, 5)$ and $B(5, 2)$.

   (d) The vector of length r which makes angle θ with the positive x-axis.

4.2   Sets and relations

A *set* is a collection of objects called *elements*. The notation $\{x: x \text{ satisfies } P\}$ is used for the set of all x which satisfy the property P. For example, the equation $R = \{x: x \text{ is real}\}$ defines R to be the set of all real numbers. We use the notation $x \in S$ to denote the fact that x is an element of the set S, and $x \notin S$ to denote the fact that x is not an element of S.

The *union* of two sets $S_1 \cup S_2$ is the set of elements in either S_1 or S_2 (or both). The *intersection* of two sets, $S_1 \cap S_2$, is the set of elements common to both S_1 and S_2. The set $\varnothing$ containing no elements is called the *empty set*. S_1 is a *subset* of S_2, denoted by $S_1 \subset S_2$, if $S_1 \cap S_2 = S_1$. Finally, S_1 and S_2 are equal, that is, $S_1 = S_2$, if $S_1 \subset S_2$ and $S_2 \subset S_1$.

Two sets S_1 and S_2 are *disjoint* or *exclusive* if $S_1 \cap S_2 = \varnothing$. If S_1, $S_2, \ldots, S_n$ are subsets of S, we say they are *exhaustive* if $\bigcup_{i=1}^{n} S_i = S$, where $\bigcup_{i=1}^{n} S_i = S_1 \cup S_2 \cup \cdots \cup S_n$.

The *Cartesian product* of two sets S_1 and S_2, denoted by $S_1 \times S_2$, is the set of ordered pairs of elements (a, b) where $a \in S_1$ and $b \in S_2$. That is, $S_1 \times S_2 = \{(a, b): a \in S_1 \text{ and } b \in S_2\}$.

Example 1   The Cartesian product $R \times R$ is the set of all pairs of real numbers. The Cartesian product $R \times C$, where C is the set of all complex numbers,[1] is the set of all pairs (a, b) where a is real and b complex. Note that (b, a) would not be in $R \times C$ nor would (a, b) be in $C \times R$.

Consider a subset B of $S_1 \times S_2$. The elements of B are pairs of the form (a, b) where $a \in S_1$ and $b \in S_2$, although not all such pairs will necessarily be included in B. For example, the set of (x, y) such that $y = mx + b$ is a subset of $R \times R$. Given a subset B of $S_1 \times S_2$, we write $a \sim b$ if the element (a, b) is contained in B. We say, then, B defines a *relation* on $S_1 \times S_2$ by $a \sim b$; that is, a *is related* to b if $(a, b) \in B$. Note that a can be related to b without b being related to a. That is, $a \sim b$ does not necessarily imply $b \sim a$.

---

[1] The student unfamiliar with the algebra of complex numbers should consult Appendix A.

Example 2 Consider a function f mapping R into R. Then the set of elements $(x, f(x))$ form a subset of $R \times R$. We have then the relation $x \sim f(x)$ although we cannot say $f(x) \sim x$. More specifically, if $f(x) = \sin x$, $(\pi/2, 1)$ is in the subset of $R \times R$ defined by the function, although $(1, \pi/2)$ is not. Hence, we write $\pi/2 \sim 1$, but not $1 \sim \pi/2$.

A particular type of relation of importance in mathematical theory is the *equivalence relation*, which is defined as follows. Let S be a set and $S \times S$ the Cartesian product of S with itself. Then the subset B of $S \times S$ defines an *equivalence relation* if the following conditions hold. For any a, b, and c in S,

1. $a \sim a$ ($\sim$ is *reflexive*)
2. $a \sim b$ implies $b \sim a$ ($\sim$ is *symmetric*)
3. $a \sim b$ and $b \sim c$ implies $a \sim c$ ($\sim$ is *transitive*)

In the event that two elements of S, say a and b, are related by an equivalence relation we shall denote this by $a \approx b$.

Note that condition (1) states every element of S is related to itself, implying $\{(a, a)\} \in B$ for every $a \in S$. We now show that an equivalence relation partitions S into a collection of mutually exclusive and exhaustive sets. Let $\approx$ denote an equivalence relation and let S_a denote the set of all elements in S which are equivalent to a. (Note that S_a is a subset of S, not of $S \times S$.) Then if $a \in S$, $b \in S$ either $S_a = S_b$ or $S_a \cap S_b = \varnothing$. Since every element in S is equivalent to itself, each element of S belongs to at least one of these subsets. Hence, the subsets of equivalent elements in S are mutually exclusive and exhaustive in S. Each of these subsets defines an *equivalence class* of elements of S.

Example 3 Let S be the set of real 2-tuples, that is, $S = R^2$. Define the relation $\sim$ such that $(x_1, x_2) \sim (y_1, y_2)$ if $x_1 + x_2 = y_1 + y_2$. That is, two elements are "related" if they lie on the same line $x_1 + x_2 = c$ for some c. Then $\sim$ is an equivalence relation.

Exercises

1. Prove that, for any sets R and S, $(R \cap S) \subset S$ and $(R \cap S) \subset R$.
2. Prove that, for any sets R and S, $R \subset (R \cup S)$ and $S \subset (R \cup S)$.

In Exercises 3 through 6 let J be the set of all integers, E the set of all even integers, O the set of all odd integers, P the set of positive integers, and N the set of negative integers.

3. Describe each of the following sets.
   (a) $E \cap P$
   (b) $E \cup O$
   (c) $J \cap N$
   (d) $(N \cap J) \cup P$
   (e) $N \cap (J \cup P)$
   (f) $(E \cap N) \cup (O \cap P)$

4. Describe each of the following sets.

(a) $J \times J$

(b) $E \times E$

(c) $P \times P$

(d) $(E \cap P) \times J$

(e) $E \cap (P \times J)$

5. The set $B = E \times E$ is a subset of $J \times J$. Describe verbally the relation $a \sim b$ defined by B. Is this an equivalence relation?

6. If a and b are integers, define $a \sim b$ to mean that their difference $a - b$ is an even integer. Show that this relation corresponds to a subset B of $J \times J$. Describe this subset. Is this an equivalence relation?

7. Let S be the set of all lines in a plane, so that $S \times S$ is the set of all ordered pairs of lines. Let B be the subset consisting of pairs of parallel lines. If a and b are lines, what is the relation $a \sim b$ defined by B? Is this an equivalence relation?

4.3   Groups and fields

In our definition of a vector space we will make use of the algebraic structures called groups and fields. This section is devoted to defining these structures and developing certain of their properties.

DEFINITION 4.2   An *additive group*[2] V is a set V together with an operation called addition $(+)$ which satisfies the following:

1. To each pair of elements v_1 and $v_2 \in V$ there is a unique element $(v_1 + v_2) \in V$
2. $(v_1 + v_2) + v_3 = v_1 + (v_2 + v_3)$ for v_1, v_2, and $v_3 \in V$
3. There exists an element $\mathbf{0} \in V$ such that $v + \mathbf{0} = v = \mathbf{0} + v$ for every $v \in V$.
4. For each $v \in V$ there is an element $v' \in V$ such that $v + v' = \mathbf{0}$.

The element $\mathbf{0}$ is called the *additive identity* element. An element v' for which $v + v' = \mathbf{0}$ is called an *additive inverse* of v. Property (2) is called the *associative* property, and (1) is called the *closure* property.

---

[2] It is usual to define a *group* as a set V with a binary operation satisfying conditions (1) to (4) of Definition 4.2. When the binary operation is denoted by the symbol $+$, the group may be called an additive group. Frequently the operation is denoted by a multiplication symbol and the group may then be called a multiplicative group.

DEFINITION 4.3 An additive group **V** is *commutative*, or *Abelian*, if $v_1 + v_2 = v_2 + v_1$ for every v_1 and $v_2 \in$ **V**.

Before developing additional properties of commutative additive groups which are implied by Definition 4.3, we will give several examples of commutative additive groups.

Example 1 Denote by **J** the set of all integers together with the usual addition. Then **J** is a commutative additive group. Note that if we had restricted **J** to just positive integers we would not have had a group as the additive inverse would not have been in our set. On the other hand, the restriction of **J** to just the number 0 (zero) would leave us with a group.

Example 2 Consider the set of all square arrays of real numbers of the form

$$A = \begin{pmatrix} a_{11} & a_{12} \\ a_{21} & a_{22} \end{pmatrix}$$

Such an array is called a 2 by 2 (or 2×2) *matrix* (plural: *matrices*). We define addition of two such matrices to be the result of numerical addition of numbers in corresponding positions within the matrices, that is

$$A + B = \begin{pmatrix} a_{11} & a_{12} \\ a_{21} & a_{22} \end{pmatrix} + \begin{pmatrix} b_{11} & b_{12} \\ b_{21} & b_{22} \end{pmatrix} = \begin{pmatrix} a_{11} + b_{11} & a_{12} + b_{12} \\ a_{21} + b_{21} & a_{22} + b_{22} \end{pmatrix}$$

The set of all such matrices together with the above defined addition forms a commutative additive group which we will denote by **M**. Note that just the set of 2×2 matrices does not form a group until the addition operation is defined.

To verify that **M** is a commutative additive group requires showing that the conditions of Definitions 4.2 and 4.3 are met. The associative and commutative properties are satisfied since each element of the array satisfies the laws of arithmetic. We take as the zero element of the group the matrix

$$\mathbf{0} = \begin{pmatrix} 0 & 0 \\ 0 & 0 \end{pmatrix}$$

which satisfies the condition $A + \mathbf{0} = A = \mathbf{0} + A$ for each $A \in$ **M**. Finally, for any matrix

$$A = \begin{pmatrix} a_{11} & a_{12} \\ a_{21} & a_{22} \end{pmatrix}$$

the matrix

$$B = \begin{pmatrix} -a_{11} & -a_{12} \\ -a_{21} & -a_{22} \end{pmatrix}$$

is the additive inverse required.

Example 3   Consider $\mathbf{R}^m$, the set of ordered m-tuples together with the usual associated addition. We now show that $\mathbf{R}^m$ is a commutative additive group. (The scalar multiplication previously defined is ignored for purposes of this example.) Again, the associative and commutative properties are satisfied by virtue of the properties of elementary arithmetic, that is, the associative and commutative properties of addition of real numbers. We define $\mathbf{0}$ to be the m-tuple of all zeros, that is, $\mathbf{0} = \{0, 0, \ldots, 0\}$. Finally, for any element $v = \{x_1, x_2, \ldots, x_m\}$ the element v' in $\mathbf{R}^m$ given by $v' = \{-x_1, -x_2, \ldots, -x_m\}$ is the required additive inverse.

We now proceed to give further properties of additive groups which follow from the requirements contained in Definitions 4.2 and 4.3.

THEOREM 4.1   The zero element $\mathbf{0}$ of an additive group is unique.

PROOF: By property (3) of Definition 4.2 the zero element is such that $v + \mathbf{0} = v = \mathbf{0} + v$ for every $v \in \mathbf{V}$. Suppose there is another zero $\mathbf{0}' \in \mathbf{V}$ with the same property. Replacing v by $\mathbf{0}'$ we have $\mathbf{0}' + \mathbf{0} = \mathbf{0}' = \mathbf{0} + \mathbf{0}'$. Since $v + \mathbf{0}' = v = \mathbf{0}' + v$ we replace v by $\mathbf{0}$ and obtain $\mathbf{0} + \mathbf{0}' = \mathbf{0} = \mathbf{0}' + \mathbf{0}$. But since $\mathbf{0} + \mathbf{0}' = \mathbf{0}$ and $\mathbf{0} + \mathbf{0}' = \mathbf{0}'$ and the sum of any two elements in the group is unique, we conclude $\mathbf{0}' = \mathbf{0}$.   ∎

THEOREM 4.2   Let v and v' be elements of an additive group $\mathbf{V}$ such that $v + v' = \mathbf{0}$. Then $v' + v = \mathbf{0}$. Further, for each $v \in \mathbf{V}$, the element v' such that $v + v' = \mathbf{0}$ is unique.

PROOF: Given that $v + v' = \mathbf{0}$, let v'' be any additive inverse of v' such that $v' + v'' = \mathbf{0}$. Then $v + v' + v'' = \mathbf{0} + v'' = v''$ and $v + v' + v'' = v + \mathbf{0} = v$. By the uniqueness of the zero element (Theorem 4.1) and the uniqueness property 1 of Definition 4.2, $v'' = v$. This proves that if v' is an additive inverse of v, then v is an additive inverse of v'. To prove the uniqueness of the additive inverse we proceed as follows.

Suppose there are two additive inverses, say v' and v'', for some element v. Then $v + v' = \mathbf{0}$ and $v + v'' = \mathbf{0}$. Hence $v + v' = v + v''$. Using the uniqueness property of addition, $v' + (v + v') = v' + (v + v'')$ which, by associativity, implies $\mathbf{0} + v' = \mathbf{0} + v''$ and therefore $v' = v''$.   ∎

We now state two further properties of additive groups but leave the proofs as exercises.

THEOREM 4.3   For any two elements v_1 and v_2 of an additive group $\mathbf{V}$, there is a unique element v_0 such that $v_1 + v_0 = v_2$.

THEOREM 4.4   Let v_0 be an element such that $v + v_0 = v$ for v some element of an additive group $\mathbf{V}$. Then $v_0 = \mathbf{0}$.

A further concept necessary in the definition of a vector space is that of a field. For the sake of simplicity we restrict attention here to fields with elements which are complex numbers.

DEFINITION 4.4 Consider the set **C** of all complex numbers with addition and multiplication defined as in Appendix A. Let **K** be a subset of **C**. Then **K** is a *field* provided

1. **K** is a commutative additive group.
2. If x_1 and x_2 are in **K**, then $x_1 x_2$ is in **K** (that is, **K** is closed under multiplication).
3. The element 1 is in **K**.
4. For every nonzero element $x \in$ **K** the element x^{-1} (such that $x \cdot x^{-1} = 1$) is also in **K**.

First note that **C** is a field. Further we see immediately that $\mathbf{R}^1$ is a field[3] because (1) the sum and product of any two real numbers is again real, (2) the negative of any real number is real and the inverse of any nonzero x is real, namely, $1/x$, and (3) 0 and 1 are both real. The set of all positive real numbers is not a field as it does not contain *additive inverses*, that is, if x is a positive real number, then $-x$ is not in the set. The set of integers is not a field as it does not contain the *multiplicative inverses*, that is, if n is an integer, n^{-1} is not an integer for $n \neq 1$.

Exercises

1. Which of the following sets of ordered pairs (x, y) are groups? In each group, state what the zero element is and what the additive inverse of (x, y) is.

   (a) x, y real, $x = y$, usual definition of addition.
   (b) x, y real, $x = y + 1$, usual definition of addition.
   (c) x, y real, addition defined unconventionally by the formula

   $$(x_1, y_1) + (x_2, y_2) = (x_1 + x_2 - 1, y_1 + y_2 - 1)$$

   (d) x, y real, addition defined by

   $$(x_1, y_1) + (x_2, y_2) = (y_1 + y_2, x_1 + x_2)$$

   (e) x, y rational numbers, usual definition of addition.
   (f) x, y complex numbers, usual definition of addition.

---

[3] Throughout the rest of this text we shall use **R** and $\mathbf{R}^1$ interchangeably for the real numbers considered as a group, a field, or later, as a vector space.

2.   Which of the following sets of 2×2 matrices

$$A = \begin{pmatrix} a_{11} & a_{12} \\ a_{21} & a_{22} \end{pmatrix}$$

are groups with addition defined as in Example 2? In each group indicate the zero element and the additive inverse.

(a) The set of symmetric matrices, that is, those with $a_{12} = a_{21}$.

(b) The set of diagonal matrices, that is, $a_{12} = a_{21} = 0$.

(c) The set of matrices with $a_{11} = a_{22} = 0$.

(d) The set of matrices with $a_{11} = a_{12}$ and $a_{21} = a_{22}$.

(e) The set of matrices with even integers in all four positions.

3.   Which of the following subsets of **C** are fields?

(a) The set of all numbers of the form ib, where b is real.

(b) The number zero.

(c) The set $\{-1, 0, 1\}$.

(d) The set of all rational numbers.

(e) The set of all numbers of the form $a + ia$, where a is real.

4.   Prove Theorem 4.3.

5.   Prove Theorem 4.4.

6.   Prove that if $u + v = u + w$ then $v = w$. That is, *cancellation* is always possible in a group.

7.   Prove that the inverse of $u + v$ is $v' + u'$, where u' is the inverse of u and v' is the inverse of v.

8.   Let **K** be the set of all numbers of the form $a + b\sqrt{2}$ where a and b are arbitrary rational numbers. Prove that this is a field. Is it a field if a and b are arbitrary real numbers?

4.4   Vector spaces

We shall now generalize the notion of vectors discussed in Section 4.1.

DEFINITION 4.5   A *vector space* **V** *over a field* **K** is a commutative additive group **V** upon which a scalar multiplication is defined such that

> 1. There exists a unique element cv in **V** for every $v \in$ **V** and every c in the field **K**.

2. For every v in **V** the element $(-1)v$ is such that $v + (-1)v = \mathbf{0}$ where **0** is the additive identity element of the group, or, as it is more commonly called, the *zero vector* of **V**.
3. $c(u + v) = cu + cv$ for every u and v in **V** and c in **K**.
4. $(a + b)v = av + bv$ for v in **V** and a and b in **K**.
5. $(ab)v = a(bv)$ for v in **V** and a and b in **K**.
6. $1 \cdot v = v$ for all v in **V**.

The elements of **V** are called *vectors* and the elements of **K** are called *scalars*. If **K** = **R**, **V** is called a vector space over the reals or a *real vector space*. If **K** = **C**, **V** is called a vector space over the complex field or a *complex vector space*. Vector spaces are also called *linear spaces*. For notational convenience we will sometimes express $u + (-1)v$ as $u - v$.

For brevity one often uses the same letter **V** both for the vector space and the set of its elements, and we shall usually follow this convenient practice. However, the reader must bear in mind that there is a logical distinction between a set and the set with associated algebraic structure.

Example 1   Consider the commutative additive group **M** of 2×2 matrices defined in Section 4.3. We define scalar multiplication by

$$cA = \begin{pmatrix} ca_{11} & ca_{12} \\ ca_{21} & ca_{22} \end{pmatrix}$$

where

$$A = \begin{pmatrix} a_{11} & a_{12} \\ a_{21} & a_{22} \end{pmatrix}$$

is any element of **M** and c is any real number. Let us show that the set of all such matrices is a vector space over the field of the real numbers. We have seen that **M** is a commutative additive group. Condition (1) of Definition 4.5 is obvious and conditions (2) through (6) can be verified by writing out the required relationships.

Example 2   Consider the space $\mathbf{R}^n$ of all ordered n-tuples of real numbers with addition and scalar multiplication defined as in Section 4.1. As shown in Section 4.3, this is an additive group in which the vector **0** is the n-tuple of zeros, namely $(0, 0, \ldots, 0)$. We can easily verify that $\mathbf{R}^n$ is a vector space over the field of real numbers. This space is sometimes called *n-dimensional real coordinate space*.

Example 3   Consider the set **V** of all triples (x_1, x_2, x_3) in $\mathbf{R}^3$ where x_1 and x_2 are arbitrary and $x_3 = x_1 + 2x_2 + 1$, with addition and scalar multiplication defined as in Section 4.1. **V** does not form a vector space since, for the scalar $c = 0$ and any arbitrary triple in **V**, (a_1, a_2, a_3), the scalar product $c(a_1, a_2, a_3) = (0, 0, 0)$ is not in **V**.

Example 4   Consider V to be the set of all polynomials of degree n or less with real coefficients. That is, $p(x)$, a typical element of V, is of the form $a_n x^n + a_{n-1} x^{n-1} + \cdots + a_0$. The sum of two polynomials

$$p(x) = a_n x^n + a_{n-1} x^{n-1} + \cdots + a_0$$

and

$$q(x) = b_n x^n + b_{n-1} x^{n-1} + \cdots + b_0$$

is defined to be

$$p(x) + q(x) = (a_n + b_n) x^n + (a_{n-1} + b_{n-1}) x^{n-1} + \cdots + (a_0 + b_0)$$

Scalar multiplication is defined by

$$cp(x) = ca_n x^n + ca_{n-1} x^{n-1} + \cdots + ca_0$$

We now proceed to show that **V**, the set V with addition and scalar multiplication defined as above, is a vector space over the field **R**. By noticing that the polynomial with all $a_i = 0$ satisfies the condition required for **0**, and since the sum is uniquely defined, we can conclude that **V** is an additive group. Condition (1) of Definition 4.5 is illustrated above and properties (2) through (6) follow from the properties of the real numbers.

The reader may wonder why the word *space* is used in the term *vector space*, since Definition 4.5 is entirely algebraic rather than geometric, and since the examples are not necessarily geometric in nature. The reason is that the abstract definition is motivated by and framed so as to include the principal properties of $\mathbf{R}^n$, which is, as we have seen, an algebraic version of geometrical vectors. As we progress, we shall find that much geometric language can be applied with profit to abstract vector spaces. For this reason, the elements of a vector space **V** are sometimes called *points*, even if they are not points in the ordinary sense of geometry.

We close this section with a theorem, the proof of which will be left to the reader.

THEOREM 4.5   Consider a vector space **V** over a field **K**.

1. The product of the zero scalar 0 and any vector v in **V** is the zero element **0** of **V**.
2. The product of any scalar c and the zero vector is the zero vector.

Exercises

1. Using Theorem 4.4 and properties (4) and (6) of Definition 4.5, prove statement (1) of Theorem 4.5.

2. Using the result of Exercise 1, prove that for every v in **V**, $v - v = \mathbf{0}$. This shows that property (2) in Definition 4.5 is redundant in the sense that it

follows from the other properties of a vector space. It also shows that the additive inverse of v (unique by Theorem 4.2) is $(-1)v$.

3. Prove statement (2) of Theorem 4.5.

4. Which of the sets of ordered pairs in Exercise 1, Section 4.3, are vector spaces over **R** with the notion of vector addition given there and with the usual notion of multiplication by a scalar; that is, $c(x, y) = (cx, cy)$?

5. Which of the sets of matrices in Exercise 2, Section 4.3, are vector spaces over **R** with the usual addition and multiplication by scalars?

6. Show that the set of real numbers is a vector space over the field of real numbers.

7. Let **K** be a field. Show that if the elements are considered to be both scalars and vectors, and if the vector operations are defined to be the corresponding field operations, the field becomes a vector space; that is, a field is a vector space over itself.

8. Prove that the set of all $n \times k$ matrices forms a vector space over **R**. Denote an arbitrary matrix A as

$$\begin{pmatrix} a_{11} & a_{12} & \cdots & a_{1k} \\ a_{21} & a_{22} & \cdots & a_{2k} \\ \vdots & & & \vdots \\ a_{n1} & a_{n2} & \cdots & a_{nk} \end{pmatrix}$$

where the element of the ith row and jth column is denoted a_{ij} and is in **R**. Addition and scalar multiplication are defined in a manner analogous to that in Example 1.

9. Consider the set of all infinite sequences of real numbers, $\{x_n\} = x_1, x_2, \ldots$. Define the sum of two sequences in the usual way, that is, by adding corresponding elements. Define multiplication by a scalar as usual, that is, c times a sequence is the sequence with cx_n as the nth element. Prove that this system is a vector space over **R**.

10. Can $\varnothing$ (the empty set) be a vector space over any field?

11. We have shown that $\mathbf{R}^n$ is a vector space over **R** under the usual definitions of addition and scalar multiplication. Determine whether $\mathbf{R}^1$ is a vector space if vector addition is defined to be algebraic multiplication, and scalar multiplication is defined to be algebraic addition.

12. Construct a definition of vector space by writing out all axioms needed without using the word "group."

13. Show that the set of all nth degree polynomials, with operations defined as in Example 4, is not a vector space.

14. Let W be the set of elements (x, y, z) in $\mathbf{R}^3$ which satisfy the equations

$$a_1 x + b_1 y + c_1 z = 0$$
$$a_2 x + b_2 y + c_2 z = 0$$
$$a_3 x + b_3 y + c_3 z = 0$$

for a given set of real numbers a_i, b_i, c_i ($i = 1, 2, 3$). Prove that W is a vector space.

15. Prove that the following properties hold in any vector space over $\mathbf{R}$.

(a) If $cv = \mathbf{0}$, then either $c = 0$ or $v = \mathbf{0}$.

(b) If $cu = cv$, where c is a nonzero scalar and u and v are vectors, then $u = v$.

(c) If $cv = dv$ where v is a nonzero vector and c and d are scalars, then $c = d$.

(d) If v is any vector, then $v + v = 2v$, and in general, the sum of n v's is nv.

4.5   Spaces of functions

Example 4 of the preceding section may have seemed startling or bizarre to the student, for in it the elements of the vector space are polynomial functions. This "space" whose elements are functions seems distant from our primitive notion of "space," and its vectors (the functions) seem remote from the geometrical vectors of experience. Nevertheless, it is vector spaces of functions that are most useful in the study of differential equations. The realization that such diverse situations satisfy the vector space axioms was of profound importance in the development of mathematics, since it permitted a unified approach to diverse phenomena. In this section we introduce several other spaces of functions which will be useful in this course.

Example 1   Consider the *spaces* $\mathbf{F}[a, b]$ *and* $\mathbf{F}(-\infty, \infty)$. Let a and b be fixed real numbers, $a < b$, and consider the set of all real-valued[4] functions f defined on the interval $a \leq x \leq b$. For any two functions f and g we define $f + g$ to be the function whose value at x is the sum of the numbers $f(x)$ and $g(x)$. Symbolically

$$(f + g)(x) = f(x) + g(x), \qquad a \leq x \leq b \tag{1}$$

For any function f and any real number c we define a function called cf by

$$(cf)(x) = cf(x), \qquad a \leq x \leq b \tag{2}$$

---

[4] A function is called real-valued if its value $f(x)$ at every x is a number in $\mathbf{R}$; it is called complex-valued if $f(x) \in \mathbf{C}$ for every x.

For example, if f is the function whose value at x is $-\frac{1}{2}\cos x$, and g is defined by $g(x) = \frac{1}{2}$ (say for $0 \le x \le 2\pi$), then $f + g$ is the function whose value at x is $\sin^2(x/2)$, and $2g$ is the function that has the constant value 1.

The set of all real functions on $[a, b]$ endowed with these operations of addition and multiplication by scalars is a vector space over the real field, which we denote as $\mathbf{F}[a, b]$. We verify this fact item by item. First, if f and g are both defined for $a \le x \le b$, then it is clear that $f + g$ is a function defined on $a \le x \le b$. Also $f + g = g + f$ and $(f + g) + h = f + (g + h)$, since the real numbers have these commutative and associative properties.[5] The zero element in $\mathbf{F}[a, b]$ is the function $\mathbf{z}$ which has value zero at every x in $[a, b]$ since then $f + \mathbf{z} = f$ for every $f \in \mathbf{F}[a, b]$. The additive inverse of a function f is the function $(-1)f$ since the value at x of $(-1)f + f$ is, by Eqs. (1) and (2), $-f(x) + f(x) = 0$. Thus, $\mathbf{F}[a, b]$ is a commutative additive group. We leave it to the reader to verify the remaining axioms for a vector space.

Sometimes we may wish to deal with functions defined on an open interval $a < x < b$ or on $-\infty < x < \infty$. The corresponding vector spaces $\mathbf{F}(a, b)$ and $\mathbf{F}(-\infty, \infty)$ are defined in analogous fashion.

Example 2 The *spaces* $\mathbf{C}^0[a, b]$, $\mathbf{C}^0(a, b)$, *and* $\mathbf{C}^0(-\infty, \infty)$ are defined as follows. $\mathbf{C}^0[a, b]$ is the vector space of all real-valued *continuous* functions defined on the closed interval $[a, b]$. The verification that this is a vector space over the real field is identical to that for $\mathbf{F}[a, b]$, except that it must be shown that the sum of two continuous functions is a continuous function, and the product of a scalar c and a continuous function f is a continuous function. These are standard theorems from calculus.[6]

Similarly $\mathbf{C}^0(a, b)$ is the vector space of all real-valued continuous functions defined on the open interval (a, b), and $\mathbf{C}^0(-\infty, \infty)$ is the space of all real-valued continuous functions defined on the infinite interval $(-\infty, \infty)$. The meaning of $\mathbf{C}^0[a, \infty)$ and $\mathbf{C}^0(-\infty, b]$ should be clear.

Example 3 Consider the *spaces* $\mathbf{C}^k[a, b]$, $\mathbf{C}^k(a, b)$, and $\mathbf{C}^k(-\infty, \infty)$. We define $\mathbf{C}^1(a, b)$ to be the real vector space of all real-valued functions defined on the interval (a, b) whose first derivative function exists and is continuous on the interval. That is, in order for f to belong to $\mathbf{C}^1(a, b)$ it is required that f' belong to $\mathbf{C}^0(a, b)$. (f also belongs to $\mathbf{C}^0(a, b)$ since it is known from calculus that a differentiable function is necessarily continuous.) To show that $\mathbf{C}^1(a, b)$ is a vector space we note that if f and g are in $\mathbf{C}^1(a, b)$ then f' and g' are in $\mathbf{C}^0(a, b)$, hence $f' + g' = (f + g)'$ is in $\mathbf{C}^0(a, b)$. But then $f + g$ is in $\mathbf{C}^1(a, b)$. Similarly if $f \in \mathbf{C}^1(a, b)$ and c is a real number, then $(cf)' = cf'$ is in $\mathbf{C}^0(a, b)$ and cf is in $\mathbf{C}^1(a, b)$. The other axioms are verified in the same way as before.

---

[5] Recall that functions are equal if they have the same domain of definition and the same value at every point of this domain.

[6] George B. Thomas, *Calculus and Analytic Geometry*, Reading, Mass., Addison-Wesley, 1968.

The space $\mathbf{C}^1[a, b]$ is defined in the same way, except that care is needed in considering the endpoints. If a function f is defined only for $a \leq x \leq b$, then by $f'(x)$ we mean the ordinary derivative if $a < x < b$; $f'(a)$ means the right-hand derivative and $f'(b)$ the left-hand derivative. Thus,

$$f'(a) = \lim_{h \to 0+} \frac{f(a + h) - f(a)}{h}, \qquad f'(b) = \lim_{h \to 0-} \frac{f(b + h) - f(b)}{h}$$

Then f' denotes the function defined on $[a, b]$ with these values. $\mathbf{C}^1[a, b]$ is the real vector space of all real functions on $[a, b]$ for which the derivative function f' is defined and continuous on $[a, b]$.

The meaning of $\mathbf{C}^1(-\infty, \infty)$ should be clear.

Next, if k is a positive integer, $\mathbf{C}^k(a, b)$ denotes the vector space of real functions defined on (a, b) whose kth derivative function exists and is continuous on (a, b). That is, $f \in \mathbf{C}^k(a, b)$ if and only if $f^{(k)} \in \mathbf{C}^0(a, b)$. Of course, if $f^{(k)} \in \mathbf{C}^0(a, b)$, then the lower order derivatives $f^{(k-1)}, \ldots, f', f$ all exist and are continuous on (a, b). $\mathbf{C}^k[a, b]$ is defined similarly, using right-hand derivatives at a and left-hand derivatives at b.

Example 4   The *space* $\mathbf{C}^\infty(a, b)$ is the real vector space of all real-valued functions defined on (a, b) which have continuous derivatives of every order on (a, b). It is called the space of *infinitely differentiable* functions.

By way of illustration we consider a few particular functions:

1. $f(x) = \sin x$ f is in $\mathbf{C}^\infty(-\infty, \infty)$.
2. $f(x) = 1/x$ f is in $\mathbf{C}^\infty(0, \infty)$ but not in $\mathbf{C}^k[0, 1]$ for any nonnegative integer k.
3. $f(x) = \sqrt{x}$ f is in $\mathbf{C}^\infty(0, \infty)$ and in $\mathbf{C}^0[0, \infty)$ but not in $\mathbf{C}^1[0, \infty)$.

Although other function spaces are defined in more advanced courses, those given here will provide almost all we need in this text. In Section 9.2 we will define spaces of complex-valued functions.

Exercises

In these exercises tell whether the given set of functions is a vector space over $\mathbf{R}$, if addition and multiplication by a scalar are defined in the usual way. For those that are not vector spaces, indicate which axioms fail to hold. The functions are real-valued.

1. All functions with $f(1) = f(2)$.
2. All functions with $f(1) = 3f(2)$.
3. All functions with $f(1) = f(2) + 1$.
4. All functions such that $|f(x)| \leq 1$ for all x.
5. All functions with $f(x) \geq 0$ for all x.
6. All functions with $f'(0) = 1$.

7. All functions with $f'(1) = 0$.

8. All functions with $f'(x) \leq 0$ for all x.

9. All functions differentiable on $a < x < b$.

10. All functions such that the Riemann integral $\int_a^b f(x)\, dx$ exists.

11. All polynomials, of any degree.

12. All rational functions $f(x) = P_1(x)/P_2(x)$ where P_1 and P_2 are polynomials.

13. All functions n times differentiable on $a < x < b$ (that is, $d^n f(x)/dx^n$ exists on $a < x < b$).

14. All functions with period $1 : f(x - 1) = f(x)$ for all x.

4.6   Subspaces

The preceding sections were concerned with defining and illustrating the notion of a vector space. The examples show that vectors can be other than directed line segments, and a vector space can be other than $\mathbf{R}^n$.

In certain cases interest might be centered upon some subset $\mathbf{W}$ of the vectors in a vector space $\mathbf{V}$. In such an event we might want to know whether $\mathbf{W}$ itself possesses the properties of a vector space in order that we might study $\mathbf{W}$ by using facts known about vector spaces. Hence we are led to the following definition.

DEFINITION 4.6   Let $\mathbf{V}$ be a vector space over a field $\mathbf{K}$ and let $\mathbf{W}$ be a subset of $\mathbf{V}$. Then $\mathbf{W}$ is a *subspace* of $\mathbf{V}$ if $\mathbf{W}$ is a vector space over $\mathbf{K}$.

THEOREM 4.6   Let $\mathbf{V}$ be a vector space over a field $\mathbf{K}$ and let $\mathbf{W}$ be a subset of $\mathbf{V}$. Then $\mathbf{W}$ is a subspace of $\mathbf{V}$ if and only if:

1. $\mathbf{0}$ is an element of $\mathbf{W}$.
2. If v_1 and v_2 are elements of $\mathbf{W}$, then $v_1 + v_2$ is an element of $\mathbf{W}$.
3. If v is an element of $\mathbf{W}$ and c an element of $\mathbf{K}$, then cv is in $\mathbf{W}$.

PROOF: The proof of this theorem is simply the verification of each requirement in the definition of a vector space. Associativity and commutivity of vectors carry over to the elements in the subset $\mathbf{W}$ of $\mathbf{V}$. Closure under addition and the inclusion in $\mathbf{W}$ of $\mathbf{0}$ are conditions (1) and (2) of the theorem. The remaining conditions on elements of $\mathbf{V}$ in order that $\mathbf{V}$ be a vector space carry over to the elements of $\mathbf{W}$ except those obtained through closure under scalar multiplication, which is condition (3) above.   ∎

The value of the theorem is that it gives a simple means of checking whether a subset of a known vector space constitutes a vector space. We shall illustrate this use in the following examples.

If **V** is any vector space, then the subset **W** which contains only the vector **0** is a subspace, since the three requirements of Theorem 4.6 are met. Also, **V** is itself a subspace of **V**. These two subspaces are sometimes called *trivial* subspaces of **V**. Ordinarily, interest centers on nontrivial subspaces.

Example 1   In Example 1 of Section 4.4 we saw that the set of all 2×2 matrices formed a vector space, which we denoted **M**, over the field **R**. Consider the subset **D** consisting of *diagonal* matrices, that is, matrices of the form

$$\begin{pmatrix} a_{11} & 0 \\ 0 & a_{22} \end{pmatrix}$$

for arbitrary a_{11} and a_{22}. Does **D** form a subspace of **M**?

Certainly

$$\mathbf{0} = \begin{pmatrix} 0 & 0 \\ 0 & 0 \end{pmatrix}$$

is in **D** since this is the case $a_{11} = a_{22} = 0$. Further, the sum of any two diagonal matrices is diagonal since

$$\begin{pmatrix} a_{11} & 0 \\ 0 & a_{22} \end{pmatrix} + \begin{pmatrix} b_{11} & 0 \\ 0 & b_{22} \end{pmatrix} = \begin{pmatrix} a_{11} + b_{11} & 0 \\ 0 & a_{22} + b_{22} \end{pmatrix}$$

And finally for any real scalar c,

$$c \begin{pmatrix} a_{11} & 0 \\ 0 & a_{22} \end{pmatrix} = \begin{pmatrix} ca_{11} & 0 \\ 0 & ca_{22} \end{pmatrix}$$

is again an element of **D**. Hence **D** is a subspace of **M**.

Example 2   Is $\mathbf{R}^2$ a subspace of $\mathbf{R}^3$? We showed in Example 2 of Section 4.4 that $\mathbf{R}^n$ is a vector space for n an integer ≥ 1. Hence we know $\mathbf{R}^2$ is a vector space, as is $\mathbf{R}^3$. But the elements of $\mathbf{R}^2$ are ordered pairs of numbers of the form (x_1, x_2), whereas the elements of $\mathbf{R}^3$ are ordered triples (x_1, x_2, x_3). For example $(3, 2)$ is an element of $\mathbf{R}^2$. Since $(3, 2)$ is not an ordered triple, it is not an element of $\mathbf{R}^3$ and similarly no element of $\mathbf{R}^2$ is in $\mathbf{R}^3$. Hence $\mathbf{R}^2$ is not a subspace of $\mathbf{R}^3$. (See Exercise 4 of Section 5.3 for further comment.)

Example 3   The Cartesian coordinates of a point in 3-space, $\mathbf{G}^3$, can be thought of as forming an element in $\mathbf{R}^3$. The equation of a plane in 3-space is $ax_1 + bx_2 + cx_3 = d$, where (x_1, x_2, x_3) represents a point on the plane provided it satisfies the equation. If addition and multiplication for points in 3-space are defined as for $\mathbf{R}^3$, what conditions on the scalars a, b, c, and d cause the resulting plane to be a subspace of three-dimensional space? To begin, the zero element of $\mathbf{R}^3$ must lie on the plane. Hence $(0, 0, 0)$ must satisfy the equation. This implies that $d = 0$. Hence the equation can be no more general than $ax_1 + bx_2 + cx_3 = 0$. For any a, b, and c, let $x = (x_1, x_2, x_3)$ and $y =$

(y_1, y_2, y_3) be "vectors" satisfying the equation. Then the point $x + y = (x_1 + y_1, x_2 + y_2, x_3 + y_3)$ also satisfies the equation since

$$a(x_1 + y_1) + b(x_2 + y_2) + c(x_3 + y_3)$$
$$= (ax_1 + bx_2 + cx_3) + (ay_1 + by_2 + cy_3) = 0$$

Further, for any scalar, k, $kx = (kx_1, kx_2, kx_3)$ lies on the plane since

$$a(kx_1) + b(kx_2) + c(kx_3) = k(ax_1 + bx_2 + cx_3) = 0$$

Hence any plane represented as $ax_1 + bx_2 + cx_3 = 0$ is a subspace of three-dimensional space.

A set of vectors of special interest is the set resulting from the intersection of two arbitrary subspaces, $\mathbf{W}_1$ and $\mathbf{W}_2$, of a vector space $\mathbf{V}$ over a field $\mathbf{K}$.

THEOREM 4.7 For any arbitrary subspaces $\mathbf{W}_1$ and $\mathbf{W}_2$ of a vector space $\mathbf{V}$ over a field $\mathbf{K}$, the intersection $\mathbf{W}_1 \cap \mathbf{W}_2$ is also a subspace of $\mathbf{V}$.

PROOF: The proof of this theorem requires verifying the conditions of Theorem 4.6.

1. Since both $\mathbf{W}_1$ and $\mathbf{W}_2$ are vector spaces they contain the vector $\mathbf{0}$, and hence $\mathbf{W}_1 \cap \mathbf{W}_2$ contains $\mathbf{0}$. Note that this guarantees $\mathbf{W}_1 \cap \mathbf{W}_2$ is not empty.
2. If v and v' are elements of $\mathbf{W}_1 \cap \mathbf{W}_2$ they must both be in $\mathbf{W}_1$ and both in $\mathbf{W}_2$. But $\mathbf{W}_1$ and $\mathbf{W}_2$ are vector spaces and therefore $v + v'$ must be in both. Hence $v + v'$ is in $\mathbf{W}_1 \cap \mathbf{W}_2$.
3. If v is an element of $\mathbf{W}_1$ and $\mathbf{W}_2$ and $c \in \mathbf{K}$, then v is in $\mathbf{W}_1$, which implies that cv is in $\mathbf{W}_1$. Similarly, cv will be in $\mathbf{W}_2$ which implies cv is in $\mathbf{W}_1 \cap \mathbf{W}_2$. Thus the theorem is proved. ∎

Exercises

1. In each case consider the set of all functions f on $-2 \le x \le 2$ with the indicated properties. Which are subspaces of $\mathbf{C}^0[-2, 2]$?

   (a) f continuous, $f(1) = f(2)$
   (b) f continuous, $f(1) = f(2) + 1$
   (c) f continuous, $|f(x)| \le 1$
   (d) f continuous, $f(1) = f'(0)$
   (e) f continuous, $f(x) \ge 0$
   (f) f continuous, $f'(0) = 1$
   (g) f continuous, $f'(1) = 0$

(h) f continuous, $f'(x) \leq 0$

(i) f differentiable on $-2 < x < 2$

(j) f constant on $[-2, 2]$

(k) f continuous, f periodic with period 1

2.   In each case consider the set of all 4-tuples (x_1, x_2, x_3, x_4) with the indicated properties. Which sets are subspaces of $\mathbf{R}^4$?

(a) $x_4 = 0$

(b) $x_1 = 2x_2$

(c) $x_2 - x_3 = 3$

(d) $x_4 = x_3^2$

(e) $x_1 = e^{x_2}$

(f) $|x_1| = |x_4|$

(g) $x_1 = \max(x_2, x_3, x_4)$

(h) $x_1 + 2x_2 \leq 0$

3.   A *hyperplane* through the origin in $\mathbf{R}^4$ is the set of all (x_1, x_2, x_3, x_4) satisfying an equation

$$ax_1 + bx_2 + cx_3 + dx_4 = 0$$

where a, b, c, d are fixed real numbers. Show that each such hyperplane is a subspace of $\mathbf{R}^4$.

4.   Show that the set of all polynomials of degree $<n$ form a subspace of the vector space of polynomials of degree $\leq n$.

5.   Show that polynomials of the form $ax^n + b$ with a, b real form a subspace of the vector space of polynomials of degree $\leq n$.

6.   Reduce the three conditions in Theorem 4.6 to a set of two conditions equivalent to the original three. Can these be reduced further?

4.7   Linear combinations and generators

Essential properties of a vector space are that if v_1 and v_2 are elements, then $v_1 + v_2$ is an element, and if c_1 and c_2 are scalars, then $c_1 v_1 + c_2 v_2$ is an element. If v_1 and v_2 are held fixed while c_1 and c_2 are allowed to vary in $\mathbf{K}$, we obtain a subset of $\mathbf{V}$. It is very important in the theory of vector spaces to discuss the nature of subsets obtained in this way. This discussion will lead us to the concepts of *basis* and *dimension* of a vector space. First, we introduce some helpful terminology.

DEFINITION 4.7 Let **V** be a vector space over a field **K**, and let $\{v_1, \ldots, v_n\}$ be a finite subset of **V**. A vector v in **V** is called a *linear combination* of the vectors $v_1, \ldots, v_n$ if there are scalars $c_1, \ldots, c_n$ in **K** such that

$$v = c_1 v_1 + c_2 v_2 + \cdots + c_n v_n = \sum_{i=1}^{n} c_i v_i \qquad (1)$$

More generally, let S be any subset of vectors in **V**. A vector v in **V** is called a *linear combination* of vectors in S if v is a linear combination of some finite subset of vectors in S.

DEFINITION 4.8 Let S be a subset of vectors in **V**. The set of all vectors in **V** which are linear combinations of vectors in S is called the set *generated* or *spanned* by S, or the *linear span* of S, and is denoted here by the symbol $[S]$. The vectors in the set S are called *generators* of $[S]$. In particular if $S = \{v_1, v_2, \ldots, v_n\}$ is finite, the set generated by S is denoted by $[v_1, v_2, \ldots, v_n]$. The span of the empty set is considered to be empty.

Thus v is in $[v_1, v_2, \ldots, v_n]$ if and only if there are scalars $c_1, c_2, \ldots, c_n$ such that Eq. (1) holds.

Example 1 Consider **V** to be $\mathbf{R}^n$, a vector space over the field **R**. Let

$$v_1 = (1, 0, \ldots, 0), v_2 = (0, 1, 0, \ldots, 0), \ldots, v_{n-1} = (0, 0, \ldots, 0, 1, 0)$$

That is, v_i is an n-tuple of zeros with a 1 in the ith position. Let x be any vector of the form $(x_1, x_2, \ldots, x_{n-1}, 0)$. Clearly $x = x_1 v_1 + x_2 v_2 + \cdots + x_{n-1} v_{n-1}$. Hence the vectors $v_1, v_2, \ldots, v_{n-1}$ generate the set of all ordered n-tuples with last element zero. Note that each element of the set is an element of $\mathbf{R}^n$, and hence $[v_1, v_2, \ldots, v_{n-1}]$ is a subset of $\mathbf{R}^n$.

THEOREM 4.8 Let $\{v_1, v_2, \ldots, v_n\}$ be a finite set of vectors in a vector space **V** over a field **K**. Then the set **W** generated by $\{v_1, \ldots, v_n\}$ is a subspace of **V**.

PROOF: We must show that the conditions of Theorem 4.6 are satisfied. (1) The linear combination $0v_1 + 0v_2 + \cdots + 0v_n$ is the zero vector **0**, which is therefore in **W**. (2) If x and y are in **W**, then

$$x = c_1 v_1 + c_2 v_2 + \cdots + c_n v_n$$

and

$$y = d_1 v_1 + d_2 v_2 + \cdots + d_n v_n$$

for some elements $c_1, c_2, \ldots, c_n, d_1, d_2, \ldots, d_n$ of **K**. Further,

$$c_1 + d_1, c_2 + d_2, \ldots, c_n + d_n$$

are elements of **K** since **K** is a field. Hence

$$(c_1 + d_1)v_1 + (c_2 + d_2)v_2 + \cdots + (c_n + d_n)v_n$$

is in **W**. But this is just

$$
\begin{aligned}
c_1 v_1 &+ d_1 v_1 + c_2 v_2 + d_2 v_2 + \cdots + c_n v_n + d_n v_n \\
&= c_1 v_1 + c_2 v_2 + \cdots + c_n v_n + d_1 v_1 + d_2 v_2 + \cdots + d_n v_n \\
&= x + y
\end{aligned}
$$

Hence $x + y$ is in **W**. (3) Similarly, for x in **W**, then $x = c_1 v_1 + c_2 v_2 + \cdots + c_n v_n$. For any scalar d in **K**,

$$dx = d(c_1 v_1) + d(c_2 v_2) + \cdots + d(c_n v_n) = (dc_1)v_1 + (dc_2)v_2 + \cdots + (dc_n)v_n$$

Since $(dc_1, dc_2, \ldots, dc_n)$ is an n-tuple of elements in **K**, dx is a linear combination of the generators $v_1, \ldots, v_n$ and hence is in **W**. Therefore **W** is a subspace of **V**.   ∎

Example 2   Consider the vector space **M** of 2×2 matrices (see Example 1 of Section 4.4). Let

$$v_1 = \begin{pmatrix} 1 & 0 \\ 0 & 0 \end{pmatrix}, \quad v_2 = \begin{pmatrix} 0 & 0 \\ 0 & 1 \end{pmatrix}, \quad v_3 = \begin{pmatrix} 0 & 0 \\ 1 & 1 \end{pmatrix}, \quad v_4 = \begin{pmatrix} 1 & 0 \\ 1 & 1 \end{pmatrix}$$

We wish to determine $[v_1, v_2, v_3, v_4]$, the subspace generated by these four matrices. Since $v_4 = v_1 + v_3$, we see that $[v_1, v_2, v_3, v_4] = [v_1, v_2, v_3]$ and therefore we need only consider those matrices which can be generated by v_1, v_2, and v_3. Next, note that any linear combination of v_1, v_2, and v_3 will have 0 as the element in the upper right place. Therefore, any matrix generated must be of the form

$$\begin{pmatrix} x_1 & 0 \\ x_2 & x_3 \end{pmatrix} \tag{2}$$

Conversely, every such matrix is a combination of v_1, v_2, v_3, for consider any matrix

$$x = \begin{pmatrix} x_1 & 0 \\ x_2 & x_3 \end{pmatrix}$$

We now proceed to find c_1, c_2, and c_3 such that $c_1 v_1 + c_2 v_2 + c_3 v_3 = x$, that is,

$$\begin{pmatrix} c_1 & 0 \\ c_3 & c_2 + c_3 \end{pmatrix} = \begin{pmatrix} x_1 & 0 \\ x_2 & x_3 \end{pmatrix}$$

Clearly, $c_1 = x_1$, $c_3 = x_2$ and $c_2 = x_3 - x_2$ give the desired result. Therefore, $[v_1, v_2, v_3]$ is the subspace consisting of all matrices of the form (2). (These matrices are called *lower triangular*.)

Example 3   Let **V** be the space $C^0(-\infty, \infty)$ of all continuous real-valued functions on **R**. Consider the functions $v_1 = e^{2x}$, $v_2 = e^{-3x}$. Then $[v_1, v_2]$ is the subspace containing all functions of the form $ae^{2x} + be^{-3x}$, where a and b are real numbers. Note that

$$\frac{d}{dx}(av_1 + bv_2) = 2av_1 - 3bv_2$$

is also in $[v_1, v_2]$.

Exercises

1. In the vector space of all real polynomials $P(x)$ give a simple description or characterization of the subspace spanned by each of the following subsets S.

    (a) $S = \{x, x^3\}$
    (b) $S = \{1, x^2, x^4\}$
    (c) $S = \{x, x^3, \ldots, x^{2n+1}, \ldots\}$
    (d) $S = \{x - 1, (x - 1)^2\}$

2. Describe the subspace spanned by each of the following subsets S of the given space **V**.

    (a) $S = \{\ln |x|, \ln |2x|\}$, $V = F(-\infty, 0)$
    (b) $S = \{x, |x|\}$, $V = F(-\infty, \infty)$
    (c) $S = \{x, e^x\}$, $V = F(-\infty, \infty)$
    (d) $S = \{x, 1, 1/x\}$, $V = F(0, \infty)$

3. Prove that the set of all complex numbers is a vector space over the reals (that is, over **R**). What is the subspace generated by the set $\{i\}$?

4. In G^3, describe the subspace generated by the indicated subset S of directed arrows.

    (a) S contains the three vectors whose endpoints have Cartesian coordinates $(1, -1, 1)$, $(1, 1, 1)$, $(1, 0, 1)$.
    (b) S contains the vectors whose endpoints have coordinates $(1, 0, 0)$, $(1, 1, 1)$, $(0, 0, 1)$.
    (c) S contains the vectors whose endpoints have coordinates $(1, 0, 0)$, $(1, 1, 1)$, $(0, 0, 0)$.

5. Let **V** be a vector space and S a finite subset. Prove that $[S]$ contains S, and that there is no other subspace **W** of **V** with the property that **W** contains S and **W** is contained in $[S]$. That is, $[S]$ is the "smallest" subspace containing S.

4.8   Linear independence

Consider two vectors v_1 and v_2 in $\mathbf{G}^2$ which have corresponding coordinates in $\mathbf{R}^2$ of (2, 1) and (4, 2), respectively, relative to a given coordinate system. The subspace $[v_1, v_2]$, that is, the subspace of vectors generated by v_1 and v_2, is the set of all vectors of the form $c_1 v_1 + c_2 v_2$, where c_1 and c_2 are in $\mathbf{R}$. But since the coordinates of v_2 are (4, 2) = 2(2, 1) = twice the coordinates of v_1, we have $v_2 = 2v_1$. Hence

$$c_1 v_1 + c_2 v_2 = c_1 v_1 + c_2(2v_1) = (c_1 + 2c_2)v_1$$

Therefore the subspace $[v_1, v_2]$ evidently consists of all multiples of v_1 and is thus the same as the subspace generated by v_1 alone. The vectors v_1 and v_2 do not in this case generate a "larger" subspace than the vector v_1 alone, since v_2 is already in the subspace generated by v_1. Since this kind of situation occurs often in arbitrary vector spaces, it is helpful to formulate the following concepts.

DEFINITION 4.9   Let $\mathbf{V}$ be a vector space over a field $\mathbf{K}$ and let $S = \{v_1, \ldots, v_n\}$ be a finite subset of $\mathbf{V}$. S is said to be *linearly dependent* over $\mathbf{K}$ if there exist scalars $c_1, c_2, \ldots, c_n$ in $\mathbf{K}$, not all of which are zero, such that

$$c_1 v_1 + \cdots + c_n v_n = \mathbf{0} \tag{1}$$

If such scalars do not exist, S is said to be *linearly independent*.

The definition of linear independence is equivalent to this alternative.

DEFINITION 4.9' S is linearly independent if and only if from the equation

$$c_1 v_1 + \cdots + c_n v_n = \mathbf{0}$$

it follows that $c_1 = \cdots = c_n = 0$. That is, if a linear combination of $v_1, \ldots, v_n$ is $\mathbf{0}$, the scalars must all be zero.

The equivalence of these definitions may seem obvious, but it will probably be worthwhile for the reader to write out a careful explanation as requested in the exercises.

We adopt the convention that the *null set* $\varnothing$ is linearly independent. (One may plausibly say that if S contains no vectors, no linear combination can be the zero vector.)

DEFINITION 4.10   Let $\mathbf{V}$ be a vector space over a field $\mathbf{K}$, let $S = \{v_1, v_2, \ldots, v_n\}$ be a finite subset of $\mathbf{V}$, and v an arbitrary vector in $\mathbf{V}$. The vector v is said to be *linearly dependent on* S (over $\mathbf{K}$) if it is a linear combination of the vectors in S. If not, v is said to be *linearly independent of* S.

These definitions can be extended to arbitrary subsets S of **V** as in Definition 4.7 of Section 4.7.

Example 1   Consider the meaning of linear dependence for geometric vectors in the plane. If v_1 and v_2 are two such vectors, dependence means that $c_1 v_1 + c_2 v_2 = \mathbf{0}$ for scalars c_1 and c_2, not both zero. If, for example, $c_1 \neq 0$, this means that $v_1 = -(c_2/c_1)v_2$, or in other words that v_1 and v_2 have the same (or opposite) direction. If v_1, v_2, and v_3 are three vectors in the plane, then dependence means $c_1 v_1 + c_2 v_2 + c_3 v_3 = \mathbf{0}$ with c_1, c_2, c_3 not all zero. If, for example, $c_1 \neq 0$, this is equivalent to $v_1 = -(c_2 v_2 + c_3 v_3)/c_1$. If v_2 and v_3 are collinear, this implies that v_1, v_2, and v_3 are all collinear. If v_2 and v_3 are not collinear, this means that v_1 is expressible as a linear combination of v_2 and v_3, as is indeed always possible. It therefore follows that three vectors in $\mathbf{G}^2$ are always linearly dependent. Two vectors in $\mathbf{G}^2$ are dependent if and only if they are collinear.

Example 2   Consider v_1 and v_2 defined as above with coordinates $(2, 1)$ and $(4, 2)$, respectively. We have seen that $v_2 = 2v_1$ and hence v_2 is linearly dependent on v_1. Note also that $v_1 = \frac{1}{2}v_2$ and hence v_1 is linearly dependent upon v_2.

Example 3   Let $\mathbf{V} = \mathbf{R}^n$ and let e_i be the n-tuple with 1 in the ith position and 0 in the other $n - 1$ positions $(i = 1, 2, \ldots, n)$. For example, $e_1 = (1, 0, \ldots, 0)$, $e_2 = (0, 1, 0, \ldots, 0)$, and so on. Then $c_1 e_1 + \cdots + c_n e_n = (c_1, c_2, \ldots, c_n)$. This is equal to the zero vector $\mathbf{0} = (0, 0, \ldots, 0)$ if and only if $c_1 = c_2 = \cdots = c_n = 0$. Therefore the set $E = \{e_1, e_2, \ldots, e_n\}$ is linearly independent.

Example 4   Let **V** be the vector space of all functions defined on an interval $[x_0, x_1]$, over **R**. Let $S = \{f_1, f_2, \ldots, f_n\}$ be a set of n functions. To say that S is linearly dependent is to say that there are real numbers $c_1, c_2, \ldots, c_n$, not all zero, such that $c_1 f_1 + c_2 f_2 + \cdots + c_n f_n$ is the zero vector of **V**, that is,

$$\sum_{i=1}^{n} c_i f_i(x) = 0 \quad \text{for } all \quad x \in [x_0, x_1]$$

As an illustration we shall show that $S = \{e^x, e^{2x}\}$ is linearly independent on any interval (x_0, x_1). To prove this, suppose that

$$c_1 e^x + c_2 e^{2x} = 0, \qquad x_0 \le x \le x_1$$

Since $e^x \neq 0$, this is equivalent to

$$c_1 + c_2 e^x = 0, \qquad x_0 \le x \le x_1$$

Since this must hold for all x in the interval, it follows that the derivative $c_2 e^x$ must also be zero for all x. Hence $c_2 = c_1 = 0$.

THEOREM 4.9

1. Any set which contains the zero vector is dependent.
2. Any set which consists of a single nonzero vector is independent.
3. Any subset of an independent set is independent.
4. Any set containing a dependent subset is dependent.

PROOF: This is left as an exercise.   ∎

The notion of linear independence can be extended to infinite subsets S of a vector space **V**. We say that S is *linearly independent* if every *finite* subset of S is linearly independent in the previous sense. We shall seldom have occasion for this extension.

It is important to be able to test a given set of vectors in a space **V** for independence. The following theorem is often useful in this connection.

THEOREM 4.10   Let **V** be the space of ordered n-tuples over a field **K**, that is, $\mathbf{V} = \mathbf{K}^n$. Let

$$v_i = (a_{1i}, a_{2i}, \ldots, a_{ni}) \qquad i = 1, 2, \ldots, r$$

The set $S = \{v_1, v_2, \ldots, v_r\}$ is dependent over **K** if and only if the system of n equations in r unknowns

$$
\begin{aligned}
a_{11}x_1 + a_{12}x_2 + \cdots + a_{1r}x_r &= 0 \\
a_{21}x_1 + a_{22}x_2 + \cdots + a_{2r}x_r &= 0 \\
&\cdots \\
a_{n1}x_1 + a_{n2}x_2 + \cdots + a_{nr}x_r &= 0
\end{aligned}
\qquad (2)
$$

can be satisfied by scalars $x_1, x_2, \ldots, x_r$ in **K** other than the "trivial solution" $x_1 = x_2 = \cdots = x_r = 0$.

PROOF: The statement that S is dependent is equivalent to the statement that there are scalars $x_1, x_2, \ldots, x_n$, not all zero, such that

$$x_1v_1 + x_2v_2 + \cdots + x_rv_r = \mathbf{0} \qquad (3)$$

This means that each of the n components of $x_1v_1 + \cdots + x_rv_r$ must be zero, which is exactly the meaning of (2).   ∎

A full discussion of systems of equations of the form (2) will be given in Chapter 10. For the moment we are content to look at simple examples.

Example 5   Let $v_1 = (2, -1, 1)$, $v_2 = (0, 2, -1)$, $v_3 = (1, -1, 0)$. To test these for independence over **R**, we consider the vector equation $x_1v_1 + x_2v_2 + x_3v_3 = \mathbf{0}$, or

$$x_1(2, -1, 1) + x_2(0, 2, -1) + x_3(1, -1, 0) = (0, 0, 0)$$

This is equivalent to the system

$$2x_1 \qquad\quad + x_3 = 0$$
$$-x_1 + 2x_2 - x_3 = 0$$
$$x_1 - \quad x_2 \qquad\;\; = 0$$

Adding the first two equations yields $x_1 + 2x_2 = 0$, and from this and the third equation it follows that x_1 and x_2 must be zero. Hence $x_3 = 0$, and there is no solution except the trivial one $x_1 = x_2 = x_3 = 0$. Consequently the three given vectors form an independent set.

Exercises

1. Show that the following sets of vectors in $\mathbf{R}^2$ or $\mathbf{R}^3$ are linearly independent over $\mathbf{R}$.

(a) $S = \{(-2, 1, 0), (0, 1, 2)\}$
(b) $S = \{(1, 2), (3, 4)\}$
(c) $S = \{(\sqrt{2}, 1), (1, \sqrt{3})\}$

2. Show that the following sets of vectors are linearly dependent over $\mathbf{R}$.

(a) $\{(1, -1, 0), (-2, 2, 0)\}$
(b) $\{(0, 3, 3), (1, 1, 3), (-10, 11, -9)\}$
(c) $\{(2, 3, 5), (5, 8, 11), (1, 2, 1)\}$

3. Are the following sets of vectors linearly independent in the space of ordered pairs of complex numbers over the field of complex numbers?

(a) $S = \{(1, i), (i, 1)\}$
(b) $S = \{(1 + i, 2 - i), (2 + 2i, 4 - i)\}$
(c) $S = \{(1, i), (i^2, i^3)\}$

4. Find all k in $\mathbf{R}$ such that the following set is linearly dependent in $\mathbf{R}^3$.

$$S = \{(1, 1, 0), (0, 1, 1), (k, 0, 2)\}$$

5. Consider the vector space $\mathbf{F}(-\infty, \infty)$ of all functions with domain $(-\infty, \infty)$ over the field $\mathbf{R}$. Prove independence or dependence, as the case may be, of each of the following sets.

(a) $S = \{x, |x|\}$
(b) $S = \{x, e^x\}$
(c) $S = \{e^{2x}, e^{3x}\}$
(d) $S = \{\sin x, \cos x\}$
(e) $S = \{x, \sin x\}$
(f) $S = \{\sin 2x, \sin x, \cos x\}$
(g) $S = \{x^2, (x + 1)^2, (x + 2)^2\}$

6. Find all functions f in $\mathbf{C}^1(-\infty, \infty)$, the space of real continuously differentiable functions, for which the set $\{f, f'\}$ is a linearly dependent subset over the reals (where f' means the derivative of f).

7. Prove Theorem 4.9.

8. In each case, either prove the stated proposition or disprove it by giving a specific example in which it is not true (a "counterexample").

   (a) If S_1 is a linearly independent subset of $\mathbf{V}$ and so is S_2, then the intersection of S_1 and S_2 is a linearly independent subset.

   (b) Same as (a) for the union of S_1 and S_2.

9. Prove that the vectors (x_1, y_1) and (x_2, y_2) are linearly dependent in $\mathbf{R}^2$ (or $\mathbf{C}^2$) if and only if $x_1 y_2 = x_2 y_1$.

10. Prove that Definitions 4.9 and 4.9′ are equivalent.

4.9 Bases and dimension

DEFINITION 4.11 Let $\mathbf{V}$ be a vector space over a field $\mathbf{K}$. A *basis* of $\mathbf{V}$ is a linearly independent set of elements of $\mathbf{V}$ which generates $\mathbf{V}$.

Certain questions which are immediate are: Does every vector space have a basis, and if a basis exists, is it unique? We have already seen that $E = \{e_1, e_2, \ldots, e_n\}$ is an independent set of $\mathbf{R}^n$, and further for any element

$$x = (x_1, x_2, \ldots, x_n) \quad \text{of} \quad \mathbf{R}^n, \quad x = \sum_{i=1}^{n} x_i e_i$$

Hence E is a basis for $\mathbf{R}^n$. However, other bases can be found such as $(1, 0, 0, \ldots, 0)$, $(1, 1, 0, \ldots, 0)$, $\ldots$, $(1, 1, 1, \ldots, 1)$. Hence we see that bases are not necessarily unique.

We proceed to derive properties of bases and defer the question of existence. The following theorem and corollary show that the number of elements in a basis, if finite, is unique.

THEOREM 4.11 Let $\mathbf{V}$ be a vector space over a field $\mathbf{K}$. Suppose that $\mathbf{V}$ has a basis $\{v_1, v_2, \ldots, v_m\}$ containing a finite number of elements. Let $w_1, w_2, \ldots, w_n$ be elements of $\mathbf{V}$ where $n > m$. Then the set $S = \{w_1, w_2, \ldots, w_n\}$ is linearly dependent.

PROOF: If any $w_i = \mathbf{0}$, the set S is dependent. Therefore assume all the $w_i \neq \mathbf{0}$. Since $\{v_1, \ldots, v_m\}$ is a basis, there exist scalars $c_1, \ldots, c_m$ in $\mathbf{K}$ such that

$$w_1 = c_1 v_1 + \cdots + c_m v_m$$

Since $w_1 \neq \mathbf{0}$, there must be at least one nonzero c_i. By reordering the vectors $v_1, \ldots, v_m$, if necessary, we can take this nonzero scalar to be c_1. Then it is possible to solve for

$$v_1 = (w_1 - c_2 v_2 - \cdots - c_m v_m)/c_1$$

Now let $\mathbf{V}_1 = [w_1, v_2, \ldots, v_m]$; that is, $\mathbf{V}_1$ is the space generated by $w_1, v_2, \ldots, v_m$. Since v_1 is a linear combination of these vectors, and v_1, $v_2, \ldots, v_m$ generate $\mathbf{V}$, it follows that $\mathbf{V}_1 \supset \mathbf{V}$. By Theorem 4.8, $\mathbf{V}_1 \subset \mathbf{V}$, and hence $\mathbf{V}_1 = \mathbf{V}$.

It has thus been shown that when v_1 is replaced by w_1 in the basis, the resulting set still generates $\mathbf{V}$. The above procedure is now repeated. There exist scalars $b_1, b_2, \ldots, b_m$ such that

$$w_2 = b_1 w_1 + b_2 v_2 + \cdots + b_m v_m$$

If $b_2 = b_3 = \cdots = b_m = 0$, the set $\{w_1, w_2\}$ is dependent, hence by Theorem 4.9, the set S is dependent, as was to be proved. If $b_2, b_3, \ldots, b_m$ are not all zero, we can take $b_2 \neq 0$, by reordering the vectors if necessary. Then

$$v_2 = (w_2 - b_1 w_1 - b_3 v_3 - \cdots - b_m v_m)/b_2$$

Let $\mathbf{V}_2 = [w_1, w_2, v_3, \ldots, v_m]$. Since $w_1, v_2, \ldots, v_m$ generates $\mathbf{V}$, and v_2 is a combination of $w_1, w_2, v_3, \ldots, v_m$, it follows that $\mathbf{V}_2 = \mathbf{V}$.

It is clear that this process can be continued.[7] At each step we show either that a subset of $\{w_1, \ldots, w_n\}$ is dependent or else we can replace a vector v_i by a vector w_i to obtain a new set which generates $\mathbf{V}$. Finally, all the v_i's have been replaced, and at this point it has been shown that $\mathbf{V} = [w_1, \ldots, w_m]$. But then w_n is a linear combination

$$w_n = c_1 w_1 + \cdots + c_m w_m$$

whence

$$c_1 w_1 + \cdots + c_m w_m - w_n = \mathbf{0}$$

Thus $\{w_1, \ldots, w_m, w_n\}$ is dependent, and so is S. The proof is complete. ∎

COROLLARY Let $\mathbf{V}$ be a vector space. If $\mathbf{V}$ has a finite basis, every basis has the same finite number of elements.

PROOF: Let $\{v_1, \ldots, v_m\}$ and $\{w_1, \ldots, w_n\}$ be two bases. It is impossible to have $n > m$ since then $\{w_1, \ldots, w_n\}$ would be dependent and could not be a basis. Likewise it is impossible to have $m > n$. Therefore $m = n$. ∎

This result makes it possible to define the dimension of $\mathbf{V}$ in an unambiguous way as follows.

---

[7] A more formal discussion using mathematical induction is possible here.

DEFINITION 4.12 A vector space **V** which has a basis consisting of a finite number of vectors is called a *finite dimensional* vector space. Other spaces are called *infinite dimensional*. The *dimension* of a finite dimensional space is the number of vectors in a basis.

A space **V** over **K** may consist of a zero vector **0** alone. This space has no basis. The convention is adopted that this space has *zero dimension*.

Example 1 The four matrices

$$\begin{pmatrix} 1 & 0 \\ 0 & 0 \end{pmatrix}, \quad \begin{pmatrix} 0 & 1 \\ 0 & 0 \end{pmatrix}, \quad \begin{pmatrix} 0 & 0 \\ 1 & 0 \end{pmatrix}, \quad \begin{pmatrix} 0 & 0 \\ 0 & 1 \end{pmatrix}$$

form a basis for the vector space **M** of 2×2 matrices over **R**. Hence **M** is a four-dimensional vector space.

Example 2 The $n + 1$ polynomials $1, x, x^2, \ldots, x^n$ form a basis for the space of all polynomials of degree $\leq n$. To prove linear independence, suppose that a linear combination of $1, x, \ldots, x^n$ is the zero polynomial, say

$$a_0 + a_1 x + \cdots + a_n x^n = 0, \quad \text{for all } x$$

Since this is an identity, it is well known that all coefficients must be zero, that is, $a_0 = a_1 = \cdots = a_n = 0$. (See Appendix B for a review of polynomials.)

It is possible to give an alternative definition of the concept of basis, using the following idea.

DEFINITION 4.13 A subset S of a vector space **V** is called a *maximal linearly independent subset of* **V** if (1) S is linearly independent and (2) for any vector w in **V**, but not in S, the set S' consisting of w and the elements of S is linearly dependent.

THEOREM 4.12 If a vector space **V** has a maximal linearly independent subset $\{v_1, \ldots, v_n\}$ with n elements, then $\{v_1, \ldots, v_n\}$ is a basis of **V**. Conversely, a basis is a maximal linearly independent subset of **V**.

PROOF: Let $S = \{v_1, \ldots, v_n\}$ be a maximal linearly independent subset. It must be shown that S generates **V**. Let w be any vector in **V**, but not in S. The set $\{w, v_1, \ldots, v_n\}$ is dependent, since S is maximal. Hence there are scalars $a_0, a_1, \ldots, a_n$, not all zero, such that $a_0 w + a_1 v_1 + \cdots + a_n v_n = 0$. Moreover, $a_0 \neq 0$, since $a_0 = 0$ would imply dependence of S. Hence $w = -(a_1 v_1 + \cdots + a_n v_n)/a_0$. This proves that every vector is a linear combination of vectors in S, which is to say that S generates **V**.

The converse is valid since any basis $\{v_1, \ldots, v_n\}$ is a maximal linearly independent subset by Theorem 4.11. ∎

Exercises

1. Give two bases for $\mathbf{R}^3$ that have no elements in common.

2. Find a basis for each subspace of $\mathbf{R}^4$ in Exercise 2 of Section 4.6.

3. Give two bases for the vector space of polynomials of degree 2 or less, one containing $x + 1$, and not $x^2 + x$ and one containing $x^2 + x$, but not $x + 1$.

4. Under what conditions on the scalar k do the vectors $(1, 1, 0)$, $(0, k, 2k)$, and $(0, k, k^2)$ form a basis for $\mathbf{R}^3$?

5. Let
$$A = \{(1, 0, 0, 1), (0, 1, 2, 1), (3, 2, 4, 5), (-1, 1, 2, 0)\}$$

    A has several different linearly independent subsets. List all of these. Show that none is a maximal linearly independent subset of $\mathbf{R}^4$.

6. A subset S of a set T contained in a vector space $\mathbf{V}$ is called a *maximal linearly independent subset of T* if S is linearly independent but $S \cup \{t\}$ is linearly dependent for any t which is an element of T but not of S. Prove that if $\mathbf{V} = [T]$ and S is a maximal linearly independent subset of T, then S is a maximal linearly independent subset of $\mathbf{V}$.

7. A subset S of a vector space $\mathbf{V}$ is called a *minimal generating set* of $\mathbf{V}$ if (a) S generates $\mathbf{V}$ and (b) if any vector is deleted from S, the resulting set S' does not generate $\mathbf{V}$. Prove that if a space $\mathbf{V}$ has a minimal generating set S with a finite number of elements, S is a basis. Conversely, prove any basis is a minimal generating set.

4.10   Properties of bases

This section will include further important ramifications of the notion of basis.

THEOREM 4.13   Let $S = \{v_1, \ldots, v_m\}$ be a linearly independent subset of a vector space $\mathbf{V}$, let $v \in \mathbf{V}$, and let $S' = S \cup \{v\}$. Then S' is linearly independent if and only if v does not lie in the subspace generated by S.

PROOF: If v lies in the subspace $[S]$, there are scalars $a_1, \ldots, a_m$ such that

$$v = a_1 v_1 + \cdots + a_m v_m, \quad a_1 v_1 + \cdots + a_m v_m + (-1)v = 0$$

Therefore S' is not independent. This proves the "only if" part of the lemma. On the other hand, suppose S' is dependent. Then there are scalars, $c_0, c_1, \ldots,$ c_m such that

$$c_0 v + c_1 v_1 + \cdots + c_m v_m = 0$$

Since S is independent, $c_0 \neq 0$, and

$$v = -(c_1 v_1 + \cdots + c_m v_m)/c_0$$

Therefore $v \in [S]$. This proves the "if" part of the lemma.   ∎

THEOREM 4.14 If $S = \{v_1, \ldots, v_m\}$ generates a vector space **V** and **V** is not the space containing only **0**, there is a linearly independent subset of S which generates **V**.

PROOF: Let v_p be the first nonzero vector in S and let $\mathbf{V}_1 = [v_p]$. Then $\mathbf{V}_1$ is a subspace of **V**. If $\mathbf{V}_1 = \mathbf{V}$, $\{v_p\}$ is an independent subset which generates **V**. Otherwise, there is a vector in S which is not in $\mathbf{V}_1$. Let v_q be the next nonzero vector in S which is not in $\mathbf{V}_1$. Then by Theorem 4.13 the set $\{v_p, v_q\}$ is independent. Let $\mathbf{V}_2 = [v_p, v_q]$. $\mathbf{V}_2$ is a subspace of **V**. If $\mathbf{V}_2 = \mathbf{V}$, the theorem is proved. If not, the argument can be repeated. Since S is finite, after a finite number of repetitions of the argument we obtain a linearly independent subset of S, we call it S', and cannot repeat the argument again. That is, we stop when every vector in S not yet in S' is in $[S']$. We must now show S' generates **V**. Since S generates **V** every element in **V** is a linear combination of S. Hence if $v \in \mathbf{V}$, then $v = \sum_{i=1}^{n} a_i v_i$. Consider those v_i in S but not in S'. These too are linear combinations of the elements in S'. Hence $a_i v_i$ is a linear combination of elements of S' for all $i = 1, \ldots, n$. Hence v is in $[S']$.   ∎

Theorem 4.14 may be succinctly stated thus: In a finite dimensional space, any set which generates the space contains a basis.

THEOREM 4.15 Let **V** be a vector space of dimension $n \geq 1$ and let $S = \{v_1, \ldots, v_n\}$ be a subset of n vectors. Then

1. If S is linearly independent, S is a basis for **V**.
2. If S generates **V**, S is a basis for **V**.

PROOF: First suppose S is linearly independent. Since by Theorem 4.11 any set of more than n vectors is dependent, S is a maximal linearly independent subset of **V**. By Theorem 4.12, S is a basis.

Next suppose S generates **V**. By Theorem 4.14 there is a linearly independent subset S' of S which generates **V**. Then S' is a basis, and must have n elements. Therefore $S' = S$, and S is a basis.   ∎

The point of Theorem 4.15 is that in a space **V** of known finite dimension n, a set S of n elements can be shown to be a basis by showing that S is independent or that S generates **V**. It is not necessary to prove both.

Example 1 The proof of Theorem 4.14 suggests an *algorithm* for extracting a linearly independent set from a given set. As an example, consider the vectors

$v_1 = (1, 1, 0)$, $v_2 = (2, 2, 0)$, $v_3 = (0, 1, 1)$, $v_4 = (3, 4, 1)$. We begin by choosing v_1 since $v_1 \neq \mathbf{0}$. Since $v_2 \in [v_1]$, $v_3 \notin [v_1]$, we do not choose v_2 but do choose v_3. Now $\{v_1, v_3\}$ is independent. It can be shown that $\{v_1, v_3, v_4\}$ is dependent; in fact, $v_4 = 3v_1 + v_3$. Therefore, $\{v_1, v_3\}$ is a maximal linearly independent subset, and it spans the same subspace of $\mathbf{R}^3$ as the dependent set $\{v_1, v_2, v_3, v_4\}$.

As the final topic in this section, we consider the problem of enlarging a linearly independent set to obtain a basis.

THEOREM 4.16   Let $\mathbf{V}$ be a vector space of finite dimension n and let $S = \{v_1, \ldots, v_r\}$ be a linearly independent subset. Then there exist $v_{r+1}, \ldots, v_n$ in $\mathbf{V}$ such that $\{v_1, \ldots, v_r, v_{r+1}, \ldots, v_n\}$ is a basis of $\mathbf{V}$.

PROOF: If $r = n$, S is a basis of $\mathbf{V}$ by Theorem 4.15 and S does not have to be enlarged. If $r < n$, S cannot generate $\mathbf{V}$. Hence there exists a vector v_{r+1} not in $[S]$. By Theorem 4.13 $S_1 = \{v_1, \ldots, v_r, v_{r+1}\}$ is linearly independent. If $r + 1 = n$, S_1 is a basis of $\mathbf{V}$. If $r + 1 < n$, S_1 cannot be a basis and there is a vector v_{r+2} not in $[S_1]$. By repeating this argument we can find a linearly independent set $\{v_1, \ldots, v_n\}$, which must be a basis for $\mathbf{V}$. ∎

Example 3   Let $v_1 = (1, 1, 0)$, $v_2 = (0, 1, 1)$. To extend the set $\{v_1, v_2\}$ to a basis for $\mathbf{R}^3$, we merely have to find a vector not in $[v_1, v_2]$. For example, $v_3 = (1, 0, 0)$ is suitable since a relation $(1, 0, 0) = a(1, 1, 0) + b(0, 1, 1)$ is impossible. Hence $\{(1, 1, 0), (0, 1, 1), (1, 0, 0)\}$ forms a basis for $\mathbf{R}^3$.

Exercises

1.   Let $S = \{v_1, \ldots, v_n\}$ be a finite set of nonzero vectors in $\mathbf{V}$. Prove that S is dependent if and only if for some integer k, $2 \leq k \leq n$, $\{v_1, \ldots, v_{k-1}\}$ is independent and v_k lies in the subspace $[v_1, \ldots, v_{k-1}]$.

2.   Let $\mathbf{V}$ be a finite dimensional vector space and let $\mathbf{W}$ be a subspace of $\mathbf{V}$. Prove that if $\dim \mathbf{V} = \dim \mathbf{W}$, then $\mathbf{V} = \mathbf{W}$.

3.   Select a maximal linearly independent subset of each set. (See Section 4.9, Exercise 6.)

   (a) $\{(1, 0, 1, 0), (0, 1, 0, 1), (2, 2, 2, 2,) (-1, 0, 1, 2)\}$

   (b) $\{(1, -1, 1, -1), (-1, 1, -1, 1), (0, 0, 0, 1), (1, -1, 1, -4)\}$

4.   Extend each set to form a basis for $\mathbf{R}^4$.

   (a) $\{(1, 0, 0, 0), (0, 1, 0, 0), (1, 1, 1, 0)\}$

   (b) $\{(1, -1, 1, -1), (-1, 1, -1, -1)\}$

5.   Give a proof of Theorem 4.14 based on Exercise 7 of Section 4.9.

4.11   Coordinates of vectors

In Section 4.9 we defined a basis of a vector space to be a set of linearly independent vectors that generate the vector space. The corollary to Theorem 4.11
proved that if a finite basis to a vector space exists, then all bases to that vector
space will have the same number of elements. The uniqueness of this number led
us to the concept of dimension of a vector space. Also, by considering the
standard bases of $\mathbf{R}^3$, namely, e_1, e_2, and e_3, and the basis given in Example 3
of Section 4.10, we see that more than one basis can be constructed for a given
vector space. This section is concerned with representation of vectors in a space
V with respect to a particular basis.

THEOREM 4.17   Consider a vector space **V** over a field **K** with basis $\{v_1,$
$v_2, \ldots, v_n\}$. Then any vector v in **V** can be represented as $v = \sum_{i=1}^n a_i v_i$ where
the scalars a_i are uniquely determined.

PROOF: That v can be represented as a linear combination of $\{v_1, \ldots, v_n\}$
follows from the fact that a basis generates the vector space. To show that the
set $\{a_1, a_2, \ldots, a_n\}$ is unique for each v we proceed as follows. Suppose two
representations exist. Then

$$v = \sum_{i=1}^n a_i v_i = \sum_{i=1}^n b_i v_i$$

It follows that

$$\sum_{i=1}^n a_i v_i - \sum_{i=1}^n b_i v_i = v - v = \mathbf{0} \quad \text{or} \quad \sum_{i=1}^n (a_i - b_i) v_i = \mathbf{0}$$

Since $\{v_1, \ldots, v_n\}$ forms a basis the set is linearly independent. Hence

$$\sum_{i=1}^n c_i v_i = \mathbf{0}$$

implies $c_i = 0$ for all i. Therefore, $a_i - b_i = 0$ or $a_i = b_i$ for all i. Hence the
set $\{a_1, a_2, \ldots, a_n\}$ of scalars is unique.   ∎

Theorem 4.17 shows that each vector in a vector space **V** over a field **K** has
a unique representation with respect to a given basis $B = \{v_1, v_2, \ldots, v_n\}$.
Each vector v in **V** can be described by the n-tuple of elements of **K**, $(a_1, a_2, \ldots,$
$a_n)$, where

$$v = \sum_{i=1}^n a_i v_i$$

DEFINITION 4.14   Let $B = \{v_1, v_2, \ldots, v_n\}$ be a basis and v an arbitrary
vector of the vector space **V** over the field **K**. The unique ordered n-tuple

$(a_1, a_2, \ldots, a_n)$ of elements of **K** such that $v = \sum_{i=1}^{n} a_i v_i$ is called the *coordinate vector* of v with respect to the basis B. We call a_i the ith *coordinate* of v with respect to B.

Note that the coordinate vector of any point (that is, vector) v in **V** with respect to the basis B is a unique vector in **K**^{n}. On the other hand, given any element $(a_1, a_2, \ldots, a_n)$ in **K**^{n}, the vector $\sum_{i=1}^{n} a_i v_i$ is a unique vector in **V** (it is in **V** since any linear combination of elements of **V** is in **V**) which has $(a_1, a_2, \ldots, a_n)$ as its coordinate vector. These two observations lead us to the following interesting and important result.

THEOREM 4.18   Given an n-dimensional vector space **V** over a field **K** and an arbitrary basis $B = \{v_1, v_2, \ldots, v_n\}$ of **V**, any element of **K**^{n} is the coordinate vector with respect to B of a unique element in **V** and conversely.

PROOF: The proof is essentially Theorem 4.17 and the discussion above.   ∎

We have seen that we can construct a one-to-one relationship between the elements of **V** and the elements of **K**^{n}; namely, take any arbitrary basis B and match the coordinates of the vectors in **V** with respect to the basis B to the corresponding elements in **K**^{n}. Further, the properties of addition and scalar multiplication survive the correspondence, for let $B = \{v_1, \ldots, v_n\}$ be a basis and let w_1 and w_2 be arbitrary vectors in **V** where

$$w_1 = \sum_{i=1}^{n} a_i v_i \quad \text{and} \quad w_2 = \sum_{i=1}^{n} b_i v_i$$

The elements in **K**^{n} corresponding to w_1 and w_2 are $k_1 = (a_1, a_2, \ldots, a_n)$ and $k_2 = (b_1, b_2, \ldots, b_n)$, respectively. The sum vector

$$w_1 + w_2 = \sum_{i=1}^{n} (a_i + b_i) v_i$$

corresponds to the element $(a_1 + b_1, a_2 + b_2, \ldots, a_n + b_n)$ of **K**^{n}, which is just $k_1 + k_2$. For any scalar c in **K**,

$$cw_1 = \sum_{i=1}^{n} ca_i v_i$$

with corresponding coordinates $(ca_1, ca_2, \ldots, ca_n)$, which is $c(a_1, a_2, \ldots, a_n)$, that is c times the element in **K**^{n} corresponding to w_1. Such a relationship is called an *isomorphism*. The vector spaces between which an isomorphism is defined are said to be *isomorphic*.[8]

---

[8] A more formal treatment of these concepts will be given in Section 5.3.

Example 1   Consider the plane P^2 with specified coordinate axes. In Section 4.1 we stated that to each vector in G^2 there corresponds a unique element in R^2, namely the coordinates of the vector. Further, if v is in G^2 with corresponding coordinates (x_1, x_2) then cv has coordinates (cx_1, cx_2) and if v and v^* have coordinates (x_1, x_2) and (x_1^*, x_2^*) respectively, then $v + v^*$ has coordinates $(x_1 + x_1^*, x_2 + x_2^*)$. By noting that to each element in R^2 there corresponds a unique element in G^2, namely, the geometrical vector with those coordinates, we see that there is an isomorphism between G^2 and R^2.

Example 2   Consider the vector space V of all polynomials with real coefficients of degree $\leq n$. In Example 2 of Section 4.9, we discussed the fact that $\{1, x, x^2, \ldots, x^n\}$ forms a basis for the space. Hence, for any arbitrary polynomial $a_0 + a_1 x + a_2 x^2 + \cdots + a_n x^n = P_n$, we have coordinates $(a_0, a_1, \ldots, a_n)$. In particular consider $n = 3$ and the polynomial $3 + 2x + 5x^2$. This polynomial has representation $(3, 2, 5, 0)$ which is the corresponding element of R^4 with respect to the particular basis given for V. Conversely, the element $(1, 2, 0, 1)$ of R^4 corresponds to the polynomial $1 + 2x + x^4$.

Example 3   Consider the space M of all 2×2 matrices and as a basis for M the vectors

$$\begin{pmatrix} 1 & 0 \\ 0 & 0 \end{pmatrix}, \quad \begin{pmatrix} 0 & 1 \\ 0 & 0 \end{pmatrix}, \quad \begin{pmatrix} 0 & 0 \\ 1 & 0 \end{pmatrix}, \quad \begin{pmatrix} 0 & 0 \\ 0 & 1 \end{pmatrix}$$

Then an arbitrary vector in M

$$\begin{pmatrix} a & b \\ c & d \end{pmatrix}$$

has, as its coordinates with respect to the given basis, (a, b, c, d), which is the corresponding vector in R^4. Consider an alternative basis for M, namely,

$$v_1 = \begin{pmatrix} 1 & 1 \\ 1 & 1 \end{pmatrix}, \quad v_2 = \begin{pmatrix} 0 & 1 \\ 1 & 1 \end{pmatrix}, \quad v_3 = \begin{pmatrix} 0 & 0 \\ 1 & 1 \end{pmatrix}, \quad \text{and} \quad v_4 = \begin{pmatrix} 0 & 0 \\ 0 & 1 \end{pmatrix}$$

Then the vector

$$v = \begin{pmatrix} a & b \\ c & d \end{pmatrix} = av_1 + (b - a)v_2 + (c - b)v_3 + (d - c)v_4$$

Hence, the coordinate vector of v with respect to the basis $\{v_1, v_2, v_3, v_4\}$ is $(a, b - a, c - b, d - c)$.

Exercises

1.   Find the coordinates of the vector v with respect to the basis $\{v_1, v_2, v_3\}$ in each case.

(a) $v = (5, -1, 3)$ $v_1 = (1, 0, 0)$ $v_2 = (0, 1, 0)$ $v_3 = (0, 0, 1)$

(b) $v = (5, -1, 3)$ $v_1 = (1, 1, 1)$ $v_2 = (1, 1, 0)$ $v_3 = (1, 0, 0)$

(c) $v = (5, -1, 3)$ $v_1 = (1, 0, 0)$ $v_2 = (1, 1, 1)$ $v_3 = (1, 1, 0)$

(d) $v = (6, -2, 0)$ $v_1 = (1, 0, 2)$ $v_2 = (1, 1, 0)$ $v_3 = (0, 2, 4)$

(e) $v = t^2$ $v_1 = 1$ $v_2 = t - 1$ $v_3 = (t - 1)^2$

2. The coordinates of a vector v in $\mathbf{R}^4$ are $1, 3, -2, 7$ relative to the basis

$$v_1 = e_1, \qquad v_2 = e_2 - e_1, \qquad v_3 = e_3 - e_2, \qquad v_4 = e_4 - e_3$$

Find the 4-tuple v.

3. The vectors $v = \sin^2 t$ and $w = \cos^2 t$ are in the space of functions spanned by $v_1 = 1$ and $v_2 = \cos 2t$. Find their coordinates.

4. Find the coordinates of $\sin^3 t$ relative to the independent set $\{\sin t, \sin 2t, \sin 3t\}$.

5. Prove that $\{e^{ix}, e^{-ix}\}$ is linearly independent over the complex field. What are the coordinates of $\sin x$ relative to e^{ix} and e^{-ix}?

6. $\{1, x, x^2, x^3\}$ is a basis for the space of polynomials of degree at most three. Find the coordinate vector relative to this basis of $f'(x)$ and $\int_1^x f(t)\, dt$, where $f(x) = x^2 - 2x + 3$.

4.12   Sums and direct sums

The final section of this chapter deals with a particular type of subspace.

DEFINITION 4.15   Let $\mathbf{U}$ and $\mathbf{W}$ be two subspaces of a vector space $\mathbf{V}$ over a field $\mathbf{K}$. Then the sum of $\mathbf{U}$ and $\mathbf{W}$, denoted $\mathbf{U} + \mathbf{W}$, is the set of all elements $u + w$ where $u \in \mathbf{U}$ and $w \in \mathbf{W}$.

By virtue of the following theorem, $\mathbf{U} + \mathbf{W}$ is a subspace of $\mathbf{V}$, not merely a subset.

THEOREM 4.19   Let $\mathbf{U}$ and $\mathbf{W}$ be subspaces of a vector space $\mathbf{V}$ over a field $\mathbf{K}$. Then $\mathbf{U} + \mathbf{W}$ is a subspace of $\mathbf{V}$.

PROOF: Let v_1 and v_2 be elements of $\mathbf{U} + \mathbf{W}$. Then $v_1 = u_1 + w_1$ and $v_2 = u_2 + w_2$, where u_1, u_2 are in $\mathbf{U}$ and w_1, w_2 are in $\mathbf{W}$. Further,

$$v_1 + v_2 = (u_1 + w_1) + (u_2 + w_2) = (u_1 + u_2) + (w_1 + w_2)$$

where $u_1 + u_2$ is in $\mathbf{U}$ and $w_1 + w_2$ is in $\mathbf{W}$ since $\mathbf{U}$ and $\mathbf{W}$ are vector spaces. Hence $(u_1 + u_2) + (w_1 + w_2)$ is in $\mathbf{U} + \mathbf{W}$. Similarly let $v = u + w$ where $u \in \mathbf{U}$ and $w \in \mathbf{W}$. Then for any $c \in \mathbf{K}$, $cu \in \mathbf{U}$ and $cw \in \mathbf{W}$ which implies $c(u + w) \in \mathbf{U} + \mathbf{W}$. Finally, $0 \in \mathbf{U}$ and $0 \in \mathbf{W}$ since $\mathbf{U}$ and $\mathbf{W}$ are subspaces of $\mathbf{V}$. Hence $0 + 0 = 0$ is in $\mathbf{U} + \mathbf{W}$.   ∎

The sum of more than two subspaces can be similarly defined. Let $\mathbf{U}_1$, $\mathbf{U}_2, \ldots, \mathbf{U}_n$ be n subspaces of $\mathbf{V}$. Then the sum, denoted $\sum_{i=1}^{n} \mathbf{U}_i$, is the set of all vectors $\sum_{i=1}^{n} u_i$, where $u_i \in \mathbf{U}_i$, $i = 1, 2, \ldots, n$. We leave as an exercise the result that the sum of n subspaces is a vector space.

Example 1   Let $\mathbf{V} = \mathbf{C}^0[1, 2]$ and let $\mathbf{U}$ be the subspace of all polynomials on $[1, 2]$ and $\mathbf{W} = [1, x^{-1}]$. Then $\mathbf{U} + \mathbf{W}$ consists of all functions of the type $f(x) = c/x + $ polynomial, $1 \leq x \leq 2$, where c is a scalar.

Example 2   Let $\mathbf{U} = [(1, 1, 0), (0, 1, 2)]$ and $\mathbf{W} = [(2, 5, 6), (-1, 2, 6)]$. Then $\mathbf{U} + \mathbf{W} = [(1, 1, 0), (0, 1, 2), (2, 5, 6,), (-1, 2, 6)]$. But the last two vectors in the generating set are linearly dependent on the first two. Hence $\mathbf{U} + \mathbf{W} = \mathbf{U}$.

Example 3   Let $\mathbf{V} = \mathbf{R}^3$, $\mathbf{U} = [e_1, e_2]$, $\mathbf{W} = [e_3]$. Then, clearly, $\mathbf{V} = \mathbf{U} + \mathbf{W}$ and moreover every vector v in $\mathbf{V}$ has exactly one representation $u + w$ where $u \in \mathbf{U}$ and $w \in \mathbf{W}$. That is, v has exactly one representation $c_1 e_1 + c_2 e_2 + c_3 e_3$. On the other hand, suppose $\mathbf{U} = [e_1, e_2]$ and $\mathbf{W} = [e_2, e_3]$. Then many vectors v have an infinity of such representations. Indeed if $v = (a, b, c)$ and a and c are not both zero, then $v = ae_1 + be_2 + ce_3 = a(e_1 + ke_2) + c(le_2 + e_3)$ provided k and l are chosen so that $ak + cl = b$. Hence we are led to the following definition.

DEFINITION 4.16   We say that the sum $\mathbf{U} + \mathbf{W}$ is the *direct sum* of $\mathbf{U}$ and $\mathbf{W}$ if for every element $v \in \mathbf{U} + \mathbf{W}$, there are *unique* elements $u \in \mathbf{U}$ and $w \in \mathbf{W}$ such that $v = u + w$. We denote the direct sum by $\mathbf{U} \oplus \mathbf{W}$.

THEOREM 4.20   Let $\mathbf{V}$ be a finite dimensional vector space over a field $\mathbf{K}$ and let $\mathbf{U}$ and $\mathbf{W}$ be subspaces of $\mathbf{V}$. Let $\mathbf{S} = \mathbf{U} + \mathbf{W}$. If $\mathbf{U} \cap \mathbf{W} = \{\mathbf{0}\}$, then $\mathbf{S}$ is the direct sum of $\mathbf{U}$ and $\mathbf{W}$, and conversely.

PROOF: Assume that $\mathbf{U} \cap \mathbf{W} = \{\mathbf{0}\}$. $\mathbf{S}$ is given as the sum of $\mathbf{U}$ and $\mathbf{W}$ so all that remains to be shown is that for any $v \in \mathbf{S}$ there is a unique $u \in \mathbf{U}$ and $w \in \mathbf{W}$ such that $u + w = v$. Let us assume any two representations. Suppose that u and u' are elements in $\mathbf{U}$ and w and w' in $\mathbf{W}$, such that $u + w = u' + w' = v \in \mathbf{S}$. Then $u + w = u' + w'$ implies $u - u' = w' - w$. But since $\mathbf{U}$ and $\mathbf{W}$ are vector spaces (subspaces of $\mathbf{V}$) then $u - u' \in \mathbf{U}$ and $w' - w \in \mathbf{W}$. The only element common to $\mathbf{U}$ and $\mathbf{W}$ is $\mathbf{0}$ which implies $u - u' = \mathbf{0}$ or $u = u'$, and similarly $w = w'$. Hence the representation of v is unique. The proof of the converse is left as an exercise. ∎

THEOREM 4.21   Let $\mathbf{V}$ be a finite dimensional vector space over a field $\mathbf{K}$ and let $\mathbf{U}$ be a subspace of $\mathbf{V}$. Then there exists a subspace $\mathbf{W}$ of $\mathbf{V}$ such that $\mathbf{U} \oplus \mathbf{W} = \mathbf{V}$.

PROOF: The proof is by construction of the subspace **W**. Let **V** have dimension n and let $\{u_1, u_2, \ldots, u_r\}$ be a basis for **U**. By the method of Theorem 4.16 extend this set to form a basis for **V**, by adding $\{w_1, \ldots, w_{n-r}\}$. Let **W** be the space generated by $\{w_1, \ldots, w_{n-r}\}$. Then $V = U + W$ since every vector v in **V** is a linear combination of the vectors $\{u_1, \ldots, u_r, w_1, \ldots, w_{n-r}\}$, which means $v = u + w$ where

$$u = \sum_{i=1}^{r} c_i u_i \in U \quad \text{and} \quad w = \sum_{i=1}^{n-r} d_i w_i \in W$$

Moreover, $U \oplus W = V$ because $U \cap W = \{0\}$ since their basis vectors are linearly independent.   ∎

DEFINITION 4.17   If $V = U \oplus W$, we say that **U** and **W** are *complementary* and that **W** is the *complementary subspace* of **U** and **U** is the *complementary subspace* of **W**.

Theorem 4.21 states, in this language, that every subspace of a finite dimensional space has a complementary subspace.

THEOREM 4.22   If **V** is a finite dimensional vector space over a field **K** and if $V = U \oplus W$, then the union of a basis for **U** and a basis for **W** is a basis for **V**. Hence, the dimension of **V** is equal to the sum of the dimension of **U** and the dimension of **W**.

PROOF: Let $\{u_1, \ldots, u_r\}$ be a basis for **U** and $\{w_1, \ldots, w_s\}$ be a basis for **W**. Then every vector v in **V** has a unique representation as $u + w$ where $u \in U$ and $w \in W$. Further, u has a unique representation by Theorem 4.17 as $\sum_{i=1}^{r} c_i u_i$ and w has a unique representation as $\sum_{i=1}^{s} d_i w_i$. Hence, every v has a unique representation as $\sum_{i=1}^{r} c_i u_i + \sum_{i=1}^{s} d_i w_i$, which implies that the set $S = \{u_1, u_2, \ldots, u_r, w_1, w_2, \ldots, w_s\}$ generates **V**. Further S is linearly independent since $U \cap W = \{0\}$. Therefore the dimension of **V** is $r + s$.   ∎

Exercises

1. In each example, describe $U + W$.
    (a) $V = R^2$, **U** is the subspace generated by $(2, 1)$, and **W** is the subspace generated by $(0, 1)$.
    (b) $V = R^4$, $U = \{(x_1, x_2, x_3, x_4) : 2x_1 - x_4 = 0\}$
       $W = \{(x_1, x_2, x_3, x_4) : 3x_2 - 2x_3 = 0\}$
    (c) **V** is the space of 2×2 real matrices over **R**, **U** and **W** are the subspaces generated, respectively, by the matrices

$$\begin{pmatrix} 1 & 1 \\ 0 & 0 \end{pmatrix} \quad \text{and} \quad \begin{pmatrix} 1 & 0 \\ 1 & 0 \end{pmatrix}$$

(d) $V = C^0[0, 1]$, $U = \{f \in V : f(0) - f(1) = 0\}$
   $W = \{f \in V : f(0) + f(1) = 0\}$

2. Prove that $U + W$ is the smallest subspace of V that contains both U and W. That is, prove that if X is a subspace of V and $U \subset X$, $W \subset X$, then

$$U + W \subset X$$

3. Using mathematical induction, prove that the sum of any finite number of subspaces is a subspace.

4. Prove that if U and W are subspaces of V, then $U + W$ is equal to the subspace generated by the set consisting of all elements in U or in W, that is, the union of U and W.

5. Find and prove a necessary and sufficient condition in order that $U + W = U$, where U and W are subspaces of a vector space V.

6. In each part of Exercise 1, show that the sum $U + W$ is direct or is not direct, as the case may be.

7. In which of the following cases is the sum $U + W$ direct?

   (a) $U = [(1, 1, 0), (0, 1, 1)]$, $W = [(1, 0, 0)]$
   (b) $U = [(1, 0, 1), (0, 1, 1)]$, $W = [(2, 1, 1)]$
   (c) $U = [(1, 1, 1), (0, 1, 2)]$, $W = [(5, 3, 1)]$
   (d) U is the space of polynomials of degree one or less,
       W is the space of polynomials of degree two or less.

8. Complete the proof of Theorem 4.20. That is, show that if $S = U \oplus W$, then $U \cap W = \{0\}$.

9. Let u, v be two vectors in R^2, and assume neither of them is 0. If there is no number c such that $cu = v$, show that u, v form a basis of R^2, and that R^2 is a direct sum of the subspaces generated by u and v, respectively.

10. Define the sum of n subspaces $U_1, U_2, \ldots, U_n$ to be the direct sum $\left(\sum\right)_{i=1}^{n} U_i$ if for any $v \in \left(\sum\right)_{i=1}^{n} U_i$ there are unique $u_1 \in U_1, u_2 \in U_2, \ldots, u_n \in U_n$ such that $v = \sum_{i=1}^{n} u_i$. State an analogue to Theorem 4.20. Why can we not require only $\bigcap_{i=1}^{n} U_i = \{0\}$?

11. Give an example in R^2 to show that the complementary subspace to a subspace U is not unique.

COMMENTARY

Section 4.1   Some of the reasons for the abstractness of linear algebra are: to coordinate various subjects, to study at once many systems insofar as they possess common properties in sufficient degree, and to discover hitherto unsuspected relationships.

Section 4.3   Definition 4.2 gives an example of what is called an *abstract system* or *axiomatic system* in mathematics. Such a system is composed of:

1. A *set* of elements.
2. Relations among the elements, including a definition of what is meant by *equality* of elements.
3. One or more *operations*, ways of combining elements to form new elements.
4. A list of *axioms* or *postulates*.

From these ingredients, theorems can be proved using the deductive laws of logic. These theorems can also be regarded as part of the abstract system.

Readers interested in the nature of abstract systems and their role in the fabric of mathematics may wish to read the article:

WILDER, R. L., "The axiomatic method," *The World of Mathematics*, vol. 3, New York, Simon and Schuster, 1965, pp. 1647–1667.

For a readable account on groups see:

KEYSER, C. J., "The group concept," *The World of Mathematics*, vol. 3, New York, Simon and Schuster, 1956, pp. 1538–1557.

Section 4.9   It may be proved that every vector space has a basis. The proof, however, requires the use of an axiom of the general theory of sets called the Axiom of Choice. In this elementary text we have elected not to introduce this somewhat difficult axiom. The reader may refer, for example, to:

STOLL, R., and E. WONG, *Linear Algebra*, New York, Academic Press, 1968.

Section 4.12   The usefulness of direct sums is that they allow us to break down a vector space into smaller spaces which we may be able to study more easily—or conversely, to build up a new space out of old ones. We will make use of these techniques in our treatment of eigenspaces in Chapter 11.

More advanced books show that Theorem 4.21 remains true for any vector space **V**. That is, if **U** is a subspace of **V**, then **U** has a complementary subspace.

Chapter 5

Linear transformations

5.1 Mappings

We assume that the reader is familiar with the notion of a real-valued function of one (or several) real variables (as presented, for example, in elementary calculus courses). Recall that such a function has a domain D, which is a set of real numbers, and that to each $x \in D$ the function f assigns a unique number $f(x)$, called the value of f at x. The set of such numbers $f(x)$, for all $x \in D$, is called the range of f. The following definition is an easy generalization of these concepts.

DEFINITION 5.1 Let A and B be two sets of elements (or "points"). A *function* or *transformation* or *mapping F from A into B* is a rule of correspondence such that to each element a in A there is associated one and only one element b in B.

The element b is often denoted by $F(a)$, or merely Fa, and is often called the *value* of F at a even though it is no longer necessarily a number. The symbols $F: A \to B$ are often used to indicate that F is a mapping from A into B, and the symbol $a \to F(a)$ is used to indicate that $F(a)$ is the element corresponding to a under F. Sometimes the symbol $a \mapsto F(a)$ is used for the latter. The word *map* is sometimes used as an abbreviation for mapping. Also, some authors reserve the word "function" for a numerical-valued function.

DEFINITION 5.2 Two mappings $F: A \to B$ and $G: A \to B$ are *equal*, denoted by $F = G$, if $F(a) = G(a)$ for each $a \in A$. More generally, two mappings $F: A_1 \to B$ and $G: A_2 \to B$ are *equal on the set C* if $F(a) = G(a)$ for each $a \in C$ where $C \subset A_1$ and $C \subset A_2$.

Sometimes a function F is not defined on all of a given set A, but only on a subset D. That is, F is a mapping from D into B. In this case, we adopt the following terminology.

DEFINITION 5.3   The *domain D* of a mapping F defined on a subset of a set A is the set of those elements in A for which F is defined. That is,

$$D = \{a: a \in A \text{ and } F(a) \text{ is defined}\}$$

Of course, if F is defined on all of A, A is the domain of F. We will reserve the notation $F: A \rightarrow B$ for the case when the whole of A is the domain of F.

As the student is aware, the graphs of numerical functions are helpful in visualizing their properties. Pictures can also be useful in visualizing properties of other types of mappings, as illustrated in the following example.

Example 1   Consider a circular disk subjected to a compressive force around the edge. Under this force, each point of the disk is moved slightly. Let us consider the disk before compression to lie in the (x, y) plane, say with center at the origin and radius 1. The point (x, y) of the disk is moved to some point (x', y') by the compression. Let A be the disk, let B be the set of points occupied after the compression, and let $F: A \rightarrow B$ be the function which assigns to (x, y) the point (x', y'). For example, under a uniform compression, it might be that $x' = rx$, $y' = ry$ where $0 < r < 1$. Then the map F can be described by the equation

$$(x, y) \rightarrow (rx, ry)$$

The domain of F is the subset of the plane consisting of points in the disk. Note that a graph of this function in the ordinary sense would require a four-dimensional space with x, y, x', y' axes. On the other hand, there is an alternative geometrical way of conceiving this function, suggested by Figure 5.1. The arrow indicates that the point (x, y) goes into the point (rx, ry) under the map F.

A somewhat similar picture is sometimes useful even when A and B are not sets of points, in the same way that geometric pictures are helpful in the study of operations on abstract sets. In Figure 5.2, we draw figures to represent the sets A and B and arrows to indicate that F sets up a correspondence from A into B. This picture enables us to adopt a geometric style of language, as in the following definition.

DEFINITION 5.4   Let $F: A \rightarrow B$. The *range* of F, denoted R_F, is the set

$$R_F = \{b: b = F(a) \text{ for some } a \in A\}$$

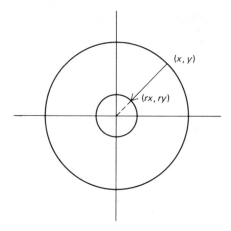

FIGURE 5.1

If $a \in A$, then $F(a)$ is called the *image* of a under the mapping F. If Z is a subset of A, the set denoted by $F(Z)$ given by

$$F(Z) = \{b: b = F(a) \text{ for some } a \in Z\}$$

is the *image of the set Z* under F.

Example 2   Let $A = \mathbf{R}^1$, $B = \mathbf{R}^2$, and let F be defined by

$$F: x \to (x, x)$$

Figure 5.3 shows a geometric representation of this function.

Example 3   Let $A = \mathbf{R}^3$, $B = \mathbf{R}^1$ and let

$$F: (x, y, z) \to x - y + 2z$$

The reader will recognize that F takes each vector (x, y, z) into the ordinary dot product of (x, y, z) with the vector $(1, -1, 2)$; that is,

$$F(x, y, z) = (x, y, z) \cdot (1, -1, 2)$$

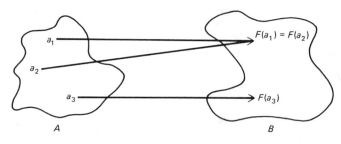

FIGURE 5.2

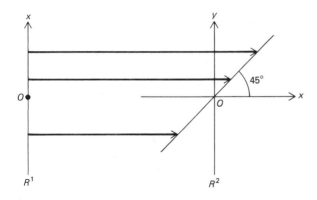

FIGURE 5.3

Example 4 Let A be the set of all functions $f(t)$ which are differentiable for $-1 < t < 1$, and let B be the set of all functions defined on $-1 < t < 1$. Let $D: A \to B$ be the map which assigns to each $f \in A$ the derivative f'. For example, in an obvious notation $D: t^2 \to 2t$ and $D: t^{3/2} \to \frac{3}{2}t^{1/2}$ so that we can speak of the function $2t$ as the image of t^2 under D, and so on. Note that it would not be correct to say that D maps A into A, since the image function need not be differentiable (for example, $\frac{3}{2}t^{1/2}$ has no derivative at $t = 0$).

Example 5 Let A be $\mathbf{C}^0[0, 1]$, that is, the set of continuous functions f defined on $[0, 1]$; and let B be the set of differentiable functions on that interval. Let $S: A \to B$ be the map which assigns to f the function g such that

$$g(t) = \int_0^t f(u)\, du, \qquad 0 \le t \le 1$$

Thus $Sf = g$. The function g is differentiable (its derivative is f) and is therefore an element of the set B.

The fact that differentiation and integration operations can be regarded as mappings on sets of functions is of great significance in modern treatments of differential equations, as we shall see.

It is important to observe that although a function $F: A \to B$ assigns to each element a in its domain a unique image $F(a)$, it may well be possible that an element in the range is the image of more than one element in A. For example, the numerical function $F(x) = x^2$ takes both x and $-x$ into the same image x^2. Thus mappings are in general *many-to-one* in the sense that many points in the domain can be mapped into one point in the image. This is illustrated in Figure 5.2 in which $a_1 \ne a_2$ but $F(a_1) = F(a_2)$. Mappings that take distinct points into distinct points are particularly important and so we give them a special name.

DEFINITION 5.5 Let $F: A \to B$ be a mapping of A into B. F is called *injective*, or *one-to-one* on A, if whenever x_1 and x_2 are distinct elements of A, $F(x_1)$ and $F(x_2)$ are distinct elements of B.

If F is one-to-one from A into B, it is possible to define an "inverse" mapping from a subset of B back into A in the following way.

DEFINITION 5.6 Let $F: A \to B$ be one-to-one and let R be the range of F. The *inverse mapping*[1] F^* is the mapping from R into A such that $F^*y = x$ where x is the unique element in A such that $Fx = y$. Thus the domain of F^* is the range of F and the range of F^* is the domain of F.

The idea of inverse mapping is illustrated in Figure 5.4.

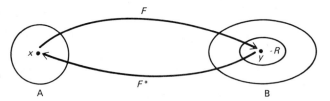

FIGURE 5.4

Example 6 Let S be the mapping defined in Example 5 above. Let A be the set $\mathbf{C}^0[0, 1]$ and B the set of differentiable functions on that interval. The range R of S is the set of functions which equal zero at $t = 0$ and have a continuous first derivative. Now S is one-to-one on A, since from

$$\int_0^t f_1(u) \, du = \int_0^t f_2(u) \, du, \qquad \text{for each } t, \quad 0 \le t \le 1$$

it follows by differentiation that $f_1(t) = f_2(t), 0 \le t \le 1$. The inverse mapping S^* therefore exists, and it is easily seen that it is the operator of differentiation as in Example 4. That is, for any function f in the range R of S, $S^*f = f'$. Evidently S^* maps the range R into A. It would be customary here to write D for S^*, but it should be kept in mind that the mapping here has a domain different from that of the mapping D in Example 4.

In Example 6, the range R of S is a proper subset of the set of differentiable functions. For this reason S is said to map A *into* the set of differentiable

---

[1] In the sequel we shall usually speak of an inverse mapping only when the range R is all of B but for the moment we do not impose this restriction.

functions, but *onto* the set of functions with a continuous derivative and with value 0 at $t = 0$. The general definition for this idea is as follows.

DEFINITION 5.7   Let $F: A \rightarrow B$ be a mapping of A into B. F is called *surjective* or *onto B*, if the range of F is all of B.

DEFINITION 5.8   Let $F: A \rightarrow B$ be one-to-one and onto. Then F is said to establish a *one-to-one correspondence* between A and B.

In Example 2, F establishes a one-to-one correspondence between $\mathbf{R}^1$ and the set $\{(x, y) : x = y\}$ in $\mathbf{R}^2$. This definition makes precise the notion of one-to-one correspondence which was previously used in discussing the concept of isomorphism.

Exercises

1.   For each mapping F on the domain $\mathbf{R}^2$ and the given subset Z of the domain, describe $F(Z)$.

(a) $F: (x, y) \rightarrow (y, x)$ $Z = \{(x, y): x^2 + y^2 \leq 4\}$

(b) $F: (x, y) \rightarrow (x, -y)$ $Z = \{(x, y): x^2 + y^2 = a^2\}$

(c) $F: (x, y) \rightarrow (x, y^2)$ $Z = \{(x, y): 0 \leq x \leq 1, 0 \leq y \leq 1\}$

(d) $F: (x, y) \rightarrow (x - 1, y - 1)$ $Z = $ arbitrary set in $\mathbf{R}^2$

(e) $F: (x, y) \rightarrow (x, 0)$ $Z = $ arbitrary set in $\mathbf{R}^2$

2.   For each mapping F and the given set S in the range of F find a set Z in the domain of F for which $S = F(Z)$.

(a) $F: x \rightarrow (x, x)$ $S = \{(u, v): 0 \leq u = v \leq 2\}$

(b) $F: (x, y) \rightarrow (2x - y)$ $S = \{u: -1 \leq u \leq 3\}$

(c) $F: (x, y) \rightarrow (2x - y, x + y)$ $S = \{(u, v): u^2 + v^2 = 1\}$

(d) $F(x_1, x_2, x_3) = (x_1, x_2, 0)$ $S = \{(u, v, 0): |u| + |v| \leq 1\}$

(e) $F(x_1, x_2, x_3) = (x_1 + x_2, 0, x_2 + x_3)$

$$S = \{(u, v, w): \max (u, v, w) \leq 1\}$$

3.   Consider each mapping F on $\mathbf{R}^2$ in Exercise 1. Is F one-to-one on $\mathbf{R}^2$? Is it onto all of $\mathbf{R}^2$?

4.   Each map in Exercise 2 is from $\mathbf{R}^n$ into $\mathbf{R}^m$ for suitable values of n and m. In each case, give the values of n and m, and tell whether F is one-to-one on its domain $\mathbf{R}^n$ and/or onto all of $\mathbf{R}^m$.

5.   Let S be the set of functions having derivatives of all orders on the interval $0 < t < 1$. Then the derivative $D = d/dt$ is a mapping from S into S.

Indeed, our rule D associates the function $df/dt = Df$ with the function f. According to our terminology, Df is the value of the mapping D at f. What is Df if:

(a) $f(x) = e^{ax}$

(b) $f(x) = $ arc tan x

(c) $f(x) = \ln (1 + x^2)$

6. Let A, B, and S be defined as in Example 5. Give $S(f)$ when f is the function:

(a) $f(x) = x^2$

(b) $f(x) = 1/(1 + x^2)$

(c) $f(x) = \sin 2x$

7. For each transformation T compute the image under T of $f_1(x) = e^{ax}$, $f_2(x) = x^n$.

(a) Tf is the function g, where $g(x) = d[xf(x)]/dx$. The domain of T is $\mathbf{C}^1(-\infty, \infty)$.

(b) Tf is the function $g(x)$, where $g(x) = \int_0^x tf(t)\, dt$. The domain of T is $\mathbf{C}^0(-\infty, \infty)$.

(c) $Tf = g$, where $g(x) = f(x - 3)$. The domain of T is the set of all real functions on $(-\infty, \infty)$.

(d) $Tf = g$, where $g(x) = xf(x)$. The domain of T is as in part (c).

5.2   Linear transformations

We shall henceforth be interested in mappings from one vector space into another vector space. Of primary interest is a special type of mapping, called a linear mapping or linear transformation.

DEFINITION 5.9 Let $\mathbf{V}$ and $\mathbf{V}'$ be vector spaces over a field $\mathbf{K}$. A *linear transformation*, or *linear operator*, or *linear mapping* from $\mathbf{V}$ into $\mathbf{V}'$ is a mapping $T: \mathbf{V} \to \mathbf{V}'$ such that

1. For any elements u, v in $\mathbf{V}$

$$T(u + v) = Tu + Tv$$

2. For any $u \in \mathbf{V}$ and any $c \in \mathbf{K}$,

$$T(cu) = cT(u)$$

Thus a linear transformation is a mapping from one vector space into another such that the image of a sum of vectors is the sum of their images, and the image of a multiple of a vector is that multiple of the image of the vector.

The properties in Definition 5.9 are sometimes referred to as the requirement that a linear transformation must be *compatible* with the algebraic operations defined on the vector spaces, or that the linear transformation must *preserve* the algebraic operations.

It should be pointed out that there is nothing to prevent **V** and **V'** from being the same vector space. In this case, T is often called a linear transformation of **V** into itself, or a linear operator on **V**. Some authors restrict the term *linear operator* to mean a linear transformation of a space into itself, but we shall use linear operator and linear transformation synonymously.

Example 1 Consider the transformation

$$T: (x, y) \rightarrow (rx, ry)$$

or more briefly

$$T(x, y) = (rx, ry)$$

of $\mathbf{R}^2$ into $\mathbf{R}^2$. This is the same transformation as in Example 1 of Section 5.1, except that the domain is now all of $\mathbf{R}^2$ instead of the unit circle. For any two vectors $u = (x_1, y_1)$ and $v = (x_2, y_2)$,

$$
\begin{aligned}
T(u + v) = T\{(x_1, y_1) + (x_2, y_2)\} &= T(x_1 + x_2, y_1 + y_2) \\
&= [r(x_1 + x_2), r(y_1 + y_2)] \\
&= (rx_1, ry_1) + (rx_2, ry_2) \\
&= T(x_1, y_1) + T(x_2, y_2) = Tu + Tv
\end{aligned}
$$

This proves that the image of a sum is the sum of the images. Similarly, if c is a real number (scalar) and $u = (x, y)$ is any vector,

$$
\begin{aligned}
T(cu) = T[c(x, y)] &= T(cx, cy) = (rcx, rcy) \\
&= c(rx, ry) = cT(x, y) = cTu
\end{aligned}
$$

Thus T has the properties required in Definition 5.9 and is a linear transformation.

Example 2 Let $T: \mathbf{R}^2 \rightarrow \mathbf{R}^2$ be defined by

$$T(x, y) = (x, -y)$$

It is easy to verify that T is a linear transformation. Geometrically, T takes each point of $\mathbf{R}^2$ into its reflection across the x-axis.

Example 3 Let $T: \mathbf{R}^2 \rightarrow \mathbf{R}^2$ be defined by

$$T(x, y) = (x \cos \theta - y \sin \theta, x \sin \theta + y \cos \theta)$$

This is a linear transformation which takes $\mathbf{R}^2$ into itself by rotation about the origin through the angle θ. (See Figure 5.5.)

Example 4 Let $T: \mathbf{R}^2 \rightarrow \mathbf{R}^2$ be defined by

$$T: (x, y) \rightarrow (x + 1, y)$$

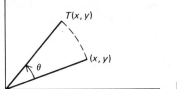

FIGURE 5.5

This mapping is not linear. For example, property (2) is not satisfied since
$$T[2(x, y)] = T(2x, 2y) = (2x + 1, 2y) \neq 2T(x, y)$$

Example 5   Let $T: \mathbf{R}^3 \to \mathbf{R}^1$ be defined as in Example 3 of Section 5.1, that is

$$T(x, y, z) = (x - y + 2z)$$

Then T is a linear transformation.

Example 6   Let $\mathbf{V} = \mathbf{C}^0[0, 1]$ be the vector space of real continuous functions on $[0, 1]$ and let $S: \mathbf{V} \to \mathbf{V}$ be defined by

$$S: f \to g \quad \text{where} \quad g(t) = \int_0^t f(u)\, du, \qquad 0 \le t \le 1$$

More briefly, S takes f into a function Sf such that

$$(Sf)(t) = \int_0^t f(u)\, du, \qquad 0 \le t \le 1$$

Then S is a transformation of $\mathbf{V}$ into itself (and more precisely into the subset of $\mathbf{V}$ consisting of functions with a continuous derivative and with value zero at $t = 0$; see Examples 5 and 6 of Section 5.1). For $f_1 \in \mathbf{V}$ and $f_2 \in \mathbf{V}$,

$$S(f_1 + f_2)(t) = \int_0^t (f_1 + f_2)(u)\, du = \int_0^t [f_1(u) + f_2(u)]\, du$$

$$= \int_0^t f_1(u)\, du + \int_0^t f_2(u)\, du$$

Therefore, $S(f_1 + f_2) = Sf_1 + Sf_2$. Similarly,

$$S(cf)(t) = \int_0^t (cf)(u)\, du = c \int_0^t f(u)\, du = c(Sf)(t)$$

and thus $S(cf) = cS(f)$. This proves that S is a linear operator on $\mathbf{C}^0[0, 1]$. Actually the statement that S is a linear operator is no more than the statement of well-known properties of the integral.

Example 7   We again consider Example 4 of Section 5.1. The set of all differentiable functions on $-1 < t < 1$ is a vector space which we now call $\mathbf{V}$

and the set of all functions on $-1 < t < 1$ is a vector space which we call **V'**. The mapping $D: f \to f'$ or $(Df)(t) = f'(t)$ for $-1 < t < 1$ is a linear operator from **V** into **V'** by virtue of the familiar differentiation rules

$$D(f_1 + f_2) = Df_1 + Df_2 \quad \text{and} \quad D(cf) = cD(f)$$

The next definition introduces a special transformation.

DEFINITION 5.10 Let **V** be a vector space over a field **K**. The mapping **Z** of **V** into **V** such that $\mathbf{Z}v = 0$ for each $v \in \mathbf{V}$ is called the *zero mapping* on **V**.

THEOREM 5.1 The zero mapping is a linear transformation.

PROOF: We leave the proof as an exercise. ∎

We close this section with two theorems which will play an important role in our development of matrix algebra.

THEOREM 5.2 Consider a linear transformation $T: \mathbf{V} \to \mathbf{V'}$, and let $\{v_1, \ldots, v_n\}$ be any basis of **V**. The linear transformation T is uniquely determined by $\{Tv_1, Tv_2, \ldots, Tv_n\}$, that is, if T^* is a linear transformation such that $Tv_i = T^*v_i$ $(i = 1, \ldots, n)$, then $T = T^*$.

PROOF: Suppose there exist two linear transformations T and T^* such that $Tv_1 = T^*v_1, Tv_2 = T^*v_2, \ldots, Tv_n = T^*v_n$. For $v \in \mathbf{V}$, $v = \sum_{i=1}^{n} a_i v_i$, where the a_i are uniquely determined by Theorem 4.17. Hence

$$T(v) = T\left(\sum_{i=1}^{n} a_i v_i\right) = \sum_{i=1}^{n} a_i T(v_i) = \sum_{i=1}^{n} a_i T^*(v_i)$$

$$= T^*\left(\sum_{i=1}^{n} a_i v_i\right) = T^*(v)$$

Since this holds for any $v \in \mathbf{V}$, $T = T^*$. ∎

Note in the above proof that no restrictions were placed upon the elements $Tv_1, \ldots, Tv_n$ other than that they be elements of **V'**. Hence the following theorem.

THEOREM 5.3 Let **V** and **V'** be vector spaces and $\{v_1, \ldots, v_n\}$ be a basis of **V**. Let $\{v'_1, \ldots, v'_n\}$ be any n elements in **V'**. Then there exists one and only one linear transformation T on **V** into **V'** such that $Tv_i = v'_i$ $(i = 1, \ldots, n)$. This transformation maps

$$\sum_{j=1}^{n} a_j v_j \quad \text{into} \quad \sum_{j=1}^{n} a_j v'_j$$

PROOF: Let T be the mapping such that an arbitrary element $\sum_{j=1}^{n} a_j v_j$ has image $\sum_{j=1}^{n} a_j v'_j$, that is,

$$T\left(\sum_{j=1}^{n} a_j v_j\right) = \sum_{j=1}^{n} a_j v'_j$$

Then clearly $Tv_i = v'_i$ $(i = 1, \ldots, n)$. Also T is linear since

$$T\left(\sum_{j=1}^{n} a_j v_j + \sum_{j=1}^{n} b_j v_j\right) = T\left(\sum_{j=1}^{n} (a_j + b_j)v_j\right) = \sum_{j=1}^{n} (a_j + b_j)v'_j$$

$$= T\left(\sum_{j=1}^{n} a_j v_j\right) + T\left(\sum_{j=1}^{n} b_j v_j\right)$$

Finally, T is the only such linear transformation, by Theorem 5.2.   ∎

Example 8   Consider the basis $\{v_1, v_2, v_3\}$ of $\mathbf{R}^3$ given by $v_1 = (1, 0, 1)$, $v_2 = (0, 1, 0)$, and $v_3 = (0, 0, 1)$. Let T be a linear mapping of $\mathbf{R}^3 \to \mathbf{R}^2$ such that $Tv_1 = (0, 1)$, $Tv_2 = (1, 0)$, and $Tv_3 = (0, 1)$. Let (a, b, c) be any vector in $\mathbf{R}^3$. Then $(a, b, c) = av_1 + bv_2 + (c - a)v_3$. Hence,

$$T(a, b, c) = aTv_1 + bTv_2 + (c - a)Tv_3$$
$$= (0, a) + (b, 0) + (0, c - a) = (b, c)$$

Hence by Theorem 5.2, the only linear mapping of $\mathbf{R}^3 \to \mathbf{R}^2$ such that $Tv_1 = (0, 1)$, $Tv_2 = (1, 0)$, and $Tv_3 = (0, 1)$ is the mapping $T(a, b, c) = (b, c)$.

Exercises

1. In each case, determine whether the mapping T is linear.

(a) $T: \mathbf{R}^2 \to \mathbf{R}^2$ defined by $T(x, y) = (y, x)$

(b) $T: \mathbf{R}^2 \to \mathbf{R}^2$ defined by $T(x, y) = (x, -y)$

(c) $T: \mathbf{R}^2 \to \mathbf{R}^2$ defined by $T(x, y) = (x, y^2)$

(d) $T: \mathbf{R}^2 \to \mathbf{R}^2$ defined by $T(x, y) = (x - 1, y - 1)$

(e) $T: \mathbf{R}^2 \to \mathbf{R}^2$ defined by $T(x, y) = (x, 0)$

(f) $T: \mathbf{R}^1 \to \mathbf{R}^2$ defined by $T(x) = (x, x)$

(g) $T: \mathbf{R}^2 \to \mathbf{R}^1$ defined by $T(x, y) = 2x - y$

(h) $T: \mathbf{R}^3 \to \mathbf{R}^3$ defined by $T(x, y, z) = (x, y, 0)$

(i) $T: \mathbf{R}^1 \to \mathbf{R}^1$ defined by $T(y) = -y$

(j) $T: \mathbf{R}^2 \to \mathbf{R}^1$ defined by $T(x, y) = xy$

(k) $T: \mathbf{R}^n \to \mathbf{R}^1$ defined by $T(x_1, \ldots, x_n) = \max(x_1, \ldots, x_n)$

(l) $T: \mathbf{R}^2 \to \mathbf{R}^1$ defined by $T(x_1, x_2) = |x_1| + |x_2|$

(m) $T: \mathbf{R}^2 \to \mathbf{R}^2$ where T maps each point onto its reflection in the line $y = -x$

(n) $T: \mathbf{R}^2 \to \mathbf{R}^2$ defined by $T(x, y) = (rx, r^2y)$ where r is a fixed number, $0 < r < 1$

2. Let $\mathbf{V}$ be the vector space of all polynomials. Which of the following are linear transformations of $\mathbf{V}$ into $\mathbf{V}$?

(a) $T: p \to p^2$

(b) $T: p \to q,$ where $q(x) = p(x + 1)$

(c) $T: p \to \dfrac{dp}{dx}$

(d) $T: p \to q,$ where $q(x) = p(0)$ for all x

(e) $T: p \to q,$ where $q(x) = \int p(x)\, dx$ and $q(0) = 0$

(f) $T: p \to q,$ where $q(x) = p(2x)$

(g) $T: p \to q,$ where $q(x) = p(x^2)$

(h) $T: p \to q,$ where $q(x) = \{p(\sqrt{x})\}^2$

3. Let T be a linear transformation from $\mathbf{V}$ into $\mathbf{V}'$. Prove that the image under T of the zero vector in $\mathbf{V}$ is the zero vector in $\mathbf{V}'$.

4. Prove Theorem 5.1.

5. Let T be a linear transformation from $\mathbf{V}$ into $\mathbf{V}'$. Prove that the image of a subspace $\mathbf{U}$ of $\mathbf{V}$ is a subspace of $\mathbf{V}'$.

6. Let T be a linear transformation from $\mathbf{V}$ into $\mathbf{V}'$. Let $\mathbf{U}'$ be a subspace of $\mathbf{V}'$. Prove that the set of all vectors in $\mathbf{V}$ whose images lie in $\mathbf{U}'$ is a subspace of $\mathbf{V}$.

7. In each case, determine whether the given information determines a unique linear transformation T on $\mathbf{R}^3$, and if so, find $T(x, y, z)$ for arbitrary (x, y, z).

(a) $T(1, 0, 0) = (1, 1, 1),$ $T(1, 1, 0) = (1, 1, 0),$ $T(1, 1, 1) = (1, 0, 0)$

(b) $T(1, 2, 3) = (3, 2, 1),$ $T(0, 0, 1) = (1, 0, 1),$ $T(1, 2, 0) = (2, 2, 0)$

(c) $T(1, 0, 0) = (1, 0),$ $T(0, 1, 0) = (1, 1),$ $T(1, 4, 1) = (3, -3)$

(d) $T(1, 0, 0) = \begin{pmatrix} 1 & 0 \\ 0 & 0 \end{pmatrix},$ $T(0, 1, 0) = \begin{pmatrix} 2 & 0 \\ 0 & 1 \end{pmatrix},$ $T(0, 0, 1) = \begin{pmatrix} 3 & 0 \\ 1 & 2 \end{pmatrix}$

8. Let $\mathbf{V} = \mathbf{C}^0[0, 1]$ and let $S: \mathbf{V} \to \mathbf{V}$ be defined by $S: f \to g$ where

$$g(x) = \int_0^{x^2} e^{-u} f(u)\, du$$

Is S a linear transformation?

9. Let $V = C^1(-\infty, \infty)$ and let S be the transformation which takes f into g where

$$g(x) = f'(x) - \int_1^x uf(u)\, du$$

Show that S is linear on V into V' for a suitable space V'.

10. Let $V = C^2(-\infty, \infty)$, and

$$Sf(x) = f''(x) - 2f'(x) - 3f(x)$$

Is S linear? What image space V' should be used?

11. Let V be the space of continuous functions $f(x, y)$ of two real variables on $-\infty < x < \infty$, $-\infty < y < \infty$. Let $T: V \to V$ be defined by $Tf = g$ where

$$g(x, y) = \int_2^x ds \int_1^y f(s, t)\, dt$$

Is T linear?

12. Let $V = C^0[-1, 1]$ and let $T: V \to V$ be defined by $Tf = g$ where

$$g(x) = \int_{-1}^1 e^{-xu} f(u)\, du$$

Is T linear?

5.3   Isomorphisms and the coordinate map

The notion of isomorphism was presented in Section 4.11. However, at that time we did not give a formal definition. In this section, we give such a definition and two examples of isomorphisms which play important roles in the development of the theory of linear mappings.

DEFINITION 5.11   Let V and V' be vector spaces over a field K and let T be a one-to-one mapping of the vector space V onto the vector space V'. Then T is an *isomorphism* if

1. $T(v_1 + v_2) = T(v_1) + T(v_2)$
2. $T(cv_1) = cT(v_1)$

   for every v_1 and v_2 in V and c in K.

The above definition describes an isomorphism between two vector spaces simply as a one-to-one and onto linear transformation.

DEFINITION 5.12   Two vector spaces V and V' are said to be *isomorphic* if there exists an isomorphism $T: V \to V'$.

We will show in Section 5.7 that if an isomorphism T exists such that $T: \mathbf{V} \to \mathbf{V}'$, there will also exist an isomorphism $T^*: \mathbf{V}' \to \mathbf{V}$. Consequently, the phrases, "$\mathbf{V}$ is isomorphic to $\mathbf{V}'$," "$\mathbf{V}'$ is isomorphic to $\mathbf{V}$," and "$\mathbf{V}$ and $\mathbf{V}'$ are isomorphic" have the same meaning. To show that two vector spaces are not isomorphic one need only show lack of existence of an isomorphism in one direction.

An isomorphism that we will use in our development of the theory of linear transformations is given in the following definition.

DEFINITION 5.13   Let $\mathbf{V}$ be a vector space over the field $\mathbf{K}$ and let $\{v_1, v_2, \ldots, v_n\}$ be a basis for $\mathbf{V}$. Let $T: \mathbf{V} \to \mathbf{K}^n$ be the mapping such that Tv is the coordinate vector of v with respect to the basis $\{v_1, v_2, \ldots, v_n\}$, for each $v \in \mathbf{V}$. Then T is called the *coordinate map* of $\mathbf{V}$, relative to the basis $\{v_1, v_2, \ldots, v_n\}$.

We observed in Section 4.11 that every vector $v \in \mathbf{V}$ has a unique representation, with respect to the specified basis,

$$v = \sum_{i=1}^{n} a_i v_i$$

and that we can associate v with the vector $(a_1, a_2, \ldots, a_n) \in \mathbf{K}^n$. In fact the mapping given by $\sigma v = (a_1, a_2, \ldots, a_n)$ is just the coordinate map. [*Note:* We will use σ to denote the coordinate map.]

THEOREM 5.4   Let $\mathbf{V}$ be a vector space over a field $\mathbf{K}$ with basis $(v_1, v_2, \ldots, v_n)$. The coordinate map of $\mathbf{V}$ relative to $(v_1, v_2, \ldots, v_n)$ is an isomorphism of $\mathbf{V}$ onto $\mathbf{K}^n$.

PROOF: By Theorem 4.18 the coordinate map is one-to-one and onto. Hence we need only verify that the coordinate map is linear. Let v and v' be any two vectors in $\mathbf{V}$ with coordinates $(a_1, a_2, \ldots, a_n)$ and $(b_1, b_2, \ldots, b_n)$, respectively, relative to $(v_1, v_2, \ldots, v_n)$. Then, denoting the coordinate map by σ, we obtain

$$\sigma(v + v') = \sigma\left(\sum_{i=1}^{n} a_i v_i + \sum_{i=1}^{n} b_i v_i\right) = \sigma\left(\sum_{i=1}^{n} (a_i + b_i)v_i\right)$$

$$= (a_1 + b_1, a_2 + b_2, \ldots, a_n + b_n) = \sigma(v) + \sigma(v')$$

Further, for any $c \in \mathbf{K}$,

$$\sigma(cv) = \sigma\left(\sum_{i=1}^{n} ca_i v_i\right) = (ca_1, ca_2, \ldots, ca_n) = c\sigma(v)$$

Hence, σ is linear and therefore an isomorphism of $\mathbf{V}$ onto $\mathbf{K}^n$.   ∎

COROLLARY   Let $\mathbf{V}$ be an n-dimensional vector space over a field $\mathbf{K}$. Then $\mathbf{V}$ and $\mathbf{K}^n$ are isomorphic.

A second isomorphism is the identity transformation.

DEFINITION 5.14   Let **V** be a vector space over a field **K**. The mapping $I_V: V \rightarrow V$ such that $I_V u = u$ for each $u \in V$ is the *identity mapping*.

In cases where no ambiguity should arise we will use I rather than I_V to denote the identity mapping of $V \rightarrow V$.

THEOREM 5.5   The identity mapping on a vector space **V** is an isomorphism of **V** onto **V**.

PROOF: The proof is left as an exercise.   ∎

Exercises

1.   Consider the mapping T on $C^1(-\infty, \infty)$ into $C^0(-\infty, \infty)$ which takes f into df/dx. Is T an isomorphism? Explain.

2.   Let **V** be the subspace of $C^1(-\infty, \infty)$ consisting of functions with $f(0) = 0$. Let $T: f \rightarrow df/dx$. Is T an isomorphism of **V** onto $C^0(-\infty, \infty)$?

3.   Let T be the mapping on $C^0[0, 1]$ into itself defined by

$$(Tf)(x) = xf(x), \qquad 0 \le x \le 1$$

Show that T is linear but is not an isomorphism. Let **W** be the subspace

$$W = \{g \in C^0[0, 1]: g(0) = 0\}$$

Is T an isomorphism of $C^0[0, 1]$ onto **W**?

4.   Prove that R^2 is isomorphic to the subspace of R^3 consisting of triples (x, y, z) with $z = 0$.

5.   Prove that the vector space of real polynomials of degree at most 3 is isomorphic to the space of 2×2 matrices with real entries ($K = R$).

6.   To each complex number $a + ib$ associate the real number a. Is this an isomorphism between R^1 (the real numbers over the real field) and the vector space of complex numbers over the complex field?

5.4   Operations on linear transformations

In Section 5.2 we defined a linear transformation T of a vector space **V** into a vector space **V**′ in a way which preserved the algebraic operations of vector addition and multiplication. That is, for u and v in **V**, $T(v + u) = T(v) + T(u)$ in **V**′, and $T(cu) = cT(u)$ in **V**′. The multiplication and addition in the above case were with respect to vectors. In this section we introduce the notion of addition and multiplication of linear transformations. The usefulness of these concepts is indicated in part by the examples in this section.

DEFINITION 5.15 Let S and T be two linear transformations mapping the vector space $\mathbf{V}$ into the vector space $\mathbf{V}'$. The sum $S + T$ is the transformation of $\mathbf{V}$ into $\mathbf{V}'$ such that $(S + T)(v) = S(v) + T(v)$ for all $v \in \mathbf{V}$.

Example 1 Let $\mathbf{V} = \mathbf{C}^0[0, 1]$, the set of all continuous functions on $[0, 1]$. Further, let $S: \mathbf{V} \to \mathbf{V}$ be defined as $Sf = g$, where

$$g(t) = \int_0^t e^u f(u)\, du \quad \text{for} \quad 0 \le t \le 1$$

Similarly, define $T: \mathbf{V} \to \mathbf{V}$ by $Tf = g$, where

$$g(t) = \int_0^t e^{-u} f(u)\, du \quad \text{for} \quad 0 \le t \le 1$$

That S and T are linear transformations is immediate. Further, $(S + T)f = g$, where

$$g(t) = \int_0^t e^u f(u)\, du + \int_0^t e^{-u} f(u)\, du = \int_0^t (e^u + e^{-u})\, f(u)\, du$$

Example 2 Let S and T map $\mathbf{R}^2$ into $\mathbf{R}^2$ where $S(x, y) = (x \cos \theta, x \sin \theta)$ and $T(x, y) = (-y \sin \theta, y \cos \theta)$. Then $S + T$ is the linear transformation $(S + T)(x, y) = (x \cos \theta - y \sin \theta, x \sin \theta + y \cos \theta)$ of Example 3 in Section 5.2. In the following, when discussing more than one vector space, we will assume that the spaces are defined over the same field.

DEFINITION 5.16 The *scalar multiple* of a linear transformation $T: \mathbf{V} \to \mathbf{V}'$ by a scalar c in $\mathbf{K}$ is the transformation $[cT]$ of $\mathbf{V}$ into $\mathbf{V}'$ such that $[cT](u) = c(T(u))$, for all $u \in \mathbf{V}$. One usually writes cT rather than $[cT]$.

This gives us the result that for any linear transformation T, $T(cu) = [cT](u)$.

We are now in a position to derive some interesting properties concerning linear transformations. These are given in the theorems below.

THEOREM 5.6 Let $\mathbf{V}$ and $\mathbf{V}'$ be vector spaces over a field $\mathbf{K}$ and S and T be two linear transformations of $\mathbf{V}$ into $\mathbf{V}'$. Then $S + T$ is a linear transformation. Further, for any scalar $c \in \mathbf{K}$, $[cT]$ is a linear transformation of $\mathbf{V}$ into $\mathbf{V}'$.

PROOF: The proof is left as an exercise. ∎

THEOREM 5.7 The set of all linear transformations mapping $\mathbf{V}$ into $\mathbf{V}'$ is a vector space.

PROOF: Theorem 5.6 shows that the sum and scalar product of linear transformations are again linear transformations. We define the additive inverse of a transformation $T: \mathbf{V} \to \mathbf{V}'$ as $(-1)T$. The zero element of the set of linear transformations mapping $\mathbf{V}$ into $\mathbf{V}'$ is the zero mapping (see Definition 5.10). It is now easy to verify that the conditions required for a vector space are met (see Exercise 2). ∎

DEFINITION 5.17   Let $S: \mathbf{V} \to \mathbf{V}'$ and $T: \mathbf{V}' \to \mathbf{V}''$ where $\mathbf{V}$, $\mathbf{V}'$, and $\mathbf{V}''$ are vector spaces, S and T are linear transformations. The *product* or *composition* of the linear transformations, TS, is the mapping defined by $(TS)(u) = T(S(u))$ for all $u \in \mathbf{V}$. [For the sake of abbreviation, one often writes $TS(u)$ instead of $(TS)(u)$.]

A pictorial representation for the product of two linear transformations is given in Figure 5.6. This idea of product is analogous to that of composite functions (function of a function) which is treated in elementary calculus texts.

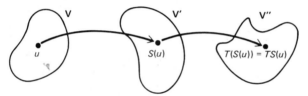

FIGURE 5.6

THEOREM 5.8   The product TS of two linear transformations $S: \mathbf{V} \to \mathbf{V}'$ and $T: \mathbf{V}' \to \mathbf{V}''$ is a linear transformation of $\mathbf{V} \to \mathbf{V}''$.

PROOF: Consider any two vectors u and v in $\mathbf{V}$. Then

$$TS(u + v) = T(S(u + v)) = T(S(u) + S(v)) = T(u' + v')$$

where $S(u) = u'$ and $S(v) = v'$ are in $\mathbf{V}'$. But $T(u' + v') = T(u') + T(v')$ since T is linear. Hence

$$TS(u + v) = T(S(u)) + T(S(v)) = TS(u) + TS(v)$$

Similarly, $TS(cu) = T(S(cu)) = T(cS(u)) = cTS(u)$. ∎

Example 3   Let $S: \mathbf{R}^3 \to \mathbf{R}^2$ with $S(x, y, z) = (x - z, y + 2z)$, and let $T: \mathbf{R}^2 \to \mathbf{R}^1$ with $T(x, y) = (x + y)$. Both S and T are linear. The product mapping $TS: \mathbf{R}^3 \to \mathbf{R}^1$ is given by

$$TS(x, y, z) = T(x - z, y + 2z) = (x - z + y + 2z) = (x + y + z)$$

Example 4   Let **V** be $\mathbf{C}^1[0, 1]$, the set of continuously differentiable functions on $[0, 1]$, and $\mathbf{V}' = \mathbf{C}^0[0, 1]$. Let $S: \mathbf{V} \to \mathbf{V}'$ with $Sf = g$ where $g(t) = df(t)/dt$. Let $T: \mathbf{V}' \to \mathbf{V}$ with $Tf(t) = \int_0^t f(u)\,du$, $0 \le t \le 1$. Then $TS: \mathbf{V} \to \mathbf{V}$ where

$$TS(f) = \int_0^t \left(\frac{df(u)}{du}\right) du$$

Note that TS is not an identity mapping, that is, $TS(f)$ does not necessarily equal f for $f \in \mathbf{V}$. For example, if $f(t) = t + 3$, then $S(f) = g$, where $g(t) = 1$, $0 \le t \le 1$, and hence $TS(f) = h$, where $h(t) = t$, $0 \le t \le 1$.

Example 5   Let S and T be defined as in Example 3. We have seen that the transformation TS maps $\mathbf{R}^3 \to \mathbf{R}^1$. Consider the product ST. From our definition of product of transformations $ST(u) = S(T(u))$. If we consider an element $u \in \mathbf{R}^2$, then T maps u into u', an element of $\mathbf{R}^1$. But S maps $\mathbf{R}^3$ into $\mathbf{R}^1$ and hence $S(u')$ is not defined. It is therefore meaningless to speak of a mapping ST.

Example 5 proves that multiplication of transformations is not, in general, commutative. This result is worth remembering as we will be encountering it in various forms throughout the remainder of the text. However, most other algebraic properties of multiplication are satisfied.

THEOREM 5.9   Assume the linear transformations R, S, and T are such that in each case the composite mappings are defined. Then

$R(ST) = (RS)T$	multiplication is associative
$(R + S)T = RT + ST$	multiplication is distributive over
$R(S + T) = RS + RT$	addition
$c(ST) = (cS)T = S(cT)$	multiplication commutes with scalar multiplication

Further, if $S: \mathbf{V} \to \mathbf{V}'$, $I_\mathbf{V}$ is the identity mapping on $\mathbf{V}$, and $I_{\mathbf{V}'}$ is the identity mapping on $\mathbf{V}'$, then $S = I_{\mathbf{V}'}S = SI_\mathbf{V}$.

PROOF: We prove here that multiplication is associative and leave the remaining proofs as exercises. Let $T: \mathbf{V} \to \mathbf{V}'$, $S: \mathbf{V}' \to \mathbf{V}''$, and $R: \mathbf{V}'' \to \mathbf{V}'''$. For any $u \in \mathbf{V}$,

$$R[S(T(u))] = (RS)(T(u)) = ((RS)T)(u)$$

and on the other hand

$$R[S(T(u))] = R[(ST)(u)] = (R(ST))(u)$$

Hence $(RS)T = R(ST)$, which is the desired result. The reader should supply the justification for each step given above. ∎

Exercises

1. Prove Theorem 5.6.

2. Prove the following properties, which are necessary for a complete proof of Theorem 5.7.

   (a) Addition of linear transformations is associative:
   $$(S + T) + U = S + (T + U)$$

   (b) $T + (-1)T = Z$ (zero transformation)
   (c) Addition is commutative: $S + T = T + S$
   (d) If c is any scalar, $c(S + T) = cS + cT$
   (e) If a and b are scalars, $(a + b)T = aT + bT$ and $a(bT) = (ab)T$
   (f) $1 \cdot T = T$

3. Let $\mathbf{V} = \mathbf{R}^2$. Consider the transformations S and T, where $S(x, y) = (y, x)$ and $T(x, y) = (-x, y)$. Prove that

   (a) S and T are linear.
   (b) $ST \neq TS$, that is, multiplication of linear transformations is not commutative in general.

4. Let S and T be transformations defined as below. In each case, compute the image of an arbitrary element under $S + T$, ST, and TS, if defined.

   (a) $T(x, y) = (y, x)$ $S(x, y) = (x, -y)$
   (b) $T(x, y) = (x, 0)$ $S(x, y) = (x - y, y)$
   (c) $T(x, y) = (x, 0)$ $S(x, y) = (0, y)$
   (d) $T(x) = (x, x)$ $S(x, y) = x + y$
   (e) $T(x) = (x, x)$ $S(x, y) = (x + y, 0)$
   (f) $T(x, y, z) = (z, y, x)$ $S(x, y, z) = (x, y - x, z - x)$

5. S and T are linear transformations on $\mathbf{V}$ into $\mathbf{V}$ where $\mathbf{V}$ is the space of all polynomials. Let
   $$p(x) = a + bx + cx^2$$
   Compute STp and TSp in each case.

   (a) $S: p(x) \rightarrow p(x + 1)$, $T: p(x) \rightarrow dp/dx$
   (b) $S: p(x) \rightarrow p(x + 1)$, $T: p(x) \rightarrow p(0)$
   (c) $S: p(x) \rightarrow \int_0^x p(u)\, du$, $T: p(x) \rightarrow dp/dx$
   (d) $S: p(x) \rightarrow p(2x)$, $T: p(x) \rightarrow p(0)$

6. Let R, S, T be defined by

$$Tf(x) = \int_0^x uf(u) \, du$$

$$Sf(x) = xf(x)$$

$$Rf(x) = f(0)x$$

Verify the formulas

$$(R + S)T = RT + ST$$
$$R(S + T) = RS + RT$$

for the particular function $f(x) = x^2 + 2x - 1$.

7. Let R, S, T be defined on the space of polynomials by

$$R: a_0 + a_1x + \cdots + a_nx^n \to a_0$$
$$S: a_0 + a_1x + \cdots + a_nx^n \to a_0x + a_1x^2 + \cdots + a_nx^{n+1}$$
$$T: a_0 + a_1x + \cdots + a_nx^n \to a_0 + a_nx^n$$

Find the image of $p(x) = x^3 - 3x + 5$ under the transformations

$$R, S, T, \quad RST, \quad RTS, \quad STR, \quad SRT$$

8. Prove that multiplication of linear mappings is distributive with respect to addition.

9. Prove that multiplication of linear transformations commutes with scalar multiplication.

10. Let S be a fixed linear mapping of $\mathbf{V} \to \mathbf{V}$. Show that the set of all linear mappings T of $\mathbf{V} \to \mathbf{V}$ such that $ST = Z$ is a subspace of the vector space of all linear transformations mapping $\mathbf{V} \to \mathbf{V}$. What is the subspace if S is the zero mapping; that is, $Sv = \mathbf{0}$ for all $v \in \mathbf{V}$? What is the subspace if $S = I_\mathbf{V}$?

5.5   Polynomials in linear transformations

In this section we develop the notion of power of a linear transformation, which leads to some interesting examples. Previously we have defined the product of two linear transformations $S: \mathbf{V} \to \mathbf{V}'$ and $T: \mathbf{V}' \to \mathbf{V}''$ as the mapping $TS: \mathbf{V} \to \mathbf{V}''$ and suggested the analogy with composite functions as presented in elementary calculus courses. The reader must be careful with the notational differences in carrying the analogy over to power of linear transformations. For, given a function f defined on some interval (a, b) of the real line, f^2 is used to denote the function defined by $f^2(x) = [f(x)]^2$ for each x in (a, b), which is not analogous to squaring a linear transformation. For example, let $f(x) = x + 1$ for all real x. The function f^2 is defined as $f^2(x) = (x + 1)^2$. This function is not the same as the composite mapping, which we denote $f \cdot f$, where

$f \cdot f(x) = f[f(x)] = f(x + 1) = (x + 1) + 1 = x + 2$. With this warning we proceed to the following definition.

DEFINITION 5.18   Let S be a linear transformation $S: V \rightarrow V$. Then S^2 is the product $S \cdot S$. In general S^n is the product $S \cdot S^{n-1}$ for any integer $n \geq 1$. Finally $S^0 = I$, the identity mapping on V.

Certain observations are immediate. To begin, S^2 is again a linear transformation, as is S^n for any positive integer n. Further, while S^n is defined to be $S^n = S \cdot S^{n-1}$ it is easy to verify that $S^n = S^{n-1} \cdot S$. In fact, $S^n = S^{n-k}S^k$ for any integer k such that $0 \leq k \leq n$. Finally, note in the definition that S is a mapping of V back into itself. If S were to map V into V', where V' is not a subset of V, the powers S^2, and so on, would not make sense.

The idea of power of a linear transformation leads naturally to the notion of polynomial in the linear transformation, with scalar coefficients. Since for all nonnegative integers n, S^n is a linear transformation, and further the sum of linear transformations is linear, we obtain the following theorem.

THEOREM 5.10   Let S be a linear transformation mapping the vector space V into itself. Then for any nonnegative integer n and set of $n + 1$ scalars $\{c_0, c_1, \ldots, c_n\}$, the polynomial in S given by $c_0 I + c_1 S + c_2 S^2 + \cdots + c_n S^n$ is a linear transformation of V into V.

PROOF: The proof is essentially contained in the above discussion.   ∎

Note in Theorem 5.10 that I plays the role of S^0 in the polynomial, which is the reason for defining S^0 as such in Definition 5.18.

Example 1   Let $V = C^\infty[0, 1]$, the set of all infinitely differentiable functions on $[0, 1]$. Let $D: V \rightarrow V$ where $D(f) = g = f'$. That is, $D(f) = df/dt = g(t)$ for $0 \leq t \leq 1$. In Example 7 of Section 5.2 we showed D to be a linear operator, which implies $D^2, D^3, \ldots, D^n$ are linear transformations also. D^2 is the composite mapping given by $D^2 f(t) = D(D(f(t))) = D(df(t)/dt) = d^2 f(t)/dt^2$ for each t in the interval $0 \leq t \leq 1$. Similarly, $D^k f = f^{(k)}$, where $f^{(k)} = d^k f(t)/dt^k$ for $0 \leq t \leq 1$. Note that $D^2 f$ is the second derivative f'', whereas $(Df)^2$ denotes the algebraic square of the first derivative.

Next consider the polynomial $D^2 + 3D + 2I$. The resulting transformation applied to f yields

$$(D^2 + 3D + 2I)f = D^2 f + 3Df + 2If = f'' + 3f' + 2f$$

By asking what set of functions f cause the polynomial to equal zero, that is, for what functions f does $(D^2 + 3D + 2I)f = 0$, we are asking for the set of solutions of the second order linear differential equation $f'' + 3f' + 2f = 0$. In this example we begin to glimpse the usefulness of the notation and language of linear transformations in the study of differential equations.

One final property of polynomials in linear transformations, which will be used later in deriving solutions for nth order linear differential equations, is that they can be factored in the usual fashion.[2] We illustrate this property below.

Example 2 Consider the polynomial in D given in Example 1. If we were to consider instead the polynomial $x^2 + 3x + 2$ we would factor it as $(x + 2)(x + 1)$. We therefore consider $(D + 2I)(D + I)$. By the distributive properties of linear transformations given in Section 5.4, we have

$$(D + 2I)(D + I) = (D + 2I)D + (D + 2I)I$$
$$= D^2 + 2ID + DI + 2I^2 = D^2 + 3D + 2I$$

Example 3 Consider the operator S defined by

$$Sf(t) = \int_0^t f(u)\, du$$

Then

$$S^2 f(t) = S(Sf(t)) = \int_0^t \left(\int_0^v f(s)\, ds \right) dv$$

That is, $S^2 f$ is an iterated integral of f. Incidentally, this iterated integral is equal to the single integral

$$S^2 f(t) = \int_0^t (t - u)f(u)\, du$$

The proof of this remark is left as an exercise. Also, the equation $(S - I)f = 0$ means

$$\int_0^t f(u)\, du - f(t) = 0$$

This is a *linear integral equation*. Its solutions are those functions mapped into the zero function by the linear transformation $S - I$. In this example and Example 1 we see that some aspects of the study of both differential equations and integral equations can be regarded in a unified fashion by using the abstract concepts of linear algebra.

Exercises

1. In each of the following, find the function obtained by applying the given polynomial operator to each given function.

   (a) $D^2 - I$ to $2e^x$; to $e^x + e^{-x}$
   (b) $D^2 + I$ to $\sin x$; to $\cos x$
   (c) $(D + I)(D - 2I)$ to x^2; to e^x; to $\sin x + e^{2x}$

---

[2] We will consider later an extended concept of polynomial in a linear transformation for which this will no longer be true.

2. Show that

$$(D + 3I)(2D - I)f = (2D - I)(D + 3I)f$$

for arbitrary twice-differentiable functions f.

3. Factor each of these polynomial operators. Use the distributive properties as in Example 2, to verify that the factorization is valid.

(a) $6D^2 + 2D - 20I$ (b) $D^3 - I$

(c) $D^4 + I$ (d) $D^2 - 2D + 2I$

4. For each transformation T, compute T^2 and more generally T^n, if defined.

(a) $T(x, y) = (y, x)$ (b) $T(x, y) = (x, 0)$

(c) $T(x, y) = (0, x)$ (d) $T(x, y) = (x, x)$

(e) $Tx = (x, x)$ (f) $T(x, y, z) = (z, y, x)$

5. For each linear transformation S in Exercise 5 of Section 5.4, compute S^n ($n = 2, 3, \ldots$) if defined.

6. Let $S = tD - I$ and $T = tD + 2I$, where D denotes differentiation. That is, for each function f,

$$(Sf)(t) = tf'(t) - f(t)$$
$$(Tf)(t) = tf'(t) + 2f(t)$$

(a) Are S and T linear? Substantiate your answer.

(b) What is the result of applying ST to the functions

$$t^2, \quad (t^3 + 1)/t, \quad e^t$$

7. Establish the formula

$$\int_0^t \left(\int_0^v f(s) \, ds \right) dv = \int_0^t (t - u) f(u) \, du$$

used in Example 3.

5.6   Null space and range of a linear transformation

Consider the problem of finding all solutions of the differential equation $f'' + 3f' + 2f = 0$. As was seen in the last section, this equation can be written in the form $(D^2 + 3D + 2I)f = 0$. Therefore, in the language of operators, we wish to find all functions f which are mapped by the linear operator $D^2 + 3D + 2I$ into the zero function (the function which is everywhere zero). The problem of finding the set of elements in a vector space mapped by a given linear transformation into the zero element of the image space occurs quite frequently in diverse parts of mathematics. We are therefore led to the following development.

THEOREM 5.11 Let $T: V \to V'$ be a linear transformation. Further let N_T be the set of elements in V mapped by T into the zero vector in V'. Then N_T is a subspace of V.

PROOF: That N_T is a subset of V is obvious. To show that N_T is a subspace we make use of Theorem 4.6. Clearly 0, the zero element of V, is in N_T. Suppose v_1 and v_2 are in N_T, that is, $Tv_1 = 0'$ and $Tv_2 = 0'$, where $0'$ denotes the zero vector of V'. Then

$$T(v_1 + v_2) = Tv_1 + Tv_2 = 0' + 0' = 0'$$

Hence $v_1 + v_2 \in N_T$. Let c be any scalar and v_1 be an element of N_T. Then $T(cv_1) = cTv_1 = c0' = 0'$. Hence N_T is a subspace of V. ∎

We are now able to make the following definition.

DEFINITION 5.19 Let $T: V \to V'$ be a linear transformation. The *null space* or *kernel* of T, denoted by N_T, or $\ker (T)$, is the subspace of V consisting of all elements v in V such that $Tv = 0'$, where $0'$ is the zero vector in V'. The *nullity* of T, denoted by $\nu(T)$, is the dimension of the null space. If the null space consists of the zero vector alone, $\nu(T)$ is defined to be zero.

In a similar vein, we can consider the range of a linear transformation.

THEOREM 5.12 Let $T: V \to V'$ be a linear transformation. The range of T, denoted R_T, is a vector space.

PROOF: The proof is left as an exercise. ∎

DEFINITION 5.20 Consider the linear transformation $T: V \to V'$. The *rank* of T, denoted by $\rho(T)$, is the dimension of the range of T.

Example 1 We reconsider an example from Section 5.1. Let $V = R^1$, $V' = R^2$, and $Tx = (x, x)$. Since $Tx = 0'$ implies $(x, x) = (0, 0)$, it follows that N_T consists of the single point $x = 0$ of R^1. That is, N_T is the trivial subspace. On the other hand, R_T is the set of points on the straight line $y = x$ in R^2. Thus the rank is one and the nullity is zero.

Example 2 Let V be any vector space and let I be the identity operator on V. Since $Iv = v$ for every v, $Iv = 0$ if and only if $v = 0$. That is, $N_I = \{0\}$, the trivial subspace. Also clearly the range is V.

Example 3 Let $V = C^1(-\infty, \infty)$, $V' = C^0(-\infty, \infty)$, and let $T = D - 2I$ be a linear operator on V into V', where D is the derivative operator. The null space N_T is the subspace of all functions $y(x)$ in V such that $(D - 2I)y = 0$, that is,

$$\frac{dy}{dx} - 2y = 0 \tag{1}$$

By the methods of Chapter 1 it can be seen that the solution of this equation is $y = ce^{2x}$, where c is an arbitrary constant. Hence N_T is the vector space generated by the element e^{2x} of V. The range R_T is the set of all functions $h(x)$ such that there is a function y in V for which $(D - 2I)y = h$. In other words, R_T is the set of all h in V' for which the equation

$$\frac{dy}{dx} - 2y = h \tag{2}$$

can be solved for y. In Chapter 1 we showed that there is a solution y for every continuous h, namely

$$y(x) = ce^{2(x-x_0)} + e^{2x} \int_{x_0}^{x} h(u)e^{-2u}\, du \tag{3}$$

Therefore R_T is the whole space V' in this case.

In the future we will dispense with making a notational distinction between the zero vectors of V and V', denoting both by 0.

THEOREM 5.13   A linear transformation $T: V \rightarrow V'$ is one-to-one (injective) if and only if N_T is the zero vector only.

PROOF: Suppose T is one-to-one. Then $Tv_1 = Tv_2$ implies $v_1 = v_2$. Let $v \in N_T$, that is, $Tv = 0$. This equation can be written $Tv = T0$. Thus $N_T = \{0\}$. Conversely, if $N_T = \{0\}$ and $Tv_1 = Tv_2$, then $T(v_1 - v_2) = 0$ and hence $v_1 = v_2$. Thus T is one-to-one.   ∎

It can be verified in Examples 1 and 2 that the transformation is one-to-one. In Example 3 this is not true. In fact, Eq. (3) provides an infinite family of solutions of Eq. (2) depending on c, so that $y_1' - 2y_1 = y_2' - 2y_2$ does not imply $y_1 = y_2$.

THEOREM 5.14   Let $T: V \rightarrow V'$ be a linear transformation with $N_T = \{0\}$. If $v_1, \ldots, v_n$ are linearly independent elements of V, then $Tv_1, \ldots, Tv_n$ are linearly independent elements of V'.

If $Tv_1, Tv_2, \ldots, Tv_n$ are linearly independent in V', then $v_1, \ldots, v_n$ must be linearly independent in V, whether or not N_T consists of the zero vector only.

PROOF: The equation

$$a_1(Tv_1) + \cdots + a_n(Tv_n) = 0$$

is, by the linearity of T, equivalent to

$$T(a_1v_1 + \cdots + a_nv_n) = 0$$

This implies that $\mathbf{N}_T$ contains $a_1v_1 + \cdots + a_nv_n$, and therefore by hypothesis

$$a_1v_1 + \cdots + a_nv_n = 0$$

Since $v_1, \ldots, v_n$ are linearly independent, it follows that $a_1 = \cdots = a_n = 0$. Therefore $Tv_1, \ldots, Tv_n$ are linearly independent. It is easy to see that if $Tv_1, \ldots, Tv_n$ are linearly independent in $\mathbf{V}'$, then $v_1, \ldots, v_n$ must be linearly independent in $\mathbf{V}$, whether or not $\mathbf{N}_T$ consists of the zero vector only. ∎

COROLLARY 1   Let $T: \mathbf{V} \to \mathbf{V}'$ be linear and one-to-one and let $\{v_1, v_2, \ldots, v_n\}$ be a basis of $\mathbf{V}$. Then $\{Tv_1, Tv_2, \ldots, Tv_n\}$ is a basis of the range $\mathbf{R}_T$.

PROOF: It only remains to show that $\{Tv_1, \ldots, Tv_n\}$ generates all of $\mathbf{R}_T$. This is left as an exercise. ∎

COROLLARY 2   If $T: \mathbf{V} \to \mathbf{V}'$ is linear and one-to-one and $\mathbf{V}$ is finite dimensional, then $\rho(T) = \dim \mathbf{V}$.

PROOF: The proof is left as an exercise. ∎

THEOREM 5.15   Consider the linear transformation $T: \mathbf{V} \to \mathbf{V}'$. Let $\{v_1, v_2, \ldots, v_q\}$ be a basis for $\mathbf{N}_T$ (the null space of T) and $\{v_1, v_2, \ldots, v_q, v_{q+1}, \ldots, v_n\}$ be a basis for $\mathbf{V}$. Then $\{Tv_{q+1}, Tv_{q+2}, \ldots, Tv_n\}$ is a basis for $\mathbf{R}_T$.

PROOF: Note that q is the nullity of T. Further, the dimension of $\mathbf{V}$ is assumed to be the finite integer n. Consider the case $q = 0$. It is then understood that $\{v_1, \ldots, v_q\}$ is the empty set, in which case the theorem asserts that the images of basis vectors of $\mathbf{V}$ comprise a basis for $\mathbf{R}_T$. This is the substance of Corollary 1. Assume, then, $q \geq 1$. Let $\{v_1, \ldots, v_q\}$ be a basis for $\mathbf{N}_T$ and $\{v_1, \ldots, v_q, v_{q+1}, \ldots, v_n\}$ be a basis for $\mathbf{V}$. Now let $v' \in \mathbf{R}_T$. Then $v' = Tv$ for some $v \in \mathbf{V}$. Let

$$v = \sum_{i=1}^{n} a_iv_i$$

Then

$$v' = Tv = \sum_{i=1}^{n} a_iTv_i = \sum_{i=q+1}^{n} a_iTv_i$$

since $Tv_i = 0$ for $i = 1, \ldots, q$. Hence $\{Tv_{q+1}, \ldots, Tv_n\}$ generates $\mathbf{R}_T$.

We now show that this set is linearly independent. Suppose that $c_{q+1}, \ldots, c_n$ are scalars, not all zero, such that

$$c_{q+1}Tv_{q+1} + \cdots + c_nTv_n = 0$$

Then

$$T(c_{q+1}v_{q+1} + \cdots + c_n v_n) = 0$$

and therefore

$$c_{q+1}v_{q+1} + \cdots + c_n v_n \text{ is in } \mathbf{N}_T$$

Hence $\sum_{i=q+1}^{n} c_i v_i$ is a linear combination of $v_1, \ldots, v_q$, which contradicts the linear independence of the set $\{v_1, \ldots, v_q, v_{q+1}, \ldots, v_n\}$. This proves that $\{Tv_{q+1}, \ldots, Tv_n\}$ is independent and therefore a basis for $\mathbf{R}_T$.   ∎

THEOREM 5.16 Let $\mathbf{V}$ be a finite dimensional vector space and T a linear transformation from $\mathbf{V}$ into $\mathbf{V}'$. Then $\rho(T) + v(T) = \dim \mathbf{V}$ where $\dim \mathbf{V}$ denotes the dimension of $\mathbf{V}$.

PROOF: From the last theorem, $\rho(T) = n - q = n - v(T)$.   ∎

COROLLARY The range of a linear transformation T on a space $\mathbf{V}$ can never have dimension greater than $\dim \mathbf{V}$.

PROOF: Since $v(T) \geq 0$, it follows that $\rho(T) \leq \dim \mathbf{V}$.   ∎

If the transformation is one-to-one, its nullity is zero and therefore $\dim \mathbf{R}_T = \dim \mathbf{V}$. Also, clearly $\dim \mathbf{R}_T \leq \dim \mathbf{V}'$ since the range is a subspace of $\mathbf{V}'$.

We close this section with theorems concerning the range and null space of composite mappings.

THEOREM 5.17 Let $S: \mathbf{V} \to \mathbf{V}'$ and $T: \mathbf{V}' \to \mathbf{V}''$ be linear transformations such that the composite transformation TS is defined. Then

$$\rho(TS) + \dim (\mathbf{R}_S \cap \mathbf{N}_T) = \rho(S)$$

PROOF: (See Figure 5.7.) Let T' denote the mapping defined on $\mathbf{R}_S$ into $\mathbf{V}''$ such that $T'v' = Tv'$ for every v' in $\mathbf{R}_S$. That is, T' is the same as T except that T' is regarded as defined only[3] on $\mathbf{R}_S$, not on all of $\mathbf{V}'$. The null space of T'

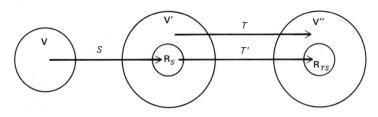

FIGURE 5.7

---

[3] The mapping T' is sometimes called the *restriction* of T to the domain $\mathbf{R}_S$.

consists of those vectors in the null space of T which are in $\mathbf{R}_S$. That is, $\mathbf{N}_{T'} = \mathbf{N}_T \cap \mathbf{R}_S$. Hence $v(T') = \dim (\mathbf{R}_S \cap \mathbf{N}_T)$. On the other hand, the range of T' is the image of $\mathbf{V}$ under the mapping TS. Therefore, $\rho(T') = \rho(TS)$. Finally, by Theorem 5.16 applied to T', we have $\rho(T') + v(T') = \dim \mathbf{R}_S = \rho(S)$. Substituting, we have

$$\rho(TS) + \dim (\mathbf{R}_S \cap \mathbf{N}_T) = \rho(S)$$

This completes the proof. ∎

COROLLARY Under the same hypothesis as in Theorem 5.17,

1. $\rho(S) \le \rho(TS) + v(T)$
2. $\rho(S) + \rho(T) \le \rho(TS) + \dim \mathbf{V}'$

PROOF: Since $\mathbf{R}_S \cap \mathbf{N}_T \subset \mathbf{N}_T$, we have $\dim (\mathbf{R}_S \cap \mathbf{N}_T) \le v(T)$. When this is applied to the equality in Theorem 5.17 we get (1). Then (2) follows since $\rho(T) + v(T) = \dim \mathbf{V}'$. ∎

THEOREM 5.18 Under the same hypothesis as in Theorem 5.17,

1. $\mathbf{R}_{TS} \subseteq \mathbf{R}_T$ and $\rho(TS) \le \min (\rho(T), \rho(S))$
2. $\mathbf{N}_S \subseteq \mathbf{N}_{TS}$ and $v(S) \le v(TS)$

PROOF: The proof is left as an exercise for the reader. ∎

Exercises

1. Find the nullity and rank of each linear transformation and describe the null space.

   (a) $T(x_1, x_2) = (x_1 + x_2, -x_2)$
   (b) $T(x_1, x_2) = (x_1, -x_2)$
   (c) $T(x_1, x_2) = (x_1, 0)$
   (d) $T(x_1, x_2, x_3) = (x_2 - x_3, 2x_1 + x_2, 0)$
   (e) $T(x_1, x_2, x_3) = (x_1, x_2)$ (T maps $\mathbf{R}^3$ into $\mathbf{R}^2$)
   (f) $T(x_1, x_2, x_3) = x_1 - x_2 + 2x_3$ (T maps $\mathbf{R}^3$ into $\mathbf{R}^1$)

2. Let $\mathbf{V} = \mathbf{C}^0[0, 1]$, and let $T: \mathbf{V} \to \mathbf{V}$ be defined by

$$Tf(t) = \int_0^t f(u) \, du$$

   Find the null space of T.

3. Let $\mathbf{V}$ be the vector space of functions which have derivatives of all orders, and let $D: \mathbf{V} \to \mathbf{V}$ be the derivative. What is the kernel of D? of D^2? of D^n?

4. Find the null space of the differential operator $D - 3I$.

5. Let $T: V \rightarrow V'$ be a linear transformation. Prove that $\mathbf{R}_T$ is a subspace of V' (that is, prove Theorem 5.12).

6. Prove the Corollaries to Theorem 5.14.

7. For each transformation in Exercise 1, give a basis for $\mathbf{N}_T$. Then extend this to a basis for all of V, and from this find a basis of $\mathbf{R}_T$.

8. Prove Theorem 5.18.

9. Let $T: V \rightarrow V'$ be a linear transformation with $\mathbf{N}_T = \{0\}$, and assume that $\dim V = \dim V' < \infty$. Then prove that $\mathbf{R}_T = V'$. In other words, if T is injective and V and V' have the same finite dimension, then T is surjective.

10. Find the rank and nullity of each linear transformation T in Exercise 2 of Section 5.2.

11. $\{1, x - 1, (x + 1)^2, x^3\}$ is a basis for the space V of polynomials of degree three or less. The transformation

$$T(a_0 + a_1 x + a_2 x^2 + a_3 x^3) = \begin{pmatrix} a_0 & a_1 \\ a_2 & a_3 \end{pmatrix}$$

takes V into V', the space of 2×2 matrices.

(a) Show that T is one-to-one and onto.

(b) Find the image of the given basis for V and express each of the following 2×2 matrices in terms of this basis for V'.

$$\begin{pmatrix} 1 & 0 \\ 0 & 1 \end{pmatrix}, \quad \begin{pmatrix} a & b \\ b & a \end{pmatrix}$$

12. Let V be the space of all continuous functions, $V = \mathbf{C}^0(-\infty, \infty)$. Describe the null space of the operator $S: V \rightarrow V$ defined by

$$(Sf)(x) = \int_0^{x^2} e^{-u} f(u)\, du$$

13. Let $T: V \rightarrow V'$ be a linear transformation. Prove that if $\dim V = \infty$, then either the null space or the range of T must be infinite dimensional. (Thus Theorem 5.16 remains true in a certain sense.)

5.7 Inverse transformations

In Section 5.1, we defined a mapping F of S into S' to be one-to-one (or injective) if distinct elements in S are mapped by F into distinct elements in S'. The mapping F of S into S' is onto (or surjective) if every element of S' is the image of some element of S. Hence, if F maps S both one-to-one and onto S' it follows that to each element $x \in S$, there is a unique element $x' \in S'$ such that $F(x) = x'$, and to each element $y' \in S'$ there corresponds a unique $y \in S$ such that $F(y) = y'$. The relationship F^* such that $F^*(y') = y$ for $F(y) = y'$, $y \in S$, $y' \in S'$ is called the inverse mapping of F.

It is immediately clear that only one-to-one mappings of S onto S' have unique inverse mappings since if F is not onto, that is, the range of F is a proper subset of S', there must be an element $y' \in S'$ such that there is no $y \in S$ for which $F(y) = y'$. In contrast to the definition in Section 5.1, we shall henceforth say that the inverse mapping of F exists only when the range of F is all of S'. Further, if F is not one-to-one there must be two elements y_0 and y_1 in S such that $F(y_0) = F(y_1) = y'$ for $y' \in S'$. Hence the inverse relationship is not a mapping since y' is associated with more than one element in S.

With the above review we turn to the topic of this section, invertible linear transformations.

DEFINITION 5.21 A linear transformation T which maps $\mathbf{V}$ one-to-one onto $\mathbf{V}'$ is said to be *nonsingular* or *invertible*. There is then a transformation T^* of $\mathbf{V}'$ onto $\mathbf{V}$ such that $T^*Tv = v$ for every $v \in \mathbf{V}$ and $TT^*v' = v'$ for every $v' \in \mathbf{V}'$. T^* is called the *inverse* of T.

If we let $I_{\mathbf{V}}$ be the identity mapping on $\mathbf{V}$ and $I_{\mathbf{V}'}$ be the identity mapping on $\mathbf{V}'$, then if T^* is the inverse of T, we have

$$T^*T = I_{\mathbf{V}}, \qquad TT^* = I_{\mathbf{V}'} \tag{1}$$

If T is invertible, the inverse is unique. In fact, suppose that T^* and T_2^* both satisfy Eq. (1). Then by multiplying the first equation on the right by T_2^* we get $(T^*T)T_2^* = I_{\mathbf{V}}T_2^* = T_2^*$. Using the associative law we obtain $T^*(TT_2^*) = T_2^*$. But $TT_2^* = I_{\mathbf{V}'}$. Hence $T^*I_{\mathbf{V}'} = T_2^*$ and $T^* = T_2^*$.

It is customary to use the symbol T^{-1} rather than T^* for the unique inverse of T, and we shall sometimes follow this custom.

THEOREM 5.19 If it exists, the inverse mapping T^* of a linear transformation $T: \mathbf{V} \to \mathbf{V}'$ is linear.

PROOF: Let v'_1 and v'_2 be two elements of $\mathbf{V}'$. Since T is one-to-one, and onto, there exist two elements v_1 and v_2 of $\mathbf{V}$ such that $T(v_1) = v'_1$ and $T(v_2) = v'_2$. Hence $T^*(v'_1) = v_1$ and $T^*(v'_2) = v_2$. And since T is linear, $T(v_1 + v_2) = v'_1 + v'_2$, which implies $T^*(v'_1 + v'_2) = v_1 + v_2$. Hence

$$T^*(v'_1 + v'_2) = T^*(v'_1) + T^*(v'_2)$$

Also, for any scalar c, $T(cv_1) = cT(v_1) = cv'_1$. Therefore $T^*(cv'_1) = cv_1 = cT^*(v'_1)$, which proves T^* is linear. ∎

Example 1   Consider the mapping $T: \mathbf{R}^2 \to \mathbf{R}^2$, where $T(x, y) = (x + y, x - y)$. To determine whether T is nonsingular we try to construct an inverse transformation. Given any (x', y') is there any (x, y) such that $T(x, y) = (x', y')$? Setting $x + y = x'$ and $x - y = y'$ and solving for x and y, we obtain $x = (x' + y')/2$ and $y = (x' - y')/2$ as the only solutions for x and y. Hence

$$T^*(x', y') = \left(\frac{x' + y'}{2}, \frac{x' - y'}{2} \right)$$

is the inverse mapping of T.

Example 2   Let $T: \mathbf{R}^3 \to \mathbf{R}^2$ where $T(x, y, z) = (y, z)$. Since $T(x_0, 0, 0) = T(x_1, 0, 0)$ for any x_0 and x_1, we have that T is not one-to-one and hence not invertible.

Example 3   Let $S: \mathbf{C}^0[0, 1] \to \mathbf{C}^1[0, 1]$ be such that $S(f) = g$, where $g(t) = \int_0^t f(u) \, du$, $0 \le t \le 1$. We are now faced with a mapping which is one-to-one but not onto since for any element of $\mathbf{C}^0[0, 1]$, $S(f) = g$ must satisfy $g(0) = 0$. There is, for example, no function $f \in \mathbf{C}^0[0, 1]$ such that $\int_0^t f(u) \, du = t + 3$ for $0 \le t \le 1$. If we consider S as the mapping of $\mathbf{C}^0[0, 1] \to \mathbf{V}'$ where $\mathbf{V}'$ is the subset of functions g in $\mathbf{C}^1[0, 1]$ for which $g(0) = 0$, then S could be shown to be onto and hence invertible.

THEOREM 5.20   Let T be a linear mapping of $\mathbf{V}$ *onto* $\mathbf{V}'$. Then the following statements are equivalent:

1. T is invertible
2. $N_T = \{0\}$
3. $v(T) = 0$
4. $\rho(T) = \dim \mathbf{V}$ (in case $\mathbf{V}$ is finite dimensional)
5. If $\{v_1, \ldots, v_n\}$ is a basis of $\mathbf{V}$, then $\{Tv_1, \ldots, Tv_n\}$ is a basis of $\mathbf{V}'$ (in case $\mathbf{V}$ is finite dimensional).
6. For any v_1 and v_2 in $\mathbf{V}$, $T(v_1) = T(v_2)$ implies $v_1 = v_2$.

PROOF: By hypothesis T maps $\mathbf{V}$ onto $\mathbf{V}'$. Hence, T is invertible if and only if T is one-to-one. Hence (1) and (6) are equivalent by definition of one-to-one. By Theorem 5.13, $N_T = \{0\}$ if and only if T is one-to-one and hence statements (1) and (2) are equivalent. Statement (3) is equivalent to statement (2) [and hence to (1)] by definition of nullity. By Theorem 5.16, if $\mathbf{V}$ is finite dimensional, $\rho(T) + v(T) = \dim \mathbf{V}$ so that $\rho(T) = \dim \mathbf{V}$ if and only if $v(T) = 0$. Thus (4) and (3) are equivalent.

Assume that $\mathbf{V}$ is finite dimensional, $\{v_1, \ldots, v_n\}$ is a basis of $\mathbf{V}$, and $\{Tv_1, \ldots, Tv_n\}$ is a basis of $\mathbf{V}'$. Then $n = \dim \mathbf{V} = \dim \mathbf{V}' = \rho(T)$, so (5) implies (4). Conversely, if $\rho(T) = \dim \mathbf{V}$ then $v(T) = 0$ and T is one-to-one. Thus by Corollary 1 to Theorem 5.14 and the hypothesis that T is onto $\mathbf{V}'$, (4) implies (5).   ∎

Note that if **V** is finite dimensional and $T: \mathbf{V} \to \mathbf{V}'$ is invertible, it follows that dim $\mathbf{V}' = $ dim **V**. Thus if **V** is finite dimensional and dim $\mathbf{V} \neq$ dim $\mathbf{V}'$, T cannot be invertible. If dim $\mathbf{V} > $ dim $\mathbf{V}'$, T cannot be one-to-one, and if dim $\mathbf{V} < $ dim $\mathbf{V}'$, T cannot be onto. It is shown in Exercise 7 that Theorem 5.20 is valid without the hypothesis that T be onto if **V** and $\mathbf{V}'$ are of equal finite dimension. In that case then, to prove T invertible it suffices either to prove that T is one-to-one or that T is onto.

Example 4   Consider the linear transformations D and S on $\mathbf{C}^{\infty}[0, 1]$ such that $D(f) = f'$ on $0 \leq t \leq 1$ and $Sf(t) = \int_0^t f(u)\, du$. Then $DS(f) = f$. Hence $DS = I$ and a first inclination might be to assume that S and D are invertible. However, consider $SD(f)$, which we denote by g. Then

$$g(t) = \int_0^t f'(u)\, du = f(t) - f(0) \neq f(t)$$

in general. Hence $SD \neq I$. Note that S maps $\mathbf{V} = \mathbf{C}^{\infty}[0, 1]$ onto the subspace $\mathbf{V}'$ of $\mathbf{C}^{\infty}[0, 1]$ where $f \in \mathbf{V}'$ implies $f(0) = 0$, and hence S is not a one-to-one mapping of $\mathbf{C}^{\infty}[0, 1]$ onto $\mathbf{C}^{\infty}[0, 1]$ and therefore is not invertible. If we were to consider $S: \mathbf{C}^{\infty}[0, 1] \to \mathbf{V}'$, as defined in Example 3, and $D: \mathbf{V}' \to \mathbf{C}^{\infty}[0, 1]$, we could show that S and D are one-to-one and hence are invertible. This example suggests the possibility of distinguishing between left side inverses and right side inverses, as in the following definition.

DEFINITION 5.22   Let T be a linear transformation on **V** into $\mathbf{V}'$. Then a linear transformation S is said to be a *left inverse* of T if S is defined on $\mathbf{R}_T$, has range in **V**, and

$$ST = I_\mathbf{V} \tag{2}$$

A linear transformation U is said to be a *right inverse* of T if U is defined on $\mathbf{V}'$, has range in **V**, and

$$TU = I_{\mathbf{V}'} \tag{3}$$

In Example 4 with **V** and $\mathbf{V}'$ equal to $\mathbf{C}^{\infty}[0, 1]$, D is a left inverse of S and S is a right inverse of D, but D is not a right inverse of S.

THEOREM 5.21   Let $T: \mathbf{V} \to \mathbf{V}'$ have a left inverse S and a right inverse U. Then T is invertible and $S = U = T^{-1}$.

PROOF: If $Tv_1 = Tv_2$, then $STv_1 = STv_2$ and by Eq. (2), $v_1 = v_2$. Hence T is one-to-one. Moreover, let v' be any vector in $\mathbf{V}'$. Then Uv' is defined. Let $v = Uv'$. Then $Tv = TUv' = v'$ be Eq. (3). This shows that T is onto. Hence T is invertible. Also, since T is onto $\mathbf{V}'$, S is defined on $\mathbf{V}'$. From Eq. (3) we get $T^{-1}TU = T^{-1}$ and therefore $U = T^{-1}$, and from Eq. (2) we get $STT^{-1} = T^{-1}$ or $S = T^{-1}$. ∎

COROLLARY 1 $T: V \to V'$ has both a left inverse and a right inverse if and only if T is invertible.

COROLLARY 2 If $T: V \to V'$ is invertible and T^{-1} is its inverse, then T^{-1} is invertible and $(T^{-1})^{-1} = T$.

PROOF: By hypothesis $TT^{-1} = I_{V'}$ and $T^{-1}T = I_V$. Thus Eqs. (2) and (3) hold with T replaced by T^{-1}, S by T, and U by T (with V and V' interchanged). ∎

As shown in Example 4, neither Eq. (2) by itself nor Eq. (3) by itself is sufficient to ensure that T is invertible, in general. On the other hand, the next theorem shows that either one by itself is sufficient if V and V' are of equal finite dimension.

THEOREM 5.22 Assume that dim V = dim V' = $m < \infty$. Let T be a linear transformation of V into V' and suppose that either

1. there exists a linear transformation S defined on R_T with range in V such that $ST = I_V$ or
2. there exists a linear transformation U with domain V' and range in V such that $TU = I_{V'}$.

Then T is invertible. In case (1), $S = T^{-1}$ and in case (2), $U = T^{-1}$.

PROOF: Consider case (1). Let $v \in N_T$. That is, $Tv = 0$. Hence $STv = S0 = 0$. Since $ST = I_V$, it follows that $v = 0$. This shows that $N_T = \{0\}$. Since Theorem 5.20 is valid without the hypothesis that T is onto but with the assumption that dim V = dim V' = $m < \infty$ (see Exercise 7), it follows that T is invertible. Then $STT^{-1} = I_V T^{-1}$ or $S = T^{-1}$.

Now consider case (2). Since $TU = I_{V'}$, T maps R_U *onto* V' (For every $v' \in V'$, T maps Uv' into v'.) Hence the range of T is all of V' and $\rho(T) = $ dim V' = dim V. From Theorem 5.20 it follows that T is invertible. Then $T^{-1}TU = T^{-1}I_{V'} = T^{-1}$ and $U = T^{-1}$. ∎

The remaining theorems in this section give conditions for a composition TS of mappings to be nonsingular.

THEOREM 5.23 Let V, V', and V'' be vector spaces, and let S and T be linear transformations, $S: V \to V'$ and $T: V' \to V''$. Then if T is nonsingular, $\rho(TS) = \rho(S)$.

PROOF: Since T is invertible, it is one-to-one and $N_T = \{0\}$. Hence by Theorem 5.17, $\rho(TS) = \rho(S)$. ∎

THEOREM 5.24 Let V, V', and V'' be vector spaces and let S and T be linear transformations, $S: V \to V'$ and $T: V' \to V''$. Then if S is nonsingular, $\rho(TS) = \rho(T)$.

PROOF: This is left as an exercise. ∎

We can summarize the last two theorems by saying that multiplication by a nonsingular linear transformation does not change rank.

THEOREM 5.25 Let S be a linear transformation of V into V' and T be a linear transformation of V' into V''. If S and T are nonsingular then TS is non-singular and $(TS)^{-1} = S^{-1}T^{-1}$.

PROOF: Suppose S and T are nonsingular. Then

$$(S^{-1}T^{-1})(TS) = S^{-1}(T^{-1}T)S = S^{-1}I_{V'}S = S^{-1}S = I_V$$

and

$$(TS)(S^{-1}T^{-1}) = T(SS^{-1})T^{-1} = TI_{V'}T^{-1} = TT^{-1} = I_{V''}$$

Therefore $S^{-1}T^{-1}$ is both a left and right inverse of TS. By Theorem 5.21, TS is invertible and $(TS)^{-1} = S^{-1}T^{-1}$. ∎

Example 5 The converse of Theorem 5.25 is not true. For example, let $V = V'' = \mathbf{R}^1$ and $V' = \mathbf{R}^2$ and let $Sx = (x, 0)$ for $x \in \mathbf{R}^1$, $T(x, y) = x$ for $(x, y) \in \mathbf{R}^2$. Then $TSx = x$ for all x and TS is the identity map of $\mathbf{R}^1$ into itself. Although TS is invertible, neither T nor S is itself invertible in the sense of our definition. However, the converse is true in the special case in which V, V', and V'' are of equal finite dimension, as we now show.

THEOREM 5.26 Let S be a linear transformation of V into V' and T be a linear transformation of V' into V''. Assume that V, V', and V'' are of equal finite dimension. Then if TS is nonsingular, both S and T are nonsingular.

PROOF: Assume that TS is nonsingular. Then there is some transformation U on V'' such that $U(TS) = I_V$. But then UT is defined on $\mathbf{R}_S$ and $(UT)S = I_V$. By Theorem 5.22, case (1), S is invertible. Similarly, $(TS)U = I_{V''}$. So $T(SU) = I_{V''}$ and by Theorem 5.22, case (2), T is invertible. ∎

Exercises

1. For each part of Exercise 1, Section 5.6, determine whether T is one-to-one and onto V' where V' is $\mathbf{R}^k$ for appropriate k. If so, find the inverse T^{-1}; that is, set $(x, y) = T^{-1}(u, v)$ and give formulas for x and y in terms of u and v.

2. In each part, determine whether the given linear transformation on $\mathbf{R}^3$ into $\mathbf{R}^3$ is one-to-one and onto. If so, find the inverse T^{-1}; that is, set $(x, y, z) = T^{-1}(u, v, w)$ and give formulas for x, y, and z in terms of u, v, and w.

    (a) $T(x, y, z) = (x, x + y, x + y + z)$

    (b) $T(x, y, z) = (x - y, y - z, z - x)$

    (c) $T(x, y, z) = (2x + y, 2y - x, x + y + z)$

3. For what values of A is the linear transformation T not invertible if

    (a) $T(x, y) = (2x + y, x + Ay)$

    (b) $T(x, y, z) = (2x + y + z, 2y - x, x + y + Az)$

4. In each part, T is a linear transformation on $\mathbf{V}$ into $\mathbf{V}$ where $\mathbf{V}$ is the space of all polynomials. Determine whether T is invertible, and if so give a formula for $T^{-1}q$ for an arbitrary polynomial q.

    (a) $T: p(x) \rightarrow p(x + 1)$

    (b) $T: p(x) \rightarrow \dfrac{dp}{dx}$

    (c) $T: p(x) \rightarrow p(0)$

    (d) $T: p(x) \rightarrow \displaystyle\int_0^x p(u)\,du$

    (e) $T: p(x) \rightarrow xp(x)$

    (f) $T: p(x) \rightarrow (p(x) - p(0))/x$

5. In each part, D represents the derivative operator. Determine whether the given linear transformation T is invertible.

    (a) D^2 on $\mathbf{C}^2(-\infty, \infty)$ into $\mathbf{C}^0(-\infty, \infty)$

    (b) D^2 on $\mathbf{V}$ into $\mathbf{C}^0(-\infty, \infty)$ where $\mathbf{V} = \{f \in \mathbf{C}^2(-\infty, \infty): f(0) = 0, f'(0) = 0\}$

    (c) $D - 4I$ on $\mathbf{V}$ into $\mathbf{C}^0(-\infty, \infty)$ where $\mathbf{V} = \{f \in \mathbf{C}^1(-\infty, \infty): f(0) = f(2)\}$

6. In Theorem 5.20 suppose that the hypothesis that T is onto $\mathbf{V}'$ is discarded. Give an example to show that statement (2) does not necessarily imply (1) and an example to show that (4) does not necessarily imply (5).

7. In Theorem 5.20, drop the hypothesis that T is onto $\mathbf{V}'$ but add the hypothesis that $\mathbf{V}$ and $\mathbf{V}'$ be of equal finite dimension. Prove that the theorem remains correct, that is, that statements (1) through (6) are equivalent.

8. Prove Theorem 5.24.

9. What modifications are needed in Theorem 5.21 and its Corollaries if in Definition 5.22 the right inverse is defined on $\mathbf{R}_T$ rather than $\mathbf{V}'$ and Eq. (3) is replaced by $TUv' = v'$ for all v' in $\mathbf{R}_T$?

COMMENTARY

Section 5.5 It is possible to define functions of an operator for functions other than polynomials. An example of this occurs in Chapter 12. See also:

GANTMACHER, F. R., *The Theory of Matrices*, New York, Chelsea Publishing, 1959.

BELLMAN, R., *Introduction to Matrix Analysis*, New York, McGraw-Hill, 1960.

Chapter 6

Matrices and linear transformations

6.1 Matrices

In several examples, we have used 2 by 2 arrays of numbers which were called 2 by 2 matrices. The purpose of this chapter is to generalize this definition and then develop an algebra of matrices in a way that will illustrate the close relationship between the theory of matrices and the theory of linear transformations.

DEFINITION 6.1 A *matrix* is a rectangular array of elements of a field **K**. The array will be written in the form

$$\begin{pmatrix} a_{11} & a_{12} & \cdots & a_{1m} \\ a_{21} & a_{22} & \cdots & a_{2m} \\ \vdots & \vdots & \vdots & \vdots \\ a_{n1} & a_{n2} & \cdots & a_{nm} \end{pmatrix}$$

A matrix with n rows and m columns, such as the one shown above, is said to be an n by m or $n \times m$ matrix. If $m = n$, the matrix is called a *square* matrix and said to be of *order n*.

For convenience we often use the symbol (a_{ij}) to denote the matrix above, where a_{ij} typifies the element in the ith row and jth column. Also, upper-case Latin letters are frequently used to denote matrices, as $A = (a_{ij})$ or $M = (m_{ij})$. The a_{ij} are called the *elements* of the matrix.

Example 1 The array

$$\begin{pmatrix} 3 & -2 & 1 \\ 1 & 0 & 1 \end{pmatrix}$$

is a 2 × 3 matrix. The matrix

$$\begin{pmatrix} 3 - i & 2i \\ -3 & 3 + i \end{pmatrix}$$

is a 2 × 2 matrix of elements of the complex field. The array

$$\begin{pmatrix} 1 & 1 & 3 \\ 0 & 1 \end{pmatrix}$$

is not a matrix as it is not rectangular. Finally

$$(1 \quad 0 \quad 1 \quad 3) \quad \text{and} \quad \begin{pmatrix} 1 \\ 0 \\ 1 \\ 3 \end{pmatrix}$$

are 1 × 4 and 4 × 1 matrices, respectively.

A matrix of dimension 1 × n is often referred to as a *row vector* of elements of $\mathbf{K}$, and a matrix of dimension n × 1 is referred to as a *column vector*. The use of the word *vector* suggests that we can think of these as elements in a vector space. As we shall see below, this is indeed the case.

DEFINITION 6.2 Two matrices $M = (m_{ij})$ and $N = (n_{ij})$ are called *equal* if they consist of the same elements in the same positions. That is, they must have the same number of rows and they must have the same number of columns, and $m_{ij} = n_{ij}$ for all i and j.

Example 2

$$\begin{pmatrix} \sin 30° & 2 \\ i^2 & 3 \end{pmatrix} = \begin{pmatrix} 0.5 & \sqrt{4} \\ -1 & 3 \end{pmatrix}$$

DEFINITION 6.3 The m × n *zero matrix*, denoted O, is the matrix with m rows and n columns in which all the elements are zero.

6.2 The matrix of a linear transformation

Before developing a theory of matrices, it will be helpful to establish a connection between matrices and linear transformations. Consider the following example.

Example 1 Let $\mathbf{V} = \mathbf{R}^3$ and let $v_1 = (1, 0, 1)$, $v_2 = (0, 1, 0)$, $v_3 = (0, 0, 1)$. Then $B = \{v_1, v_2, v_3\}$ is a basis for $\mathbf{V}$. Let T be the linear transformation on $\mathbf{V}$ into $\mathbf{V}$ defined by $T(a, b, c) = (b, a, 0)$ for all (a, b, c) in $\mathbf{R}^3$. Then in particular

$$\begin{aligned} Tv_1 &= (0, 1, 0) = 0v_1 + v_2 + 0v_3 \\ Tv_2 &= (1, 0, 0) = v_1 + 0v_2 - v_3 \\ Tv_3 &= (0, 0, 0) = 0v_1 + 0v_2 + 0v_3 \end{aligned}$$

Recall Theorem 5.2 which states that a linear transformation is uniquely determined by the images under the transformation of the vectors in a basis. Here, for example, the fact that

$$Tv_1 = v_2, \quad Tv_2 = v_1 - v_3, \quad Tv_3 = \mathbf{0} \tag{1}$$

is enough to specify T uniquely. On the other hand, all the information in (1) is conveyed by giving the coordinate vectors of Tv_1, Tv_2, and Tv_3 relative to the given basis. The coordinates relative to B of Tv_1 are 0, 1, and 0, the coordinates of Tv_2 are 1, 0, -1, and the coordinates of Tv_3 are 0, 0, 0. We shall display the coordinates as 3×1 matrices or "column vectors,"

$$\begin{pmatrix} 0 \\ 1 \\ 0 \end{pmatrix}, \quad \begin{pmatrix} 1 \\ 0 \\ -1 \end{pmatrix}, \quad \text{and} \quad \begin{pmatrix} 0 \\ 0 \\ 0 \end{pmatrix}$$

Combining these into a 3×3 matrix we have

$$M = \begin{pmatrix} 0 & 1 & 0 \\ 1 & 0 & 0 \\ 0 & -1 & 0 \end{pmatrix}$$

In this matrix, the first column contains the coordinates 0, 1, 0 of Tv_1 relative to the basis B, or in other words, it tells us that $Tv_1 = 0v_1 + v_2 + 0v_3 = v_2$. The second column tells us that Tv_2 has coordinates 1, 0, -1, that is, $Tv_2 = v_1 - v_3$, and the third column tells us that $Tv_3 = \mathbf{0}$. This is enough to determine Tv for arbitrary v in V since if $v = a_1v_1 + a_2v_2 + a_3v_3$, then, by linearity of T,

$$Tv = a_1 Tv_1 + a_2 Tv_2 + a_3 Tv_3 = a_1v_2 + a_2(v_1 - v_3)$$

Any other linear transformation on $\mathbf{R}^3$ into $\mathbf{R}^3$ could similarly be associated with a 3×3 matrix. Provided we deal with a fixed basis and provided we adopt the above conventional way of constructing a matrix, each linear transformation has a unique matrix associated with it, and each matrix is associated with a unique linear transformation.

After these preliminaries, we can now give a general and formal definition.

DEFINITION 6.4   Let V and V' be finite dimensional vector spaces with bases $B = \{v_1, \ldots, v_p\}$ and $B' = \{v'_1, \ldots, v'_q\}$, respectively. Let T be a linear transformation of V into V'. The *matrix of T* or *matrix representative of T relative to the bases B and B'* is the $q \times p$ matrix in which the elements in the jth column are the coordinates of Tv_j relative to B' ($j = 1, \ldots, p$).

Suppose that the elements $Tv_1, \ldots, Tv_p$ of V' are expressed in terms of the basis B' by the equations

$$Tv_1 = \sum_{i=1}^{q} m_{i1}v'_i, \quad Tv_2 = \sum_{i=1}^{q} m_{i2}v'_i, \ldots, Tv_p = \sum_{i=1}^{q} m_{ip}v'_i$$

Then the matrix of T relative to B and B' is

$$\begin{pmatrix} m_{11} & m_{12} & \cdots & m_{1p} \\ m_{21} & m_{22} & \cdots & m_{2p} \\ \vdots & \vdots & \vdots & \vdots \\ m_{q1} & m_{q2} & \cdots & m_{qp} \end{pmatrix}$$

Example 2   Consider the space of geometrical 2-vectors $\mathbf{G}^2$ and let the elements of $\mathbf{V} = \mathbf{R}^2$ be the coordinates of these 2-vectors relative to the usual rectangular coordinate system. Take as basis for $\mathbf{V}$, $B = B' = \{e_1, e_2\}$, where $e_1 = (1, 0)$ and $e_2 = (0, 1)$. Let T be the transformation on $\mathbf{R}^2$ corresponding to the rotation of vectors in $\mathbf{G}^2$ 90 degrees counterclockwise. Then clearly $Te_1 = e_2$ and $Te_2 = -e_1$. Therefore, relative to the basis $\{e_1, e_2\}$, T is represented by the matrix

$$\begin{pmatrix} 0 & -1 \\ 1 & 0 \end{pmatrix}$$

More generally, let T correspond to rotation counterclockwise through an angle θ. Then $Te_1 = (\cos \theta, \sin \theta) = (\cos \theta)e_1 + (\sin \theta)e_2$ and $Te_2 = (-\sin \theta, \cos \theta) = (-\sin \theta)e_1 + (\cos \theta)e_2$. Therefore the matrix representing T is

$$\begin{pmatrix} \cos \theta & -\sin \theta \\ \sin \theta & \cos \theta \end{pmatrix}$$

Example 3   Let $\mathbf{V} = \mathbf{R}^3$, $\mathbf{V}' = \mathbf{R}^2$, and let T be the linear transformation such that $T(a, b, c) = (a + b, a - b)$. Choose the standard bases for $\mathbf{V}$ and $\mathbf{V}'$. That is, $B = \{e_1, e_2, e_3\}$ where $e_1 = (1, 0, 0)$, $e_2 = (0, 1, 0)$, $e_3 = (0, 0, 1)$; and $B' = \{e'_1, e'_2\}$ where $e'_1 = (1, 0)$, $e'_2 = (0, 1)$. Then

$$Te_1 = (1, 1) = e'_1 + e'_2$$
$$Te_2 = (1, -1) = e'_1 - e'_2$$
$$Te_3 = (0, 0)$$

The matrix of T is

$$\begin{pmatrix} 1 & 1 & 0 \\ 1 & -1 & 0 \end{pmatrix}$$

It must be emphasized that the matrix representing T depends on the choice of bases for $\mathbf{V}$ and $\mathbf{V}'$, as well as on T itself.

Example 4   Suppose that in the above example we choose the bases $B = \{v_1, v_2, v_3\}$, $B' = \{v'_1, v'_2\}$, where

$$v_1 = (1, 0, 1), \quad v_2 = (0, 1, 0), \quad v_3 = (0, 0, 1)$$
$$v'_1 = (1, 1), \quad v'_2 = (0, 1).$$

Then

$$Tv_1 = (1, 1) = v'_1$$
$$Tv_2 = (1, -1) = v'_1 - 2v'_2$$
$$Tv_3 = (0, 0)$$

and the matrix of T is

$$\begin{pmatrix} 1 & 1 & 0 \\ 0 & -2 & 0 \end{pmatrix}$$

Example 5   Let **V** be the space of polynomials of degree less than or equal to 2 and **V'** the space of polynomials of degree less than or equal to 3. Let $B = \{v_1, v_2, v_3\}$ be a basis for **V** where $v_1 = x^2$, $v_2 = x^2 - x$, $v_3 = x - 1$; and $B' = \{v'_1, v'_2, v'_3, v'_4\}$ be a basis for **V'** where $v'_1 = 1$, $v'_2 = x - 1$, $v'_3 = x^2 - x + 1$ and $v'_4 = x^3$. Consider the linear transformation $T: \mathbf{V} \to \mathbf{V'}$ where $T(ax^2 + bx + c) = ax^3 + bx^2 + cx$. Then $Tv_1 = T(x^2) = x^3 = v'_4$ has coordinates with respect to B' of

$$\begin{pmatrix} 0 \\ 0 \\ 0 \\ 1 \end{pmatrix}$$

$Tv_2 = T(x^2 - x) = x^3 - x^2 = v'_4 - v'_3 - v'_2$ has coordinates with respect to B' of

$$\begin{pmatrix} 0 \\ -1 \\ -1 \\ 1 \end{pmatrix}$$

and $Tv_3 = T(x - 1) = x^2 - x = v'_3 - v'_1$ has coordinates with respect to B' of

$$\begin{pmatrix} -1 \\ 0 \\ 1 \\ 0 \end{pmatrix}$$

Thus the matrix of T relative to B and B' is

$$\begin{pmatrix} 0 & 0 & -1 \\ 0 & -1 & 0 \\ 0 & -1 & 1 \\ 1 & 1 & 0 \end{pmatrix}$$

The above discussion leads to the following theorem. .

THEOREM 6.1   Let **V** and **V′** be any finite dimensional vector spaces with bases $B = \{v_1, \ldots, v_p\}$ of **V** and $B' = \{v'_1, \ldots, v'_q\}$ of **V′**. Then there is a one-to-one correspondence between the set of all linear transformations T of **V** into **V′** and the set of $q \times p$ matrices.

PROOF: We have shown above that for any transformation T there exists a unique matrix M with columns representing the coordinates of the images of the elements of B under T with respect to the elements of B'. But by Theorem 5.3, to each set of images of the basis vectors in B, there corresponds a unique linear transformation. Hence to each set of p columns, that is, to each $q \times p$ matrix, there is a unique transformation.   ∎

For a linear transformation which maps **V** into itself, $T: \mathbf{V} \to \mathbf{V}$, only one basis is needed. The associated matrix is then square, which is to say that it has the same number of rows as columns. Note that the matrix representing $T: \mathbf{V} \to \mathbf{V}$ relative to the bases B and B', where $B \neq B'$, is also square as is any matrix representing any linear transformation of **V** into **V′** as long as **V** and **V′** are of the same dimension.

Exercises

1. In each part of Exercise 1, Section 5.6, find the matrix of T relative to the standard bases for the spaces involved.

2. In each part, a linear transformation T from $\mathbf{R}^p$ into $\mathbf{R}^2$ has the given matrix representative relative to the standard bases. Find p and T; that is, find $T(x_1, \ldots, x_p)$ for arbitrary $(x_1, \ldots, x_p)$.

(a) $\begin{pmatrix} 0 & 1 \\ 1 & 0 \end{pmatrix}$

(b) $\begin{pmatrix} 2 & 1 \\ 3 & -1 \end{pmatrix}$

(c) $\begin{pmatrix} 3 \\ 2 \end{pmatrix}$

(d) $\begin{pmatrix} 1 & 0 & 4 \\ 0 & 1 & 2 \end{pmatrix}$

(e) $\begin{pmatrix} 1 & 0 & 0 & 0 \\ 0 & 1 & 0 & 0 \end{pmatrix}$

3. Find the matrix representing $T: \mathbf{R}^3 \to \mathbf{R}^2$, where $T(x_1, x_2, x_3) = (x_1 + x_2, x_1 - x_3)$, relative to each pair of bases.

(a) B and B' are the standard bases:
$B = \{e_1, e_2, e_3\}, \quad B' = \{e'_1, e'_2\}$

(b) $B = \{e_1, e_2, e_3\}, \quad B' = \{(1, -1), (-1, -1)\}$

(c) $B = \{(1, 1, 0), (1, 0, 1), (0, 1, 1)\}, \quad B' = \{(2, 1), (-1, -2)\}$

4. Find the matrix representing the identity map of $\mathbf{R}^3$ onto itself relative to each basis.

(a) The standard basis $\{e_1, e_2, e_3\}$

(b) The basis $\{(1, 0, 0), (1, 1, 0), (1, 1, 1)\}$

(c) An arbitrary basis B

5. In each part, T is a linear transformation of $\mathbf{V}$ into $\mathbf{R}^q$ for some q where $\mathbf{V}$ is the space of polynomials of degree at most 3. Find the matrix representing T relative to the bases

$$B = \{1, x, x^2, x^3\}, \qquad B' = \{e_1, \ldots, e_q\}$$

(a) $T: p(x) \to p(0)$

(b) $T: p(x) \to (p(0), p'(0))$

(c) $T: p(x) \to \left(\int_0^2 p(x)\, dx,\ p(0),\ p'(0) \right)$

6. In each part, T is a linear transformation of $\mathbf{V}$ into $\mathbf{V}$ where $\mathbf{V}$ is the space of polynomials of degree at most two. Find the matrix of T relative to the basis $\{1, x, x^2\}$.

(a) $T: p(x) \to p(x + 1)$

(b) $T: p(x) \to dp/dx$

(c) $T: p(x) \to p(0)x$

(d) $T: p(x) \to (p(x) - p(0))/x$

7. Repeat Exercise 6 if the basis is $\{1, x - 1, (x - 1)^2\}$.

8. Let $\mathbf{V}$ be of finite dimension n, and B a basis of $\mathbf{V}$. Suppose that the matrix of a linear operator T on $\mathbf{V}$ into $\mathbf{V}$ with respect to a basis B is the $n \times n$ matrix D where $d_{ii} = 1$, and $d_{ij} = 0$ for $i \neq j$. Prove that T is the identity operator.

9. Let $B = \{e^x, e^{2x}\}$ and let $\mathbf{V}$ be the space of functions generated by B. Find the matrix of each of the following transformations relative to B.

(a) D where $D = d/dx$

(b) D^2

(c) $D - I$

(d) $(D - I)(D - 2I)$

10. Let T be the transformation of $\mathbf{R}^2$ into $\mathbf{R}^2$ defined by $T(x_1, x_2) = (x_1 + x_2, x_1 - x_2)$. Find the matrix of T relative to bases B and B', where $B = \{e_1, e_2\}$ and $B' = \{(1, 2), (-1, 1)\}$.

6.3 Addition and scalar multiplication for matrices

According to Theorem 6.1, given bases B and B' for spaces $\mathbf{V}$ and $\mathbf{V}'$, there is a one-to-one correspondence between the set of all linear transformations from $\mathbf{V}$ into $\mathbf{V}'$ and the set of all $q \times p$ matrices, where $q = \dim \mathbf{V}'$ and $p = \dim \mathbf{V}$. However, there is still an imbalance in our discussion of linear transformations and matrices, since for the former we have defined algebraic operations of addition and multiplication by a scalar, whereas for the latter this has been done only in a special case (see Example 2 of Section 4.3 and Example 1 of Section 4.4).

There are many ways in which one could define addition of two $q \times p$ matrices, but only one way which corresponds naturally to the addition of linear transformations as previously defined. In order to see this we take two matrices

$$M = \begin{pmatrix} m_{11} & m_{12} & \cdots & m_{1p} \\ m_{21} & m_{22} & \cdots & m_{2p} \\ & & \cdots & \\ m_{q1} & m_{q2} & \cdots & m_{qp} \end{pmatrix} \qquad N = \begin{pmatrix} n_{11} & n_{12} & \cdots & n_{1p} \\ n_{21} & n_{22} & \cdots & n_{2p} \\ & & \cdots & \\ n_{q1} & n_{q2} & \cdots & n_{qp} \end{pmatrix} \qquad (1)$$

Let $\mathbf{V}$ be any p-dimensional vector space and $\mathbf{V}'$ any q-dimensional vector space. Let $B = \{v_1, v_2, \ldots, v_p\}$ and $B' = \{v'_1, v'_2, \ldots, v'_q\}$ be fixed bases for these spaces. Then, as we have seen, there is a unique linear transformation $S \colon \mathbf{V} \to \mathbf{V}'$ which has M as its representative relative to B and B', and a unique linear transformation $T \colon \mathbf{V} \to \mathbf{V}'$ which has N as its representative. Thus,

$$Sv_i = \sum_{k=1}^{q} m_{ki}v'_k, \qquad Tv_i = \sum_{k=1}^{q} n_{ki}v'_k \qquad (i = 1, 2, \ldots, p) \qquad (2)$$

By definition, $S + T$ is that linear transformation U of $\mathbf{V}$ into $\mathbf{V}'$ such that $Uv = Sv + Tv$ for every v in $\mathbf{V}$. Hence

$$Uv_i = Sv_i + Tv_i = \sum_{k=1}^{q} (m_{ki} + n_{ki})v'_k \qquad (i = 1, 2, \ldots, p) \qquad (3)$$

This equation states that the matrix which represents $S + T = U$, relative to bases B and B', is

$$\begin{pmatrix} m_{11} + n_{11} & m_{12} + n_{12} & \cdots & m_{1p} + n_{1p} \\ m_{21} + n_{21} & m_{22} + n_{22} & \cdots & m_{2p} + n_{2p} \\ & & \cdots & \\ m_{q1} + n_{q1} & m_{q2} + n_{q2} & \cdots & m_{qp} + n_{qp} \end{pmatrix} \qquad (4)$$

We therefore adopt the following definition.

DEFINITION 6.5 Let M and N be two $q \times p$ matrices with elements in a field $\mathbf{K}$, as in Eq. (1). The *sum* $M + N$ is the $q \times p$ matrix obtained by adding corresponding elements of M and N. That is, $M + N$ is the matrix in (4).

By adopting this definition we ensure that if a matrix M represents a transformation $S: \mathbf{V} \rightarrow \mathbf{V}'$ and a matrix N represents $T: \mathbf{V} \rightarrow \mathbf{V}'$, relative to the same bases for $\mathbf{V}$ and $\mathbf{V}'$, then $M + N$ represents $S + T$ relative to these bases.

If c is any scalar, cS is that linear transformation U of $\mathbf{V}$ into $\mathbf{V}'$ such that $Uv = c(Sv)$ for every v in $\mathbf{V}$. Hence

$$Uv_i = cSv_i = \sum_{k=1}^{q} (cm_{ki})v'_k \qquad (i = 1, 2, \ldots, p) \tag{5}$$

This equation states that the matrix which represents $cS = U$ is the matrix

$$\begin{pmatrix} cm_{11} & cm_{12} & \cdots & cm_{1p} \\ cm_{21} & cm_{22} & \cdots & cm_{2p} \\ & & \cdots & \\ cm_{q1} & cm_{q2} & \cdots & cm_{qp} \end{pmatrix} \tag{6}$$

Therefore we make the following definition.

DEFINITION 6.6   Let M be a $q \times p$ matrix with elements in a field $\mathbf{K}$. The *product* of M by a scalar c in $\mathbf{K}$ is defined to be the matrix (6), that is, the matrix obtained by multiplying each element of M by c.

This definition ensures that if S is represented by a matrix M relative to given bases, then cS is represented relative to the same bases by cM.

Example 1   Let

$$M = \begin{pmatrix} 2 & 1 & 3 \\ 4 & 0 & 6 \end{pmatrix} \qquad N = \begin{pmatrix} 4 & 1 & 0 \\ 0 & 0 & -2 \end{pmatrix} \qquad P = \begin{pmatrix} 2 & 1 \\ -4 & 3 \end{pmatrix}$$

Then

$$M + N = \begin{pmatrix} 6 & 2 & 3 \\ 4 & 0 & 4 \end{pmatrix} = N + M$$

$$7M = \begin{pmatrix} 14 & 7 & 21 \\ 28 & 0 & 42 \end{pmatrix}$$

$$-3P = \begin{pmatrix} -6 & -3 \\ 12 & -9 \end{pmatrix}$$

whereas $M + P$ and $N + P$ are undefined.

THEOREM 6.2   The set of $q \times p$ matrices with elements in a field $\mathbf{K}$ is a vector space over $\mathbf{K}$ with addition and scalar multiplication defined as above.

PROOF: It is easy to see that the set of $q \times p$ matrices forms a commutative additive group, with the $q \times p$ zero matrix O being the zero for the group. The verification of the other axioms for a vector space (Definition 4.5) is left as an exercise. ∎

According to Theorem 5.7, the set of all linear transformations on V into V' is also a vector space. The one-to-one correspondence between linear transformations and matrices established in Theorem 6.1 is therefore an isomorphism between vector spaces, as we shall prove.

DEFINITION 6.7 If V and V' are vector spaces, $\mathscr{L}(V, V')$ denotes the vector space of all linear transformations mapping V into V'.

THEOREM 6.3 Let V and V' be any vector spaces over K with dim $V = p$, dim $V' = q$. Then the vector space of all linear transformations of V into V' is isomorphic to the vector space of all $q \times p$ matrices with elements in K.

PROOF: Let $\mathbf{M}_{qp}$ be the vector space of all $q \times p$ matrices.[1] We must show that there is an isomorphism of $\mathscr{L}(V, V')$ onto $\mathbf{M}_{qp}$. Let B and B' be any fixed bases of V and V', respectively. By Theorem 6.1, there is a one-to-one correspondence between $\mathscr{L}(V, V')$ and $\mathbf{M}_{qp}$. Moreover, the above discussion shows that this correspondence is preserved under the operations. That is, if S corresponds to M and T corresponds to N (which is to say that the matrices M and N represent S and T, respectively, relative to the given bases), then $S + T$ corresponds to $M + N$ and cS corresponds to cM. Alternatively, we can say that there is a one-to-one mapping τ of $\mathscr{L}(V, V')$ onto $\mathbf{M}_{qp}$ such that

$$\tau(S + T) = \tau(S) + \tau(T)$$

$$\tau(cS) = c\tau(S)$$

Thus the correspondence is an isomorphism. ∎

This result is extremely important because it says that every fact or theorem concerning the vector space operations on linear transformations has a counterpart for matrices, and vice versa. This sometimes enables one to give simple proofs of results which might otherwise be tedious. For example, we have the following theorem.

---

[1] *Warning on notation:* Do not confuse the symbol $\mathbf{M}_{qp}$ which is used for this space with symbols such as m_{ij} which are used for elements of a matrix called M.

THEOREM 6.4 If **V** and **V**′ are finite dimensional vector spaces, then so is $\mathscr{L}(\mathbf{V}, \mathbf{V}')$, the space of all linear transformations on **V** into **V**′, and

$$\dim \mathscr{L}(\mathbf{V}, \mathbf{V}') = (\dim \mathbf{V})(\dim \mathbf{V}')$$

PROOF: Let $\dim \mathbf{V} = p$, $\dim \mathbf{V}' = q$. Then $\mathscr{L}(\mathbf{V}, \mathbf{V}')$ and $\mathbf{M}_{qp}$ have the same dimension since they are isomorphic. Now there are qp distinct matrices with q rows and p columns, each having all entries 0 except for one entry that is 1. It is left as an exercise to show that these matrices form a basis for $\mathbf{M}_{qp}$, and hence that $\dim \mathbf{M}_{qp} = qp$. This gives the desired result. (See Example 1 in Section 4.9.)   ∎

We shall conclude this section by pointing out that the vector space of $q \times 1$ matrices is isomorphic to $\mathbf{R}^q$, as is easily seen from Definitions 6.5 and 6.6. Therefore when we consider the coordinate map of some space **V** relative to a given basis $\{v_1, \ldots, v_q\}$, we can if we wish represent the coordinates as elements of the space of $q \times 1$ matrices or the space of $1 \times q$ matrices rather than as elements of $\mathbf{R}^q$.

Exercises

1. Let

$$A = \begin{pmatrix} 2 & 3 \\ 1 & 0 \\ 5 & -2 \end{pmatrix} \qquad B = \begin{pmatrix} 8 & 2 \\ 7 & 0 \\ -6 & 3 \end{pmatrix}$$

$$C = \begin{pmatrix} 1 & 4 & \sqrt{2} \\ \frac{1}{2} & 0 & -\frac{5}{2} \end{pmatrix} \qquad D = \begin{pmatrix} \dfrac{1}{\sqrt{2}} & \dfrac{2}{\sqrt{2}} & 1 \\ \dfrac{1}{\sqrt{2}} & 1 & -\sqrt{2} \end{pmatrix}$$

Compute, if possible,

(a) $2A + B$ (b) $2C - \sqrt{2}D$
(c) $B - C$

2. If a linear transformation S has matrix

$$\begin{pmatrix} 3 & 0 & 2 \\ -1 & 4 & 1 \\ 2 & 1 & -3 \end{pmatrix}$$

find the matrix of (a) $3S$ and (b) $I - S$ relative to the same bases (where I is the identity map).

3. Let T and S be linear transformations on $\mathbf{R}^p$ into $\mathbf{R}^q$ as defined below. In each case, find the matrix representing $2S + T$ relative to the given bases.

(a) $p = q = 2$
    Bases $B = B' = \{(1, 0), (0, 1)\}$
    $S(x, y) = (2x + y, 3x - y)$
    T is represented by

$$\begin{pmatrix} 4 & 0 \\ 1 & 5 \end{pmatrix}$$

(b) $p = 2, q = 3$
    Bases $B = \{(1, 6), (1, 1)\}$, $B' = \{(1, 0, 0), (0, 1, 0), (0, 0, 1)\}$
    $S(x, y) = (2x, y - x, y)$
    $T(x, y) = (x + 2y, x - 2y, 0)$

4. T and S are linear transformations of the space of all polynomials of degree 2 or less into $\mathbf{R}^2$. Relative to the bases

$$B = \{1, x, x^2\}, \qquad B' = \{(1, 0), (0, 1)\}$$

the matrix representing $S + T$ is

$$\begin{pmatrix} 1 & 2 & 3 \\ -1 & 0 & 1 \end{pmatrix}$$

and the matrix representing $2S - T$ is

$$\begin{pmatrix} 2 & 1 & -3 \\ 4 & -6 & -1 \end{pmatrix}$$

Compute $Sp(x)$ and $Tp(x)$ where $p(x) = a_0 + a_1 x + a_2 x^2$.

5. Exhibit a basis for $\mathbf{M}_{32}$.

6. Write out a demonstration that $\mathbf{M}_{23}$ is a commutative additive group.

7. Write out a proof that the matrices described in the proof of Theorem 6.4 form a basis for $\mathbf{M}_{qp}$.

8. Prove Theorem 6.2.

6.4   Matrix multiplication

There is another operation which has been introduced for linear transformations, the product or composition of mappings. Recall that if $S: \mathbf{V} \to \mathbf{V}'$ and $T: \mathbf{V}' \to \mathbf{V}''$, then TS is the mapping $U: \mathbf{V} \to \mathbf{V}''$ defined by $Uv = T(Sv)$, for v in $\mathbf{V}$. We now define an operation on matrices which corresponds to this operation on transformations. Consider the following simple example.

Example 1   Let $V = V' = V'' = \mathbf{R}^2$, $S(x, y) = (x + y, x - y)$,   and $T(x, y) = (2x, 3y)$. Let the basis chosen be the standard basis $\{e_1, e_2\} = \{(1, 0), (0, 1)\}$. Then $(TS)(x, y) = (2x + 2y, 3x - 3y)$. Since

$$
\begin{array}{ll}
Se_1 = (1, 1) = e_1 + e_2, & Se_2 = (1, -1) = e_1 - e_2, \\
Te_1 = (2, 0) = 2e_1, & Te_2 = (0, 3) = 3e_2, \\
TSe_1 = (2, 3) = 2e_1 + 3e_2, & TSe_2 = (2, -3) = 2e_1 - 3e_2,
\end{array}
$$

the matrices representing T, S, and TS are

$$
N = \begin{pmatrix} 2 & 0 \\ 0 & 3 \end{pmatrix} \qquad M = \begin{pmatrix} 1 & 1 \\ 1 & -1 \end{pmatrix} \qquad P = \begin{pmatrix} 2 & 2 \\ 3 & -3 \end{pmatrix}
$$

respectively. The question is: What is the relation of P to M and N? Can P be constructed directly from N and M, without computing TSe_1 and TSe_2?

Let us show how this can be done, in general. Let $B = \{v_1, v_2, \ldots, v_p\}$, $B' = \{v'_1, v'_2, \ldots, v'_q\}$, and $B'' = \{v''_1, v''_2, \ldots, v''_r\}$ be bases for $\mathbf{V}$, $\mathbf{V}'$, and $\mathbf{V}''$, respectively. Let $M = (m_{kj})$ represent S relative to B and B' and let $N = (n_{ik})$ represent T relative to B' and B''. That is,

$$
Sv_j = \sum_{k=1}^{q} m_{kj} v'_k \qquad j = 1, \ldots, p
$$

$$
Tv'_k = \sum_{i=1}^{r} n_{ik} v''_i \qquad k = 1, \ldots, q
$$

Then

$$
(TS)(v_j) = T(Sv_j) = T\left(\sum_{k=1}^{q} m_{kj} v'_k \right)
$$

$$
= \sum_{k=1}^{q} m_{kj} (Tv'_k) = \sum_{k=1}^{q} m_{kj} \sum_{i=1}^{r} n_{ik} v''_i
$$

This is a finite sum containing qr terms which can be written in any order, in particular

$$
(TS)(v_j) = \sum_{i=1}^{r} \left[\sum_{k=1}^{q} n_{ik} m_{kj} \right] v''_i
$$

The last formula expresses the image of v_j under TS in terms of the basis B'' for $\mathbf{V}''$. The coefficients in this expression are the numbers $\sum_{k=1}^{q} n_{ik} m_{kj}$; that is, the coordinate vector of TSv_j relative to the basis in $\mathbf{V}''$ is

$$
\sum_{k=1}^{q} n_{1k} m_{kj}, \qquad \sum_{k=1}^{q} n_{2k} m_{kj}, \ldots, \sum_{k=1}^{q} n_{rk} m_{kj}
$$

It follows that this coordinate vector makes up the jth column of the matrix which represents TS relative to B and B''. In other words, the entry in row i, column j, of this matrix is

$$\sum_{k=1}^{q} n_{ik}m_{kj}$$

Since TS is called the product of the transformations T and S, it is natural to refer to this matrix as the "product" of the matrices N and M. Therefore we shall adopt the following definition.

DEFINITION 6.8   Let N be an $r \times q$ matrix and M be a $q \times p$ matrix of elements from a field **K**. The *product NM* is the $r \times p$ matrix

$$NM = \left(\sum_{k=1}^{q} n_{ik}m_{kj} \right) \tag{1}$$

It is important to notice that the product NM is defined *only when the number of columns of N is equal to the number of rows of M*. This corresponds to the fact that the transformation TS is defined only when the range of S is contained in the domain of T.

With the definition of matrix multiplication given above, it is apparent that if a matrix M represents a transformation $S: \mathbf{V} \to \mathbf{V}'$ and a matrix N represents a transformation $T: \mathbf{V}' \to \mathbf{V}''$, relative to bases B, B', and B'', then the matrix NM represents the transformation $TS: \mathbf{V} \to \mathbf{V}''$ relative to the bases B and B''.

Example 2   In Example 1 we have

$$n_{11} = 2, \quad n_{12} = 0, \quad n_{21} = 0, \quad n_{22} = 3,$$
$$m_{11} = 1, \quad m_{12} = 1, \quad m_{21} = 1, \quad m_{22} = -1$$

By Eq. (1) the elements of the product $P = NM$ are

$$\begin{aligned}
p_{11} &= n_{11}m_{11} + n_{12}m_{21} = 2 \cdot 1 + 0 \cdot 1 = 2 \\
p_{12} &= n_{11}m_{12} + n_{12}m_{22} = 2 \cdot 1 + 0(-1) = 2 \\
p_{21} &= n_{21}m_{11} + n_{22}m_{21} = 0 \cdot 1 + 3 \cdot 1 = 3 \\
p_{22} &= n_{21}m_{12} + n_{22}m_{22} = 0 \cdot 1 + 3(-1) = -3
\end{aligned}$$

This agrees with the matrix P found in Example 1 by computing TSe_1 and TSe_2.

The definition of matrix multiplication can most conveniently be recalled by the spatial pattern

$$\text{row } i \to \begin{pmatrix} \vdots & \vdots & & \vdots \\ n_{i1} & n_{i2} & \cdots & n_{iq} \\ \vdots & \vdots & & \vdots \end{pmatrix} \begin{pmatrix} \cdots & m_{1j} & \cdots \\ \cdots & m_{2j} & \cdots \\ & \vdots & \\ \cdots & m_{qj} & \cdots \\ & \uparrow & \\ & \text{column } j & \end{pmatrix} = \text{row } i \to \begin{pmatrix} & \vdots & \\ \cdots & p_{ij} & \cdots \\ & \vdots & \\ & \uparrow & \\ & \text{column } j & \end{pmatrix}$$

which is intended to indicate that the element p_{ij} of $P = NM$ is obtained from row i of N and column j of M. Another arrangement is:

$$\begin{pmatrix} \vdots & \vdots & & \vdots & \\ n_{i1} & n_{i2} & \cdots & n_{iq} & \\ \vdots & \vdots & & \vdots & \end{pmatrix} \begin{pmatrix} \cdots & m_{1j} & \cdots \\ \cdots & m_{2j} & \cdots \\ & \vdots & \\ \cdots & m_{qj} & \cdots \\ & \downarrow & \\ & \longrightarrow & p_{ij} \end{pmatrix}$$

Example 3 If

$$A = \begin{pmatrix} 2 & 1 & 4 \\ 0 & 2 & -1 \end{pmatrix}, \quad B = \begin{pmatrix} 3 & 1 \\ 2 & 6 \\ 4 & 3 \end{pmatrix}, \quad C = \begin{pmatrix} 1 \\ 0 \\ 5 \end{pmatrix}$$

Then

$$AB = \begin{pmatrix} 2 \cdot 3 + 1 \cdot 2 + 4 \cdot 4 & 2 \cdot 1 + 1 \cdot 6 + 4 \cdot 3 \\ 0 \cdot 3 + 2 \cdot 2 - 1 \cdot 4 & 0 \cdot 1 + 2 \cdot 6 - 1 \cdot 3 \end{pmatrix} = \begin{pmatrix} 24 & 20 \\ 0 & 9 \end{pmatrix}$$

Similarly,

$$BA = \begin{pmatrix} 6 & 5 & 11 \\ 4 & 14 & 2 \\ 8 & 10 & 13 \end{pmatrix}$$

and

$$AC = \begin{pmatrix} 22 \\ -5 \end{pmatrix}$$

but CA is not defined. Note that $AB \neq BA$, showing that in multiplication of matrices the order of the factors is important, or as we may say, *matrix multiplication is not commutative*.

THEOREM 6.5   Matrix multiplication is associative whenever it is defined. That is,

$$(AB)C = A(BC)$$

provided all the indicated products are permissible. Also, multiplication is distributive over addition,

$$(A + B)C = AC + BC$$

$$C(A + B) = CA + CB$$

and multiplication commutes with scalar multiplication,

$$c(AB) = (cA)B = A(cB)$$

whenever the indicated operations are permissible.

PROOF: Suppose A is $k \times m$, B is $m \times n$, and C is $n \times p$. Then AB and BC are defined, and of dimensions $k \times n$ and $m \times p$, respectively. Hence both $(AB)C$ and $A(BC)$ are defined and $k \times p$. We now regard A, B, and C as representatives of R, S, and T, respectively, where $R: \mathbf{V}'' \to \mathbf{V}'''$, $S: \mathbf{V}' \to \mathbf{V}''$, $T: \mathbf{V} \to \mathbf{V}'$ are linear transformations on vector spaces of appropriate dimensions. Then $(AB)C$ represents $(RS)T$ and $A(BC)$ represents $R(ST)$. Since we know from Chapter 5 that $(RS)T = R(ST)$, it follows that $(AB)C = A(BC)$. The distributive properties likewise follow at once from the distributive properties of linear transformations. ∎

Direct computational proofs of these results are notationally very awkward (see the exercises).

This theorem shows that the set $\mathbf{M}_{nn}$ of $n \times n$ square matrices with the addition, multiplication, and scalar multiplication defined above, is a linear algebra in the language introduced in Exercise 18.

We conclude this section with useful applications of matrix multiplication. Suppose that M is the matrix of a linear transformation $T: \mathbf{V} \to \mathbf{V}'$ relative to fixed bases. Let $v \in \mathbf{V}$, $v' \in \mathbf{V}'$, and $v' = Tv$. Then it is almost evident, and will now be proved, that the coordinates of v' can be obtained from the coordinates of v by matrix multiplication of M times the latter.

THEOREM 6.6 Let M be the matrix of a linear transformation $T: \mathbf{V} \to \mathbf{V}'$ relative to bases B and B'. Let $v \in \mathbf{V}$, $v' \in \mathbf{V}'$, and $v' = Tv$. Let

$$X = \begin{pmatrix} x_1 \\ x_2 \\ \vdots \\ x_p \end{pmatrix} \qquad Y = \begin{pmatrix} y_1 \\ y_2 \\ \vdots \\ y_q \end{pmatrix}$$

be the $p \times 1$ and $q \times 1$ matrices of coordinates of v and v' with respect to B and B', respectively. Then $Y = MX$, that is,

$$y_i = \sum_{j=1}^{p} m_{ij} x_j \qquad (i = 1, \ldots, q) \tag{2}$$

PROOF: Let $B = \{v_1, v_2, \ldots, v_p\}$ be the basis for $\mathbf{V}$ and $B' = \{v'_1, v'_2, \ldots, v'_q\}$ be the basis for $\mathbf{V}'$. Then

$$Tv_j = \sum_{i=1}^{q} m_{ij} v'_i \qquad (j = 1, \ldots, p)$$

Since X is the representation of v relative to B, we have

$$v = x_1 v_1 + x_2 v_2 + \cdots + x_p v_p$$

Therefore

$$Tv = \sum_{j=1}^{p} x_j Tv_j = \sum_{j=1}^{p} x_j \sum_{i=1}^{q} m_{ij} v'_i$$

$$= \sum_{i=1}^{q} \left[\sum_{j=1}^{p} m_{ij} x_j \right] v'_i$$

It follows that the coordinate vector of Tv relative to B' is

$$Y = \begin{pmatrix} y_1 \\ \vdots \\ y_q \end{pmatrix} \qquad \text{where } y_i = \sum_{j=1}^{p} m_{ij} x_j$$

This completes the proof.   ∎

According to this theorem the effect of a linear transformation on a vector v can be calculated by finding the coordinate vector of v (regarded as a column matrix) and multiplying it on the left by the matrix which represents the transformation.

Example 4   Let

$$M = \begin{pmatrix} 1 & -2 & 0 \\ 2 & 3 & 1 \\ 1 & -1 & 2 \end{pmatrix}$$

be the matrix of a transformation $T: V \rightarrow V'$ relative to bases $\{v_1, v_2, v_3\}$ and $\{v'_1, v'_2, v'_3\}$ of V and V', respectively.

Let $v = 2v_1 - v_2 + v_3$. The coordinate vector of v is

$$X = \begin{pmatrix} 2 \\ -1 \\ 1 \end{pmatrix}$$

Therefore the coordinate vector of Tv is, by the above theorem,

$$MX = \begin{pmatrix} 1 & -2 & 0 \\ 2 & 3 & 1 \\ 1 & -1 & 2 \end{pmatrix} \begin{pmatrix} 2 \\ -1 \\ 1 \end{pmatrix} = \begin{pmatrix} 4 \\ 2 \\ 5 \end{pmatrix}$$

Hence $Tv = 4v'_1 + 2v'_2 + 5v'_3$. In the same way, for an arbitrary $v = a_1 v_1 + a_2 v_2 + a_3 v_3$,

$$MX = \begin{pmatrix} 1 & -2 & 0 \\ 2 & 3 & 1 \\ 1 & -1 & 2 \end{pmatrix} \begin{pmatrix} a_1 \\ a_2 \\ a_3 \end{pmatrix} = \begin{pmatrix} a_1 - 2a_2 \\ 2a_1 + 3a_2 + a_3 \\ a_1 - a_2 + 2a_3 \end{pmatrix}$$

Example 5   Let $V = V'$ be the space of polynomials of degree at most two. Choose the basis $B = B' = \{1, x - 1, (x - 1)^2\}$. Let D be the differentiation operator and let $v_1 = 1, v_2 = x - 1, v_3 = (x - 1)^2$. Since $D(1) = 0$,

$D(x - 1) = 1 = v_1$, $D(x - 1)^2 = 2(x - 1) = 2v_2$, the operator $D: \mathbf{V} \to \mathbf{V}$ is represented by the matrix

$$\begin{pmatrix} 0 & 1 & 0 \\ 0 & 0 & 2 \\ 0 & 0 & 0 \end{pmatrix}$$

The polynomial $p(x) = x^2 - 3x + 4 = (x - 1)^2 - (x - 1) + 2$ has coordinate vector with entries $2, -1, 1$. Hence the image of this polynomial has coordinate vector

$$Y = \begin{pmatrix} 0 & 1 & 0 \\ 0 & 0 & 2 \\ 0 & 0 & 0 \end{pmatrix} \begin{pmatrix} 2 \\ -1 \\ 1 \end{pmatrix} = \begin{pmatrix} -1 \\ 2 \\ 0 \end{pmatrix}$$

The image polynomial is thus $-1 + 2(x - 1) = 2x - 3$, which is $p'(x)$ as it should be.

Example 6 Let $I_{\mathbf{V}}: \mathbf{V} \to \mathbf{V}$ be the identity operator on $\mathbf{V}$ and let dim $\mathbf{V} = p$. Then the matrix

$$I_p = \begin{pmatrix} 1 & 0 & 0 & \cdots & 0 \\ 0 & 1 & 0 & \cdots & 0 \\ 0 & 0 & 1 & \cdots & 0 \\ & & \cdots & & \\ 0 & 0 & 0 & \cdots & 1 \end{pmatrix}$$

with p rows and p columns represents $I_{\mathbf{V}}$ relative to any basis for $\mathbf{V}$. This matrix is called the *identity matrix* of dimension p. We shall use the letter I to denote both identity operators and identity matrices. Subscripts will be used only if needed to avoid confusion.

Exercises

1. Let $S(x, y) = (x, y, x + y)$
   $T(x, y, z) = (z, y, x)$
   Find the matrices M, N, and P of S, T, and TS, respectively, relative to the given bases, and verify that $NM = P$ in each case.

   (a) $B = \{(1, 0), (0, 1)\}$ $B' = \{(1, 0, 0), (0, 1, 0), (0, 0, 1)\}$
   (b) $B = \{(1, 1), (1, -1)\}$ $B' = \{(1, 1, 2), (1, 0, 0), (1, 1, 0)\}$

2. Let S and T be linear operators on the space of polynomials of degree at most two defined by

   $$S: p(x) \to p(x + 1) \qquad T: p(x) \to xp'(x) + p(0)$$

   Find the matrices of S, T, and TS relative to the basis

   $$B = \{1, x, x^2\}$$

   and verify that the product of the first two is the last.

3. Compute each matrix product

(a) $\begin{pmatrix} 1 & 0 & \frac{1}{2} \\ 2 & 4 & 1 \\ 2 & -5 & 3 \end{pmatrix} \begin{pmatrix} 3 & 7 \\ -1 & 0 \\ -2 & 4 \end{pmatrix}$

(b) $\begin{pmatrix} 5 & -1 & 0 \\ 2 & 6 & 3 \end{pmatrix} \begin{pmatrix} 1 \\ 4 \\ 7 \end{pmatrix}$

(c) $(6 \quad 0 \quad 3) \begin{pmatrix} 1 \\ 4 \\ 7 \end{pmatrix}$

(d) $\begin{pmatrix} 1 \\ 4 \\ 7 \end{pmatrix} (6 \quad 0 \quad 3)$

4. It is natural to define $A^2 = AA$. Under what conditions is this defined?

5. Compute A^n if

(a) $A = \begin{pmatrix} 1 & 1 \\ 1 & 1 \end{pmatrix}$

(b) $A = \begin{pmatrix} 0 & 1 \\ 1 & 0 \end{pmatrix}$

6. Find all 2×2 matrices A which commute with

$$B = \begin{pmatrix} 1 & 1 \\ 0 & 2 \end{pmatrix}$$

7. Show that the $n \times n$ identity matrix commutes with every $n \times n$ matrix.

8. Are there other 2×2 matrices A which commute with every 2×2 matrix?
   Hint: Such a matrix A would in particular have to commute with

$$\begin{pmatrix} 1 & 0 \\ 0 & 0 \end{pmatrix}, \quad \begin{pmatrix} 0 & 1 \\ 0 & 0 \end{pmatrix}, \quad \begin{pmatrix} 0 & 0 \\ 1 & 0 \end{pmatrix}, \quad \text{and} \quad \begin{pmatrix} 0 & 0 \\ 0 & 1 \end{pmatrix}$$

9. A matrix A is called a *diagonal matrix* if $a_{ij} = 0$ for $i \neq j$. That is the only nonzero elements are on the *main diagonal* ($i = j$). Show that the product of two diagonal matrices is diagonal.

10. Prove the associative law $(AB)C = A(BC)$ for 2×2 matrices directly from the definition of matrix multiplication (without using associativity of the composition of mappings).

11. Prove the distributive law $A(B + C) = AB + AC$ for 2×2 matrices without using the distributive law for mappings.

12. Let

$$A = \begin{pmatrix} 1 & 2 & 3 \\ 0 & -1 & -2 \\ 4 & -3 & 0 \end{pmatrix}$$

represent T relative to bases $\{v_1, v_2, v_3\}$ and $\{u_1, u_2, u_3\}$. Use A to find Tv where

(a) $v = v_1 - v_2 + 2v_3$

(b) $v = 3v_1 + 4v_2$

13. Let

$$A = \begin{pmatrix} 2 & 1 & 0 \\ -1 & 2 & 3 \\ 4 & 0 & -2 \end{pmatrix}$$

be the matrix of $T: \mathbf{R}^3 \to \mathbf{V}$, where $\mathbf{V}$ is the space of polynomials of degree at most 2, relative to the bases $B = \{e_1, e_2, e_3\}$, $B' = \{1, x, x^2\}$. Find the image under T of each of the following:

(a) $(6, 1, 4)$ (b) $(1, 1, 1)$ (c) $(0, 1, 2)$

14. Let $\mathbf{V} = \mathbf{V}'$ be the space of polynomials of degree ≤ 2 and choose the basis $B = \{1, x, x^2\}$. Find the matrix A which represents the operator $D - 2I$ on $\mathbf{V}$. Find the image of $p(x) = x^2 - 3x + 4$

(a) directly, and

(b) by using A.

15. Let $T: \mathbf{R}^3 \to \mathbf{R}^2$ be the transformation such that $Tv_1 = Tv_2 = u'_1$ and $Tv_3 = u'_2$ where $v_1 = (1, 0, 0)$, $v_2 = (0, 1, 0)$, $v_3 = (0, 0, 1)$, $u'_1 = (1, 0)$, $u'_2 = (0, 1)$. Find the matrix A of T relative to $B = \{v_1, v_2, v_3\}$ and $B' = \{u'_1, u'_2\}$ and use A to find $T(a, b, c)$.

16. A matrix such as

$$M = \begin{pmatrix} 1 & 0 & 2 \\ 0 & 1 & -1 \\ 0 & 0 & 4 \end{pmatrix}$$

can sometimes conveniently be considered in the form

$$\left(\begin{array}{c|c} A & B \\ \hline C & D \end{array} \right)$$

where A, B, C, D represent the submatrices

$$A = \begin{pmatrix} 1 & 0 \\ 0 & 1 \end{pmatrix}, \quad B = \begin{pmatrix} 2 \\ -1 \end{pmatrix}, \quad C = (0 \;\; 0), \quad D = (4)$$

In this form the original matrix is said to be *partitioned*, or in *block form.* Show that if

$$M_1 = \left(\begin{array}{c|c} A_1 & B_1 \\ \hline C_1 & D_1 \end{array}\right), \quad M_2 = \left(\begin{array}{c|c} A_2 & B_2 \\ \hline C_2 & D_2 \end{array}\right)$$

are any two partitioned matrices with the same block dimensions, then

$$M_1 + M_2 = \left(\begin{array}{c|c} A_1 + A_2 & B_1 + B_2 \\ \hline C_1 + C_2 & D_1 + D_2 \end{array}\right)$$

and

$$cM_1 = \left(\begin{array}{c|c} cA_1 & cB_1 \\ \hline cC_1 & cD_1 \end{array}\right)$$

17. Let

$$A = \begin{pmatrix} A_{11} & A_{12} \\ A_{21} & A_{22} \end{pmatrix}, \quad B = \begin{pmatrix} B_{11} & B_{12} \\ B_{21} & B_{22} \end{pmatrix}$$

be partitioned matrices. Show that

$$AB = \begin{pmatrix} A_{11}B_{11} + A_{12}B_{21} & A_{11}B_{12} + A_{12}B_{22} \\ A_{21}B_{11} + A_{22}B_{21} & A_{21}B_{12} + A_{22}B_{22} \end{pmatrix}$$

provided all the indicated products and sums of matrices are defined.

18. A *linear algebra* **A** over a field **K** is a vector space over **K** on which is defined a "product" which is associative and bilinear; that is, the following properties hold:

(a) If T_1 and T_2 are in **A**, there is a product $T_1 T_2$ which is in **A** (**A** is closed under the product).

(b) If T_1, T_2, $T_3 \in$ **A**, then

$$T_1(T_2 T_3) = (T_1 T_2)T_3$$

(The product is associative.)

(c) If T_1, T_2, $T_3 \in$ **A** and $a, b \in$ **K**, then

$$T_1(aT_2 + bT_3) = aT_1 T_2 + bT_1 T_3$$
$$(aT_1 + bT_2)T_3 = aT_1 T_3 + bT_2 T_3$$

(The product is *bilinear*.)

Prove that $\mathscr{L}(\mathbf{V}, \mathbf{V})$ is a linear algebra.

19. Two linear algebras are called isomorphic if there is an isomorphism between them as vector spaces, which also preserves the product. Let **V** be an m-dimensional vector space over **K**. Prove that the set $\mathbf{M}_{mm}$ of all m by m matrices with elements in **K** forms a linear algebra which is isomorphic to the linear algebra of all linear operators on **V** into **V**.

20. Prove that matrix multiplication is bilinear, that is, equations such as in Exercise 18c are satisfied.

21. Show that a product is bilinear in the sense of Exercise 18 if and only if it is distributive on the left and on the right, and multiplication commutes with scalar multiplication, that is, $c(ST) = (cS)T = S(cT)$.

6.5 Matrix transpose

In this section we shall introduce an operation on matrices which is useful in both theory and applications.

DEFINITION 6.9 Let M be a $q \times p$ matrix with typical element m_{ij}. The *transpose* of M, denoted M^T, is the $p \times q$ matrix with typical element n_{ij}, where $n_{ij} = m_{ji}$.

Example 1 If

$$M = \begin{pmatrix} 1 & 0 \\ 2 & 3 \\ 1 & 2 \end{pmatrix} \quad \text{then } M^T = \begin{pmatrix} 1 & 2 & 1 \\ 0 & 3 & 2 \end{pmatrix}$$

The operation which replaces a matrix by its transpose is in one sense unlike the operations on matrices defined previously. In matrix addition and multiplication two matrices are combined to form a new matrix, whereas the transpose operation uses a single matrix to form a new one.

The reader may also notice that the operations of matrix addition and multiplication were defined in relation to addition and multiplication of linear transformations. So far, however, we have no operation on linear transformations which corresponds to transposition of a matrix. This gap can be filled as is noted in the Commentary.

THEOREM 6.7 Let M and N be two matrices for which MN is defined. Then the transpose of the product is the product of the transposes in the reverse order, that is, $(MN)^T = N^T M^T$.

PROOF: The proof is left as an exercise. ∎

Example 2 Let

$$M = \begin{pmatrix} 1 & 2 & 0 \\ -1 & 4 & 3 \end{pmatrix} \qquad N = \begin{pmatrix} 6 \\ -3 \\ 1 \end{pmatrix}$$

Then

$$MN = \begin{pmatrix} 0 \\ -15 \end{pmatrix} \qquad (MN)^T = (0 \quad -15)$$

and

$$N^T M^T = (6 \quad -3 \quad 1) \begin{pmatrix} 1 & -1 \\ 2 & 4 \\ 0 & 3 \end{pmatrix} = (0 \quad -15)$$

DEFINITION 6.10 A square matrix M is said to be *symmetric* if $M = M^T$, and *skew-symmetric* if $M = -M^T$.

Example 3 Let

$$M = \begin{pmatrix} 2 & 4 & 1 \\ 4 & -3 & -5 \\ 1 & -5 & 6 \end{pmatrix} \qquad N = \begin{pmatrix} 0 & 7 & 3 \\ -7 & 0 & -2 \\ -3 & 2 & 0 \end{pmatrix}$$

Then M is symmetric and N is skew-symmetric.

As we shall see later, symmetric matrices are in certain ways simpler than general matrices. They occur in various applications.

Exercises

1. Compute the transposes of the matrices A, B, C, D in Exercise 1 of Section 6.3.

2. Let $\mathbf{M}_{pq}$ and $\mathbf{M}_{qp}$ be the spaces of $p \times q$ and $q \times p$ matrices, respectively. The transpose operation converts a matrix in $\mathbf{M}_{pq}$ into a matrix in $\mathbf{M}_{qp}$. Prove that this operator is linear by showing that

$$(M + N)^T = M^T + N^T \qquad \text{for all } M, N \text{ in } \mathbf{M}_{pq},$$
$$(cM)^T = cM^T \qquad \text{for all } M \text{ in } \mathbf{M}_{pq}, c \text{ in } \mathbf{K}$$

(Compare this with our general notation $S(M + N) = SM + SN$, $S(cM) = cSM$ where S is a linear transformation.)

3. Take $p = q = 2$ in Exercise 2, and choose the basis

$$B = \left\{ \begin{pmatrix} 1 & 0 \\ 0 & 0 \end{pmatrix}, \begin{pmatrix} 0 & 1 \\ 0 & 0 \end{pmatrix}, \begin{pmatrix} 0 & 0 \\ 1 & 0 \end{pmatrix}, \begin{pmatrix} 0 & 0 \\ 0 & 1 \end{pmatrix} \right\}$$

for $\mathbf{M}_{22}$. What is the matrix, relative to this basis, of the transpose operator?

4. Prove that $(M^T)^T = M$ for any $q \times p$ matrix M.

5. Prove Theorem 6.7.

6. Let A, B, C, and D be the matrices in Exercise 1 of Section 6.3.

    (a) Compute $(A + B)^T$ and verify that this is equal to $A^T + B^T$.

    (b) Compute $(3C - D)^T$ and verify that this is equal to $3C^T - D^T$.

    (c) Compute $(AC)^T$ and verify that this is equal to $C^T A^T$.

7. Prove that for any square matrix M, $M + M^T$ is symmetric and $M - M^T$ is skew-symmetric.

8. If M and N are symmetric and MN is defined, must MN be symmetric?

9. Prove that any square matrix can be expressed in a unique way as the sum of a symmetric matrix and a skew-symmetric matrix.

6.6 Rank of a matrix

Because of the isomorphism between the space of all $q \times p$ matrices over a field $\mathbf{K}$ and a space $\mathscr{L}(\mathbf{V}, \mathbf{V}')$ of linear transformations of $\mathbf{V}$ into $\mathbf{V}'$ established in Theorem 6.3, many concepts and theorems relating to linear transformations have counterparts for matrices, and vice versa. For example, in Chapter 5 the rank $\rho(T)$ of a linear transformation $T: \mathbf{V} \to \mathbf{V}'$ was defined as the dimension of $\mathbf{R}_T$, the range or image of T. We shall now introduce a corresponding concept of rank for matrices. In fact, there are several possible definitions which turn out to be equivalent.

DEFINITION 6.11 Let M be any $q \times p$ matrix with elements in $\mathbf{K}$. The *column space* of M is the space of all $q \times 1$ matrices generated by the column vectors of M. The *row space* of M is the space of all $1 \times p$ matrices generated by the row vectors of M.

DEFINITION 6.12 The *column rank* of a matrix is the dimension of the column space of the matrix. The *row rank* is the dimension of the row space.

Example 1 Let

$$M = \begin{pmatrix} 4 & 6 & -2 & 8 \\ 2 & 3 & -1 & 4 \\ 3 & 7 & 0 & 6 \end{pmatrix}$$

The column space is the span of the set

$$\begin{pmatrix} 4 \\ 2 \\ 3 \end{pmatrix}, \quad \begin{pmatrix} 6 \\ 3 \\ 7 \end{pmatrix}, \quad \begin{pmatrix} -2 \\ -1 \\ 0 \end{pmatrix}, \quad \begin{pmatrix} 8 \\ 4 \\ 6 \end{pmatrix}$$

It can be verified that

$$\begin{pmatrix} 8 \\ 4 \\ 6 \end{pmatrix} = 2 \begin{pmatrix} 4 \\ 2 \\ 3 \end{pmatrix} = \frac{1}{7} \left\{ 6 \begin{pmatrix} 6 \\ 3 \\ 7 \end{pmatrix} - 10 \begin{pmatrix} -2 \\ -1 \\ 0 \end{pmatrix} \right\}$$

Therefore, only two of the column vectors are independent and the column rank of M is 2. The row space is the span of the set

$$\{(4\ \ 6\ \ -2\ \ 8),\ (2\ \ 3\ \ -1\ \ 4),\ (3\ \ 7\ \ 0\ \ 6)\}$$

Clearly, only two of these vectors are independent and the row rank of M is also 2. As we shall see, the row and column ranks of a matrix are always equal.

THEOREM 6.8 The number of linearly independent rows in a matrix is equal to the number of linearly independent columns. In other words, the row rank of a matrix is equal to the column rank of the matrix.

We do not yet have enough machinery to prove this theorem. Two proofs will be given later, one based on properties of determinants and one on the so-called reduced echelon form of a matrix. Because of the theorem, we no longer need distinguish between row rank and column rank.

DEFINITION 6.13 The *rank* of a matrix M is the value of its row rank or column rank as defined above, and is denoted by $\rho(M)$.

The next theorem will show the relation between the rank of a matrix and the rank of a linear transformation. First let us recall that the coordinate map σ of V' relative to a basis $B' = \{v'_1, v'_2, \ldots, v'_q\}$ of V' is defined as the linear transformation of V' onto K^q which takes $c_1 v'_1 + c_2 v'_2 + \cdots + c_q v'_q$ into $(c_1, c_2, \ldots, c_q)$. This is an isomorphism (see Section 5.3).

THEOREM 6.9 Let M be a $q \times p$ matrix over K. Let V and V' be any vector spaces such that dim $V = p$, dim $V' = q$ and let T be a linear transformation of V into V'. If M is a representative of T relative to any bases of V and V' (say B and B' respectively), the range R_T is isomorphic to the column space of M. Therefore, $\rho(M) = \rho(T)$.

PROOF: Let C denote the subspace of K^q isomorphic[2] to the column space of M. Let σ be the coordinate map with respect to B' of V' into K^q. Then σ maps R_T, a subspace of V', onto a subspace W of K^q (see Figure 6.1). W is the subspace

---

[2] The isomorphism is the obvious one which maps $(a_1\ \ a_2\ \ \cdots\ \ a_q)^T$ onto $(a_1, a_2, \cdots, a_q)$.

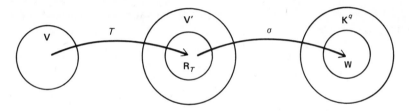

FIGURE 6.1

consisting of coordinate vectors of the vectors in $\mathbf{R}_T$. Clearly, σ establishes an isomorphism between $\mathbf{R}_T$ and $\mathbf{W}$. Therefore, we need only show that $\mathbf{W} = \mathbf{C}$.

Suppose $v' \in \mathbf{R}_T$. Then $v' = Tv$ where $v \in \mathbf{V}$. Hence there are scalars $c_1, \ldots, c_p$ such that

$$v = \sum_{i=1}^{p} c_i v_i$$

where v_i is the ith element in the basis B. It follows that

$$v' = Tv = \sum_{i=1}^{p} c_i(Tv_i) \quad \text{and} \quad \sigma v' = \sum_{i=1}^{p} c_i(\sigma Tv_i)$$

Since M represents T, the coordinate vector of Tv_i is the ith column vector of M. Therefore v', which is a linear combination of vectors Tv_i, has a coordinate vector $\sigma v'$ which is a linear combination of column vectors of M. This shows that $\mathbf{W} \subset \mathbf{C}$. Conversely, the ith column vector of M is the coordinate vector of Tv_i, which is in $\mathbf{R}_T$. Hence the column vectors of M are in $\mathbf{W}$, and $\mathbf{C} \subset \mathbf{W}$. This proves that $\mathbf{W} = \mathbf{C}$ and completes the proof of the theorem.   ∎

Remark The rank of T is an *intrinsic* property of T, in the sense that its value is independent of the choice of bases B and B'. The rank of a matrix M is an intrinsic property of M independent of how M may be associated with a linear transformation. The theorem asserts that all matrices which represent a given T, relative to any bases, have the same rank, $\rho(T)$. Also all transformations represented by a given matrix M relative to any bases have the same rank, $\rho(M)$.

The concept of rank of a matrix will be useful later when we discuss the solution of sets of linear algebraic equations. Also, Theorem 6.9 shows that we can find the dimension of the range of a linear transformation by determining the rank of any matrix representative of T. In fact, this theorem suggests an algorithm for finding a basis for the range $\mathbf{R}_T$ of a transformation T. Pick bases $B = \{v_1, \ldots, v_p\}$ and $B' = \{v'_1, \ldots, v'_q\}$, and let M be the matrix of T. Find a maximal independent set of column vectors of M. These vectors are coordinate vectors of a basis for $\mathbf{R}_T$.

Example 2   Consider the differentiation operator D on the space of poly-nomials of degree ≤ 2, with basis $B = \{1, x, x^2\}$. Then the matrix repre-sentative of D with respect to B is

$$M = \begin{pmatrix} 0 & 1 & 0 \\ 0 & 0 & 2 \\ 0 & 0 & 0 \end{pmatrix}$$

Clearly the second and third columns form a maximal independent set. There-fore $\{1, 2x\}$ is a basis for $\mathbf{R}_D$. This checks with the known result that the most general element in $\mathbf{R}_D$ is $D(ax^2 + bx + c) = 2ax + b = b \cdot 1 + a \cdot 2x$.

We conclude this section with two theorems which will be useful later.

THEOREM 6.10   Let M and N be $q \times p$ matrices. Then $\rho(M + N) \leq \rho(M) + \rho(N)$.

PROOF: The proof is left as an exercise.   ∎

THEOREM 6.11 (Sylvester's Law)   Let N be a $q \times p$ matrix and M a $p \times r$ matrix. Then

$$\rho(M) + \rho(N) - p \leq \rho(NM) \leq \min\,[\rho(M), \rho(N)]$$

PROOF: According to Theorem 6.9, it is sufficient to prove the corresponding inequalities for linear transformations T and S representing N and M, respec-tively; that is

$$\rho(S) + \rho(T) - p \leq \rho(TS) \leq \min\,[\rho(S), \rho(T)]$$

These inequalities have already been proved in Theorem 5.18 and the Corollary to Theorem 5.17.   ∎

Exercises

1.   Find a basis for the row space and a basis for the column space of the matrix

$$A = \begin{pmatrix} 1 & 4 & -2 & 8 \\ -2 & 5 & 4 & 7 \\ -1 & 9 & 2 & 15 \end{pmatrix}$$

2.   Find the row rank and the column rank of each matrix

$$A = \begin{pmatrix} 3 & 1 \\ 4 & 2 \\ 0 & 1 \end{pmatrix} \qquad B = \begin{pmatrix} 2 & 3 & 7 \\ -5 & 8 & -2 \\ -1 & 5 & 3 \end{pmatrix}$$

3.   Let T be an operator on $\mathbf{R}^3$ into $\mathbf{R}^3$ with matrix B of Exercise 2 relative to the standard basis. Find a basis for $\mathbf{R}_T$.

4. Let M be an upper triangular matrix

$$M = \begin{pmatrix} m_{11} & m_{12} & \cdots & m_{1n} \\ 0 & m_{22} & \cdots & m_{2n} \\ & & \cdots & \\ 0 & 0 & \cdots & m_{nn} \end{pmatrix}$$

with nonzero diagonal entries m_{ii}. Prove that M has rank n.

5. For each transformation T in Exercise 5, Section 6.2, find the rank of T by finding the rank of the matrix M representing T. Also, find a basis for $\mathbf{R}_T$ by finding a basis for the column space of M.

6. Do the same as in Exercise 5 for the transformations T in Exercise 6 of Section 6.2.

7. Prove Theorem 6.10.

8. Let M be a $p \times p$ matrix of rank one. Prove that $M = CR$ where C is a $p \times 1$ matrix (column vector) and R is a $1 \times p$ matrix (row vector).

6.7 Inverse of a matrix

Our development of matrix algebra has been parallel to our development of an algebra of linear transformations. In keeping with this approach we now turn to the concepts of singular and nonsingular transformations, introduced in Section 5.7, from which we will derive the concept of a matrix inverse.

THEOREM 6.12 Let M be a $q \times q$ matrix with $\rho(M) = q$. Then there exists a unique matrix M^{-1}, such that $M^{-1}M = MM^{-1} = I$ where I is the $q \times q$ identity matrix.

PROOF: Since M is $q \times q$, we make use of an isomorphism between the space $\mathbf{M}_{qq}$ of all $q \times q$ matrices and $\mathscr{L}(\mathbf{V}, \mathbf{V}')$ where $\mathbf{V}$ and $\mathbf{V}'$ are arbitrarily chosen q-dimensional vector spaces. In the present case it is convenient to choose $\mathbf{V} = \mathbf{V}'$, and we do this. Let B be a basis of $\mathbf{V}$. Then to each matrix in $\mathbf{M}_{qq}$ there corresponds with respect to B a unique transformation in $\mathscr{L}(\mathbf{V}, \mathbf{V}')$ and conversely. Let T denote the transformation corresponding to M. Since $\rho(M) = q$, by Theorem 6.9 we know that $\rho(T) = q$. Therefore, by Exercise 7 of Section 5.7, there exists a transformation T^{-1} such that $T^{-1}T = I_{\mathbf{V}}$ and $TT^{-1} = I_{\mathbf{V}}$, where T^{-1} is a mapping of $\mathbf{V}$ into $\mathbf{V}$. Let M^{-1} be the matrix of T^{-1} with respect to B. Then, by our construction of matrix multiplication, $M^{-1}M$ is the representation of $T^{-1}T = I_{\mathbf{V}}$ with respect to B. It is immediate that $M^{-1}M$ must be I, the $q \times q$ identity. By a similar argument $MM^{-1} = I$. The proof of uniqueness of M^{-1} is left as an exercise. ∎

DEFINITION 6.14 Let M be a $q \times q$ square matrix. Then M is said to be *nonsingular* or *invertible* if the rank of M is q. If the rank of M is less than q, M is said to be *singular*.

We immediately get the following corollary to the above theorem.

COROLLARY If the $q \times q$ matrix M is invertible, then there exists a $q \times q$ matrix M^{-1} such that $M^{-1}M = MM^{-1} = I$. (M^{-1} is called the *inverse* of M.)

By an argument similar to those used above we can deduce the following.

THEOREM 6.13 If M is a $q \times q$ matrix with $\rho(M) < q$, there exists no matrix M^* such that $M^*M = I$ or $MM^* = I$ where I is the $q \times q$ identity matrix.

PROOF: This is left as an exercise. ∎

To summarize the above discussion, if we consider a $q \times q$ matrix M, the following are equivalent:

1. M is nonsingular or invertible.
2. $\rho(M) = q$.
3. There is a matrix M^{-1} such that $M^{-1}M = MM^{-1} = I$.
4. The columns of M are linearly independent.
5. The rows of M are linearly independent.

Example 2 Items (4) and (5) provide methods for determining whether a matrix is singular or nonsingular. For example, the matrix

$$\begin{pmatrix} 1 & 3 & 0 \\ 2 & 0 & 6 \\ 3 & -1 & 10 \end{pmatrix}$$

is singular, since the third column vector is three times the first column minus the second column.

Example 3 The matrix

$$A = \begin{pmatrix} 2 & 1 \\ 5 & 3 \end{pmatrix}$$

is nonsingular since its columns are independent. One way to find the inverse is as follows. Let

$$A^{-1} = \begin{pmatrix} a & b \\ c & d \end{pmatrix}$$

Then

$$AA^{-1} = \begin{pmatrix} 2 & 1 \\ 5 & 3 \end{pmatrix}\begin{pmatrix} a & b \\ c & d \end{pmatrix} = \begin{pmatrix} 2a + c & 2b + d \\ 5a + 3c & 5b + 3d \end{pmatrix}$$

Since this must be the identity matrix,

$$2a + c = 1 \qquad 2b + d = 0$$
$$5a + 3c = 0 \qquad 5b + 3d = 1$$

Solving these equations gives $a = 3$, $c = -5$, $b = -1$, $d = 2$. Therefore

$$A^{-1} = \begin{pmatrix} 3 & -1 \\ -5 & 2 \end{pmatrix}$$

Later we shall describe much more efficient methods for the computation of matrix inverses, methods which can be applied to $p \times p$ matrices where p is 3, 4, 5, or even larger.

The following theorems, the proofs of which are left as exercises, give useful properties of nonsingular matrices.

THEOREM 6.14 Let A, B, C be matrices such that BA and AC are defined. If A is nonsingular, $\rho(BA) = \rho(B)$ and $\rho(AC) = \rho(C)$.

THEOREM 6.15 Let A and B be square matrices of the same size. Then AB is nonsingular if and only if A and B are each nonsingular. In this case, $(AB)^{-1} = B^{-1}A^{-1}$.

THEOREM 6.16 If A is nonsingular, then A^{-1} is nonsingular and cA is nonsingular for any nonzero scalar c, and $(A^{-1})^{-1} = A$, $(cA)^{-1} = (1/c)A^{-1}$.

THEOREM 6.17 If A is nonsingular, then A^T is nonsingular and $(A^T)^{-1} = (A^{-1})^T$.

Exercises

1. Which of the following matrices are nonsingular?

(a)
$$M = \begin{pmatrix} 1 & 3 & 0 & 2 \\ 4 & 6 & -1 & 5 \\ 3 & 0 & 2 & 1 \end{pmatrix}$$

(b)
$$M = \begin{pmatrix} 4 & 5 & 8 \\ 2 & 0 & 1 \\ 6 & 5 & 10 \end{pmatrix}$$

(c)
$$M = \begin{pmatrix} 3 & 1 & -1 \\ 2 & 0 & -8 \\ -5 & -2 & -2 \end{pmatrix}$$

2. Use the method of Example 3 to compute A^{-1} if

$$A = \begin{pmatrix} 4 & 5 \\ 2 & 3 \end{pmatrix}$$

3. In each part, T is a linear transformation from the space of polynomials of degree at most 2 into $\mathbf{R}^3$. Compute the matrix M of T relative to the bases

$$B = \{1, x, x^2\}, \qquad B' = \{e_1, e_2, e_3\}$$

Then compute T^{-1} and the matrix of T^{-1}; verify that the latter is M^{-1}.

(a) $T: a_0 + a_1 x + a_2 x^2 \rightarrow (a_0, a_2, a_1)$

(b) $T: a_0 + a_1 x + a_2 x^2 \rightarrow (a_0 + a_1, a_1 + a_2, a_2)$

(c) $T: p(x) \rightarrow \left(\int_0^2 p(x)\, dx,\; p(0),\; p'(0) \right)$

4. Find the inverse of the diagonal matrix

$$\begin{pmatrix} a_1 & 0 & 0 & \cdots & 0 \\ 0 & a_2 & 0 & \cdots & 0 \\ 0 & 0 & a_3 & \cdots & 0 \\ & & \cdots & & \\ 0 & 0 & 0 & \cdots & a_n \end{pmatrix}$$

where the diagonal entries are not zero.

5. Find the inverse of the triangular matrix

$$\begin{pmatrix} a & b & c \\ 0 & d & e \\ 0 & 0 & f \end{pmatrix}$$

assuming that a, d, and f are nonzero.

6. A matrix of the form

$$a \begin{pmatrix} 1 & -v \\ -v/c^2 & 1 \end{pmatrix}$$

where c is the velocity of light, v is the velocity of a moving body, and $a = c/\sqrt{c^2 - v^2}$, is called a Lorentz matrix. Under what conditions is it nonsingular?

7. Complete the proof of Theorem 6.12.

8. Prove Theorem 6.13.

9. Prove Theorem 6.14.

10. Prove Theorem 6.15.

11. Prove Theorem 6.16.

12. Prove Theorem 6.17.

13. Let A be a partitioned matrix

$$A = \begin{pmatrix} A_1 & 0 \\ 0 & A_2 \end{pmatrix}$$

where A is $n \times n$, A_1 is $r \times r$, A_2 is $s \times s$, and $r + s = n$. Show that if A_1 and A_2 are nonsingular, then A is nonsingular and

$$A^{-1} = \begin{pmatrix} A_1^{-1} & 0 \\ 0 & A_2^{-1} \end{pmatrix}$$

14. Use the method of Exercise 13 to compute A^{-1} if

(a)
$$A = \begin{pmatrix} 1 & 2 & 0 & 0 \\ 0 & 5 & 0 & 0 \\ 0 & 0 & 4 & 5 \\ 0 & 0 & 2 & 3 \end{pmatrix}$$

(b)
$$A = \begin{pmatrix} 4 & 5 & 0 & 0 & 0 \\ 2 & 3 & 0 & 0 & 0 \\ 0 & 0 & 4 & 0 & 0 \\ 0 & 0 & 0 & 1 & 0 \\ 0 & 0 & 0 & 0 & 2 \end{pmatrix}$$

15. Let A be invertible.

(a) Let B be obtained from A by interchanging two columns. Prove that B is invertible.

(b) Let B be obtained from A by multiplying row i by c, $c \neq 0$. Prove that B is invertible.

(c) Let B be obtained by replacing the first column of A by the sum of the first and ith columns of $A (i \neq 1)$. Prove that B is invertible.

16. Let M be a $p \times q$ matrix. Suppose that the only column q-vector y (that is, y is $q \times 1$) for which $My = 0$ is $y = 0$. Also suppose that the only row p-vector x (that is, x is $1 \times p$) for which $xM = 0$ is $x = 0$. Prove that M is square $(p = q)$ and nonsingular.

[*Hint:* The vector My is a linear combination of the column vectors making up M, and the row vector xM is a linear combination of the row vectors making up M.]

6.8   Similar matrices

In Theorem 6.1 we established a one-to-one correspondence between the set of all $q \times p$ matrices and the set of all linear transformations of the vector space $\mathbf{V}$ to $\mathbf{V}'$ where dim $\mathbf{V} = p$ and dim $\mathbf{V}' = q$. This was done by picking bases C of $\mathbf{V}$ and C' of $\mathbf{V}'$, and showing that the columns of each $q \times p$ matrix correspond to the coordinates with respect to C' of the images under a unique $T: \mathbf{V} \to \mathbf{V}'$ of the elements of C, and to each transformation T there exists a unique $q \times p$ matrix which is such a representation.

Let us now restrict our attention to the set of transformations of $\mathbf{V}$ into itself, $\mathscr{L}(\mathbf{V}, \mathbf{V})$, and the set $\mathbf{M}_{pp}$ of $p \times p$ matrices. For a specific basis B of $\mathbf{V}$ and a specific transformation $T: \mathbf{V} \to \mathbf{V}$, there is a unique $p \times p$ matrix M representing T with respect to B, where the columns of M represent the coordinates of the images under T of the elements of B with respect to B. If we next consider a second basis B' of $\mathbf{V}$ there will be a unique matrix N in $\mathbf{M}_{pp}$ representing T with respect to B'. This motivates the following definition.

DEFINITION 6.15   Two $p \times p$ matrices M and N with elements in $\mathbf{K}$ are *similar* if there exist bases B and B' of $\mathbf{V}$ and a linear transformation $T: \mathbf{V} \to \mathbf{V}$ such that M represents T with respect to B and N represents T with respect to B'.[2]

Let us now consider a transformation T of $\mathbf{V}$ into itself and let M denote the matrix of T with respect to B, and N denote the matrix of T with respect to B'. Can we find a matrix relationship between M and N? To answer this question we proceed as follows.

Consider the identity mapping of $\mathbf{V}$ onto itself, that is, $I_{\mathbf{V}}v = v$ for every $v \in \mathbf{V}$. Let P be the matrix representative of $I_{\mathbf{V}}$ with respect to B and B'. Then by Theorem 6.6, if

$$X = \begin{pmatrix} x_1 \\ x_2 \\ \vdots \\ x_p \end{pmatrix} \quad \text{and} \quad Y = \begin{pmatrix} y_1 \\ y_2 \\ \vdots \\ y_p \end{pmatrix}$$

are the coordinates of $v \in \mathbf{V}$ with respect to B and B', respectively, we conclude $Y = PX$. But the transformation $I_{\mathbf{V}}$ is invertible and its rank is p. Therefore by

---

[2] Here $\mathbf{V}$ is any p-dimensional vector space over $\mathbf{K}$. Because all p-dimensional spaces are isomorphic, it can be proved that this definition is independent of the choice of $\mathbf{V}$. Also, see Theorem 6.19.

Theorem 6.9, $\rho(P) = \rho(I_\mathbf{V}) = p$. Hence P is nonsingular. We have proved the following lemma.

LEMMA 6.1 Let $\mathbf{V}$ be a finite dimensional vector space and let B and B' be bases of $\mathbf{V}$. Then there exists a nonsingular matrix P, called the *transition matrix* from B to B', such that if v is any vector in $\mathbf{V}$ with coordinates

$$X = \begin{pmatrix} x_1 \\ x_2 \\ \vdots \\ x_p \end{pmatrix}$$

relative to B and

$$Y = \begin{pmatrix} y_1 \\ y_2 \\ \vdots \\ y_p \end{pmatrix}$$

relative to B', then $Y = PX$ and $X = P^{-1}Y$. The matrix P is the matrix representative of the identity map on $\mathbf{V}$ relative to B and B'.

Example 1 Consider the vector space $\mathbf{R}^2$ and the bases $B = \{(1, 1), (1, -1)\}$ and $B' = \{(0, 2), (-1, 0)\}$. Since

$$I_\mathbf{V}(1, 1) = (1, 1) = \tfrac{1}{2}(0, 2) - 1(-1, 0)$$
$$I_\mathbf{V}(1, -1) = (1, -1) = -\tfrac{1}{2}(0, 2) - 1(-1, 0)$$

the matrix representing $I_\mathbf{V}$ relative to B and B' is

$$P = \begin{pmatrix} \tfrac{1}{2} & -\tfrac{1}{2} \\ -1 & -1 \end{pmatrix}$$

Now consider a vector v, for example, $v = (4, 7)$. Since $(4, 7) = \tfrac{11}{2}(1, 1) - \tfrac{3}{2}(1, -1)$, the vector v has coordinates $X = \begin{pmatrix} \tfrac{11}{2} \\ -\tfrac{3}{2} \end{pmatrix}$ relative to B. Similarly, $(4, 7)$ has coordinates $Y = \begin{pmatrix} \tfrac{7}{2} \\ -4 \end{pmatrix}$ relative to B'. It is easy to check that

$$PX = \begin{pmatrix} \tfrac{1}{2} & -\tfrac{1}{2} \\ -1 & -1 \end{pmatrix}\begin{pmatrix} \tfrac{11}{2} \\ -\tfrac{3}{2} \end{pmatrix} = \begin{pmatrix} \tfrac{7}{2} \\ -4 \end{pmatrix} = Y$$

The following theorem will be useful in our discussion of eigenvalues and eigenvectors in Chapter 11.

THEOREM 6.18 Let T be a linear transformation of $\mathbf{V}$ into itself. Let M be the matrix of T with respect to the basis B, and N the matrix of T with respect to B'. Then there exists a nonsingular matrix P such that $M = P^{-1}NP$. In fact, P is the transition matrix from B to B'.

PROOF: Let v be any vector in **V**. Denote by X the coordinates of v with respect to B, and by Y the coordinates of v with respect to B'. Since M is the matrix of T with respect to B, MX is the coordinate vector of Tv with respect to B. Similarly, NY is the coordinate vector of Tv with respect to B'. By Lemma 6.1 we know there is a transition matrix P from B to B', and that $Y = PX$. Moreover, since Tv has coordinates MX relative to B and coordinates NY relative to B', it also follows from Lemma 6.1 that $NY = PMX$. Therefore, $NPX = PMX$. Since this must hold for every X, $NP = PM$, or $M = P^{-1}NP$.   ∎

LEMMA 6.2   Let **V** be an n-dimensional vector space and let B be any basis for **V**. Let P be a nonsingular $n \times n$ matrix. Then there exists a unique basis B' for **V** such that P is the transition matrix from B to B'.

PROOF: Let $B = \{v_1, v_2, \ldots, v_n\}$. We define a set of vectors $B' = \{v'_1, v'_2, \ldots, v'_n\}$ by the equations

$$v'_1 = q_{11}v_1 + q_{12}v_2 + \cdots + q_{1n}v_n$$
$$\cdots$$
$$v'_n = q_{n1}v_1 + q_{n2}v_2 + \cdots + q_{nn}v_n \tag{1}$$

where the matrix $Q = (q_{ij})$ is defined by $Q = (P^T)^{-1}$, the inverse of the transpose of P. In order to show that B' is a basis for **V** we must show that it is a linearly independent set. Suppose that there are scalars $c_1, \ldots, c_n$ such that

$$\sum_{i=1}^{n} c_i v'_i = 0$$

Then

$$\sum_{i=1}^{n} c_i \left(\sum_{j=1}^{n} q_{ij}v_j \right) = \sum_{j=1}^{n} \left(\sum_{i=1}^{n} q_{ij}c_i \right) v_j = 0$$

Since B is a linearly independent set it follows that

$$\sum_{i=1}^{n} q_{ij}c_i = 0 \qquad (j = 1, 2, \ldots, n)$$

or equivalently

$$Q^T \begin{pmatrix} c_1 \\ \vdots \\ c_n \end{pmatrix} = 0$$

Since Q^T is nonsingular this implies $c_i = 0$ ($i = 1, \ldots, n$), proving that B' is linearly independent and hence is a basis.

Next we show that P is the transition matrix from B to B'. To do so, consider the identity transformation I_V on **V**. Equation (1) can be written

$$\begin{pmatrix} v'_1 \\ \vdots \\ v'_n \end{pmatrix} = Q \begin{pmatrix} v_1 \\ \vdots \\ v_n \end{pmatrix}, \tag{2}$$

where it is understood that the multiplication on the right is to be carried out by the usual rule for matrix multiplication, even though $v_1, \ldots, v_n$ are vectors rather than scalars. Then

$$Q^{-1}\begin{pmatrix} v'_1 \\ \vdots \\ v'_n \end{pmatrix} = P^T\begin{pmatrix} v'_1 \\ \vdots \\ v'_n \end{pmatrix} = \begin{pmatrix} v_1 \\ \vdots \\ v_n \end{pmatrix}, \tag{3}$$

that is,

$$I_V v_1 = v_1 = p_{11}v'_1 + p_{21}v'_2 + \cdots + p_{n1}v'_n$$
$$\cdots \tag{4}$$
$$I_V v_n = v_n = p_{1n}v'_1 + p_{2n}v'_2 + \cdots + p_{nn}v'_n$$

It follows that P is the matrix representing I_V relative to the bases B and B'; that is, P is the transition matrix from B to B'.

Finally, there cannot be two distinct bases B' and $B'' = \{v''_1, \ldots, v''_n\}$ such that P is the transition matrix from B to B' and also from B to B''. For then Eq. (4) holds and hence Eq. (2) with the v'_i replaced by v''_i. Equation (2) shows that B and P uniquely determine the vectors in the set B'. ∎

If we now recall our definition of similar matrices we have the following theorem

THEOREM 6.19   A necessary and sufficient condition that two $p \times p$ matrices M and N be similar over a field K is that there exist a nonsingular matrix P with elements in K such that $M = P^{-1}NP$.

PROOF: The proof is left as an exercise. ∎

Example 2   Consider the vector space $\mathbf{R}^2$ and bases B and B' given in Example 1. Let T be the transformation taking the vector (r_1, r_2) into $(3r_1, 2r_2)$. Let M be the matrix of the transformation with respect to B, and N the matrix with respect to B'. To determine N we note the first column of N contains the coordinates with respect to B' of the image under T of $(0, 2)$, and the second column of N contains the coordinates of the image of $(-1, 0)$. Therefore

$$N = \begin{pmatrix} 2 & 0 \\ 0 & 3 \end{pmatrix}$$

By Theorem 6.18, we know $M = P^{-1}NP$. Since

$$P^{-1} = \begin{pmatrix} 1 & -\frac{1}{2} \\ -1 & -\frac{1}{2} \end{pmatrix}$$

we obtain

$$M = \begin{pmatrix} 1 & -\frac{1}{2} \\ -1 & -\frac{1}{2} \end{pmatrix}\begin{pmatrix} 2 & 0 \\ 0 & 3 \end{pmatrix}\begin{pmatrix} \frac{1}{2} & -\frac{1}{2} \\ -1 & -1 \end{pmatrix} = \begin{pmatrix} \frac{5}{2} & \frac{1}{2} \\ \frac{1}{2} & \frac{5}{2} \end{pmatrix}$$

The reader can verify that this is the matrix of T relative to B.

Exercises

1.　Prove Theorem 6.19.

2.　Prove that similarity is an equivalence relation.

In Exercises 3 through 5 let $\mathbf{V}$ be $\mathbf{R}^3$. Find the transition matrix between the specified bases B and B' of $\mathbf{V}$.

3.　$B = \{(1, 0, 0), (1, 1, 0), (1, 1, 1)\}$　$B' = \{(1, 0, 0), (0, 1, 0), (0, 0, 1)\}$

4.　$B = \{(1, -1, 1), (1, 0, 1), (1, 1, 0)\}$　$B' = \{(1, 2, 0), (2, 1, 0), (0, 0, 1)\}$

5.　$B = \{(a_1, a_2, 0), (b_1, b_2, 0), (c_1, c_2, c_3)\}$
$$B' = \{(1, 0, 0), (0, 1, 0), (0, 0, 1)\}$$

In Exercises 6 and 7 compute the specified matrix.

6.　Let B and B' be as defined in Exercise 3. Let

$$M = \begin{pmatrix} 1 & 1 & 0 \\ 0 & -1 & 1 \\ 1 & 1 & 0 \end{pmatrix}$$

be the matrix of T relative to B. Find the matrix N of T relative to B'.

7.　Let B and B' be as defined in Exercise 4. Let

$$N = \begin{pmatrix} 1 & 2 & 1 \\ 1 & 3 & 1 \\ 2 & 1 & 1 \end{pmatrix}$$

be the matrix of T relative to B'. Find the matrix M of T relative to B.

6.9　Generalized inverses

Our discussion in Section 6.7 related invertible linear transformations with invertible or nonsingular matrices. We saw that in order for the inverse M^{-1} of a matrix M to exist, M had to be a representation of an invertible linear transformation. That is, if M is a representative of a transformation $T: \mathbf{V} \rightarrow \mathbf{V}'$, then in order that M^{-1} exist, T must be one-to-one and onto.

In many applications, particularly in statistical inference, situations arise in which a matrix is singular, and yet something with some of the properties of an inverse is desired. For example, suppose we are faced with a linear system of algebraic equations

$$\begin{aligned} y_1 &= m_{11}x_1 + \cdots + m_{1k}x_k \\ y_2 &= m_{21}x_1 + \cdots + m_{2k}x_k \\ &\vdots \\ y_n &= m_{n1}x_1 + \cdots + m_{nk}x_k \end{aligned} \tag{1}$$

This system can be presented in matrix notation as $Y = MX$, where

$$Y = \begin{pmatrix} y_1 \\ y_2 \\ \vdots \\ y_n \end{pmatrix}$$

$$M = \begin{pmatrix} m_{11} & m_{12} & \cdots & m_{1k} \\ m_{21} & m_{22} & \cdots & m_{2k} \\ & & \cdots & \\ m_{n1} & m_{n2} & \cdots & m_{nk} \end{pmatrix}$$

and

$$X = \begin{pmatrix} x_1 \\ x_2 \\ \vdots \\ x_k \end{pmatrix}$$

If Y and M are given, we could be interested in solving for X. If M is nonsingular (implying $n = k$), then the unique solution to system (1) is $X = M^{-1}Y$. But what if $n = k$ and M is singular? Or what if n is not equal to k? The generalized inverse is used to attack these problems.

We begin our discussion of generalized inverses by investigating the transformations associated with singular matrices. Let T be a linear transformation mapping a vector space $\mathbf{V}$ into a vector space $\mathbf{V}'$, and let B and B' be bases of $\mathbf{V}$ and $\mathbf{V}'$, respectively. Assume the dimension of $\mathbf{V}$ to be n and of $\mathbf{V}'$ to be m, and let M be the $m \times n$ matrix representation of T with respect to B and B'. If T is either not one-to-one or not onto, the matrix M will not have an inverse.

Consider the case where T is not onto. This means there exists an element v' in $\mathbf{V}'$ which is not the image under T of any element of $\mathbf{V}$. That is, v' is not in the range space of T. We cannot define an inverse mapping to T since there is no inverse image for v'. In the event T is not one-to-one, the problem encountered in trying to define an inverse mapping is that there are several distinct elements, say v_0, v_1 in $\mathbf{V}$ mapped by T to the same image v' in $\mathbf{V}'$. Since a mapping must have a unique image, no mapping T^* can be defined which would allow $T^*(v') = v_0$ and $T^*(v') = v_1$. (See Figure 6.2.)

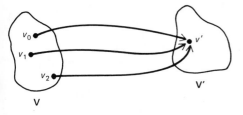

FIGURE 6.2

In the case that T is not one-to-one we see that T establishes equivalence classes on $\mathbf{V}$, each class consisting of elements mapped by T to the same image point. With the above review, we now define a generalized inverse.

DEFINITION 6.16   Let T be a linear transformation $T: \mathbf{V} \to \mathbf{V}'$. The transformation $T^-: \mathbf{V}' \to \mathbf{V}$ is a *generalized inverse* of T if T^- is linear and $TT^-T = T$.

In the event that T is not one-to-one, T^- must map each element in the range space of T back to an element in the corresponding equivalence class. In Figure 6.2, v' may be mapped by T^- to either v_0, v_1, or v_2. If T is not onto, then T^- must map the range space of $\mathbf{V}$ under T in the fashion described above, but may be defined arbitrarily on the complement of the range space, provided of course it remains a linear transformation. One particular example would be for T^- to map the complement of the range space into the zero vector of $\mathbf{V}$.

THEOREM 6.20   Let $\mathbf{V}$ and $\mathbf{V}'$ be finite dimensional vector spaces. There exists a generalized inverse for every linear transformation $T: \mathbf{V} \to \mathbf{V}'$.

PROOF: If T is the zero mapping, that is, $Tv = \mathbf{0}$ for every v in $\mathbf{V}$, define T^- to be the zero mapping from $\mathbf{V}'$ to $\mathbf{V}$. Then $TT^-T = \mathbf{Z} = T$. If T is not the zero mapping, let $\{v'_1, v'_2, \ldots, v'_k\}$ be a basis for the range space $\mathbf{R}_T$ of T. Extend this to a basis $B' = \{v'_1, v'_2, \ldots, v'_k, \ldots, v'_q\}$ of $\mathbf{V}'$. We now construct a linear transformation T^- which is the generalized inverse of T. Since $v'_1, v'_2, \ldots, v'_k$ are elements of $\mathbf{R}_T$, we can find elements $v_1, v_2, \ldots, v_k$ in $\mathbf{V}$ such that $Tv_1 = v'_1$, $Tv_2 = v'_2, \ldots$, $Tv_k = v'_k$. (Note that our choice is not necessarily unique. However, any set $\{v_1, v_2, \ldots, v_k\}$ satisfying our condition will suffice.) We define

$$T^-v'_1 = v_1, \ T^-v'_2 = v_2, \ldots, \ T^-v'_k = v_k, \ T^-v'_{k+1} = \mathbf{0}, \ldots, \ T^-v'_q = \mathbf{0}$$

Since T^- is now defined for each element of the basis B', we know there is a unique extension which will make T^- linear on $\mathbf{V}'$. All that remains is to verify that $TT^-Tv = Tv$ for each element in $\mathbf{V}$. This is left as Exercise 1.   ∎

THEOREM 6.21   A generalized inverse of a linear transformation T is unique if and only if T is invertible, in which case $T^- = T^{-1}$.

PROOF: We leave this for the reader (Exercise 2).   ∎

In keeping with our attempt to relate matrix properties with the properties of linear transformations, we now define the generalized inverse of a matrix.

DEFINITION 6.17   A *generalized inverse of a matrix* M is a matrix, denoted by M^-, satisfying $MM^-M = M$.

As an immediate consequence of Theorems 6.20 and 6.21 we have the following.

THEOREM 6.22 Every matrix M has an associated generalized inverse. Further, the generalized inverse is unique if and only if M is nonsingular, in which case $M^- = M^{-1}$.

PROOF: The proof is required in Exercise 3. ∎

Example 1 Consider the matrix $M = \begin{pmatrix} 1 & 2 \\ 2 & 4 \end{pmatrix}$. Since the second column of M is twice the first we see immediately that M is singular. Therefore there is no matrix M^{-1} such that $M^{-1}M = I$. We therefore will search for a generalized inverse satisfying

$$\begin{pmatrix} 1 & 2 \\ 2 & 4 \end{pmatrix} M^- \begin{pmatrix} 1 & 2 \\ 2 & 4 \end{pmatrix} = \begin{pmatrix} 1 & 2 \\ 2 & 4 \end{pmatrix} \tag{2}$$

Letting $M^- = \begin{pmatrix} a_{11} & a_{12} \\ a_{21} & a_{22} \end{pmatrix}$ we obtain from Eq. (2),

$$\begin{pmatrix} a_{11} + 2a_{21} & a_{12} + 2a_{22} \\ 2a_{11} + 4a_{21} & 2a_{12} + 4a_{22} \end{pmatrix} \begin{pmatrix} 1 & 2 \\ 2 & 4 \end{pmatrix}$$

$$= \begin{pmatrix} a_{11} + 2a_{21} + 2a_{12} + 4a_{22} & 2a_{11} + 4a_{21} + 4a_{12} + 8a_{22} \\ 2a_{11} + 4a_{21} + 4a_{12} + 8a_{22} & 4a_{11} + 8a_{21} + 8a_{12} + 16a_{22} \end{pmatrix}$$

$$= \begin{pmatrix} 1 & 2 \\ 2 & 4 \end{pmatrix}$$

We see that the second row of these equations is twice the first and therefore we need only choose the elements of M^- to satisfy

$$\begin{aligned} a_{11} + 2a_{21} + 2a_{12} + 4a_{22} &= 1 \\ 2a_{11} + 4a_{21} + 4a_{12} + 8a_{22} &= 2 \end{aligned} \tag{3}$$

But any solution of the first equation will be a solution of the second also. For a solution we can arbitrarily pick three of the elements. Let us choose $a_{11} = a_{12} = a_{21} = 1$. Then substituting into Eq. (3) we obtain $a_{22} = -1$ and hence

$$M^- = \begin{pmatrix} 1 & 1 \\ 1 & -1 \end{pmatrix}$$

The reader should verify that M^- does satisfy $MM^-M = M$.

As an alternative solution we could choose $a_{11} = 1$, $a_{12} = 1$, $a_{21} = -1$ which, when substituted into Eq. (3), would require $a_{22} = 0$. Therefore,

$$M_1^- = \begin{pmatrix} 1 & 1 \\ -1 & 0 \end{pmatrix}$$ is also a generalized inverse of M.

Exercises

1. Complete the proof of Theorem 6.20.

2. Prove Theorem 6.21.

3. Prove Theorem 6.22.

4. Compute a generalized inverse for each of the following matrices.

(a) $\begin{pmatrix} 1 & 0 \\ 0 & 1 \\ 0 & 0 \end{pmatrix}$
(b) $\begin{pmatrix} 1 & 2 \\ 1 & 0 \\ 2 & 1 \end{pmatrix}$
(c) $\begin{pmatrix} 1 & 1 & 2 \\ 2 & 0 & 1 \end{pmatrix}$

(d) $\begin{pmatrix} 1 & 1 & 1 \\ 2 & 0 & 1 \\ 0 & 2 & 1 \end{pmatrix}$
(e) $\begin{pmatrix} 2 & 0 & 0 \\ 0 & 1 & 2 \\ 0 & 2 & 5 \end{pmatrix}$

COMMENTARY

A complete survey of the theory of matrices may be found in:

M. Marcus and H. Minc, *A Survey of Matrix Theory and Matrix Inequalities*, Boston, Allyn and Bacon, 1964.

Section 6.4   A word about notation is in order here. In this text we have used TS to represent composition of transformations in the order S, then T, and Sv for the image of v under S. We have chosen to represent a transformation $T: \mathbf{V} \to \mathbf{V}'$ by the matrix N having in its columns the coordinates of the vectors Tv_i relative to a basis for $\mathbf{V}'$, where the $\{v_i\}$ form a basis for $\mathbf{V}$. With this choice, we have seen that if N represents T and M represents S, then NM represents TS, provided matrix multiplication is defined as "row times column."

An alternative notation is also in widespread use. In it, ST is used to represent composition in the order S, then T. This has the advantage that one lists the transformations as they occur in our natural left-to-right order of writing. On the other hand, the image of v under S must be written vS [so that $v(ST) = (vS)T$]. In order to retain the fact that if M represents S, and N represents T, then MN represents ST, one must either change the definition of matrix multiplication to "column times row" or change the definition of matrix representative of a transformation. Usually, the latter is done: If $\mathbf{V}$ has basis $\{v_1, \ldots, v_p\}$ and $\mathbf{V}'$ has basis $\{v'_1, \ldots, v'_q\}$, the matrix representing T has as its ith row the coordinates of v_iT relative to $\{v'_1, \ldots, v'_q\}$. Finally, in this notation, if $v' = vT$ and T is represented by N, then coordinates X of v and Y of v' are related by $Y = XN$.

Section 6.5   The notion of the transpose of a matrix is related to the concept of the *adjoint* of a linear transformation. See:

NOMIZU, K., *Fundamentals of Linear Algebra*, New York, McGraw-Hill, 1966.

SHIELDS, P. C., *Linear Algebra*, Reading, Mass., Addison-Wesley, 1964, pp. 139–143.

Section 6.9   For further discussion and application of generalized inverses see:

RAO, C. R., *Linear Statistical Inference and Its Applications*, New York, Wiley, 1967.

BOULLION, T. L., and P. L. ODELL, *Generalized Inverse Matrices*, New York, Wiley, 1971.

RAO, C. R., and S. K. MITRA, *Generalized Inverse of Matrices and Its Applications*, New York, Wiley, 1971.

7

Determinants

7.1 Determinants defined

One of the most useful characteristics of a square matrix is its determinant. We use this characteristic later in our discussions of solving systems of algebraic equations and in our treatment of linear differential equations. However, before discussing applications of determinants, we must define them and investigate their properties.

Most readers learned to compute the determinant of a 2 × 2 matrix in elementary mathematics courses. The determinant of

$$M = \begin{pmatrix} m_{11} & m_{12} \\ m_{21} & m_{22} \end{pmatrix}$$

is given as

$$\det M = m_{11}m_{22} - m_{21}m_{12} \tag{1}$$

We begin our general discussion of determinants by noting properties possessed in the 2 × 2 case.

Observe that the determinant is really a function mapping 2 × 2 matrices into scalars; that is, to each 2 × 2 matrix M there is a scalar which we denote $\det M$. This function has certain properties, which we shall now describe. Let us regard the ith column of M as a vector and denote it by M_i ($i = 1, 2$). Thus,

$$M_1 = \begin{pmatrix} m_{11} \\ m_{21} \end{pmatrix}, \qquad M_2 = \begin{pmatrix} m_{12} \\ m_{22} \end{pmatrix}$$

We can write $M = (M_1 \ M_2)$. Moreover, we can now regard the determinant as a function which maps each pair $(M_1 \ M_2)$ into a scalar $\det M$. The special properties to which we direct attention are as follows:

1. $\det (M_1 \quad M_2 + M_2^*) = \det (M_1 \quad M_2) + \det (M_1 \quad M_2^*)$ for any column vectors M_1, M_2, and M_2^*

2. $\det (cM_1 \quad M_2) = \det (M_1 \quad cM_2) = c \det (M_1 \quad M_2)$ for any scalar c and any column vectors M_1 and M_2

3. If $M_1 = M_2$, $\det M = 0$

4. $\det \begin{pmatrix} 1 & 0 \\ 0 & 1 \end{pmatrix} = 1$

That the determinant defined in Eq. (1) has the properties above can be verified by direct expansion as follows:

1. $\det (M_1 \quad M_2 + M_2^*) = \det \begin{pmatrix} m_{11} & m_{12} + m_{12}^* \\ m_{21} & m_{22} + m_{22}^* \end{pmatrix}$

$$= m_{11}(m_{22} + m_{22}^*) - m_{21}(m_{12} + m_{12}^*)$$
$$= m_{11}m_{22} - m_{21}m_{12} + m_{11}m_{22}^* - m_{21}m_{12}^*$$
$$= \det (M_1 \quad M_2) + \det (M_1 \quad M_2^*)$$

2. $\det (cM_1 \quad M_2) = \det \begin{pmatrix} cm_{11} & m_{12} \\ cm_{21} & m_{22} \end{pmatrix}$

$$= cm_{11}m_{22} - cm_{12}m_{21}$$
$$= c(m_{11}m_{22} - m_{12}m_{21})$$
$$= c \det (M_1 \quad M_2)$$

That $\det (M_1 \quad cM_2) = c \det (M_1 \quad M_2)$ can similarly be verified.

3. Let $M_1 = M_2 = \begin{pmatrix} m_1 \\ m_2 \end{pmatrix}$. Then

$$\det M = \det \begin{pmatrix} m_1 & m_1 \\ m_2 & m_2 \end{pmatrix} = m_1 m_2 - m_1 m_2 = 0$$

4. $\det \begin{pmatrix} 1 & 0 \\ 0 & 1 \end{pmatrix} = 1 \cdot 1 - 0 \cdot 0 = 1$

In defining determinants, we must be careful to be consistent with the above example. And, in fact, our definition is based on a generalization of properties (1) to (4).

DEFINITION 7.1 Let $\mathbf{M}_{pp}$ be the space of all $p \times p$ matrices with elements in $\mathbf{K}$. The *determinant* is a mapping of $\mathbf{M}_{pp}$ into $\mathbf{K}$ defined by the requirement that if $M_1, M_2, \ldots, M_p$ are the columns of M, then:

1. For any j $(1 \leq j \leq p)$ and any column vector M_j^*,

$$\det (M_1 \quad M_2 \quad \cdots \quad M_j + M_j^* \quad \cdots \quad M_p)$$
$$= \det (M_1 \quad M_2 \quad \cdots \quad M_j \quad \cdots \quad M_p)$$
$$+ \det (M_1 \quad M_2 \quad \cdots \quad M_j^* \quad \cdots \quad M_p)$$

2. For any $c \in \mathbf{K}$, and any j $(1 \leq j \leq p)$,

$$\det (M_1 \quad M_2 \quad \cdots \quad cM_j \quad \cdots \quad M_p) = c \det M$$

3. If $M_j = M_{j+1}$ for any j, $1 \leq j \leq p - 1$, then $\det M = 0$
4. $\det I_p = 1$ where I_p is the $p \times p$ identity matrix

For any matrix M in $\mathbf{M}_{pp}$, the image of M is called the *value of the determinant*[1] of M and is denoted $\det M$ or $|M|$. Thus

$$\det \begin{pmatrix} m_{11} & m_{12} & \cdots & m_{1p} \\ m_{21} & m_{22} & \cdots & m_{2p} \\ & & \cdots & \\ m_{p1} & m_{p2} & \cdots & m_{pp} \end{pmatrix} = \begin{vmatrix} m_{11} & m_{12} & \cdots & m_{1p} \\ m_{21} & m_{22} & \cdots & m_{2p} \\ & & \cdots & \\ m_{p1} & m_{p2} & \cdots & m_{pp} \end{vmatrix}$$

Properties (1) and (2) mean that the determinant is a linear function of the vector in the jth column $(j = 1, \ldots, n)$. Property (3) states that if a matrix has two adjacent columns which are identical, the value of its determinant is zero.

We have already verified that for 2×2 matrices a mapping exists satisfying these conditions,[2] and have therefore illustrated the existence of a determinant mapping in the special case $p = 2$. In the next several sections we will be leading up to a proof that for each positive integer p there does exist a determinant over the space $\mathbf{M}_{pp}$, and further, this determinant is unique. Moreover, we will develop a technique for evaluating determinants.

We conclude this section with a series of theorems which give additional properties of a determinant function. Note that each theorem does not depend on the uniqueness of the determinant, but gives a property which any determinant function must satisfy.

THEOREM 7.1 Let M be a $p \times p$ matrix and let det be a determinant function on $\mathbf{M}_{pp}$. Then the value of det at the matrix obtained by interchanging two adjacent columns of M is $(-1) \det M$.

PROOF: Let $M_1, \ldots, M_p$ be the columns of M. Consider the matrix which has $M_j + M_{j+1}$ in its jth column and its $(j + 1)$st column, but is otherwise the same as M. By property (3), the value of any determinant of this matrix is zero.

---

[1] It is common practice to use the term *determinant* in referring to both the determinant function and the value of the determinant function for a specific matrix. We will follow this loose convention, and will distinguish either the function or the value of the function only when necessary for clarity of the discussion.

Additionally note that for each p there is a determinant function. Hence, to be strictly correct we should use different notation for the determinant on, say, $\mathbf{M}_{22}$ and $\mathbf{M}_{33}$. However, we follow the standard practice of assuming the reader will not be confused by lack of distinction.

[2] Readers unsure of the above definitions should work Exercises 5 and 6 before proceeding.

Thus, $0 = \det (M_1 \quad M_2 \quad \cdots \quad (M_j + M_{j+1}) \quad (M_j + M_{j+1}) \quad \cdots \quad M_p)$. By property (1) we can rewrite this as

$$
\begin{aligned}
0 &= \det (M_1 \quad M_2 \quad \cdots \quad M_j \quad (M_j + M_{j+1}) \quad \cdots \quad M_p) \\
&\quad + \det (M_1 \quad M_2 \quad \cdots \quad M_{j+1} \quad (M_j + M_{j+1}) \quad \cdots \quad M_p) \\
&= \det (M_1 \quad M_2 \quad \cdots \quad M_j \quad M_j \quad \cdots \quad M_p) \\
&\quad + \det (M_1 \quad M_2 \quad \cdots \quad M_j \quad M_{j+1} \quad \cdots \quad M_p) \\
&\quad + \det (M_1 \quad M_2 \quad \cdots \quad M_{j+1} \quad M_j \quad \cdots \quad M_p) \\
&\quad + \det (M_1 \quad M_2 \quad \cdots \quad M_{j+1} \quad M_{j+1} \quad \cdots \quad M_p) \\
&= 0 + \det M + \det (M_1 \quad M_2 \quad \cdots \quad M_{j+1} \quad M_j \quad \cdots \quad M_p) + 0
\end{aligned}
$$

Hence $\det M = - \det (M_1 \quad M_2 \quad \cdots \quad M_{j+1} \quad M_j \quad \cdots \quad M_p)$. ∎

COROLLARY 1 If two columns of M are equal, that is, $M_i = M_j$ for $i \neq j$, then $\det M = 0$.

PROOF: The absolute value of $\det M$, denoted $|\det M|$, is left unchanged by interchanging two adjacent columns of M. If $i < j$, interchange M_i with M_{i+1}, then with M_{i+2}, and so on, until M_i is adjacent to M_j. Since the absolute value of $\det M$ is left unchanged and the value of the determinant of the resulting matrix is zero by (3) of Definition 7.1, $\det M = 0$. ∎

COROLLARY 2 Let M be a $p \times p$ matrix. The value of a determinant of the matrix obtained by interchanging any two columns of M is $(-1) \det M$.

PROOF: The proof follows from Corollary 1 using the method of proof for Theorem 7.1. The details are left as an exercise. ∎

THEOREM 7.2 The addition of a scalar multiple of one column of a matrix to a second column leaves the value of a determinant unchanged. That is,

$$
\det (M_1 \quad M_2 \quad \cdots \quad M_j + cM_k \quad \cdots \quad M_k \quad \cdots \quad M_p)
$$
$$
= \det (M_1 \quad M_2 \quad \cdots \quad M_j \quad \cdots \quad M_k \quad \cdots \quad M_p)
$$

for any j, k ($j \neq k$) and any scalar c.

PROOF: The proof is left as an exercise. ∎

Exercises

1. Using Eq. (1), evaluate

$$
\det \begin{pmatrix} 1 & 2 \\ 3 & 5 \end{pmatrix} \qquad \det \begin{pmatrix} 1 & -2 \\ 3 & 4 \end{pmatrix} \qquad \det \begin{pmatrix} 1 & 0 \\ 3 & 9 \end{pmatrix}
$$

and thus verify property (1) for this special case.

2.   Let (m) be an element of $\mathbf{M}_{11}$. Define det $(m) = m$.

(a) Prove det (m) is a determinant function on $\mathbf{M}_{11}$.

(b) Prove that there is no other determinant function on $\mathbf{M}_{11}$, that is, that det (m) is unique.

3.   Show that if M is any 2×2 matrix, det $M = $ det M^T.

4.   Solve these equations:

(a) $\det \begin{pmatrix} 2x - 1 & x + 16 \\ x - 1 & x + 11 \end{pmatrix} = 0$

(b) $\det \begin{pmatrix} 2^x & x - 1 \\ \sin^2 x & \ln 1 \end{pmatrix} + \det \begin{pmatrix} x^{13} & 1 - x \\ x^2 & 0 \end{pmatrix} = 0$

5.   Given that

$$\begin{vmatrix} 1 & 3 & 0 \\ 0 & 2 & 1 \\ 5 & 0 & -1 \end{vmatrix} = 13, \quad \begin{vmatrix} 1 & 0 & 0 \\ 0 & 3 & 1 \\ 5 & 2 & -1 \end{vmatrix} = -5$$

find the value of the determinant of each of these matrices, and explain which parts of Definition 7.1 are used:

(a) $\begin{pmatrix} 1 & 3 & 0 \\ 0 & 5 & 1 \\ 5 & 2 & -1 \end{pmatrix}$ (b) $\begin{pmatrix} 1 & 0 & 0 \\ 0 & -6 & 1 \\ 5 & -4 & -1 \end{pmatrix}$

6.   Let

$$M = \begin{pmatrix} m_{11} & m_{12} & m_{13} \\ m_{21} & m_{22} & m_{23} \\ m_{31} & m_{32} & m_{33} \end{pmatrix}$$

Most readers have learned to expand the determinant of M as

$$m_{11}m_{22}m_{33} + m_{12}m_{23}m_{31} + m_{21}m_{32}m_{13} - m_{13}m_{22}m_{31}$$
$$- m_{23}m_{32}m_{11} - m_{12}m_{21}m_{33} = \det M$$

Show that this function satisfies all the conditions of Definition 7.1.

7.   Prove Corollary 2 to Theorem 7.1.

8.   Prove Theorem 7.2.

9.   Given $c \in \mathbf{K}$ and let M be a $p \times p$ matrix. Show from Definition 7.1 that

$$\det (cM) = c^p(\det M)$$

7.2   Minors and cofactors

In Section 7.1 we asserted that there is one and only one mapping of $\mathbf{M}_{pp}$ into $\mathbf{K}$ with all the properties (1) to (4) of Definition 7.1. Neither existence nor uniqueness was proved. In the next two sections we will assume a unique determinant for each p and will present a method for computing these determinants. An easy extension of this method will later lead to a method for computing the inverse of a square matrix M, when it exists.

DEFINITION 7.2   Let M be a $p \times p$ matrix, $p > 1$. The i,j *minor* of M, denoted M_{ij}, is the value of the determinant of the $(p - 1) \times (p - 1)$ matrix formed by removing the ith row and jth column from M.

For each i and j, $1 \le i \le p$ and $1 \le j \le p$, there is an associated minor. Therefore, there are p^2 minors associated with each matrix.

Example 1   Consider the 3×3 matrix

$$M = \begin{pmatrix} 3 & 1 & 2 \\ 1 & 1 & 0 \\ 0 & 1 & 2 \end{pmatrix}$$

Then using Eq. (1) of Section 7.1 we get

$$M_{11} = \det \begin{pmatrix} 1 & 0 \\ 1 & 2 \end{pmatrix} = 2 \qquad M_{12} = \det \begin{pmatrix} 1 & 0 \\ 0 & 2 \end{pmatrix} = 2 \qquad M_{13} = \det \begin{pmatrix} 1 & 1 \\ 0 & 1 \end{pmatrix} = 1$$

$$M_{21} = \det \begin{pmatrix} 1 & 2 \\ 1 & 2 \end{pmatrix} = 0 \qquad M_{22} = \det \begin{pmatrix} 3 & 2 \\ 0 & 2 \end{pmatrix} = 6 \qquad M_{23} = \det \begin{pmatrix} 3 & 1 \\ 0 & 1 \end{pmatrix} = 3$$

$$M_{31} = \det \begin{pmatrix} 1 & 2 \\ 1 & 0 \end{pmatrix} = -2 \quad M_{32} = \det \begin{pmatrix} 3 & 2 \\ 1 & 0 \end{pmatrix} = -2 \quad M_{33} = \det \begin{pmatrix} 3 & 1 \\ 1 & 1 \end{pmatrix} = 2$$

We represent these results as

$$\text{minor } M = \begin{pmatrix} 2 & 2 & 1 \\ 0 & 6 & 3 \\ -2 & -2 & 2 \end{pmatrix}$$

and call this array the *matrix of minors* of M.

Example 2   Consider the 2×2 matrix $M = \begin{pmatrix} 4 & 7 \\ 1 & 3 \end{pmatrix}$. In this case each minor is the determinant of a 1×1 matrix. For example $M_{21} = \det (7) = 7$. The matrix of minors is

$$\text{minor } M = \begin{pmatrix} 3 & 1 \\ 7 & 4 \end{pmatrix}$$

Example 3 Let M be the 4×4 matrix

$$M = \begin{pmatrix} 1 & 2 & 0 & 1 \\ 2 & 2 & 3 & 1 \\ 0 & 3 & 1 & 2 \\ 1 & 1 & 2 & 1 \end{pmatrix}$$

By expanding the determinants of the appropriate 3×3 matrix in the manner described in Exercise 6 of Section 7.1, we obtain

$$M_{11} = \det \begin{pmatrix} 2 & 3 & 1 \\ 3 & 1 & 2 \\ 1 & 2 & 1 \end{pmatrix} = -4 \qquad M_{12} = \det \begin{pmatrix} 2 & 3 & 1 \\ 0 & 1 & 2 \\ 1 & 2 & 1 \end{pmatrix} = -1$$

$$M_{13} = \det \begin{pmatrix} 2 & 2 & 1 \\ 0 & 3 & 2 \\ 1 & 1 & 1 \end{pmatrix} = 3 \qquad M_{14} = \det \begin{pmatrix} 2 & 2 & 3 \\ 0 & 3 & 1 \\ 1 & 1 & 2 \end{pmatrix} = 3$$

$$M_{21} = \det \begin{pmatrix} 2 & 0 & 1 \\ 3 & 1 & 2 \\ 1 & 2 & 1 \end{pmatrix} = -1$$

and so on. The total matrix of minors is

$$\text{minor } M = \begin{pmatrix} -4 & -1 & 3 & 3 \\ -1 & -4 & 2 & 7 \\ 3 & 2 & -1 & -1 \\ 3 & 7 & -1 & -11 \end{pmatrix}$$

The reader might note that the original matrix is symmetric, that is, $M = M^T$, and that this property carries over to minor M.

DEFINITION 7.3 Let M be a $p \times p$ matrix with i,j minor M_{ij}. Then $(-1)^{i+j} M_{ij}$ is the i,j *cofactor* of M, and is denoted by M^{ij}. The $p \times p$ matrix (M^{ij}) is called the *matrix of cofactors*.

Example 4 Consider the matrix

$$M = \begin{pmatrix} 3 & 1 & 2 \\ 1 & 1 & 0 \\ 0 & 1 & 2 \end{pmatrix} \quad \text{with} \quad \text{minor } M = \begin{pmatrix} 2 & 2 & 1 \\ 0 & 6 & 3 \\ -2 & -2 & 2 \end{pmatrix}$$

as derived in Example 1. Then the matrix of cofactors is given by

$$\text{cofactor } M = \begin{pmatrix} 2 & -2 & 1 \\ 0 & 6 & -3 \\ -2 & 2 & 2 \end{pmatrix}$$

Note that the element in the ijth position for $i + j$ odd is the negative of the corresponding minor element, whereas the i, j cofactor equals the i, j minor for $i + j$ even.

Example 5 Let M be the 2×2 matrix of Example 2. Then

$$\text{cofactor } M = \begin{pmatrix} 3 & -1 \\ -7 & 4 \end{pmatrix}$$

Example 6 Let M be the 4×4 matrix of Example 3. Then

$$\text{cofactor } M = \begin{pmatrix} -4 & 1 & 3 & -3 \\ 1 & -4 & -2 & 7 \\ 3 & -2 & -1 & 1 \\ -3 & 7 & 1 & -11 \end{pmatrix}$$

Again, note that symmetry still obtains in the cofactor matrix.

7.3 The cofactor expansion

In Definitions 7.2 and 7.3, the minors and cofactors of a $p \times p$ matrix were defined in terms of the determinants of certain $(p - 1) \times (p - 1)$ matrices. In order for these definitions to make sense in all cases, it is necessary that the determinant exist and be unique. We have already shown in Section 7.1 that

$$\begin{vmatrix} m_{11} & m_{12} \\ m_{21} & m_{22} \end{vmatrix} = m_{11}m_{22} - m_{12}m_{21} \tag{1}$$

has the properties required of a determinant, thus verifying the existence of det M if $p = 2$. We shall now exhibit a formula for computing the determinant function on $\mathbf{M}_{pp}$. We will show later that this function satisfies the four required properties of Definition 7.1.

Let p and i be positive integers for which $1 \leq i \leq p$ and $2 \leq p$. Then

$$\det M = \sum_{j=1}^{p} m_{ij}M^{ij} \tag{2}$$

where M^{ij} is the i, j cofactor of M. This formula is called the *expansion of* $|M|$ in *cofactors of the ith row*, or simply the expansion of $|M|$ about the ith row.

Before proving that det M uniquely exists and provides a determinant function, we wish to show how the equation can be used to compute determinants.

Example 1 Consider the matrix

$$Q = \begin{pmatrix} 2 & 3 & 1 \\ 3 & 1 & 2 \\ 1 & 2 & 1 \end{pmatrix}$$

encountered in Example 3 of Section 7.2. According to Eq. (2), with $i = 1$,

$$\det Q = q_{11}Q^{11} + q_{12}Q^{12} + q_{13}Q^{13}$$

$$= 2\begin{vmatrix} 1 & 2 \\ 2 & 1 \end{vmatrix} - 3\begin{vmatrix} 3 & 2 \\ 1 & 1 \end{vmatrix} + 1\begin{vmatrix} 3 & 1 \\ 1 & 2 \end{vmatrix}$$

$$= 2(1 \cdot 1 - 2 \cdot 2) - 3(3 \cdot 1 - 1 \cdot 2) + 1(3 \cdot 2 - 1 \cdot 1) = -4$$

On the other hand, using $i = 3$, we get

$$\det Q = q_{31}Q^{31} + q_{32}Q^{32} + q_{33}Q^{33}$$

$$= 1\begin{vmatrix} 3 & 1 \\ 1 & 2 \end{vmatrix} - 2\begin{vmatrix} 2 & 1 \\ 3 & 2 \end{vmatrix} + 1\begin{vmatrix} 2 & 3 \\ 3 & 1 \end{vmatrix} = -4$$

It is easily seen that the same value of $\det Q$ is obtained for $i = 2$. We leave it to the reader to use Eq. (2) to verify the values for M_{12}, M_{13}, M_{14}, and so on, stated in Example 3 of Section 7.2.

Example 2   Now let us evaluate the determinant of the matrix

$$M = \begin{pmatrix} 1 & 2 & 0 & 1 \\ 2 & 2 & 3 & 1 \\ 0 & 3 & 1 & 2 \\ 1 & 1 & 2 & 1 \end{pmatrix}$$

The matrix of cofactors is

$$\text{cofactor } M = \begin{pmatrix} -4 & 1 & 3 & -3 \\ 1 & -4 & -2 & 7 \\ 3 & -2 & -1 & 1 \\ -3 & 7 & 1 & -11 \end{pmatrix}$$

as stated in Example 6 of Section 7.2. Using Eq. (2) with the four possible values for i we obtain

$$
\begin{aligned}
i = 1: \ \det M &= 1(-4) + 2(1) + 0(3) + 1(-3) = -5 \\
i = 2: \ \det M &= 2(1) + 2(-4) + 3(-2) + 1(7) = -5 \\
i = 3: \ \det M &= 0(3) + 3(-2) + 1(-1) + 2(1) = -5 \\
i = 4: \ \det M &= 1(-3) + 1(7) + 2(1) + 1(-11) = -5
\end{aligned}
$$

The value obtained is again independent of i.

Equation (2) provides a method for computing the value of the determinant of an arbitrary $p \times p$ matrix. It should be pointed out that while this method leads to the determinant, there are other methods which are computationally more efficient, some of which we shall present later. However, the fact (now to be proved) that the above method leads to a determinant of a matrix provides us with a method for proving the existence of a determinant function. The fact that in the above examples the same value was obtained regardless of which row was chosen for the expansion illustrates (but does not prove) the uniqueness of the determinant.

Exercises

1.   Let $M = \begin{pmatrix} m_{11} & m_{12} \\ m_{21} & m_{22} \end{pmatrix}$

   (a) Find the matrix of minors and the matrix of cofactors of M.

   (b) Apply Eq. (2) with $i = 1$ and then with $i = 2$ to compute det M and compare with Eq. (1).

2.   For each matrix M compute minor M, cofactor M, and det M.

   (a)
   $$M = \begin{pmatrix} 3 & 7 & 4 \\ -1 & 2 & 3 \\ -4 & 0 & -1 \end{pmatrix}$$
   (b) $M = \begin{pmatrix} 0 & 6 & 1 \\ 2 & -1 & -2 \\ 3 & -2 & -4 \end{pmatrix}$

   (c)
   $$M = \begin{pmatrix} 1 & 0 & 2 & 3 \\ 3 & 0 & 1 & -4 \\ -5 & 3 & -2 & 2 \\ 1 & 2 & 0 & 3 \end{pmatrix}$$

3.   Let

   $$M = \begin{pmatrix} m_{11} & m_{12} & m_{13} \\ m_{21} & m_{22} & m_{23} \\ m_{31} & m_{32} & m_{33} \end{pmatrix}$$

   (a) Compute the cofactors M^{11}, M^{12}, M^{13}

   (b) Write the expansion of det M in cofactors of the first row, and verify that the result agrees with Exercise 6 of Section 7.1.

7.4   Uniqueness of the determinant

In this section we will prove the uniqueness of the determinant. That is, if $\mathbf{M}_{pp}$ is a space upon which a determinant function det exists, then this is the only determinant function on $\mathbf{M}_{pp}$.

   In Section 7.1 we established the existence of a determinant function on $\mathbf{M}_{22}$. However the proof of existence in higher dimensions will be postponed until the next section.

   Since the proof of the uniqueness theorem is fairly intricate, we will insert between the theorem and its proof an example which is the proof for the special case of 2×2 matrices. It is recommended that the reader work through this example.

THEOREM 7.3   There is at most one determinant on $\mathbf{M}_{pp}$.

Example 1   We have already shown that a determinant exists on $\mathbf{M}_{22}$. We will now show it is unique. Let Δ be any function on $\mathbf{M}_{22}$ satisfying the conditions of Definition 7.1. Property (4) of Definition 7.1 requires $\Delta(I_2) = 1$. Denoting the ith column of I by e_i, that is, $e_1 = \begin{pmatrix} 1 \\ 0 \end{pmatrix}$ and $e_2 = \begin{pmatrix} 0 \\ 1 \end{pmatrix}$, and letting $H = (e_2 \quad e_1) = \begin{pmatrix} 0 & 1 \\ 1 & 0 \end{pmatrix}$, we have, by Corollary 2 of Theorem 7.1, that $\Delta(H) = -1$.

Let M be any matrix in $\mathbf{M}_{22}$ and denote the columns of M by M_i. Then $M_1 = m_{11}e_1 + m_{21}e_2$ and $M_2 = m_{12}e_1 + m_{22}e_2$. Hence

$$
\begin{aligned}
\Delta(M) &= \Delta(m_{11}e_1 + m_{21}e_2 \quad m_{12}e_1 + m_{22}e_2) \\
&= \Delta(m_{11}e_1 \quad m_{12}e_1) + \Delta(m_{21}e_2 \quad m_{12}e_1) + \Delta(m_{11}e_1 \quad m_{22}e_2) \\
&\quad + \Delta(m_{21}e_2 \quad m_{22}e_2)
\end{aligned}
$$

by repeated applications of property (1) of Definition 7.1. Next we have

$$
\begin{aligned}
\Delta(M) &= m_{11}m_{12}\,\Delta(e_1 \quad e_1) + m_{21}m_{12}\,\Delta(e_2 \quad e_1) \\
&\quad + m_{11}m_{22}\,\Delta(e_1 \quad e_2) + m_{21}m_{22}\,\Delta(e_2 \quad e_2)
\end{aligned}
$$

by application of property (2). Then

$$
\Delta(M) = 0 + m_{21}m_{12}\,\Delta(e_2 \quad e_1) + m_{11}m_{22}\,\Delta(e_1 \quad e_2) + 0
$$

by property (3). But $(e_2 \quad e_1)$ is the matrix H. Hence

$$
\Delta(M) = m_{11}m_{22}\,\Delta(I_2) + m_{12}m_{21}\,\Delta(H) = m_{11}m_{22} - m_{12}m_{21}
$$

Thus the value of the determinant is unique; it is, of necessity, $m_{11}m_{22} - m_{12}m_{21}$.

PROOF: This proof is a generalization of Example 1. Consider any $p \times p$ matrix $M = (M_1 \quad M_2 \quad \cdots \quad M_p)$ with typical element m_{ij}. Suppose two functions, Δ_1 and Δ_2, are determinants. By condition (4) of the definition,

$$
\Delta_1(I_p) = \Delta_2(I_p) = 1.
$$

If we let H denote any matrix which is a reordering of the columns of I_p, then $\Delta_1(H) = \Delta_2(H)$, since interchanging two columns of a matrix causes both determinant values to be multiplied by minus one, and H can be obtained from I_p by interchanging columns of I_p.

Next consider

$$
M_i = m_{1i}e_1 + m_{2i}e_2 + \cdots + m_{pi}e_p = \sum_{j=1}^{p} m_{ji}e_j
$$

where e_j is the jth column of I_p. Then

$$\Delta_1(M) = \Delta_1(M_1 \quad M_2 \quad \cdots \quad M_p)$$

$$= \Delta_1\left(\sum_{j_1=1}^{p} m_{j_1 1} e_{j_1} \quad \sum_{j_2=1}^{p} m_{j_2 2} e_{j_2} \quad \cdots \quad \sum_{j_p=1}^{p} m_{j_p p} e_{j_p}\right)$$

$$= \sum_{j_1=1}^{p} \sum_{j_2=1}^{p} \cdots \sum_{j_p=1}^{p} \Delta_1(m_{j_1 1} e_{j_1} \quad m_{j_2 2} e_{j_2} \quad \cdots \quad m_{j_p p} e_{j_p})$$

by repeated applications of property (1), and therefore

$$\Delta_1(M) = \sum_{j_1=1}^{p} \sum_{j_2=1}^{p} \cdots \sum_{j_p=1}^{p} m_{j_1 1} m_{j_2 2} \cdots m_{j_p p} \Delta_1(e_{j_1} \quad e_{j_2} \quad \cdots \quad e_{j_p}) \quad (1)$$

by repeated applications of property (2).
By exactly the same argument we have

$$\Delta_2(M) = \sum_{j_1=1}^{p} \sum_{j_2=1}^{p} \cdots \sum_{j_p=1}^{p} m_{j_1 1} m_{j_2 2} \cdots m_{j_p p} \Delta_2(e_{j_1} \quad e_{j_2} \quad \cdots \quad e_{j_p})$$

By applying property (3) of Definition 7.1, we know that

$$\Delta_1(e_{j_1} \quad e_{j_2} \quad \cdots \quad e_{j_p}) = \Delta_2(e_{j_1} \quad e_{j_2} \quad \cdots \quad e_{j_p}) = 0$$

if any two columns are equal, that is, if $e_{j_k} = e_{j_i}$ for $k \neq i$. Further, if all the columns are different, that is, $e_{j_k} \neq e_{j_i}$ for $k \neq i$, then $(e_{j_1} \quad e_{j_2} \quad \cdots \quad e_{j_p})$ is a matrix of the form H described above and hence

$$\Delta_1(e_{j_1} \quad e_{j_2} \quad \cdots \quad e_{j_p}) = \Delta_2(e_{j_1} \quad e_{j_2} \quad \cdots \quad e_{j_p})$$

Therefore

$$\sum_{j_1=1}^{p} \sum_{j_2=1}^{p} \cdots \sum_{j_p=1}^{p} m_{j_1 1} m_{j_2 2} \cdots m_{j_p p} \Delta_1(e_{j_1} \quad e_{j_2} \quad \cdots \quad e_{j_p})$$

$$= \sum_{j_1=1}^{p} \sum_{j_2=1}^{p} \cdots \sum_{j_p=1}^{p} m_{j_1 1} m_{j_2 2} \cdots m_{j_p p} \Delta_2(e_{j_1} \quad e_{j_2} \quad \cdots \quad e_{j_p})$$

and hence $\Delta_1(M) = \Delta_2(M)$ for any $p \times p$ matrix. Hence the value of the determinant at each M is unique and therefore the determinant function is unique. ∎

Exercises

1. Work through the proof of Theorem 7.3 for the special case $p = 3$.

2. Prove that if M is $p \times q$ and N is such that det MN is defined, then N is $q \times p$.

7.5   Existence of the determinant

So far we have established that unique determinants exist on $\mathbf{M}_{11}$ and $\mathbf{M}_{22}$. (See Exercise 2 of Section 7.1.) Additionally, we know by Theorem 7.3 that if a determinant exists on $\mathbf{M}_{pp}$, $p > 2$, then it is unique.

We asserted in Section 7.3 that the cofactor expansion defines the determinant for any space of square matrices. In this section we will verify this assertion, thus proving the existence of a determinant on $\mathbf{M}_{pp}$ for any $p \geq 1$.

THEOREM 7.4   Let p be any positive integer. A unique determinant exists on $\mathbf{M}_{pp}$. Further, if i is any integer $1 \leq i \leq p$ and $p \geq 2$, the function defined by

$$\det M = \sum_{j=1}^{p} m_{ij}M^{ij} \tag{1}$$

where M^{ij} is the i, j cofactor, is the determinant function.

PROOF: We have already proved the theorem for $p \leq 2$. We now proceed with an inductive argument to show that, for each p Eq. (1) defines a determinant which, by Theorem 7.3, must be unique.

Assume that Eq. (1) defines the unique determinant for matrices of order up to and including $(p - 1) \times (p - 1)$. Let M be a $p \times p$ matrix $M = (M_1 \ M_2 \ \cdots \ M_p)$.

1. It must be shown that if Eq. (1) is adopted as the determinant for $p \times p$ matrices, property (1) will hold. We therefore consider the matrices

$$\begin{aligned} M &= (M_1 \ \cdots \ M_j \ \cdots \ M_p) \\ M^* &= (M_1 \ \cdots \ M_j^* \ \cdots \ M_p) \\ N &= (M_1 \ \cdots \ M_j + M_j^* \ \cdots \ M_p) \end{aligned}$$

which differ only in their jth columns. By Eq. (1),

$$\det N = m_{i1}N^{i1} + m_{i2}N^{i2} + \cdots + (m_{ij} + m_{ij}^*)N^{ij} + \cdots + m_{ip}N^{ip}$$

For $k \neq j$, $N^{ik} = (-1)^{i+k}N_{ik}$ where the minor N_{ik} is the determinant of a $(p - 1) \times (p - 1)$ matrix, one column of which is a vector equal to $M_j + M_j^*$ except that the ith row element is missing. But for a matrix of order $(p - 1) \times (p - 1)$ we know the linearity property (1) holds. Hence for $k \neq j$, $N^{ik} = M^{ik} + M^{*ik}$. For $k = j$, $N^{ij} = M^{ij} = M^{*ij}$. Hence

$$\begin{aligned} \det N &= m_{i1}(M^{i1} + M^{*i1}) + m_{i2}(M^{i2} + M^{*i2}) + \cdots \\ &\quad + (m_{ij} + m_{ij}^*)M^{ij} + \cdots + m_{ip}(M^{ip} + M^{*ip}) \\ &= \det M + \det M^* \end{aligned}$$

Hence this linearity property holds for $p \times p$ matrices.

2. Next, define $N = (M_1 \quad M_2 \quad \cdots \quad cM_j \quad \cdots \quad M_p)$, that is N is the matrix obtained by multiplying the jth column of M by c. By Eq. (1),

$$\det N = m_{i1}N^{i1} + m_{i2}N^{i2} + \cdots + cm_{ij}N^{ij} + \cdots + m_{ip}N^{ip} \qquad (2)$$

For $k \neq j$, N^{ik} is $(-1)^{i+k}$ times the value of the determinant of a matrix of order $(p-1) \times (p-1)$, where this matrix differs from the matrix used to obtain M^{ik} only in that one column of the first is c times that of the second. Hence $N^{ik} = cM^{ik}$ for $k \neq j$. Further $N^{ij} = M^{ij}$. Hence, from Eq. (2) we obtain

$$\det N = m_{i1}cM^{i1} + m_{i2}cM^{i2} + \cdots + cm_{ij}M^{ij} + \cdots + m_{ip}cM^{ip}$$
$$= c \det M$$

3. Let M be a $p \times p$ matrix such that $M_j = M_{j+1}$ for some j, $1 \leq j \leq p-1$. Then

$$\det M = m_{i1}M^{i1} + m_{i2}M^{i2} + \cdots + m_{ij}M^{ij}$$
$$+ m_{ij+1}M^{ij+1} + \cdots + m_{ip}M^{ip} \qquad (3)$$

For $k \neq j$ and $k \neq j+1$, M^{ik} is the determinant of a $(p-1) \times (p-1)$ matrix with two equal columns and hence $M^{ik} = 0$. Since the jth and $(j+1)$st columns of M are equal, the minors $M_{ij} = M_{ij+1}$. Hence

$$M^{ij} = (-1)^{i+j}M_{ij} = -(-1)^{i+j+1}M_{ij+1} = -M^{ij+1}$$

Since $m_{ij} = m_{ij+1}$, Eq. (3) gives

$$\det M = m_{i1} \cdot 0 + m_{i2} \cdot 0 + \cdots + m_{ij}M^{ij} - m_{ij}M^{ij} + \cdots + m_{ip} \cdot 0 = 0$$

4. That $\det I_p = 1$ is left as an exercise and the proof is complete. ∎

COROLLARY 1 Let D be a $p \times p$ diagonal matrix. That is, if $D = (d_{ij})$, then $d_{ij} = 0$ for $i \neq j$. Then $\det D$ is the product of the diagonal elements, that is, $\det D = \prod_{i=1}^{p} d_{ii}$.[3]

PROOF: We use an inductive argument. If $p = 1$, $\det D = d_{11}$. Suppose the corollary true for $(p-1) \times (p-1)$ matrices. Then, expanding $\det D$ about the pth row,

$$\det D = 0 \cdot D^{p1} + 0 \cdot D^{p2} + \cdots + 0 \cdot D^{p\,p-1} + d_{pp}D^{pp}$$

---

[3] The symbol $\prod$ is used to denote a product in a manner analogous to the use of the symbol Σ for summation. Hence,

$$\prod_{i=1}^{p} d_{ii} = d_{11} \times d_{22} \times \cdots \times d_{pp}$$

But D^{pp} is the determinant of a $(p - 1) \times (p - 1)$ diagonal matrix and is therefore $\prod_{i=1}^{p-1} d_{ii}$. Hence det $D = \prod_{i=1}^{p} d_{ii}$. ∎

COROLLARY 2   Let T be a $p \times p$ upper triangular matrix. That is, if $T = (t_{ij})$, $t_{ij} = 0$ for $i > j$. Then det $T = \prod_{i=1}^{p} t_{ii}$.

PROOF: The proof is left as an exercise. ∎

Exercises

1.   Prove Corollary 2 to Theorem 7.4. Note that

$$T = \begin{pmatrix} t_{11} & t_{12} & t_{13} & \cdots & t_{1p} \\ 0 & t_{22} & t_{23} & \cdots & t_{2p} \\ 0 & 0 & t_{33} & \cdots & t_{3p} \\ \vdots & \vdots & \vdots & & \vdots \\ 0 & 0 & 0 & \cdots & t_{pp} \end{pmatrix}$$

2.   Show that det $I_p = 1$ when Eq. (1) is used to evaluate the determinant. This verifies condition (4) needed in the proof of Theorem 7.4.

3.   Taking $p = 4$ and $i = 1$, write the matrices N^{ik}, $(k \neq j)$, and N^{ij} in full array form. Verify the relations $N^{ik} = M^{ik} + M^{*ik}$ and $N^{ij} = M^{ij} = M^{*ij}$ used in part (1) of the proof of Theorem 7.4.

4.   Taking $p = 4$, $j = 2$, and $i = 2$, write out in full array form part (3) of the proof of Theorem 7.4.

5.   Evaluate each determinant.

(a) $\begin{vmatrix} 1 & 0 & 0 & \cdots & 0 \\ 0 & 2 & 0 & \cdots & 0 \\ 0 & 0 & 4 & \cdots & 0 \\ \vdots & \vdots & \vdots & & \vdots \\ 0 & 0 & 0 & \cdots & 2^{p-1} \end{vmatrix}$
(b) $\begin{vmatrix} 5 & 7 & 5 & 2 \\ 0 & -2 & e & \pi \\ 0 & 0 & 3 & \sqrt{2} \\ 0 & 0 & 0 & 4 \end{vmatrix}$

7.6   Relating determinants to singularity

In this section we will treat another property of determinants, and then make use of this property in classifying matrices according to the values of their determinants. We begin with a theorem, the proof of which parallels that of Theorem 7.3. We will outline the proof, but leave the details as an exercise for the ambitious reader.

THEOREM 7.5   Let M and N be arbitrary $p \times p$ matrices. Then

$$\det NM = (\det N)(\det M).$$

PROOF: As mentioned above, the proof parallels that of Theorem 7.3. If we denote the columns of N by N_1, N_2, and so on, then by definition of matrix product, the ith column of NM is

$$m_{1i}N_1 + m_{2i}N_2 + \cdots + m_{pi}N_p$$

Therefore

$$\det NM = \sum_{j_1=1}^{p} \sum_{j_2=1}^{p} \cdots \sum_{j_p=1}^{p} m_{j_11}m_{j_22} \cdots m_{j_pp} \det (N_{j_1} \quad N_{j_2} \quad \cdots \quad N_{j_p}) \quad (1)$$

Note that the right-hand side corresponds to Eq. (1) of Section 7.4 with e_{j_k} replaced by N_{j_k}. As argued in Section 7.4, $\det (N_{j_1} \quad N_{j_2} \quad \cdots \quad N_{j_p}) = 0$ whenever some $j_k = j_i$ for $k \neq i$, that is, $\det (N_{j_1} \quad N_{j_2} \quad \cdots \quad N_{j_p}) = 0$ if the same column of N is included more than once.

Next, consider $\det (N_{j_1} \quad N_{j_2} \quad \cdots \quad N_{j_p})$ when $j_k \neq j_i$ when $k \neq i$, that is, each column of $(N_{j_1} \quad N_{j_2} \quad \cdots \quad N_{j_p})$ is a different column of N. (Note that two columns of $(N_{j_1} \quad N_{j_2} \quad \cdots \quad N_{j_p})$ might still be equal if N had two equal columns.) The matrix of which the determinant is being taken is that of N with columns interchanged. But this gives us $\det (N_{j_1} \quad N_{j_2} \quad \cdots \quad N_{j_p}) = \det N$ times plus or minus one, depending upon how many columns of N need interchanging to obtain $(N_{j_1} \quad N_{j_2} \quad \cdots \quad N_{j_p})$ from N. But this is the same as the product $(\det N) \det (e_{j_1} \quad e_{j_2} \quad \cdots \quad e_{j_p})$.

Returning to Eq. (1) we have

$$\det NM = \sum_{j_1=1}^{p} \sum_{j_2=1}^{p} \cdots \sum_{j_p=1}^{p} m_{j_11}m_{j_22} \cdots m_{j_pp}(\det N) \det (e_{j_1} \quad e_{j_2} \quad \cdots \quad e_{j_p})$$

$$= \det N \sum_{j_1=1}^{p} \sum_{j_2=1}^{p} \cdots \sum_{j_p=1}^{p} m_{j_11}m_{j_22} \cdots m_{j_pp} \det (e_{j_1} \quad e_{j_2} \quad \cdots \quad e_{j_p})$$

which by Eq. (1) of Section 7.4 is $(\det N)(\det M)$. ∎

Although the above proof is rather intricate in general it would be worthwhile for the reader to work through the details for the case $p = 2$ or 3.

Example 1   Consider the matrices

$$M = \begin{pmatrix} 2 & 1 & 1 \\ 1 & 2 & 1 \\ 0 & 0 & 3 \end{pmatrix} \quad \text{and} \quad N = \begin{pmatrix} 1 & 4 & 2 \\ -1 & 1 & -2 \\ 2 & 1 & 3 \end{pmatrix}$$

$\det M = 9$, and $\det N = -5$ and hence, by Theorem 7.5, $\det MN = -45$. To verify this, we compute the product

$$MN = \begin{pmatrix} 3 & 10 & 5 \\ 1 & 7 & 1 \\ 6 & 3 & 9 \end{pmatrix}$$

By direct calculation the determinant of this matrix is -45.

Example 2   Consider

$$M = \begin{pmatrix} 1 & 2 & 1 \\ 1 & 1 & 3 \end{pmatrix} \quad \text{and} \quad N = \begin{pmatrix} 1 & 1 \\ 1 & 0 \\ 0 & 1 \end{pmatrix}$$

While det M and det N are undefined since neither M nor N is square, the matrices

$$MN = \begin{pmatrix} 3 & 2 \\ 2 & 4 \end{pmatrix} \quad NM = \begin{pmatrix} 2 & 3 & 4 \\ 1 & 2 & 1 \\ 1 & 1 & 3 \end{pmatrix}$$

are square. Calculation shows that det $MN = 8$ and det $NM = 0$.

The above example leads to some interesting discussion. First, if M and N both have determinants defined, then by Theorem 7.5, det $MN =$ det $NM =$ det M det N. But in those cases where M and N are not square but are such that MN and NM do have determinants, it is not necessarily true that det $MN =$ det NM. In fact, we will be able to show that one of these must always be zero.

We now come to a most important property of determinants.

THEOREM 7.6   Let M be a $p \times p$ matrix. Then det $M = 0$ if and only if M is singular.

PROOF: First suppose M is nonsingular. Then there exists a matrix M^{-1} such that $MM^{-1} = I_p$. Therefore, det $(MM^{-1}) =$ det $I_p = 1$. But det $MM^{-1} =$ det M det M^{-1} and hence, since the product is one, neither factor can be zero. In particular, det $M \neq 0$. Thus the "only if" is proven.

Next, suppose M is singular. Then the columns of M are linearly dependent. Therefore, there is one column, say M_k, which is a linear combination of the remaining. Hence

$$M_k = \sum_{\substack{j=1 \\ j \neq k}}^{p} c_j M_j$$

We therefore write

$$\det M = \det \left(M_1 \quad M_2 \quad \cdots \quad \sum_{\substack{j=1 \\ j \neq k}}^{p} c_j M_j \quad \cdots \quad M_p \right)$$

Applying properties (1) and (2) gives

$$\det M = \sum_{\substack{j=1 \\ j \neq k}}^{p} c_j \det (M_1 \quad M_2 \quad \cdots \quad M_j \quad \cdots \quad M_p)$$

But in each term of the sum, column M_j appears twice, once in the jth position and once in the kth position (note that the sum is over all $j \neq k$), and hence

each determinant in the sum is zero. Thus det $M = 0$ and the theorem is
proved.   ∎

COROLLARY   Let M be a nonsingular matrix. Then det $M^{-1} = 1/\text{det } M$.

PROOF: The proof follows directly from the fact that det M det $M^{-1} = 1$.   ∎

Before closing this section we note that the proof of Theorem 7.6 used the
fact that M is singular only if the columns of M are linearly dependent. No
assumption was made concerning the rows. This is important to keep in mind
because we will make use of this theorem in Section 7.9 to prove that the rows
of a matrix are linearly independent if and only if the columns are linearly
independent.

Exercises

1. Give a direct proof of Theorem 7.5 for $p = 2$ based directly on Eq. (1) of
    Section 7.1.

2. Write a detailed proof of Theorem 7.5 in the case $p = 3$.

3. Compute det M, det N, det NM, and det MN in each case.

    (a)
    $$M = \begin{pmatrix} 1 & -1 & 2 \\ -2 & 3 & -3 \\ 4 & -4 & 5 \end{pmatrix} \qquad N = \begin{pmatrix} 0 & 1 & 2 \\ -1 & 0 & 3 \\ 1 & 1 & 4 \end{pmatrix}$$

    (b)
    $$M = \begin{pmatrix} 1 & -1 & 0 & 1 \\ 0 & 2 & 1 & 2 \\ 0 & 0 & -1 & 3 \\ 0 & 0 & 0 & -2 \end{pmatrix} \qquad N = \begin{pmatrix} 4 & 1 & 0 & 0 \\ 0 & 2 & 0 & 0 \\ 0 & 0 & 3 & 2 \\ 0 & 0 & 0 & 6 \end{pmatrix}$$

4. Find all solutions of the equation det $(AB) = 0$ if
    $$A = \begin{pmatrix} x + 2 & 3x \\ 3 & x + 2 \end{pmatrix} \qquad B = \begin{pmatrix} x & 0 \\ 5 & x + 2 \end{pmatrix}$$

5. For each matrix in Exercise 2 of Section 7.3, determine whether M is singu-
    lar. If nonsingular, compute det M^{-1}.

6. In each case, determine whether M is singular. If nonsingular, compute
    det M^{-1}.

    (a) M is the matrix M in Exercise 3a.

    (b) M is the matrix M in Exercise 3b.

    (c)
    $$M = \begin{pmatrix} 1 & 3 & 0 & 5 \\ 2 & -1 & 2 & -2 \\ 5 & 0 & 3 & -1 \\ -6 & 5 & -4 & 10 \end{pmatrix}$$

7. Let M be any 1×2 matrix and N any 2×1 matrix. Show that det $NM = 0$.

8. Prove that if M is $p \times q$, $q \neq p$, and if det MN is defined, then either det MN or det NM must be zero. [*Hint:* Consider the linear transformations associated with M and N.]

9. Let

$$M = \left(\begin{array}{c|c} A & 0 \\ \hline 0 & B \end{array}\right)$$

Prove that det $M = (\det A)(\det B)$.

7.7 Row and column expansion

In this section we will show that the determinant of a matrix can be obtained by expanding in cofactors of a column as well as of a row. This result will play a central role in showing that row rank equals column rank. We will also use the column expansion in deriving a computational procedure for obtaining the inverses of nonsingular matrices.

Example 1 Let M be the 2×2 matrix

$$M = \begin{pmatrix} m_{11} & m_{12} \\ m_{21} & m_{22} \end{pmatrix}$$

We obtain the transpose of M by interchanging rows and columns, that is, $M^T = \begin{pmatrix} m_{11} & m_{21} \\ m_{12} & m_{22} \end{pmatrix}$. Then det $M = m_{11}m_{22} - m_{21}m_{12} = \det M^T$. Since the determinant of M^T can be obtained by a row expansion on M^T, and the rows of M^T are the columns of M, we see that the determinant of M can be obtained by a column expansion on M.

Example 2 In this example we will illustrate the method of proof of the following theorem. Let M be 3×3, that is,

$$M = \begin{pmatrix} m_{11} & m_{12} & m_{13} \\ m_{21} & m_{22} & m_{23} \\ m_{31} & m_{32} & m_{33} \end{pmatrix}$$

Denote by M_{ij} the i,j minor of M. Then, by Theorem 7.4,

$$\det M = \sum_{j=1}^{3} (-1)^{1+j} M_{1j}m_{1j} = m_{11}M_{11} - m_{12}M_{12} + m_{13}M_{13}$$

Next we expand the determinant M_{1j} by a column expansion, which we know we can do by Example 1 (M_{1j} is the determinant of a 2×2 matrix). Hence

$$
\begin{aligned}
\det M &= m_{11}(m_{22}m_{33} - m_{23}m_{32}) - m_{12}(m_{21}m_{33} - m_{23}m_{31}) \\
&\quad + m_{13}(m_{21}m_{32} - m_{22}m_{31}) \\
&= m_{11}M_{11} - m_{21}(m_{12}m_{33} - m_{32}m_{13}) + m_{31}(m_{12}m_{23} - m_{22}m_{13}) \\
&= m_{11}M_{11} - m_{21}M_{21} + m_{31}M_{31}
\end{aligned}
$$

which is the column expansion of the determinant around the first column of M, or alternatively, the determinant of M^T.

We generalize the above examples in the following theorem.

THEOREM 7.7 Let M be a $p \times p$ matrix and M^T the transpose of M. Then $\det M = \det M^T$, and $\det M$ is given either by row or column expansion. That is, for any integer j, $1 \le j \le p$,

$$
\det M = \sum_{k=1}^{p} m_{kj}M^{kj} \tag{1}
$$

where M^{kj} is the k,j cofactor of M.

PROOF: We have seen in Examples 1 and 2 that the theorem holds for $p = 2$ and 3. We will use an inductive argument. Assume the theorem true for $p - 1$. Let M be a $p \times p$ matrix with M_{ij} the i,j minor. By Theorem 7.4,

$$
\det M = \sum_{j=1}^{p} m_{1j}(-1)^{1+j}M_{1j} = m_{11}M_{11} + \sum_{j=2}^{p} (-1)^{1+j}m_{1j}M_{1j} \tag{2}
$$

For $j > 1$ we expand M_{1j}, the determinant of a $(p-1) \times (p-1)$ matrix, about its first column and denote by N_{1j}^{k1} the determinant of the matrix obtained by omitting the first and kth rows and first and jth columns from M. We obtain

$$
M_{1j} = \sum_{k=2}^{p} m_{k1}(-1)^{k}N_{1j}^{k1}
$$

Substituting into Eq. (2) gives

$$
\begin{aligned}
\det M &= m_{11}M_{11} + \sum_{j=2}^{p} (-1)^{1+j}m_{1j} \sum_{k=2}^{p} (-1)^{k}m_{k1}N_{1j}^{k1} \\
&= m_{11}M_{11} + \sum_{k=2}^{p} (-1)^{k+1}m_{k1} \sum_{j=2}^{p} (-1)^{j}m_{1j}N_{1j}^{k1} \tag{3}
\end{aligned}
$$

But $\sum_{j=2}^{p} (-1)^{j}m_{1j}N_{1j}^{k1}$ is the first row expansion for M_{k1}. Hence, from Eq. (3)

$$
\det M = m_{11}M_{11} + \sum_{k=2}^{p} (-1)^{k+1}m_{k1}M_{k1} = \sum_{k=1}^{p} (-1)^{k+1}m_{k1}M_{k1} \tag{4}
$$

which is the column expansion for det M about the first column. The last sum in Eq. (4) is the row expansion about the first row for M^T. Thus from Eq. (4) we have det $M = $ det M^T, and since expansion about any row of M^T gives the (unique) determinant, it follows that the determinant of M can be obtained by expansion about any column of M, as in Eq. (1).   ∎

Exercises

1. For the given matrix M, write det M as a linear combination of the minors of the first column. Leave the minors in determinant form. Repeat for the second or third columns.

$$M = \begin{pmatrix} 6 & 2 & 0 & -5 \\ 3 & 1 & -5 & 7 \\ 0 & -3 & -4 & 6 \\ 4 & -2 & -1 & 2 \end{pmatrix}$$

2. For the matrix in Exercise 2c of Section 7.3, compute det M by expansion about

  (a) the second row.

  (b) the third column.

  (c) the fourth column.

3. Write out the proof of Theorem 7.7 for the case $p = 4$, writing all the determinants M_{1j}, N_{1j}^{k1}, and so on, in full array.

4. In hand calculation of determinants of small order, it is sometimes expedient to utilize Theorem 7.2 before using the expansion in cofactors. The objective is to replace the given matrix by one having an equal determinant but many zeros in one row or column. For example,

$$\begin{vmatrix} 1 & 2 & -3 \\ -2 & 4 & 6 \\ -1 & 5 & 7 \end{vmatrix} = \begin{vmatrix} 1 & 2 & -3 \\ -2 & 4 & 6 \\ 0 & 7 & 4 \end{vmatrix} = \begin{vmatrix} 1 & 2 & -3 \\ 0 & 8 & 0 \\ 0 & 7 & 4 \end{vmatrix}$$

The first equality is obtained by adding row one to row three, the second by adding twice row one to row two. Expansion of the last determinant by the first column gives

$$1 \cdot \begin{vmatrix} 8 & 0 \\ 7 & 4 \end{vmatrix} - 0 \cdot \begin{vmatrix} 2 & -3 \\ 7 & 4 \end{vmatrix} + 0 \cdot \begin{vmatrix} 2 & -3 \\ 8 & 0 \end{vmatrix} = \begin{vmatrix} 8 & 0 \\ 7 & 4 \end{vmatrix} = 32$$

Using this method compute

  (a) $\begin{vmatrix} 1 & 2 & -2 \\ 2 & 6 & 2 \\ 5 & 7 & 3 \end{vmatrix}$ by getting two zeros in the first row;

(b) $\begin{vmatrix} 3 & 2 & -1 & 4 \\ 1 & 4 & -2 & 0 \\ 0 & 2 & 3 & -2 \\ 5 & 0 & 5 & 1 \end{vmatrix}$ by getting three zeros in the second column (note that a factor 2 can be removed).

5. Show that this determinant vanishes for any a, b, c, d:

$$\begin{vmatrix} a+d & 3a & b+2a & b+d \\ 2b & b+d & c-b & c-d \\ a+c & c-2d & d & a+3d \\ b-d & c-d & a+c & a+b \end{vmatrix}$$

6. Let $M(x)$ be a $p \times p$ matrix with elements which are differentiable functions of x. Let $M_1(x), M_2(x), \cdots, M_p(x)$ denote the columns of $M(x)$. Prove that

$$d[\det M(x)]/dx = \det [M_1'(x) \quad M_2(x) \quad \cdots \quad M_p(x)]$$
$$+ \det [M_1(x) \quad M_2'(x) \quad \cdots \quad M_p(x)]$$
$$+ \cdots + \det [M_1(x) \quad M_2(x) \quad \cdots \quad M_p'(x)]$$

where $M_j'(x)$ is the column vector with elements $dm_{ij}(x)/dx$.
[*Hint:* Use induction beginning with $p = 2$.]

7.8 Calculation of matrix inverse

We are now ready to present a computational procedure for obtaining the inverse of a nonsingular matrix. This procedure, like that of row and column expansions for the computation of determinants, is computationally inefficient. However, it will suffice for our purposes in the next few chapters. A more efficient scheme, based on Gaussian elimination, will be presented later.

DEFINITION 7.4 Let M be a $p \times p$ matrix with cofactors M^{ij}. Let (M^{ij}) be the cofactor matrix of M, where M^{ij} is in the ith row and jth column of (M^{ij}). Then the *adjugate* or *adjoint* of M, denoted adj M, is the transpose of (M^{ij}).

Example 1 Consider again the matrix

$$M = \begin{pmatrix} 3 & 1 & 2 \\ 1 & 1 & 0 \\ 0 & 1 & 2 \end{pmatrix}$$

with cofactor $M = (M^{ij}) = \begin{pmatrix} 2 & -2 & 1 \\ 0 & 6 & -3 \\ -2 & 2 & 2 \end{pmatrix}$. Then

$$\text{adj } M = \begin{pmatrix} 2 & 0 & -2 \\ -2 & 6 & 2 \\ 1 & -3 & 2 \end{pmatrix}$$

LEMMA 7.1 Let M be a $p \times p$ matrix. Let i and j be integers such that $i \neq j$, $1 \leq i \leq p$, $1 \leq j \leq p$. Then

$$\sum_{k=1}^{p} m_{ik} M^{jk} = 0 \quad \text{and} \quad \sum_{k=1}^{p} m_{ki} M^{kj} = 0 \tag{1}$$

PROOF: The sum $S = \sum_{k=1}^{p} m_{ik} M^{jk}$ can be considered to be the cofactor expansion about the jth row of the matrix $\hat{M}$ obtained by replacing the jth row of M by the ith row,

$$\hat{M} = \begin{pmatrix} m_{11} & m_{12} & \cdots & m_{1p} \\ \vdots & \vdots & & \vdots \\ m_{i1} & m_{i2} & \cdots & m_{ip} \\ \vdots & \vdots & & \vdots \\ m_{i1} & m_{i2} & \cdots & m_{ip} \\ \vdots & \vdots & & \vdots \\ m_{p1} & m_{p2} & \cdots & m_{pp} \end{pmatrix} \begin{matrix} \\ \\ \leftarrow i\text{th row} \\ \\ \leftarrow j\text{th row} \\ \\ \\ \end{matrix}$$

since $\hat{M}^{jk} = M^{jk}$. Since $\hat{M}$ has two equal rows, $\hat{M}^T$ has two equal columns and by Corollary 1 to Theorem 7.1, we have

$$S = \det \hat{M} = \det \hat{M}^T = 0$$

A similar argument can be applied to the other sum in Eq. (1). ∎

THEOREM 7.8 Let M be a $p \times p$ nonsingular matrix. The inverse of M, denoted by M^{-1}, is given by

$$M^{-1} = \frac{1}{\det M} \text{adj } M$$

PROOF: Before beginning let us recall that, since M is nonsingular, $\det M \neq 0$ and hence the above division is possible. Also, the matrix adj M is the transpose of the cofactor matrix of M, that is, adj $M = (M^{ij})^T$. The proof now follows by considering the product $M(\text{adj } M/\det M)$ and verifying that this is indeed the identity matrix. Let A denote the adjoint of M. Then, if $N = M(\text{adj } M/\det M)$, we have

$$n_{ii} = \sum_{k=1}^{p} \frac{m_{ik} a_{ki}}{\det M}$$

But the adjoint is the transpose of the cofactor matrix. Hence $a_{ki} = M^{ik}$. Therefore

$$n_{ii} = \sum_{k=1}^{p} \frac{m_{ik} M^{ik}}{\det M}$$

But this sum is the expansion about the ith row of the determinant of M. Hence $n_{ii} = \det M/\det M = 1$. Next consider n_{ij} where $i \neq j$. By the preceding lemma,

$$n_{ij} = \frac{1}{\det M} \sum_{k=1}^{p} m_{ik}a_{kj} = \sum_{k=1}^{p} \frac{m_{ik}M^{jk}}{\det M} = 0 \qquad \blacksquare$$

Example 2 Consider the matrix

$$M = \begin{pmatrix} 1 & 2 & 0 & 1 \\ 2 & 2 & 3 & 1 \\ 0 & 3 & 1 & 2 \\ 1 & 1 & 2 & 1 \end{pmatrix}$$

treated in Examples 3 and 6 of Section 7.2. There we computed the cofactor matrix from which we obtain

$$\text{adj } M = \begin{pmatrix} -4 & 1 & 3 & -3 \\ 1 & -4 & -2 & 7 \\ 3 & -2 & -1 & 1 \\ -3 & 7 & 1 & -11 \end{pmatrix}$$

The determinant of M is -5 and therefore

$$M^{-1} = \begin{pmatrix} \frac{4}{5} & -\frac{1}{5} & -\frac{3}{5} & \frac{3}{5} \\ -\frac{1}{5} & \frac{4}{5} & \frac{2}{5} & -\frac{7}{5} \\ -\frac{3}{5} & \frac{2}{5} & \frac{1}{5} & -\frac{1}{5} \\ \frac{3}{5} & -\frac{7}{5} & -\frac{1}{5} & \frac{11}{5} \end{pmatrix}$$

The reader can verify by computing MM^{-1} that this is indeed the inverse of M.

Exercises

1. Compute adj M and M^{-1} if M is nonsingular and $M = \begin{pmatrix} m_{11} & m_{12} \\ m_{21} & m_{22} \end{pmatrix}$.

2. Compute adj M and M^{-1} for each matrix in Exercise 2 of Section 7.3.

3. Compute A^{-1} if

$$A = \begin{pmatrix} 1 & 0 & 3 & -2 \\ 2 & 3 & 6 & -3 \\ 1 & 26 & 7 & 5 \\ -3 & 5 & -8 & 7 \end{pmatrix}$$

4. Use Theorem 7.8 to prove that $(A^T)^{-1} = (A^{-1})^T$ if A is a nonsingular matrix.

5. Let M be a nonsingular diagonal matrix. What is M^{-1}?

6.   Let M be a matrix with complex elements m_{ij} and let $\overline{M} = (\overline{m}_{ij})$, where the bar denotes the complex conjugate. That is, the entries in $\overline{M}$ are the conjugates of the entries in M. If M is nonsingular, show that the inverse of $\overline{M}$ is the conjugate of the inverse of M, that is, $(\overline{M})^{-1} = (\overline{M^{-1}})$.

7.   Prove that if M is any $p \times p$ matrix, $M(\text{adj } M) = (\det M)I$.

7.9   Matrix rank from determinants

This section is devoted to the use of determinants in computing the rank of a matrix, and to an important theorem concerning rank. In Section 6.6 we defined column rank and row rank of a matrix, and asserted that these were equal. We defined the rank of matrix M, denoted $\rho(M)$, to be this common integer. In this section we will prove the above assertion, using the information we have built up concerning determinants.

The column rank of a $p \times q$ matrix M was defined as the dimension of the space of $q \times 1$ matrices generated by the columns of M. This is equivalent to defining the column rank to be the number of vectors in the largest linearly independent subset of the columns. We now proceed to the basic theorem needed to prove equality of row and column ranks.

THEOREM 7.9   Let M be a $p \times q$ matrix with column rank ρ. Let r denote the order (that is, dimension) of the largest (square) submatrix of M with nonzero determinant. Then $\rho = r$.

PROOF: If M is the zero matrix ($m_{ij} = 0$ for every i and j) then it follows immediately that $\rho = r = 0$. Suppose this is not the case. Then $r > 0$ and we define N as follows. Let

$$N = \begin{pmatrix} m_{11} & m_{12} & \cdots & m_{1r} \\ m_{21} & m_{22} & \cdots & m_{2r} \\ \vdots & \vdots & & \\ m_{r1} & m_{r2} & \cdots & m_{rr} \end{pmatrix}$$

be such that $\det N \neq 0, r \geq 1$, but any $(r + 1) \times (r + 1)$ submatrix of M will have determinant zero. Although N represents the first r rows and columns of M, we have actually chosen this submatrix without loss of generality since interchanging rows and/or columns does not affect the linear dependence relationships. (The verification of this is left as an exercise.) Let $M_1, M_2, \cdots, M_r$ be the first r columns of M, and let $N_1, N_2, \cdots, N_r$ be the first r columns of N. Then $\{M_1, M_2, \cdots, M_r\}$ is independent for if not, then certainly $\{N_1, N_2, \cdots, N_r\}$ would be dependent. This would imply $\det N = 0$ by Theorem 7.6, which contradicts our assumption. Therefore we can conclude that $\rho \geq r$.

If $r = q$, the number of columns of M, then $q = r \leq \rho \leq q$ implies $r = \rho$ and we are through. Therefore suppose $r < q$. We now create a set of matrices by adding certain elements of M to our previously defined N. Let $N[s, t]$ be the matrix

$$N[s, t] = \begin{pmatrix} m_{11} & m_{12} & \cdots & m_{1r} & m_{1t} \\ m_{21} & m_{22} & \cdots & m_{2r} & m_{2t} \\ \vdots & & & \vdots & \vdots \\ m_{r1} & m_{r2} & \cdots & m_{rr} & m_{rt} \\ m_{s1} & m_{s2} & \cdots & m_{sr} & m_{st} \end{pmatrix}$$

That is, $N[s, t]$ is obtained by augmenting N with elements of the sth row and tth column of M. $N[s, t]$ is of dimension $(r + 1) \times (r + 1)$. We add the further restriction that $1 \leq s \leq p$ and $r + 1 \leq t \leq q$. It follows immediately that det $N[s, t] = 0$ since if $s \geq r + 1$, we have $N[s, t]$ a submatrix of M of order $r + 1$ which contradicts the fact that N has the order of the largest submatrix with nonzero determinant. On the other hand, if $s \leq r$, then $N[s, t]$ has two identical rows. Hence, the transpose of $N[s, t]$ will have two identical columns implying $N^T[s, t]$ singular. Therefore det $N^T[s, t] = 0$, and by Theorem 7.7, det $N[s, t] = 0$.

Next, denote by N^{sj} the corresponding cofactor of $N[s, t]$, for $j = 1, 2, \ldots, r, t$. Note that these cofactors are independent of our choice of s. Expanding det $N[s, t]$ about the last row we obtain

$$0 = m_{s1}N^{s1} + m_{s2}N^{s2} + \cdots + m_{sr}N^{sr} + m_{st}N^{st}$$

Since $N^{s1}, N^{s2}, \ldots, N^{sr}, N^{st}$ are independent of our choice of s, call these $c_1, c_2, \ldots, c_r, c_t$. Since N^{st} is just the determinant of N, we know $N^{st} = c_t \neq 0$ for any t or s. Therefore, we have shown that there exist scalars $c_1, c_2, \ldots, c_r, c_t$, not all zero, such that $c_1 m_{s1} + c_2 m_{s2} + \cdots + c_r m_{sr} + c_t m_{st} = 0$ for all s, $1 \leq s \leq p$. Hence $c_1 M_1 + c_2 M_2 + \cdots + c_r M_r + c_t M_t = 0$ where M_j is the jth column of the original M matrix. Hence these $r + 1$ columns are dependent for any choice of t and therefore $\rho \leq r$. Combining this with the inequality $\rho \geq r$ obtained above, we obtain $\rho = r$. ∎

Theorem 7.9 tells us that the column rank of a matrix equals the order of the largest submatrix with nonzero determinant. We now have the tools for the following important theorem.

THEOREM 7.10 Let M be a $p \times q$ matrix. Then the row rank of M is equal to the column rank of M.

PROOF: Let ρ denote the column rank of M. Then there exists a submatrix N of order ρ such that det $N \neq 0$ but any larger order submatrix of M will have zero determinant. Consider M^T. Then N^T is a submatrix of M^T and it must similarly be a largest submatrix of M^T with nonzero determinant, for if not,

N could not be the largest such submatrix of M. That N^T has nonzero determinant follows from Theorem 7.7. Therefore, the column rank of M^T is the order of N^T, that is the order of N, which is p. But the column rank of M^T is the row rank of M and our theorem is proved. ∎

We close this section with an example to illustrate the use of Theorem 7.9 in obtaining the rank of a matrix, and hence of any linear transformation for which the matrix is a representation.

Example 1   Consider

$$M = \begin{pmatrix} 1 & 1 & 0 & 1 \\ 3 & 2 & 1 & 1 \\ 2 & 1 & 1 & 0 \end{pmatrix}$$

Since the determinant is defined only for square matrices and the largest square submatrix is 3×3, the rank of M is at most 3. Denoting the columns of M by M_1, M_2, M_3, and M_4, respectively, we obtain

$$\det (M_1 \ M_2 \ M_3) = \det (M_1 \ M_2 \ M_4) = \det (M_1 \ M_3 \ M_4)$$
$$= \det (M_2 \ M_3 \ M_4) = 0$$

Consider the 2×2 array $\begin{pmatrix} m_{11} & m_{12} \\ m_{21} & m_{22} \end{pmatrix} = \begin{pmatrix} 1 & 1 \\ 3 & 2 \end{pmatrix}$. This array has determinant -1 which implies that $\rho(M) = 2$.

Example 2   Consider

$$M = \begin{pmatrix} 1 & 2 & 1 \\ 2 & 4 & 2 \\ 1 & 3 & 2 \end{pmatrix}$$

Since $\det M = 0$, we conclude $\rho(M) < 3$. If we consider

$$\begin{pmatrix} m_{11} & m_{12} \\ m_{21} & m_{22} \end{pmatrix} = \begin{pmatrix} 1 & 2 \\ 2 & 4 \end{pmatrix}$$

we have again a matrix with determinant zero. But

$$\det \begin{pmatrix} m_{21} & m_{22} \\ m_{31} & m_{32} \end{pmatrix} = \det \begin{pmatrix} 2 & 4 \\ 1 & 3 \end{pmatrix} = 2$$

which indicates $\rho(M) = 2$.

Exercises

1.   Find the rank of each matrix.

(a) $\begin{pmatrix} 1 & -1 & 2 & 1 \\ 2 & 1 & -1 & 1 \\ 3 & 0 & 1 & 2 \end{pmatrix}$ (b) $\begin{pmatrix} 2 & -3 & 1 \\ 1 & 1 & -1 \\ 5 & 1 & 1 \\ 3 & 4 & 0 \end{pmatrix}$

(c)
$$\begin{pmatrix} 1 & 2 & -4 & 0 \\ -1 & -2 & 4 & 0 \\ 3 & 6 & -12 & 0 \\ -2 & -4 & 8 & 0 \end{pmatrix}$$

2. Find the rank of M as a function of x.

$$M = \begin{pmatrix} 2 & 2 & -6 & 8 \\ 3 & 3 & -9 & 8 \\ 1 & 1 & x & 4 \end{pmatrix}$$

3. Verify that the choice of the matrix N in the proof of Theorem 7.9 does not cause a loss of generality. That is, prove that neither ρ nor r is changed if two columns of a matrix are interchanged, or if two rows are interchanged.

4. If M is a $p \times q$ matrix, where $p \le q$, how many $(p - 1) \times (p - 1)$ submatrices are there?

5. Prove that if $r = 0$, then M is the zero matrix. (This is used in the first paragraph in the proof of Theorem 7.9.)

6. Write out a proof of Theorem 7.9 for the special case $p = q = 3, r = 2$. Follow the general lines of the proof in the text, but write explicitly the matrices such as $N[s, t]$.

7.10   Cramer's rule

This section is devoted to a classical method for solving systems of linear algebraic equations. We present Cramer's rule only for completeness and as an application of the determinant function. We will give a more efficient method of solution later in our discussion of Gaussian elimination.

An example of a system of linear algebraic equations is

$$2x_1 + 4x_2 + x_3 = 7$$
$$x_1 - 3x_2 + x_3 = 4$$

This is a system of two equations in three unknowns. Cramer's rule treats systems of p equations (p an arbitrary positive integer) in p unknowns, that is, the same number of equations as unknowns. Such a system can be represented in matrix notation as

$$MX = Y \tag{1}$$

where Y is p by 1, M is $p \times p$, X is p by 1, and where M and Y are given and X is unknown. The problem is to solve for an X satisfying Eq. (1).

Example 1   The system

$$3x_1 + 4x_2 + x_3 = 7$$
$$x_1 - 3x_2 + x_3 = 4$$
$$x_1 \qquad + 5x_3 = 12$$

is of the form of Eq. (1) with

$$M = \begin{pmatrix} 3 & 4 & 1 \\ 1 & -3 & 1 \\ 1 & 0 & 5 \end{pmatrix} \qquad X = \begin{pmatrix} x_1 \\ x_2 \\ x_3 \end{pmatrix} \qquad Y = \begin{pmatrix} 7 \\ 4 \\ 12 \end{pmatrix}$$

In order to apply Cramer's rule to a system given by Eq. (1), one further condition must be imposed upon M, namely, M must be nonsingular. To understand the effect of this last constraint we need only notice that if $MX = Y$ where M is nonsingular, then multiplying both sides of Eq. (1) by M^{-1} gives the unique result $X = M^{-1}Y$. Conversely, if X is defined as $M^{-1}Y$, then clearly $MX = Y$. Recall that M can be thought of as the matrix of a transformation of $\mathbf{K}^p$ onto $\mathbf{K}^p$ assuming M is nonsingular, and hence MX gives the coordinates of the image of X. Since M is nonsingular, there is a unique vector X whose image MX has the given coordinate Y, and this X is just $M^{-1}Y$. We conclude that if $\det M \neq 0$, there is one and only one solution of Eq. (1), $X = M^{-1}Y$.

Cramer's rule is given in the following theorem. It provides a way to compute the solution X without first computing M^{-1}.

THEOREM 7.11 (Cramer's Rule)   Let Y be a column vector in $\mathbf{K}^p$ and let M be a $p \times p$ matrix, $\det M \neq 0$. Then the system of linear equations $MX = Y$ has one and only one solution X in $\mathbf{K}^p$. Moreover, the ith element of X, denoted by x_i, is given by

$$x_i = \frac{\det (M_1 \quad M_2 \quad \cdots \quad M_{i-1} \quad Y \quad M_{i+1} \quad \cdots \quad M_p)}{\det M}, \qquad 1 \leq i \leq p \quad (2)$$

where M_j is the jth column of M.

PROOF: The above discussion has already shown that there is exactly one solution, $X = M^{-1}Y$. All that is left is to prove that Eq. (2) gives this solution.
The system $MX = Y$ can be rewritten as

$$x_1 M_1 + x_2 M_2 + \cdots + x_p M_p = Y$$

Therefore, if X is the solution

$$\det [M_1 \quad M_2 \quad \cdots \quad M_{i-1} \quad Y \quad M_{i+1} \quad \cdots \quad M_p]$$
$$= \det [M_1 \quad M_2 \quad \cdots \quad M_{i-1} \quad (x_1 M_1 + x_2 M_2 + \cdots$$
$$+ x_p M_p) \quad M_{i+1} \quad \cdots \quad M_p]$$

Expanding this determinant according to Property (1) of Definition 7.1 gives

$$
\begin{aligned}
\det [M_1 \quad M_2 \quad \cdots \quad M_{i-1} \quad (x_1 M_1) \quad M_{i+1} \quad \cdots \quad M_p] \\
+ \det [M_1 \quad M_2 \quad \cdots \quad M_{i-1} \quad (x_2 M_2) \quad M_{i+1} \quad \cdots \quad M_p] \\
+ \cdots + \det [M_1 \quad M_2 \quad \cdots \quad M_{i-1} \quad (x_i M_i) \quad M_{i+1} \quad \cdots \quad M_p] \\
+ \cdots + \det [M_1 \quad M_2 \quad \cdots \quad M_{i-1} \quad (x_p M_p) \quad M_{i+1} \quad \cdots \quad M_p]
\end{aligned}
$$

which by condition (3) can be written

$$
\begin{aligned}
x_1 \det [M_1 \quad M_2 \quad \cdots \quad M_{i-1} \quad M_1 \quad M_{i+1} \quad \cdots \quad M_p] + \cdots \\
+ x_i \det [M_1 \quad M_2 \quad \cdots \quad M_{i-1} \quad M_i \quad M_{i+1} \quad \cdots \quad M_p] + \cdots \\
+ x_p \det [M_1 \quad M_2 \quad \cdots \quad M_{i-1} \quad M_p \quad M_{i+1} \quad \cdots \quad M_p]
\end{aligned}
$$

However $\det [M_1 \, M_2 \cdots M_{i-1} \, M_j \, M_{i+1} \cdots M_p] = 0$ for $j \neq i$ by Corollary 1 of Theorem 7.1. Hence

$$
\begin{aligned}
\det [M_1 \quad M_2 \quad \cdots \quad M_{i-1} \quad Y \quad M_{i+1} \quad \cdots \quad M_p] \\
= x_i \det [M_1 \quad M_2 \quad \cdots \quad M_{i-1} \quad M_i \quad M_{i+1} \quad \cdots \quad M_p] \\
= x_i \det M
\end{aligned}
$$

Dividing through by $\det M$, which is given to be nonzero, verifies that x_i must have the form in Eq. (2). ∎

Example 2 Solve the system of equations

$$
\begin{aligned}
x + 2y + z &= 1 \\
x - y + z &= 3 \\
x + y + z &= 0
\end{aligned}
$$

First note that this system is given by

$$
\begin{pmatrix} 1 & 2 & 1 \\ 1 & -1 & 1 \\ 1 & 1 & 1 \end{pmatrix} \begin{pmatrix} x \\ y \\ z \end{pmatrix} = \begin{pmatrix} 1 \\ 3 \\ 0 \end{pmatrix} \quad \text{where} \quad \det \begin{pmatrix} 1 & 2 & 1 \\ 1 & -1 & 1 \\ 1 & 1 & 1 \end{pmatrix} = 0
$$

Hence the conditions of Cramer's rule are not satisfied and another method of solution must be found. Subtracting twice the first equation from three times the third yields $x - y + z = -2$. Since the second equation requires $x - y + z = 3$, a contradiction occurs and hence there are no values (x, y, z) satisfying all three equations simultaneously.

Example 3 Solve the system of equations

$$
\begin{aligned}
x + 2y + z &= 2 \\
x - y + z &= 1 \\
x + y - z &= 3
\end{aligned}
$$

Again expressing the system as a matrix equation gives equivalently $MX = Y$ where

$$M = \begin{pmatrix} 1 & 2 & 1 \\ 1 & -1 & 1 \\ 1 & 1 & -1 \end{pmatrix} \quad X = \begin{pmatrix} x \\ y \\ z \end{pmatrix} \quad \text{and} \quad Y = \begin{pmatrix} 2 \\ 1 \\ 3 \end{pmatrix}$$

Since $\det M = 6$, we can solve for (x, y, z). By Cramer's rule

$$x = \frac{1}{6} \det \begin{pmatrix} 2 & 2 & 1 \\ 1 & -1 & 1 \\ 3 & 1 & -1 \end{pmatrix} = \frac{12}{6} = 2$$

$$y = \frac{1}{6} \det \begin{pmatrix} 1 & 2 & 1 \\ 1 & 1 & 1 \\ 1 & 3 & -1 \end{pmatrix} = \frac{2}{6} = \frac{1}{3}$$

$$z = \frac{1}{6} \det \begin{pmatrix} 1 & 2 & 2 \\ 1 & -1 & 1 \\ 1 & 1 & 3 \end{pmatrix} = -\frac{4}{6} = -\frac{2}{3}$$

Exercises

1. Solve by Cramer's rule, if a solution exists.

(a) $2x - 3y = 7$
    $3x - 2y = -2$

(b) $2x - y + z = 2$
    $3x - 2y + 2z = 3$
    $x - 4y + z = -2$

(c) $4x + y - z = 3$
    $2y + 5z = 1$
    $-8x + 7z = 5$

(d) $2x + 4y - z = -3$
    $x + y + 2z = 9$
    $3x - 3y - z = 2$

(e) $x_1 + 3x_2 - 2x_3 - 3x_4 = -5$
    $2x_1 + 4x_2 + 3x_3 + x_4 = 7$
    $4x_2 - x_3 - x_4 = 5$
    $x_1 - 2x_2 - 5x_3 = 0$

2. Derive Eq. (2) from the formula $M^{-1} = (\text{adj } M)/\det M$ (see Theorem 7.8).

3. Let A be any $p \times p$ matrix and let

$$e_1 = \begin{pmatrix} 1 \\ 0 \\ \vdots \\ 0 \end{pmatrix}, \quad e_2 = \begin{pmatrix} 0 \\ 1 \\ \vdots \\ 0 \end{pmatrix}, \dots, \quad e_p = \begin{pmatrix} 0 \\ 0 \\ \vdots \\ 1 \end{pmatrix}$$

(a) Show that Ae_j is the jth column of A.

(b) Show that if A is nonsingular, the jth column of A^{-1} is the solution of the system of equations $AX = e_j$. This provides another technique for the calculation of the matrix inverse.

4. By solving the system $AX = e_j$, where e_j is given in Exercise 3, for $j = 1, 2, \ldots$, compute the inverse of each matrix A. Check your answer by computing AA^{-1}.

(a)
$$A = \begin{pmatrix} 2 & -1 & 1 \\ 3 & -2 & 2 \\ 1 & -4 & 1 \end{pmatrix}$$

(b)
$$A = \begin{pmatrix} 2 & 4 & -1 \\ 1 & 1 & 2 \\ 3 & -3 & -1 \end{pmatrix}$$

(c)
$$A = \begin{pmatrix} 1 & 3 & -2 & -3 \\ 2 & 4 & 3 & 1 \\ 0 & 4 & -1 & -1 \\ 1 & -2 & -5 & 0 \end{pmatrix}$$

5. Let $M_1, M_2, \ldots, M_p$, and Y be given $p \times 1$ matrices (column vectors). Discuss the problem of expressing Y as a linear combination, where $x_1, x_2, \ldots, x_p$ are scalars:

$$Y = x_1 M_1 + x_2 M_2 + \cdots + x_p M_p$$

6. Express Y as a linear combination of M_1, M_2, M_3 if

(a)
$$Y = \begin{pmatrix} 0 \\ 1 \\ 2 \end{pmatrix} \quad M_1 = \begin{pmatrix} 2 \\ 1 \\ 3 \end{pmatrix} \quad M_2 = \begin{pmatrix} 4 \\ 1 \\ -3 \end{pmatrix} \quad M_3 = \begin{pmatrix} -1 \\ 2 \\ -1 \end{pmatrix}$$

(b)
$$Y = \begin{pmatrix} 5 \\ 0 \\ 3 \end{pmatrix} \quad M_1 = \begin{pmatrix} 1 \\ 2 \\ 3 \end{pmatrix} \quad M_2 = \begin{pmatrix} 5 \\ -1 \\ 2 \end{pmatrix} \quad M_3 = \begin{pmatrix} 2 \\ 1 \\ 3 \end{pmatrix}$$

7. Let M be a $p \times p$ matrix and consider the system of p linear equations $MX = 0$. Show that if this system has a solution X different from 0, then determinant $M = 0$. (For further results on systems $MX = 0$, see Chapter 10.)

7.11   An alternate definition of determinant

In the previous development of the determinant we have attempted to give a presentation which emphasizes the properties of this function. Some mathematicians prefer an alternate definition and development. We shall outline this development below.[4]

---

[4] For more complete presentations see: I. N. Herstein, *Topics in Algebra*, Waltham, Mass., Blaisdell, 1964, and Katsumi Nomizu, *Fundamentals of Linear Algebra*, New York, McGraw-Hill, 1966.

We begin by treating the determinant of a 2×2 matrix. Consider

$$M = \begin{pmatrix} m_{11} & m_{12} \\ m_{21} & m_{22} \end{pmatrix}$$

Then, as we have seen, det $M = m_{11}m_{22} - m_{12}m_{21}$. By knowing what we are seeking, namely det M, we can verify that, at least in the two-dimensional case, we can compute the determinant of M as follows. We first take the product of the diagonal elements of M, that is, $m_{11}m_{22}$. From this we subtract the product of the diagonal elements of the matrix obtained by interchanging the columns of M, which is just the product of the diagonal elements of $\begin{pmatrix} m_{12} & m_{11} \\ m_{22} & m_{21} \end{pmatrix}$.

The above example illustrates the concepts necessary for the general development of *determinants by permutations*.

DEFINITION 7.5   Consider a set of n (distinguishable) elements. Every ordering of these elements is called a *permutation*.

It is known that the number of permutations of n elements is $n!$ (n factorial).

For our purpose we will be considering, as elements, the columns of a matrix. Denote the set of all such permutations by P_n. Also consider the set of the integers $\{1, 2, \ldots, n\}$. To each permutation $(i_1, i_2, \ldots, i_n)$, where i_k is any integer in the set and $i_k \neq i_j$ for $k \neq j$, we can associate the corresponding permutation of the columns of an arbitrary $n \times n$ matrix $M = (M_1 \quad M_2 \quad \cdots \quad M_n)$; that is, the permutation of M associated with $(i_1, i_2, \ldots, i_n)$ is $(M_{i_1} \quad M_{i_2} \quad \cdots \quad M_{i_n})$.

Next we explain the concept of the sign of a permutation. We have seen in Corollary 2 of Theorem 7.1 that interchanging two columns of a matrix has the effect on the determinant of multiplication by minus one. Further, in the example of the two-dimensional determinant above we subtracted $m_{12}m_{21}$ from $m_{11}m_{22}$. In accordance with these observations we have the following definition.

DEFINITION 7.6   Consider an $n \times n$ matrix $M = (M_1 \quad M_2 \quad \cdots \quad M_n)$. The *sign* of the permutation $(M_{i_1} \quad M_{i_2} \quad \cdots \quad M_{i_n})$ on the columns of M is $+1$ if $(M_{i_1} \quad M_{i_2} \quad \cdots \quad M_{i_n})$ is obtained from M by an even number of transpositions (interchanges of two columns) of the columns of M, and -1 if $(M_{i_1} \quad M_{i_2} \quad \cdots \quad M_{i_n})$ is obtained by an odd number of transpositions.

In the example of the two-dimensional determinant, $(M_2 \quad M_1)$ is obtained from $M = (M_1 \quad M_2)$ by one transposition, namely, interchanging column one with column two. As another example, suppose $M = (M_1 \quad M_2 \quad M_3 \quad M_4)$ is 4×4. Then the permutation $(M_2 \quad M_1 \quad M_4 \quad M_3)$ is obtained from M by two transpositions, first interchanging columns one and two, and then interchanging columns three and four. Since this is an even number of transpositions, the sign of the permutation is $+1$. The permutation $(M_3 \quad M_1 \quad M_4 \quad M_2)$ can be obtained from M as follows. First interchange the first and third columns of M, which gives us $(M_3 \quad M_2 \quad M_1 \quad M_4)$. Our second transposition is to interchange the second and third columns of the *resulting matrix* which gives $(M_3 \quad M_1 \quad M_2 \quad M_4)$, and finally, transpose the third and fourth columns of this matrix, giving $(M_3 \quad M_1 \quad M_4 \quad M_2)$ as was desired. Since this was accomplished in three transpositions, the sign of this permutation is -1.

We state the following theorem, needed for our alternative definition of the determinant, without proof.

THEOREM 7.13   The sign of a permutation is unique.

In general, there are many ways in which to pass from $(M_1 \quad M_2 \quad \cdots \quad M_n)$ to a certain permutation $(M_{i_1} \quad M_{i_2} \quad \cdots \quad M_{i_n})$ by a sequence of transpositions. The theorem asserts that no matter how this is done the number of interchanges used for a particular permutation is always even or always odd. We are now in a position to give our alternate definition.

DEFINITION 7.7   Let $M = (M_1 \quad M_2 \quad \cdots \quad M_n)$ be an $n \times n$ matrix. Then

$$\det M = \sum_{P_n} [\text{sign} \, (M_{i_1} \quad M_{i_2} \quad \cdots \quad M_{i_n})][\Delta(M_{i_1} \quad M_{i_2} \quad \cdots \quad M_{i_n})]$$

where $\Delta(M_{i_1} \quad M_{i_2} \quad \cdots \quad M_{i_n})$ denotes the product of the elements on the principal diagonal of $(M_{i_1} \quad M_{i_2} \quad \cdots \quad M_{i_n})$.

Before this can be accepted as a definition of a determinant, it must be shown to be consistent with Definition 7.1. The interested reader should consult the suggested references.

Example 1   Let

$$M = \begin{pmatrix} 2 & 1 & 3 \\ 2 & 3 & 2 \\ 4 & 1 & 5 \end{pmatrix}$$

Since there are 3 columns, there will be $3! = 6$ permutations of these columns. These are listed below together with the product of the corresponding diagonal elements multiplied by the sign of the permutation. The sum of these products will be the determinant.

$$\begin{pmatrix} 2 & 1 & 3 \\ 2 & 3 & 2 \\ 4 & 1 & 5 \end{pmatrix} +30 \qquad \begin{pmatrix} 2 & 3 & 1 \\ 2 & 2 & 3 \\ 4 & 5 & 1 \end{pmatrix} -4$$

$$\begin{pmatrix} 1 & 2 & 3 \\ 3 & 2 & 2 \\ 1 & 4 & 5 \end{pmatrix} -10 \qquad \begin{pmatrix} 1 & 3 & 2 \\ 3 & 2 & 2 \\ 1 & 5 & 4 \end{pmatrix} +8$$

$$\begin{pmatrix} 3 & 1 & 2 \\ 2 & 3 & 2 \\ 5 & 1 & 4 \end{pmatrix} -36 \qquad \begin{pmatrix} 3 & 2 & 1 \\ 2 & 2 & 3 \\ 5 & 4 & 1 \end{pmatrix} +6$$

Hence det $M = 30 - 4 - 10 + 8 - 36 + 6 = -6$.

Exercises

1. List the six permutations of the set $(1, \ 2, \ 3)$ and state which are odd and which are even. Use this to derive the formula:

$$\det M = m_{11}m_{22}m_{33} + m_{12}m_{23}m_{31} + m_{21}m_{32}m_{13}$$
$$- m_{13}m_{22}m_{31} - m_{23}m_{32}m_{11} - m_{12}m_{21}m_{33}$$

where

$$M = \begin{pmatrix} m_{11} & m_{12} & m_{13} \\ m_{21} & m_{22} & m_{23} \\ m_{31} & m_{32} & m_{33} \end{pmatrix}$$

2. List all permutations of the set $(1, \ 2, \ 3, \ 4)$ and give the sign of each.

3. If M is 3×3, det M is the sum of the three products indicated by the three arrows in the following scheme which point downward, less the sum of the three products indicated by the three arrows which point upward. Verify this.

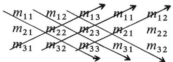

4. Show that if M is 4×4, the scheme analogous to that in Exercise 3 does not give the correct result for det M.

5. If a computer can print out one thousand permutations per second, how long will it take to print all the permutations on ten symbols? on twenty? How long will it take if the computer can output one million permutations per second?

COMMENTARY

The theory of determinants has a venerable past. The interested reader can sample this by looking at:

MUIR, SIR THOMAS, *Theory of Determinants in the Historical Order of Development*, Vols. I–IV, New York, Dover Publications, 1960.

MUIR, SIR THOMAS, *The History of Determinants 1900–1920*, Glasgow, Blackie and Son, 1930.

8

Linear differential equations

8.1 Review

We have now developed an algebraic structure which will aid us in our study of very general types of linear differential equations. To begin, let us review the material in Chapters 1 and 3 on first order linear differential equations and, at the same time, introduce a new perspective afforded by the concepts of linear algebra.

The equations to be considered are of the form

$$y' + p(x)y = 0 \tag{1}$$

and

$$y' + p(x)y = q(x) \tag{2}$$

where $p(x)$ and $q(x)$ are given continuous functions. These equations are called *homogeneous* and *nonhomogeneous* (or inhomogeneous), respectively. As stated in Theorem 3.5, the set of all solutions of the homogeneous equation is given by the formula

$$y(x) = c \exp\left[-\int_{x_0}^{x} p(s) \, ds\right] \tag{3}$$

where c is an arbitrary constant (equal to the value of y at $x = x_0$). The set of all solutions of the nonhomogeneous equation is given by

$$y(x) = c \exp\left[-\int_{x_0}^{x} p(s) \, ds\right]$$

$$+ \exp\left[-\int_{x_0}^{x} p(s) \, ds\right]\left\{\int_{x_0}^{x} q(r) \exp\left[\int_{x_0}^{r} p(s) \, ds\right] dr\right\} \tag{4}$$

From Eq. (3) we see that the solutions of Eq. (1) form a one-dimensional vector space, namely the space generated by

$$\left\{ \exp\left[-\int_{x_0}^{x} p(s)\,ds \right] \right\}$$

That is, the solutions of Eq. (1) are all scalar multiples of a single function. In the remainder of this chapter we will be dealing with linear equations of order n where $n \geq 1$. One of our principal objectives will be to show that the solutions of homogeneous linear differential equations always form a finite-dimensional vector space.

As seen from Eq. (4), all solutions of Eq. (2) are of the form

$$y(x) = y_p(x) + c \exp\left[-\int_{x_0}^{x} p(s)\,ds \right] \tag{5}$$

where $y_p(x)$ is the function

$$y_p(x) = \exp\left[-\int_{x_0}^{x} p(s)\,ds \right] \int_{x_0}^{x} q(r) \exp\left[\int_{x_0}^{r} p(s)\,ds \right] dr$$

These solutions form a family depending on one arbitrary constant c. This family is not a vector space unless it happens to contain the zero function, which can only occur, by Eq. (2), when $q(x) \equiv 0$. We can say that each solution of Eq. (2) is the sum of y_p (a particular solution) and some solution of Eq. (1). This property will remain valid for higher order linear differential equations.

Example 1 Consider the equation

$$y' + x^{-1}y = x \qquad (x > 0)$$

Here $p(x) = x^{-1}$, and (taking $x_0 = 1$)

$$\int_{1}^{x} p(s)\,ds = \int_{1}^{x} \frac{ds}{s} = \ln x$$

The solutions of the homogeneous equation $y' + x^{-1}y = 0$ are of the form

$$y(x) = ce^{-\ln x} = \frac{c}{x}$$

and from Eq. (4), the solutions of the nonhomogeneous equation are

$$y(x) = \frac{c}{x} + e^{-\ln x} \int_{1}^{x} re^{\ln r}\,dr$$

$$= \frac{c}{x} + \left(\frac{x^2}{3} - \frac{1}{3x} \right)$$

or equivalently

$$y(x) = \frac{c'}{x} + \frac{1}{3}x^2 \qquad (x > 0)$$

The function $y_p(x) = x^2/3$ is a particular or fixed solution of the nonhomogeneous equation, and c'/x is an arbitrary solution of the homogeneous equation.[1]

Equation (2) can be rewritten in terms of the operator D for differentiation which has been used often in this text. Thus, we can write Eq. (2) in any of the forms

$$Dy + p(x)y = q(x)$$
$$(D + p(x))y = q(x)$$
$$(D + p(x)I)y = q(x)$$

where I is the identity operator. The expression $D + p(x)I$ is a linear transformation or operator defined on the space of all differentiable functions. We call $D + p(x)I$ a first order linear *differential operator*. This idea can also be extended to higher order operators; many of the general ideas regarding linear transformations on vector spaces can be used in studying these differential operators. In this chapter we propose to use the concepts and theorems on vector spaces and linear transformations to illuminate the problem of solving linear differential equations.

To begin the study of higher order linear differential equations we consider the second order equation

$$y'' + a_1(x)y' + a_0(x)y = q(x) \tag{6}$$

which appears to be a natural generalization of Eq. (2). Here we are thinking of $a_0(x)$, $a_1(x)$, and $q(x)$ as known functions and $y(x)$ as unknown. A vast number of phenomena in the physical, biological, and social sciences lead to equations of precisely this type. Many others lead to equations of order 3, 4, and so on, and so we also need to consider the more general equation

$$y^{(n)} + a_{n-1}(x)y^{(n-1)} + \cdots + a_1(x)y' + a_0(x)y = q(x) \tag{7}$$

We will call Eq. (7) an nth *order linear differential equation*.

Unfortunately, the simple methods of Chapter 1 cannot readily be extended so as to obtain "nice" formulas for the solutions of Eq. (6) or (7) in terms of integrals, as was true for first order equations. In fact, the construction of solutions requires, in general, sophisticated techniques involving infinite series, integral transform methods, or numerical procedures. We will, in due course, introduce many of these methods. But before doing so we shall develop a general theory of homogeneous linear equations. The main objective of this theory is to show that the solutions form a vector space of dimension equal to the order of the equation. Armed with this information we will return later to the problem of constructing a basis for the space of solutions.

---

[1] We could just as well say that $y_p(x) = \frac{1}{3}(x^2 - x^{-1})$ is a particular solution of the nonhomogeneous equation and c/x is an arbitrary solution of the homogeneous equation. The particular solution chosen does not affect the set of all solutions.

8.2   Linear differential operators and their algebra

Various examples in Chapter 5 dealt with differential operators such as $D^2 + 2D + 2I$, $xD + 2I$, and so on, where D represents differentiation. We now wish to give a more thorough discussion of such operators. To begin, we repeat a definition. (See Section 4.5.)

DEFINITION 8.1   Let J denote an interval on the real line. Then $C^n(J)$ denotes the vector space of all real-valued functions f defined on J such that $f, f', \ldots, f^{(n)}$ exist and are continuous on J. Addition and scalar multiplication of functions in $C^n(J)$ are defined in the usual way. If J is an interval (a, b) or $[a, b]$ or $(-\infty, \infty)$, we may often write $C^n(a, b)$ or $C^n[a, b]$ or $C^n(-\infty, \infty)$. We exclude the case in which J consists of a single point or is empty.

Suppose that $a_0(x)$ and $a_1(x)$ are continuous real-valued functions defined on an interval J, and consider the expression

$$y'' + a_1(x)y' + a_0(x)y = (D^2 + a_1(x)D + a_0(x)I)y$$

where I denotes the identity operator. For any function y in $C^2(J)$, this expression is a continuous function. In other words, the expression defines a map of $C^2(J)$ into $C^0(J)$. Under this map, the image of y is $y'' + a_1(x)y' + a_0(x)y$.

Example 1   Taking $a_1(x) = 2x$, $a_0(x) = 1$, the mapping $D^2 + a_1(x)D + a_0(x)I$ is given by $y \to y'' + 2xy' + y$. For example, $y = x^2$ has image $2 + 2x(2x) + x^2 = 2 + 5x^2$.

More generally, suppose that $a_0(x), a_1(x), \ldots, a_n(x)$ are continuous real-valued functions defined on an interval J. Let I denote the identity linear transformation on $C^n(J)$. We can for any y in $C^n(J)$ form $Dy = y'$, $D^2y = y'', \ldots, D^n y = y^{(n)}$. These are functions defined and continuous on J. Furthermore we can form the products $a_1(x)y' = a_1(x)(Dy)$, and so on, and combine these to obtain

$$a_n(x)D^n y + a_{n-1}(x)D^{n-1}y + \cdots + a_1(x)Dy + a_0(x)Iy =$$
$$a_n(x)y^{(n)}(x) + a_{n-1}(x)y^{(n-1)}(x) + \cdots + a_1(x)y'(x) + a_0(x)y(x) \qquad (1)$$

(where for convenience the dependence of the derivatives on x is understood but not written in the first line). It is evident that this combination defines a continuous function on J, for each given y in $C^n(J)$. In other words, it defines a transformation from $C^n(J)$ into $C^0(J)$. We shall denote this transformation by L. Thus L maps $y \in C^n(J)$ into $z \in C^0(J)$ where

$$z(x) = (Ly)(x) = a_n(x)y^{(n)}(x) + a_{n-1}(x)y^{(n-1)}(x) + \cdots$$
$$+ a_1(x)y'(x) + a_0(x)y(x) \qquad (2)$$

or more briefly,

$$Ly = a_n(x)y^{(n)} + a_{n-1}(x)y^{(n-1)} + \cdots + a_1(x)y' + a_0(x)y \qquad (3)$$

It is convenient to write

$$L = a_n(x)D^n + a_{n-1}(x)D^{n-1} + \cdots + a_1(x)D + a_0(x)I \tag{4}$$

and refer to this expression in D as a *differential operator*. If $a_n(x)$ is not identically zero, this operator is said to be of *order n*. It has to be understood that in finding the image of a function y under this operator, one has to compute the derivatives of y first, then multiply by the functions $a_i(x)$, and add, as indicated in Eq. (2) or Eq. (3). For example, the operator $x^2D^2 + 2xD + I$ maps the function x into the function $3x$, the function e^x into the function $(x^2 + 2x + 1)e^x$ and the function y into the function $x^2y''(x) + 2xy'(x) + y(x)$.

Strictly speaking, Eq. (2) is a formula for the value attained by the image Ly of y at the point x. Often this is abbreviated to simply $Ly(x)$, and referred to as the linear transformation. For example, one may speak of the transformation $Ly(x) = x^2y''(x) + 2xy'(x) + y(x)$. To be precise, this is the value at x of the image of y under the transformation. Since such locutions are tedious and unnecessary when one understands the concepts involved, it is customary to use simpler if somewhat less precise language.

For convenience, the identity operator I is often not written explicitly in Eq. (4). For example, $D - 2$ means $D - 2I$.

Perhaps the most important fact about differential operators of the type in Eq. (4) is that they are linear transformations.

Example 2   Let $L = D^2 + 2xD + 1$. Then for any two functions y and z which are twice differentiable we have

$$Ly = y'' + 2xy' + y, \qquad Lz = z'' + 2xz' + z$$

Therefore,

$$\begin{aligned} Ly + Lz &= y'' + z'' + 2x(y' + z') + y + z \\ &= D^2(y + z) + 2xD(y + z) + (y + z) = L(y + z) \end{aligned}$$

Similarly $L(ky) = k(Ly)$ if k is a real scalar. Hence L is a linear transformation.

Example 3   The mapping $y \to (y') + x(y)^2$ takes functions in $C^1(J)$ into continuous functions. For example, the image of e^x is $e^x + xe^{2x}$ and the image of x is $1 + x^3$. However, the image of $e^x + x$ is

$$\frac{d}{dx}(e^x + x) + x(e^x + x)^2 = e^x + 1 + x(e^{2x} + 2xe^x + x^2)$$

and is not the sum of the images of e^x and x. This mapping is not linear (it is called *nonlinear*).

THEOREM 8.1   If $a_0(x), a_1(x), \ldots, a_n(x)$ are continuous real functions on an interval J, the transformation L from $C^n(J)$ into $C^0(J)$ defined by Eq. (4) is a linear transformation over the field **R**.

PROOF: It has to be shown that

$$L(y + z) = Ly + Lz \quad \text{and} \quad L(ky) = kLy$$

if y, z are in $\mathbf{C}^n(J)$ and $k \in \mathbf{R}$. Details are left as an exercise.   ∎

Since the spaces $\mathbf{C}^n(J)$ and $\mathbf{C}^0(J)$ are infinite dimensional, it is not possible to represent L by a finite matrix. In fact, we have not even proved the existence of bases for these two spaces. Fortunately, it happens that we can make great progress in the study of L by other means.

It should be remarked that not every linear transformation from $\mathbf{C}^n(J)$ into $\mathbf{C}^0(J)$ is a linear differential operator. The transformations

$$T: f \to g \quad \text{where} \quad g(x) = \int_0^x f(u)\,du, \quad 0 \le x \le 1$$

$$S: f \to g \quad \text{where} \quad g(x) = f(0), \quad 0 \le x \le 1$$

are examples.

We recall that two transformations S and T are equal on a set C if $Sy = Ty$ for all y in C. The following theorem gives a necessary and sufficient condition for two differential operators to be equal.

THEOREM 8.2   Let

$$L = a_n(x)D^n + \cdots + a_1(x)D + a_0(x)I$$
$$M = b_m(x)D^m + \cdots + b_1(x)D + b_0(x)I$$

where $a_i(x)$ $(0 \le i \le n)$ and $b_i(x)$ $(0 \le i \le m)$ are continuous on an interval J, $a_n(x)$ and $b_m(x)$ not identically zero. Then $Ly = My$ for all y for which Ly and My are defined if, and only if, $m = n$ and $a_i(x) = b_i(x)$ for $0 \le i \le n$, $x \in J$.

PROOF: The proof is left as an exercise.   ∎

As we know, if $\mathbf{V}$ and $\mathbf{V}'$ are vector spaces, linear transformations from $\mathbf{V}$ into $\mathbf{V}'$ can be added and multiplied by scalars, and indeed $\mathscr{L}(\mathbf{V}, \mathbf{V}')$ is itself a vector space. It follows that if K and M are linear differential operators on $\mathbf{C}^n(J)$ and k a real scalar, then $K + M$ and kK are linear transformations on $\mathbf{C}^n(J)$. In fact, $K + M$ and kK are linear differential operators. We illustrate this by an example.

Example 3   Let $K = D^2 + xD + I$, $M = D^3 - xD + 2xI$. Then by definition $(K + M)y = Ky + My$. Thus

$$(K + M)y = (y'' + xy' + y) + (y''' - xy' + 2xy)$$
$$= y''' + y'' + (2x + 1)y$$

Therefore $K + M = D^3 + D^2 + (2x + 1)I$.

In general, if

$$K = \sum_{i=0}^{n} a_i(x)D^i, \qquad M = \sum_{i=0}^{n} b_i(x)D^i \tag{5}$$

then

$$K + M = \sum_{i=0}^{n} (a_i(x) + b_i(x))D^i, \qquad kK = \sum_{i=0}^{n} (ka_i(x))D^i \tag{6}$$

The verification is left to the reader as an exercise.

Since the set of differential operators is closed under addition and scalar multiplication, and contains the zero operator (by taking $a_0 = a_1 = \cdots = a_n = 0$), it is a subspace of $\mathcal{L}(\mathbf{C}^n(J), \mathbf{C}^0(J))$. From the fact that the linear differential operators on $\mathbf{C}^n(J)$ form a vector space, or directly from Eq. (6), it is easy to see that if K, L, M are differential operators on $\mathbf{C}^n(J)$ and k_1, k_2 are scalars, then

$$L + M = M + L \qquad \text{(commutative property)}$$
$$K + (L + M) = (K + L) + M \qquad \text{(associative property)}$$
$$k_1(L + M) = k_1 L + k_1 M \tag{7}$$
$$(k_1 + k_2)L = k_1 L + k_2 L$$
$$(k_1 k_2)L = k_1(k_2 L)$$

Under suitable circumstances it is also possible to multiply linear transformations and therefore, in particular, linear differential operators. In Chapter 5, for example, it was shown that $(D + 2I)(D + I) = D^2 + 3D + 2I$. Before giving a general discussion of multiplication of differential operators, we consider two examples.

Example 4   The operator $D + 2I$ maps $\mathbf{C}^2(J)$ into $\mathbf{C}^1(J)$ (for any J) and the operator D is defined on $\mathbf{C}^1(J)$. Hence, using the linearity of multiplication of linear transformations (that is, $R(S + T) = RS + RT$), we get

$$D(D + 2I) = D^2 + D2I = D^2 + 2D$$

This means, of course, that

$$D(D + 2I)y = y'' + 2y'$$

for y in $\mathbf{C}^2(J)$.

Example 5   Let $L = xD$, $M = D$ be operators on $\mathbf{C}^2(J)$. Then

$$LMy = (xD)Dy = (xD)y' = xy'' = (xD^2)y$$
$$MLy = D(xDy) = D(xy') = xy'' + y' = (xD^2 + D)y$$

Thus $LM = xD^2$ and $ML = xD^2 + D$ are linear differential operators, but $ML \neq LM$.

It can be shown that the product of two linear differential operators of the form (4) is a linear differential operator. As seen in Example 5, multiplication is not commutative, in general. It is, however, associative and bilinear; that is, the equations

$$K(LM) = (KL)M \qquad \text{(associative)}$$

$$\begin{aligned} K(k_1 L + k_2 M) &= k_1 KL + k_2 KM \\ (k_1 K + k_2 L)M &= k_1 KM + k_2 LM \end{aligned} \qquad \text{(bilinear)} \qquad (8)$$

hold whenever the indicated products exist. This was proved in Chapter 5 for linear transformations in general. One of the important things to remember is that differential operators can be handled by the usual rules of algebra except that the order of operations cannot in general be reversed. One of the implications of this is explored in the next example.

Example 6   Consider the operator $L = xD - D$. We have

$$\begin{aligned} L^2 &= (xD - D)(xD - D) = xD(xD - D) - D(xD - D) \\ &= xDxD - xD^2 - D(xD) + D^2 \\ &= x(xD^2 + D) - xD^2 - (xD^2 + D) + D^2 \\ &= x^2 D^2 - 2xD^2 + D^2 + xD - D = (x^2 - 2x + 1)D^2 + (x - 1)D \end{aligned}$$

Note that if we were to expand $(xD - D)^2$ without care we would get $x^2 D^2 - 2xD^2 + D^2$, which is not correct. Another way to perform the calculation correctly is as follows. For any twice differentiable function y we have $Ly = (xD - D)y = xy' - y'$; hence

$$\begin{aligned} L^2 y &= (xD - D)(xD - D)y = (xD - D)(xy' - y') \\ &= x(xy' - y')' - (xy' - y')' \\ &= x(xy'' + y' - y'') - (xy'' + y' - y'') \\ &= (x^2 - 2x + 1)y'' + (x - 1)y' \end{aligned}$$

For differential operators with constant coefficients, multiplication is commutative (see Exercise 12). Consequently, all the usual rules of algebra apply to such operators. In particular, they can be factored in the familiar way.

Exercises

1. Evaluate the image Lf of the given function f:

   (a) $L = D^3 + 2D$, $f(x) = \sin x$
   (b) $L = D^2 - 2D + 4I$, $f(x) = e^{2x}$
   (c) $L = x^2 D^2 - 2xD + I$, $f(x) = \ln x$

2. For the operator $L = xD - xI$, compute the images of x^2 and of e^x and verify that the sum of the images is the image of the sum.

3. Prove Theorem 8.1.

4. Prove that the transformation $Sf = g$ where $g(x) = f(0)$ is not a differential operator. That is, prove that if L is any operator of the type in Eq. (4), $S \neq L$.

5. Prove Theorem 8.2. [*Hint:* Use the fact that $Ly = My$, for $y = 1, x,$ $x^2, \ldots,$ in turn.]

6. Find the indicated operators.

   (a) $L_1 + 2L_2$ if $L_1 = 3xD + 2I$, $L_2 = xD - I$
   (b) $L_1 + L_2$ if $L_1 = xD + I$, $L_2 = x^2D$
   (c) $2L_1 - L_2$ if $L_1 = e^xD^2 + xD$, $L_2 = e^{-x}D^2 - xD$

7. Prove the result in Eq. (6).

8. Find the null space (kernel) of each operator

   (a) $D - I$ (b) $D - x^2$ (c) D^n

9. Write each of the following differential operators in the polynomial form of Eq. (4).

   (a) $(xD^2 + 2D)^2$
   (b) $x^2D(xD - x)$ where $(xD - x)$ is an abbreviation for $(xD - xI)$
   (c) $D(e^xD + 2I) + \ln x$

10. The differential operator D maps $C^1(J)$ into $C^0(J)$. Is this mapping onto? Is it one-to-one? Verify your answers.

11. Let $K = k_1D^m$, $L = k_2D^n$ where k_1 and k_2 are scalars. Prove that $KL = k_1k_2D^{m+n}$.

12. Let L and M be linear differential operators with constant coefficients, that is,

$$L = \sum_{i=0}^{n} a_iD^i, \qquad M = \sum_{i=0}^{n} b_iD^i$$

where a_i and b_i are constants ($0 \leq i \leq n$). Prove that

$$LM = ML = \sum_{i=0}^{2n} c_iD^i \quad \text{where} \quad c_i = \sum_{k=0}^{i} a_kb_{i-k}$$

and where it is assumed that $a_k = 0$ and $b_k = 0$ if $k > n$. Thus such operators can be multiplied or factored as if they were ordinary polynomials.

13. Factor the following differential operators over the complex field:

    (a) $D^3 + 4D^2 - 5$ (b) $D^4 - I$

14. Let L and M be linear differential operators of order one, $L = a_1(x)D + a_0(x)I$, $M = b_1(x)D + b_0(x)I$. Assuming that the coefficients $a_i(x)$ and $b_i(x)$ have derivatives of as high order as necessary, prove that LM is a linear differential operator of order two or less. When will it be less than two?

15. Let K, L, and M be linear differential operators of order one. Give a direct computational proof of Eqs. (7) and (8) for this case.

8.3   Linear differential equations

A linear differential equation is defined to be an equation

$$Ly = f \tag{1}$$

where L is a linear differential operator and f is a function in $C^0(J)$ for an interval J. Thus Eq. (1) has the form

$$a_n(x)y^{(n)}(x) + a_{n-1}(x)y^{(n-1)}(x) + \cdots + a_1(x)y'(x) + a_0(x)y = f(x) \tag{2}$$

or

$$a_n(x)\frac{d^n y}{dx^n} + a_{n-1}(x)\frac{d^{n-1} y}{dx^{n-1}} + \cdots + a_1(x)\frac{dy}{dx} + a_0(x)y = f(x) \tag{3}$$

For example,

$$y'' + (\sin x)y = 0, \qquad x\frac{d^3 y}{dx^3} + x^2\frac{dy}{dx} = e^x$$

are equations of this kind. The basic questions connected with such equations are similar to those encountered previously for first order differential equations. Are there any solutions? How many? How can solutions be described by convenient formulas or in other ways? The foundation for answering these questions is provided by a theorem on existence and uniqueness of solutions which we shall give below. However, we first review the definition of a solution and describe some of the ideas to be developed later.

Recall that the differential operator

$$L = \sum_{i=0}^{n} a_i(x)D^i \tag{4}$$

is said to be of *order n* on the interval J if $a_n(x)$ is not identically zero on J, and the differential equation $Ly = f$ is then said to be of *order n*.

DEFINITION 8.2   Let L be the nth order operator given in Eq. (4) and let $f(x)$ and $a_i(x)$ be continuous on J ($0 \le i \le n$). A *solution* of the nth order differential equation (1) is a function $y(x)$ in $\mathbf{C}^n(J)$ which satisfies the equation; that is, $Ly(x) = f(x)$ for all x in the interval J.

DEFINITION 8.3 Equation (1) is called *homogeneous* if $f(x) \equiv 0$, and *nonhomogeneous* otherwise. The functions $a_i(x)$ $(0 \leq i \leq n)$ are called the *coefficients* in the differential equation.

THEOREM 8.3 The set of all solutions of an nth order homogeneous linear differential equation $Ly = 0$ is a subspace of the vector space $\mathbf{C}^n(J)$.

PROOF: The set of all solutions is the null space of L. Since L is a linear transformation on $\mathbf{C}^n(J)$ into $\mathbf{C}^0(J)$, its null space is a subspace of $\mathbf{C}^n(J)$, by Theorem 5.11.   ∎

The solutions of a nonhomogeneous equation $Ly = f$ do not form a vector space, as was pointed out in Section 8.1 for the first-order case.

From Theorem 8.3 it follows that if $y_1(x)$ and $y_2(x)$ are solutions of $Ly = 0$, then $c_1 y_1(x) + c_2 y_2(x)$ is also a solution. More generally, any finite linear combination of solutions is a solution. This principle of combining or super-imposing solutions is basic to linear mathematical analysis.

Theorem 8.3 can be strengthened provided that the coefficient $a_n(x)$ is never zero on J. In fact, it will be proved below that the space of solutions of $Ly = 0$ is *finite dimensional* and its dimension is equal to the order of the differential operator L. This result implies that if we can find n linearly independent solutions $y_1(x), \ldots, y_n(x)$, where n is the order of L, then $y_1(x), \ldots, y_n(x)$ is a basis for the space of all solutions. Therefore every solution will be of the form $y(x) = c_1 y_1(x) + \cdots + c_n y_n(x)$ for some choice of constants $c_1, \ldots, c_n$.

Example 1 Consider the equation $Ly = 0$ where $L = D^2 - I$. Thus the equation is $y'' - y = 0$. Since L has constant coefficients, the operator can be factored, as shown in Section 8.2. Thus the equation can be written

$$(D - I)(D + I)y = 0 \tag{5}$$

A technique which can be used to solve this equation is as follows. Let $z = (D + I)y$. Then Eq. (5) can be written as

$$(D - I)z = 0 \tag{6}$$

Equation (6) is a first order linear equation and we can use our previous methods to find that z must be of the form $z = ce^x$ where c is any constant. But now we have

$$(D + I)y = z = ce^x \tag{7}$$

This is a linear nonhomogeneous equation for y, and can be solved by our previous methods. For example,

$$(y' + y)e^x = ce^{2x}$$

$$\frac{d}{dx}(ye^x) = ce^{2x}$$

$$ye^x = \frac{c}{2}e^{2x} + c_2$$

$$y = c_1e^x + c_2e^{-x} \tag{8}$$

(where $c_1 = c/2$). It is not hard to show that e^x and e^{-x} are linearly independent functions. All solutions are given by Eq. (8) and the space of all solutions is $[e^x, e^{-x}]$. This, of course, agrees with our assertion that the solution space should be of dimension equal to the order of L. The fact that Eq. (8) gives all solutions is sometimes expressed by saying that the family of all solutions depends on two arbitrary constants. For first order equations, there was one arbitrary constant. This difference can be interpreted in terms of the number of initial conditions required to pick out a unique solution. For example, suppose that for Eq. (5) we impose the auxiliary conditions $y(0) = 0$, $y'(0) = 1$. Then we must have $y(0) = c_1 + c_2 = 0$, $y'(0) = c_1 - c_2 = 1$. Hence $c_1 = \frac{1}{2}$, $c_2 = -\frac{1}{2}$, and $y = (e^x - e^{-x})/2 = \sinh x$.

We shall now give another method for solving Eq. (5). The function $y = c_1e^{-x}$, where c_1 is a constant, is in the null-space of the operator $D + I$. Hence it is in the null space of $(D - I)(D + I)$; therefore it is a solution of Eq. (5). The function $y = c_2e^x$, where c_2 is a constant, is in the null space of the operator $D - I$. Hence it is in the null space of the operator $(D + I)(D - I)$, which equals $(D - I)(D + I)$. Therefore $y = c_2e^x$ is a solution of Eq. (5). By linearity, $y = c_1e^{-x} + c_2e^x$ is a solution. This method has an advantage over the other in that we do not have to solve a nonhomogeneous equation. On the other hand, it is not clear from this method that all solutions have been found.

In later sections, we shall discuss more completely the solution of linear differential equations using both of the methods introduced here.

Exercises

1. Classify each differential equation as linear or nonlinear. For the linear equations, classify as homogeneous or nonhomogeneous and give the order.

(a) $\left(\dfrac{dy}{dx}\right)^3 - x\left(\dfrac{dy}{dx}\right)^2 + ye^{x} = 0$

(b) $3 - y = e^x y' - y''$

(c) $\dfrac{dy}{dx} - e^{x-y} = 1$

(d) $x^2 y'' - (x^2 - 1)y' + y = \sin x$

2. Verify that the given functions form a basis for the space of solutions of the given equation, under the assumption that the dimension of the space equals the order of the equation.

(a) $y'' + y = 0$, $y_1 = \sin x$, $y_2 = \cos x$

(b) $y'' - 2y' + y = 0$, $y_1 = e^x$, $y_2 = xe^x$

(c) $y' - (\cos x)y = 0$, $y_1 = e^{\sin x}$

3. Solve each equation by factoring the operator and proceeding as in Example 1.

(a) $y'' - 2y' - 3y = 0$

(b) $(2D^2 - 3D + 1)y = 0$

4. Find the solution of the given equation satisfying the given auxiliary conditions (see Exercise 2).

(a) $y'' + y = 0$, $y(0) = 0$, $y'(0) = 2$

(b) $y'' - 2y' + y = 0$, $y(0) = 1$, $y'(0) = 2$

(c) $y' - (\cos x)y = 0$, $y\left(\dfrac{\pi}{2}\right) = 1$

8.4   Existence-uniqueness theorems for linear differential equations

We are now ready to state a theorem on the existence and uniqueness of solutions of initial value problems for linear differential equations of order n. As in Section 8.3, we suppose that the *leading coefficient*, or coefficient of the highest order derivative, is nowhere zero on the interval of interest. An equation for which this is true is sometimes said to be in *normal form*.

THEOREM 8.4   Consider the nth order linear differential equation

$$Ly = a_n(x)y^{(n)} + \cdots + a_1(x)y' + a_0(x)y = f(x) \tag{1}$$

where $a_n(x), \ldots, a_0(x), f(x)$ are given functions in $C^0(J)$ and $a_n(x)$ is nonzero at every point of J. Let x_0 be a given number in J, let $z_0, z_1, \ldots, z_{n-1}$ be any real numbers, and consider the initial conditions[2]

$$y(x_0) = z_0, \qquad y'(x_0) = z_1, \qquad \ldots, \qquad y^{(n-1)}(x_0) = z_{n-1} \tag{2}$$

Then there is one and only one function $y(x)$ in $C^n(J)$ which is a solution of the initial value problem (1) and (2).

---

[2] If J is a closed interval and x_0 is an endpoint of J, all derivatives at x_0 must be interpreted as one-sided derivatives from inside J. The theorem is then still correct.

The proof of this theorem can be accomplished by the method of successive approximations in a manner similar to the proof of Theorem 3.3 given in Sections 3.7 and 3.8. It involves the construction of a sequence of functions that can be shown to converge to a function which is a solution of the problem. It is seldom possible to find a simple formula for the limit function. We shall omit the proof of this theorem and instead illustrate it, and compare it with the similar theorems given in Chapter 3.

Example 1   Consider the problem

$$y'' + \frac{x}{1-x} y' + y = 0, \qquad y(2) = 1, \qquad y'(2) = 2$$

The coefficient function $x/(1 - x)$ is continuous everywhere except at $x = 1$. Therefore there is a unique solution $y(x)$ which exists on any interval J which contains $x = 2$ but not $x = 1$. Since the right end of J is arbitrary, we can in fact say that the solution exists on the infinite interval $(1, \infty)$. Later, methods will be given for finding this solution.

If we compare Theorem 8.4 with Theorems 3.2 and 3.3, we find the following. A first difference concerns the type of equation to which the respective theorems are applicable. Theorem 8.4 applies to equations of order n, not just order one. On the other hand, Theorem 8.4 can be applied only to linear equations, whereas the earlier theorems were for arbitrary first order equations $y' = f(x, y)$. Thus Theorem 8.4 says nothing about the existence of solutions of an equation such as $y'' - yy' + xy^2 = 0$. (A theorem which can be applied to such problems is stated in Section 12.5.)

A second difference concerns the conclusion which can be drawn about solutions of the equations. Theorem 8.4 asserts the existence of a solution over the whole interval on which all the coefficient functions $a_i(x)$ are continuous, whereas Theorems 3.2 and 3.3 asserted much less. For example, the solutions of $y'' - y = 0$ found in Section 8.3 exist for all x. The solution of $y' = y^2$, $y(0) = 100$, is $y = 1/(0.01 - x)$ which exists only up to $x = 0.01$, despite the fact that the function y^2 is continuous everywhere. The fact that the interval of existence of a solution of a nonlinear equation is not easy to predict is one of the reasons these equations are much more difficult to analyze than linear ones.

If in Theorem 8.4 we take $z_0 = z_1 = \cdots = z_{n-1} = 0$, we obtain the following important corollary.

COROLLARY   If the differential operator L satisfies the conditions of Theorem 8.4, the unique solution of the problem

$$Ly = 0, \qquad y(x_0) = 0, \qquad y'(x_0) = 0, \qquad \ldots, \qquad y^{(n-1)}(x_0) = 0$$

is the "trivial solution," $y(x) = 0$ for every $x \in J$.

In Theorem 8.4, the hypothesis that the leading coefficient $a_n(x)$ be nonzero everywhere in J is essential. Without it, the theorem is not generally true, as this example shows.

Example 2 We shall show that the initial value problem

$$x^3 y'' - y' + y = 0, \qquad y(0) = 0, \quad y'(0) = 1$$

has no solution. If there were a solution y in $C^2(J)$, where J is some interval surrounding $x = 0$, then $y''(0)$ would have some finite value which we denote by c. The differential equation then gives $0c - 1 + 0 = 0$, which is impossible. Therefore, the initial value problem has no solution. The difficulty is that $a_2(x) = x^3$ vanishes at $x = 0$, the point where the constraining conditions $y = 0$, $y' = 1$ are set. This can be seen in another way. If the equation is written

$$y'' - \frac{1}{x^3} y' + \frac{1}{x^3} y = 0,$$

the coefficient of y'' never vanishes, but the coefficients of y and y' are not continuous throughout any interval J which contains $x = 0$.

The reader may wonder whether there is a convenient geometrical interpretation of Theorem 8.4, as there was for Theorems 3.2 and 3.3. Recall that for a first order equation $y' = f(x, y)$, the equation could be regarded as specifying a direction field, and a solution curve had to have the specified direction at each of its points. The existence theorem told us that through each point (x, y) there is a solution curve, and the uniqueness theorem indicated that through each point there is only one solution curve.

For a higher order equation, this interpretation breaks down. For example, an equation such as $y'' - y = 0$ or $y'' - 3xy' + y = 0$ does not tell us what the slope of a solution should be at an arbitrary point. Rather, if y and y' are known for some x, the equation tells us what y'' must be. It is possible to have many solution curves passing through a single point (x_0, y_0).

Later, when we discuss systems of differential equations, we shall give an alternate geometrical interpretation which is quite helpful.

Exercises

1. Verify that the given functions are solutions of the given differential equation on the specified interval.

(a) $y'' + 4y = 0$; $\sin 2x, \cos 2x, \cos\left(2x + \dfrac{\pi}{8}\right)$;   on $(-\infty, \infty)$

(b) $4x^2 y'' + 4xy' + (x^2 - 1)y = 0$; $(1/\sqrt{x}) \sin \dfrac{x}{2}$;   on $(0, \infty)$

2. For each initial value problem, what information on existence, uniqueness, and interval of existence of solutions is provided by Theorems 3.2, 3.3, or 8.4?

(a) $(x + e^x)y' + 5y = 0, \quad y(0) = 1$

(b) $y' + e^{xy} = 0, \quad y(1) = 3$

(c) $4\sqrt{x - 1}\, y'' - y' + e^x y = 0, \quad y(2) = 0, \quad y'(2) = 1$

(d) $(1 - x^2)y'' + 2xy = 0, \quad y(\tfrac{1}{2}) = y'(\tfrac{1}{2}) = 1$

(e) $y'' + [\sin(ax + b)]y = 0, \quad y(0) = 1, \quad y'(0) = 2$

(f) $y'' + [\cot(\ln x)]y = e^x, \quad y(1) = y'(1) = -2$

3. Prove that the graph of a nontrivial solution of a first order linear equation (1), where $a_1(x)$ is nonzero and $f(x) \equiv 0$, cannot intersect the x-axis.

4. Let $y_1(x)$ and $y_2(x)$ be distinct solutions of a second order linear differential equation (1). Prove that their graphs cannot have a point of tangency.

5. Show that $y(x) \equiv 0$ and $y(x) \equiv x$ are two different solutions of the problem $y'' + xy' - y = 0$, $y(0) = 0$, $y''(0) = 0$. Discuss the relationship of this to Theorem 8.4.

8.5 Dimension of the space of solutions

It was proved in Theorem 8.3 that the set of all real solutions of the nth order linear homogeneous equation

$$Ly = a_n(x)y^{(n)} + \cdots + a_1(x)y' + a_0(x)y = 0 \tag{1}$$

in which $a_n(x), \ldots, a_0(x)$ are real continuous functions on an interval J is a vector space, a subspace of $C^n(J)$.

DEFINITION 8.4 The vector space of all solutions of $Ly = 0$ is called the *solution space* of the differential equation. This is the null space N_L of L.

Theorem 8.4 will now be used to show that if $a_n(x)$ is nonzero, the solution space is of *finite* dimension n, even though $C^n(J)$ itself has infinite dimension.[3] It will be helpful to recall that a set of functions $y_1, y_2, \ldots, y_n$ is linearly independent in $C^n(J)$ if a relation

$$c_1 y_1 + c_2 y_2 + \cdots + c_n y_n = 0 \tag{2}$$

---

[3] The functions $1, x, x^2, x^3$, and so on, are all in C^n and any finite set of them is linearly independent. Hence C^n cannot have finite dimension.

implies that the real scalars $c_1, c_2, \ldots, c_n$ must be all zero. In Eq. (2), of course, the zero is the zero function in $\mathbf{C}^n(J)$, so that Eq. (2) means more explicitly that

$$c_1 y_1(x) + c_2 y_2(x) + \cdots + c_n y_n(x) = 0 \quad \text{for all } x \text{ in } J \tag{3}$$

Example 1   The functions $y_1(x) = x$, $y_2(x) = x^2$ are independent on any interval. For if c_1 and c_2 are not both zero, the equation $c_1 x + c_2 x^2 = 0$ can hold for at most two values of x, $x = 0$ and $x = -c_1/c_2$ if $c_2 \neq 0$, and not for all x in an interval.

THEOREM 8.5   Let

$$Ly = a_n(x)y^{(n)} + \cdots + a_1(x)y' + a_0(x)y$$

where $a_0(x), \ldots, a_n(x)$ are in $\mathbf{C}^0(J)$ and $a_n(x) \neq 0$ on J. The solution space of $Ly = 0$ has dimension n.

PROOF: Let x_0 be a fixed point in J. By Theorem 8.4 there are n functions $y_1, y_2, \ldots, y_n$ which satisfy $Ly = 0$ and which satisfy the respective initial conditions

$$
\begin{aligned}
y_1(x_0) &= 1, & y_1{}'(x_0) &= 0, & \ldots & & y_1^{(n-1)}(x_0) &= 0 \\
y_2(x_0) &= 0, & y_2{}'(x_0) &= 1, & \ldots & & y_2^{(n-1)}(x_0) &= 0 \\
& & & & \ldots & & & \\
y_n(x_0) &= 0, & y_n{}'(x_0) &= 0, & \ldots & & y_n^{(n-1)}(x_0) &= 1
\end{aligned}
\tag{4}
$$

In other words, the vector $(y_i(x_0), y_i{}'(x_0), \ldots, y_i^{(n-1)}(x_0))$ is the ith standard basis vector e_i for $\mathbf{R}^n$ ($1 \leq i \leq n$). We will now show that the set $\{y_1, y_2, \ldots, y_n\}$ is a basis for the solution space $\mathbf{N}_L$.

Suppose that $c_1, \ldots, c_n$ are scalars such that

$$c_1 y_1(x) + c_2 y_2(x) + \cdots + c_n y_n(x) = 0, \qquad x \in J \tag{5}$$

Since each function y_i is in $\mathbf{C}^n(J)$, we can differentiate this equation $n - 1$ times, then put $x = x_0$ and obtain the equations

$$
\begin{aligned}
y_1(x_0)c_1 + y_2(x_0)c_2 + \cdots + y_n(x_0)c_n &= 0 \\
y_1{}'(x_0)c_1 + y_2{}'(x_0)c_2 + \cdots + y_n{}'(x_0)c_n &= 0 \\
\ldots & \\
y_1^{(n-1)}(x_0)c_1 + y_2^{(n-1)}(x_0)c_2 + \cdots + y_n^{(n-1)}(x_0)c_n &= 0
\end{aligned}
\tag{6}
$$

Because of Eqs. (4), these equations reduce to

$$
\begin{aligned}
1c_1 & & &= 0 \\
& 1c_2 & &= 0 \\
& & \ldots & \\
& & 1c_n &= 0
\end{aligned}
$$

and therefore $c_1 = c_2 = \cdots = c_n = 0$. Since Eq. (5) therefore holds only if all c_i are zero, the set of functions $\{y_1, y_2, \ldots, y_n\}$ is linearly independent.

Finally, we shall show that the set $\{y_1, y_2, \ldots, y_n\}$ generates the whole solution space $\mathbf{N}_L$. Let $y(x)$ be an arbitrary function in the solution space. Let x_0 be in J and let $z_0, z_1, \ldots, z_{n-1}$ be the values taken on by $y(x)$ and its derivatives at x_0; that is,

$$y(x_0) = z_0, \qquad y'(x_0) = z_1, \qquad \ldots, \qquad y^{(n-1)}(x_0) = z_{n-1} \qquad (7)$$

Consider the function $\hat{y}(x)$ defined by the equation

$$\hat{y}(x) = z_0 y_1(x) + z_1 y_2(x) + \cdots + z_{n-1} y_n(x), \qquad x \in J \qquad (8)$$

This function is in $\mathbf{N}_L$, since $y_1, y_2, \ldots, y_n$ are. Also it has the same initial values $z_0, z_1, \ldots, z_{n-1}$ as $y(x)$. It follows from the uniqueness part of Theorem 8.4 that $\hat{y}(x) = y(x)$ on J. Therefore, $y(x)$ is a linear combination of $y_1(x), \ldots, y_n(x)$. Since $y(x)$ was an arbitrary solution, this shows that the set $\{y_1, y_2, \ldots, y_n\}$ generates $\mathbf{N}_L$, and consequently is a basis. This completes the proof that there is a basis containing n elements. By the corollary to Theorem 4.11, every basis has n elements. ∎

The practical significance of this theorem is that if we can find n independent solutions of an nth order linear homogeneous differential equation, the equation is completely solved since every solution is a linear combination of these n. To illustrate this we return to the example

$$y'' - y = 0$$

We can verify that $y_1(x) = \cosh x$ satisfies the equation and $y_1(0) = 1$, $y_1'(0) = 0$, and $y_2(x) = \sinh x$ satisfies the equation and $y_2(0) = 0$, $y_2'(0) = 1$. Hence every solution has the form $y(x) = c_1 \cosh x + c_2 \sinh x$ where c_1 and c_2 are scalars. It is customary to refer to this expression as a general solution of the equation, although this term is in disrepute in some quarters because we have heretofore used the term *solution* to refer to a single function, not to a family of functions or as a symbolic expression representing a family of functions. As long as we use it only in relation to linear differential equations of the type we have considered, the danger of confusion is slight and we prefer not to discard such a handy term.

DEFINITION 8.5   Let

$$Ly = a_n(x)y^{(n)} + \cdots + a_1(x)y' + a_0(x)y$$

where $a_0(x), \ldots, a_n(x)$ are in $\mathbf{C}^0(J)$ and $a_n(x) \neq 0$ on J. Let $\{y_1, y_2, \ldots, y_n\}$ be any basis for the solution space of $Ly = 0$. An expression

$$c_1 y_1 + c_2 y_2 + \cdots + c_n y_n,$$

where $c_1, c_2, \ldots, c_n$ are arbitrary or unspecified constants, is called a *general solution* of $Ly = 0$. The term *general solution* is also sometimes used to refer to the solution space itself.

The form of a general solution is not unique. For example, the equation $y'' - y = 0$ has a general solution $y = c_1 \cosh x + c_2 \sinh x$, as shown above. Another general solution is $y = a_1 e^x + a_2 e^{-x}$, since $\{e^x, e^{-x}\}$ is also a basis for the solution space. Accordingly, we shall always refer to *a* general solution rather than *the* general solution.

8.6   Independence and the Wronskian

In Sections 8.4 and 8.5 we combined the existence-uniqueness theorem with fundamental results on vector spaces to show that an nth order linear homogeneous differential equation has a solution space of dimension n. It now seems reasonable to ask how one can go about finding a basis for the solution space, that is, n independent solutions. Note that the proof of Theorem 8.5 depended on the existence of n independent solutions, but no method for actually finding these was given. It turns out that obtaining n linearly independent solutions is often very difficult. As a first step toward resolving this problem, we shall develop a method for deciding whether a given set of solutions, by whatever method found, is linearly independent.

Example 1   It can be verified that the functions $y_1(x) = e^{2x}$, $y_2(x) = e^{-x}$ are solutions of $y'' - y' - 2y = 0$. To test these for linear independence, consider the equation $c_1 e^{2x} + c_2 e^{-x} = 0$. If this equation holds for all x, then by differentiation, the equation $2c_1 e^{2x} - c_2 e^{-x} = 0$ also holds for all x. Setting $x = 0$, we obtain

$$c_1 + c_2 = 0$$
$$2c_1 - c_2 = 0$$

It is easy to see directly that the unique solution of this system is $c_1 = 0$, $c_2 = 0$. Alternatively, we can write the equations as

$$\begin{pmatrix} 1 & 1 \\ 2 & -1 \end{pmatrix}\begin{pmatrix} c_1 \\ c_2 \end{pmatrix} = \begin{pmatrix} 0 \\ 0 \end{pmatrix} \quad \text{or} \quad c_1\begin{pmatrix} 1 \\ 2 \end{pmatrix} + c_2\begin{pmatrix} 1 \\ -1 \end{pmatrix} = \begin{pmatrix} 0 \\ 0 \end{pmatrix}$$

Since the matrix

$$\begin{pmatrix} y_1(0) & y_2(0) \\ y_1'(0) & y_2'(0) \end{pmatrix} = \begin{pmatrix} 1 & 1 \\ 2 & -1 \end{pmatrix}$$

is nonsingular, its columns are linearly independent and the only solution is $c_1 = c_2 = 0$.

In what follows, we develop these ideas into a general method for proving linear independence or dependence of functions. In Example 1, the elements of the matrix are the values of the two functions and their derivatives. Similarly, in the proof of Theorem 8.5, we encountered a set of equations with the coefficient matrix

$$\mathscr{W}(x) = \begin{pmatrix} y_1(x) & y_2(x) & \cdots & y_n(x) \\ y_1'(x) & y_2'(x) & \cdots & y_n'(x) \\ & & \cdots & \\ y_1^{(n-1)}(x) & y_2^{(n-1)}(x) & \cdots & y_n^{(n-1)}(x) \end{pmatrix} \tag{1}$$

Given any functions $y_1, y_2, \ldots, y_n$, in $\mathbf{C}^{n-1}(J)$, we can construct this matrix.

DEFINITION 8.6 The matrix given in Eq. (1) is called the *Wronski matrix* of the functions $y_1, y_2, \ldots, y_n$ in $\mathbf{C}^{n-1}(J)$. The determinant of the Wronski matrix is called the *Wronskian* of these functions and is denoted $W(x)$. It is a function in $\mathbf{C}^0(J)$.

In Example 1, the functions $y_1(x) = e^{2x}$, $y_2(x) = e^{-x}$ have Wronskian

$$W(x) = \begin{vmatrix} e^{2x} & e^{-x} \\ 2e^{2x} & -e^{-x} \end{vmatrix} = -3e^x$$

The principal usefulness of the Wronskian to us is in testing a set of functions for linear independence. Consider the following results.

LEMMA 8.1 Let $y_1, y_2, \ldots, y_n$ be functions in $\mathbf{C}^{n-1}(J)$ and suppose that the vectors

$$Y_i(x_0) = \begin{pmatrix} y_i(x_0) \\ y_i'(x_0) \\ \cdots \\ y_i^{(n-1)}(x_0) \end{pmatrix} \qquad (i = 1, 2, \ldots, n) \tag{2}$$

are linearly independent (in $\mathbf{R}^n$) for some x_0 in J. Then $y_1, y_2, \ldots, y_n$ are linearly independent in $\mathbf{C}^{n-1}(J)$.

PROOF: Suppose $c_1, c_2, \ldots, c_n$ are scalars such that

$$c_1 y_1(x) + c_2 y_2(x) + \cdots + c_n y_n(x) = 0, \qquad \text{for all } x \text{ in } J \tag{3}$$

As in the proof of Theorem 8.5, we can differentiate and deduce Eq. (6) of Section 8.5. In terms of the vectors $Y_1, \ldots, Y_n$ these equations are

$$c_1 Y_1(x_0) + c_2 Y_2(x_0) + \cdots + c_n Y_n(x_0) = 0$$

Since the vectors $Y_1(x_0), Y_2(x_0), \ldots, Y_n(x_0)$ are independent, by hypothesis, it follows that $c_1 = c_2 = \cdots = c_n = 0$. Thus Eq. (3) implies that all scalars are zero, and therefore $\{y_1, y_2, \ldots, y_n\}$ is a linearly independent set. The converse of this lemma is false (see Exercise 1). ∎

THEOREM 8.6   Let $y_1, y_2, \ldots, y_n$ be in $\mathbf{C}^{n-1}(J)$. If the Wronskian of these functions is nonzero for at least one point in J, then $\{y_1, y_2, \ldots, y_n\}$ is a linearly independent set in $\mathbf{C}^{n-1}(J)$.

PROOF: Suppose $W(x_0) \neq 0$ where x_0 is in J. Then the Wronski matrix $\mathscr{W}(x_0)$ is nonsingular and its columns are linearly independent by Theorem 7.6. Thus the vectors $Y_i(x_0)$ of Eq. (2) are independent, and it follows from Lemma 8.1 that $\{y_1, y_2, \ldots, y_n\}$ is independent.   ∎

Example 2   If $y_1(x) = xe^x$, $y_2(x) = \sin x$, $y_3(x) = \cos x$, then

$$W(x) = \begin{vmatrix} xe^x & \sin x & \cos x \\ (x+1)e^x & \cos x & -\sin x \\ (x+2)e^x & -\sin x & -\cos x \end{vmatrix} = -2[1+x]e^x$$

Since $W(x)$ is nonzero for any $x \neq -1$, the functions are linearly independent.

The converse of this theorem is, in general, false. That is, it is possible to find a set of linearly independent functions which have an identically vanishing Wronskian (see Exercise 2). On the other hand, if we consider only functions which are in the null space of a linear differential operator, this behavior cannot occur, as the next theorem will show.

THEOREM 8.7   Let $y_1, y_2, \ldots, y_n$ be solutions of the homogeneous linear differential equation

$$Ly = a_n(x)y^{(n)} + \cdots + a_1(x)y' + a_0(x)y = 0 \tag{4}$$

on an interval J where the coefficients are continuous and $a_n(x)$ is nonzero. Then the following statements are equivalent:

1. The Wronskian $W(x)$ is nonzero for every x in J.
2. There exists a point x_0 in J such that $W(x_0) \neq 0$.
3. $\{y_1, y_2, \ldots, y_n\}$ is a linearly independent set.

PROOF: From Theorem 8.6, we already know that if (2) holds then so does (3); that is, (2) implies (3). Obviously (1) implies (2). We shall now prove that (3) implies (1), thus completing the cycle, (1) implies (2) implies (3) implies (1) and proving equivalence of (1), (2), and (3).

To prove that (3) implies (1), we suppose that $\{y_1, y_2, \ldots, y_n\}$ is a linearly independent set of solutions of $Ly = 0$. We shall prove that the set $\{Y_1, Y_2, \ldots, Y_n\}$ is linearly independent at every x_0 in J where the Y_i are the vectors defined in Lemma 8.1. Suppose the contrary, that is, suppose that the Y_i are dependent at some x_0 in J. Then there are scalars $c_1, c_2, \ldots, c_n$, not all zero, such that

$$c_1 Y_1(x_0) + c_2 Y_2(x_0) + \cdots + c_n Y_n(x_0) = 0$$

Equivalently, Eq. (6) of Section 8.5 holds. Now define the function $y(x)$ by the equation

$$y(x) = c_1 y_1(x) + c_2 y_2(x) + \cdots + c_n y_n(x)$$

Clearly $y(x)$ is a solution of $Ly = 0$. Also by Eq. (6) of Section 8.5

$$y(x_0) = 0, \qquad y'(x_0) = 0, \quad \ldots, \quad y^{(n-1)}(x_0) = 0$$

It follows from the corollary to Theorem 8.4 that $y(x)$ is identically zero on J. Thus

$$c_1 y_1(x) + c_2 y_2(x) + \cdots + c_n y_n(x) = 0, \qquad \text{for all } x \text{ in } J,$$

where not all of $c_1, \ldots, c_n$ are zero. This contradicts the hypothesis that the set $\{y_1, \ldots, y_n\}$ is independent. Thus the set $\{Y_1, Y_2, \ldots, Y_n\}$ must be independent at all x_0 in J, as we set out to prove. Since the Y_i are the columns of $W(x_0)$, the Wronski matrix is nonsingular and the determinant $W(x_0) \neq 0$. This is true for any x_0, and so (1) is proved. ∎

We can summarize our principal results in the following corollary.

COROLLARY   Let $y_1, y_2, \ldots, y_n$ be solutions on J of the nth order linear homogeneous differential equation $Ly = 0$. Then exactly one of these alternatives holds:

1. The Wronskian $W(x)$ vanishes identically on J and $\{y_1, y_2, \ldots, y_n\}$ is linearly dependent.
2. The Wronskian $W(x)$ vanishes at no point on J and $\{y_1, y_2, \ldots, y_n\}$ is linearly independent and is a basis for the solution space.

Example 3   By direct differentiation and substitution it can be shown that $y_1 = \cos^3 x$ and $y_2 = \sec^2 x$ are solutions of $y'' - (\cot x)y' - 6(\tan^2 x)y = 0$ on any interval on which $\tan x$ and $\cot x$ are defined, for example, on $(0, \pi/2)$. The Wronskian of these solutions is

$$W(x) = \begin{vmatrix} \cos^3 x & \sec^2 x \\ -3 \cos^2 x \sin x & 2 \sec^2 x \tan x \end{vmatrix}$$

$$= 2 \cos x \tan x + 3 \sin x = 5 \sin x$$

Therefore $W(x)$ is nonzero on such an interval. It follows that these solutions are linearly independent and form a basis for the solution space of the equation.

Example 4   We reconsider Example 2, in which $W(x) = -2(1 + x)e^x$. Here $W(x_0) \neq 0$ for some x_0 but there is an x for which $W(x) = 0$. This does not contradict Theorem 8.7, but tells us that these functions are not simultaneously solutions of any third order homogeneous differential equation.

Exercises

1. Show that the functions $y_1(x) = x$ and $y_2(x) = x^2$ are linearly independent on $[-1, 1]$, but the vectors $(y_1(0), y_1'(0))$ and $(y_2(0), y_2'(0))$ are linearly dependent in $\mathbf{R}^2$.

2. Show that the functions $y_1(x) = x^3$ and $y_2(x) = |x|^3$ are in $\mathbf{C}^1[-1, 1]$ and are linearly independent on $\mathbf{C}^1[-1, 1]$ but that their Wronskian is identically zero.

3. Using the Wronskians, if possible, test each of these sets of functions for linear independence:

   (a) e^{ax}, e^{bx}, $J = (-\infty, \infty)$
   (b) e^{ax}, $\sin bx$, $J = (-\infty, \infty)$, $(b \neq 0)$
   (c) $e^{ax} \sin bx$, $e^{ax} \cos bx$, $J = (-\infty, \infty)$, $(b \neq 0)$
   (d) x, $\max (x, -x)$, $J = (-1, 1)$
   (e) $\max (x, -x)$, $\min (x, -x)$, $J = (-1, 1)$
   (f) $x^j e^{ax}$, $x^k e^{bx}$, $J = (-\infty, \infty)$ where j, k are nonnegative integers
   (g) e^x, $\sinh x$, $\cosh x$, $J = (0, 2)$

4. Prove that the set $\{1, x, x^2, \ldots, x^n\}$ is linearly independent in $\mathbf{C}^0$ for any interval J.

5. Let f, g be in $\mathbf{C}^1(J)$ and suppose $g(x)$ is never zero on J. Prove that if the Wronskian is identically zero on J, then f and g are linearly dependent. [*Hint:* Calculate $d/dx\ (f(x)/g(x))$.]

6. Prove that the functions $y_1(x) = x$ and $y_2(x) = x^2$ in Exercise 1 cannot be solutions of any linear differential equation $y'' + a_1(x)y' + a_0(x)y = 0$ on $[-1, 1]$ if $a_1(x)$ and $a_0(x)$ are continuous.

7. Let $y_1(x)$ and $y_2(x)$ be solutions of the equation

$$y'' + a_1(x)y' + a_0(x)y = 0$$

on an interval where a_1 and a_0 are continuous. Show that the Wronskian $W(x)$ satisfies the first order differential equation

$$W'(x) + a_1(x)W(x) = 0, \qquad x \text{ in } J$$

and therefore that $W(x) = c \exp [-\int a_1(x)\, dx]$. This is known as Abel's formula. It provides a proof, independent of the one in the text, that either $W(x)$ is identically zero $(c = 0)$ or never zero $(c \neq 0)$.

8. Let $y_1, y_2, \ldots, y_n$ be solutions of Eq. (4) on J, and let $W(x)$ be their

Wronskian. Compute $W'(x)$ and show that $W(x)$ satisfies the first order differential equation

$$a_n(x)W'(x) + a_{n-1}(x)W(x) = 0, \qquad x \text{ in } J$$

[*Hint*: See Exercise 6 of Section 7.7.]

9. Let y_1, y_2 be real functions in $C^2(J)$.

   (a) Show that if

   $$\begin{vmatrix} y_1(x_0) & y_2(x_0) \\ y_1''(x_0) & y_2''(x_0) \end{vmatrix} \neq 0$$

   for one x_0 in J, then y_1, y_2 are independent functions.

   (b) Show that the converse fails, even if y_1 and y_2 are solutions of a second order equation $Ly = 0$. [*Hint*: Consider the independent solutions $\sin x$, $\cos x$ of $y'' + y = 0$.]

10. Let $y_1, \ldots, y_n$ be functions in $C^0[a, b]$, let

    $$g_{ij} = \int_a^b y_i(x)y_j(x) \, dx \qquad (i, j = 1, 2, \ldots, n)$$

    and let G be the $n \times n$ matrix $G = (g_{ij})$. This is called the *Gram* matrix. Prove that $\{y_1, \ldots, y_n\}$ is linearly independent if, and only if, $\det G \neq 0$. [*Hint*: If $\det G \neq 0$, and $c_1 y_1 + \cdots + c_n y_n = 0$ on $[a, b]$, then

    $$\int_a^b (c_1 y_1 + \cdots + c_n y_n)y_j \, dx = 0, \qquad (j = 1, \ldots, n)$$

    Discuss solving for $c_1, \ldots, c_n$.]

11. Prove that if two linearly independent solutions of a linear second order homogeneous equation $y'' + a_1(x)y' + a_0(x)y = 0$ are known, then the coefficients are uniquely determined. Give formulas for $a_1(x)$ and $a_0(x)$. [*Outline of solution*: If ϕ_1 and ϕ_2 are known solutions on an interval J, any other solution y has the form $y = c_1\phi_1 + c_2\phi_2$. Therefore for $x \in J$ the equations

    $$y(x) - \phi_1(x)c_1 - \phi_2(x)c_2 = 0$$
    $$y'(x) - \phi_1'(x)c_1 - \phi_2'(x)c_2 = 0$$
    $$y''(x) - \phi_1''(x)c_1 - \phi_2''(x)c_2 = 0$$

    are satisfied. Therefore, the system of algebraic equations $\mathcal{W}v = 0$, where $\mathcal{W}$ is the Wronskian matrix of $y(x)$, $\phi_1(x)$, and $\phi_2(x)$, has a nontrivial solution

    $$v = \begin{pmatrix} 1 \\ -c_1 \\ -c_2 \end{pmatrix}$$

It follows that $\mathcal{W}$ is singular and det $\mathcal{W} = 0$ (see Exercise 7 in Section 7.10).]

12. Let f_1, f_2, f_3 be twice-differentiable functions on an interval J and let $\mathcal{W}$ be their Wronskian. Let

$$\mathcal{W}_1 = \begin{vmatrix} f_2 & f_3 \\ f_2' & f_3' \end{vmatrix} \qquad \mathcal{W}_2 = \begin{vmatrix} f_1 & f_3 \\ f_1' & f_3' \end{vmatrix} \qquad \mathcal{W}_3 = \begin{vmatrix} f_1 & f_2 \\ f_1' & f_2' \end{vmatrix}$$

Show that if $\mathcal{W}_3$ does not equal zero at any point of J, then

$$\frac{d}{dx}\left(\frac{\mathcal{W}_1}{\mathcal{W}_3}\right) = \frac{f_2 \mathcal{W}}{\mathcal{W}_3^2}, \qquad \frac{d}{dx}\left(\frac{\mathcal{W}_2}{\mathcal{W}_3}\right) = \frac{f_1 \mathcal{W}}{\mathcal{W}_3^2}$$

(See Thomas Muir, *The Theory of Determinants in the Historical Order of Development*, London, Macmillan, Vol. 2, 1906, p. 225.)

13. Using the result of the preceding exercise, prove the following theorem. If the Wronskian $\mathcal{W}$ of f_1, f_2, f_3 vanishes identically on J, while the Wronskian $\mathcal{W}_3$ of f_1, f_2 is nonzero on J, then f_1, f_2, f_3 are linearly dependent on J. (See G. H. Meisters, "Local linear dependence and the vanishing of the Wronskian," *Amer. Math. Monthly* **68,** pp. 847–856 (1961). This theorem was originally proved by M. Bocher in 1900.)

8.7   Using a general solution to solve an initial value problem

Suppose $Ly = 0$ is a linear homogeneous differential equation of order n. Here are two typical problems we may be asked to solve:

1. Find a general solution or a basis for the solution space.
2. Find the unique solution satisfying given initial conditions.

If (1) has been solved, then (2) can be resolved by purely algebraic methods. Consider the following initial value problem.

Example 1 $y'' - y = 0$, $y(1) = 3$, $y'(1) = -1$. We have seen before that a general solution is $y = c_1 e^x + c_2 e^{-x}$. For the initial conditions to be satisfied, it is necessary and sufficient that

$$\begin{matrix} c_1 e + c_2 e^{-1} = 3 \\ c_1 e - c_2 e^{-1} = -1 \end{matrix} \quad \text{or} \quad \begin{pmatrix} e & e^{-1} \\ e & -e^{-1} \end{pmatrix}\begin{pmatrix} c_1 \\ c_2 \end{pmatrix} = \begin{pmatrix} 3 \\ -1 \end{pmatrix}$$

These equations can be solved to yield $c_1 = e^{-1}$, $c_2 = 2e$. Therefore the required solution is $y = e^{x-1} + 2e^{1-x}$.

The procedure illustrated here is always applicable. Let $\{y_1, y_2, \ldots, y_n\}$ be a basis for the solution space and let the initial conditions which must be satisfied be

$$y(x_0) = z_0, \qquad y'(x_0) = z_1, \qquad \ldots, \qquad y^{(n-1)}(x_0) = z_{n-1}$$

Since the unique solution must be a linear combination of any basis, there must be scalars $c_1, c_2, \ldots, c_n$ such that

$$y(x) = c_1 y_1(x) + \cdots + c_n y_n(x)$$

By differentiating $n - 1$ times and then setting $x = x_0$, we obtain the equations

$$
\begin{aligned}
c_1 y_1(x_0) \quad &+ \cdots + c_n y_n(x_0) \quad = z_0 \\
c_1 y_1'(x_0) \quad &+ \cdots + c_n y_n'(x_0) \quad = z_1 \\
&\cdots \\
c_1 y_1^{(n-1)}(x_0) &+ \cdots + c_n y_n^{(n-1)}(x_0) = z_{n-1}
\end{aligned}
\tag{1}
$$

Since $\{y_1, y_2, \ldots, y_n\}$ is a linearly independent set of solutions, it follows from Theorem 8.7 that the Wronskian is nonzero at x_0. Therefore Eqs. (1) have a unique solution (which can be computed by Cramer's rule, for example). *Thus, given a basis for the solution space, the solution satisfying any initial condition can be found by solving a system of n linear algebraic equations.*

Our theory of the homogeneous equation $Ly = 0$ is now in a very satisfactory state. What remains is to find methods for constructing n independent solutions. This will be dealt with in subsequent chapters.

Exercises

For each problem, verify that the given functions are solutions of the given differential equation. Write a general solution and then find the solution which satisfies the given initial conditions.

1. $y'' + y' - 2y = 0$; e^x, e^{-2x}; $y(0) = 2, y'(0) = 0$

2. $y'' - 2y' + y = 0$; e^x, xe^x; $y(0) = 0, y'(0) = 8$

3. $x^2 y'' - x(x + 2)y' + (x + 2)y = 0$; x, xe^x; $y(1) = 1, y'(1) = 2$

4. $xy'' + 2y' + xy = 0$; $x^{-1} \sin x, x^{-1} \cos x$; $y(\pi) = 1, y'(\pi) = 0$

5. $x(x - 1)y'' + [(a + 1)x + 1]y' = 0$; $1, \displaystyle\int_2^x t(t - 1)^{-a-2}\, dt$;

$$y(2) = 1, y'(2) = 4$$

(See E. Kamke, *Differentialgleichungen Lösungsmethoden und Lösungen,* Bd.I, New York, Chelsea, 1948, p. 464.

8.8   Nonhomogeneous linear differential equations

The preceding discussion has shown that the solutions of an nth order homogeneous linear differential equation $Ly = 0$ constitute a vector space of dimension n. In this section we study nonhomogeneous equations of the form

$$Ly = a_n(x)y^{(n)} + a_{n-1}(x)y^{(n-1)} + \cdots + a_0(x)y = f(x) \qquad (1)$$

Here it is assumed that $f, a_0, \ldots, a_n$ are in $C^0(J)$ for some interval J and f is not identically zero. We can solve Eq. (1) by first finding a general solution of the homogeneous equation $Ly = 0$ and then finding just one solution of the equation $Ly = f$. This is shown by the following theorem.

THEOREM 8.8   Assume that $f, a_0, \ldots, a_n$ are in $C^0(J)$ and that $a_n(x)$ is nonzero on J. Let $\{y_1, y_2, \ldots, y_n\}$ be a basis for the solution space of the homogeneous equation $Ly = 0$, and let y_p be any one solution of the nonhomogeneous equation $Ly = f$ on J. Then the set of all solutions of $Ly = f$ on J consists of all functions of the form

$$c_1 y_1 + c_2 y_2 + \cdots + c_n y_n + y_p \qquad (2)$$

where $c_1, c_2, \ldots, c_n$ are arbitrary scalars.

PROOF: First we verify that every function y of the form in (2) is a solution of $Ly = f$. Since L is a linear operator on $C^n(J)$, we have

$$Ly = L\left(\sum_{i=1}^{n} c_i y_i + y_p\right) = \sum_{i=1}^{n} c_i(Ly_i) + Ly_p$$

Since $Ly_i = 0$ and $Ly_p = f$, the above equation reduces to $Ly = f$, as required.

Next we must show that every solution of $Ly = f$ is of the form in (2). Let y be any solution. Then since $Ly = f$, it follows that $L(y - y_p) = 0$. Therefore, $y - y_p$ is a solution of the homogeneous equation, and is therefore a linear combination of the basis $\{y_1, \ldots, y_n\}$. That is, there are scalars $c_1, \ldots, c_n$ such that

$$y - y_p = c_1 y_1 + c_2 y_2 + \cdots + c_n y_n$$

This shows that y is of the form in (2).   ∎

Example 1   Consider the equation $y'' - y = 1 + x$. We have shown by the method of factorization of operators (Section 8.3) that the corresponding homogeneous equation $y'' - y = 0$ has solution space $[e^x, e^{-x}]$. It can be directly verified that $y_p(x) = -(1 + x)$ is one solution of the nonhomogeneous equation. By Eq. (2), all solutions of the nonhomogeneous equation have the form

$$y = c_1 e^x + c_2 e^{-x} - (1 + x)$$

DEFINITION 8.7 An expression of the form (2) is called a *general solution* of the nonhomogeneous linear differential equation (1). The set of all solutions is also called the general solution.

The content of Theorem 8.8 is that a general solution of a nonhomogeneous equation $Ly = f$ can be constructed in two steps. First, construct a general solution of the corresponding homogeneous equation $Ly = 0$; second, add to this any one solution y_p of $Ly = f$. The solution y_p in Eq. (2) is often referred to as a *particular solution* of $Ly = f$. In the next chapter we shall develop systematic methods for constructing the required solutions.

Exercises

1. In each part, verify that the given function is a particular solution of the nonhomogeneous linear differential equation. Then use the solutions of the Exercises in Section 8.7 to write a general solution. Finally, solve the given initial value problem.

    (a) $y'' + y' - 2y = 2 \cos x - 6 \sin x$; $y_p = 2 \sin x$

        Initial value problem: $y(0) = 1$, $y'(0) = 1$

    (b) $y'' - 2y' + y = x^3 - 6x^2 + 6x$; $y_p = x^3$

        Initial value problem: $y(0) = 1$, $y'(0) = 7$

    (c) $x^2y'' - x(x + 2)y' + (x + 2)y = (2 - x)e^x$; $y_p = e^x$

        Initial value problem: $y(1) = 1$, $y'(1) = 2$

    (d) $xy'' + 2y' + xy = 2 \cos x$; $y_p = \sin x$

        Initial value problem: $y(\pi) = 0$, $y'(\pi) = 1$

    (e) $x(x - 1)y'' + (x + 1)y' = x + 1$; $y_p = x + 1$

        Initial value problem: $y(2) = 3$, $y'(2) = 0$

2. Three particular solutions of a second order equation $y'' + a_1(x)y' + a_0(x)y = f(x)$ are given. Find a general solution.

    (a) $e^x - x$, $e^{-x} - x$, $\sinh x - x$

    (b) $1 - \frac{1}{4}x^2$, $2 - \frac{1}{4}x^2$, $e^{2x} - \frac{1}{4}x^2$

8.9 Appanage

The remainder of this chapter will be devoted to an extension of the concept of vector space which may be helpful in the interpretation of Theorem 8.8. This discussion will not, however, be prerequisite to understanding later material in this text.

Recall that in order for a set of vectors to form a vector space the set must contain the zero vector. In Example 3 of Section 4.6 we saw that a plane in $\mathbf{G}^3$ must have an equation of the form $ax + by + cz = 0$ in order to form a subspace of $\mathbf{G}^3$, that is, the plane must contain the point represented by (0, 0, 0). There are many situations where we are not interested in a subspace of a vector space, but in a collection of vectors with many similarities to a subspace but which does not necessarily contain the zero vector. Such a collection entered our discussion of solution spaces of nonhomogeneous linear differential equations.

DEFINITION 8.8 Let $\mathbf{V}$ be a vector space, $\mathbf{U}$ a subspace of $\mathbf{V}$, and v_0 a vector in $\mathbf{V}$. Then the set W of all vectors $w = u + v_0$ for $u \in \mathbf{U}$ is called an *affine subspace* or *affine manifold* of $\mathbf{V}$ through v_0 and parallel to $\mathbf{U}$.

In general an affine subspace will not be a subspace of $\mathbf{V}$. In fact, if W is an affine subspace of $\mathbf{V}$ through v_0 (that is if every element of W can be represented as the sum $v_0 + u$ where u is an element of $\mathbf{U}$), then W will be a subspace of $\mathbf{V}$ if and only if $v_0 \in \mathbf{U}$. However, there are certain similarities between subspaces and affine subspaces which are given in Exercise 1.

Example 1   Consider the vector space $\mathbf{G}^2$ with coordinate representation in $\mathbf{R}^2$ relative to the usual Cartesian coordinate system. It is easy to verify that any line L with coordinates (x, y) satisfying $ax + by = 0$ is a subspace of $\mathbf{G}^2$, and further any line with coordinates (x, y) satisfying $ax + by = c$ is not a subspace when $c \neq 0$. Consider the particular subspace described by $x - 2y = 0$ (see Figure 8.1).

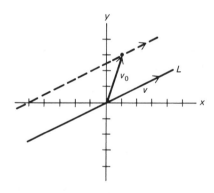

FIGURE 8.1

If we now consider the particular vector v_0 with coordinates (1, 3), we can form the affine subspace through v_0 parallel to L by adding (1, 3) to the coordinates of each vector on the line L to obtain the coordinates of vectors in the affine subspace. Hence (x_0, y_0) are coordinates of a vector in the affine subspace if $(x_0 - 1, y_0 - 3)$ satisfy $(x_0 - 1) - 2(y_0 - 3) = 0$ or $x_0 - 2y_0 = -5$.

We close this section with a theorem which is a generalization of Theorem 8.8. The proof is quite similar to that of Theorem 8.8 and we leave it as an exercise for the reader.

THEOREM 8.9   Let T be a linear transformation of $\mathbf{V} \to \mathbf{V}'$ and let v_0' be an element in the range space of T. Then the set of elements in $\mathbf{V}$ which are mapped by T into v_0' is an affine subspace of $\mathbf{V}$ parallel to the null space of T.

Example 2   We have shown previously that

$$Ly = y'' - y$$

is a linear transformation of $\mathbf{C}^2(J)$ into $\mathbf{C}^0(J)$. Further, the null space of this transformation is the solution space of the corresponding homogeneous differential equation $Ly = y'' - y = 0$. Now consider the nonhomogeneous equation $Ly = y'' - y = 1 + x$. The set of functions mapped by L into the function $1 + x$ is of course the set of all solutions of $Ly = 1 + x$. As shown in Section 8.8, Example 1, a particular solution is $y_p(x) = -(1 + x)$, and a general solution is $c_1 e^x + c_2 e^{-x} - (1 + x)$. This general solution of $Ly = 1 + x$ is the affine subspace of $\mathbf{C}^2(J)$ through $y_p = -(1 + x)$ and parallel to the null space of L, $[e^x, e^{-x}]$.

Exercises

1. Let $\mathbf{V}$ be a vector space over a field $\mathbf{K}$, $\mathbf{U}$ a subspace of $\mathbf{V}$ and v_0 an element of $\mathbf{V}$. Let W be an affine subspace of $\mathbf{V}$ through v_0 and parallel to $\mathbf{U}$. Verify that the following are true:

    (a) W is a subspace if and only if v_0 is an element of $\mathbf{U}$, in which case $W = \mathbf{U}$.

    (b) If w_1 and w_2 are elements of W, then $w_1 + w_2 - v_0$ is in W.

    (c) If w is an element of W and c an element of $\mathbf{K}$, then $cw - (c - 1)v_0$ is an element of W although cw is not necessarily an element of W.

2. Prove that any plane in $\mathbf{G}^3$ is an affine subspace. Show that the affine subspaces of $\mathbf{G}^3$ are points, lines, and planes, as well as the whole space.

3. Prove Theorem 8.9.

COMMENTARY

Section 8.2   There is an ambiguity in the definition of "differential operator" as explained here. The symbol L is used to denote the mapping from $\mathbf{C}^n(J)$ into $\mathbf{C}^0(J)$

defined by Eq. (2). It is also used to denote the expression in Eq. (4) which represents this mapping, which is not the same thing. This is analogous to the fact, known to the reader, that a function is not logically the same as a formula which defines the function. Indeed, different formulas may define the same function, as, for example, x and $(x^2)^{1/2}$ for $x > 0$. For a discussion of the concept of function, see:

MacLane, S., and G. Birkhoff, *Algebra*, New York, Macmillan, 1967.

More advanced treatments of differential operators are given in various books, for example, in:

Lanczos, C., *Linear Differential Operators*, Princeton, N.J., Van Nostrand, 1961.

Section 8.4   For a proof of Theorem 8.4, see:

Coddington, E. A., *An Introduction to Ordinary Differential Equations*, Englewood Cliffs, N.J., Prentice-Hall, 1961, Chapter 6.

Section 8.9   In Definition 8.8, the affine subspace was said to be parallel to U. The reader may notice that no concept of angle has been defined for general vector spaces. This matter will be taken up later. For now we use the term *parallel* only in the sense of Definition 8.8. For further discussion of the concept of affine subspace, refer to:

Nomizu, K., *Fundamentals of Linear Algebra*, New York, McGraw-Hill, 1966.

Chapter 9

Obtaining solutions of linear differential equations

9.1 Introduction

Chapter 8 was primarily concerned with general theoretical results for linear differential equations of arbitrary order of the form

$$Ly = a_n(x)y^{(n)} + a_{n-1}(x)y^{(n-1)} + \cdots + a_0(x)y = f \tag{1}$$

We shall now turn to the practical side of this topic and investigate various methods by which solutions can be constructed or found. Our first objective is a simple and systematic method for finding a basis of the solution space of the homogeneous equation. We know that we can determine a solution satisfying any particular set of initial conditions from such a basis. Our second objective is a similarly simple method for finding one solution of the nonhomogeneous equation. By the results of Section 8.8, we can then construct a general solution of the nonhomogeneous equation.

Unfortunately, these objectives are not easily attained. As a matter of fact, Eq. (1) is so general that no single method can be utilized in dealing with every instance of it, and we shall have to devote several chapters to the exposition of various methods. In this chapter we shall restrict most of our attention to equations with constant coefficients, that is, to Eq. (1) in the case where $a_n, a_{n-1}, \ldots, a_0$ are constants, that is, $(a_0, a_1, \ldots, a_n)$ is a point in $\mathbf{R}^{n+1}$. In this case, at least, we can attain our goal of a fairly simple and systematic procedure.

In subsequent chapters we will take up the important idea of seeking solutions in the form of integrals or infinite series or by means of Laplace transforms.

273

9.2   Complex functions and solutions

In the discussion of linear differential equations with constant coefficients it is necessary to consider complex-valued solutions.[1] To see why this might be so, consider the following example.

Example 1   Consider the equation $y'' + y = 0$, or $(D^2 + I)y = 0$. Since the operator has constant coefficients, it can be factored (as will be justified below) and the equation can be written

$$(D + iI)(D - iI)y = 0$$

where I is the identity operator and $i^2 = -1$. In order to use the method of solution by factorization of operators (see Example 1 of Section 8.3) we now let $z = (D - iI)y$, obtaining $(D + iI)z = 0$, or $z' + iz = 0$. This equation is linear, and our previous procedures lead to the solutions

$$z = ce^{-ix}$$

Clearly, z is complex-valued. Moreover, we have

$$\frac{dy}{dx} - iy = ce^{-ix}$$

$$\frac{d}{dx}(ye^{-ix}) = ce^{-2ix} \tag{1}$$

$$y = c_1 e^{ix} + c_2 e^{-ix}$$

This appears to be a practical method for solving the given equation, provided the manipulations can be rigorously justified.

In order to justify techniques of the type in Example 1, it is necessary to define complex-valued functions of a real variable x and to show how to differentiate and integrate them. In particular, the meaning of e^{ix}, e^{-ix}, and other exponential functions with complex exponents must be made clear. Also, Eq. (1) gives complex-valued solutions of $y'' + y = 0$, but so far no method has been given for deducing real-valued solutions, which was and is our main objective. The path to real solutions will presently be cleared. First, let us discuss complex functions in general.

DEFINITION 9.1   A complex function of a real variable is a mapping f from a subset of **R**, the real field, into **C**, the complex field.

---

[1] The reader is advised to review Appendix A prior to undertaking this section.

Since the value at x of f is a complex number $u + iv$, and u and v depend on x, we can write

$$f(x) = u(x) + iv(x) \qquad (2)$$

where u and v are real-valued functions of x. Thus, every complex function of x is associated with two real functions.

DEFINITION 9.2 If $f(x) = u(x) + iv(x)$, the *real part* of f is $u(x)$ and the *imaginary part* of f is $v(x)$. The symbols Re (f) and Im (f) denote the real and imaginary parts, respectively. Thus Re $(f) = u(x)$ and Im $(f) = v(x)$.

Note the peculiarity of this (commonly used) terminology. The imaginary part of f is real-valued.

Example 2 If $f(x) = \cos x + i \sin x$, then Re $(f) = \cos x$, Im $(f) = \sin x$. The *absolute value* of a complex number $z = a + ib$ is $|z| = (a^2 + b^2)^{1/2}$ and the *argument* is arg $z = \tan^{-1}(b/a)$. Therefore, if f is given by Eq. 2,

$$|f(x)| = \{[u(x)]^2 + [v(x)]^2\}^{1/2}$$

$$\arg [f(x)] = \tan^{-1} \frac{v(x)}{u(x)} \qquad (3)$$

In Example 2, $|f(x)| = 1$ and arg $[f(x)] = \tan^{-1}(\tan x)$. Because of the ambiguity in the inverse tangent function, we can only say that arg $[f(x)] = x + 2k\pi$, where k is an integer.

DEFINITION 9.3 A complex function $f(x) = u(x) + iv(x)$ is said to be *continuous* at x_0 if u and v are both continuous at x_0, and *differentiable* at x_0 if u and v are both differentiable at x_0. The *derivative* $f'(x)$ is defined as

$$f'(x) = u'(x) + iv'(x) \qquad (4)$$

Similarly, the *integral* of f is defined by

$$\int f(x)\, dx = \int u(x)\, dx + i \int v(x)\, dx$$

The definition of the spaces $C^n(J)$ can be extended to complex functions as follows.

DEFINITION 9.4 Let J denote an interval on the real line and let n be a nonnegative integer. Then $C^n(J, C)$ denotes the vector space of all complex-valued functions f defined on J such that $f, f', \ldots, f^{(n)}$ exist and are continuous on J. Addition of functions and multiplication of a function and a (complex) scalar are defined in the same way as before.

We shall now show that much of the theory of linear differential equations in Chapter 8 can be generalized very easily to include complex solutions.

The concept of a linear differential operator, introduced in Section 8.2, extends easily to the complex case. Let $a_0(x), a_1(x), \ldots, a_n(x)$ be continuous real or complex functions on an interval J of the real line. Then we write

$$Ly = a_n(x)y^{(n)} + \cdots + a_1(x)y' + a_0(x)y \tag{5}$$

and we refer to L as an nth order linear differential operator. It is easy to show that L is a linear transformation from $\mathbf{C}^n(J, \mathbf{C})$ into $\mathbf{C}^0(J, \mathbf{C})$ and to verify that the various algebraic properties (associativity, commutativity, and so on) given in Section 8.2 remain valid. In particular, if L has constant coefficients in $\mathbf{C}$, L can be factored by the ordinary procedures of algebra. (See Exercise 15.) The following proposition is new.

THEOREM 9.1   Let L be a linear differential operator with *real-valued* coefficient functions $a_n(x), \ldots, a_0(x)$. Let $y(x) = u(x) + iv(x)$ be a function in $\mathbf{C}^n(J, \mathbf{C})$. Then

$$Ly = Lu + iLv \tag{6}$$

or in other words, Re $(Ly) = Lu$ and Im $(Ly) = Lv$.

PROOF:

$$Ly = \sum_{j=0}^{n} a_j(x)D^j y(x) = \sum_{j=0}^{n} a_j(x)D^j[u(x) + iv(x)]$$

$$= \sum_{j=0}^{n} a_j(x)D^j u(x) + i \sum_{j=0}^{n} a_j(x)D^j v(x) = Lu + iLv \qquad ∎$$

We now consider complex-valued solutions of differential equations.

DEFINITION 9.5   Let L be a linear differential operator of the form in Eq. (5), and let $f \in \mathbf{C}^0(J, \mathbf{C})$. A complex function $y(x)$ in $\mathbf{C}^n(J, \mathbf{C})$ is said to be a *solution* of the equation

$$Ly = f \tag{7}$$

on J if $Ly(x) = f(x)$ for all x in J.

COROLLARY   Let L be a linear differential operator with *real* coefficients $a_n(x), \ldots, a_0(x)$. Let $y(x) = u(x) + iv(x)$ be a solution of the equation $Ly = 0$. Then Re $(y) = u$ and Im $(y) = v$ are also solutions of $Ly = 0$.

PROOF: By Theorem 9.1, $0 = Ly = Lu + iLv$, for x in J. Since a complex number is zero only if its real and imaginary parts are both zero, $Lu = 0$ and $Lv = 0$ for all x in J.   ∎

Example 2   Let $L = D^2 - 2iD$. Then for $y = x + ix^2$ we have $Ly = 4x$. Thus, the complex function $x + ix^2$ is a solution of the equation $y'' - 2iy' = 4x$, which has coefficients in $\mathbf{C}$.

Example 3   Let $L = x^2D^2 - 2xD + 2$, $y = x^2 + ix$. Then y is a solution of the equation

$$(x^2D^2 - 2xD + 2)y = 0$$

Since this equation has real coefficients, it follows from the above corollary that $\operatorname{Re}(y) = x^2$ and $\operatorname{Im}(y) = x$ are solutions of the differential equation.

Theorem 8.3 extends to the complex case; that is, if L is a linear differential operator with real or complex coefficients, the set of all complex solutions of $Ly = 0$ is a subspace of $\mathbf{C}^n(J, \mathbf{C})$. The proof, which is virtually unchanged, is left as an exercise for the reader.

Theorem 8.4 also remains valid in the complex case. That is, an initial value problem consisting of Eq. (7), with $a_n(x) \neq 0$, and initial conditions

$$y(x_0) = z_0, \qquad y'(x_0) = z_1, \qquad \ldots, \qquad y^{(n-1)}(x_0) = z_{n-1}$$

has one and only one solution y in $\mathbf{C}^n(J, \mathbf{C})$. Here, f and the coefficients in L may be complex functions in $\mathbf{C}^0(J, \mathbf{C})$ and $z_0, z_1, \ldots, z_{n-1}$ may be complex numbers. For a proof the reader may consult the reference given at the end of the chapter. The corollary to Theorem 8.4 also remains valid.

Of most importance to us is the fact that Theorem 8.5 remains valid in the complex case. Specifically, we have the following result.

THEOREM 9.2   Let L be an nth order linear differential operator with coefficients in $\mathbf{C}^0(J, \mathbf{C})$ such that $a_n(x) \neq 0$ in J. Then the complex solution space of $Ly = 0$ is an n-dimensional vector space (over $\mathbf{C}$).

PROOF: The proof of this theorem is left as an exercise.   ∎

The theorem is important because it shows that a general solution of a complex differential equation involves n linearly independent complex solutions. (The definition of general solution in Chapter 8 is easy to extend to the complex case.) For instance, we saw in Example 3 that x and x^2 are solutions of the equation $x^2y'' - 2xy' + 2y = 0$. A general solution is $y = c_1x + c_2x^2$. If we want to restrict attention to real solutions, we take c_1 and c_2 in $\mathbf{R}$ and then $c_1x + c_2x^2$ represents all real solutions; if we want to generalize to complex solutions, we take c_1 and c_2 in $\mathbf{C}$ and $c_1x + c_2x^2$ represents all complex solutions.

We note, finally, that the definition of Wronskian and related results on linear independence carry over to complex solutions. See the following exercises.

We will conclude this section by defining the exponential function e^{kx}, where $k \in C$, and deriving some of its useful properties. These properties and the theoretical results outlined above will make it possible to justify and complete the discussion in Example 1.

In Example 1, the complex functions e^{ix} and e^{-ix} appeared to be solutions of the equation $y'' + y = 0$. (By the corollary to Theorem 9.1, Re (e^{ix}), Im (e^{ix}), Re (e^{-ix}), Im (e^{-ix}) are real-valued solutions.) We shall now give precise meaning to exponential functions having complex values in the exponent. Consider the well-known Maclaurin series

$$e^x = 1 + \frac{x}{1!} + \frac{x^2}{2!} + \cdots = \sum_{k=0}^{\infty} \frac{x^k}{k!}$$

This series is known to converge to e^x for all real x. If we replace x by ix we obtain

$$e^{ix} = 1 + \frac{ix}{1!} + \frac{i^2 x^2}{2!} + \cdots = \sum_{k=0}^{\infty} \frac{(ix)^k}{k!} \qquad (8)$$

It can be proved that this series converges for all real x; that is, if

$$s_m(x) = \sum_{k=0}^{m} \frac{(ix)^k}{k!}$$

then $s_m(x)$, the mth partial sum of the series, approaches a limit as $m \to \infty$, for every real x. (See Exercise 2 for a discussion of limits of complex functions.) We define e^{ix} to be this limit.

Since $i^2 = -1$, $i^3 = -i$, $i^4 = 1$, and so on, we have

$$e^{ix} = 1 + \frac{ix}{1!} - \frac{x^2}{2!} - i\frac{x^3}{3!} + \frac{x^4}{4!} + i\frac{x^5}{5!} - \cdots$$

$$= \left(1 - \frac{x^2}{2!} + \frac{x^4}{4!} - \cdots\right) + i\left(x - \frac{x^3}{3!} + \frac{x^5}{5!} - \cdots\right)$$

It can be proved that it is legitimate to rearrange the terms of the series in this way. Recalling the Maclaurin series for sin x and cos x,

$$\sin x = x - \frac{x^3}{3!} + \frac{x^5}{5!} - \cdots$$

$$\cos x = 1 - \frac{x^2}{2!} + \frac{x^4}{4!} - \cdots \qquad (9)$$

we see that we have proved that

$$e^{ix} = \cos x + i \sin x \qquad (10)$$

for all real x. This famous equation is known as *Euler's formula*. By replacing x by $-x$ and using $\cos(-x) = \cos x$, $\sin(-x) = -\sin x$, we also get

$$e^{-ix} = \cos x - i \sin x \qquad (11)$$

Addition and subtraction of Eqs. (10) and (11) give the further formulas

$$\cos x = \frac{1}{2}(e^{ix} + e^{-ix}), \qquad \sin x = \frac{1}{2i}(e^{ix} - e^{-ix}) \qquad (12)$$

Finally

$$\frac{d}{dx}e^{ix} = \frac{d}{dx}\cos x + i\frac{d}{dx}\sin x = -\sin x + i\cos x$$

$$= i(\cos x + i\sin x) = ie^{ix}$$

We shall also consider the function $f(x) = e^{(a+ib)x}$ where $a + ib$ is a complex number. For $a = 0$, we merely replace x by bx in Eq. (10), and obtain $e^{ibx} = \cos bx + i\sin bx$. We shall define

$$e^{(a+bi)x} = e^{ax}e^{ibx} = e^{ax}(\cos bx + i\sin bx) \qquad (13)$$

For $b = 0$ this reduces to e^{ax}. For $a = 0$ this is our previous definition of e^{ibx}. It can also be shown that

$$e^{(a+ib)x} = \sum_{n=0}^{\infty} \frac{(a + ib)^n x^n}{n!} \qquad (14)$$

but we omit the proof. From Eq. (13) we have

$$\frac{d}{dx}e^{(a+ib)x} = \frac{d}{dx}(e^{ax}\cos bx) + i\frac{d}{dx}(e^{ax}\sin bx)$$

$$= e^{ax}(a\cos bx - b\sin bx) + ie^{ax}(a\sin bx + b\cos bx)$$

$$= e^{ax}[a(\cos bx + i\sin bx) + ib(\cos bx + i\sin bx)]$$

$$= e^{ax}(a + ib)(\cos bx + i\sin bx)$$

$$= (a + ib)e^{(a+ib)x}$$

Thus we have proved that

$$\frac{d}{dx}e^{kx} = ke^{kx} \qquad (15)$$

whether k be real, imaginary, or complex.

Many books on calculus give more information on complex functions, and there are more advanced books devoted entirely to the theory of complex functions of complex variables. The short treatment given here will be suitable for our purposes. We consider Example 1 again.

Example 1 (continued) For arbitrary complex scalars c_1 and c_2, the function $y = c_1 e^{ix} + c_2 e^{-ix}$ is a solution of the equation $y'' + y = 0$, as can be verified by differentiation with the aid of the rule in Eq. (15). It is shown in the next section that e^{ix} and e^{-ix} are linearly independent functions. Since, by Theorem 9.2, the complex solution space has dimension two, $y = c_1 e^{ix} + c_2 e^{-ix}$ is a complex general solution. By the corollary to Theorem 9.1, $\text{Re}(e^{ix}) = \cos x$ and $\text{Im}(e^{ix}) = \sin x$ are real solutions. (Also e^{-ix} is a solution, but its real and imaginary parts do not yield any new solutions.) In fact, $\cos x$ and $\sin x$ are linearly independent real solutions on any interval and a real general solution is, for a_1 and a_2 real,

$$y = a_1 \cos x + a_2 \sin x$$

We suggest that the reader who has not previously been exposed to complex functions preview the remainder of this chapter without concern over the distinction between real and complex functions. When the techniques of solution have been mastered, he should return for a thorough reading.

Exercises

1. Establish the following differentiation formulas for complex functions of a real variable x.

   (a) $\dfrac{d}{dx}(cf) = c\,\dfrac{df}{dx}$,   where c is a complex scalar.

   (b) $\dfrac{d}{dx}(f_1 + f_2) = \dfrac{df_1}{dx} + \dfrac{df_2}{dx}$

   (c) $\dfrac{d}{dx}(f_1 f_2) = f_1\,\dfrac{df_2}{dx} + \dfrac{df_1}{dx}\,f_2$

   (d) $\dfrac{d}{dx}\left(\dfrac{f_1}{f_2}\right) = \dfrac{f_2 f'_1 - f_1 f'_2}{f_2{}^2}$ if $f_2(x) \neq 0$

   (e) $\dfrac{d}{dx} f(g(x)) = f'(g(x))g'(x)$

2. A complex function $f(x) = u(x) + iv(x)$ is said to have limit $\alpha = a + ib$ as $x \to x_0$ if the real number $|f(x) - \alpha|$ has 0 limit as $x \to x_0$, that is

$$\lim_{x \to x_0} |f(x) - \alpha| = 0$$

We then write

$$\lim_{x \to x_0} f(x) = \alpha, \quad \text{or} \quad f(x) \to \alpha \quad \text{as} \quad x \to x_0$$

Prove that $f(x) \to \alpha = a + ib$ as $x \to x_0$ if, and only if,

$$\lim_{x \to x_0} u(x) = a \quad \text{and} \quad \lim_{x \to x_0} v(x) = b$$

3. Say that a complex function $f(x)$ is continuous at x_0 if f is defined at x_0 and $f(x) \to f(x_0)$ as $x \to x_0$. Prove that this is equivalent to Definition 9.3.

4. Show that the derivative $f'(x_0)$ can be defined by the limit

$$f'(x_0) = \lim_{x \to x_0} \frac{f(x) - f(x_0)}{x - x_0}$$

5. Using Eq. (13), prove the law of exponents $e^z e^w = e^{z+w}$ for arbitrary complex numbers $z = a + ib$, $w = c + id$.

6. Using Exercise 5, show that $(e^{ix})^m = e^{ixm}$ for every integer m. Hence deduce that

$$(\cos x + i \sin x)^m = \cos mx + i \sin mx$$

This is known as DeMoivre's formula.

7. Use Eq. (12) to prove that

$$2 \cos^2 x = 1 + \cos 2x, \quad 2 \sin^2 x = 1 - \cos 2x$$

Derive similar formulas for $\cos^3 x$ and $\sin^3 x$.

8. Solve each equation by factoring the operator, and then find a real general solution.

(a) $(D^2 + 4I)y = 0$ (b) $(D^2 - 2D + 2I)y = 0$

9. Compute the derivative and indefinite integral of each function.

(a) $(x + ix)^2$ (b) $e^{(a+ib)x}$

(c) $(\sin x + i \cos x)^n$

10. Let L be a linear differential operator with real coefficients $a_n(x), \ldots, a_0(x)$ and let $y(x) = u(x) + iv(x)$ be a complex solution of $Ly = 0$. Show that the conjugate $\bar{y}(x) = u(x) - iv(x)$ is also a solution, that is $L\bar{y} = 0$.

11. Assuming the validity of the *ratio test* for series of complex terms, prove that the series in Eq. (8) converges absolutely for all real x.

12. Assume that $\sum_{k=1}^{\infty} a_k$ is an absolutely convergent series of complex constants. Prove that

$$\sum_{k=1}^{\infty} a_k = \sum_{k \text{ even}} a_k + \sum_{k \text{ odd}} a_k$$

where each of the series on the right is convergent.

13. Is the space $C^n(J)$ a subspace of the space $C^n(J, \mathbf{C})$?

14. Consider Eqs. (7) and (8) of Section 8.2. Show that these are valid if K, L, M are linear differential operators on $C^n(J, \mathbf{C})$ with complex coefficients.

15. Show that linear differential operators with constant real or complex coefficients can be factored according to the usual rules of algebra.

16. Prove the assertion in the text that Theorem 8.3 extends to the complex case.

17. Prove Theorem 9.2. In what respects must the proof of Theorem 8.5 be altered?

18. Prove the following extension of Lemma 8.1. Let $y_1, y_2, \ldots, y_n$ be in $C^{n-1}(J, \mathbf{C})$ and suppose that the vectors $Y_1(x_0)$, $Y_2(x_0)$, $\ldots$, $Y_n(x_0)$ are linearly independent in $\mathbf{C}^n$, where $\mathbf{C}^n$ denotes the vector space of n-tuples of complex numbers. Then $y_1, y_2, \ldots, y_n$ are linearly independent in $C^{n-1}(J, \mathbf{C})$.

19. Let $y_1, y_2, \ldots, y_n$ be in $C^{n-1}(J, \mathbf{C})$ and suppose that the Wronskian of these functions is nonzero for at least one point in J. Show that $\{y_1, y_2, \ldots, y_n\}$ is linearly independent in $C^{n-1}(J, \mathbf{C})$. (This is the extension of Theorem 8.6.)

20. Formulate and prove a theorem extending Theorem 8.7 to the complex case.

21. Discuss the problem of finding all complex solutions of a nonhomogeneous equation $Ly = f$. (See Theorem 8.8.)

9.3   Method of factorization of operators

In this section we shall show that the method based on factoring operators can be used to find a basis for the solution space of any nth order homogeneous linear differential equation with real or complex constant coefficients

$$Ly = (a_n D^n + a_{n-1} D^{n-1} + \cdots + a_1 D + a_0 I)y = 0 \qquad (1)$$

where $a_n \neq 0$. As we know from Section 8.2 and Exercise 15 of Section 9.2, it is legitimate to factor L in exactly the same way as the ordinary polynomial

$$p(r) = a_n r^n + a_{n-1} r^{n-1} + \cdots + a_1 r + a_0 \qquad (2)$$

DEFINITION 9.6   The polynomial $p(r)$ is called the *characteristic polynomial* or *auxiliary polynomial* associated with L, and its roots λ_i are called the *characteristic roots* of L.

It is known that $p(r)$ can be factored in the form

$$p(r) = a_n(r - \lambda_1)^{m_1}(r - \lambda_2)^{m_2} \cdots (r - \lambda_k)^{m_k} \qquad (3)$$

where $\lambda_1, \lambda_2, \ldots, \lambda_k$ are complex numbers called the distinct *roots* of $p(r) = 0$ and $m_1, m_2, \ldots, m_k$ are positive integers called the *multiplicities* of these roots.[2] Therefore from Eq. (1) we obtain (if $a_n \neq 0$)

$$(D - \lambda_1 I)^{m_1}(D - \lambda_2 I)^{m_2} \cdots (D - \lambda_k I)^{m_k} y = 0 \qquad (4)$$

As a preliminary to solving Eq. (4) we first consider an equation of the form

$$(D - \lambda I)^m z = 0 \qquad (5)$$

where $\lambda \in \mathbf{C}$ and m is a positive integer. We shall prove by mathematical induction that any solution of Eq. (5) has the form

$$z = (c_0 + c_1 x + \cdots + c_{m-1} x^{m-1}) e^{\lambda x} \qquad (6)$$

This is clear for $m = 1$. Assuming that it is true for an integer m, we consider the equation $(D - \lambda I)^{m+1} z = 0$. Letting $w = (D - \lambda I)z$, we can write this as $(D - \lambda I)^m w = 0$. By the inductive hypothesis, every solution of this equation is of the form

$$w = (c_0 + \cdots + c_{m-1} x^{m-1}) e^{\lambda x}$$

Hence we have

$$(D - \lambda I)z = (c_0 + \cdots + c_{m-1} x^{m-1}) e^{\lambda x}$$

Multiplying this first order equation by $e^{-\lambda x}$ we have

$$\frac{d}{dx}(z e^{-\lambda x}) = c_0 + c_1 x + \cdots + c_{m-1} x^{m-1}$$

$$z e^{-\lambda x} = c_0 x + \frac{c_1}{2} x^2 + \cdots + \frac{c_{m-1}}{m} x^m + c$$

Multiplying by $e^{\lambda x}$ and changing the notation for the arbitrary constants gives

$$z = (c_0 + c_1 x + \cdots + c_m x^m) e^{\lambda x}$$

This completes the proof that every solution of Eq. (5) has the form in Eq. (6). It is also easy to prove the converse, that every function of the form in Eq. (6) is a solution of Eq. (5). In fact, we have from Eq. (6)

$$z e^{-\lambda x} = c_0 + c_1 x + \cdots + c_{m-1} x^{m-1}$$

and by differentiation m times,

$$D^m(z e^{-\lambda x}) = 0 \qquad (7)$$

---

[2] For a review of basic information on polynomials, see Appendix B.

It can be proved from this (see Exercise 1) that Eq. (5) is satisfied. Since the solution space of Eq. (5) has dimension m, our result can be summarized as follows.

LEMMA 9.1   If $\lambda \in \mathbf{C}$ and m is a positive integer, a complex general solution of $(D - \lambda I)^m z = 0$ is given by

$$z = (c_0 + c_1 x + \cdots + c_{m-1} x^{m-1}) e^{\lambda x}$$

where $c_0, \ldots, c_{m-1}$ are arbitrary complex constants. If λ is real, a real general solution is given by the same formula with $c_0, c_1, \ldots, c_{m-1}$ in $\mathbf{R}$.

Consider Eq. (4) again. We shall use another simple result.

LEMMA 9.2   If for some integer j, $1 \leq j \leq k$, y is a function in the null space of $(D - \lambda_j I)^{m_j}$, then y is in the null space of

$$L = (D - \lambda_1 I)^{m_1} \cdots (D - \lambda_k I)^{m_k}$$

PROOF: Since the order of the factors can be permuted, we can write

$$Ly = \left\{ \prod_{\substack{i=1 \\ i \neq j}}^{k} (D - \lambda_i I)^{m_i} \right\} (D - \lambda_j I)^{m_j} y$$

Since $(D - \lambda_j I)^{m_j} y = 0$, clearly $Ly = 0$.   ∎

Combining Lemmas 9.1 and 9.2, we find that the solution space of Eq. (1) contains all the functions

$$e^{\lambda_1 x}, xe^{\lambda_1 x}, \ldots, x^{m_1 - 1} e^{\lambda_1 x}, \ldots, e^{\lambda_k x}, xe^{\lambda_k x}, \ldots, x^{m_k - 1} e^{\lambda_k x}$$

That is, any function which satisfies any of the equations $(D - \lambda_j I)^{m_j} y = 0$ also satisfies $Ly = 0$. The total number of solutions found in this way is $m_1 + m_2 + \cdots + m_k = n$, since the sum of the multiplicities of the roots is n. The next theorem states that the n solutions are linearly independent.

THEOREM 9.3   Let L be the differential operator

$$L = a_n (D - \lambda_1)^{m_1} (D - \lambda_2)^{m_2} \cdots (D - \lambda_k)^{m_k}$$

where $\lambda_1, \lambda_2, \ldots, \lambda_k$ are the distinct roots of the characteristic polynomial of L and $m_1, m_2, \ldots, m_k$ are their respective multiplicities. Then

$$\{ e^{\lambda_1 x}, xe^{\lambda_1 x}, \ldots, x^{m_1 - 1} e^{\lambda_1 x}, \ldots, e^{\lambda_k x}, xe^{\lambda_k x}, \ldots, x^{m_k - 1} e^{\lambda_k x} \}$$

is a basis for the complex solution space of the equation $Ly = 0$. (If $\lambda_1, \lambda_2, \ldots, \lambda_k$ are all real, this is also a basis for the real solution space.)

Since the solution space has dimension n, it is only necessary to prove linear independence of this set of n functions. Proofs of this are somewhat complicated. One of the simpler proofs is outlined in Exercise 9.

Example 1 Consider the equation

$$y^{(iv)} + y''' + y' + y = 0$$

The characteristic polynomial is $r^4 + r^3 + r + 1 = (r + 1)^2(r^2 - r + 1)$. The characteristic roots are -1 (multiplicity 2) and $(1 \pm i\sqrt{3})/2$. Therefore

$$\{e^{-x}, xe^{-x}, e^{(1/2+i\sqrt{3}/2)x}, e^{(1/2-i\sqrt{3}/2)x}\}$$

is a basis for the complex solution space and a complex general solution is

$$y = (c_1 + c_2 x)e^{-x} + c_3 e^{(1/2+i\sqrt{3}/2)x} + c_4 e^{(1/2-i\sqrt{3}/2)x}$$

$(c_1, c_2, c_3, c_4 \in \mathbf{C})$. By the corollary to Theorem 9.1, the functions

$$\text{Re}\,\{e^{(1/2+i\sqrt{3}/2)x}\} = e^{x/2}\cos\sqrt{3}x/2 \qquad \text{Im}\,\{e^{(1/2+i\sqrt{3}/2)x}\} = e^{x/2}\sin\sqrt{3}x/2$$

are real solutions of the equation. The set of four solutions

$$\{e^{-x}, xe^{-x}, e^{x/2}\cos\sqrt{3}x/2, e^{x/2}\sin\sqrt{3}x/2\}$$

can be shown to be linearly independent, and is therefore a basis for the real solution space of the equation. A real general solution of the equation is

$$y = (c_1 + c_2 x)e^{-x} + e^{x/2}(c_3\cos\sqrt{3}x/2 + c_4\sin\sqrt{3}x/2)$$

$(c_1, c_2, c_3, c_4 \in \mathbf{R})$.

In this example, a real general solution was deduced from an independent set of complex solutions. The following theorem proves the general validity of this procedure. Recall that for an equation with real coefficients, complex solutions must occur in conjugate pairs. (See Exercise 10 of Section 9.2.)

THEOREM 9.4 Let L be a linear differential operator with real coefficients. Let the set of functions

$$\{z_1, \ldots, z_p, w_1, \bar{w}_1, \ldots, w_q, \bar{w}_q\} \tag{8}$$

be a linearly independent set (over the complex field) of solutions of $Ly = 0$, where $z_1, \ldots, z_p$ are real functions, $w_1, \bar{w}_1, \ldots, w_q, \bar{w}_q$ are complex functions, and $\bar{w}_j$ is the conjugate of w_j, that is,

$$w_j(x) = u_j(x) + iv_j(x), \quad \bar{w}_j(x) = u_j(x) - iv_j(x) \qquad (j = 1, \ldots, q)$$

Then

$$\{z_1, \ldots, z_p, u_1, v_1, \ldots, u_q, v_q\} \tag{9}$$

is a linearly independent set (over the real field) of real solutions of $Ly = 0$.

PROOF: By the corollary to Theorem 9.1, the functions $u_1, v_1, \ldots, u_q, v_q$ are solutions. The linear independence has to be proved. Suppose that there are real constants, $a_1, \ldots, a_p, b_1, c_1, \ldots, b_q, c_q$, such that

$$a_1 z_1 + \cdots + a_p z_p + b_1 u_1 + c_1 v_1 + \cdots + b_q u_q + c_q v_q = 0 \qquad (x \in J) \tag{10}$$

Let

$$\beta_j = \frac{b_j - i c_j}{2}, \qquad \gamma_j = \frac{b_j + i c_j}{2}, \qquad (j = 1, \ldots, q)$$

Then

$$\begin{aligned} b_j u_j + c_j v_j &= \beta_j(u_j + i v_j) + \gamma_j(u_j - i v_j) \\ &= \beta_j w_j + \gamma_j \overline{w}_j \end{aligned}$$

Therefore Eq. (10) can be written

$$a_1 z_1 + \cdots + a_p z_p + \beta_1 w_1 + \gamma_1 \overline{w}_1 + \cdots + \beta_q w_q + \gamma_q \overline{w}_q = 0 \tag{11}$$

Since the set of functions in Eq. (8) is linearly independent, Eq. (11) can hold only if all the coefficients are zero, $a_1 = \cdots = a_p = \beta_1 = \gamma_1 = \cdots = \beta_q = \gamma_q = 0$. But then all the a_j, b_j and c_j are zero, and this proves independence of the set (9). [*Note:* The same proof can be easily extended to show that the set (9) of real solutions is linearly independent over the complex field.]   ∎

Theorem 9.4 tells us that if we find an independent set of complex solutions of a linear equation with real coefficients, we can deduce an independent set of real solutions quite simply. We need only replace each conjugate pair of complex solutions w_j, $\overline{w}_j$ by the real pair Re (w_j), Im (w_j).

Example 2   Consider the equation

$$(D^2 + 1)^2 (D - 2)y = 0$$

By Theorem 9.3, a complex basis for the solutions is

$$\{e^{2x}, e^{ix}, e^{-ix}, xe^{ix}, xe^{-ix}\}$$

By Theorem 9.4, a basis for real solutions is

$$\{e^{2x}, \cos x, \sin x, x \cos x, x \sin x\}$$

A real general solution is

$$y = c_1 e^{2x} + (c_2 + c_3 x) \cos x + (c_4 + c_5 x) \sin x$$

We can now write the most general form possible for a basis for the solution space of an equation with real constant coefficients.

THEOREM 9.5   Let L be a linear differential operator of order n with real constant coefficients. Let its real characteristic roots be $\lambda_1, \ldots, \lambda_p$ with multiplicities $m_1, \ldots, m_p$, respectively. Let the complex characteristic roots be $\mu_1 + iv_1, \mu_1 - iv_1, \ldots, \mu_q + iv_q, \mu_q - iv_q$ with multiplicities $n_1, n_1, \ldots, n_q, n_q$. Then the following functions comprise a basis for the real solution space N_L:

$$e^{\lambda_j x}, xe^{\lambda_j x}, \ldots, x^{m_j - 1}e^{\lambda_j x} \qquad (j = 1, \ldots, p)$$

$$e^{\mu_j x} \cos v_j x, \ e^{\mu_j x} \sin v_j x, \ xe^{\mu_j x} \cos v_j x, \ xe^{\mu_j x} \sin v_j x, \ldots,$$

$$x^{n_j - 1}e^{\mu_j x} \cos v_j x, \ x^{n_j - 1}e^{\mu_j x} \sin v_j x \qquad (j = 1, \ldots, q)$$

PROOF: To prove this theorem, start with the n linearly independent solutions listed in Theorem 9.3. Those with λ_j real are not changed. Those with complex roots are replaced by their real and imaginary parts. For example, $e^{(\mu_1 + iv_1)x}$ and $e^{(\mu_1 - iv_1)x}$ are replaced by $e^{\mu_1 x} \cos v_1 x$ and $e^{\mu_1 x} \sin v_1 x$, and so on.   ∎

In summary, we can assert that we have achieved our goal of a systematic procedure for solving any linear homogeneous equation with real constant coefficients. The steps in this procedure are as follows.

1. Find all roots of the characteristic polynomial $p(r)$ with their multiplicities.
2. Construct linearly independent solutions as in Theorem 9.3.
3. If desired, replace complex exponentials by their real and imaginary parts to obtain a real basis.

The reader will notice that steps (2) and (3) are easy, whereas step (1) is a problem in algebra which can in some cases be onerous. However, the solution has been reduced to a problem of a more elementary sort, finding the roots of polynomials.

Exercises

1. (a) Prove that $D(ze^{-\lambda x}) = e^{-\lambda x}(D - \lambda I)z$ for any differentiable function z and any real or complex scalar λ.

   (b) Prove that

   $$D^m(ze^{-\lambda x}) = e^{-\lambda x}(D - \lambda I)^m z, \qquad m = 1, 2, \ldots.$$

   Note that $(D - \lambda I)^m z = e^{\lambda x} D^m(ze^{-\lambda x})$.

2. Find three linearly independent complex solutions of each equation.

   (a) $(D - 3i)(D^2 - 2iD + 1)y = 0$

   (b) $[D^3 + (1 + 2i)D^2 - (1 - 2i)D - 1]y = 0$. Note that $D + 1$ is a factor.

3. In each part, find a complex general solution and then a real general solution.

(a) $y'' - 5y' + 4y = 0$

(b) $(D^2 + D + I)y = 0$

(c) $4\dfrac{d^2z}{dx^2} - 7\dfrac{dz}{dx} + 3z = 0$

(d) $y''' + y = 0$

(e) $(D^2 + D + 1)^2 y = 0$

(f) $(D^2 - 2D + 1)(D^2 + 1)y = 0$

(g) $z^{(iv)} - 3z'' + 2z = 0$

(h) $(D^2 - 4D + 13)^2 y = 0$

(i) $(D^4 + 1)y = 0$

(j) $(D^3 - 1.5D^2 - 2.5D + 3)y = 0$

4. Find the solution, in real form, of each of the following initial value problems (see Exercise 3).

(a) $y'' - 5y' + 4y = 0$ $y(1) = 2$ $y'(1) = 1$

(b) $(D^2 + D + I)y = 0$ $y(0) = 1$ $y'(0) = 0$

(c) $4\dfrac{d^2z}{dx^2} - 7\dfrac{dz}{dx} + 3z = 0,$ $y(8) = 10,$ $y'(8) = 1$

(d) $y''' + y = 0,$ $y(0) = 2,$ $y'(0) = 0,$ $y''(0) = 1$

5. In each part find a linear differential equation with real constant coefficients which has the given general solution.

(a) $c_1 e^{-2x} + c_2 e^{3x}$

(b) $(c_1 + c_2 x)e^{10x}$

(c) $e^{-x}(c_1 \cos 2x + c_2 \sin 2x)$

(d) $(c_1 + c_2 x)e^x \cos \dfrac{x}{2} + (c_3 + c_4 x)e^x \sin \dfrac{x}{2}$

6. Consider the second order equation $y'' + 2by' + cy = 0$ and assume that $b^2 - c < 0$.

(a) Show that a real general solution is

$$y = e^{-bx}(c_1 \cos wx + c_2 \sin wx)$$

where $w^2 = c - b^2$.

(b) Show that a general solution is

$$y = k_1 e^{-bx} \cos (wx + k_2)$$

where k_1 and k_2 are arbitrary real constants. This is often called the *phase-amplitude* form of solution. What is the significance of k_1, w, and k_2 relative to the graph of y?

7. Prove that $\{e^{\lambda x}, xe^{\lambda x}, \ldots, x^{m-1}e^{\lambda x}\}$ is a linearly independent set on the interval $J = (-\infty, \infty)$. [*Hint:* If

$$c_0 e^{\lambda x} + c_1 xe^{\lambda x} + \cdots + c_{m-1}x^{m-1}e^{\lambda x} = 0$$

then

$$c_0 + c_1 x + \cdots + c_{m-1}x^{m-1} = 0]$$

8. Using Exercise 1b, prove that

$$(D - \lambda)^m(x^j e^{\lambda x}) = 0 \quad \text{if} \quad m > j$$
$$= j(j-1)\cdots(j - m + 1)x^{j-m}e^{\lambda x} \quad \text{if} \quad m \le j$$

In particular, show that

$$(D - \lambda)^m(x^m e^{\lambda x}) = m!e^{\lambda x}$$

9. (a) Let $L = (D - \lambda_1)^{m_1}(D - \lambda_2)^{m_2}$ where $\lambda_1 \ne \lambda_2$. Prove that the set

$$\{e^{\lambda_1 x}, xe^{\lambda_1 x}, \ldots, x^{m_1 - 1}e^{\lambda_1 x}, e^{\lambda_2 x}, xe^{\lambda_2 x}, \ldots, x^{m_2 - 1}e^{\lambda_2 x}\}$$

is linearly independent on any interval J. [*Hint:* Suppose that

$$\sum_{j=1}^{m_1} a_j x^{j-1}e^{\lambda_1 x} + \sum_{j=1}^{m_2} b_j x^{j-1}e^{\lambda_2 x} = 0, \quad x \in J$$

Applying the operator $(D - \lambda_2)^{m_2}(D - \lambda_1)^{m_1 - 1}$ to this equation yields

$$(D - \lambda_2)^{m_2} \sum_{j=1}^{m_1} a_j(D - \lambda_1)^{m_1 - 1}(x^{j-1}e^{\lambda_1 x}) = 0$$

Using Exercise 8, deduce that $a_{m_1} = 0$. Now apply the operators $(D - \lambda_2)^{m_2}(D - \lambda_1)^{m_1 - 2}$, $(D - \lambda_2)^{m_2}(D - \lambda_1)^{m_1 - 3}$, and so on, and show that all $a_j = 0$. Similarly show that all $b_j = 0$.]

(b) Generalize the result in (a) to operators L of the form

$$L = (D - \lambda_1)^{m_1}(D - \lambda_2)^{m_2} \ldots (D - \lambda_k)^{m_k}$$

10. (a) Show that the Wronskian of the functions $e^{\lambda_1 x}, \ldots, e^{\lambda_k x}$ is $W(x) = \exp\left[(\lambda_1 + \cdots + \lambda_k)x\right] V(\lambda_1, \ldots, \lambda_k)$ where V is the determinant

$$V(\lambda_1, \lambda_2, \ldots, \lambda_k) = \det \begin{pmatrix} 1 & 1 & \cdots & 1 \\ \lambda_1 & \lambda_2 & \cdots & \lambda_k \\ \lambda_1^2 & \lambda_2^2 & \cdots & \lambda_k^2 \\ & & \cdots & \\ \lambda_1^{k-1} & \lambda_2^{k-1} & \cdots & \lambda_k^{k-1} \end{pmatrix}$$

This is known as the Vandermonde determinant.

(b) Prove that

$$V(\lambda_1, \lambda_2, \ldots, \lambda_k) = \prod_{\substack{m, n = 1 \\ m > n}}^{k} (\lambda_m - \lambda_n);$$

that is, V is the product of all distinct differences $(\lambda_m - \lambda_n)$ for m and n between 1 and k. [*Hint:* Think of $\lambda_2, \ldots, \lambda_k$ as fixed and λ_1 as variable. Then show that V is a polynomial of degree $k - 1$ in λ_1 with roots $\lambda_2, \ldots, \lambda_k$, so that

$$V(\lambda_1, \ldots, \lambda_k) = a(\lambda_2, \ldots, \lambda_k)(\lambda_1 - \lambda_2)(\lambda_1 - \lambda_3) \cdots (\lambda_1 - \lambda_k)$$

where a is independent of λ_1. Now regard V as a function of λ_2, and so on.]

(c) Deduce that the set $\{e^{\lambda_1 x}, \ldots, e^{\lambda_k x}\}$ is linearly independent if the λ_j are distinct.

9.4   Variation of parameters

In the preceding section, a method was given for determining a general solution of an nth order homogeneous linear equation $Ly = 0$, provided the coefficients in the operator L are constants. As was shown in Chapter 8 (and Exercise 21 of Section 9.2) a general solution of a nonhomogeneous equation $Ly = f$ can be constructed by adding any one solution to a general solution of $Ly = 0$. We turn now to methods for finding a solution of the nonhomogeneous equation. There are two principal methods. The one presented in this section, called the method of *variation of parameters* or *variation of constants*, is the more general and powerful, but the other, called the method of *undetermined coefficients*, is easier to use in some circumstances.

We consider the nth order differential operator

$$Ly = y^{(n)} + a_{n-1}(x)y^{(n-1)} + \cdots + a_1(x)y' + a_0(x)y \qquad (1)$$

and the equation

$$Ly = f \qquad (2)$$

Since the method of variation of parameters applies in principle to equations with either constant or variable coefficients, it is assumed here that $a_{n-1}(x), \ldots, a_0(x)$, and $f(x)$ are given continuous complex functions on an interval J. Let us recall a few results. Consider n functions $y_1(x), y_2(x), \ldots, y_n(x)$ which are in $\mathbf{C}^n(J, \mathbf{C})$. The Wronski matrix is

$$\mathscr{W}(x) = \begin{pmatrix} y_1(x) & y_2(x) & \cdots & y_n(x) \\ y_1'(x) & y_2'(x) & \cdots & y_n'(x) \\ & & \cdots & \\ y_1^{(n-1)}(x) & y_2^{(n-1)}(x) & \cdots & y_n^{(n-1)}(x) \end{pmatrix}$$

and the Wronskian $W(x)$ is the determinant of $\mathscr{W}(x)$. If the functions $y_1, y_2, \ldots, y_n$ are solutions of $Ly = 0$ on J, then by the corollary to Theorem 8.7 and Exercise 20 in Section 9.2 either (1) $W(x) = 0$ for all x in J and $\{y_1, y_2, \ldots, y_n\}$ is a linearly dependent set or (2) $W(x) \neq 0$ for all x in J and $\{y_1, y_2, \ldots, y_n\}$ is a basis for the complex solution space of the homogeneous equation $Ly = 0$.

The method of variation of parameters can be used only when the homogeneous equation has already been completely solved. That is, n linearly independent solutions $y_1, y_2, \ldots, y_n$ of $Ly = 0$ must be known. That we can find n linearly independent solutions (or even one solution) to every homogeneous nth order linear differential equation is a gross assumption. However, in some cases, such as with constant coefficients, solutions are obtainable.

We know that Eq. (2) has solutions. In fact, by the existence theorem, there is a unique solution corresponding to any set of initial conditions. What we want is a method or algorithm for constructing one of these solutions. To begin our search for a particular solution of Eq. (2), we let $\{y_1, y_2, \ldots, y_n\}$ be a basis for the space of complex solutions of $Ly = 0$, and we *guess* that we can find a solution of Eq. (2) in the form

$$y_p = b_1(x)y_1 + b_2(x)y_2 + \cdots + b_n(x)y_n \tag{3}$$

where the $b_i(x)$ are in $\mathbf{C}^1(J, \mathbf{C})$. We shall prove that functions $b_i(x)$ always exist such that y_p as defined by Eq. (3) satisfies Eq. (2). Taking the derivative of both sides of Eq. (2) yields

$$\begin{aligned} y_p' = &[b_1'(x)y_1 + b_2'(x)y_2 + \cdots + b_n'(x)y_n] \\ &+ [b_1(x)y_1' + b_2(x)y_2' + \cdots + b_n(x)y_n'] \end{aligned}$$

If we place the restriction on $b_1, b_2, \ldots, b_n$ that

$$b_1'(x)y_1 + b_2'(x)y_2 + \cdots + b_n'(x)y_n = 0 \tag{4}$$

we obtain

$$y_p' = b_1(x)y_1' + b_2(x)y_2' + \cdots + b_n(x)y_n'$$

Differentiating again yields

$$\begin{aligned} y_p'' = &[b_1'(x)y_1' + b_2'(x)y_2' + \cdots + b_n'(x)y_n'] \\ &+ [b_1(x)y_1'' + b_2(x)y_2'' + \cdots + b_n(x)y_n''] \end{aligned}$$

We impose a second restriction

$$b_1'(x)y_1' + b_2'(x)y_2' + \cdots + b_n'(x)y_n' = 0 \tag{5}$$

which results in

$$y_p'' = b_1(x)y_1'' + b_2(x)y_2'' + \cdots + b_n(x)y_n''$$

Continuing in this fashion we compute the first $n - 1$ derivatives of y_p with restrictions of the form of Eqs. (4) and (5), which, in general, are given by

$$b_1'(x)y_1^{(k-1)} + b_2'(x)y_2^{(k-1)} + \cdots + b_n'(x)y_n^{(k-1)} = 0$$

$$k = 1, \ldots, n - 1 \qquad (6)$$

and obtain

$$y_p^{(k)} = b_1(x)y_1^{(k)} + b_2(x)y_2^{(k)} + \cdots + b_n(x)y_n^{(k)} \qquad k = 0, \ldots, n - 1 \qquad (7)$$

From Eq. (7) we can obtain

$$y_p^{(n)} = [b_1'(x)y_1^{(n-1)} + b_2'(x)y_2^{(n-1)} + \cdots + b_n'(x)y_n^{(n-1)}]$$
$$+ [b_1(x)y_1^{(n)} + b_2(x)y_2^{(n)} + \cdots + b_n(x)y_n^{(n)}] \qquad (8)$$

Substituting the results of Eqs. (7) and (8) into $Ly = f(x)$ yields

$$[b_1'(x)y_1^{(n-1)} + b_2'(x)y_2^{(n-1)} + \cdots + b_n'(x)y_n^{(n-1)}]$$
$$+ [b_1(x)y_1^{(n)} + b_2(x)y_2^{(n)} + \cdots + b_n(x)y_n^{(n)}]$$
$$+ a_{n-1}(x)[b_1(x)y_1^{(n-1)} + b_2(x)y_2^{(n-1)} + \cdots + b_n(x)y_n^{(n-1)}]$$
$$+ \cdots + a_0(x)[b_1(x)y_1 + b_2(x)y_2 + \cdots + b_n(x)y_n] = f(x)$$

which can be rewritten as

$$[b_1'(x)y_1^{(n-1)} + b_2'(x)y_2^{(n-1)} + \cdots + b_n'(x)y_n^{(n-1)}]$$
$$+ b_1(x)[y_1^{(n)} + a_{n-1}(x)y_1^{(n-1)} + \cdots + a_0(x)y_1]$$
$$+ \cdots + b_n(x)[y_n^{(n)} + a_{n-1}(x)y_n^{(n-1)} + \cdots + a_0(x)y_n] = f(x) \qquad (9)$$

But since $y_1, y_2, \ldots, y_n$ are solutions of the homogeneous equation, Eq. (9) reduces to

$$b_1'(x)y_1^{(n-1)} + b_2'(x)y_2^{(n-1)} + \cdots + b_n'(x)y_n^{(n-1)} = f(x) \qquad (10)$$

Combining the results of Eq. (10) with the restriction of Eq. (6) yields a system of n equations in n unknowns, namely $b_1', b_2', \ldots, b_n'$. The system can be represented in matrix notation as

$$\mathscr{W}(x) \begin{pmatrix} b_1'(x) \\ b_2'(x) \\ \vdots \\ b_n'(x) \end{pmatrix} = \begin{pmatrix} 0 \\ 0 \\ \vdots \\ f(x) \end{pmatrix} \qquad (11)$$

The above discussion began with the consideration of a function

$$y_p = b_1(x)y_1 + b_2(x)y_2 + \cdots + b_n(x)y_n$$

where $y_1, y_2, \ldots, y_n$ are given and known to be linearly independent. Restrictions were imposed upon $b_1, b_2, \ldots, b_n$ which are given in Eq. (11). If we can find a set $b_1, b_2, \ldots, b_n$ satisfying Eq. (11) we will have located a particular solution of the differential equation $Ly = f(x)$.

Since the $\{y_i\}$ are known to be linearly independent, $W(x)$ is nonzero for all x in J, which implies $\mathcal{W}(x)$ is nonsingular at each x in J. Hence $\mathcal{W}^{-1}(x)$ exists at each point, and Eq. (11) implies

$$
\begin{pmatrix} b_1{}'(x) \\ b_2{}'(x) \\ \vdots \\ b_n{}'(x) \end{pmatrix} = \mathcal{W}^{-1}(x) \begin{pmatrix} 0 \\ 0 \\ \vdots \\ f(x) \end{pmatrix} \tag{12}
$$

Each of the functions $b_i{}'(x)$ is determined by Eq. (12), and $b_i(x)$ is then found by integration. Denote by $W^{ij}(x)$ the i,j cofactor of $\mathcal{W}(x)$. Recall that $\mathcal{W}^{-1}(x) =$ adj $\mathcal{W}/W(x)$. Therefore

$$
b_i{}'(x) = \frac{W^{ni}(x)f(x)}{W(x)} \qquad (i = 1, 2, \ldots, n) \tag{13}
$$

and

$$
b_i(x) = \int^x \frac{W^{ni}(z)f(z)}{W(z)} \, dz \qquad (i = 1, 2, \ldots, n) \tag{14}
$$

Any indefinite integral can be selected in Eq. (14).

The following theorem has been established.

THEOREM 9.6 Let $a_{n-1}(x), \ldots, a_0(x)$, $f(x)$ be in $C^0(J, C)$ and let $y_1, y_2, \ldots, y_n$ be a basis for the complex solution space of the equation $Ly = 0$ where L is given in Eq. (1). Then a particular solution of Eq. (2) is given by

$$
y_p(x) = b_1(x)y_1(x) + \cdots + b_n(x)y_n(x) \tag{15}
$$

where $b_1(x), \ldots, b_n(x)$ are chosen to satisfy Eqs. (12), (13), or (14). A complex general solution of Eq. (2) is

$$
y = c_1 y_1 + \cdots + c_n y_n + b_1(x)y_1 + \cdots + b_n(x)y_n \tag{16}
$$

where $c_1, \ldots, c_n$ are arbitrary constants in C.

If L has real coefficients, f is real, and $\{y_1, \ldots, y_n\}$ is a basis for real solutions, then $b_1, \ldots, b_n$ are real functions, $y_p(x)$ is real, and Eq. (16) is a real general solution for $c_1, \ldots, c_n \in R$.

Example 1 Solve $y'' - y = e^{2x}$ by the method of variation of parameters. Here $y_1 = e^x$ and $y_2 = e^{-x}$ are linearly independent solutions of $y'' - y = 0$. The Wronski matrix is

$$
\mathcal{W}(x) = \begin{pmatrix} e^x & e^{-x} \\ e^x & -e^{-x} \end{pmatrix}
$$

and the Wronskian is $W(x) = -2$. Since

$$W^{11} = -e^{-x}, \quad W^{12} = -e^x, \quad W^{21} = -e^{-x}, \quad W^{22} = e^x$$

we get

$$\mathscr{W}^{-1}(x) = \tfrac{1}{2}\begin{pmatrix} e^{-x} & e^{-x} \\ e^x & -e^x \end{pmatrix}$$

Therefore

$$b_1'(x) = \frac{1}{2}e^x, \quad b_2'(x) = \frac{-1}{2}e^{3x}$$

We can take any indefinite integral to get b_1 and b_2. A simple choice is

$$b_1(x) = \frac{1}{2}e^x, \quad b_2(x) = \frac{-1}{6}e^{3x}$$

A particular solution is

$$y_p = \tfrac{1}{2}e^x \cdot e^x - \tfrac{1}{6}e^{3x} \cdot e^{-x} = \tfrac{1}{3}e^{2x}$$

and a general solution is

$$y = c_1 e^x + c_2 e^{-x} + \tfrac{1}{3}e^{2x} \tag{17}$$

If we had chosen some other integral, for example,

$$b_1(x) = \frac{1}{2}e^x + 5, \quad b_2(x) = \frac{-1}{6}e^{3x}$$

then we would have obtained $y_p(x) = \tfrac{1}{3}e^{2x} + 5e^x$, $y = c_1 e^x + c_2 e^{-x} + \tfrac{1}{3}e^{2x} + 5e^x$, which is equivalent to the result in Eq. (17).

Example 2 Solve $y''' - y' = 2 \sin x$ by the method of variation of parameters. A general solution to the homogeneous equation is $c_1 + c_2 e^x + c_3 e^{-x}$. Setting $y_1 = 1$, $y_2 = e^x$, $y_3 = e^{-x}$, we obtain

$$\mathscr{W}(x) = \begin{pmatrix} 1 & e^x & e^{-x} \\ 0 & e^x & -e^{-x} \\ 0 & e^x & e^{-x} \end{pmatrix}, \qquad W(x) = 2$$

$$W^{31}(x) = -2, \quad W^{32}(x) = e^{-x}, \quad \text{and} \quad W^{33}(x) = e^x$$

Solving for $b_i(x)$ yields

$$b_1(x) = \int^x (-2 \sin z)\, dz = 2 \cos x$$

$$b_2(x) = \int^x (e^{-z}/2)2 \sin z\, dz$$

$$b_2(x) = \int^x (e^z/2)2 \sin z\, dz$$

Hence a general solution to the given equation is

$$y = c_1 + 2 \cos x + \left(c_2 + \int^x e^{-z} \sin z \; dz \right) e^x$$

$$+ \left(c_3 + \int^x e^z \sin z \; dz \right) e^{-x}$$

The integrals can be worked out explicitly.

Example 3 Solve $(D^2 + 1)y = 1$. A complex general solution to the homogeneous equation is $c_1 e^{ix} + c_2 e^{-ix}$. Let $y_1 = e^{ix}$, $y_2 = e^{-ix}$. Then

$$\mathcal{W}(x) = \begin{pmatrix} e^{ix} & e^{-ix} \\ ie^{ix} & -ie^{-ix} \end{pmatrix}, \qquad W(x) = -2i$$

$$W^{11} = -ie^{-ix}, \quad W^{12} = -ie^{ix}, \quad W^{21} = -e^{-ix}, \quad W^{22} = e^{ix}$$

Hence

$$2i\mathcal{W}^{-1}(x) = \begin{pmatrix} ie^{-ix} & e^{-ix} \\ ie^{ix} & -e^{ix} \end{pmatrix}$$

$$b_1(x) = \tfrac{1}{2} e^{-ix}, \qquad b_2(x) = \tfrac{1}{2} e^{ix}$$

A particular solution is

$$y_p(x) = \tfrac{1}{2} e^{-ix} e^{ix} + \tfrac{1}{2} e^{ix} e^{-ix} = 1$$

Therefore a complex general solution to the real differential equation is

$$y = 1 + c_1 e^{ix} + c_2 e^{-ix}$$

and a real general solution is

$$y = 1 + c_1 \sin x + c_2 \cos x$$

In this example, the same $y_p(x)$ could have been obtained starting from the real solutions $\sin x$ and $\cos x$ of the homogeneous equation.

The reader should note that the method of variation of parameters is a perfectly general method for solving every equation $Ly = f$, provided only that n independent solutions of $Ly = 0$ are known.

Exercises

1. Use variation of parameters to obtain a particular solution of each equation. Write a real general solution.

(a) $(D^2 + 1)y = \tan x$
(b) $y'' - y = e^x$

(c) $(4D^2 + 4D + 1)y = e^{-x/2}$

(d) $(D^3 - D)y = \ln x, \quad x > 0$

(e) $\dfrac{d^2y}{dx^2} - 4y = \sin^2 x$

(f) $y''' - 3y'' = 9(x + 1)$

(g) $\dfrac{d^4y}{dx^4} = x$

2. (a) Find two linearly independent polynomial solutions of

$$x^2y'' - 2xy' + 2y = 0$$

(b) Use variation of parameters to find a general solution of

$$x^2y'' - 2xy' + 2y = x$$

3. Show that $y_1 = x$ and $y_2 = x \ln x$ are solutions of

$$Ly = x^2\dfrac{d^2y}{dx^2} - x\dfrac{dy}{dx} + y = 0 \qquad x > 0$$

Find a particular solution of the nonhomogeneous equation

$$Ly = x^2 + x$$

4. Using results from Exercises 1, 2, and 3, solve each initial value problem.

(a) $(D^2 + 1)y = \tan x$, $y(0) = 1$, $y'(0) = 2$

(b) $y'' - y = e^x$, $y(0) = -1$, $y'(0) = -1$

(c) $(4D^2 + 4D + 1)y = e^{-x/2}$, $y(0) = 0$, $y'(0) = 0$

(d) $(D^3 - D)y = \ln x$, $y(1) = 1$, $y'(1) = 0$ $y''(1) = 0$

(e) $x^2y'' - 2xy' + 2y = x$, $y(1) = 4$, $y'(1) = 0$

(f) $x^2y'' - xy' + y = x^2 + x$, $y(1) = 0$, $y'(1) = 0$

5. (a) Using $y_1 = e^{(1+i)x}$, $y_2 = e^{(1-i)x}$, and variation of parameters, find a particular solution of the equation $(D^2 - 2D + 2)y = e^x$.

(b) Using $y_1 = e^x \cos x$, $y_2 = e^x \sin x$, find a real particular solution of the same equation.

6. Consider the equation $(D - \lambda)(D - \bar\lambda)y = f(x)$, where $\lambda = a + ic$, $\bar\lambda = a - ic$, and f is real. Take $y_1 = e^{\lambda x}$, $y_2 = e^{\bar\lambda x}$ and compute $b_1(x)$, $b_2(x)$, and $y_p(x)$. Show that $y_p(x)$ is real and in fact

$$y_p(x) = \dfrac{1}{c}\int_c^x e^{a(x-z)}\sin c(x - z)f(z)\,dz$$

9.5   Operator inverse and the one-sided Green's function

The results of the preceding section can be stated in a way which leads to new insight and which will later lead to important generalizations. The functions $b_i(x)$ of Theorem 9.6 were determined by Eq. (14), that is, as indefinite integrals of certain functions. To remove the ambiguity in the choice of the $b_i(x)$ due to the arbitrary constants of integration, we choose a point x_0 in the interval of interest and define

$$b_i(x) = \int_{x_0}^{x} \frac{W^{ni}(z)f(z)}{W(z)}\, dz \qquad (i = 1, \ldots, n) \tag{1}$$

When x_0 is selected, the functions $b_1(x), \ldots, b_n(x)$ are uniquely determined by Eq. (1). The particular solution of $Ly = f$ constructed in Theorem 9.6 can now be written as follows:

$$
\begin{aligned}
y_p(x) &= b_1(x)y_1(x) + \cdots + b_n(x)y_n(x) \\
&= y_1(x) \int_{x_0}^{x} \frac{W^{n1}(z)f(z)}{W(z)}\, dz + \cdots + y_n(x) \int_{x_0}^{x} \frac{W^{nn}(z)f(z)}{W(z)}\, dz \\
&= \int_{x_0}^{x} \frac{y_1(x)W^{n1}(z) + \cdots + y_n(x)W^{nn}(z)}{W(z)} f(z)\, dz
\end{aligned}
$$

Here, $y_1(x), \ldots, y_n(x)$ can be brought under the integral sign since they are independent of the integration variable z. The result of Theorem 9.6 can thus be restated as follows.

THEOREM 9.7   Let L be a linear differential operator with leading coefficient 1 and continuous coefficients on J, let f be continuous on J, and let $y_1, \ldots, y_n$ be a basis for the null space of L. Then the function

$$y_p(x) = \int_{x_0}^{x} G(x, z)f(z)\, dz \tag{2}$$

where G is given by

$$G(x, z) = \frac{y_1(x)W^{n1}(z) + \cdots + y_n(x)W^{nn}(z)}{W(z)} \tag{3}$$

is a solution of the equation $Ly = f$, for $x \in J$.

DEFINITION 9.7   The function $G(x, z)$ defined by Eq. (3) is called the *one-sided Green's function* or *Green's function for the initial value problem* associated with the linear differential operator L.

The function G is independent of the choice of the point x_0 and also independent of the function f. It can be constructed by Eq. (3) from n independent solutions $y_1, \ldots, y_n$ of the equation $Ly = 0$. When G has been constructed, Eq. (2) gives a particular solution of $Ly = f$, for any continuous function f.

Example 1 Consider the equation $y'' - y = f$. As shown in Example 1 of Section 9.4, we can take $y_1 = e^x$, $y_2 = e^{-x}$, and

$$W^{11} = -e^{-x}, \quad W^{12} = -e^x, \quad W^{21} = -e^{-x}, \quad W^{22} = e^x$$

Therefore

$$G(x, z) = \frac{-e^x e^{-z} + e^{-x} e^z}{-2} = \tfrac{1}{2}[e^{x-z} - e^{-(x-z)}] = \sinh(x - z)$$

A particular solution is (taking $x_0 = 0$),

$$y_p(x) = \int_0^x f(z) \sinh(x - z)\, dz$$

and a general solution is

$$y(x) = c_1 e^x + c_2 e^{-x} + \int_0^x f(z) \sinh(x - z)\, dz \tag{4}$$

Equation (4) makes it possible to solve not just one differential equation, but rather all equations of the form $y'' - y = f$, merely by inserting the given function f into the integral. In the same way, Eq. (2) provides a solution of $Ly = f$ for all functions f.

Equation (2) assigns a unique function y_p to each function f. This suggests that it would be interesting to interpret Theorem 9.7 as a statement about inverting the linear transformation L of $C^n(J)$ into $C^0(J)$. (We shall state this for the real case, but our remarks are valid for the complex case too.) For every f in $C^0(J)$, the equation $Ly = f$ has a solution. Therefore the mapping L is onto $C^0(J)$. However, it is not one-to-one, since for given f there are many solutions of the differential equation. Therefore the differential operator L is not invertible. On the other hand, suppose that we impose a particular set of initial conditions, for example,

$$y(x_0) = 0, \quad y'(x_0) = 0, \quad \ldots, \quad y^{(n-1)}(x_0) = 0 \tag{5}$$

Then there is a unique solution of $Ly = f$ satisfying Eqs. (5). Consequently, if the domain of L is restricted to just those functions in $C^n(J)$ which satisfy Eqs. (5), then L is one-to-one and onto $C^0(J)$, and L^{-1} exists. The next theorem shows that Eq. (2) gives an explicit formula for L^{-1}.

THEOREM 9.8 Let L be a linear differential operator with coefficients in $C^0(J)$ and let $y_1, \ldots, y_n$ be a basis for the null space of L. Let $G(x, z)$ be the one-sided Green's function of L. If the domain of L is restricted to functions in $\mathbf{C}''(J)$ which satisfy the zero initial conditions in Eq. (5), the resulting operator L_0 is invertible, and

$$L_0^{-1} f(x) = \int_{x_0}^x G(x, z) f(z) \, dz \tag{6}$$

PROOF: It is known that there is a unique y satisfying $Ly = f$ and Eqs. (5). It is only necessary to verify that the y_p defined by Eq. (2) satisfies the differential equation and Eqs. (5). The former was proved in Theorem 9.7. A proof of the latter is sketched in the Exercises. ∎

Example 2 Find the inverse of the restricted operator L_0 if $L = y'' - y'$. Here $y_1 = 1$, $y_2 = e^x$ are linearly independent functions in the null space of L. We have

$$\mathcal{W}(x) = \begin{pmatrix} 1 & e^x \\ 0 & e^x \end{pmatrix}, \quad W(x) = e^x$$

$$W^{21} = -e^x, \quad W^{22} = 1$$

$$G(x, z) = \frac{-1 \cdot e^z + 1 \cdot e^x}{e^z} = e^{x-z} - 1$$

$$L_0^{-1} f(x) = \int_{x_0}^x (e^{x-z} - 1) f(z) \, dz$$

Additional information concerning Green's function is given in the exercises.

Exercises

1. Find the one-sided Green's function for each operator in Exercise 1 of Section 9.4.

2. Let $G(x, z)$ be defined by Eq. (3). Show that

$$W(z)G(x, z) = \begin{vmatrix} y_1(z) & y_2(z) & \cdots & y_n(z) \\ y_1'(z) & y_2'(z) & \cdots & y_n'(z) \\ & & \cdots & \\ y_1^{(n-2)}(z) & y_2^{(n-2)}(z) & \cdots & y_n^{(n-2)}(z) \\ y_1(x) & y_2(x) & \cdots & y_n(x) \end{vmatrix}$$

3. Leibniz's formula for differentiating integrals[3] is

$$\frac{d}{dx} \int_{g(x)}^{h(x)} f(x, z) \, dz = \int_{g(x)}^{h(x)} \frac{\partial f(x, z)}{\partial x} \, dz + f(x, h(x)) \frac{dh}{dx} - f(x, g(x)) \frac{dg}{dx}$$

---

[3] See any advanced calculus text, for example R. C. Buck, *Advanced Calculus*, 2nd ed., New York, McGraw-Hill, 1965.

Using this formula, verify that the function

$$y(x) = \int_0^x f(z) \sinh (x - z) \, dz$$

satisfies $y'' - y = f$ and $y(0) = y'(0) = 0$.

4. Verify that

$$y(x) = \int_0^x (e^{x-z} - 1)f(z) \, dz$$

satisfies $y'' - y' = f$ and $y(0) = y'(0) = 0$.

5. Show that the solution of $y'' + a^2 y = f$ with $y(0) = 0$, $y'(0) = 0$, is

$$y = \frac{1}{a} \int_0^x f(z) \sin a(x - z) \, dz$$

6. Let $L = x^2 D^2 - 2xD + 2$, as in Exercise 2 of Section 9.4. Let L_0 be the operator $x^{-2} L$ restricted to functions with $y(1) = 0$, $y'(1) = 0$. Find L_0^{-1}.

7. Repeat Exercise 6 for the operator $L = x^2 D^2 - xD + 1$ of Exercise 3, Section 9.4.

8. Using the rule for differentiating determinants and the result of Exercise 2, calculate

$$\frac{\partial}{\partial x} G(x, z), \ldots, \frac{\partial^{(n-2)}}{\partial x^{n-2}} G(x, z), \frac{\partial^{(n-1)}}{\partial x^{n-1}} G(x, z)$$

Show that all but the last of these reduces to zero for $x = z$, but

$$\left. \frac{\partial^{(n-1)}}{\partial x^{(n-1)}} G(x, z) \right|_{z=x} = 1$$

9. Show that $G(x, z)$ satisfies

$$[D_x^n + a_{n-1}(x)D_x^{n-1} + \cdots + a_0(x)]G(x, z) = 0$$

where $D_x = \partial/\partial x$, for x in J and z in J. Also

$$G = D_x G = \cdots = D_x^{n-2} G = 0 \quad \text{for} \quad x = z$$
$$D_x^{n-1} G = 1 \quad \text{for} \quad x = z$$

Thus G as a function of x is the particular solution of $LG = 0$ with the indicated "initial values" at $x = z$.

[*Remark:* By the existence-uniqueness theorem it follows that $G(x, z)$ is uniquely determined. In particular, G is independent of the choice of basis $\{y_1, \ldots, y_n\}$.]

10. Using the results in Exercises 8 and 9, and Leibniz's formula from Exercise 3, show that the function

$$y(x) = \int_{x_0}^{x} G(x, z)f(z) \, dz$$

satisfies $Ly = f$ and $y(x_0) = 0, \ldots, y^{(n-1)}(x_0) = 0$.

9.6   Undetermined coefficients

In this section we shall describe the second principal method for finding a particular solution y_p of a linear nonhomogeneous equation of the form

$$Ly = a_n y^{(n)} + a_{n-1} y^{(n-1)} + \cdots + a_0 y = f(x) \tag{1}$$

This method, known as the *method of undetermined coefficients*, is less general than the method of variation of constants in that it is applicable only when Ly is a differential operator with constant coefficients and only for special types of functions $f(x)$ in Eq. (1). However, under appropriate conditions, the method enables one to predict the form of a particular solution and then, with relatively little work, to actually compute one. Also, the method reappears in various guises throughout mathematical analysis and so is worth having in one's bag of tricks.

Begin by considering any particular solution of the differential equation $Ly = f(x)$. For most functions y the operations involved in the linear operator L would masticate the identity of y to the point that $f(x)$ would give no hint as to y's identity. However there are a few functions on which the effect of L is somewhat predictable. In particular, consider y of the form of a polynomial in x, an exponential $e^{\alpha x}$, $\sin \beta x$, or $\cos \beta x$. We treat these cases separately.

Case 1 Polynomial f   Consider first the case where $f(x)$ is a polynomial, that is, the equation considered is of the form

$$Ly = a_n y^{(n)} + a_{n-1} y^{(n-1)} + \cdots + a_0 y = b_m x^m + b_{m-1} x^{m-1} + \cdots + b_0$$

where $b_m \neq 0$. We will assume that $a_0, a_1, \ldots, a_n, b_0, b_1, \ldots, b_m$ are real. The condition $b_m \neq 0$ is really no restriction, serving only to establish the true order of the polynomial $f(x)$. The question arises, what sort of function y_p, when operated on by L, leads to a polynomial of the form of f? And the answer is evident—a polynomial. Consider the following example.

Example 1   Solve the differential equation

$$Ly = y''' - y'' + 4y' - 4y = 3x^2 + x$$

A real general solution of the corresponding homogeneous equation is (by methods of Section 9.3) $k_1 e^x + k_2 \cos 2x + k_3 \sin 2x$. Since f is a polynomial, to find a particular solution we take y_p of the form

$$y_p = c_2 x^2 + c_1 x + c_0$$

where c_0, c_1, and c_2 are constants to be determined. Then

$$y_p''' = 0, \quad y_p'' = 2c_2, \quad \text{and} \quad y_p' = 2c_2 x + c_1$$

Substituting these results into the differential equation yields

$$L y_p = -2c_2 + 8c_2 x + 4c_1 - 4c_2 x^2 - 4c_1 x - 4c_0 = 3x^2 + x$$

Since two polynomials are equal only if like powers of x have equal coefficients, this gives

$$-4c_2 = 3$$
$$8c_2 - 4c_1 = 1$$
$$-2c_2 + 4c_1 - 4c_0 = 0$$
$$c_2 = -\tfrac{3}{4}, \ c_1 = -\tfrac{7}{4}, \quad \text{and} \quad c_0 = -\tfrac{11}{8}$$

Therefore, a real general solution to the nonhomogeneous equation is

$$y = k_1 e^x + k_2 \cos 2x + k_3 \sin 2x - \tfrac{3}{4} x^2 - \tfrac{7}{4} x - \tfrac{11}{8}$$

In order to complete the case where $f(x)$ is a polynomial, we must consider the situation where $a_0 = a_1 = \cdots = a_{k-1} = 0$ for some $k \le n$, or in other words, where $\lambda = 0$ is a root of multiplicity k of the characteristic polynomial $p(r)$ associated with L.

Example 2   Solve $L y = y''' - 2y'' = x^2 - 2x$.

It is easy to see that $k_1 e^{2x} + k_2 x + k_3$ is a general solution of $L y = 0$. If we choose $y_p = c_0 + c_1 x + c_2 x^2$ for the particular solution, we obtain $L y = -4c_2$ and this cannot equal $x^2 - 2x$. The difficulty here is that $L = (D - 2)D^2$ so that L acting on a polynomial reduces the degree of the polynomial by two. This suggests choosing $y_p = x^2(c_0 + c_1 x + c_2 x^2)$ which gives

$$L y_p = 6c_1 + 24c_2 x - 4c_0 - 12c_1 x - 24c_2 x^2 = x^2 - 2x$$

Therefore, $c_2 = -\tfrac{1}{24}$, $c_1 = \tfrac{1}{12}$, and $c_0 = \tfrac{1}{8}$, which leads to the general solution

$$y = k_1 e^{2x} + k_2 x + k_3 + x^2(\tfrac{1}{8} + \tfrac{1}{12} x - \tfrac{1}{24} x^2)$$

On the basis of these examples, we conjecture that the following rule will always lead to a particular solution. Suppose that

$$f(x) = b_m x^m + b_{m-1} x^{m-1} + \cdots + b_0, \qquad b_m \ne 0.$$

If 0 is not a root of the characteristic polynomial, there is a solution y_p of $Ly = f$ of the form $y_p = c_m x^m + c_{m-1} x^{m-1} + \cdots + c_0$. If 0 is a root of multiplicity k of the characteristic polynomial, there is a solution y_p of the form

$$y_p = x^k(c_m x^m + c_{m-1} x^{m-1} + \cdots + c_0)$$

The validity of this rule is proved in Section 9.8.

Case 2 Polynomial times $e^{\alpha x}$ Suppose that

$$f(x) = e^{\alpha x}(b_m x^m + b_{m-1} x^{m-1} + \cdots + b_0)$$

Then a rule for determining the form of a particular solution is:

(a) $y_p = e^{\alpha x}(c_m x^m + c_{m-1} x^{m-1} + \cdots + c_0)$ provided α is not a root of $p(r)$;

(b) $y_p = e^{\alpha x} x^k(c_m x^m + c_{m-1} x^{m-1} + \cdots + c_0)$ if α is a k-fold root of $p(r)$.

We shall illustrate by examples. The demonstration that this rule *always* leads to a solution is given in Section 9.8. Note that $g(x) = b_0 e^{\alpha x}$ is the special case where $m = 0$.

Example 3 $Ly = y'' - 4y = xe^x$. Since $\alpha = 1$ is not a root of the characteristic polynomial $p(r) = r^2 - 4$, we try $y_p = e^x(c_0 + c_1 x)$. Then

$$y_p'' - 4y_p = e^x(-3c_1 x + 2c_1 - 3c_0)$$

In order that this be equal to xe^x it is necessary and sufficient that $-3c_1 = 1$, $2c_1 - 3c_0 = 0$, which yields $c_1 = -\frac{1}{3}$, $c_0 = -\frac{2}{9}$. Therefore,

$$y_p = -e^x(\tfrac{1}{3}x + \tfrac{2}{9})$$

and a general solution of the equation is

$$y = k_1 e^{2x} + k_2 e^{-2x} - \left(\frac{2}{9} + \frac{x}{3}\right) e^x$$

Example 4 $Ly = y'' - 4y = xe^{2x}$. Now $\alpha = 2$ and $p(\alpha) = 0$. We are in Case 2 (b) above, and accordingly try $y_p = xe^{2x}(c_0 + c_1 x)$. Calculation yields

$$y_p'' - 4y_p = e^{2x}(8c_1 x + 4c_0 + 2c_1)$$

Setting this equal to xe^{2x} we obtain $c_1 = \frac{1}{8}$, $c_0 = -\frac{1}{16}$. A general solution is

$$y = k_1 e^{2x} + k_2 e^{-2x} + \tfrac{1}{16} xe^{2x}(-1 + 2x)$$

Case 3 Polynomial times $e^{\alpha x} \sin \beta x$ or $e^{\alpha x} \cos \beta x$ Suppose that $f(x)$ is a real function of the form

$$f(x) = e^{\alpha x}(b_m x^m + b_{m-1} x^{m-1} + \cdots + b_0) \sin \beta x \tag{2}$$

or a similar expression with $\cos \beta x$ instead of $\sin \beta x$ or a sum of both. The appropriate choice in this situation is as follows:

(a) If $\alpha \pm i\beta$ is not a root of $p(r)$,

$$y_p = e^{\alpha x}(c_m x^m + c_{m-1}x^{m-1} + \cdots + c_0) \sin \beta x$$
$$+ e^{\alpha x}(d_m x^m + d_{m-1}x^{m-1} + \cdots + d_0) \cos \beta x \qquad (3)$$

(b) If $\alpha \pm i\beta$ is a root of multiplicity k of $p(r)$, y_p is of the form in Eq. (3) multiplied by x^k.

In Case 3 it is perhaps easier to use the "method of complex arithmetic," which we shall now explain. Recall that

$$e^{i\beta x} = \cos \beta x + i \sin \beta x$$

and therefore if c is any real number,

$$\mathrm{Re}\,(ce^{i\beta x}) = c\,\mathrm{Re}\,(e^{i\beta x}) = c \cos \beta x$$
$$\mathrm{Im}\,(ce^{i\beta x}) = c\,\mathrm{Im}\,(e^{i\beta x}) = c \sin \beta x$$

Therefore $f(x)$, as given by Eq. (2), can be written

$$f(x) = \mathrm{Im}\,[(b_m x^m + b_{m-1}x^{m-1} + \cdots + b_0)e^{\alpha x}e^{i\beta x}]$$
$$= \mathrm{Im}\,[(b_m x^m + b_{m-1}x^{m-1} + \cdots + b_0)e^{(\alpha + i\beta)x}]$$

assuming that $b_m, b_{m-1}, \ldots, b_0$ are real constants. We now solve the equation

$$Ly = (b_m x^m + b_{m-1}x^{m-1} + \cdots + b_0)e^{(\alpha + i\beta)x}$$

using the rule for Case 2 and obtaining a complex function as a particular solution. (The validity of the rule in this situation is shown in the exercise in Section 9.8.) The imaginary part of this solution will then be a solution of the original equation (see Exercise 9). We illustrate the method by an example.

Example 5 $Ly = y'' - 4y = e^x \sin 2x$. Since $e^x \sin 2x = \mathrm{Im}\,(e^{(1+2i)x})$, we consider the equation

$$y'' - 4y = e^{(1+2i)x}$$

Since $1 + 2i$ is not a root of $p(r) = r^2 - 4$, we have Case 2(a) and accordingly try

$$y_p = c_0 e^{(1+2i)x}$$

We readily obtain

$$y_p'' - 4y_p = c_0[(1+2i)^2 - 4]e^{(1+2i)x} = e^{(1+2i)x},$$

hence

$$c_0 = \frac{1}{(1+2i)^2 - 4} = \frac{1}{-7+4i}$$

Thus

$$y_p = \frac{e^{(1+2i)x}}{-7+4i} = e^x \frac{\cos 2x + i \sin 2x}{-7+4i}$$

It remains to take the imaginary part of this function. To do this, we must obtain an equal expression with real denominator (since it is not true that the imaginary part of the quotient is the quotient of the imaginary parts of the numerator and denominator). We write

$$y_p = e^x \frac{(\cos 2x + i \sin 2x)(-7 - 4i)}{(-7 + 4i)(-7 - 4i)}$$

$$= \frac{e^x}{65} [(-7 \cos 2x + 4 \sin 2x) + i(-4 \cos 2x - 7 \sin 2x)]$$

Hence

$$\text{Im}\,(y_p) = \frac{-e^x}{65}(4 \cos 2x + 7 \sin 2x)$$

Thus, a general solution of the original equation is

$$y = k_1 e^{2x} + k_2 e^{-2x} - \frac{e^x}{65}(4 \cos 2x + 7 \sin 2x)$$

Exercises

1. Prove: If y_1 and y_2 are particular solutions to $Ly = g_1$, $Ly = g_2$, respectively, then $y_1 + y_2$ is a particular solution to $Ly = g_1 + g_2$.

2. Find a particular solution of each equation by the method of undetermined coefficients.

   (a) $y''' - 3y'' + 2y' + y = x + 1$
   (b) $y''' - 3y'' = x + 1$
   (c) $D'' y = x + 5$
   (d) $(D^2 - 1)y = e^x$
   (e) $(4D^2 + 4D + 1)y = e^{-x/2}$
   (f) $\dfrac{d^2 y}{dx^2} - 4y = \sin^2 x$
   (g) $(D - 1)^2 y = e^x$
   (h) $(D - 1)^2 y = e^x \sin x$
   (i) $(D^2 + D + 1)y = e^{-x/2}\left(\sin \dfrac{\sqrt{3}}{2} x - 2 \cos \dfrac{\sqrt{3}}{2} x\right)$
   (j) $(D^2 + 4)y = \sin 2x$

3. Find y such that $y^{(4)} + 2y^{(2)} + y = 3x + 4$, $y(0) = y'(0) = 0$, $y^{(2)}(0) = y^{(3)}(0) = 1$.

4. Using the result of Exercise 1, find a general solution for

$$y^{(4)} - 2y^{(3)} + 2y^{(2)} = 3e^x + 2xe^{-x} + e^{-x}\sin x$$

5. Prove that $y_p = e^{ax}/p(a)$ is a solution of $Ly = e^{ax}$, if p is the characteristic polynomial of L and $p(a) \neq 0$.

6. If a is a root of p of multiplicity m, then $p(r) = (r - a)^m q(r)$ where $q(r) \neq 0$. Show that in this case,

$$y_p = \frac{x^m e^{ax}}{m!\, q(a)}$$

is a solution of $Ly = e^{ax}$.

7. The method of factorization of operators can be used to obtain a general solution of $Ly = f$ if L has constant coefficients, even if f is not of one of the forms discussed in this section. For example, letting $z = (D - 1)y$ reduces the equation $(D^2 - D)y = \ln x$, $x > 0$, to $Dz = \ln x$. Hence, $z = c_1 + x \ln x - x$. Then $D(ye^{-x}) = e^{-x}(c_1 + x \ln x - x)$ and one more integration yields y. Complete this solution.

8. Use the method of factorization of operators as in Exercise 7 to find a general solution of each equation.

(a) $(D^2 - 1)y = e^x$
(b) $(4D^2 + 4D + 1)y = e^{-x/2}$
(c) $(D^2 + 4)y = \sin 2x$
(d) $(D^2 - 1)y = \tan x$
(e) $y'' - y' = \dfrac{1}{x}$, $x > 0$

(f) $y'' + y = \dfrac{1}{x}$, $x > 0$

9. Let L be a linear differential operator with real coefficients $a_n(x), \ldots, a_0(x)$. Let $f(x) = \phi(x) + i\psi(x)$ where ϕ and ψ are real and let $y(x)$ be a solution of $Ly = f$. Prove that $u(x) = \text{Re}\,(y(x))$ is a solution of $Ly = \phi$ and $v(x) = \text{Im}\,(y(x))$ is a solution of $Ly = \psi$.

10. If p is the characteristic polynomial of L and $p(r)$ contains only even powers of r, we can write $p(r) = q(r^2)$ where q is a polynomial. Show that a solution of $Ly = \cos ax$ is $y = (\cos ax)/q(-a^2)$ and a solution of $Ly = \sin ax$ is $y = (\sin ax)/q(-a^2)$, provided $q(-a^2) \neq 0$.

9.7   Reduction and substitution

The methods described in Section 9.3 for solving a linear homogeneous equation

$$Ly = y^{(n)} + a_{n-1}(x)y^{(n-1)} + \cdots + a_0(x)y = 0 \tag{1}$$

apply only when L has constant coefficients. Methods applicable to the general case (non-constant coefficients) depend mostly on the use of infinite series, numerical methods, and methods of approximation. In some special cases, however, it may be possible to replace a given equation by one of a more amenable type. In this section we shall suggest two such possibilities.

The first method to be discussed is the method of *reduction of order*, and can be used if one solution of Eq. (1) is known. Suppose that $y = u_1(x)$ is such a solution. Then make the substitution

$$y(x) = u_1(x)z(x) \tag{2}$$

Then $y' = u_1 z' + u_1' z$, $y'' = u_1 z'' + 2u_1' z' + u_1'' z$, and in general

$$y^{(r)} = \sum_{j=0}^{r} \binom{r}{j} u_1^{(j)} z^{(r-j)} \qquad (r = 1, 2, \ldots, n) \tag{3}$$

Substitution into Eq. (1) yields

$$u_1 z^{(n)} + b_{n-1}(x)z^{(n-1)} + \cdots + b_1(x)z' = 0 \tag{4}$$

where $b_{n-1}, \ldots, b_1$ are new coefficients depending on u_1. The term in z has dropped out, since its coefficient is

$$u_1^{(n)} + a_{n-1}u_1^{(n-1)} + \cdots + a_0 u_1 = 0$$

If z' is replaced by a new variable w in Eq. (4), the result is

$$u_1 w^{(n-1)} + b_{n-1}(x)w^{(n-2)} + \cdots + b_1(x)w = 0 \tag{5}$$

This is a linear homogeneous equation of order one less than the original, and as such may possibly be easier to solve.

Example 1   Consider the equation

$$x^2 y'' - 2(1 + x^2)xy' + 2(1 + x^2)y = 0$$

After some experimentation (guessing), one may find that $y = x$ is a solution. Therefore make the substitution $y = xz$. This gives $y' = xz' + z$, $y'' = xz'' + 2z'$, and the equation reduces to

$$x^2(xz'' + 2z') - 2(1 + x^2)x(xz' + z) + 2(1 + x^2)xz = 0$$
$$x^3(z'' - 2xz') = 0$$

Put $z' = w$ and we get the first order equation

$$w' - 2xw = 0$$

A general solution is $w = c_1 e^{x^2}$. Therefore, $z = c_1 \int e^{x^2} \, dx$, and $y = c_1 x \int e^{x^2} \, dx$ is a solution of the original equation. The set $\{x, x \int e^{x^2} \, dx\}$ is a basis for the solution space (on any interval not containing $x = 0$).

This method can be applied in the more general case in which two or more independent solutions of Eq. (1) are known. The reader should see the exercises and the references listed at the end of the chapter for more information.

The second method to be presented is the use of a change of variables which will convert the differential equation to one which has constant coefficients. We shall find that the method is successful only under certain special conditions, but is effective in these cases.

In order to simplify the presentation, we shall consider the second order equation

$$y'' + a_1(x)y' + a_0(x)y = 0 \tag{6}$$

where we assume that $a_1(x)$ and $a_0(x)$ are real and continuous on some interval J. Without loss of generality we can assume that $a_0(x)$ is nonzero on this interval, since otherwise we could consider a subinterval on which $a_0(x)$ is nonzero. We shall try to simplify Eq. (6) by means of a transformation of the independent variable of the form

$$\xi = \int^x f(x_1) \, dx_1 \quad \text{for} \quad x \text{ in } J, \tag{7}$$

where the function f remains to be chosen, and where the integral is any convenient choice of the indefinite integral.

Under this change of variable, the function $y(x)$ goes into a function η of $\xi; \eta(\xi) = y(x)$. We shall assume that Eq. (7) defines a one-to-one correspondence between the x-interval J and a ξ-interval J_ξ, so that to each ξ there is a unique value of η. Now we have

$$\frac{dy}{dx} = \frac{d\eta}{d\xi} \frac{d\xi}{dx} = f(x) \frac{d\eta}{d\xi} \tag{8}$$

$$\frac{d^2 y}{dx^2} = f(x)^2 \frac{d^2 \eta}{d\xi^2} + f'(x) \frac{d\eta}{d\xi}$$

(assuming the existence of f'). Hence Eq. (6) is transformed into

$$[f(x)]^2 \frac{d^2 \eta}{d\xi^2} + [f'(x) + a_1(x)f(x)] \frac{d\eta}{d\xi} + a_0(x)\eta = 0 \tag{9}$$

If Eq. (9) is to be reducible to one with constant coefficients, its coefficients must be proportional; that is, there must be constants β_1 and β_2 and a function $f(x)$ such that

$$[f(x)]^2 = \beta_2 a_0(x), \qquad f'(x) + a_1(x)f(x) = \beta_1 a_0(x) \tag{10}$$

We may suppose that β_2 is ± 1, since multiplication of $f(x)$ by a constant is irrelevant to our purpose. Taking $\beta_2 = 1$ if $a_0(x) > 0$ and $\beta_2 = -1$ if $a_0(x) < 0$, we obtain (taking the positive square root for f) the equations

$$f(x) = |a_0(x)|^{1/2}, \qquad f'(x) + a_1(x)f(x) = \beta_1 a_0(x) \tag{11}$$

Thus, if the transformation is to succeed, the coefficient functions $a_1(x)$ and $a_0(x)$ must be related by Eqs. (11), wherein β_1 is some constant. On the other hand, if Eqs. (11) are satisfied, the transformation will succeed as stated in the next theorem.

THEOREM 9.9 Consider the second order equation $y'' + a_1(x)y' + a_0(x)y = 0$, where $a_0(x)$ and $a_1(x)$ are real and continuous and $a_0(x)$ is nonzero on an interval J. Suppose there is a constant β_1 such that the functions $a_0(x)$, $a_1(x)$ satisfy Eqs. (11) on J. Then Eq. (7) defines a one-to-one mapping of the interval J onto an interval J_ξ under which Eq. (5) is transformed into an equation with constant coefficients.

PROOF: Since $f(x)$ as defined by Eq. (11) is positive on J, Eq. (7) defines a one-to-one map of J onto an interval J_ξ. Under this map, Eq. (6) is transformed into Eq. (9), that is, into

$$|a_0(x)| \frac{d^2\eta}{d\xi^2} + \beta_1 a_0(x) \frac{d\eta}{d\xi} + a_0(x)\eta = 0$$

Dividing out $a_0(x)$, we obtain an equation with constant coefficients. ∎

Equations (11) can be rewritten in the form

$$a_0'(x) + 2a_0(x)a_1(x) = 2\beta_1 |a_0(x)|^{3/2} \tag{12}$$

Of course, such a relation is satisfied only for very special equations. Generally, we cannot expect a change of variable of the type in Eq. (7) to yield an equation with constant coefficients.

Example 2 Consider the equation

$$xy'' + (x^2 - 1)y' + x^3y = 0, \qquad \text{for } x > 0$$

Dividing by x we obtain an equation of the form in Eq. (6) with $a_0(x) = x^2$, $a_1(x) = x - x^{-1}$. Taking $f(x) = |x^2|^{1/2} = x$, we see that Eqs. (11) are satisfied with $\beta_1 = 1$. Hence we let

$$\xi = \int^x x_1 \, dx_1 = \frac{x^2}{2}, \qquad x = \sqrt{2\xi}$$

This maps $x > 0$ onto $\xi > 0$. Now

$$\frac{dy}{dx} = x\frac{d\eta}{d\xi}, \qquad \frac{d^2y}{dx^2} = x^2\frac{d^2\eta}{d\xi^2} + \frac{d\eta}{d\xi}$$

and the resulting equation, after dividing by x^3, is

$$\frac{d^2\eta}{d\xi^2} + \frac{d\eta}{d\xi} + \eta = 0$$

The characteristic roots are $\lambda = (-1 \pm i\sqrt{3})/2$ and a real general solution is

$$\eta = e^{-\xi/2}\left(c_1 \cos\frac{\xi\sqrt{3}}{2} + c_2 \sin\frac{\xi\sqrt{3}}{2}\right)$$

Transforming back to the original variables we get

$$y = e^{-x^2/4}\left(c_1 \cos\frac{x^2\sqrt{3}}{4} + c_2 \sin\frac{x^2\sqrt{3}}{4}\right), \qquad x > 0$$

Many other possibilities exist for simplifying a given differential equation by a change of variables. See the references at the end of the chapter.

Exercises

1. Show that each equation has a solution of the form x^m for some integer m. Then use the method of reduction of order to obtain a general solution.

   (a) $(x^2 \ln x)y'' - (x + 4x \ln x)y' + (2 + 6 \ln x)y = 0$
   (b) $xy'' - (x + 1)y' + 2y = 0$
   (c) $x^2y'' + x(2x^2 + 1)y' + (2x^2 - 1)y = 0$

2. Show that each equation has a solution of the form e^{ax} and then use the method of reduction of order to obtain a general solution.

   (a) $xy'' + (2x^2 - 4x - 1)y' - 2(2x^2 - 2x - 1)y = 0$
   (b) $y'' - (2 - \sec x \csc x)y' + (1 - \sec x \csc x)y = 0$

3. Let $u_1(x)$ be a solution of $y'' + a_1(x)y' + a_0(x)y = 0$.

   (a) By use of the method of reduction of order, deduce that another solution is

   $$y = u_1(x)\int^x [u_1(t)]^{-2} \exp\left[-\int^t a_1(s)\,ds\right]dt$$

   on any interval where u_1 does not vanish.

   (b) Prove that these solutions are linearly independent.

4. Use the formula in Exercise 3 to find a second solution in each part of Exercise 1.

5. If one solution of $Ly = 0$ is known, the order of a corresponding non-homogeneous equation can be reduced in the same way as before. For example, consider the equation $xy'' + 2y' = 2x$. The homogeneous equation $xy'' + 2y' = 0$ has a solution $y = x^{-1}$. Therefore, we put $y = x^{-1}z$ and obtain $z'' = 2x$. Hence

$$z = c_1 x + c_2 + x^3/3 \quad \text{and} \quad y = c_1 + c_2 x^{-1} + x^2/3$$

Use this method to find a general solution for each of the following equations. (Refer to Exercises 1 and 2.)

(a) $(x^2 \ln x)y'' - (x + 4x \ln x)y' + (2 + 6 \ln x)y = x^2 \ln x \quad (x > 1)$
(b) $x^2 y'' + x(2x^2 + 1)y' + (2x^2 - 1)y = x^3$
(c) $xy'' + (2x^2 - 4x - 1)y' - 2(2x^2 - 2x - 1)y = x^2 e^{-x^2}$
(d) $y'' - (2 - \sec x \csc x)y' + (1 - \sec x \csc x)y = 1$

6. Using mathematical induction, prove the validity of Eq. (3).

7. Find a general solution for each equation by reducing to an equation with constant coefficients.

(a) $y'' + (2e^x - 1)y' + e^{2x}y = 0$
(b) $(x \ln x)y'' + [x (\ln x)^2 - 1]y' + x (\ln x)^3 y = 0, \qquad x > 1$
(c) $xy'' + (\beta x^{\alpha+1} - \alpha)y' + x^{2\alpha+1}y = 0, \quad x > 0, \quad \alpha \neq -1$

8. Suppose that Eq. (6) has two linearly independent solutions of the form

$$y_1(x) = [\phi(x)]^{\alpha_1}, \qquad y_2(x) = [\phi(x)]^{\alpha_2}$$

where α_1 and α_2 are distinct constants and $\phi(x)$ is some sufficiently smooth positive function on $a \leq x \leq b$. Prove that the transformation given in Eq. (7) with

$$f(x) = \frac{\phi'(x)}{\phi(x)}$$

carries Eq. (6) into an equation that is proportional to an equation with constant coefficients. [*Hint:* $\ln \phi(x) = \int f(x_1) \, dx_1$.]

9. Derive conditions under which a third order equation

$$y''' + a_2(x)y'' + a_1(x)y' + a_0(x)y = 0$$

can be reduced to an equation with constant coefficients by a transformation such as in Eq. (7).

10. Show that Eqs. (11) can be rewritten in the form in Eq. (12).

11. Show that the substitution

$$y = v \exp\left[-\tfrac{1}{2} \int a_1(x)\, dx\right]$$

reduces the equation

$$y'' + a_1(x)y' + a_0(x)y = f$$

to the form

$$v'' + (a_0 - \tfrac{1}{4} a_1{}^2 - \tfrac{1}{2}a_1')v = f \exp\left[\tfrac{1}{2} \int a_1(x)\, dx\right]$$

which contains no term in v'.

12. Solve the equation

$$y'' + \frac{1}{x}\, y' - \frac{1}{4x^2}\, y = \frac{1}{\sqrt{x}}, \qquad (x > 0)$$

9.8   Validity of the method of undetermined coefficients

The purpose of this section is to prove the following theorem.

THEOREM 9.10   Let

$$f(x) = e^{\alpha x} p(x) \sin \beta x + e^{\alpha x} q(x) \cos \beta x$$

where $p(x) = b_s x^s + \cdots + b_0$ and $q(x) = a_s x^s + \cdots + a_0$, and where α, β, a_j, b_j $(j = 0, 1, \ldots, s)$ are real. Let the operator L have real constant coefficients and let

$$L = (D - \lambda_1)^{m_1} \cdots (D - \lambda_k)^{m_k}$$

where $\lambda_1, \ldots, \lambda_k$ are the characteristic roots and $m_1, \ldots, m_k$ are their multiplicities. Then if $\alpha + i\beta$ is not equal to any λ, the equation $Ly = f$ has a solution of the form

$$e^{\alpha x}\left(\sum_{j=0}^{s} c_j x^j\right) \sin \beta x + e^{\alpha x}\left(\sum_{j=0}^{s} d_j x^j\right) \cos \beta x \tag{1}$$

where $c_j, d_j \in \mathbf{R}$ $(j = 0, 1, \ldots, s)$. If $\alpha + i\beta$ is equal to a characteristic root, say $\alpha + i\beta = \lambda_1$, then there is a solution of the form

$$x^{m_1} e^{\alpha x}\left(\sum_{j=0}^{s} c_j x^j\right) \sin \beta x + x^{m_1} e^{\alpha x}\left(\sum_{j=0}^{s} d_j x^j\right) \cos \beta x \tag{2}$$

where $c_j, d_j \in \mathbf{R}$ $(j = 0, 1, \ldots, s)$.

PROOF: The given function $f(x)$ is a particular solution of a linear homogeneous differential equation

$$\begin{aligned} L_1 y &= [D - (\alpha + i\beta)]^{s+1}[D - (\alpha - i\beta)]^{s+1} y \\ &= [(D - \alpha)^2 + \beta^2]^{s+1} y = 0 \end{aligned}$$

as can be seen by writing out a general solution of $L_1 y = 0$ by the method of Section 9.3. Now let y_p be a particular solution of $L y_p = f$. Then $L_1 L y_p = L_1 f = 0$. Therefore, y_p is also a solution of the equation $L_1 L y = 0$, that is, of the equation

$$[D - (\alpha + i\beta)]^{s+1}[D - (\alpha - i\beta)]^{s+1}(D - \lambda_1)^{m_1} \cdots (D - \lambda_k)^{m_k} y_p = 0 \qquad (3)$$

If $\alpha + i\beta$ is different from any λ_j, a general solution of this equation is the sum of a general solution of $L y = 0$ and a general solution of $L_1 y = 0$. Hence, $y_p = y_0 + y_1$, where y_0 is in the null space of L and y_1 is in the null space of L_1. Therefore, y_p must be of the form

$$y_p(x) = p_1(x)e^{\lambda_1 x} + \cdots + p_k(x)e^{\lambda_k x} + r_1(x)e^{(\alpha + i\beta)x} + r_2(x)e^{(\alpha - i\beta)x}$$

where $p_1, \ldots, p_k, r_1, r_2$ are polynomials, possibly complex, with r_1 and r_2 of degree at most s. However, $L y_p = f$ and

$$L[p_j(x)e^{\lambda_j x}] = 0, \qquad j = 1, \ldots, k \qquad (4)$$

and therefore we must have

$$L[r_1(x)e^{(\alpha + i\beta)x} + r_2(x)e^{(\alpha - i\beta)x}] = f(x)$$

that is, there is a particular solution of $L y = f$ of the form $r_1(x)e^{(\alpha + i\beta)x} + r_2(x)e^{(\alpha - i\beta)x}$. Taking the real part, we obtain a solution of the form in Eq. (1). If $\alpha + i\beta = \lambda_1, \alpha - i\beta = \lambda_2$, Eq. (3) becomes

$$(D - \lambda_1)^{m_1 + s + 1}(D - \bar{\lambda}_1)^{m_1 + s + 1} \cdots (D - \lambda_k)^{m_k} y_p = 0 \qquad (5)$$

Therefore, y_p must be of the form

$$y_p(x) = r_1(x)e^{\lambda_1 x} + r_2(x)e^{\bar{\lambda}_1 x} + \cdots + p_k(x)e^{\lambda_k x}$$

where r_1 and r_2 are polynomials of degree at most $m_1 + s$. Since Eq. (4) is still valid, we deduce that

$$L[r_1(x)e^{\lambda_1 x} + r_2(x)e^{\bar{\lambda}_1 x}] = f(x) \qquad (6)$$

Since λ_1 is a characteristic root of L of multiplicity m_1,

$$L(x^j e^{\lambda_1 x}) = 0 \quad \text{for } j = 0, 1, \ldots, m_1 - 1$$

Therefore, from Eq. (6) we get

$$L[\hat{r}_1(x)e^{\lambda_1 x} + \hat{r}_2(x)e^{\bar{\lambda}_1 x}] = f(x)$$

where $\hat{r}_1(x)$ and $\hat{r}_2(x)$ are polynomials containing powers of x from m_1 to $m_1 + s$; that is, there is a particular solution of the form

$$x^{m_1}(\gamma_0 + \gamma_1 x + \cdots + \gamma_s x^s)e^{\lambda_1 x} + x^{m_1}(\delta_0 + \delta_1 x + \cdots + \delta_s x^s)e^{\bar{\lambda}_1 x}$$

(where the γ_j and δ_j are constants, possibly complex). Taking the real part, we see that there is a particular solution of the form in Eq. (2). This completes the proof.   ∎

It is also possible to prove that the constants c_j, d_j in Eq. (1) or Eq. (2) are uniquely determined; that is, there is exactly one solution of the indicated form.[4]

Exercise

1. Formulate and prove a theorem which establishes the form of a complex particular solution of the equation $Ly = f$ where

$$f(x) = e^{(\alpha + i\beta)x}p(x)$$

and where α and β are real and $p(x)$ is a polynomial with coefficients in **C**.

Miscellaneous Exercises

1. In Section 2.5, it was indicated that in an electric circuit with generator, resistor, inductor, and capacitor in series, the equation

$$L\ddot{q} + R\dot{q} + \frac{1}{C}q = E(t) \tag{1}$$

describes the charge $q(t)$ on the capacitor at time t. Consider an LC circuit, that is, a circuit in which $R = 0$. Assume that $E(t) = E_0$ is a constant, and that at time $t = 0$ there is no charge on the capacitor and no current flowing in the circuit. Find $q(t)$ and $i(t)$, and sketch their graphs.

2. Repeat Exercise 1 if $E(t) = E_0 \cos \omega t$, where ω is a constant and $\omega \neq 1/\sqrt{LC}$. Sketch $q(t)$ and $i(t)$ for $L = C = 1$ and for $\omega = 2$ and $\omega = \pi$.

3. Find the solution of $L\ddot{q} + q/C = E_0 \cos \omega t$, $\omega \neq 1/\sqrt{LC}$, with arbitrary initial conditions $i(0) = i_0$, $q(0) = q_0$.

4. (a) Show that every solution of $L\ddot{q} + R\dot{q} + q/C = 0$ tends to zero as $t \to \infty$ if L, R, and C are positive constants.

---

[4] See E. A. Coddington, *An Introduction to Ordinary Differential Equations*, Englewood Cliffs, N. J., Prentice-Hall, 1961, p. 96.

(b) Show that every solution of Eq. (1), where $E(t) = E_0$, is the sum of $E_0 C$ and a term which tends to zero as $t \to \infty$. The latter is called the *transient*, and $E_0 C$ is called the *steady-state*.

5. Find a general solution of $L\ddot{q} + q/C = E_0 \cos \omega t$ when $\omega = 1/\sqrt{LC}$. Show that every solution oscillates with increasing amplitude as t increases, that is, successive maxima of $[q(t)]$ tend to ∞ as $t \to \infty$. This phenomenon is called *resonance*. It occurs when the applied or generated voltage oscillates at the *natural frequency*, that is, the frequency of oscillation of the circuit with no applied voltage.

6. The problem

$$y'' + y = 0 \qquad \text{on } 0 \le x \le 1$$
$$y(0) = 0 \qquad y(1) = 0$$

is called a *boundary value problem* because a solution is sought which satisfies conditions at the two boundaries, $x = 0$ and $x = 1$. These conditions are called *boundary conditions*. Show that $y \equiv 0$ is the only solution of this problem.

7. Show that there are infinitely many solutions of the boundary value problem

$$y'' + \pi^2 y = 0, \qquad 0 \le x \le 1$$
$$y(0) = 0, \qquad y(1) = 0$$

8. A long straight rod subjected to axial compressive forces P at each end is compressed and retains its initial shape as long as P does not exceed a certain critical value. When P reaches this critical value, the rod buckles suddenly and becomes curved. It is known that for small deflections y

$$\frac{d^2 y}{dx^2} = - \frac{Py}{EI}$$

where E and I are constants depending on the physical characteristics of the rod. Also,

$$y(0) = 0, \qquad y(l) = 0$$

where l is the length of the rod, since the ends of the rod must remain on the x-axis. Show that the only solution of this boundary value problem is $y \equiv 0$ unless P has one of the values

$$P_n = \frac{EI\pi^2 n^2}{l^2}, \qquad n = 1, 2, \ldots$$

Thus, $P_1 = EI\pi^2/l^2$ may be considered to be the smallest force at which buckling occurs.

COMMENTARY

Section 9.2   There are numerous texts on the theory of complex functions. An elementary approach is taken in:

HAMILTON, H., *A Primer of Complex Variables*, Belmont, Calif., Brooks/Cole, 1966.
One of the standard textbooks at a more advanced level is:
AHLFORS, L., *Complex Analysis*, New York, McGraw-Hill, 1953.

An extensive discussion of differential equations from the point of view of functions of a complex variable can be found in:

INCE, E. L., *Ordinary Differential Equations*, London, Longmans, Green, 1927. Reprint, New York, Dover, 1944 and 1953.

The proof of the existence–uniqueness theorem (Theorem 8.4) in the complex case is given in:

CODDINGTON, E. A., *An Introduction to Ordinary Differential Equations*, Englewood Cliffs, N.J., Prentice-Hall, 1961, Section 9 of Chapter 5 and Chapter 6.

Section 9.7   For further discussion of the method of reduction of order, see:

INCE, E. L., *op. cit.*, Chapter 5.
FORD, L. R., *Differential Equations*, New York, McGraw-Hill, 1955.

This method of reduction to an equation with constant coefficients was published (in more general form) in:

BREUER, S. and D. GOTTLIEB, "The reduction of linear ordinary differential equations to equations with constant coefficients," *J. Math. Anal. Appl.*, vol. 32, 67–76 (1970).

Many other special transformations are listed in the compendious work:

KAMKE, E., *Differentialgleichungen, Lösungsmethoden und Lösungen*. Band 1, *Gewöhnliche Differentialgleichungen*, New York, Chelsea, 1948.

Section 9.8   An algebraic approach to these results is given in:

GOLOMB, M., "An algebraic method in differential equations," *Amer. Math. Monthly*, vol. 72, 1107–1110 (1965).

Elementary matrix operations and systems of linear equations[1]

10.1 Gaussian elimination method

In Chapter 7, the theory of determinants was developed and used to calculate matrix inverses and solutions of systems of linear algebraic equations. It was pointed out there that other methods are better from the computational standpoint. In this chapter we shall explain some of these methods and their theoretical ramifications.

The starting point of our investigation is the method of solving a set of linear equations by successive elimination of variables, a method which has probably been encountered previously by most readers. This is illustrated in the next example.

Example 1   Consider the system

$$
\begin{aligned}
x + 2y + z &= 3 \\
x - y + 2z &= 9 \\
x + y - z &= -2
\end{aligned}
\tag{1}
$$

If the first equation is subtracted from the second and from the third, the result is the system

$$
\begin{aligned}
x + 2y + z &= 3 \\
-3y + z &= 6 \\
-y - 2z &= -5
\end{aligned}
\tag{2}
$$

---

[1] Chapters 4 through 6 provide sufficient background for reading this chapter.

Now if one third of the second equation is subtracted from the third equation, the resulting system is

$$
\begin{aligned}
x + 2y + \ z &= \ \ \ 3 \\
- 3y + \ z &= \ \ \ 6 \\
-\tfrac{7}{3} z &= -7
\end{aligned}
\tag{3}
$$

System (3) is easy to solve because it is in triangular form. Starting from the bottom we get $z = 3$, then $-3y + 3 = 6$ or $y = -1$, and finally $x - 2 + 3 = 3$ or $x = 2$. It is easy to check that $x = 2$, $y = -1$, $z = 3$ satisfies System (1).

Example 2 Consider

$$
\begin{aligned}
x + 2y + z &= 1 \\
x - \ y + z &= 3 \\
x + \ y + z &= 0
\end{aligned}
\tag{4}
$$

Subtracting the first equation from the second and third equations we get

$$
\begin{aligned}
x + 2y + z &= \ \ \ 1 \\
- 3y \ \ \ \ &= \ \ \ 2 \\
- y \ \ \ \ &= -1
\end{aligned}
\tag{5}
$$

The latter two equations are contradictory or inconsistent since y cannot equal both 1 and $-\tfrac{2}{3}$. Therefore, System (4) has no solution.

Example 3 Consider the system

$$
\begin{aligned}
x + 2y + z &= \ \ \ 1 \\
x - \ y + z &= -2 \\
x + \ y + z &= \ \ \ 0
\end{aligned}
\tag{6}
$$

This time subtraction of the first equation from the others yields

$$
\begin{aligned}
x + 2y + z &= \ \ \ 1 \\
- 3y \ \ \ \ &= -3 \\
- y \ \ \ \ &= -1
\end{aligned}
\tag{7}
$$

From the last two equations we find $y = 1$. The first equation is then satisfied by any x and z for which $x + z = -1$. This shows that the System (6) has infinitely many solutions.

The method illustrated in these examples is called the method of Gaussian elimination[2] or reduction. In this method the original system is replaced by a sequence of systems, all equivalent to the original system in the sense that they have the same solutions. The last system in the sequence is in a special form from which the solutions can easily be deduced. There are many variants of the

---

[2] Named after K. F. Gauss (1777–1855), a famous German mathematician.

basic scheme, but all are designed to provide a systematic procedure for generating the sequence of systems. We shall explain a method which depends on repeated applications of three kinds of operations:

Type 1 The interchange of two equations of the system.
Type 2 The replacement of one equation of the system by a nonzero multiple of itself.
Type 3 The replacement of one equation by the sum of that equation and any other equation.

The reader should convince himself that a new system resulting from the application of any one of these operations has exactly the same solutions as the original system. Consequently, any finite sequence of such operations must result in a system having the same solutions as the original. Many other simple operations can be generated by a sequence of operations of Types 1, 2, and 3. In Example 1, subtraction of equation one from equation two can be accomplished by first multiplying equation one by -1 (Type 2), then replacing equation two by the sum of equations one and two (Type 3), and finally multiplying equation one by -1 (Type 2).

We know that a system of linear equations can be written conveniently in matrix-vector form. For example, System (1) can be written in the more flexible notation

$$
\begin{aligned}
x_1 + 2x_2 + x_3 &= 3 \\
x_1 - x_2 + 2x_3 &= 9 \\
x_1 + x_2 - x_3 &= -2
\end{aligned}
\tag{8}
$$

and then in the form

$$MX = Y \tag{9}$$

where

$$
M = \begin{pmatrix} 1 & 2 & 1 \\ 1 & -1 & 2 \\ 1 & 1 & -1 \end{pmatrix}, \quad
X = \begin{pmatrix} x_1 \\ x_2 \\ x_3 \end{pmatrix}, \quad
Y = \begin{pmatrix} 3 \\ 9 \\ -2 \end{pmatrix}
\tag{10}
$$

The matrix M is called the *matrix of coefficients* of the system. The steps in the reduction from (1) to (2) and (2) to (3) correspond to certain operations performed on M and Y. For example, subtraction of the first equation from the second and third equations in (1) corresponds to subtracting the first row of M from the second and third rows and subtracting the first row of Y from the second and third rows. In general, the operations of Types 1, 2, and 3 on systems of equations correspond to the following operations on matrices:

Type 1 The interchange of two rows.
Type 2 The replacement of one row by a nonzero multiple of itself.
Type 3 The replacement of one row by the sum of that row and any other row.

In the next sections, we shall discuss the effect of these operations on matrices, introduce the associated concept of elementary matrices, and show that by such operations a given matrix can be transformed into a matrix of a simple structure (analogous to the triangular linear systems obtained in the above examples). The concept of equivalence of matrices will be presented. Finally, we will derive theorems on the solvability of linear systems and on invertibility of matrices, and discuss practical numerical methods for solving systems, inverting matrices, and computing determinants.

Exercises

1. In each of the following, verify that the given X is a solution of the system $MX = Y$.

   (a)
   $$M = \begin{pmatrix} 2 & 1 \\ 4 & 3 \\ -1 & 1 \end{pmatrix}, \quad X = \begin{pmatrix} 2 \\ 5 \end{pmatrix}, \quad Y = \begin{pmatrix} 9 \\ 23 \\ 3 \end{pmatrix}$$

   (b)
   $$M = \begin{pmatrix} 1 & 3 & 5 & 7 \\ 0 & 2 & 4 & 6 \end{pmatrix}, \quad X = \begin{pmatrix} 0 \\ -1 \\ -2 \\ 3 \end{pmatrix}, \quad Y = \begin{pmatrix} 8 \\ 8 \end{pmatrix}$$

   (c)
   $$M = \begin{pmatrix} 1 & 2 & 1 \\ 1 & -1 & 2 \\ 2 & 7 & 1 \end{pmatrix}, \quad X = \begin{pmatrix} -3 \\ 0 \\ 6 \end{pmatrix} + c\begin{pmatrix} -5 \\ 1 \\ 3 \end{pmatrix}, \quad Y = \begin{pmatrix} 3 \\ 9 \\ 0 \end{pmatrix}$$

   where c is arbitrary.

2. Using the method of elimination of variables, convert the system

   $$\begin{aligned} x_1 + 2x_2 + x_3 &= 3 \\ x_1 - x_2 + 2x_3 &= 9 \\ 2x_1 + 7x_2 + x_3 &= 0 \end{aligned}$$

   to the system

   $$\begin{aligned} x_1 + 2x_2 + x_3 &= 3 \\ -3x_2 + x_3 &= 6 \\ 0 &= 0 \end{aligned}$$

   Taking $x_2 = c$, derive the solution given in Exercise 1c.

3. Using the method of elimination, show that the following system has no solutions.

   $$\begin{aligned} 2x_1 - x_2 + x_3 &= 1 \\ 4x_1 + x_2 - 2x_3 &= 2 \\ 6x_1 - 6x_2 + 7x_3 &= 0 \end{aligned}$$

10.2 Elementary operations on matrices

In this section, we will define and study the elementary matrix operations and certain related matrices called elementary matrices.

DEFINITION 10.1 The three *elementary row operations* on a matrix are as follows:

1. The interchange (permutation) of two rows.
2. The replacement of a single row by a nonzero scalar multiple of itself.
3. The replacement of a row by the sum of that row and any other row. If row number r is replaced by the sum of row r and row number s, we say that we have added row s to row r.

DEFINITION 10.2 The three *elementary column operations* on a matrix are as follows:

1. The interchange (permutation) of two columns.
2. The replacement of a column by a nonzero scalar multiple of itself.
3. The replacement of a column by the sum of that column and any other column.

The collection of elementary row and column operations is referred to as the *elementary matrix operations*.

Example 1 Consider the matrices

$$M = \begin{pmatrix} 1 & 3 & 2 & 2 \\ 2 & 1 & 4 & 3 \\ 3 & 3 & 2 & 1 \end{pmatrix} \qquad N_1 = \begin{pmatrix} 1 & 2 & 2 & 3 \\ 2 & 3 & 4 & 1 \\ 3 & 1 & 2 & 3 \end{pmatrix}$$

The matrix N_1 is obtainable from M by an elementary column operation, namely, the permutation of columns 2 and 4. The matrix

$$N_2 = \begin{pmatrix} 2 & 6 & 4 & 4 \\ 2 & 1 & 4 & 3 \\ 3 & 3 & 2 & 1 \end{pmatrix}$$

is obtainable from M by the elementary row operation of multiplying the first row of M by the scalar 2. The matrix

$$N_3 = \begin{pmatrix} 4 & 7 & 8 & 7 \\ 2 & 1 & 4 & 3 \\ 3 & 3 & 2 & 1 \end{pmatrix}$$

is obtainable from N_2 by the elementary operation of addition of row 2 of N_2 to row 1 of N_2. However, N_3 is not obtainable from M by an elementary row operation, but rather a sequence of such operations.

It would be possible to dispense with the elementary operations of Type (1), since an interchange of two rows (or columns) can be effected by a sequence of six operations of Types (2) and (3). However, it is not our purpose to work with a minimal list of operations, but rather to use a set of operations which is convenient for theory and computational practice.

DEFINITION 10.3   A matrix L is an *elementary matrix* if it is the result of performing a single elementary matrix operation on an identity matrix.

Example 2   The following illustrate elementary matrices. Note that each elementary matrix can be the result of either an elementary row or column operation.

1.
$$L = \begin{pmatrix} 1 & 0 & 0 \\ 0 & 0 & 1 \\ 0 & 1 & 0 \end{pmatrix}$$

is obtained by interchanging rows (columns) 2 and 3 of the 3×3 identity matrix I.

2.
$$L = \begin{pmatrix} 1 & 0 & 0 \\ 0 & c & 0 \\ 0 & 0 & 1 \end{pmatrix}$$

is obtained by multiplying the second row (column) of I by the scalar c.

3.
$$L = \begin{pmatrix} 1 & 0 & 0 \\ 0 & 1 & 1 \\ 0 & 0 & 1 \end{pmatrix}$$

is obtained by adding the third row (second column) to the second row (third column) of I.

On the other hand,

4.
$$\begin{pmatrix} 1 & 2 & 0 \\ 0 & 1 & 0 \\ 0 & 0 & 1 \end{pmatrix}$$

is not an elementary matrix, since no single elementary operation on the identity matrix gives this result.

THEOREM 10.1   Let A and B be $m \times n$ matrices. If A is obtained from B by an elementary row operation, then $A = EB$ where E is an $m \times m$ elementary matrix. Similarly, if A is obtained from B by an elementary column operation, then $A = BE$ where E is an $n \times n$ elementary matrix.

PROOF: We will only prove the part of the theorem dealing with row operations. The proof with regard to column operations is analogous and is left to the reader. We treat each type of operation separately.

1. Suppose A is obtainable from B by interchanging the pth and qth rows of B. Then let E be the $m \times m$ matrix (e_{ij}) where

$$e_{ii} = 1 \qquad \text{for} \quad i \neq p \text{ and } i \neq q$$
$$e_{qp} = e_{pq} = 1$$
$$e_{ij} = 0 \qquad \text{otherwise}$$

In other words, E is obtained from the $m \times m$ identity matrix by interchanging rows p and q.

If we let $A^* = (a_{ij}) = EB$, we have

$$a_{ij} = \sum_{k=1}^{m} e_{ik} b_{kj} = \begin{cases} b_{ij} & \text{for} \quad i \neq p, i \neq q \\ b_{qj} & \text{for} \quad i = p \\ b_{pj} & \text{for} \quad i = q \end{cases} \qquad (j = 1, \ldots, n)$$

which shows $A^* = EB$ to be the desired matrix A.

2. Suppose A is obtainable from B by the elementary operation of multiplying the pth row of B by a scalar c. The corresponding elementary matrix is $E = (e_{ij})$ where

$$e_{ij} = \begin{cases} 1 & i = j \neq p \\ c & i = j = p \\ 0 & i \neq j \end{cases}$$

In other words, E is obtained from the $m \times m$ identity matrix by multiplying the pth row by c.

Defining $A^* = (a_{ij}) = EB$ we have

$$a_{ij} = \sum_{k=1}^{m} e_{ik} b_{kj} = \begin{cases} b_{ij} & i \neq p \\ c b_{ij} & i = p \end{cases} \qquad (j = 1, \ldots, n)$$

which proves $A^* = EB$ is the desired matrix A.

3. Suppose A is derived by addition of the qth row to the pth row of B. The corresponding elementary matrix is $E = (e_{ij})$ where

$$e_{ij} = \begin{cases} 1 & i = j \\ 1 & i = p \text{ and } j = q \\ 0 & \text{otherwise} \end{cases}$$

In other words, E is obtained from the $m \times m$ identity matrix by addition of the qth row to the pth row.

Then $A^* = EB = (a_{ij})$ where

$$a_{ij} = \begin{cases} b_{ij} & i \neq p \\ b_{pj} + b_{qj} & i = p \end{cases} \qquad (j = 1, \ldots, n)$$

Thus A^* is the desired matrix A. Hence, the theorem is proved with respect to row operations. ∎

Example 3   Let

$$A = \begin{pmatrix} 1 & 2 \\ 3 & 0 \\ 4 & -1 \end{pmatrix} \qquad B = \begin{pmatrix} 3 & 0 \\ 1 & 2 \\ 4 & -1 \end{pmatrix}$$

Then A is obtainable from B by interchanging the first two rows. Therefore, if we let E be the elementary matrix

$$E = \begin{pmatrix} 0 & 1 & 0 \\ 1 & 0 & 0 \\ 0 & 0 & 1 \end{pmatrix}$$

then $A = EB$. The reader can verify that this is true.

A useful property of elementary matrices is given in the following theorem.

THEOREM 10.2   Elementary matrices are invertible.

PROOF: We leave it to the reader to prove (Exercise 2) that every $m \times m$ elementary column matrix is also an $m \times m$ elementary row matrix and conversely. Hence, we can restrict our attention to elementary column matrices. Again we break the proof into three parts.

1. Let E be the $m \times m$ elementary matrix for interchanging two columns, say the pth and qth. This is just the matrix resulting from interchanging the pth and qth columns of an identity matrix I, and hence the set of column vectors of E is the same as the set of column vectors of I. Since the columns of I, and hence of E, are linearly independent, we know that E is a matrix representation of an invertible transformation, and hence is invertible.

2. Consider the elementary matrix E associated with scalar multiplication of a column, say the pth. This is the matrix obtained from I by multiplying the pth column of I by a nonzero scalar c. It is obvious that the columns of this matrix are linearly independent and hence E is invertible.

3. Finally, consider the matrix E associated with addition of the qth column of a matrix to the pth column. The column vectors of E are those of I with the pth column replaced by the vector containing a 1 in both the pth and qth position. That is, $E = (e_1 \ e_2 \ \cdots \ e_{p-1} \ e_p + e_q \ e_{p+1} \ \cdots \ e_n)$, where e_i is the vector of all zeros except a 1 in the ith position. Since $(e_p + e_q) - e_q = e_p$, the column vectors of E generate the same n-dimensional vector space as the columns of I (namely, the set of all $n \times 1$ vectors). Since there are n columns of E, these columns are linearly independent, which implies that E is invertible. ∎

It is worth noting the nature of the inverse matrix of an elementary matrix. If E is the matrix associated with interchanging two rows or columns, then $E = E^{-1}$. This follows since interchanging the same two rows twice will return the matrix to its original form. If E is the matrix associated with multiplication of a row (or column) by a scalar c, then the inverse is the matrix associated with multiplying the row by the scalar $1/c$. And finally, if the elementary operation of E is addition of the qth column to the pth, then the inverse operation is subtraction of the qth column from the pth and hence

$$E^{-1} = (e_1 \quad e_2 \quad \cdots \quad e_p - e_q \quad \cdots \quad e_n).$$

Note that in the first two cases the inverse matrix is also an elementary matrix. In the other case, the inverse matrix is a product of three elementary matrices (see Exercise 3).

Exercises

1.   Show how to interchange two rows of a matrix by a sequence of six row operations of Types 2 and 3.

2.   Prove that every elementary matrix which is obtained from an identity by performing an elementary row operation can also be obtained by an elementary column operation, and vice versa.

3.   Verify the statement in the text that if E is an elementary matrix obtained from I by adding the qth column to the pth column, E^{-1} is the product of three elementary matrices.

4.   Identify the elementary row operations and the elementary column operations associated with each of the following elementary matrices.

(a) $\begin{pmatrix} 1 & 0 & 0 \\ 0 & 1 & 0 \\ 1 & 0 & 1 \end{pmatrix}$ (b) $\begin{pmatrix} 0 & 0 & 1 \\ 0 & 1 & 0 \\ 1 & 0 & 0 \end{pmatrix}$

(c) $\begin{pmatrix} 1 & 0 & 0 \\ 0 & 1 & 0 \\ 0 & 0 & -2 \end{pmatrix}$

5.   In each case $A = EB$ where E is an elementary matrix. What is E?

(a) $A = \begin{pmatrix} 1 & -2 & -5 \\ 3 & 1 & 2 \end{pmatrix}$ $B = \begin{pmatrix} 3 & 1 & 2 \\ 1 & -2 & -5 \end{pmatrix}$

(b) $A = \begin{pmatrix} 6 & 2 & 4 \\ 1 & -2 & -5 \end{pmatrix}$ $B = \begin{pmatrix} 3 & 1 & 2 \\ 1 & -2 & -5 \end{pmatrix}$

(c) $A = \begin{pmatrix} 3 & 1 & 2 \\ 4 & -1 & -3 \end{pmatrix}$ $B = \begin{pmatrix} 3 & 1 & 2 \\ 1 & -2 & -5 \end{pmatrix}$

6. In each case $A = BE$, where E is an elementary matrix. What is E?

(a)
$$A = \begin{pmatrix} 2 & 1 & 3 \\ -5 & -2 & 1 \end{pmatrix} \quad B = \begin{pmatrix} 3 & 1 & 2 \\ 1 & -2 & -5 \end{pmatrix}$$

(b)
$$A = \begin{pmatrix} 3 & 1 & -2 \\ 1 & -2 & 5 \end{pmatrix} \quad B = \begin{pmatrix} 3 & 1 & 2 \\ 1 & -2 & -5 \end{pmatrix}$$

(c)
$$A = \begin{pmatrix} 3 & 3 & 2 \\ 1 & -7 & -5 \end{pmatrix} \quad B = \begin{pmatrix} 3 & 1 & 2 \\ 1 & -2 & -5 \end{pmatrix}$$

10.3   Echelon forms

The purpose of this section is to study the form which a matrix will take when we perform a sequence of elementary operations on it.

DEFINITION 10.4   A matrix is said to be in (row) *echelon form*[3] if the following are satisfied:

1. The first nonzero element of every row is a one and this one appears to the right of the first nonzero element of every preceding row.

2. If there is a zero row, then every row following will also be zero. (A row will be called a *zero row* if every element in the row is zero.)

Example 1   The following matrix is in echelon form:

$$\begin{pmatrix} 0 & 1 & 2 & 3 & 0 & 4 & 6 & 2 & 1 \\ 0 & 0 & 0 & 1 & 0 & 2 & 1 & 1 & 3 \\ 0 & 0 & 0 & 0 & 0 & 1 & 2 & 2 & 1 \\ 0 & 0 & 0 & 0 & 0 & 0 & 0 & 0 & 1 \\ 0 & 0 & 0 & 0 & 0 & 0 & 0 & 0 & 0 \end{pmatrix}$$

DEFINITION 10.5   A matrix is in *reduced echelon form* if the two following conditions are satisfied:

1. It is in echelon form.

2. The first nonzero element in each nonzero row is the only nonzero element in its column.

Example 2   The following matrix is in reduced echelon form

$$\begin{pmatrix} 0 & 1 & 2 & 0 & 0 & 0 & 6 & 2 & 0 \\ 0 & 0 & 0 & 1 & 0 & 0 & 1 & 1 & 0 \\ 0 & 0 & 0 & 0 & 0 & 1 & 2 & 2 & 0 \\ 0 & 0 & 0 & 0 & 0 & 0 & 0 & 0 & 1 \\ 0 & 0 & 0 & 0 & 0 & 0 & 0 & 0 & 0 \end{pmatrix}$$

---

[3] An analogous theory can be developed based on column echelon forms.

Notice that the second column contains the first nonzero element of row 1 and is therefore e_1, the fourth column contains the first nonzero element of the second row and is e_2, and so on.

THEOREM 10.3   Every $m \times n$ matrix A can be reduced to a matrix in echelon form by a sequence of elementary row operations. That is, there exist an integer k and elementary matrices $L_1, L_2, \ldots, L_k$ such that

$$L_k L_{k-1} \ldots L_2 L_1 A = B$$

where B is in echelon form.

PROOF: We prove the theorem by illustrating how to go about finding the L_i. The reader may wish to study Example 3 before reading this proof.

Consider the matrix A and find the first column containing a nonzero element. Pick any row with a nonzero element in this column and interchange that row with row 1. The matrix of this elementary row operation is L_1. Next, multiply every element of row 1 of the new matrix, $L_1 A$, by the reciprocal of the first nonzero element in row 1. The matrix of this elementary row operation is L_2. The first nonzero element in row 1 of $L_2 L_1 A$ is now 1, and $L_2 L_1 A$ looks like this:

$$L_2 L_1 A = \begin{pmatrix} \cdots & 0 & 1 & \cdots & * \\ \cdots & 0 & * & \cdots & * \\ & & \cdots & & \\ \cdots & 0 & * & \cdots & * \end{pmatrix}$$

where the asterisks denote unspecified elements. Next, subtract an appropriate scalar multiple of row 1 from every other row so that the elements under the 1 are replaced by zeros. As we know, this can be accomplished by a sequence of row operations. The resulting matrix B_1 looks like this:

$$B_1 = L_{k_1} \cdots L_2 L_1 A = \begin{pmatrix} \cdots & 0 & 1 & * & \cdots \\ \cdots & 0 & 0 & * & \cdots \\ & & \cdots & & \\ \cdots & 0 & 0 & * & \cdots \end{pmatrix}$$

In this matrix the first nonzero column is

$$e_1^m = \begin{pmatrix} 1 \\ 0 \\ \cdots \\ 0 \end{pmatrix}$$

the standard basis vector with m elements and 1 as first element.

Now consider the last $m - 1$ rows of B_1. Repeat the operations above on B_1 so as to reduce the last $m - 1$ rows to an $(m - 1) \times n$ matrix with first nonzero column e_1^{m-1}. This can again be done by a sequence of elementary row operations. Repeating this procedure on the last $m - 2$, then $m - 3$, and so on, rows will lead eventually to a matrix B with the desired property of being in echelon form and such that

$$B = L_k L_{k-1} \cdots L_1 A$$

where the L_i are elementary matrices. ∎

THEOREM 10.4 Every $m \times n$ matrix A can be transformed into a matrix in reduced echelon form by a sequence of elementary row operations. In other words, there is a sequence $L_k, L_{k-1}, \ldots, L_2, L_1$ such that $L_k L_{k-1} \cdots L_1 A = B$ where B is in reduced echelon form.

PROOF: This is left as an exercise. ∎

Example 3   Consider the matrix

$$M = \begin{pmatrix} 0 & 4 & 2 & 1 & 3 \\ 1 & 1 & 2 & 2 & 2 \\ 0 & 8 & 4 & 2 & 6 \\ 1 & 5 & 4 & 3 & 5 \end{pmatrix}$$

We begin by placing a one in the upper left position. This is done by interchanging rows 2 and 1, yielding

$$M_1 = \begin{pmatrix} 1 & 1 & 2 & 2 & 2 \\ 0 & 4 & 2 & 1 & 3 \\ 0 & 8 & 4 & 2 & 6 \\ 1 & 5 & 4 & 3 & 5 \end{pmatrix}$$

Next, we eliminate one from every other element in column 1. This is done by subtracting row 1 from row 4. (Note this requires three elementary row operations, first, multiplication of row 1 by minus 1, then addition of that row to row 4, and finally multiplication of the new row 1 by minus one). This gives

$$M_2 = \begin{pmatrix} 1 & 1 & 2 & 2 & 2 \\ 0 & 4 & 2 & 1 & 3 \\ 0 & 8 & 4 & 2 & 6 \\ 0 & 4 & 2 & 1 & 3 \end{pmatrix}$$

We now make the first nonzero element of row 2 equal to one by multiplication of row 2 by $\frac{1}{4}$. This gives

$$M_3 = \begin{pmatrix} 1 & 1 & 2 & 2 & 2 \\ 0 & 1 & \frac{1}{2} & \frac{1}{4} & \frac{3}{4} \\ 0 & 8 & 4 & 2 & 6 \\ 0 & 4 & 2 & 1 & 3 \end{pmatrix}$$

By subtracting 8 times row 2 from row 3, and 4 times row 2 from row 4 (each requiring three elementary row operations) we are led to

$$M_4 = \begin{pmatrix} 1 & 1 & 2 & 2 & 2 \\ 0 & 1 & \frac{1}{2} & \frac{1}{4} & \frac{3}{4} \\ 0 & 0 & 0 & 0 & 0 \\ 0 & 0 & 0 & 0 & 0 \end{pmatrix}$$

M_4 is in echelon form. Subtraction of row 2 from row 1 yields

$$M_5 = \begin{pmatrix} 1 & 0 & \frac{3}{2} & \frac{7}{4} & \frac{5}{4} \\ 0 & 1 & \frac{1}{2} & \frac{1}{4} & \frac{3}{4} \\ 0 & 0 & 0 & 0 & 0 \\ 0 & 0 & 0 & 0 & 0 \end{pmatrix}$$

The matrix M_5 is in reduced echelon form.

DEFINITION 10.6 If a matrix A can be reduced to a matrix B in echelon form by a sequence of elementary row operations, B is called an *echelon form* of A. If B is in reduced echelon form, B is called the *reduced echelon form* of A or the *Hermite normal form* of A.

The use of the definite article in defining *the* reduced echelon form of A is justified by the fact (to be proved below) that the reduced echelon form of a matrix is unique.

THEOREM 10.5 A matrix A has row rank r if and only if its echelon forms all have row rank r and have r nonzero rows.

PROOF: It is clear that each of the elementary row operations leaves the row space of a matrix unchanged, and hence the row rank. Therefore, all echelon forms of A have the same row rank as A. Further, the nonzero rows of an echelon matrix are easily seen to be linearly independent (Exercise 2). Therefore, the row rank of an echelon matrix equals the number of nonzero rows. ∎

THEOREM 10.6 The reduced echelon form of a matrix is unique.

PROOF: Let A be a given $m \times n$ matrix and suppose that by one sequence of row operations we obtain a reduced echelon form B, whereas by another sequence of row operations we obtain a reduced echelon form C. We must show that $B = C$. Since the inverse of an elementary row operation is either an elementary row operation or a product of several such, it is possible to pass from B to C by a sequence of elementary row operations. We shall show that this is impossible if $B \neq C$.

First, B and C have the same number r of nonzero rows, where r is the row rank. Let the leading 1 in row i appear in column k_i in B and in column k_i'

in C $(i = 1, \ldots, r)$. We shall show that $k_i' = k_i$ $(i = 1, \ldots, r)$. Suppose on the contrary that $k_1' > k_1$. Then, clearly, no linear combination of the rows of C can yield the first row of B. But then B and C cannot have the same row space, which is a contradiction. Hence $k_1' = k_1$. Similarly, suppose $k_2' > k_2$. Then B and C look like this:

$$
\text{Column}
$$

$$
\begin{array}{ccc}
k_1 & k_2 & k_2'
\end{array}
$$

$$
B = \begin{pmatrix}
\cdots & 1 & \cdots & 0 & \cdots & * & \cdots \\
\cdots & 0 & \cdots & 1 & \cdots & * & \cdots \\
\cdots & 0 & \cdots & 0 & \cdots & * & \cdots \\
\cdots & 0 & \cdots & 0 & \cdots & * & \cdots
\end{pmatrix}
$$

$$
C = \begin{pmatrix}
\cdots & 1 & \cdots & * & \cdots & 0 & \cdots \\
\cdots & 0 & \cdots & 0 & \cdots & 1 & \cdots \\
\cdots & 0 & \cdots & 0 & \cdots & 0 & \cdots \\
\cdots & 0 & \cdots & 0 & \cdots & 0 & \cdots
\end{pmatrix}
$$

The asterisks denote unspecified entries. It is clear that no linear combination of rows of C can yield the second row of B, since the first row of C must be used to produce the 1 in row 2, column k_2 of B, but then row 2, column k_1 will contain a nonzero entry. A similar argument shows that $k_i' = k_i$ $(i = 1, \ldots, r)$.

We now know that columns $k_1, \ldots, k_r$ of B and C are identical. Any column before column k_1 consists entirely of zeros and is the same in B and C. Suppose that there is a k with $k_i < k < k_{i+1}$ such that column k of B differs from column k of C, whereas all columns to the left are identical. It will be enough to illustrate with the case $k_2 < k < k_3$, in which case B and C look like this:

$$
\begin{array}{cccc}
k_1 & k_2 & k & k_3
\end{array}
$$

$$
B = \begin{pmatrix}
1 & \cdots & 0 & \cdots & b_1 & \cdots & 0 & \cdots \\
0 & \cdots & 1 & \cdots & b_2 & \cdots & 0 & \cdots \\
0 & \cdots & 0 & \cdots & 0 & \cdots & 1 & \cdots \\
0 & \cdots & 0 & \cdots & 0 & \cdots & 0 & \cdots
\end{pmatrix}
$$

$$
C = \begin{pmatrix}
1 & \cdots & 0 & \cdots & c_1 & \cdots & 0 & \cdots \\
0 & \cdots & 1 & \cdots & c_2 & \cdots & 0 & \cdots \\
0 & \cdots & 0 & \cdots & 0 & \cdots & 1 & \cdots \\
0 & \cdots & 0 & \cdots & 0 & \cdots & 0 & \cdots
\end{pmatrix}
$$

Since B and C have the same row space, the first row of B must be a linear combination of the rows of C. Denoting the rows of B by $b^{(1)}, b^{(2)}, \ldots$, and the rows of C by $c^{(1)}, c^{(2)}, \ldots$, we therefore have $b^{(1)} = x_1 c^{(1)} + x_2 c^{(2)} + \cdots$. In order to obtain the 1 in column k_1 it is necessary to take $x_1 = 1$, and in order to obtain the 0 in column k_2 it is necessary to take $x_2 = 0$. Then, clearly, the element in column k of $b^{(1)}$ is just c_1, that is, $b_1 = c_1$. Similarly, $b^{(2)}$ is a linear combination $b^{(2)} = y_1 c^{(1)} + y_2 c^{(2)} + \cdots$ and clearly, $y_1 = 0$, $y_2 = 1$. Hence, $b_2 = c_2$. It has been shown that column k is the same in B and C. The same argument is valid if $k_r < k$. ∎

Example 4   The method of row reduction provides an easy technique for determining the row rank of a matrix. For example, the matrix M in Example 3 has M_4 as an echelon form. Since M_4 has two nonzero rows, the row rank of M is 2.

It was shown by the theory of determinants that the row rank and column rank of a matrix are the same, and this common value is called simply the rank of the matrix. We shall give an independent proof of this important result below. Therefore, in Theorem 10.5, the term *row rank* may be replaced by *rank*.

Exercises

1.   Prove Theorem 10.4. [*Hint:* Begin with the matrix in echelon form and work on preceding rows.]

2.   Prove that the nonzero rows of an echelon matrix are linearly independent.

3.   Determine the row rank of each matrix by finding an echelon form.

(a) $\begin{pmatrix} 1 & -1 & 0 & 3 \\ 1 & 2 & -1 & 1 \\ 2 & 1 & -1 & 4 \\ 1 & -7 & 2 & 7 \end{pmatrix}$
(b) $\begin{pmatrix} 0 & 0 & 3 & 6 \\ 0 & 1 & 3 & 0 \\ 0 & 2 & 6 & 0 \end{pmatrix}$

(c) $\begin{pmatrix} 2 & 7 & 7 & 2 & 3 \\ 1 & 3 & 2 & 4 & 1 \\ -1 & -2 & 1 & -10 & 0 \\ 1 & 5 & 8 & -8 & 3 \\ 0 & 0 & 0 & 0 & 1 \end{pmatrix}$

4.   Find the Hermite normal form of each matrix in Exercise 3.

10.4   Row rank and column rank

The purpose of this section is to prove that the row rank and column rank of any matrix are the same. The proof uses the properties of the echelon form, and is independent of determinant theory given in Chapter 7. Recall that the column rank of a matrix is the dimension of the space spanned by its columns.

THEOREM 10.7   If A is any $m \times n$ matrix, the column rank of A is unchanged when A is subjected to an elementary row operation.

PROOF: Suppose that the column rank of A is s and let columns with indices $k_1, k_2, \ldots, k_s$ comprise a basis for the column space of A. Let A_j denote any other column of A. Then there are constants $c_1, c_2, \ldots, c_s$ such that

$$A_j = c_1 A_{k_1} + c_2 A_{k_2} + \cdots + c_s A_{k_s}$$

or

$$\begin{pmatrix} a_{1j} \\ \cdots \\ a_{mj} \end{pmatrix} = c_1 \begin{pmatrix} a_{1k_1} \\ \cdots \\ a_{mk_1} \end{pmatrix} + c_2 \begin{pmatrix} a_{1k_2} \\ \cdots \\ a_{mk_2} \end{pmatrix} + \cdots + c_s \begin{pmatrix} a_{1k_s} \\ \cdots \\ a_{mk_s} \end{pmatrix} \tag{1}$$

Now we consider in turn the three types of row operations. Suppose that two rows of A are interchanged, say rows p and q. Then rows p and q are reversed in A_j and in $A_{k_1}, A_{k_2}, \ldots, A_{k_s}$. The new columns with indices $k_1, k_2, \ldots, k_s$ are clearly still independent. Also, the relation (1) will remain valid, with the same constants $c_1, \ldots, c_s$, the only difference being that the pth and qth entries in each column are interchanged. Therefore, an operation of this kind does not change the number of linearly independent columns.

Next, consider what happens if row p is multiplied by a nonzero scalar c. Again it is obvious that the new columns $k_1, k_2, \ldots, k_s$ will still be independent. Moreover, from Eq. (1) we get

$$\begin{pmatrix} a_{1j} \\ \cdots \\ ca_{pj} \\ \cdots \\ a_{mj} \end{pmatrix} = c_1 \begin{pmatrix} a_{1k_1} \\ \cdots \\ ca_{pk_1} \\ \cdots \\ a_{mk_1} \end{pmatrix} + \cdots + c_s \begin{pmatrix} a_{1k_s} \\ \cdots \\ ca_{pk_s} \\ \cdots \\ a_{mk_s} \end{pmatrix} \tag{2}$$

showing that the jth column of the new matrix is a linear combination of the new columns $k_1, \ldots, k_s$. Thus, an operation of this kind does not affect the column rank.

We leave it to the reader to complete the proof by considering the third type of row operation.   ∎

The above proof was based strictly on matrix algebraic arguments, which is in keeping with the spirit of this chapter. An alternative proof, using the isomorphism between matrices and linear transformations, is outlined as follows.

ALTERNATIVE PROOF: Let V and V' be vector space with bases

$$B = \{v_1, v_2, \ldots, v_n\} \quad \text{and} \quad B' = \{v_1', v_2', \ldots, v_m'\}$$

respectively. Then there exists a unique linear transformation $T: V \to V'$ with matrix representative A with respect to B and B'. Let $C = EA$, where

E is an elementary matrix. E is $m \times m$ and invertible. Therefore, E is the transition matrix corresponding to the map on V' from the basis B' to a new basis B''. Hence C is the matrix of T with respect to B and B''. But the column rank of C is just the rank of T, as is the column rank of A. Hence, column rank $C =$ column rank A and the theorem is proved.   ∎

THEOREM 10.8 If A is any $m \times n$ matrix, the row rank and the column rank of A are equal.

PROOF: Let B be the reduced echelon form of A. By Theorem 10.5, A and B have the same row rank and by Theorem 10.7 they have the same column rank. Let r denote the row rank. Then B contains r nonzero rows. By the nature of the reduced echelon form, there are r columns of B which are the vectors $e_1, e_2, \ldots, e_r$, in turn (see the proof of Theorem 10.3). Moreover, every other column of B has nonzero entries only in the first r rows, and is therefore a linear combination of $e_1, e_2, \ldots, e_r$; that is, the column space of B has basis $\{e_1, e_2, \ldots, e_r\}$, and so the column rank of B is r. Consequently, the column rank of A equals r, the row rank of A.   ∎

The result in Theorem 10.5 can now be rephrased as follows:

COROLLARY The rank of a matrix A is the number of nonzero rows in any of its (row-reduced) echelon forms.

10.5 Solution of linear systems

We began this chapter by giving examples of solutions of linear systems of equations, and were led to the notion of elementary operations and echelon forms. Now we want to return to a more thorough discussion of linear systems of equations of the form

$$AX = Y \tag{1}$$

where A is an $m \times n$ matrix, called the *matrix of coefficients*, X is an $n \times 1$ column vector, and Y is an $m \times 1$ column vector. We suppose that A and Y are given and we wish to solve for X. Written out in full, the system (1) has the form

$$a_{11}x_1 + a_{12}x_2 + \cdots + a_{1n}x_n = y_1$$
$$\cdots$$
$$a_{m1}x_1 + a_{m2}x_2 + \cdots + a_{mn}x_n = y_m$$

Note that we do not require that m and n be equal. We have already pointed out that three kinds of elementary operations used to simplify the system (1) correspond to the three kinds of elementary row operations performed on A and Y. This suggests the following definition.

DEFINITION 10.7 The $m \times (n + 1)$ matrix

$$
\begin{pmatrix}
a_{11} & a_{12} & \cdots & a_{1n} & y_1 \\
 & & \cdots & & \\
a_{m1} & a_{m2} & \cdots & a_{mn} & y_m
\end{pmatrix}
$$

is called the *augmented matrix* for the system (1), and is denoted by the symbol $(A : Y)$.

If any elementary row operation is performed on $(A : Y)$ to yield a new $m \times (n + 1)$ matrix $(B : Z)$, then $(B : Z)$ is the augmented matrix of the system $BX = Z$. By what we have said, the systems $AX = Y$ and $BX = Z$ have exactly the same solutions.

The method of solution by Gaussian elimination can be carried out by reducing $(A : Y)$ to echelon form and solving the resulting system. We illustrate this by examples.

Example 1 For the system

$$
\begin{aligned}
x_1 + 2x_2 + x_3 &= 3 \\
x_1 - x_2 + 2x_3 &= 9 \\
x_1 + x_2 - x_3 &= -2
\end{aligned}
$$

we have

$$
(A : Y) = \begin{pmatrix}
1 & 2 & 1 & 3 \\
1 & -1 & 2 & 9 \\
1 & 1 & -1 & -2
\end{pmatrix}
$$

Row operations yield successively

$$
\begin{pmatrix}
1 & 2 & 1 & 3 \\
0 & -3 & 1 & 6 \\
0 & -1 & -2 & -5
\end{pmatrix},
\qquad
\begin{pmatrix}
1 & 2 & 1 & 3 \\
0 & -1 & -2 & -5 \\
0 & -3 & 1 & 6
\end{pmatrix}
$$

$$
\begin{pmatrix}
1 & 2 & 1 & 3 \\
0 & 1 & 2 & 5 \\
0 & -3 & 1 & 6
\end{pmatrix},
\qquad
\begin{pmatrix}
1 & 2 & 1 & 3 \\
0 & 1 & 2 & 5 \\
0 & 0 & 7 & 21
\end{pmatrix}
$$

and finally,

$$
(B : Z) = \begin{pmatrix}
1 & 2 & 1 & 3 \\
0 & 1 & 2 & 5 \\
0 & 0 & 1 & 3
\end{pmatrix}
$$

The system with augmented matrix $(B : Z)$ is

$$
\begin{aligned}
x_1 + 2x_2 + x_3 &= 3 \\
x_2 + 2x_3 &= 5 \\
x_3 &= 3
\end{aligned}
$$

and has solution $x_3 = 3$, $x_2 = -1$, $x_1 = 2$. These values satisfy the original system also.

With a little practice, one can learn to make a few short cuts in the reduction process.

Example 2   This example illustrates the case where $m \neq n$. Consider the system

$$
\begin{array}{rcl}
x_1 + 3x_2 + 2x_3 + 4x_4 &=& 1 \\
-x_1 - 2x_2 + x_3 - 10x_4 &=& 0 \\
2x_1 + 7x_2 + 7x_3 + 2x_4 &=& 3
\end{array}
$$

containing three equations and four unknowns. The augmented matrix is

$$
(A:Y) = \begin{pmatrix} 1 & 3 & 2 & 4 & 1 \\ -1 & -2 & 1 & -10 & 0 \\ 2 & 7 & 7 & 2 & 3 \end{pmatrix}
$$

Row operations yield

$$
\begin{pmatrix} 1 & 3 & 2 & 4 & 1 \\ 0 & 1 & 3 & -6 & 1 \\ 0 & 1 & 3 & -6 & 1 \end{pmatrix}
$$

and then the echelon form

$$
(B:Z) = \begin{pmatrix} 1 & 3 & 2 & 4 & 1 \\ 0 & 1 & 3 & -6 & 1 \\ 0 & 0 & 0 & 0 & 0 \end{pmatrix}
$$

The corresponding system is

$$
\begin{array}{rcl}
x_1 + 3x_2 + 2x_3 + 4x_4 &=& 1 \\
x_2 + 3x_3 - 6x_4 &=& 1 \\
0 &=& 0
\end{array}
$$

It is evident that x_3 and x_4 can be assigned any values whatever, and then x_2 and x_1 will be determined. That is, the totality of solutions is given by

$$
x_3 = c_1, \qquad x_4 = c_2, \qquad x_2 = 1 - 3c_1 + 6c_2
$$
$$
x_1 = 1 - 3(1 - 3c_1 + 6c_2) - 2c_1 - 4c_2 = -2 + 7c_1 - 22c_2
$$

where c_1 and c_2 are arbitrary constants. In vector notation,

$$
X = \begin{pmatrix} -2 \\ 1 \\ 0 \\ 0 \end{pmatrix} + c_1 \begin{pmatrix} 7 \\ -3 \\ 1 \\ 0 \end{pmatrix} + c_2 \begin{pmatrix} -22 \\ 6 \\ 0 \\ 1 \end{pmatrix}
$$

We may refer to this last formula as a *general solution* of the system.

Example 2 indicates the need for a general theory for systems in which $m \neq n$, and we shall now develop such a theory.

DEFINITION 10.8   A system $AX = Y$ is called *homogeneous* if $Y = 0$ and *nonhomogeneous* if $Y \neq 0$. A *solution* of such a system is a vector X in $\mathbf{R}^n$ (or $\mathbf{C}^n$) for which the matrix product AX equals the vector Y.

THEOREM 10.9   Let A be an $m \times n$ real (or complex) matrix of rank r. Then the solutions of the homogeneous system $AX = 0$ form a subspace of $\mathbf{R}^n$ (or $\mathbf{C}^n$) of dimension $n - r$.

PROOF: We can regard A as the matrix of a linear transformation T from a space $\mathbf{V}$ of dimension n into a space $\mathbf{V}'$ of dimension m. The solutions of the equation $AX = 0$ therefore form a vector space isomorphic to the null space of T. By Theorem 5.16, the dimension of this null space is dim $\mathbf{V} - \rho(T) = n - \rho(T)$. By Theorem 6.9, $\rho(T) = r$, which completes the proof. ∎

The reduced echelon form of A can be used to construct a basis[4] for the space of solutions of the system $AX = 0$. This is illustrated in the next example.

Example 3   Consider the system

$$
\begin{aligned}
x_1 + 3x_2 + 2x_3 + 4x_4 &= 0 \\
-x_1 - 3x_2 + x_3 - 10x_4 &= 0 \\
2x_1 + 6x_2 + 7x_3 + 2x_4 &= 0
\end{aligned}
$$

We easily see that an echelon form of A is

$$
\begin{pmatrix}
1 & 3 & 2 & 4 \\
0 & 0 & 1 & -2 \\
0 & 0 & 0 & 0
\end{pmatrix}
$$

and so the reduced form is

$$
B = \begin{pmatrix}
1 & 3 & 0 & 8 \\
0 & 0 & 1 & -2 \\
0 & 0 & 0 & 0
\end{pmatrix}
$$

This corresponds to the system

$$
\begin{aligned}
x_1 + 3x_2 + 8x_4 &= 0 \\
x_3 - 2x_4 &= 0 \\
0 &= 0
\end{aligned}
$$

---

[4] This can also be used to give a direct proof of Theorem 10.9, not using the theorems in Chapters 5 and 6.

We see that x_2 and x_4 can be treated as parameters; that is, any solution has the form

$$X = \begin{pmatrix} -3x_2 - 8x_4 \\ x_2 \\ 2x_4 \\ x_4 \end{pmatrix} = x_2 X_2 + x_4 X_4$$

$$X_2 = \begin{pmatrix} -3 \\ 1 \\ 0 \\ 0 \end{pmatrix}, \quad X_4 = \begin{pmatrix} -8 \\ 0 \\ 2 \\ 1 \end{pmatrix}$$

where x_2 and x_4 are arbitrary scalars. The solution space is the span of the vectors X_2 and X_4.

The following corollary is of frequent use.

COROLLARY   A homogeneous system $AX = 0$ containing m equations in n unknowns always has nontrivial (nonzero) solutions X if $m < n$.

PROOF: If $m < n$, then the rank r of A is at most m, and so $r < n$. Hence, the solution space has dimension $n - r \geq 1$.   ∎

THEOREM 10.10   Let A be an $m \times n$ matrix of rank r. The nonhomogeneous system $AX = Y$ possesses a solution if and only if the augmented matrix $(A : Y)$ has rank r. In case there is a solution, X_p, the family of all solutions is obtained by adding to X_p any solution of the homogeneous system $AX = 0$.

PROOF: Suppose that $AX = Y$ has a solution, X_p, and let X_0 be any solution of $AX = 0$. Let $X = X_p + X_0$. Then since $AX = AX_p + AX_0 = Y$, X is a solution of the nonhomogeneous equation. Conversely, if X is any solution of the nonhomogeneous equation, and if we let $X_0 = X - X_p$, then $AX_0 = AX - AX_p = Y - Y = 0$; therefore, X must be the sum of X_p and a solution of the homogeneous equation. Thus we have proved the second part of the theorem.

In order to prove the first part of the theorem, we observe that the equation $AX = Y$ can be written as

$$x_1 A_1 + \cdots + x_n A_n = Y \tag{2}$$

where $A_1, \ldots, A_n$ are the n columns of A. If the system $AX = Y$ has a solution X, the components $x_1, \ldots, x_n$ satisfy Eq. (2). Hence, Y is a linear combination of the columns of A, or in other words, Y is in the column space of A. Therefore the matrices $(A : Y)$ and A have the same number of linearly independent columns and consequently the same rank (using the fact that column rank is equal to rank). Conversely, suppose that $(A : Y)$ and A have the same rank r. Then both matrices have r linearly independent columns. There-fore, Y can be expressed as a linear combination of the columns of A, as in Eq. (2); that is, the system $AX = Y$ has a solution.   ∎

According to Theorem 10.10, a nonhomogeneous system can be solved in two steps: (1) find a general solution of the homogeneous system; (2) find one solution of the nonhomogeneous system. The sum of these is a general solution of the nonhomogeneous system. In practice, however, it is usually sufficient to combine the two parts, as was illustrated in Example 2.

The method of row reduction or Gaussian elimination has given us the important computational method for solving linear systems illustrated in the examples, as well as a simple proof of the equality of row rank and column rank. In certain applications it is necessary to deal with very large systems, containing perhaps several hundred equations and variables. For many of these problems, the method of Gaussian elimination is the most efficient (quickest) known method. Many computer programs have been written for this purpose, based on modified and systematized variants of the method we have described. For a discussion of numerical methods, see the references given at the end of the chapter. Also, see the exercises below, in which we sketch a method of comparing the efficiency of Gaussian elimination and Cramer's rule, and also give an example of possible pitfalls in the numerical solution of systems.

Exercises

1. Find the dimension of the solution space of the system $AX = 0$ for each matrix A. Give a basis for each solution space.

(a) $\begin{pmatrix} 1 & 0 & 2 & -3 & 5 \\ 2 & 1 & 0 & 1 & 2 \end{pmatrix}$ (b) $\begin{pmatrix} 1 & 0 & 2 & -3 & 5 \\ 2 & 1 & 0 & 1 & 2 \\ -1 & 1 & -6 & 10 & -13 \end{pmatrix}$

(c) $\begin{pmatrix} 1 & 4 & 1 \\ 0 & 2 & 2 \\ 1 & 6 & 2 \end{pmatrix}$ (d) $\begin{pmatrix} 1 & 2 & 3 \\ 1 & 0 & 1 \\ -1 & 1 & 1 \\ 2 & 2 & 1 \end{pmatrix}$

2. For each system $AX = Y$, find the rank of A and of the augmented matrix, and find a general solution of the system.

(a) $\begin{pmatrix} 1 & 0 & 2 & -3 & 5 \\ 2 & 1 & 0 & 1 & 2 \end{pmatrix}$ $Y = \begin{pmatrix} 2 \\ 1 \end{pmatrix}$

(b) $A = \begin{pmatrix} 1 & 0 & 2 & -3 & 5 \\ 2 & 1 & 0 & 1 & 2 \\ -1 & 1 & -6 & 10 & -13 \end{pmatrix}$ $Y = \begin{pmatrix} 2 \\ 1 \\ 4 \end{pmatrix}$

(c)

Same A as in (b)

$$Y = \begin{pmatrix} 2 \\ 1 \\ -5 \end{pmatrix}$$

(d)

$$A = \begin{pmatrix} 1 & 2 & 3 \\ 1 & 0 & 1 \\ -1 & 1 & 1 \\ 2 & 2 & 1 \end{pmatrix} \qquad Y = \begin{pmatrix} 1 \\ 0 \\ 0 \\ 0 \end{pmatrix}$$

(e)

Same A as in (d)

$$Y = \begin{pmatrix} 3 \\ 1 \\ 0 \\ 4 \end{pmatrix}$$

3. Write out a proof of Theorem 10.9 which uses the reduced echelon form to construct a basis for the space of solutions.

4. Let $f(n)$ denote the number of multiplications (of two numbers) required to evaluate an $n \times n$ determinant by the method of expansion in cofactors. For example, $f(2) = 2$, and since a 3×3 determinant A can be expanded as $a_{11}A^{11} + a_{12}A^{12} + a_{13}A^{13}$, $f(3) = 3f(2) + 3 = 9$. Here, of course, we assume general conditions, meaning that the number of multiplications is not reduced by special circumstances such as the presence of zero elements. Also, we assume that the operations are performed as indicated, that is, A^{11} is evaluated and the result is multiplied by a_{11}, and so on.

    (a) Show that $f(n) = n + nf(n - 1)$, $n = 2, 3, \dots$ where $f(1) = 0$.
    (b) The equation $f(n) - nf(n - 1) = n$ is a *difference equation* (see Chapter 13). To solve it, it is helpful to let $g(n) = f(n)/n!$. Show that $g(n)$ satisfies the difference equation $g(n) - g(n - 1) = 1/(n - 1)!$, $g(1) = 0$. Hence, show that the solution is

$$g(n) = \sum_{k=1}^{n-1} \frac{1}{k!}$$

    Therefore, $\lim_{n \to \infty} g(n) = e - 1$, and for large n, $f(n)$ is approximately equal to $(e - 1)n!$
    (c) How many multiplications are required to solve a system of n equations in n unknowns by Cramer's rule?

5. Using the results in Exercise 4, estimate how long it will take an electronic computer to do the necessary multiplications to solve a system of ten equations in ten unknowns if the computer performs 10^6 multiplications per second. How about 20 equations in 20 unknowns? (There are 3.15×10^7

seconds in a year). (By way of comparison, "register to storage" multiplication of fixed point numbers in the IBM 360/40 computer takes an average of 86.4×10^{-6} seconds.)

6. (a) Show that the reduction of an $n \times (n + 1)$ augmented matrix $(A: Y)$ to echelon form requires

$$\sum_{k=2}^{n+1} k(k - 1) = \sum_{k=1}^{n} (k + 1)k$$

multiplications. We again assume "general conditions." Also, there may be slight variations in the result depending on the exact way in which the reduction is performed.

(b) Use the formulas

$$\sum_{k=1}^{n} k = \frac{n(n + 1)}{2}, \qquad \sum_{k=1}^{n} k^2 = \frac{n(n + 1)(2n + 1)}{6}$$

to express the result as $n(n + 1)(n + 2)/3$.

(c) After $(A: Y)$ is reduced to echelon form $(B: Z)$, show that the solution X of $AX = Y$ can be found with an additional $n(n - 1)/2$ multiplications. Thus, the total number of multiplications for the process is $n^3/3 + 3n^2/2 + n/6$, or approximately $n^3/3$ for large n.

7. Repeat Exercise 5, assuming that Gaussian elimination is used rather than Cramer's rule.

8. The system

$$x_1 + 0.99x_2 = 1.99$$
$$0.99x_1 + 0.98x_2 = 1.97$$

has exact solution $x_1 = 1$, $x_2 = 1$. However, if Gaussian elimination is used and if only four significant digits are retained in all calculations, the reduced system is

$$x_1 + 0.99x_2 = 1.99$$
$$- 0.0001x_2 = 1.97 - 1.970 = 0$$

The computed solution is therefore $x_1 = 1.99$, $x_2 = 0$, which is vastly in error. A system of this sort is called ill-conditioned, which we may define loosely as meaning that the errors caused by calculating with a limited number of digits, as in a computer, may be large.

Show that the system

$$x_1 + 0.99x_2 = 1.9903$$
$$0.99x_1 + 0.98x_2 = 1.9706$$

has exact solution $x_1 = 4$, $x_2 = -2.03$. Systems like this are also sometimes called *nearly singular* because the determinant of the coefficients

$$\begin{vmatrix} 1 & 0.99 \\ 0.99 & 0.98 \end{vmatrix} = -0.0001$$

is very small relative to the size of the individual coefficients.

9. How many multiplications are required to evaluate a determinant by reducing the matrix to echelon form and then multiplying the diagonal elements? How many if the determinant is evaluated as a sum of the $n!$ products formed by taking an element from each row and column?

10.6   Matrix inverse

In this section we shall show how the method of row operations can be used to decide whether a given square matrix has an inverse, and to compute the inverse when it does exist. The method requires many fewer operations than the method of the adjugate matrix (see Section 7.8), at least for matrices with more than three rows.

THEOREM 10.11   An $n \times n$ matrix is nonsingular if and only if its reduced echelon form is the identity matrix.

PROOF: By Definition 6.14, we know an $n \times n$ matrix A is nonsingular if and only if its rank is n. Then by Theorem 10.5, A is nonsingular if and only if the reduced echelon form of A has n nonzero rows. Since the first element in every row of a matrix in reduced echelon form is a one, and every other element in the corresponding column is a zero, this is true if and only if the echelon form of A contains the columns $e_1, e_2, \ldots, e_n$. Since the reduced echelon form is $n \times n$, this holds if and only if it is the identity matrix.   ∎

THEOREM 10.12   A matrix is nonsingular if and only if it is the product of elementary matrices.

PROOF: Since elementary matrices L_i are nonsingular the product $\prod_{i=1}^{k} L_i$ is nonsingular. In fact, $(\prod_{i=1}^{k} L_i)^{-1} = L_k^{-1} L_{k-1}^{-1} \cdots L_1^{-1}$. Hence, if $A = \prod_{i=1}^{k} L_i$ where the L_i are elementary matrices, A is nonsingular.

Next suppose A is nonsingular. Then by Theorem 10.11, the reduced echelon form of A is the identity I. By the definition of reduced echelon forms, $I = L_k \cdots L_2 L_1 A$ for a sequence $L_1, L_2, \ldots, L_k$ of elementary matrices. This formula implies that $L_k \cdots L_2 L_1$ is the inverse of A, since its product with A is I. Thus

$$A^{-1} = L_k \cdots L_2 L_1 \tag{1}$$

Hence

$$A = L_1^{-1}L_2^{-1} \cdots L_k^{-1}. \qquad \blacksquare$$

Equation (1) indicates a simple method for computing A^{-1} for a nonsingular matrix A. The equation $L_k \cdots L_2 L_1 A = I$ can be interpreted as saying that the sequence of row operations corresponding to $L_1, L_2, \ldots, L_k$ reduces A to the identity. Then the equation $L_k \cdots L_2 L_1 I = A^{-1}$, which follows from Eq. (1), says that the same sequence of row operations applied to I results in A^{-1}. This suggests the following procedure or algorithm for computing A^{-1}:

1. Form the partitioned matrix $(A : I)$.
2. Apply a sequence of row operations to reduce $(A : I)$ to the form $(B : C)$, where the matrix B is the reduced echelon form of A.
3. If $B = I$, then $C = A^{-1}$. If B is not I, it has zero rows and A is singular.

Example 1 Let

$$A = \begin{pmatrix} 2 & 4 & 4 \\ 1 & 3 & 4 \\ 1 & 1 & 1 \end{pmatrix}$$

Then

$$(A : I) = \begin{pmatrix} 2 & 4 & 4 & 1 & 0 & 0 \\ 1 & 3 & 4 & 0 & 1 & 0 \\ 1 & 1 & 1 & 0 & 0 & 1 \end{pmatrix}$$

By row operations we obtain

$$\begin{pmatrix} 1 & 2 & 2 & \frac{1}{2} & 0 & 0 \\ 0 & 1 & 2 & -\frac{1}{2} & 1 & 0 \\ 0 & -1 & -1 & -\frac{1}{2} & 0 & 1 \end{pmatrix}, \qquad \begin{pmatrix} 1 & 2 & 2 & \frac{1}{2} & 0 & 0 \\ 0 & 1 & 2 & -\frac{1}{2} & 1 & 0 \\ 0 & 0 & 1 & -1 & 1 & 1 \end{pmatrix}$$

and finally

$$(I : A^{-1}) = \begin{pmatrix} 1 & 0 & 0 & -\frac{1}{2} & 0 & 2 \\ 0 & 1 & 0 & \frac{3}{2} & -1 & -2 \\ 0 & 0 & 1 & -1 & 1 & 1 \end{pmatrix}$$

Thus

$$A^{-1} = \begin{pmatrix} -\frac{1}{2} & 0 & 2 \\ \frac{3}{2} & -1 & -2 \\ -1 & 1 & 1 \end{pmatrix}$$

Exercises

1. For each matrix A, compute A^{-1} if it exists, and if not find the rank of A.

(a) $\begin{pmatrix} 1 & 4 & 1 \\ 0 & 2 & 2 \\ 1 & 6 & 2 \end{pmatrix}$ 　　　　　　　(b) $\begin{pmatrix} 1 & 2 & 3 \\ 2 & 0 & 1 \\ 0 & 8 & 10 \end{pmatrix}$

(c) $\begin{pmatrix} 1 & 2 & 3 & 3 \\ 1 & 0 & 1 & 3 \\ -1 & 1 & 1 & -2 \\ 2 & 2 & 1 & 3 \end{pmatrix}$ (d) $\begin{pmatrix} 4 & 3 & 2 & 0 \\ 5 & 4 & 3 & 0 \\ -2 & -2 & -1 & 0 \\ 11 & 6 & 4 & 1 \end{pmatrix}$

2. Let A be a nonsingular $n \times n$ matrix and let $e_1, \ldots, e_n$ be the standard basis vectors. Let X_j be the vector solution of $AX_j = e_j$ ($j = 1, \ldots, n$). Prove that the matrix $B = (X_1 X_2 \cdots X_n)$ which has X_j in its jth column ($j = 1, \ldots, n$) is A^{-1}.

3. How many multiplications (and divisions) are needed to calculate the inverse of a nonsingular matrix by the method given in this section?

4. How many multiplications are required to solve an $n \times n$ system $AX = Y$ by computing A^{-1} and then multiplying $A^{-1}Y$? Compare this with Gaussian elimination; see the Exercises in Section 10.5.

10.7 Equivalence of matrices

In Section 10.2, elementary row operations and elementary column operations were both defined, but in subsequent sections row operations played the principal role. For example, it was proved in Theorem 10.4 that every $m \times n$ matrix can be transformed into a matrix in (row) reduced echelon form by a sequence of elementary row operations. In this section we shall see what more can be said if elementary column operations are also allowed. This will lead us to the concept of equivalence of matrices, a concept similar in nature to that of similarity, which was discussed in Section 6.8. Finally, the equivalence concept will be given a geometric interpretation in terms of change of bases.

THEOREM 10.13 Let A be an $m \times n$ matrix of rank r. There exist nonsingular matrices P and Q, where P is $m \times m$ and Q is $n \times n$, such that

$$PAQ = \left(\begin{array}{c|c} I_r & 0 \\ \hline 0 & 0 \end{array} \right) \qquad (1)$$

where I_r denotes the $r \times r$ identity matrix and the zeros denote zero matrices of appropriate size; that is, the $m \times n$ matrix PAQ has 1's in the first r positions on the main diagonal, and all other entries are zeros.

PROOF: We already know that a series of elementary row operations transform A into reduced echelon form. Equivalently, there is a nonsingular matrix P such that $PA = B$ is in reduced echelon form. Since A has rank r, B must have r non-zero rows and $m - r$ zero rows, and must have $e_1, \ldots, e_r$ in r of its columns, say in columns $k_1, \ldots, k_r$, respectively. By column operations of the second and third types, namely, by adding suitable multiples of column k_1 to other columns, we can obtain a matrix in which all entries in the first row are 0 except the 1 in column k_1. Then by adding multiples of column k_2 to other columns, we can obtain a matrix in which the only nonzero entry in the second row is the 1 in column k_2. Continuing in this way, we obtain a matrix in which $e_1, \ldots, e_r$ appear in columns $k_1, \ldots, k_r$ and all other columns are zero columns. Now by interchanging columns, we obtain a matrix in which $e_1, \ldots, e_r$ appear in the first r columns. This matrix has the form

$$\left(\begin{array}{c|c} I_r & 0 \\ \hline 0 & 0 \end{array} \right)$$

Since all this has been done with elementary column operations, there is a nonsingular matrix Q, a product of elementary matrices, such that PAQ has the indicated form.   ∎

DEFINITION 10.9   Two $m \times n$ matrices A and B with elements in a field **K** are called *equivalent* over **K** if there exist two nonsingular matrices P and Q, with elements in **K**, such that $PAQ = B$.

COROLLARY   If A is an $m \times n$ matrix of rank r, with elements in **K**, then A is equivalent over **K** to a matrix of the form in (1). The latter matrix is called the *canonical form of A under equivalence*.

THEOREM 10.14   A necessary and sufficient condition in order that two $m \times n$ matrices with elements in **K** be equivalent over **K** is that they have the same rank.

PROOF: Let A and B be $m \times n$ matrices. Suppose that A and B have the same rank r. Then by Theorem 10.13, there exist nonsingular matrices P_A, Q_A, P_B, Q_B, such that

$$P_A A Q_A = \left(\begin{array}{c|c} I_r & 0 \\ \hline 0 & 0 \end{array} \right), \qquad P_B B Q_B = \left(\begin{array}{c|c} I_r & 0 \\ \hline 0 & 0 \end{array} \right)$$

Hence, $P_A A Q_A = P_B B Q_B$ and $P_B^{-1} P_A A Q_A Q_B^{-1} = B$. If we define $P = P_B^{-1} P_A$, $Q = Q_A Q_B^{-1}$, we have $PAQ = B$, with P and Q nonsingular, which shows that A and B are equivalent.

Conversely, suppose that A and B are equivalent. Then $PAQ = B$ for certain nonsingular matrices P and Q. This means that A may be transformed to B by a sequence of elementary row and column operations. Such operations do not alter rank (Theorems 6.14 and 10.2). Therefore, A and B have the same rank. This completes the proof.   ∎

In the remainder of this section, a geometrical interpretation is given to the concept of equivalence of matrices, much as the concept of similarity of matrices was related to geometric ideas in Section 6.8. Recall that two square matrices M and N are called similar if $M = P^{-1}NP$ for some nonsingular matrix P, or, equivalently, if M and N are representatives of the same linear transformation $T: \mathbf{V} \to \mathbf{V}$ relative to different bases.

In order to obtain an analogous theory for equivalence it is necessary to consider two vector spaces $\mathbf{V}$ and $\mathbf{V}'$ where the dimension of $\mathbf{V}$ is n and of $\mathbf{V}'$ is m. We know by Theorem 6.1 that given any basis B of $\mathbf{V}$ and any basis B' of $\mathbf{V}'$, and given any transformation $T: \mathbf{V} \to \mathbf{V}'$, there exists an $m \times n$ matrix M which is the matrix of T with respect to the bases B of $\mathbf{V}$ and B' of $\mathbf{V}'$ (see Figure 10.1).

If we consider a second set of bases C of $\mathbf{V}$ and C' of $\mathbf{V}'$, then there will exist a matrix N of T with respect to the bases C and C'. Since both B and C are bases of the n-dimensional vector space $\mathbf{V}$ we know by Lemma 6.1 that a non-singular $n \times n$ matrix P exists which is the transition matrix from B to C. Further, there exists a nonsingular $m \times m$ matrix Q which is the transition matrix from C' to B'. Hence the following theorem is obtained.

THEOREM 10.15 Let T be a transformation of an n-dimensional vector space $\mathbf{V}$ into an m-dimensional vector space $\mathbf{V}'$. Further, let M be the matrix of T with respect to the bases B of $\mathbf{V}$ and B' of $\mathbf{V}'$, and N the matrix of T with respect to the bases C of $\mathbf{V}$ and C' of $\mathbf{V}'$. Then there exist nonsingular matrices $P\ (n \times n)$ and $Q\ (m \times m)$ such that $M = QNP$. The matrix P can be chosen to be the transition matrix from B to C and the matrix Q to be the transition matrix from C' to B'.

PROOF: Let P and Q be the transition matrices from B to C and C' to B', respectively. Let X be the coordinate vector of an arbitrary vector $v \in \mathbf{V}$ with respect to B. Then $PX = Y$ is the coordinate vector of v with respect to C. By definition of N, $NY = NPX$ is the coordinate vector of Tv with respect to C'. Since Q is the transition matrix from C' to B', $QNY = QNPX$ is the coordinate vector of Tv with respect to B'. Since the above is true for every $v \in \mathbf{V}$, QNP is the matrix of T with respect to B and B'. Hence $M = QNP$ where Q and P are the (non-singular) transition matrices. ∎

Figure 10.2 gives a diagrammatic presentation of the above theorem. The dotted lines represent coordinate maps of vectors in $\mathbf{V}$ and $\mathbf{V}'$.

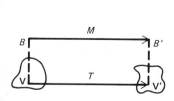

FIGURE 10.1

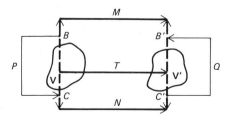

FIGURE 10.2

COROLLARY 1 Let M and N be $m \times n$ matrices. Let **V** be any vector space of dimension n and **V'** any vector space of dimension m. Then M is equivalent to N if and only if there are bases B and C for **V** and B' and C' for **V'** such that M is the matrix of T relative to B and B' and N is the matrix of T relative to C and C'.

PROOF: This is left as an exercise. ∎

COROLLARY 2 Similar matrices are equivalent.

PROOF: This is left as an exercise. ∎

It must be noted that the converse of Corollary 2 does not hold in general for square matrices. For M and N to be similar requires $M = QNP$ where $Q = P^{-1}$. For M and N to be equivalent all that is required is that P and Q be nonsingular.

It may be helpful to examine Figure 10.3, which is the specialization of Figure 10.2 to the case of similar matrices in which $\mathbf{V'} = \mathbf{V}$.

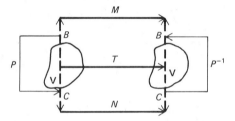

FIGURE 10.3

Example 1 Consider the transformation $T: \mathbf{R}^2 \rightarrow \mathbf{R}^3$ where $T(1, 0) = (4, 3, 6)$ and $T(0, 1) = (3, 2, 6)$. If we consider bases $C = \{(1, 0), (0, 1)\}$ of $\mathbf{R}^2$ and $C' = \{(1, 1, 1), (1, 0, 1), (0, 0, 1)\}$ of $\mathbf{R}^3$, then the matrix of T with respect to C and C' is

$$N = \begin{pmatrix} 3 & 2 \\ 1 & 1 \\ 2 & 3 \end{pmatrix}$$

This can be verified by noting that the coordinates of the image of $(1, 0)$ are

$$N \begin{pmatrix} 1 \\ 0 \end{pmatrix} = \begin{pmatrix} 3 & 2 \\ 1 & 1 \\ 2 & 3 \end{pmatrix} \begin{pmatrix} 1 \\ 0 \end{pmatrix} = \begin{pmatrix} 3 \\ 1 \\ 2 \end{pmatrix}$$

and hence the image of $(1, 0)$ is

$$3(1, 1, 1) + 1(1, 0, 1) + 2(0, 0, 1) = (4, 3, 6)$$

and the coordinates of the image of (0, 1) are

$$N\begin{pmatrix} 0 \\ 1 \end{pmatrix} = \begin{pmatrix} 2 \\ 1 \\ 3 \end{pmatrix}$$

and hence the image of (0, 1) is

$$2(1, 1, 1) + 1(1, 0, 1) + 3(0, 0, 1) = (3, 2, 6)$$

If we next consider bases $B = \{(1, 1), (0, 1)\}$ and $B' = \{1, 0, 0), (0, 1, 0),$ (0, 0, 1)\}$ of $\mathbf{R}^2$ and $\mathbf{R}^3$, respectively, we can derive the matrix of T with respect to B and B' as follows. The transition matrices from B to C and C' to B' are

$$P = \begin{pmatrix} 1 & 0 \\ 1 & 1 \end{pmatrix}, \qquad Q = \begin{pmatrix} 1 & 1 & 0 \\ 1 & 0 & 0 \\ 1 & 1 & 1 \end{pmatrix},$$

respectively. Hence, the matrix of the transformation T with respect to B and B' is given by

$$M = QNP = \begin{pmatrix} 1 & 1 & 0 \\ 1 & 0 & 0 \\ 1 & 1 & 1 \end{pmatrix}\begin{pmatrix} 3 & 2 \\ 1 & 1 \\ 2 & 3 \end{pmatrix}\begin{pmatrix} 1 & 0 \\ 1 & 1 \end{pmatrix} = \begin{pmatrix} 7 & 3 \\ 5 & 2 \\ 12 & 6 \end{pmatrix}$$

The coordinates of the vector (1, 0) with respect to B are $\begin{pmatrix} 1 \\ -1 \end{pmatrix}$ since (1, 0) =
(1, 1) − (0, 1), and the coordinates of (0, 1) with respect to B are $\begin{pmatrix} 0 \\ 1 \end{pmatrix}$.

Multiplying M by the coordinates with respect to B of (1, 0) and (0, 1) we get the coordinates of $T(1, 0)$ and $T(0, 1)$ with respect to B' as

$$\begin{pmatrix} 7 & 3 \\ 5 & 2 \\ 12 & 6 \end{pmatrix}\begin{pmatrix} 1 \\ -1 \end{pmatrix} = \begin{pmatrix} 4 \\ 3 \\ 6 \end{pmatrix} \quad \text{and} \quad \begin{pmatrix} 7 & 3 \\ 5 & 2 \\ 12 & 6 \end{pmatrix}\begin{pmatrix} 0 \\ 1 \end{pmatrix} = \begin{pmatrix} 3 \\ 2 \\ 6 \end{pmatrix}$$

which verifies that M is the matrix of T with respect to B and B'. The matrices M and N are equivalent, by Corollary 1.

Exercises

1. Are the following matrices equivalent?

$$M = \begin{pmatrix} 1 & 0 & 0 \\ 0 & 0 & 0 \\ 0 & 0 & 0 \end{pmatrix}, \qquad N = \begin{pmatrix} 1 & 1 & 0 \\ 0 & 1 & 1 \\ 0 & 0 & 0 \end{pmatrix}$$

2. By Theorem 10.14, two $m \times n$ matrices with the same rank are equivalent. Give an example of two matrices with the same rank which do not have the same Hermite normal form (row reduced echelon form).

3.   Characterize the class of all 3×3 matrices equivalent to the 3×3 identity matrix.

4.   Let $T(1, 0) = (2, 0, 1)$, $T(0, 1) = (1, 1, 0)$. Find the matrix N representing T, relative to the bases $C = \{(1, 0), (0, 1)\}$, $C' = \{(1, 1, 1), (0, 1, 0), (0, 0, 1)\}$ and the matrix M representing T relative to the bases $B = \{(1, 1), (1, 0)\}$, $B' = \{(1, 0, 0), (0, 1, 1), (1, 0, 1)\}$. Find the transition matrices P and Q and verify that $M = QNP$.

5.   If M and N are equivalent, show that

(a) M^T and N^T are equivalent

(b) M^2 and N^2 are not necessarily equivalent

(c) M^{-1} and N^{-1} are equivalent, if they exist.

6.   Prove Corollary 1 to Theorem 10.15.

7.   Prove Corollary 2 to Theorem 10.15.

COMMENTARY

The elementary operations of Types 1, 2, and 3 are closely related to the defining properties of determinants (Definition 7.1) and some of the results of this chapter can be deduced easily from the theory of determinants given in Chapter 7. However, we have chosen to develop our theory here in an independent manner.

Section 10.5   For discussion of numerical techniques for solving linear systems, and related matters, see:

FORSYTHE, G. E., and C. B. MOLER, *Computer Solution of Linear Algebraic Systems*, Englewood Cliffs, N.J., Prentice-Hall, 1967.

WILKINSON, J. H., *Rounding Errors in Algebraic Processes*, Englewood Cliffs, N.J., Prentice-Hall, 1963.

FADDEEV, D. K., and V. N. FADDEEVA, *Computational Methods of Linear Algebra* (translated from the Russian by R. C. Williams), San Francisco, W. H. Freeman, 1963.

Chapter 11

Eigenvectors and eigenvalues[1]

11.1 Motivation and definition

Some of the most useful properties of linear transformations and their corresponding matrix representatives can be deduced from a knowledge of their eigenvectors (and associated eigenvalues). We shall see one application of these concepts in the next chapter when we discuss systems of linear equations. Eigenvectors and eigenvalues play important roles in the mathematical development of numerous other areas of pure and applied mathematics, such as the theory of integral equations, Markov chains in probability theory, multivariate statistical inference, and so on.

To facilitate our discussion of eigenvectors and eigenvalues let us briefly reexamine the effects of a linear transformation. In particular, let us restrict our attention to elements of $\mathscr{L}(\mathbf{V}, \mathbf{V})$; that is, transformations of $\mathbf{V}$ into itself. First let us consider an arbitrary singular transformation T. Obviously, any vector v_0 in the null space of T will be mapped into a scalar multiple of itself, namely, $Tv_0 = \mathbf{0} = 0 \cdot v_0$. We might ask whether there are linear transformations under which vectors are mapped into scalar multiples of themselves for scalars other than zero; that is, vectors not in the null space of the transformation. Again an example is immediate, namely, the identity transformation $I_\mathbf{V}$ under which $I_\mathbf{V}v = v = 1 \cdot v$ for every v in $\mathbf{V}$. We shall later give examples of linear transformations, where certain vectors are mapped into scalar multiples of themselves for scalars other than zero or one.

DEFINITION 11.1 Let $\mathbf{V}$ be a vector space over a field $\mathbf{K}$ and let T be a linear transformation of $\mathbf{V}$ into $\mathbf{V}$. A nonzero vector v_0 in $\mathbf{V}$ is called an

---

[1] Eigenvectors and eigenvalues are also commonly referred to as characteristic vectors and characteristic values.

349

eigenvector of T if there exists a scalar λ in $\mathbf{K}$ such that $Tv_0 = \lambda v_0$; in which case λ is called an *eigenvalue* of T, or more specifically, the eigenvalue of T associated with the eigenvector v_0.

It is strongly recommended that the reader review the material in Chapter 6, particularly in Section 6.8, before proceeding.

Exercise

1. Show that if v_0 is an eigenvector of T, the eigenvalue associated with v_0 is unique.

11.2 Eigenvectors and eigenvalues of matrices

In Definition 11.1 we defined the eigenvectors and eigenvalues of a linear transformation T. The purpose of this section is to define the eigenvectors and eigenvalues of a matrix.

THEOREM 11.1   Consider the linear transformation $T: \mathbf{V} \to \mathbf{V}$, a basis B of $\mathbf{V}$, and the matrix M representative of T with respect to B. Suppose v_0 is an eigenvector of T with corresponding eigenvalue λ. Then if X is the coordinate vector of v_0 with respect to B, $MX = \lambda X$.

PROOF: Denote the elements of B by $\{v_1, v_2, \ldots, v_p\}$. Since

$$X = \begin{pmatrix} x_1 \\ x_2 \\ \vdots \\ x_p \end{pmatrix}$$

is the coordinate vector of v_0, that is, $v_0 = x_1 v_1 + x_2 v_2 + \cdots + x_p v_p$, we have $Tv_0 = \lambda v_0 = \lambda x_1 v_1 + \lambda x_2 v_2 + \cdots + \lambda x_p v_p$, which confirms λX as the coordinate vector of Tv_0. But since M is the matrix representative of T, MX is the coordinate vector of Tv_0 and we can conclude that $MX = \lambda X$.   ∎

We have just established a relationship between eigenvectors of linear transformations and a property of their coordinate vectors. We now formalize this property.

DEFINITION 11.2   Let M be a $p \times p$ matrix with elements in a scalar field $\mathbf{K}$. A nonzero vector X in $\mathbf{K}^p$ such that $MX = \lambda X$, for some scalar λ, is called an *eigenvector of M*, and λ is called the *eigenvalue* associated with the eigenvector X.

Theorem 11.1 has shown that if v_0 is an eigenvector of T with eigenvalue λ, the coordinate vector X of T is an eigenvector of M, the matrix representative of T, with eigenvalue λ. If we now recall that similar matrices are matrix representatives of the same linear transformation but with respect to different bases, we have the following theorem.

THEOREM 11.2 Let M and N be two similar $p \times p$ matrices, and let P be the transition matrix such that $M = P^{-1}NP$. If X is an eigenvector of M with eigenvalue λ then $PX = Y$ is an eigenvector of N with the same eigenvalue.

PROOF: We leave the proof of this as an exercise. See Exercises 1 and 2 of this section. ∎

Theorem 11.2 establishes a one-to-one correspondence between eigenvectors (and eigenvalues) of similar matrices, since to an eigenvector X of M with eigenvalue λ, there corresponds an eigenvector $Y = PX$ of N with eigenvalue λ, and conversely to an eigenvector Y of N, there corresponds an eigenvector $P^{-1}Y = X$ of M.

To complete the relationship we must have the following theorem, which is the converse of Theorem 11.1.

THEOREM 11.3 Let T be a linear transformation $T: \mathbf{V} \to \mathbf{V}$ and let M be the matrix representation of T with respect to some basis B of $\mathbf{V}$. If X is an eigenvector of M, with eigenvalue λ, and X is the coordinate vector with respect to B of the vector v_0, then v_0 is an eigenvector of T with associated eigenvalue λ.

PROOF: We are given that X satisfies the relationship $MX = \lambda X$. Further, if $B = \{v_1, v_2, \ldots, v_p\}$, $v_0 = x_1 v_1 + x_2 v_2 + \cdots + x_p v_p$. Since M is the matrix of T with respect to B, the coordinates of Tv_0 are λX; that is,

$$Tv_0 = \lambda x_1 v_1 + \lambda x_2 v_2 + \cdots + \lambda x_p v_p$$
$$= \lambda(x_1 v_1 + x_2 v_2 + \cdots + x_p v_p) = \lambda v_0$$

Hence, v_0 is an eigenvector of T with associated eigenvalue λ. ∎

Example 1 Consider the set $\mathbf{C}^\infty(J)$, where J is the real line. Let T be the linear transformation $Tf = df(x)/dx$. Then $Te^{\lambda x} = \lambda e^{\lambda x}$ and hence for any λ, real or complex, $e^{\lambda x}$ is an eigenvector with λ the associated eigenvalue. This transformation has infinitely many eigenvalues.

Example 2 Consider the matrix $M = \begin{pmatrix} 1 & 0 \\ 0 & 2 \end{pmatrix}$. If we consider

$$X = \begin{pmatrix} 1 \\ 0 \end{pmatrix} \quad \text{and} \quad Y = \begin{pmatrix} 0 \\ 1 \end{pmatrix}$$

we see that

$$MX = \begin{pmatrix} 1 \\ 0 \end{pmatrix} = X \quad \text{and} \quad MY = \begin{pmatrix} 0 \\ 2 \end{pmatrix} = 2Y$$

Hence, X is an eigenvector with associated eigenvalue 1, and Y is an eigenvector with associated eigenvalue 2. Note that cX and cY, for any scalar c, are also eigenvectors.

Example 3   In Example 2 of Section 6.8 we considered a transformation $T: \mathbf{R}^2 \to \mathbf{R}^2$ where $T(r_1, r_2) = (3r_1, 2r_2)$. The matrix of T with respect to the basis $B = \{(1, 1), (1, -1)\}$ was found to be

$$M = \begin{pmatrix} \frac{5}{2} & \frac{1}{2} \\ \frac{1}{2} & \frac{5}{2} \end{pmatrix}$$

and the matrix of T with respect to $B' = \{(0, 2), (-1, 0)\}$ was found to be

$$N = \begin{pmatrix} 2 & 0 \\ 0 & 3 \end{pmatrix}$$

It is easy to see that $\begin{pmatrix} c_1 \\ 0 \end{pmatrix}$ and $\begin{pmatrix} 0 \\ c_2 \end{pmatrix}$ are eigenvectors of N for arbitrary scalars

c_1 and c_2, and the respective eigenvalues are 2 and 3. By Theorem 11.2 we know 2 and 3 are eigenvalues of M also. Further, in Section 6.8, we found the transition matrix from B' to B to be

$$P^{-1} = \begin{pmatrix} 1 & -\frac{1}{2} \\ -1 & -\frac{1}{2} \end{pmatrix}$$

Since

$$Y_1 = \begin{pmatrix} c_1 \\ 0 \end{pmatrix}$$

is an eigenvector of N,

$$X_1 = P^{-1}Y_1 = \begin{pmatrix} c_1 \\ -c_1 \end{pmatrix}$$

is an eigenvector of M with eigenvalue 2. Similarly, since $Y_2 = \begin{pmatrix} 0 \\ c_2 \end{pmatrix}$ is an eigenvector of N,

$$X_2 = P^{-1}Y_2 = \begin{pmatrix} -c_2/2 \\ -c_2/2 \end{pmatrix} = \begin{pmatrix} k \\ k \end{pmatrix}$$

is, for arbitrary scalar k, an eigenvector of M with eigenvalue 3. The reader can verify that X_1 and Y_1 are the coordinates with respect to B and B', respectively, of the same vector $c_1(0, 2)$, and similarly X_2 and Y_2 are coordinates of the same vector $c_2(-1, 0)$. Further, these vectors in $\mathbf{R}^2$ are eigenvectors of T.

We close this section with a theorem giving the equivalence of various common definitions of eigenvalues of a matrix.

THEOREM 11.4   The following statements are equivalent:

1. λ is an eigenvalue of the matrix M.
2. The null space of $M - \lambda I$ is not $\{0\}$.
3. $\det [M - \lambda I] = 0$.

PROOF: The proof is left as an exercise. See the next section for further discussion of this theorem. ∎

Exercises

1.   Prove Theorem 11.2 by direct matrix manipulations; that is, prove the following. If $M = P^{-1}NP$ and $MX = \lambda X$, then $Y = PX$ implies $NY = \lambda Y$.

2.   Prove Theorem 11.2 by using the relationships between M and N and the transformation T. Do not prove the theorem by matrix manipulations.

3.   A matrix A is called *idempotent* if $A^2 = A$.

   (a) Prove that the eigenvalues of A can assume only the values 0 and 1.

   (b) Describe geometrically, in terms of eigenvectors, the transformation associated with A.

4.   Let D be a $p \times p$ diagonal matrix

$$D = \begin{pmatrix} d_{11} & 0 & \cdots & 0 \\ 0 & d_{22} & \cdots & 0 \\ & & \cdots & \\ 0 & 0 & \cdots & d_{pp} \end{pmatrix}$$

Prove that d_{ii} is an eigenvalue of D for each $i = 1, 2, \ldots, p$. Characterize the eigenvectors to which d_{ii} is associated.

5.   Prove Theorem 11.4.

11.3   Eigenspaces

Let us now turn to the questions involved in solving for eigenvalues and eigenvectors of a given $p \times p$ matrix M. If we are interested in finding all the eigenvectors and eigenvalues of M, we should know how many of each exist, or how we can tell when we have obtained every one. We begin by example.

Example 1   Consider the matrix

$$M = \begin{pmatrix} 0 & -1 \\ 1 & 0 \end{pmatrix}$$

associated with a transformation $T: \mathbf{R}^2 \to \mathbf{R}^2$. By Theorem 11.4, λ is an eigenvalue of M if and only if det $[M - \lambda I] = 0$ or

$$\det \begin{pmatrix} -\lambda & -1 \\ 1 & -\lambda \end{pmatrix} = \lambda^2 + 1 = 0$$

The roots of this equation are $\lambda = \pm\, i$. If X is an eigenvector of M, then either $MX = iX$ or $MX = -iX$. But all the entries in M are real, and hence no real eigenvectors exist. The reader might note that this matrix is representative of the transformation in $\mathbf{G}^2$ which rotates the coordinate axes 90°. Hence, we see that no real vector except the zero vector is mapped into a scalar multiple of itself.

This example sheds some light on a possible confusion regarding the meaning of Theorem 11.4, for we see that if we take the basic scalar field to be the real field $\mathbf{R}$, then M has no eigenvalues and the equation det $(M - \lambda I) = 0$ has no solutions. On the other hand, if we take the complex field, then M has eigenvalues $\pm i$ and these are the solutions of the equation det $(M - \lambda I) = 0$. Thus, in applying Theorem 11.4, we must consider what is the underlying scalar field.

Example 1 illustrates the fact that no real eigenvectors are associated with complex eigenvalues (when M is a real matrix). Hence, real eigenvectors do not always exist. Let us consider two examples involving only real eigenvalues.

Example 2   Consider the matrix

$$M = \begin{pmatrix} 1 & 1 \\ 0 & 1 \end{pmatrix}$$

The eigenvalues satisfy

$$\det \begin{pmatrix} 1 - \lambda & 1 \\ 0 & 1 - \lambda \end{pmatrix} = 0$$

or

$$(1 - \lambda^2) = 0$$

Here we see that $\lambda = 1$ is the one and only eigenvalue. It is a double root, or root of order (or multiplicity) 2. To be an eigenvector, X must satisfy $MX = \lambda X$, or $(M - \lambda I)X = 0$. Since $M - \lambda I$ is singular for $\lambda = 1$, the system of equations has a nontrivial solution, as follows.

$$(M - \lambda I)X = \begin{pmatrix} 1 - 1 & 1 \\ 0 & 1 - 1 \end{pmatrix}\begin{pmatrix} x_1 \\ x_2 \end{pmatrix} = \begin{pmatrix} 0 & 1 \\ 0 & 0 \end{pmatrix}\begin{pmatrix} x_1 \\ x_2 \end{pmatrix} = \begin{pmatrix} x_2 \\ 0 \end{pmatrix} = \begin{pmatrix} 0 \\ 0 \end{pmatrix}$$

Hence, any vector of the form $\begin{pmatrix} x_1 \\ 0 \end{pmatrix}$ will be an eigenvector. On the other hand, a vector of the form $\begin{pmatrix} 0 \\ x_2 \end{pmatrix}$ is not an eigenvector of M and hence the nullity of

$(M - \lambda I)$ is 1, which means we can find only one linearly independent eigenvector associated with the eigenvalue 1.

DEFINITION 11.3   The set of all eigenvectors of a transformation $T: \mathbf{V} \to \mathbf{V}$ associated with the eigenvalue λ, together with the vector $\mathbf{0}$, is called the *eigenspace* associated with λ, and will be denoted $\mathbf{V}_\lambda$.

THEOREM 11.5   An eigenspace $\mathbf{V}_\lambda$ of $T: \mathbf{V} \to \mathbf{V}$ is a subspace of $\mathbf{V}$.

PROOF: If v_1 and v_2 are any eigenvectors associated with λ, then

$$T(c_1 v_1 + c_2 v_2) = c_1 T v_1 + c_2 T v_2 = c_1 \lambda v_1 + c_2 \lambda v_2 = \lambda(c_1 v_1 + c_2 v_2)$$

Hence $c_1 v_1 + c_2 v_2$ is an eigenvector too. Since $\mathbf{0}$ is included as an element of $\mathbf{V}_\lambda$, $\mathbf{V}_\lambda$ is a subspace.   ∎

THEOREM 11.6   Eigenvectors associated with distinct eigenvalues are linearly independent.

PROOF: Suppose $v_1, v_2, \ldots, v_k$ are eigenvectors of a transformation T associated with eigenvalues $\lambda_1, \lambda_2, \ldots, \lambda_k$, respectively. Assume that $\lambda_1, \lambda_2, \ldots, \lambda_k$ are distinct. If the set $\{v_1, v_2, \ldots, v_k\}$ is not linearly independent, there is a largest integer $j < k$ such that $\{v_1, \ldots, v_j\}$ is independent but $\{v_1, \ldots, v_j, v_{j+1}\}$ is dependent. Thus, there are constants $c_1, \ldots, c_j$ not all zero, such that

$$v_{j+1} = c_1 v_1 + \cdots + c_j v_j$$

Then

$$T v_{j+1} = c_1 T v_1 + \cdots + c_j T v_j$$

or

$$\lambda_{j+1} v_{j+1} = c_1 \lambda_1 v_1 + \cdots + c_j \lambda_j v_j \tag{1}$$

Also, clearly

$$\lambda_{j+1} v_{j+1} = \lambda_{j+1}(c_1 v_1 + \cdots + c_j v_j) \tag{2}$$

Subtracting Eq. (1) from Eq. (2) we get

$$0 = (\lambda_{j+1} - \lambda_1) c_1 v_1 + \cdots + (\lambda_{j+1} - \lambda_j) c_j v_j$$

Since $\lambda_{j+1} - \lambda_1, \ldots, \lambda_{j+1} - \lambda_j$ are not zero, this shows that $\{v_1, \ldots, v_j\}$ are dependent, which is a contradiction.   ∎

COROLLARY   If λ_1 and λ_2 are distinct eigenvalues of T, then

$$\mathbf{V}_{\lambda_1} \cap \mathbf{V}_{\lambda_2} = \{\mathbf{0}\}.$$

PROOF: The proof is left as an exercise.   ∎

Example 3   Consider the matrix

$$M = \begin{pmatrix} 1 & 0 \\ 0 & 1 \end{pmatrix}$$

To solve for the eigenvalues, we obtain

$$\det \begin{pmatrix} 1 - \lambda & 0 \\ 0 & 1 - \lambda \end{pmatrix} = (1 - \lambda)^2 = 0$$

As in Example 2, we have $\lambda = 1$ as a double root. To find the eigenvectors of M, we must solve the equation $MX = \lambda X$ or $(M - \lambda I)X = 0$ where $\lambda = 1$. If we appeal to the relationships between matrices and linear transformations we know there is a transformation T for which $(M - \lambda I)$ is a representative matrix (Theorem 6.1). Further, the X satisfying $(M - \lambda I)X = 0$ are the co-ordinates of vectors in the null space of T. Since $\rho(T) = \rho(M - \lambda I)$, by Theorem 6.9, we can conclude that the number of linearly independent X satisfying $(M - \lambda I)X = 0$ is the dimension of M minus the rank of $M - \lambda I$. In this case $M - \lambda I = \begin{pmatrix} 0 & 0 \\ 0 & 0 \end{pmatrix}$ so $\rho(M - \lambda I) = 0$. Since M has dimension 2, we conclude there are two linearly independent eigenvectors of M with associated eigenvalue 1. Hence every vector of the form $\begin{pmatrix} x_1 \\ x_2 \end{pmatrix}$ is an eigenvector. This is not surprising as M is the identity matrix.

Before embarking on our final theorem of this section, which will formalize some of the discussion in the above example, we introduce the following definition.

DEFINITION 11.4   The *eigenspace* $\mathbf{X}_\lambda$ of a $p \times p$ matrix M associated with the eigenvalue λ of M is the set of all X in $\mathbf{K}^p$ satisfying $(M - \lambda I)X = 0$.

The reader should verify that $\mathbf{X}_\lambda$ will be a subspace of $\mathbf{K}^p$.

THEOREM 11.7   The dimension of the eigenspace $\mathbf{X}_\lambda$ of a $p \times p$ matrix M is p minus the rank of $(M - \lambda I)$.

PROOF: The proof (essentially contained in the discussion of Example 3) is left as an exercise.   ∎

According to the discussion in Section 11.2, similar matrices M and N have the same eigenvalues. However, if λ is one of these, the eigenspace of M associated with λ will, in general, be different from the eigenspace of N associated with λ, because if X is an eigenvector of M, then $Y = PX$ is an eigenvector of N (where P is a transition matrix). In other words, X and Y are coordinate vectors, relative to different bases, of an eigenvector v of the transformation T for which M and N are representatives.

Exercises

1. Prove the corollary of Theorem 11.6.

2. Prove Theorem 11.7.

3. Verify the statement following Definition 11.4.

   In Exercises 4 through 7, compute all eigenvalues and then find a basis for the eigenspace associated with each eigenvalue.

4. $M = \begin{pmatrix} 5 & -6 \\ 2 & -2 \end{pmatrix}$

   5. $M = \begin{pmatrix} 2 & 1 \\ 2 & 3 \end{pmatrix}$

6. $M = \begin{pmatrix} 2 & 0 & 0 \\ 0 & 1 & 0 \\ 0 & 0 & 1 \end{pmatrix}$

   7. $M = \begin{pmatrix} 3 & 0 & 0 \\ 0 & 1 & 1 \\ 0 & 0 & 1 \end{pmatrix}$

8. Let $T: \mathbf{R}^3 \to \mathbf{R}^3$ be defined by $T(x_1, x_2, x_3) = (2x_1, x_3, x_2 - x_1)$.

   (a) Find the three eigenvalues of T and corresponding eigenvectors by solving

   $$(2x_1, x_3, x_2 - x_1) = (\lambda x_1, \lambda x_2, \lambda x_3)$$

   (b) The matrix

   $$M = \begin{pmatrix} 3 & -1 & 3 \\ 0 & 0 & 1 \\ -1 & 1 & -1 \end{pmatrix}$$

   is the representative of T relative to $B = \{(1, 0, 0), (0, 1, 0), (1, 0, 1)\}$. Find the eigenvalues and eigenvectors of M. Show that the eigenvectors are the coordinates relative to B of the eigenvectors found in (a).

   (c) Find the matrix N of T relative to $B' = \{(1, 1, 1), (1, 2, 0), (2, 0, 0)\}$. Compute its eigenvalues and eigenvectors. Show that the eigenvectors of N are the coordinates relative to B' of the eigenvectors found in (a).

   (d) Show that each eigenvector of N is P times an eigenvector of M, where P is the transition matrix from B to B'.

9. Let S be the linear operator on $C^0(-\infty, \infty)$ defined by $Sf(x) = \int_0^x f(t)\, dt$. Prove that S has no eigenvalues.

10. Let M and N be similar matrices, $M = P^{-1}NP$, and let λ be an eigenvalue of M and N. Show that the eigenspace $\mathbf{X}_\lambda$ relative to M has the same dimension as the eigenspace $\mathbf{Y}_\lambda$ relative to N.

11.4   The characteristic polynomial

We have discussed previously how to find eigenvalues and eigenvectors of a given matrix or linear transformation. As we saw, the eigenvalues of a matrix M satisfy $\det (M - \lambda I) = 0$. Therefore, we turn to an analysis of $\det (M - \lambda I)$, considered as a function of the variable λ. If $M = (m_{ij})$ is a 2×2 matrix,

$$\det (M - \lambda I) = \begin{pmatrix} m_{11} - \lambda & m_{12} \\ m_{21} & m_{22} - \lambda \end{pmatrix} = (m_{11} - \lambda)(m_{22} - \lambda) - m_{12}m_{21}$$

Clearly, $\det (M - \lambda I)$ is a polynomial of degree 2 in λ, and the coefficient of λ^2 is $+1$. If M is a 3×3 matrix, it can be verified that $\det (M - \lambda I)$ is a polynomial of degree 3 and the coefficient of λ^3 is -1. The following general result is easily inferred.

THEOREM 11.8   Let M be a $p \times p$ matrix of scalars in a field **K**. Then $\det [M - \lambda I]$ is a pth degree polynomial in λ. The coefficient of λ^p in this polynomial is $(-1)^p$.

PROOF: This theorem can be proved formally by mathematical induction, but this is omitted here (see the exercises).   ∎

DEFINITION 11.5   Let M be a $p \times p$ matrix with elements in **K**. The pth degree polynomial

$$p(\lambda) = \det [M - \lambda I] = (-\lambda)^p + c_{p-1}\lambda^{p-1} + \cdots + c_0$$

is called the *characteristic polynomial of the matrix* M. The equation $p(\lambda) = 0$ is called the *characteristic equation* of M. Any number λ in **K** satisfying the characteristic equation is called a *characteristic root* of M.

By Theorem 11.4, the eigenvalues of a matrix M satisfy the equation

$$\det [M - \lambda I] = (-\lambda)^p + c_{p-1}\lambda^{p-1} + \cdots + c_0 = 0$$

Hence, λ is an eigenvalue if and only if λ is a root of the characteristic polynomial, or equivalently if λ satisfies the characteristic equation. Since **K** is a subfield of the complex field, we know from the theory of equations that there are exactly p roots. However, the roots might not all be distinct and some might be complex but not in **K**. If M is a real matrix and λ is a complex root, its complex conjugate is also a root; that is, if $\lambda = \alpha + i\beta$ is a complex root, then $\bar{\lambda} = \alpha - i\beta$ is also a root.

In the last section there was an example in which the characteristic roots were the complex pair i and $-i$. There were also two examples in which 1 was a root of order 2. Let us now consider the following.

Example 1   Find the eigenvalues of the diagonal matrix

$$D = \begin{pmatrix} d_{11} & 0 & 0 & \cdots & 0 \\ 0 & d_{22} & 0 & \cdots & 0 \\ 0 & 0 & d_{33} & \cdots & 0 \\ & & \cdots & & \\ 0 & 0 & 0 & \cdots & d_{pp} \end{pmatrix}$$

$$\det[D - \lambda I] = \begin{vmatrix} d_{11} - \lambda & 0 & 0 & \cdots & 0 \\ 0 & d_{22} - \lambda & 0 & \cdots & 0 \\ 0 & 0 & d_{33} - \lambda & \cdots & 0 \\ & & \cdots & & \\ 0 & 0 & 0 & \cdots & d_{pp} - \lambda \end{vmatrix}$$

$$= \prod_{i=1}^{p} (d_{ii} - \lambda)$$

The roots of this polynomial are $d_{11}, d_{22}, \ldots, d_{pp}$, the elements on the diagonal of the matrix. Hence, the following theorem has been partly proved.

THEOREM 11.9   Let D be a $p \times p$ diagonal matrix with elements in **K**. Then D has p eigenvalues (not necessarily distinct) which are the diagonal elements of D. Further, the dimension of the eigenspace associated with any eigenvalue λ_0 is the number of times λ_0 appears on the diagonal and hence a basis for $\mathbf{K}^p$ can be formed of eigenvectors of D.

PROOF: Above we have shown that the characteristic polynomial of a diagonal matrix is

$$p(\lambda) = \prod_{i=1}^{p} (d_{ii} - \lambda)$$

where d_{ii} is the ith diagonal element. Suppose λ_0 appears in the $i_1, i_2, \ldots, i_k$ diagonal positions of D. Then λ_0 is a k-fold root of $p(\lambda)$ or an eigenvalue of order k. We can construct a basis for the eigenspace associated with λ_0 as follows. Define e_j to be the jth column of the $p \times p$ identity matrix. Then e_j is a column vector with a one in the jth position and zeros elsewhere. Multiplying D by e_j gives $d_{jj}e_j$. Hence, $e_{j_1}, e_{j_2}, \ldots, e_{j_k}$ are vectors which are mapped into λ_0 times themselves by D. Further, since they are distinct columns of the identity matrix they are linearly independent. Since the number of linearly independent columns of $D - \lambda_0 I$ is $p - k$ we know by Theorem 11.7 that the dimension of the eigenspace of λ_0 is k. Hence, $\{e_{j_1}, e_{j_2}, \ldots, e_{j_k}\}$ is a basis of this eigenspace. Finally, it is clear that when we combine the bases for the eigenspaces of all the different eigenvalues, we obtain the complete set of columns, $\{e_1, e_2, \ldots, e_p\}$. This is a basis for $\mathbf{K}^p$.   ∎

We now generalize the above theorem to establish necessary and sufficient conditions for an arbitrary $p \times p$ matrix to have p linearly independent eigenvectors. This theorem will also lead us into the topic of the next section.

THEOREM 11.10   A $p \times p$ matrix M with elements in a field $\mathbf{K}$ is similar over $\mathbf{K}$ to a diagonal matrix if and only if M has p linearly independent eigenvectors in $\mathbf{K}^p$. In this case, $Y^{-1}MY$ is diagonal where Y is a matrix in which the columns are independent eigenvectors of M.

PROOF: Suppose M to be similar to a diagonal matrix D; that is, $D = P^{-1}MP$ for some nonsingular matrix P with elements in $\mathbf{K}$. By Theorem 11.9, p linearly independent eigenvectors exist for D, say $\{X_1, X_2, \ldots, X_p\}$. By Theorem 11.2, $Y_i = PX_i$ is an eigenvector of M associated with the same eigenvalue as X_i. Hence, $\{Y_1, Y_2, \ldots, Y_p\}$ are eigenvectors of M and $Y_i \in \mathbf{K}^p$. To show the Y_i linearly independent, suppose $\sum_{i=1}^{p} c_i Y_i = 0$. Applying P^{-1} to this equation, we get $c_1 = c_2 = \cdots = c_p = 0$ because the X_i are linearly independent. Hence the Y_i are linearly independent and M has p linearly independent eigenvectors in $\mathbf{K}^p$.

Next suppose M has p linearly independent eigenvectors $\{Y_1, Y_2, \ldots, Y_p\}$ in $\mathbf{K}^p$ associated with eigenvalues $d_{11}, d_{22}, \ldots, d_{pp}$, respectively, where the d_{ii} are not necessarily distinct. Recall that each Y_i is a $p \times 1$ column vector satisfying $MY_i = d_{ii}Y_i$. If we define the matrix Y to be the $p \times p$ array $Y = [Y_1 \quad Y_2 \quad \cdots \quad Y_p]$, we obtain

$$MY = [d_{11}Y_1 \quad d_{22}Y_2 \quad \cdots \quad d_{pp}Y_p]$$

$$= [Y_1 \quad Y_2 \quad \cdots \quad Y_p] \begin{pmatrix} d_{11} & 0 & \cdots & 0 \\ 0 & d_{22} & \cdots & 0 \\ & & \ddots & \\ 0 & 0 & \cdots & d_{pp} \end{pmatrix} \tag{1}$$

Multiplying both sides of Eq. (1) by Y^{-1}, which exists since the columns of Y are linearly independent, gives us

$$Y^{-1}MY = \begin{pmatrix} d_{11} & 0 & \cdots & 0 \\ 0 & d_{22} & \cdots & 0 \\ & & \ddots & \\ 0 & 0 & \cdots & d_{pp} \end{pmatrix}$$

Hence M is similar to a diagonal matrix.   ∎

DEFINITION 11.6   Let $\mathbf{V}$ be a vector space over a field $\mathbf{K}$. A linear transformation $T: \mathbf{V} \rightarrow \mathbf{V}$ is *diagonable* (or *diagonalizable*) over $\mathbf{K}$ if a basis B of $\mathbf{V}$ exists such that the matrix representation of T with respect to B is a diagonal matrix. A matrix M is *diagonable* (*diagonalizable*) over a field **K** if M is similar over **K** to a diagonal matrix.

The above theorem can be restated as follows:

COROLLARY Let T be a linear transformation $T: \mathbf{V} \to \mathbf{V}$ where $\mathbf{V}$ is p-dimensional. Then T is diagonable if and only if T has p linearly independent eigenvectors in $\mathbf{V}$.

We next consider one further case in which computation of eigenvalues is relatively simple.

DEFINITION 11.7 A matrix T is called *upper triangular* if $t_{ij} = 0$ for $i > j$, *lower triangular* if $t_{ij} = 0$ for $i < j$, and *triangular* if it is either upper triangular or lower triangular.

THEOREM 11.11 If T is a $p \times p$ triangular matrix, then the eigenvalues of T are the diagonal entries t_{ii}.

PROOF: The proof is left as an exercise.   ∎

We close this section with a theorem relating characteristic polynomials to the linear transformation $T: \mathbf{V} \to \mathbf{V}$.

THEOREM 11.12 Similar matrices have the same characteristic polynomial.

PROOF: Let M and N be two similar matrices. Then there exists a nonsingular P such that $M = PNP^{-1}$. The characteristic polynomial of M is $p_M(\lambda) = $ det $[M - \lambda I]$. Since $M = PNP^{-1}$

$$
\begin{aligned}
p_M(\lambda) &= \det \left[PNP^{-1} - \lambda PP^{-1} \right] \\
&= \det \left[P(N - \lambda I)P^{-1} \right] = \det P \det (N - \lambda I) \det P^{-1} \\
&= \det PP^{-1} \det (N - \lambda I) = \det (N - \lambda I) = p_N(\lambda)
\end{aligned}
$$ ∎

COROLLARY If λ is an eigenvalue of similar matrices M and N, the multiplicity of λ with respect to M is the same as the multiplicity of λ with respect to N.

As a direct consequence of the above theorem, we can give the following definition.

DEFINITION 11.8 Let $T: \mathbf{V} \to \mathbf{V}$ be a linear transformation and let M be any matrix representation of T. Then if $p(\lambda)$ is the characteristic polynomial of M, we define $p(\lambda)$ to be the *characteristic polynomial of the transformation T*.

Example 2   It was shown previously that

$$M = \begin{pmatrix} \frac{5}{2} & \frac{1}{2} \\ \frac{1}{2} & \frac{5}{2} \end{pmatrix} \quad \text{and} \quad N = \begin{pmatrix} 2 & 0 \\ 0 & 3 \end{pmatrix}$$

are similar. Their characteristic polynomials are

$$\det (M - \lambda I) = \det \begin{vmatrix} \frac{5}{2} - \lambda & \frac{1}{2} \\ \frac{1}{2} & \frac{5}{2} - \lambda \end{vmatrix} = (\tfrac{5}{2} - \lambda)^2 - \tfrac{1}{4} = \lambda^2 - 5\lambda + 6$$

$$\det (N - \lambda I) = \det \begin{vmatrix} 2 - \lambda & 0 \\ 0 & 3 - \lambda \end{vmatrix} = (2 - \lambda)(3 - \lambda) = \lambda^2 - 5\lambda + 6$$

Exercises

1. Prove Theorem 11.11.

   In Exercises 2 through 5 compute the characteristic polynomial and from it determine the eigenvalues of the given matrix.

2. $M = \begin{pmatrix} 3 & 2 \\ 1 & 5 \end{pmatrix}$ 3. $M = \begin{pmatrix} 1 & 2 \\ 7 & 3 \end{pmatrix}$

4. $M = \begin{pmatrix} 1 & 2 & 7 \\ 3 & 3 & 1 \\ 4 & 5 & 8 \end{pmatrix}$ 5. $M = \begin{pmatrix} 2 & 7 & 4 \\ 1 & 1 & 3 \\ 1 & 6 & 1 \end{pmatrix}$

6. Prove that zero is an eigenvalue of a singular $p \times p$ matrix M, and the dimension of the eigenspace $\mathbf{X}_0$ is $p - \rho(M)$.

   In Exercises 7 through 9 find the characteristic polynomial of each matrix M. Also give the eigenvalues in the complex number field, as well as the eigenvectors. Construct a matrix Y such that $Y^{-1}MY$ is a diagonal matrix D, and find D.

7. $\begin{pmatrix} 2 & -1 \\ 1 & 2 \end{pmatrix}$ 8. $\begin{pmatrix} 1 & 5 \\ -1 & -1 \end{pmatrix}$ 9. $\begin{pmatrix} 3 & 2 & 1 \\ 0 & 1 & 2 \\ 0 & 1 & -1 \end{pmatrix}$

10. Prove that a transformation of $\mathbf{R}^p$ into itself is diagonable if it has p distinct real eigenvalues.

11. Let M be a $p \times p$ matrix of scalars. Let N_λ be obtained from M by subtracting the variable λ from at most one element in each row and each column.

    (a) Using mathematical induction and expansion in cofactors, prove that the determinant of N_λ is a polynomial in λ of degree at most p.

(b) Prove by induction that $\det (M - \lambda I)$ is a polynomial of degree exactly p in λ and the coefficient of λ^p is $(-1)^p$.

12. Prove that if M is nonsingular then the eigenvalues of M^{-1} are the reciprocals of the eigenvalues of M. Describe the relation between the eigenvectors of M and the eigenvectors of M^{-1}.

13. Describe the relation between the eigenvalues and eigenvectors of M and of M^k where $k = 2, 3, \ldots$.

14. Show that the constant term in the characteristic polynomial is $\det M$.

15. Prove that a matrix M is nonsingular if and only if zero is not an eigenvalue of M.

16. (a) Let A and B be $q \times p$ and $p \times q$ matrices respectively, and let I_p and I_q be the identity matrices of orders p and q. Then prove that

$$\lambda^p \det (\lambda I_q - AB) = \lambda^q \det (\lambda I_p - BA)$$

(b) Deduce that if $p = q$, the matrices AB and BA have the same characteristic polynomial and, in particular, the same characteristic roots. [*Hint:* Compute $\det CD$ and $\det DC$ where

$$C = \begin{pmatrix} \lambda I_q & A \\ B & I_p \end{pmatrix}, \qquad D = \begin{pmatrix} I_q & 0 \\ -B & \lambda I_p \end{pmatrix}$$

See J. Schmid, "A Remark on Characteristic Polynomials," *Amer. Math. Monthly*, vol. 77, 998 (1970).]

11.5 Polynomials in a matrix

The extraction of eigenvalues and eigenvectors for diagonal matrices was shown in Theorem 11.9 to be simple. It therefore seems reasonable to investigate further the conditions under which matrices are similar to diagonal matrices (that is, when a transformation $T: V \to V$ has a diagonal representation). At present such a representation seems only aesthetically satisfying, but we will see later that it has great practical value.

Consider any nth degree polynomial in a variable x, $p(x) = c_n x^n + c_{n-1} x^{n-1} + \cdots + c_0$. We now introduce the notion of a polynomial in a $q \times q$ matrix M. Such a polynomial would be represented as

$$p(M) = c_n M^n + c_{n-1} M^{n-1} + \cdots + c_0 M^0 \tag{1}$$

In order for this to represent a matrix, the addition on the right-hand side must be defined. First note that M^k is again a $q \times q$ matrix, for $k = 1, 2, \ldots, n$.

Further, we agree to interpret M^0 as the $q \times q$ identity matrix I. This can be justified by the fact that if M represents a transformation $T: \mathbf{V} \to \mathbf{V}$, then M^k represents the composite transformation of T applied k times. Hence, M^0 represents applying T zero times which is equivalent to leaving $\mathbf{V}$ unchanged, or the identity transformation. Since the set $\mathbf{M}_{qq}$ of all $q \times q$ matrices is a vector space, $p(M)$ is an element of $\mathbf{M}_{qq}$. Note that since $M^0 = I$, we can write Eq. (1) as $p(M) = c_n M^n + c_{n-1} M^{n-1} + \cdots + c_0 I$.

Example 1 Consider the polynomial $p(x) = 3x^2 + 2x - 4$, and the matrix $M = \begin{pmatrix} 2 & 1 \\ 1 & -1 \end{pmatrix}$. Then $M^2 = \begin{pmatrix} 5 & 1 \\ 1 & 2 \end{pmatrix}$. Hence,

$$p(M) = 3\begin{pmatrix} 5 & 1 \\ 1 & 2 \end{pmatrix} + 2\begin{pmatrix} 2 & 1 \\ 1 & -1 \end{pmatrix} - 4\begin{pmatrix} 1 & 0 \\ 0 & 1 \end{pmatrix} = \begin{pmatrix} 15 & 5 \\ 5 & 0 \end{pmatrix}$$

We are now in a position to state and prove a somewhat surprising theorem.

THEOREM 11.13 Let $r(\lambda)$ be a polynomial of degree n in λ, and Q a $q \times q$ matrix with elements which are polynomials of degree $\leq n - 1$ in λ such that $(M - \lambda I)Q \equiv r(\lambda)I$. Then $r(M) = 0$; that is, $r(M)$ is the $q \times q$ zero matrix.

PROOF: Since $r(\lambda)$ is a polynomial of degree n we have

$$r(\lambda)I = (b_0 + b_1\lambda + b_2\lambda^2 + \cdots + b_n\lambda^n)I$$

where $b_0, b_1, \ldots, b_n$ are scalars. Since Q has entries which are polynomials of degree $\leq n - 1$, that is, Q is an array of such polynomials, we have

$$Q = Q_0 + Q_1\lambda + Q_2\lambda^2 + \cdots + Q_{n-1}\lambda^{n-1}$$

where Q_i is a $q \times q$ matrix. Hence, since $(M - \lambda I)Q$ is given as equal to $r(\lambda)I$, we have

$$MQ_0 + MQ_1\lambda + \cdots + MQ_{n-1}\lambda^{n-1} - Q_0\lambda - Q_1\lambda^2 - \cdots - Q_{n-1}\lambda^n$$
$$= r(\lambda)I = (b_0 + b_1\lambda + \cdots + b_n\lambda^n)I \quad (2)$$

Equating coefficients of like powers of λ in Eq. (2) we get the following equalities.

$$
\begin{aligned}
MQ_0 &= b_0 I \\
MQ_1 - Q_0 &= b_1 I \\
&\vdots \\
MQ_{n-1} - Q_{n-2} &= b_{n-1} I \\
-Q_{n-1} &= b_n I
\end{aligned}
\qquad (3)
$$

If we multiply the equalities in Eq. (3) by I, M, $M^2, \ldots, M^{n-1}$, M^n, respectively, we obtain

$$MQ_0 = b_0 I$$
$$M^2 Q_1 - MQ_0 = b_1 M$$
$$M^3 Q_2 - M^2 Q_1 = b_2 M^2$$
$$\vdots \qquad \qquad \qquad (4)$$
$$M^n Q_{n-1} - M^{n-1} Q_{n-2} = b_{n-1} M^{n-1}$$
$$-M^n Q_{n-1} = b_n M^n$$

The sum of the left-hand sides of Eqs. (4) is the $q \times q$ zero matrix, and the sum of the right-hand sides is $r(M)$. Hence, $r(M) = 0$. ∎

Theorem 11.13 can be thought of as a generalization of the factor theorem of elementary algebra, which states that if a polynomial $r(x)$ has a linear factor $m - x$, then $r(m) = 0$.

As a direct consequence of Theorem 11.13 we get the famous Cayley–Hamilton Theorem.

THEOREM 11.14 (Cayley–Hamilton) If $p(\lambda)$ is the characteristic polynomial of an $n \times n$ matrix M, then $p(M) = 0$. In other words, a matrix satisfies its characteristic equation.

PROOF: If $n = 1$, the theorem is obvious. We now assume $n \geq 2$. Define the matrix $R = M - \lambda I$. By Theorem 7.4, $\det R = \sum_{j=1}^{n} r_{ij} R^{ij}$, where R^{ij} is the i, j cofactor of R. Further, $\sum_{j=1}^{n} r_{ij} R^{kj} = 0$, $k \neq i$, by Lemma 7.1. Next consider $R(\text{adj } R)$. Since the adjugate (or adjoint) of R is the transpose of the cofactor matrix, the i, kth element of $R(\text{adj } R)$ is $\sum_{j=1}^{n} r_{ij} R^{kj}$. Hence

$$(\det R)I = R \cdot (\text{adj } R) \qquad (5)$$

But $\det R = \det (M - \lambda I) = p(\lambda)$, the characteristic polynomial of M, and is therefore a polynomial of degree n in λ. Further, the elements in adj R are cofactors of R and (see Exercise 11, Section 11.4) are polynomials of degree at most $n - 1$ in λ. If we denote adj R by Q and replace R by $(M - \lambda I)$ in Eq. (5), we obtain

$$(\det R)I = p(\lambda)I = (M - \lambda I)Q$$

satisfying the conditions of Theorem 11.13. Hence, $p(M) = 0$ and the theorem is proved. ∎

We can restate the above theorem in terms of linear transformations as follows.

THEOREM 11.15 If $p(\lambda)$ is the characteristic polynomial of a linear transformation $T: \mathbf{V} \to \mathbf{V}$, then $p(T) = 0$.

PROOF: We must recall the definition of polynomials in transformations (Section 5.5) and the definition of characteristic polynomial of a transformation. The details of the proof are left as an exercise. ∎

Example 2 In Example 2 of Section 11.4, we found the characteristic polynomial of $M = \begin{pmatrix} \frac{5}{2} & \frac{1}{2} \\ \frac{1}{2} & \frac{5}{2} \end{pmatrix}$ to be $p(\lambda) = \lambda^2 - 5\lambda + 6$. Since $M^2 = \begin{pmatrix} \frac{13}{2} & \frac{5}{2} \\ \frac{5}{2} & \frac{13}{2} \end{pmatrix}$ we get

$$p(M) = M^2 - 5M + 6I = \begin{pmatrix} \frac{13}{2} & \frac{5}{2} \\ \frac{5}{2} & \frac{13}{2} \end{pmatrix} - \begin{pmatrix} \frac{25}{2} & \frac{5}{2} \\ \frac{5}{2} & \frac{25}{2} \end{pmatrix} + \begin{pmatrix} 6 & 0 \\ 0 & 6 \end{pmatrix} = \begin{pmatrix} 0 & 0 \\ 0 & 0 \end{pmatrix}$$

Exercises

In Exercises 1 through 4 evaluate the polynomial at the specified matrix.

1. $p(x) = 3x^2 + 2x + 1,$ $M = \begin{pmatrix} 2 & 1 \\ 1 & 3 \end{pmatrix}$

2. $p(x) = 2x^4 - x^2 + 4,$ $M = \begin{pmatrix} 1 & 3 \\ 3 & 2 \end{pmatrix}$

3. $p(x) = x^3 - 3x^2 + 3,$ $M = \begin{pmatrix} 1 & 2 & 1 \\ -1 & 0 & 1 \\ 1 & -1 & 1 \end{pmatrix}$

4. $p(x) = x^2 - x + 1,$ $M = \begin{pmatrix} 1 & 1 & 0 & 0 \\ 0 & 1 & -1 & 2 \\ 0 & 0 & 1 & 3 \\ 0 & 0 & 0 & 2 \end{pmatrix}$

5–8. Determine the characteristic equation of each matrix in Exercises 1 through 4. Then verify the Cayley–Hamilton theorem.

9. Prove Theorem 11.15.

11.6 The minimal polynomial and diagonable transformations

We introduce the concepts of this section with an example.

Example 1 Consider the matrix

$$M = \begin{pmatrix} 3 & 0 & -1 \\ 0 & 3 & 1 \\ 0 & 0 & 2 \end{pmatrix}$$

The characteristic polynomial associated with M is $p(\lambda) = (3 - \lambda)^2(2 - \lambda)$. By the Cayley–Hamilton theorem we know $p(M) = (3I - M)^2(2I - M) = 0$. Let us define $p^*(\lambda) = (3 - \lambda)(2 - \lambda)$. If we evaluate this polynomial at M, we obtain

$$p^*(M) = \begin{pmatrix} 0 & 0 & 1 \\ 0 & 0 & -1 \\ 0 & 0 & 1 \end{pmatrix} \begin{pmatrix} -1 & 0 & 1 \\ 0 & -1 & -1 \\ 0 & 0 & 0 \end{pmatrix} = 0$$

Hence, $p^*(\lambda)$ is a lower degree polynomial than the characteristic polynomial $p(\lambda)$ yet $p^*(M) = p(M) = 0$. This leads to the following definition.

DEFINITION 11.9 A polynomial $r(\lambda) = b_n\lambda^n + b_{n-1}\lambda^{n-1} + \cdots + b_0$ with leading coefficient $b_n = 1$ is called a *monic polynomial*. For a given square matrix M, the lowest degree monic polynomial $p^*(\lambda)$ satisfying $p^*(M) = 0$ is called the *minimal polynomial* of M.

We already know that for each $n \times n$ matrix M a monic polynomial $p(\lambda)$ exists which satisfies $p(M) = 0$. This follows directly from the Cayley–Hamilton theorem. However, we now must show that there is a unique monic polynomial of minimal degree.

THEOREM 11.16 The minimal polynomial of an $n \times n$ matrix is unique.

PROOF: Suppose there are two monic polynomials $r(\lambda) = \lambda^k + a_{k-1}\lambda^{k-1} + \cdots + a_0$ and $s(\lambda) = \lambda^k + b_{k-1}\lambda^{k-1} + \cdots + b_0$ such that $r(M) = s(M) = 0$, and these are both minimal polynomials. If $k = 0$, clearly $r(\lambda)$ and $s(\lambda)$ are the same, and so we suppose $k \geq 1$. Since $r(M) = 0 = s(M)$, we obtain

$$M^k = -a_{k-1}M^{k-1} - a_{k-2}M^{k-2} - \cdots - a_0I \tag{1}$$

and

$$M^k = -b_{k-1}M^{k-1} - b_{k-2}M^{k-2} - \cdots - b_0I \tag{2}$$

Subtracting Eq. (1) from Eq. (2) gives us

$$0 = (a_{k-1} - b_{k-1})M^{k-1} + (a_{k-2} - b_{k-2})M^{k-2} + \cdots + (a_0 - b_0)I$$

If for any i, $1 \leq i \leq k - 1$, a_i is not equal to b_i, we will be able to obtain a monic polynomial of degree less than k, which yields the zero matrix when evaluated at M. But this contradicts the fact that $r(\lambda)$ and $s(\lambda)$ are minimal. Hence, our theorem is proved. ∎

Our next objective is Theorem 11.17, which gives a test, based on the minimal polynomial, which can be used to determine whether a linear transformation is diagonable. In the proof of this theorem, we make use of the following two lemmas.

LEMMA 11.1 The minimal polynomial p^* of a matrix M divides every polynomial s such that $s(M) = 0$; that is, if $p^*(\lambda)$ is the minimal polynomial of M and if $s(\lambda)$ is any polynomial such that $s(M) = 0$, then $s(\lambda) = p^*(\lambda)q(\lambda)$ for some polynomial $q(\lambda)$.

PROOF: Given $s(\lambda)$ and $p^*(\lambda)$, there exist polynomials $q(\lambda)$ and $r(\lambda)$ (the quotient and remainder when s is divided by p^*) such that $s(\lambda) = p^*(\lambda)q(\lambda) + r(\lambda)$ and such that $r(\lambda)$ has lower degree than $p^*(\lambda)$. It follows that $s(M) = p^*(M)q(M) + r(M)$. Since $s(M)$ and $p^*(M)$ both equal zero, $r(M)$ must equal zero. But since $r(M)$ is of degree less than that of p^*, r is identically zero or else p^* could not be the minimal polynomial. ∎

In particular, the minimal polynomial of a matrix divides the characteristic polynomial of the matrix.

DEFINITION 11.10 A nonconstant polynomial p is said to be *irreducible* over a field $\mathbf{K}$ if $p(x) = r(x)s(x)$ (where r and s are polynomials) implies either $r(x)$ or $s(x)$ is a constant $a \in \mathbf{K}$ and the other is $a^{-1}p(x)$. Otherwise, p is called *reducible* over $\mathbf{K}$; that is, p is irreducible over $\mathbf{K}$ if it has no non-trivial factorization.

For example, $\lambda^2 - 1$ is reducible over the real field since $\lambda^2 - 1 = (\lambda - 1)(\lambda + 1)$, but $\lambda^2 + 1$ is irreducible over the real field. However, $\lambda^2 + 1$ is reducible over the complex field since $\lambda^2 + 1 = (\lambda - i)(\lambda + i)$.

LEMMA 11.2 Every irreducible factor of the characteristic polynomial p of an $n \times n$ matrix M is a factor of its minimal polynomial p^*.

PROOF: If $r(\lambda)$ is any polynomial, $r(x) - r(y)$ is a polynomial in x for fixed y and a polynomial in y for fixed x. Since $y = x$ is a root of $r(x) - r(y) = 0$, $x - y$ is a factor of $r(x) - r(y)$. Hence, $r(x) - r(y) = (y - x)g(x, y)$, where g is a polynomial in x for fixed y and in y for fixed x. Replacing r by p^*, x by λI, and y by M we obtain

$$p^*(\lambda I) - p^*(M) = (M - \lambda I)g(\lambda I, M)$$

But $p^*(M) = 0$. Also if $p^*(\lambda) = \lambda^k + b_{k-1}\lambda^{k-1} + \cdots + b_0$, then

$$p^*(\lambda I) = \lambda^k I^k + b_{k-1}\lambda^{k-1}I^{k-1} + \cdots + b_0 I = p^*(\lambda)I$$

Therefore,

$$p^*(\lambda)I = (M - \lambda I)g(\lambda I, M)$$

Taking determinants, we get $[p^*(\lambda)]^n = p(\lambda)q(\lambda)$ where $q(\lambda) = \det[g(\lambda I, M)]$ is a polynomial in λ. Hence, every factor of $p(\lambda)$ is also a factor of $[p^*(\lambda)]^n$, and thus every irreducible factor of $p(\lambda)$ is a factor of $p^*(\lambda)$. ∎

LEMMA 11.3 Similar matrices have the same minimal polynomial.

PROOF: Let $N = P^{-1}MP$ and let $p_N^*(\lambda)$ and $p_M^*(\lambda)$ be the minimal polynomials of N and M, respectively. Suppose that $p_N^*(\lambda) = \lambda^k + b_{k-1}\lambda^{k-1} + \cdots + b_0$. Then

$$0 = p_N^*(N) = N^k + b_{k-1}N^{k-1} + \cdots + b_0I$$

$$= (P^{-1}MP)^k + b_{k-1}(P^{-1}MP)^{k-1} + \cdots + b_0I$$

Since $(P^{-1}MP)^j = P^{-1}M^jP$, we get

$$0 = P^{-1}(M^k + b_{k-1}M^{k-1} + \cdots + b_0I)P$$

Multiplying on the left by P and on the right by P^{-1} yields

$$0 = M^k + b_{k-1}M^{k-1} + \cdots + b_0I = p_N^*(M)$$

This shows that $p_N^*(\lambda)$ is a polynomial which is zero at $\lambda = M$. By Lemma 11.1 it follows that $p_M^*(\lambda)$ divides $p_N^*(\lambda)$. Clearly, the roles of M and N in this argument can be interchanged, showing that $p_N^*(\lambda)$ divides $p_M^*(\lambda)$ also. Therefore,[2] $p_N^*(\lambda) = p_M^*(\lambda)$. ∎

LEMMA 11.4 Let

$$D = \begin{pmatrix} d_{11} & 0 & \cdots & 0 \\ 0 & d_{22} & \cdots & 0 \\ & & \cdots & \\ 0 & 0 & \cdots & d_{nn} \end{pmatrix}$$

be a diagonal matrix. Let the distinct numbers among the diagonal elements $d_{11}, d_{22}, \ldots, d_{nn}$ be called $\lambda_1, \lambda_2, \ldots, \lambda_k$. Then the minimal polynomial of D is $(\lambda - \lambda_1)(\lambda - \lambda_2) \cdots (\lambda - \lambda_k)$.

PROOF: The characteristic polynomial of D is

$$p(\lambda) = (d_{11} - \lambda)(d_{22} - \lambda) \cdots (d_{nn} - \lambda)$$

By Lemma 11.1, $p^*(\lambda)$ divides $p(\lambda)$, and by Lemma 11.2, each distinct factor in this product is a factor of $p^*(\lambda)$, and therefore,

$$p^*(\lambda) = \prod_{i=1}^{k} (\lambda - \lambda_i)^{n_i}$$

---

[2] This conclusion can be derived from the prime factorization theorem for polynomials over a field, which states that a nonscalar monic polynomial can be factored into a product of monic prime polynomials in a unique way (except for the order of the factors). See the references in Appendix B.

where $n_1, \ldots, n_k$ are positive integers. We want to show that each n_i is one. Consider the polynomial

$$s(\lambda) = \prod_{i=1}^{k} (\lambda - \lambda_i)$$

We have

$$s(D) = \prod_{i=1}^{k} (D - \lambda_i I)$$

For any m, $1 \leq m \leq n$, one of the factors $D - \lambda_i I$ has all elements of the mth row zero. Therefore, $s(D) = 0$. By Lemma 11.1, $p^*(\lambda)$ must divide $s(\lambda)$. Consequently, every $n_i = 1$, $p^*(\lambda) = s(\lambda)$, and the minimal polynomial is the product of the linear factors $\lambda - \lambda_i$.   ∎

The main result of this section is as follows.

THEOREM 11.17 An $n \times n$ matrix M is diagonable over a field **K** if and only if its minimal polynomial $p^*(\lambda)$ factors in **K** into distinct linear factors. In this case, $p^*(\lambda) = \prod_{i=1}^{k} (\lambda - \lambda_i)$ where $\lambda_1, \lambda_2, \ldots, \lambda_k$ are the distinct eigenvalues of M and $\mathbf{K}^n$ is the direct sum $\mathbf{K}_1^n \oplus \mathbf{K}_2^n \oplus \cdots \oplus \mathbf{K}_k^n$ of the eigenspaces $\mathbf{K}_1^n, \mathbf{K}_2^n, \ldots, \mathbf{K}_k^n$ associated with the eigenvalues $\lambda_1, \lambda_2, \ldots, \lambda_k$, respectively.

PROOF: First suppose that M is diagonable. Then there exists a matrix P such that $P^{-1}MP = D$ where D is a diagonal matrix with entries $d_{11}, d_{22}, \ldots, d_{nn}$ on the main diagonal. By Lemma 11.3 and 11.4, the minimal polynomial of M is $p^*(\lambda) = (\lambda - \lambda_1) \cdots (\lambda - \lambda_k)$, where $\lambda_1, \ldots, \lambda_k$ are the distinct members among $d_{11}, \ldots, d_{nn}$, that is, the distinct eigenvalues of D or of M.

Conversely, suppose p^* factors into distinct linear factors; that is, $p^*(\lambda) = \prod_{i=1}^{k} (\lambda - \lambda_i)$, where the λ_i are distinct numbers, which must be eigenvalues. Denote by $\mathbf{K}_i^n$ the eigenspace associated with λ_i and let $r_i = \dim \mathbf{K}_i^n$. Hence, by Theorem 11.7, $\rho(M - \lambda_i I) = n - r_i$. By the Corollary to Theorem 11.6, we know $\mathbf{K}_i^n \cap \mathbf{K}_j^n = \{0\}$ for $i \neq j$. Further, $\sum_{i=1}^{k} \mathbf{K}_i^n = \bigoplus_{i=1}^{k} \mathbf{K}_j^n$ (where $\bigoplus_{i=1}^{k} \mathbf{K}_i^n$ denotes the direct sum of the $\mathbf{K}_i^n$. (See Section 4.12.) For suppose v is in $\sum_{i=1}^{k} \mathbf{K}_i^n$, and suppose there are two distinct representations

$$v = u_1 + u_2 + \cdots + u_k = v_1 + v_2 + \cdots + v_k$$

where u_1 and v_1 are in $\mathbf{K}_1^n$, u_2 and v_2 are in $\mathbf{K}_2^n$, and so on. Then

$$(u_1 - v_1) + (u_2 - v_2) + \cdots + (u_k - v_k) = v - v = 0$$

and therefore the set $\{(u_1 - v_1), (u_2 - v_2), \ldots, (u_k - v_k)\}$ is linearly dependent. But since each $(u_i - v_i)$ is in $\mathbf{K}_i^n$, this dependence contradicts Theorem 11.6.

Hence, by Theorem 4.22, dim $\sum_{i=1}^{k} K_i^n = \sum_{i=1}^{k} r_i$. But since $\sum_{i=1}^{k} K_i^n \subset$ K^n we have $\sum_{i=1}^{k} r_i \leq n$. Next

$$0 = p^*(M) = \prod_{i=1}^{k} (M - \lambda_i I)$$

so

$$0 = \rho[\prod_{i=1}^{k} (M - \lambda_i I)] \geq n - \sum_{i=1}^{k} r_i$$

by repeated applications of Theorem 6.11. Hence, $\sum_{i=1}^{k} r_i \geq n$. Since we have both $\sum_{i=1}^{k} r_i \leq n$ and $\sum_{i=1}^{k} r_i \geq n$ we know that $\sum_{i=1}^{k} r_i = n$. But this means that there exists a set of n linearly independent eigenvectors and hence, by Theorem 11.10, M is diagonable. ∎

The results in this section can be of assistance in computing the minimal polynomial of a given matrix and in determining whether a matrix is diagonable.

Example 2 Consider again the matrix M in Example 1. As shown there, the characteristic polynomial of M is $\rho(\lambda) = (3 - \lambda)^2 (2 - \lambda)$. By Lemma 11.2, the factors $\lambda - 3$ and $\lambda - 2$ must divide the minimal polynomial $p^*(\lambda)$. By Lemma 11.1, $p^*(\lambda)$ cannot contain these factors with powers higher than in $p(\lambda)$, nor can $p^*(\lambda)$ contain any different irreducible factors. Therefore, either

$$p^*(\lambda) = (\lambda - 2)(\lambda - 3) \quad \text{or} \quad p^*(\lambda) = (\lambda - 2)(\lambda - 3)^2 = -p(\lambda)$$

There are no other possibilities. As shown in Example 1, $(M - 2I)(M - 3I) = 0$, hence, in fact, $p^*(\lambda) = (\lambda - 2)(\lambda - 3)$. Finally, since $p^*(\lambda)$ factors into distinct linear factors, M is diagonable. A diagonal matrix to which M is similar is

$$D = \begin{pmatrix} 3 & 0 & 0 \\ 0 & 3 & 0 \\ 0 & 0 & 2 \end{pmatrix}$$

since D must have the same characteristic polynomial as M.

Example 3 Consider

$$M = \begin{pmatrix} 1 & 1 \\ 0 & 1 \end{pmatrix}$$

The characteristic polynomial is $(\lambda - 1)^2$ and the minimal polynomial must be either $\lambda - 1$ or $(\lambda - 1)^2$. Since, clearly, $M - I \neq 0$, $p^*(\lambda) = (\lambda - 1)^2$. Therefore, M is not diagonable over the real field.

Example 4 Consider

$$M = \begin{pmatrix} 0 & -1 \\ 1 & 0 \end{pmatrix}$$

The characteristic polynomial is $p(\lambda) = \lambda^2 + 1$. This polynomial is irreducible over the real field, and so by Lemmas 11.1 and 11.2 the minimal polynomial is $p^*(\lambda) = \lambda^2 + 1$. By Theorem 11.17, M is not diagonable over the real field. On the other hand, $p(\lambda) = p^*(\lambda) = (\lambda - i)(\lambda + i)$ in the complex field. Therefore, M is diagonable by means of eigenvectors in $\mathbf{C}^2$ (the complex analogue of $\mathbf{R}^2$).

In Example 2, we ascertained that M is similar to the diagonal matrix D, but did not exhibit a matrix P such that $D = P^{-1}MP$. In order to do so, we must find the eigenvectors of M. If P is a matrix with these eigenvectors as columns, it follows from Theorem 11.10 that $P^{-1}MP$ is diagonal. Thus, there are two ways to test a given matrix to see whether it is diagonable. The first requires computing the eigenvalues and eigenvectors. If there is a full set of linearly independent eigenvectors, Theorem 11.10 shows that the matrix is diagonable (and shows how to construct the diagonalizing matrix P). The other method is to compute the minimal polynomial and then to apply Theorem 11.17.

In conclusion we note that the concept of minimal polynomial can be applied to linear transformations.

DEFINITION 11.11   Let T be a linear transformation on $\mathbf{V}$ into $\mathbf{V}$. The lowest degree monic polynomial $p^*(\lambda)$ satisfying $p^*(T) = \mathbf{Z}$, the zero transformation, is called the minimal polynomial of T.

The minimal polynomial of T is the same as the minimal polynomial of all the matrix representatives of T.

Exercises

1. Compute the minimal polynomial for the matrices in Exercises 2 through 5, and 7 through 9 of Section 11.4. In each case decide whether the matrix is diagonable over the real field and over the complex field.

2. Let M be an upper triangular matrix with entries $d_{11}, \ldots, d_{nn}$ on the main diagonal. Is it true that the minimal polynomial is $(\lambda - \lambda_1), \ldots, (\lambda - \lambda_k)$, where $\lambda_1, \ldots, \lambda_k$ are the distinct numbers among the diagonal entries?

3. Prove that a matrix and its transpose have the same characteristic polynomial and the same minimal polynomial.

4. Are two matrices necessarily similar if they have (a) the same minimal polynomial, or (b) the same characteristic polynomial? Give examples.

5. Let M be a 2×2 symmetric matrix with real entries; that is, $M = M^T$. Prove that M is diagonable over the real field.

6. Let $T(x, y, z) = (x - y, y - x, x + y + 2z)$.

    (a) Find the matrix M which represents T relative to the standard basis.

(b) Show that T is diagonable by finding the minimal polynomial of M.

(c) Find a matrix P such that $P^{-1}MP$ is a diagonal matrix.

7. Suppose that M is a diagonable matrix having only one eigenvalue. Show that $M = cI$ for some scalar c.

11.7   Generalized null spaces

Suppose that T is a diagonable transformation over a field $\mathbf{K}$ and let M be an $n \times n$ diagonable matrix representative of T. Let $\lambda_1, \ldots, \lambda_k$ be its distinct eigenvalues. According to Theorem 11.17, the space $\mathbf{K}^n$ can then be decomposed into a direct sum

$$\mathbf{K}^n = \mathbf{K}_1^n \oplus \mathbf{K}_2^n \oplus \cdots \oplus \mathbf{K}_k^n \tag{1}$$

where $\mathbf{K}_i^n$ is the eigenspace of λ_i. If we choose a basis for each $\mathbf{K}_i^n$, $i = 1, \ldots, k$, it follows that the union of all these vectors is a basis for $\mathbf{K}^n$. Furthermore, the linear transformation T has a diagonal matrix representation relative to this basis of the form

$$\begin{pmatrix} \lambda_1 & & & & & \\ & \lambda_1 & & \mathbf{O} & & \\ & & \ddots & & & \\ & \mathbf{O} & & \lambda_k & & \\ & & & & \ddots & \\ & & & & & \lambda_k \end{pmatrix} \tag{2}$$

If T is not diagonable, it is not possible to find a matrix representation of T as simple as that in Eq. (2), but it happens that a fairly simple result is still possible. In this section, we shall develop a generalization of the decomposition in (1) which is valid for arbitrary T.

DEFINITION 11.12   Let T be a linear transformation on a space $\mathbf{V}$. A subspace $\mathbf{S}$ of $\mathbf{V}$ is called *invariant under T* if Tv is in $\mathbf{S}$ whenever v is in $\mathbf{S}$, or more briefly, if $T(\mathbf{S}) \subset \mathbf{S}$.

If $\mathbf{S}$ is an invariant subspace under T, it is possible to study the effect of T on vectors of $\mathbf{S}$, ignoring what happens to vectors not in $\mathbf{S}$, since vectors in $\mathbf{S}$ are mapped into vectors in $\mathbf{S}$. In fact, let us define $T_\mathbf{S}$ by

$$T_\mathbf{S}x = Tx \quad \text{for } x \in \mathbf{S}$$

Then $T_\mathbf{S}$ is a linear transformation of $\mathbf{S}$ into $\mathbf{S}$ which agrees with T on $\mathbf{S}$. $T_\mathbf{S}$ is called the *restriction* of T to $\mathbf{S}$. Let $\{v_1, \ldots, v_k\}$ be a basis for the invariant subspace S and let $\{v_1, v_2, \ldots, v_n\}$ be a basis for **V**. Then the matrix representing T relative to $\{v_1, v_2, \ldots, v_n\}$ has the form

$$\left(\begin{array}{c|c} A_1 & A_2 \\ \hline 0 & A_3 \end{array}\right)$$

where A_1 is a $k \times k$ matrix which represents T_S relative to $\{v_1, \ldots, v_k\}$. In particular, if T is diagonable, then each eigenspace $\mathbf{K}_i^n$ is invariant under T (since any vector v in $\mathbf{K}_i^n$ is a combination of eigenvectors of λ_i and therefore, $Tv = \lambda_i v$).

Following this line of thought, we will now show that for any given linear transformation T, it is possible to decompose the whole space into a direct sum of invariant subspaces. This result, Theorem 11.20, not only gives a satisfactory geometric concept, but will be very useful in the study of differential equations in the next chapter. Theorem 11.20 also leads us to the so-called Jordan canonical form of a matrix, discussed below.

LEMMA 11.5   Let T be a linear transformation on a vector space **V** of finite dimension n. Then each of the spaces $\mathbf{N}_{T^k}$ and $\mathbf{R}_{T^k}$ $(k = 1, 2, \ldots)$, where T^k is the kth power of T, is invariant under T and under any polynomial in T. Moreover, there is a positive integer p such that

$$\mathbf{N}_T \subset \mathbf{N}_{T^2} \subset \cdots \subset \mathbf{N}_{T^p} = \mathbf{N}_{T^{p+1}} = \cdots \tag{3}$$

$$\mathbf{R}_T \supset \mathbf{R}_{T^2} \supset \cdots \supset \mathbf{R}_{T^p} = \mathbf{R}_{T^{p+1}} = \cdots \tag{4}$$

where the inclusions are strict.

PROOF: Let $v \in \mathbf{N}_{T^k}$. Since $T^k v = 0$, then $T^{k+1} v = 0$, or $T^k(Tv) = 0$. Therefore $v \in \mathbf{N}_{T^{k+1}}$ and $(Tv) \in \mathbf{N}_{T^k}$. Thus $\mathbf{N}_{T^k} \subset \mathbf{N}_{T^{k+1}}$ (not necessarily strictly) and $\mathbf{N}_{T^k}$ is invariant under T.

Next, let $v \in \mathbf{R}_{T^k}$. Then there is some u in **V** such that $T^k u = v$. For $k > 1$, $T^{k-1}(Tu) = v$ so $v \in \mathbf{R}_{T^{k-1}}$ and thus $\mathbf{R}_{T^k} \subset \mathbf{R}_{T^{k-1}}$. Moreover, if $v \in \mathbf{R}_{T^k}$ $(k \geq 1)$, then $T^k u = v$, $T^{k+1} u = Tv$, and it follows that Tv is in $\mathbf{R}_{T^{k+1}}$ and therefore in $\mathbf{R}_{T^k}$. Thus, $\mathbf{R}_{T^k}$ is invariant under T.

Next we shall prove the existence of the integer p in Eqs. (3) and (4). Since each of the subspaces in (3) has integer dimension and is a subspace of a space of finite dimension n, equality must occur somewhere in the chain. Let p be the smallest integer for which $\mathbf{N}_{T^p} = \mathbf{N}_{T^{p+1}}$ but $\mathbf{N}_T \subset \mathbf{N}_{T^2} \subset \cdots \subset \mathbf{N}_{T^p}$ in the strict sense. Let $v \in \mathbf{N}_{T^{p+2}}$. Then $T^{p+1} Tv = 0$, so $(Tv) \in \mathbf{N}_{T^{p+1}} = \mathbf{N}_{T^p}$. This shows that $T^{p+1} v = 0$, and thus $\mathbf{N}_{T^{p+2}} \subset \mathbf{N}_{T^{p+1}}$. Since we already have $\mathbf{N}_{T^{p+1}} \subset \mathbf{N}_{T^{p+2}}$, it follows that $\mathbf{N}_{T^{p+1}} = \mathbf{N}_{T^{p+2}}$. The same reasoning shows that $\mathbf{N}_{T^{p+2}} = \mathbf{N}_{T^{p+3}}$, and so on, and the proof of Eq. (3) is complete.

Because $\mathbf{R}_{T^k} \supset \mathbf{R}_{T^{k+1}}$, the equality $\mathbf{R}_{T^k} = \mathbf{R}_{T^{k+1}}$ holds if and only if $\rho(T^k) = \rho(T^{k+1})$. Since the sum of the rank and nullity of any transformation is n, we can say that $\mathbf{R}_{T^k} = \mathbf{R}_{T^{k+1}}$ if and only if $\nu(T^k) = \nu(T^{k+1})$. But this is true if and only if $\mathbf{N}_{T^k} = \mathbf{N}_{T^{k+1}}$. Hence, equality can occur in the chain in Eq. (4) at and only at places where equality occurs in the chain in Eq. (3). Therefore, Eq. (4) follows from Eq. (3).

Finally, we know that each $\mathbf{N}_{T^k}$ and each $\mathbf{R}_{T^k}$ is invariant under T, hence under any power of T, and so under any polynomial in T. ∎

Now we apply Lemma 11.5 to $T - \lambda I$, rather than T, where λ is an eigenvalue of T.

DEFINITION 11.13 Let T be a linear transformation on a space $\mathbf{V}$ of finite dimension n. Let λ be an eigenvalue of T and let p be the integer such that

$$\mathbf{N}_{T-\lambda I} \subset \mathbf{N}_{(T-\lambda I)^2} \subset \cdots \subset \mathbf{N}_{(T-\lambda I)^p} = \mathbf{N}_{(T-\lambda I)^{p+1}} = \cdots$$

The space $\mathbf{N}_{(T-\lambda I)^p}$ is called the *generalized eigenspace* of λ (for the transformation T) and will be denoted by $\mathbf{M}_\lambda(T)$. We will use $\mathbf{Q}_\lambda(T)$ to denote the space $\mathbf{R}_{(T-\lambda I)^p}$. We use the abbreviations $\mathbf{M}_\lambda$ and $\mathbf{Q}_\lambda$ for $\mathbf{M}_\lambda(T)$ and $\mathbf{Q}_\lambda(T)$, respectively, if no confusion will result. The integer p is the *index* of λ.

Generally speaking, $\mathbf{M}_\lambda(T)$ is the largest subspace of $\mathbf{V}$ annihilated by a power of $T - \lambda I$. We note that $\mathbf{M}_\lambda(T)$ includes the ordinary eigenspace of λ. Recall that if the underlying scalar field is the real field, some linear transformations fail to have any (real) eigenvalues. In such a case, there will be no generalized eigenspaces to talk about and no hope of decomposing the space in terms of them. Consequently, we shall, throughout the remainder of this section, assume that $\mathbf{V}$ is a space over the complex field $\mathbf{C}$.

THEOREM 11.18 Let T be a linear transformation on $\mathbf{V}$ into $\mathbf{V}$, where $\mathbf{V}$ is an n-dimensional space over $\mathbf{C}$. Let λ be an eigenvalue of T. Then $\mathbf{V}$ is the direct sum of the invariant spaces $\mathbf{M}_\lambda$ and $\mathbf{Q}_\lambda$. Also, the restriction of $T - \lambda I$ to $\mathbf{Q}_\lambda$ is invertible.

PROOF: We already know that $\mathbf{M}_\lambda$ and $\mathbf{Q}_\lambda$ are invariant under T. Since

$$(T - \lambda I)\mathbf{Q}_\lambda = (T - \lambda I)^{p+1}\mathbf{V} = \mathbf{R}_{(T-\lambda I)^{p+1}} = \mathbf{R}_{(T-\lambda I)^p} = \mathbf{Q}_\lambda$$

we see that $T - \lambda I$ maps $\mathbf{Q}_\lambda$ onto $\mathbf{Q}_\lambda$. Therefore, the restriction of $T - \lambda I$ to $\mathbf{Q}_\lambda$ is invertible (see Theorem 5.20). Now let v be any vector in $\mathbf{V}$. Then $(T - \lambda I)^p v = w$ is a vector in $\mathbf{Q}_\lambda$. Since $T - \lambda I$ maps $\mathbf{Q}_\lambda$ onto $\mathbf{Q}_\lambda$, so do $(T - \lambda I)^2, \ldots, (T - \lambda I)^p$. Therefore, $(T - \lambda I)^p$ is invertible on $\mathbf{Q}_\lambda$ and it follows that there is a unique vector u in $\mathbf{Q}_\lambda$ for which $(T - \lambda I)^p u = w$. We have $(T - \lambda I)^p(v - u) = \mathbf{0}$, so $z = v - u$ is in $\mathbf{M}_\lambda$. We have proved that every v in $\mathbf{V}$ has a decomposition $v = u + z$ where $u \in \mathbf{Q}_\lambda$ and $z \in \mathbf{M}_\lambda$; that is, $\mathbf{V} = \mathbf{M}_\lambda + \mathbf{Q}_\lambda$. Finally, since rank plus nullity of $(T - \lambda I)^p$ is n, we have $\dim \mathbf{M}_\lambda + \dim \mathbf{Q}_\lambda = n$. Therefore, the sum is direct; that is, $\mathbf{V} = \mathbf{M}_\lambda \oplus \mathbf{Q}_\lambda$. ∎

Our objective should now become clear. We will first decompose $\mathbf{V} = \mathbf{M}_{\lambda_1} \oplus \mathbf{Q}_{\lambda_1}$. Letting λ_2 be another eigenvalue, $\lambda_2 \neq \lambda_1$, we then decompose $\mathbf{Q}_{\lambda_1}$ into a direct sum by the same procedure. By continuing in this way we hope to reach a complete decomposition of $\mathbf{V}$. The next theorem will be needed in establishing what we desire.

THEOREM 11.19   If λ_i and λ_j are distinct eigenvalues of T, then $\mathbf{M}_{\lambda_i} \subset \mathbf{Q}_{\lambda_j}$.

PROOF: Our first objective is to show that

$$\mathbf{M}_{\lambda_i} \cap \mathbf{N}_{T-\lambda_j I} = \{\mathbf{0}\} \tag{5}$$

Let v be a vector in the intersection. Let p_i be the index of λ_i, and p_j the index of λ_j, so that

$$\mathbf{M}_{\lambda_i} = \mathbf{N}_{(T-\lambda_i I)^{p_i}}, \quad \mathbf{Q}_{\lambda_j} = \mathbf{R}_{(T-\lambda_j I)^{p_j}} \tag{6}$$

Then

$$(\lambda_j - \lambda_i)^{p_i} v = [(T - \lambda_i I) - (T - \lambda_j I)]^{p_i} v$$

$$= (T - \lambda_i I)^{p_i} v + \sum_{k=1}^{p_i} (-1)^k \binom{p_i}{k} (T - \lambda_i I)^{p_i - k} (T - \lambda_j I)^k$$

The first term on the right is zero since $v \in \mathbf{M}_{\lambda_i}$. All the other terms are zero because $(T - \lambda_j I)v = \mathbf{0}$. Since $\lambda_i \neq \lambda_j$, we deduce that $v = \mathbf{0}$, as asserted.

Next we consider the restriction of $T - \lambda_j I$ to $\mathbf{M}_{\lambda_i}$. We know that every space $\mathbf{N}_{(T-\lambda_i I)^k}$ and every space $\mathbf{R}_{(T-\lambda_i I)^k}$ is invariant under any polynomial in $T - \lambda_i I$. Therefore, they are invariant under any polynomial in T, and in particular under $T - \lambda_j I$. Thus $T - \lambda_j I$ maps $\mathbf{M}_{\lambda_i}$ into $\mathbf{M}_{\lambda_i}$. Moreover, by Eq. (5), the only vector in $\mathbf{M}_{\lambda_i}$ which is mapped by $T - \lambda_j I$ into $\mathbf{0}$ is $\mathbf{0}$ itself. Therefore the restriction of $T - \lambda_j I$ to $\mathbf{M}_{\lambda_i}$ is invertible and maps $\mathbf{M}_{\lambda_i}$ onto $\mathbf{M}_{\lambda_i}$. Hence, $(T - \lambda_j I)^{p_j}$ also maps $\mathbf{M}_{\lambda_i}$ onto $\mathbf{M}_{\lambda_i}$; that is, every vector in $\mathbf{M}_{\lambda_i}$ is in the range of $(T - \lambda_j I)^{p_j}$, which is $\mathbf{Q}_j$. ∎

THEOREM 11.20   Let T be a linear transformation on a space $\mathbf{V}$, of finite dimension, over the complex field. Let $\lambda_1, \ldots, \lambda_s$ be the distinct eigenvalues of T, and let $\mathbf{M}_{\lambda_1}, \ldots, \mathbf{M}_{\lambda_s}$ be their respective generalized eigenspaces. Then

$$\mathbf{V} = \mathbf{M}_{\lambda_1} \oplus \mathbf{M}_{\lambda_2} \oplus \cdots \oplus \mathbf{M}_{\lambda_s}$$

PROOF: By Theorem 11.18, $\mathbf{V} = \mathbf{M}_{\lambda_2} \oplus \mathbf{Q}_{\lambda_2}$, and also $\mathbf{V} = \mathbf{M}_{\lambda_1} \oplus \mathbf{Q}_{\lambda_1}$. By Theorem 11.19, $\mathbf{M}_{\lambda_2} \subset \mathbf{Q}_{\lambda_1}$ from which it follows that $\mathbf{Q}_{\lambda_1} = \mathbf{M}_{\lambda_2} \oplus (\mathbf{Q}_{\lambda_1} \cap \mathbf{Q}_{\lambda_2})$. (See Figure 11.1.) Therefore $\mathbf{V} = \mathbf{M}_{\lambda_1} \oplus \mathbf{M}_{\lambda_2} \oplus (\mathbf{Q}_{\lambda_1} \cap \mathbf{Q}_{\lambda_2})$. Repeating this argument, we finally obtain

$$\mathbf{V} = \mathbf{M}_{\lambda_1} \oplus \mathbf{M}_{\lambda_2} \oplus \cdots \oplus \mathbf{M}_{\lambda_s} \oplus \mathbf{Q}$$

where

$$\mathbf{Q} = \mathbf{Q}_{\lambda_1} \cap \mathbf{Q}_{\lambda_2} \cap \cdots \cap \mathbf{Q}_{\lambda_s}$$

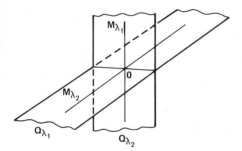

FIGURE 11.1

It only remains to prove that $\mathbf{Q}$ is the trivial space consisting of the zero vector. But we know from Theorem 11.18 that $T - \lambda_i I$ maps $\mathbf{Q}_{\lambda_i}$ onto $\mathbf{Q}_{\lambda_i}$ for each i. Hence,

$$(T - \lambda_1 I) \cdots (T - \lambda_s I)$$

maps $\mathbf{Q}$ onto $\mathbf{Q}$. Let

$$U_m(T) = [(T - \lambda_1 I) \cdots (T - \lambda_s I)]^m, \qquad m = 1, 2, \ldots$$

Then $U_m(T)$ is invertible on $\mathbf{Q}$ for every m. However, $U_m(x)$ contains every factor of the characteristic polynomial $p(x)$. Therefore, if m is large enough, $U_m(x)$ is a multiple of $p(x)$. By the Cayley–Hamilton Theorem, $p(T) = \mathbf{Z}$. Therefore, $U_m(T) = \mathbf{Z}$, for large enough m. Because $U_m(T)$ is invertible on $\mathbf{Q}$, this implies that $\mathbf{Q}$ contains only the zero vector. ∎

COROLLARY Let the minimal polynomial of T be

$$p^*(\lambda) = (\lambda - \lambda_1)^{n_1} \cdots (\lambda - \lambda_s)^{n_s}$$

Let the index of λ_i be p_i ($i = 1, \ldots, s$). Then $n_i = p_i$ ($i = 1, \ldots, s$).

PROOF: We know that $(T - \lambda_i)^{p_i} \mathbf{M}_{\lambda_i} = \mathbf{0}$. Hence,

$$(T - \lambda_1 I)^{p_1} \cdots (T - \lambda_s I)^{p_s}$$

takes every vector of $\mathbf{V}$ into $\mathbf{0}$, because every vector is a sum of vectors in the $\mathbf{M}_{\lambda_i}$, by Theorem 11.20. We have

$$(T - \lambda_1 I)^{p_1} \cdots (T - \lambda_s I)^{p_s} = \mathbf{0}$$

and therefore the minimal polynomial $p^*(\lambda)$ must divide $(\lambda - \lambda_1)^{p_1} \cdots (\lambda - \lambda_s)^{p_s}$, by Lemma 11.1. Thus, $n_i \le p_i$ for each i.

Now suppose that $n_i < p_i$ for some i. Since $\mathbf{N}_{(T - \lambda_i I)^{n_i}}$ is strictly smaller than $\mathbf{M}_{\lambda_i}$, by definition of p_i, there exists a vector v in $\mathbf{M}_{\lambda_i}$ such that $(T - \lambda_i I)^{n_i} v \ne \mathbf{0}$. Now since

$$(T - \lambda_1 I)^{n_1} \cdots (T - \lambda_s I)^{n_s} v = \mathbf{0}$$

it follows that $(T - \lambda_i I)^n v = 0$, because each $T - \lambda_j I$ is invertible on $\mathbf{M}_{\lambda_i}$ for $j \neq i$. Since this is a contradiction, we cannot have $n_i < p_i$.   ∎

Theorem 11.20 is adequate for the applications to differential equations which we have in mind. However, it can be refined still further; that is, the structure of the subspaces $\mathbf{M}_{\lambda_i}$ can be described more completely in the sense that we can tell how to construct bases for the $\mathbf{M}_{\lambda_i}$ in a rather simple way. Relative to the basis of $\mathbf{V}$ found in this way, the matrix of T has a special form, called the *Jordan canonical form*. This shows, then, that every matrix is similar to a matrix in Jordan form. The precise result is as follows.

THEOREM 11.21   Let A be an $n \times n$ matrix with characteristic polynomial

$$p(\lambda) = (-1)^n (\lambda - \lambda_1)^{m_1} \cdots (\lambda - \lambda_s)^{m_s}$$

and minimum polynomial

$$p^*(\lambda) = (\lambda - \lambda_1)^{n_1} \cdots (\lambda - \lambda_s)^{n_s}$$

$(\lambda_1, \lambda_2, \ldots, \lambda_s \in \mathbf{C})$. Then A is similar to a block diagonal matrix

$$\begin{pmatrix} J_1 & & \\ & J_2 & \\ & & \ddots \end{pmatrix}$$

where each J_k has the form

$$J_k = \begin{pmatrix} \lambda_i & 1 & 0 & \cdots & 0 & 0 \\ 0 & \lambda_i & 1 & \cdots & 0 & 0 \\ 0 & 0 & \lambda_i & \cdots & 0 & 0 \\ & & & \cdots & & \\ 0 & 0 & 0 & \cdots & \lambda_i & 1 \\ 0 & 0 & 0 & \cdots & 0 & \lambda_i \end{pmatrix}$$

For each eigenvalue λ_i, the number of blocks corresponding to λ_i is equal to the geometric multiplicity of λ_i (that is, the dimension of the eigenspace). At least one of these has order $n_i (= p_i$, the index of $\lambda_i)$, and the others have order no greater than n_i. The sum of the orders of all block matrices containing a given λ_i is m_i.

For the proof of this theorem, and additional discussion, see the references given in the Commentary.

Exercises

1.   For each of the following matrices, compute the eigenvalues, the index of each eigenvalue, and give a basis for the generalized eigenspace of each eigenvalue.

(a)
$$A = \begin{pmatrix} 0 & -1 & 2 \\ 0 & 1 & 0 \\ 1 & 1 & -1 \end{pmatrix}$$

(b)
$$A = \begin{pmatrix} -2 & 4 & 1 & 1 \\ 0 & 1 & 0 & 0 \\ 0 & 0 & 1 & 0 \\ -1 & 0 & -1 & 0 \end{pmatrix}$$

2. Let J_k be a Jordan block matrix; that is,

$$J_k = \begin{pmatrix} \lambda_i & 1 & \cdots & 0 & 0 \\ 0 & \lambda_i & \cdots & 0 & 0 \\ & & \cdots & & \\ 0 & 0 & \cdots & \lambda_i & 1 \\ 0 & 0 & \cdots & 0 & \lambda_i \end{pmatrix}$$

Compute J_k^r, where r is a positive integer.

3. With the aid of the Jordan canonical form, prove the following theorem, which is of basic importance in certain problems of numerical analysis. Let A be an $n \times n$ matrix. Then

$$\lim_{r \to \infty} A^r = 0$$

if and only if every eigenvalue of A satisfies $|\lambda_i| < 1$.

COMMENTARY

Section 11.2 and 11.4 Some texts use the term *characteristic value* for any complex solution of the characteristic equation, and the term *eigenvalue* as we have defined it for any λ in the field **K** for which $Tv = \lambda v$ or $MX = \lambda X$. If **K** is the complex field, there is no difference, but if **K** is, say, the real field, then not all characteristic values in this sense need be eigenvalues.

Section 11.3 In abstract algebra, a field is called *algebraically closed* if every polynomial with coefficients in the field factors into linear factors in the field. The real field is not algebraically closed, but according to the *Fundamental Theorem of Algebra*, the complex field is. Thus, in Example 1, the characteristic polynomial $\lambda^2 + 1$ factors in the complex field, but not in the real field.

Section 11.7 For a proof of Theorem 11.21, and additional related discussion, see:
HERSTEIN, I., *Topics of Algebra*, Boston, Mass., Ginn and Co., 1964.
SMIRNOV, V., *Linear Algebra and Group Theory*, New York, McGraw-Hill, 1961.

Chapter 12

Linear systems
of differential equations

12.1 Systems of differential equations

The purpose of this chapter is to study the theory of systems of simultaneous differential equations and methods for solving them. Such systems frequently arise in applied problems, as the following two examples illustrate.

Example 1 Newton's laws for the motion of a mass particle in space have the form

$$m \frac{d^2x}{dt^2} = F_1(x, y, z, t)$$

$$m \frac{d^2y}{dt^2} = F_2(x, y, z, t) \qquad (1)$$

$$m \frac{d^2z}{dt^2} = F_3(x, y, z, t)$$

where m is the mass of the particle, x, y, and z are the rectangular coordinates of the particle at time t, and F_1, F_2, and F_3 are the components of the external force at the point (x, y, z) at time t.[1] The determination of the path of the particle when the force components F_1, F_2, F_3 are given depends on solving the system of differential Eqs. (1); that is, on finding functions $x(t)$, $y(t)$, $z(t)$ which satisfy the equations.

---

[1] In some problems, the force components may also be functions of the velocity components dx/dt, dy/dt, dz/dt. Also, nonrectangular coordinates are frequently employed.

Example 2 In the study of the transport of a chemical compound (drug) through biological tissue, one sometimes uses a so-called *compartment model* to simulate the actual process. In our model, it is supposed that there is a chain of alternate aqueous (water) and lipid (fatty) type compartments. The lipid compartments correspond to membranes which the drug molecule must penetrate, macromolecules to which it must absorb, and so on, so that it is possible to have different concentrations of the compound in different compartments. In Figure 12.1 we illustrate this for a model with four compartments. The arrows indicate the transport of the compound through the system. As a first approximation, assume the volumes and surface areas of the compartments to be equal, and that the drug concentration throughout each compartment is uniform. (The solution is instantaneously "stirred.") Let

$$C_i(t) = \text{drug concentration in the } i\text{th compartment at time } t \qquad (2)$$

The concentration $C_1(t)$ is being decreased by transport from compartment 1 to compartment 2, and increased by transport back from 2 to 1. If we let α denote the rate of transport from an aqueous to a lipid compartment, and β the rate from lipid to aqueous, the change in concentration in compartment 1 in a short time interval Δt will be approximately

$$C_1(t + \Delta t) - C_1(t) = \beta C_2(t)\, \Delta t - \alpha C_1(t)\, \Delta t$$

Dividing by Δt and then letting Δt tend to zero, we find that the instantaneous rate of change of C_1 is given by

$$\frac{dC_1}{dt} = -\alpha C_1 + \beta C_2$$

In applying the same reasoning to compartment 2, we must take account of inflow and outflow over both of the boundaries. The complete set of equations is found to be

$$\frac{dC_1}{dt} = -\alpha C_1 + \beta C_2$$
$$\frac{dC_2}{dt} = \alpha C_1 - 2\beta C_2 + \alpha C_3$$
$$\frac{dC_3}{dt} = \beta C_2 - 2\alpha C_3 + \beta C_4 \qquad (3)$$
$$\frac{dC_4}{dt} = \alpha C_3 - \beta C_4$$

This is a system of simultaneous differential equations. Suppose that compartment 1 represents an injection site where an initial quantity of the compound is placed, say c_0. This means that the following initial conditions hold:

$$C_1(0) = c_0, \quad C_2(0) = 0, \quad C_3(0) = 0, \quad C_4(0) = 0 \qquad (4)$$

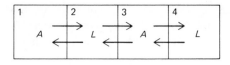

A aqueous   L lipid   **FIGURE 12.1**

The problem is now to solve Eqs. (3) subject to (4), and thus to predict the concentration in each compartment as a function of time. For example, in cancer chemotherapy, the compound is an anticancer drug and we might wish to know what concentration is present in the last compartment, the *activity site*, in order to know how large the initial amount c_0 should be. This in turn will depend on α and β, which are numbers characteristic of the compound being used. In a more realistic model, we might have a large number N of compartments rather than just 4, leading to a system of N simultaneous differential equations. One of many interesting mathematical questions associated with such a model is the question of determining how sensitive the results are to a change in the value of N.

It is evident from these examples that there is practical value in being able to solve systems of differential equations. Moreover, the theory of linear systems which we shall give is built on the theory of vector spaces, operators, and matrices. It thus provides an opportunity to see these algebraic concepts in use. We shall also find that the theory of the nth order linear differential equations given in Chapters 8 and 9 can be viewed as a special instance of the theory to be given now. This is illustrated in the examples below.

DEFINITION 12.1   A system of n equations of the form

$$
\begin{aligned}
y_1' &= f_1(x, y_1, \ldots, y_n) \\
y_2' &= f_2(x, y_1, \ldots, y_n) \\
&\cdots \\
y_n' &= f_n(x, y_1, \ldots, y_n)
\end{aligned}
\tag{5}
$$

(where y_i' denotes dy_i/dx) will be called a *first order system in normal form*.

We note that each of the equations in (5) contains only one derivative, which is of first order, and moreover that the equation is solved for this derivative. As we shall see, this kind of a system is sufficiently general to include most cases of interest.

The system in Example 2 is a first order system in normal form. (The independent variable was called t rather than x.)

Example 3   Consider the second order equation

$$
\frac{d^2 z}{dx^2} = g\left(x, z, \frac{dz}{dx}\right)
\tag{6}
$$

where g is a given function. Let us introduce new unknowns y_1 and y_2 defined by

$$y_1 = z, \quad y_2 = \frac{dz}{dx}$$

Then

$$\frac{dy_1}{dx} = \frac{dz}{dx} = y_2 \quad \text{and} \quad \frac{dy_2}{dx} = \frac{d^2z}{dx^2} = g(x, y_1, y_2)$$

Thus Eq. (6) appears to be equivalent to the first order normal system

$$\frac{dy_1}{dx} = y_2, \quad \frac{dy_2}{dx} = g(x, y_1, y_2)$$

The meaning of "equivalent" will be made more precise later.

Example 4   The system in Example 1 is not of first order. By introducing new unknowns much as in Example 3, it can be shown to be "equivalent" to a normal first order system containing six equations. (See Exercise 2.)

Example 5   Consider an arbitrary nth order equation of the form

$$z^{(n)} = g(x, z, z', z'', \ldots, z^{(n-1)}) \tag{7}$$

We can always "reduce" such an equation to a normal first order system. Define $y_1, \ldots, y_n$ by $y_1 = z$, $y_2 = z'$, $y_3 = z'', \ldots, y_n = z^{(n-1)}$. Equation (7) then yields $y_n' = g(x, y_1, y_2, \ldots, y_n)$. Hence, the required normal system is

$$
\begin{aligned}
y_1' &= y_2 \\
y_2' &= y_3 \\
&\cdots \\
y_{n-1}' &= y_n \\
y_n' &= g(x, y_1, y_2, \ldots, y_n)
\end{aligned}
\tag{8}
$$

Exercises

1.   Reduce each equation to a normal first order system:

(a) $\dfrac{d^2z}{dx^2} - 3\dfrac{dz}{dx} + xz = e^x$

(b) $\dfrac{d^2z}{dx^2} = \left(\dfrac{dz}{dx}\right)^2 + x$

(c) $\dfrac{d^3z}{dx^3} - x\dfrac{d^2z}{dx^2} + z^3 = 0$

2.   Reduce the system (1) to a normal first order system.

3.   Reduce the general nth order linear equation

$$y^{(n)} + a_1(x)y^{(n-1)} + \cdots + a_n(x)y = g(x)$$

to a first order system.

4.   Write the equations analogous to the system (3) in case there are five compartments rather than four.

5.   How is the system (3) modified in case no transport is possible from the last compartment back to the next-to-last?

12.2   Vector functions of a real variable

Before describing what is meant by a solution of a system of differential equations, we shall discuss the concept of a vector-valued function of a real variable.

DEFINITION 12.2   A function f with domain a subset of $\mathbf{R}^1$ and range a subset of $\mathbf{M}_{n1}$ (where n is a positive integer) is called a *vector function* of a real variable. For each x in the domain of f, $f(x)$ is a vector in $\mathbf{M}_{n1}$. Let $f_i(x)$ denote the ith component of this vector, that is,

$$f(x) = \begin{pmatrix} f_1(x) \\ \vdots \\ f_n(x) \end{pmatrix}$$

Then f_i is a function with domain in $\mathbf{R}^1$ and range in $\mathbf{C}^1$ called the ith *coordinate function* of f.

In what follows, we will often refer to the range of f as a subset of $\mathbf{C}^n$ (or $\mathbf{R}^n$), rather than $\mathbf{M}_{n1}$. Since $\mathbf{C}^n$ and $\mathbf{M}_{n1}$ are isomorphic, this convenient abuse of notation will lead to no harm.

Example 1   Consider the motion of a particle on a curve in space. (See Figure 12.2.) Its position at time t can be described by the position vector $R(t) = (x(t), y(t), z(t))$. (For convenience, the coordinates are written as a row rather than a column.) The coordinate functions x, y, z are real-valued functions of the real variable t, whereas R is a vector function.

DEFINITION 12.3   A vector function $f: \mathbf{R}^1 \to \mathbf{C}^n$ is said to be *continuous* at x_0 if each coordinate function is continuous at x_0. It is said to be *differentiable* at x_0 if each coordinate function is differentiable at x_0, and its *derivative* at x_0 is the vector

$$f'(x_0) = \begin{pmatrix} f_1'(x_0) \\ \vdots \\ f_n'(x_0) \end{pmatrix}$$

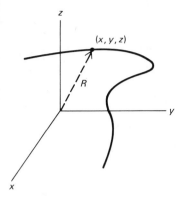

FIGURE 12.2

It is said to be *integrable* over an interval (a, b) if each coordinate function is integrable, and its *integral* is

$$\int_a^b f(x)\, dx = \begin{pmatrix} \int_a^b f_1(x)\, dx \\ \cdots \\ \int_a^b f_n(x)\, dx \end{pmatrix}$$

By means of this definition the concepts of differential calculus can be carried over to vector functions. More detailed discussions of these concepts can be found in textbooks on multivariable calculus.

Example 2   Let $R(t)$ be the position vector defined in Example 1. Then

$$V(t) = R'(t) = \begin{pmatrix} x'(t) \\ y'(t) \\ z'(t) \end{pmatrix}$$

is the velocity vector and

$$A(t) = R''(t) = \begin{pmatrix} x''(t) \\ y''(t) \\ z''(t) \end{pmatrix}$$

is the acceleration vector. Newton's law as formulated in Eq. (1) of Section 12.1 can be written

$$mA = mR'' = F(x, y, z, t)$$

where F is the vector

$$\begin{pmatrix} F_1 \\ F_2 \\ F_3 \end{pmatrix}$$

Exercises

1.   Compute $f'(x_0)$ for the indicated vector functions f and real numbers x_0.

(a) $f(x) = \begin{pmatrix} e^{ix} \\ e^{-ix} \end{pmatrix}$, $x_0 = 0$ and π

(b) $f(x) = \begin{pmatrix} x \\ x + ix^2 \\ x + i \\ \dfrac{1}{x + i} \end{pmatrix}$, $x_0 = 0$ and 1

2.   Compute the integral of each function in Exercise 1 over the indicated interval.

(a) Over $(0, \pi)$ (b) Over $(0, 1)$

3.   Let $f(x)$ be a continuous vector function. Prove that

$$\frac{d}{dx} \int_a^x f(u)\, du = f(x)$$

4.   In Example 2, suppose that $F = g(|R|)u$, where $|R|$ represents the length of the vector R, g is a function of this length, and u is a vector of unit length in the direction opposite to R. Find the coordinates of u, in terms of x, y, z, and thus express F_1, F_2, and F_3 in terms of x, y, z, and g.

12.3   The concept of solution of a system

Consider the normal first order system

$$y_i' = f_i(x, y_1, \ldots, y_n) \qquad (i = 1, 2, \ldots, n) \tag{1}$$

By a *solution* of this system we mean a vector function y defined on some interval $a < x < b$ such that the coordinate functions $y_1, \ldots, y_n$ are differentiable and satisfy the n equations of the system; that is,

$$y_i'(x) = f_i(x, y_1(x), \ldots, y_n(x)), \qquad a < x < b, \qquad i = 1, 2, \ldots, n \tag{2}$$

For the time being, we shall restrict attention to real-valued functions $y_1, \ldots, y_n$.

Example 1

$$\begin{aligned} y_1' &= y_1 + 2y_2 \\ y_2' &= -3y_1 - 4y_2 \end{aligned} \tag{3}$$

The vector function of y given by

$$y(x) = \begin{pmatrix} e^{-x} \\ -e^{-x} \end{pmatrix}$$

has $\mathbf{R}^1$ as its domain and has its range in $\mathbf{R}^2$. It is easy to see that the coordinate functions $y_1(x) = e^{-x}$, $y_2(x) = -e^{-x}$ satisfy the equations in (3), since

$$-e^{-x} = e^{-x} - 2e^{-x}$$
$$e^{-x} = -3e^{-x} + 4e^{-x}$$

The reader can verify that

$$y(x) = \begin{pmatrix} 2e^{-2x} \\ -3e^{-2x} \end{pmatrix}$$

is another solution.

It is important to keep in mind that the word *solution* applied to a system always refers to a vector function y and not to one of the individual coordinate functions y_i. This is analogous to our terminology for linear algebraic systems in n unknowns,

$$\sum_{j=1}^{n} m_{ij}x_j = y_i \qquad (i = 1, \ldots, n)$$

where the word *solution* refers to a vector $(x_1, \ldots, x_n)$ in $\mathbf{K}^n$. (See Section 10.5.)

It is customary to replace Eq. (1) by a more abbreviated notation, writing

$$y' = f(x, y_1, \ldots, y_n) \tag{4}$$

Here f denotes the vector function (of $n + 1$ scalar variables) defined by

$$f(x, y_1, \ldots, y_n) = \begin{pmatrix} f_1(x, y_1, \ldots, y_n) \\ f_2(x, y_1, \ldots, y_n) \\ \cdots \\ f_n(x, y_1, \ldots, y_n) \end{pmatrix} \tag{5}$$

and y' denotes the derivative of the vector function y,

$$y' = \begin{pmatrix} y_1' \\ y_2' \\ \vdots \\ y_n' \end{pmatrix} \tag{6}$$

The vector equality in Eq. (4) is valid if and only if respective components are equal; that is, if and only if Eq. (1) holds.

In more concise form, Eq. (4) becomes

$$y' = f(x, y) \tag{7}$$

In this form, the system has the same *appearance* as the single first order equation discussed in Chapter 3, but, of course, y and f must now be regarded as vector functions.

Let us write a formal definition of solution.

DEFINITION 12.4   Consider the system

$$y_i' = f_i(x, y_1, \ldots, y_n) \qquad (i = 1, 2, \ldots, n)$$

or more concisely

$$y' = f(x, y)$$

A *solution of the system* is a vector function y defined on an interval J such that

1. $y'(x)$ exists for each x in the interval,
2. The point $(x, y_1(x), \ldots, y_n(x))$ is in the domain of the function f, for each x in the interval, and
3. $y'(x) = f(x, y(x))$ for $x \in J$.

Exercises

1. Verify that

$$y = \begin{pmatrix} 2e^{-2x} \\ -3e^{-2x} \end{pmatrix}$$

is a solution of the system in Example 1.

2. Verify that

$$z = \begin{pmatrix} t^2 \\ t \end{pmatrix}$$

is a solution of the system below for all t.

$$z_1'' - z_1'z_2 + z_2'' + 2z_1 = 2$$
$$z_1 - z_2^2 + z_2' - z_1'' = -1$$

3. Verify that

$$y = \begin{pmatrix} (1 - x)^{-1} \\ x \end{pmatrix}$$

is a solution of

$$(y_2 - y_2')^{-2} = y_1'$$
$$y_1 y_2 + y_2' = y_1$$

on $x < 1$ and on $x > 1$.

12.4   Direction field and initial conditions

The definition of solution of a system in Definition 12.4 reduces, for $n = 1$, to the one given in Section 1.3. The notion of solution curve given there can be extended to systems with $n = 2$. Indeed, the graph of the vector solution

$y(x) = (y_1(x), y_2(x))$ is by definition the set of points $(x, y_1(x), y_2(x))$, $x \in J$. (For convenience we will write vectors in rows rather than columns here.) This is a curve in three-dimensional space which we can call a *solution curve* or *integral curve*. A tangent vector to this curve at the point corresponding to a certain x is a vector in the direction (dx, dy_1, dy_2) or $(1, y_1'(x), y_2'(x))$. Since $y_1'(x) = f_1(x, y_1(x), y_2(x))$ and $y_2'(x) = f_2(x, y_1(x), y_2(x))$ on a solution curve, we see that the tangent vector has components $1, f_1(x, y_1, y_2), f_2(x, y_1, y_2)$ at the point $(x, y_1(x), y_2(x))$.

These considerations lead us once again to the notion of the direction field defined by the system. In the case $n = 2$, at each point (x, y_1, y_2) in space, a direction is determined, namely, the direction of the vector $(1, f_1(x, y_1, y_2), f_2(x, y_1, y_2))$. The *direction field* determined by the system $y_1' = f_1(x, y_1, y_2)$, $y_2' = f_2(x, y_1, y_2)$ is the function which assigns to each point (x, y_1, y_2) the direction $(1, f_1(x, y_1, y_2), f_2(x, y_1, y_2))$. A solution of the system determines a solution curve which has the property that at each point on the curve, the tangent has the direction of the direction field at that point. The solution curves can be thought of as threading their way through the field.

Example 1 Consider the system

$$\frac{dy_1}{dx} = y_2$$

$$\frac{dy_2}{dx} = -y_1$$

From these equations it follows that $y_1'' = -y_1$, and so $y_1 = c_1 \sin x + c_2 \cos x$. Hence, $y_2 = y_1' = c_1 \cos x - c_2 \sin x$. Thus, we have found a solution,

$$y_1 = c_1 \sin x + c_2 \cos x$$
$$y_2 = c_1 \cos x - c_2 \sin x$$

depending on two arbitrary constants. It is easy to verify that $y_1^2 + y_2^2 = c_1^2 + c_2^2$. Therefore, for each choice of c_1 and c_2, $y_1^2 + y_2^2$ is constant, and the corresponding solution curve must therefore lie on a circular cylinder in (x, y_1, y_2) space. As x increases, the curve winds around the cylinder. (See Figure 12.3.)

To graph a solution for $n > 2$ a geometric space of dimension greater than three is required. Nevertheless, we shall continue to speak of direction fields and solution curves, since this terminology is a substantial aid to our intuition. All theorems will be proved by analytical arguments which are independent of geometrical interpretations (although geometrical interpretations may have suggested what is needed in the way of an analytical proof).

DEFINITION 12.5 A *solution curve* of the system $y' = f(x, y)$ is the set of points $(x, y(x)) = (x, y_1, \ldots, y_n)$ in $\mathbf{R}^{n+1}$ for $x \in J$, where $y(x)$ is a solution.

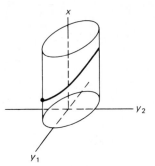

FIGURE 12.3

The *direction field* is the function which assigns to each point (x, y) in the domain of f the value $(1, f(x, y))$.

In Chapter 3, we found that a given differential equation has many solutions generally, but that a unique solution is selected (under suitable hypotheses) by specifying an initial condition. We might expect the same thing to be true here, provided we correctly interpret the phrase "initial condition." In Chapter 3, an appropriate initial condition had the form $y(x_0) = y_0$, x_0 and y_0 being given numbers; that is, we had to specify one point on a solution curve. For a system of equations, it remains true (under appropriate conditions on f) that a unique solution curve is determined by specifying one point on the curve; that is, if the values $y_1(x_0), \ldots, y_n(x_0)$ are specified for a given $x = x_0$, then there will be a unique solution $y_1(x), \ldots, y_n(x)$. This is the content of the fundamental theorems on existence and uniqueness of solutions, which we can now state.

12.5   Existence-uniqueness theorems

We wish to consider the *initial value problem* consisting of the vector differential equation (that is, system of differential equations)

$$y' = f(x, y) \tag{1}$$

and the initial condition

$$y(x_0) = \eta \tag{2}$$

where η is a given vector. Equation (2) specifies the value of the vector function y at the single point x_0 and is equivalent to the equations

$$y_1(x_0) = \eta_1, \ldots, y_n(x_0) = \eta_n \tag{3}$$

where

$$\eta = \begin{pmatrix} \eta_1 \\ \eta_2 \\ \vdots \\ \eta_n \end{pmatrix}$$

Example 1 The set of equations

$$y_1' = y_1 + 2y_2$$
$$y_2' = -3y_1 - 4y_2$$
$$y_1(0) = 1, \qquad y_2(0) = -1$$

defines an initial value problem. A solution of this problem is

$$y(x) = \begin{pmatrix} e^{-x} \\ -e^{-x} \end{pmatrix}$$

If the differential equations are not altered but the initial conditions are changed to $y_1(0) = 3$, $y_2(0) = -4$, we have a new initial value problem. A solution is

$$y(x) = \begin{pmatrix} 2e^{-2x} + e^{-x} \\ -3e^{-2x} - e^{-x} \end{pmatrix}$$

THEOREM 12.1 Suppose that each coordinate function of $f(x, y)$ is real and continuous in the "box" or $(n + 1)$-dimensional rectangle

$$B = \{(x, y): a \le x \le b, \quad c_i \le y_i \le d_i, \quad i = 1, \dots, n\} \qquad (4)$$

and that the point (x_0, η) is an interior point of B. Then there is at least one solution $y(x)$ to the initial value problem (1) and (2). Moreover, $y(x)$ exists on any interval J containing x_0 for which the points $(x, y(x))$ remain in B.

THEOREM 12.2 Suppose that each coordinate function $f_i(x, y)$ is real and continuous in B and also each partial derivative

$$\frac{\partial f_i(x, y_1, \cdots, y_n)}{\partial y_j} \qquad (i = 1, \dots, n; j = 1, \dots, n)$$

is continuous in B and (x_0, η) is an interior point of B. Then there is a unique solution $y(x)$ of the initial value problem (1) and (2).

These theorems are direct generalizations of Theorems 3.2 and 3.3. The proofs are beyond the scope of this text.

Example 2 Consider the nonlinear initial value problem

$$\frac{dy_1}{dx} = -2y_2^3$$

$$\frac{dy_2}{dx} = -y_1$$

$$y_1(0) = -1, \qquad y_2(0) = 1$$

Here we have $f_1(x, y_1, y_2) = -2y_2^3$, $f_2(x, y_1, y_2) = -y_1$. The functions and their partial derivatives with respect to y_1 and y_2 exist and are continuous in any box B containing the initial point $(0, -1, 1)$. The above theorems therefore indicate that the problem has a unique solution. The reader can verify that the solution is $y_1(x) = -(x - 1)^{-2}$, $y_2(x) = -(x - 1)^{-1}$. Note that the solution exists on any interval $J = (b, 1)$ where $b < 0$, but "blows up" at $x = 1$. As often happens when the differential equations are nonlinear, the solution cannot be extended arbitrarily far from the initial point.

Exercises

1. Verify that

$$y(x) = \begin{pmatrix} e^{-x} \\ -e^{-x} \end{pmatrix} \quad \text{and} \quad y(x) = \begin{pmatrix} 2e^{-2x} + e^{-x} \\ -3e^{-2x} - e^{-x} \end{pmatrix}$$

are solutions of the respective initial value problems in Example 1.

2. Verify that

$$y(x) = \begin{pmatrix} -(x - 1)^{-2} \\ -(x - 1)^{-1} \end{pmatrix}$$

is a solution of the problem in Example 2.

3. For each system, find a box B surrounding the initial point such that the hypotheses of Theorem 12.1 are fulfilled, and a box (the same or different) such that the hypotheses of Theorem 12.2 are fulfilled.

    (a) $y_1' = 2x^2 y_1 - e^{y_2}$
       $y_2' = y_1 \ln (1 - x^2)$ $y_1(0) = 1, \qquad y_2(0) = 0$

    (b) $(x - 2)y_1' = 3y_1 - 4y_2$
       $(3 + x)y_2' = y_1 + y_2$ $y_1(0) = 1, \qquad y_2(0) = 2$

    (c) $y_1' = \sqrt{y_1} - (y_2 - 3)^{-1}$
       $y_2' = \sqrt{y_1} - 47 + 29y_2^{-1}$ $y_1(0) = 100, \qquad y_2(0) = 5$

4. Find the solution of

$$y_1' = -y_1, \qquad y_2' = y_1 + y_2,$$
$$y_1(0) = 2, \qquad y_2(0) = 1$$

5. Find a solution containing two arbitrary constants of

$$y_1' = -y_1, \qquad y_2' = y_1 - xy_2$$

12.6   Linear systems

A (real) system of differential equations $y' = f(x, y)$ will be called a linear system if f is a linear function so far as its dependence on y is concerned. This means that for each fixed x, f defines a linear transformation from $\mathbf{R}^n$ into $\mathbf{R}^n$.

As we know, f is then representable by an $n \times n$ matrix A. Of course, the elements of A may depend on x. We therefore are led to the following definition.

DEFINITION 12.6 A first order normal system of differential equations is called a *linear homogeneous system* if it has the form

$$y' = A(x)y \tag{1}$$

where $A(x)$ is an $n \times n$ matrix, the elements of which are scalar functions of x.

Equation (1) when rewritten in full is

$$y_1' = a_{11}(x)y_1 + a_{12}(x)y_2 + \cdots + a_{1n}(x)y_n$$
$$y_2' = a_{21}(x)y_1 + a_{22}(x)y_2 + \cdots + a_{2n}(x)y_n$$
$$\cdots$$
$$y_n' = a_{n1}(x)y_1 + a_{n2}(x)y_2 + \cdots + a_{nn}(x)y_n$$

Example 1 The system

$$\frac{dy_1}{dx} = xy_1 - (\sin x)y_2$$

$$\frac{dy_2}{dx} = 2y_1 + x^2 y_2$$

can be written in the form (1) with

$$y = \begin{pmatrix} y_1 \\ y_2 \end{pmatrix}, \qquad A(x) = \begin{pmatrix} x & -\sin x \\ 2 & x^2 \end{pmatrix}$$

DEFINITION 12.7 A first order normal system is called a *linear inhomogeneous* (or *nonhomogeneous*) *system* if it has the form

$$y' = A(x)y + b(x) \tag{2}$$

where $A(x)$ is an $n \times n$ matrix and $b(x)$ is an $n \times 1$ matrix or column vector.

Equation (2) is of the form in Eq. (1) of Section 12.5 with $f(x, y) = A(x)y + b(x)$. Strictly speaking, this function f is not a linear function but rather an affine function (of y). However, the system (2) is customarily called linear nonhomogeneous.

Example 2 The following system is linear nonhomogeneous.

$$\frac{dy_1}{dx} = xy_1 - (\sin x)y_2 + \ln (x^2 + 1)$$

$$\frac{dy_2}{dx} = 2y_1 + x^2 y_2 + e^x \sinh x$$

Let us look at systems (1) and (2) from the point of view of linear operator theory. We want to think of the equations as expressing a linear operation on elements y in some suitable space, as we did in Chapter 8 for scalar equations. The space, of course, must contain vector functions y which have derivatives. We therefore make the following definition.

DEFINITION 12.8   Let J denote an interval on the real line. Then $C_n^0(J)$ denotes the vector space of all continuous functions y with domain J and range in R^n (complex values will sometimes be allowed later). $C_n^1(J)$ denotes the vector subspace of $C_n^0(J)$ consisting of vector functions with a continuous derivative. $C_n^\infty(J)$ is the space of vector functions with continuous derivatives of every order.

Addition of elements in $C_n^0(J)$ is defined in an obvious way; if y and z are in $C_n^0(J)$ then $y + z$ is the vector function such that $(y + z)(x) = y(x) + z(x)$ for all x. This definition makes sense since for each x, the values $y(x)$ and $z(x)$ are vectors in R^n and can be added component-by-component. Multiplication by a scalar has an equally obvious meaning.

DEFINITION 12.9   Let $\mathscr{D}$ be the operator on $C_n^1(J)$ into $C_n^0(J)$ such that $\mathscr{D}y = y'$. Thus

$$\mathscr{D}y = \mathscr{D}\begin{pmatrix} y_1 \\ \vdots \\ y_n \end{pmatrix} = \begin{pmatrix} y_1' \\ \vdots \\ y_n' \end{pmatrix}$$

For each y in $C_n^0(J)$, we can form the product $A(x)y(x)$ for x in J. This defines a vector function with domain J and range in R^n. The combination $\mathscr{D}y - A(x)y$ can therefore be construed as the image of y under a mapping which we denote $\mathscr{D} - A(x)$ or $\mathscr{D} - A(x)I$. This is something of an abuse of notation but is exactly analogous to the situation described in Section 8.2 in which, for example, $D + a_0(x)I$ was regarded as an operator.[2] Equations (1) and (2) can now be written

$$(\mathscr{D} - A(x))y = 0 \tag{3}$$

$$(\mathscr{D} - A(x))y = b(x) \tag{4}$$

respectively. It is left to the reader to show that $\mathscr{D} - A(x)$ is a linear operator on the space $C_n^1(J)$ into $C_n^0(J)$ or on $C_n^\infty(J)$ into $C_n^\infty(J)$, if $A(x)$ satisfies suitable conditions.

---

[2] On reflection, we see that the symbol $A(x)$ can now have three meanings:

1. $A(x)$ may be a matrix of functions;

2. $A(x)$ may be considered to be a linear transformation which takes each vector in R^n into a vector in R^n;

3. $A(x)$ may be considered to be an operator on $C_n^0(J)$ into itself.

It would be possible to consider higher order operators such as $\mathscr{D}^2 y$, but as we have shown in Section 12.1, we can comfortably restrict attention to Eqs. (3) and (4) for most purposes.

Exercises

1. Find $(\mathscr{D} - A(x))y$ for each given $A(x)$ and y.

   (a) $A = \begin{pmatrix} 1 & 2 \\ -3 & -4 \end{pmatrix}$ $\quad y = \begin{pmatrix} e^{-x} \\ -e^{-x} \end{pmatrix}$

   (b) $A = \begin{pmatrix} x & 0 \\ 0 & x \end{pmatrix}$ $\quad y = \begin{pmatrix} e^x \\ xe^x \end{pmatrix}$

2. Prove that $\mathscr{D} - A(x)$ is a linear operator on $\mathbf{C}_n{}^1(J)$ into $\mathbf{C}_n{}^0(J)$. What hypothesis on $A(x)$ is required? Under what hypothesis is $\mathscr{D} - A(x)$ a linear operator on $\mathbf{C}_n{}^\infty(J)$ into $\mathbf{C}_n{}^\infty(J)$?

3. Consider the set of operators $\mathscr{D} - A(x)$ for all continuous matrices $A(x)$. Is this a *subspace* of the space of all linear operators on $\mathbf{C}_n{}^1(J)$ into $\mathbf{C}_n{}^0(J)$? Why?

12.7 General theory for the linear system

For linear systems one can prove the following existence theorem, which gives more information than the theorems in Section 12.5.

THEOREM 12.3 Assume that the $n \times n$ matrix $A(x)$ and the $n \times 1$ vector $b(x)$ have components which are real and continuous for all x in an open interval J. Let x_0 lie in this interval and let η be any $n \times 1$ vector in $\mathbf{R}^n$. Then the initial value problem

$$y' = A(x)y + b(x) \tag{1}$$

$$y(x_0) = \eta \tag{2}$$

has a unique solution $y = \phi(x)$. Moreover, this solution $\phi(x)$ exists over the entire interval J.

We note that the matrix $A(x)$ is called continuous if each of its elements $a_{ij}(x)$ is a continuous function in the ordinary sense.

For a proof of this theorem, the reader may consult one of the references given at the end of the chapter. This theorem is stronger than Theorems 12.1 and 12.2 in two respects. First, the box B of the earlier theorems has been replaced by the infinite strip

$$\{(x, y): x \in J, y \text{ arbitrary}\}$$

Consequently, η is entirely arbitrary. Second, the solution of the linear system exists over the whole interval of continuity of $A(x)$ and $b(x)$, whereas for the more general equation $y' = f(x, y)$, the solution may possibly exist only on a very small x-interval.

We remark that if $A(x)$ and $b(x)$ are continuous for all x, $-\infty < x < \infty$, we can be sure that the solution y exists for all x, since it must exist on *every* finite interval J. We also point out the following corollary, obtained by taking $\eta = 0$.

COROLLARY   If $A(x)$ is continuous for x in J, the only solution of the initial value problem

$$y' = A(x)y, \qquad y(x_0) = 0$$

is the trivial solution $y(x) \equiv 0$.

The next theorem describes the null space of the operator $\mathscr{D} - A(x)$. Before giving the theorem, let us recall the meaning of a few terms. First, the null space of $\mathscr{D} - A(x)$ is the set of vectors ϕ such that $(\mathscr{D} - A(x))\phi = 0$; that is, it is the set of vector functions $\phi(x)$ for which $\phi'(x) - A(x)\phi(x)$ is the zero vector over the whole interval J. Next, a set of elements $\{\phi_1, \ldots, \phi_k\}$ in $C_n^1(J)$ is linearly dependent if there are scalars $c_1, \ldots, c_k$, not all zero, such that $c_1\phi_1 + \cdots + c_k\phi_k$ is the zero element. This means that

$$c_1\phi_1(x) + \cdots + c_k\phi_k(x) = 0, \qquad x \in J$$

In this equation, each $\phi_j(x)$ is a vector with n components and the 0 on the right is the vector with n zero components.

Example 1   The system in Example 1 of Section 12.5 can be written $y' = Ay$ where

$$A = \begin{pmatrix} 1 & 2 \\ -3 & -4 \end{pmatrix}$$

Two of its solutions are

$$\phi_1(x) = \begin{pmatrix} e^{-x} \\ -e^{-x} \end{pmatrix}, \qquad \phi_2(x) = \begin{pmatrix} -2e^{-x} \\ 2e^{-x} \end{pmatrix}$$

since

$$(\mathscr{D} - A)\phi_1 = \begin{pmatrix} \dfrac{d}{dx}e^{-x} \\[2mm] -\dfrac{d}{dx}e^{-x} \end{pmatrix} - \begin{pmatrix} 1 & 2 \\ -3 & -4 \end{pmatrix}\begin{pmatrix} e^{-x} \\ -e^{-x} \end{pmatrix} = \begin{pmatrix} 0 \\ 0 \end{pmatrix}$$

and similarly, $(\mathscr{D} - A)\phi_2 = 0$. However, ϕ_1 and ϕ_2 are linearly dependent in $C_2{}^1(-\infty, \infty)$ because

$$2\phi_1(x) + \phi_2(x) = 2\begin{pmatrix} e^{-x} \\ -e^{-x} \end{pmatrix} + \begin{pmatrix} -2e^{-x} \\ 2e^{-x} \end{pmatrix} = \begin{pmatrix} 0 \\ 0 \end{pmatrix}$$

DEFINITION 12.10 The *real solution space* of the system $(\mathscr{D} - A(x))y = 0$ is the vector space consisting of all solutions $y = y(x)$ for which all the components $y_1(x), \ldots, y_n(x)$ are real-valued continuously differentiable functions on J.

THEOREM 12.4 Let $A(x)$ be an $n \times n$ matrix the elements of which are continuous real-valued functions for x in an interval J. Then the real solution space of

$$y' = A(x)y \tag{3}$$

is an n-dimensional subspace of $\mathbf{C}_n{}^1(J)$.

PROOF: The solution space is the null space of the operator $\mathscr{D} - A(x)$ on $\mathbf{C}_n{}^1(J)$ into $\mathbf{C}_n{}^0(J)$ and is therefore a vector space. The proof that this space has dimension n, similar to the proof of Theorem 8.5, is as follows. Let x_0 be a fixed point in J. Let

$$e_1 = \begin{pmatrix} 1 \\ 0 \\ 0 \\ \vdots \\ 0 \end{pmatrix}, \qquad e_2 = \begin{pmatrix} 0 \\ 1 \\ 0 \\ \vdots \\ 0 \end{pmatrix}, \qquad \ldots, \qquad e_n = \begin{pmatrix} 0 \\ 0 \\ 0 \\ \vdots \\ 1 \end{pmatrix}$$

denote the respective columns of I_n. Let the vector functions $\phi_1, \phi_2, \ldots, \phi_n$ be the n solutions of Eq. (3) determined by the respective initial conditions

$$\phi_i(x_0) = e_i, \qquad i = 1, 2, \ldots, n \tag{4}$$

Now consider the equation

$$c_1\phi_1(x) + \cdots + c_n\phi_n(x) = 0, \qquad x \in J$$

Putting $x = x_0$, we get

$$c_1 e_1 + \cdots + c_n e_n = 0$$

Since $\{e_1, \ldots, e_n\}$ is a linearly independent set in $\mathbf{R}^n$, we conclude that $c_1, \ldots, c_n$ must all be zero, and this proves that $\{\phi_1, \ldots, \phi_n\}$ is a linearly independent set in $\mathbf{C}_n{}^1(J)$.

To complete the proof we shall show that every solution of Eq. (3) is a linear combination of $\phi_1, \ldots, \phi_n$. Let y be any solution in the real solution space and suppose that η is the value of y at x_0; that is, $y(x_0) = \eta$. The vector η is in $\mathbf{R}^n$ and hence we can write

$$\eta = \eta_1 e_1 + \cdots + \eta_n e_n,$$

where $\eta_1, \ldots, \eta_n$ are real numbers. We can define the vector function $z = \eta_1 \phi_1 + \cdots + \eta_n \phi_n$. Clearly z is a solution of Eq. (3) since it is a linear combination of solutions, and clearly also $z(x_0) = \sum \eta_i \phi_i(x_0) = \sum \eta_i e_i = \eta$. Therefore y and z are solutions with the same initial condition and, by Theorem 12.3, the functions y and z must be identical. Hence every solution y is a linear combination of $\phi_1, \ldots, \phi_n$. Thus $\{\phi_1, \ldots, \phi_n\}$ is a basis for the real solution space, which consequently has dimension n.   ∎

Example 2   The system

$$y_1' = y_1 + 2y_2$$
$$y_2' = -3y_1 - 4y_2$$

has solutions

$$\phi_1(x) = \begin{pmatrix} e^{-x} \\ -e^{-x} \end{pmatrix} \quad \text{and} \quad \phi_2(x) = \begin{pmatrix} 2e^{-2x} \\ -3e^{-2x} \end{pmatrix}$$

If it can be shown that these solutions are independent, it will follow that every solution has the form

$$y(x) = c_1 \phi_1(x) + c_2 \phi_2(x) = c_1 \begin{pmatrix} e^{-x} \\ -e^{-x} \end{pmatrix} + c_2 \begin{pmatrix} 2e^{-2x} \\ -3e^{-2x} \end{pmatrix}$$

or equivalently

$$y_1(x) = c_1 e^{-x} + 2c_2 e^{-2x}$$
$$y_2(x) = -c_1 e^{-x} - 3c_2 e^{-2x}$$

where c_1 and c_2 are arbitrary real numbers. To show that ϕ_1 and ϕ_2 are linearly independent we suppose that $c_1 \phi_1 + c_2 \phi_2 = 0$. This means

$$c_1 e^{-x} + 2c_2 e^{-2x} = 0$$
$$-c_1 e^{-x} - 3c_2 e^{-2x} = 0 \qquad (-\infty < x < \infty)$$

From this follows $c_2 e^{-2x} = 0$, and since this is to hold for all x, $c_2 = 0$. Then $c_1 e^{-x}$ must be zero; hence $c_1 = 0$. In the next section we will give a general criterion for deciding when n solutions are independent.

As in the discussion of nth order linear differential equations, it will be helpful to consider solutions whose components are complex-valued functions of the real variable x, even when $A(x)$ and $b(x)$ have real components.

DEFINITION 12.11   Let J denote an interval on the real line and let n be a nonnegative integer. Then $\mathbf{C}_n{}^0(J, \mathbf{C})$ denotes the vector space of all continuous vector functions with domain J and range in $\mathbf{C}^n$, and $\mathbf{C}_n{}^1(J, \mathbf{C})$ denotes the vector subspace of $\mathbf{C}_n{}^0(J, \mathbf{C})$ consisting of functions with a continuous derivative. Let

$A(x)$ and $b(x)$ have components which are continuous real or complex functions for $x \in J$. The *complex solution space* of the system $(\mathscr{D} - A(x))y = 0$ is the vector space consisting of all solutions in $C_n^1(J, C)$.

It can be shown that Theorem 12.3 remains valid when A, b, and y have complex components, and that the following result holds.

THEOREM 12.5 Let $A(x)$ be an $n \times n$ matrix, the elements of which are continuous real or complex functions of the real variable x for x in an interval J. Then the complex solution space of $(\mathscr{D} - A(x))y = 0$ is an n-dimensional vector space, a subspace of $C_n^1(J, C)$.

PROOF: The proof is left as Exercise 4. ∎

Exercises

1. What information does Theorem 12.3 give concerning the existence of a solution of each initial value problem? Over what interval does the theorem guarantee existence of the solution?

    (a) $y_1' = (\sin x)y_1 - (\csc x)y_2$
    $y_2' = xy_1 - xy_2$
    $y_1(1) = 1, y_2(1) = 2$

    (b) $y_1' = y_1 \ln x - e^x y_2$

    $y_2' = xy_1 - y_2 + \dfrac{1}{\ln x}$

    $y_1(\tfrac{1}{2}) = 0, y_2(\tfrac{1}{2}) = 3$

2. *Reduction of order.* Suppose that $u(x)$ is one known solution of the $n \times n$ linear system $y' = A(x)y$. Show that the substitution $y = fu + z$, where f is an unknown scalar function and z is an unknown vector function, leads to the system

$$f'u + z' = Az$$

    Taking the first component of z to be zero, show that this system can be written

$$u_1(x)f'(x) = \sum_{j=2}^{n} a_{1j}(x)z_j(x)$$

$$z_i'(x) = \sum_{j=2}^{n} a_{ij}(x)z_j(x) - f'(x)u_i(x), \qquad i = 2, \ldots, n$$

    or

$$z_i'(x) = \sum_{j=2}^{n} \left[\left(a_{ij}(x) - \frac{u_i(x)}{u_1(x)} a_{1j}(x)\right) z_j(x)\right], \qquad i = 2, \ldots, n$$

3.   The system $y' = A(x)y$ where

$$A(x) = \begin{pmatrix} x & 0 & -1 \\ 0 & x & -1 \\ 2 & -1 & 0 \end{pmatrix}$$

has a solution

$$u(x) = \begin{pmatrix} 1 \\ 1 \\ x \end{pmatrix}$$

Using the method of the previous exercise, reduce to a system of order two, and thereby find a general solution of the original system.

4.   Prove Theorem 12.5.

12.8   Linear independence and fundamental matrices

DEFINITION 12.12   A set of n linearly independent solutions of the $n \times n$ homogeneous system

$$y' = A(x)y \tag{1}$$

is called a *fundamental set of solutions*.

To say that a set of solutions is fundamental is equivalent to saying that the set is a basis for the solution space. In any given case, we must specify whether it is the real or complex solution space to which reference is made.

DEFINITION 12.13   An $n \times n$ matrix is called a *solution matrix* if each of its n columns defines a vector function which is a solution of Eq. (1). It is called a *fundamental matrix* if its n columns form a basis for the solution space of Eq. (1).

Example 1   The matrix

$$\Phi(x) = \begin{pmatrix} e^{-x} & 2e^{-2x} \\ -e^{-x} & -3e^{-2x} \end{pmatrix}$$

is a solution matrix and a fundamental matrix for the system in Example 2 of the preceding section, since its columns are linearly independent solutions.

We shall now develop tests for deciding when a set of n solutions is a basis, or equivalently when a solution matrix is fundamental.

LEMMA 12.1   Let $\phi_1(x)$, $\phi_2(x)$, ..., $\phi_k(x)$ be any k functions in the space $C_n^0(J)$. Let $x_0 \in J$. If the set of k constant vectors

$$\{\phi_1(x_0), \phi_2(x_0), \ldots, \phi_k(x_0)\}$$

is linearly independent in $\mathbf{R}^n$, then the set of k functions

$$\{\phi_1(x),\ \phi_2(x),\ \ldots,\ \phi_k(x)\}$$

is linearly independent in $C_n^0(J)$. A similar result holds in the complex case.

PROOF: This is left as an exercise. ▌

The converse of this lemma is false; that is, from linear independence of $\{\phi_1(x), \ldots, \phi_k(x)\}$ over an interval, we cannot in general deduce linear independence of $\{\phi_1(x_0), \ldots, \phi_k(x_0)\}$. (See Exercise 1 of Section 8.6.)

In the discussion of linear independence for the nth order scalar equation in Chapter 8, the Wronski matrix was important. We can employ a similar concept here.

DEFINITION 12.14 Let $\phi_1, \phi_2, \ldots, \phi_n$ be continuous vector functions on an interval J. The *Wronski matrix* of these functions is the matrix Φ in which the jth column contains the n elements of ϕ_j $(j = 1, 2, \ldots, n)$. The determinant of Φ is called the *Wronskian* of $\phi_1, \ldots, \phi_n$ and is denoted by $W(x)$ or $\det \Phi(x)$.

If we introduce a double subscript notation for the components of the vector functions $\phi_1, \phi_2, \ldots, \phi_n$, we can write

$$\phi_1 = \begin{pmatrix} \phi_{11} \\ \phi_{21} \\ \cdots \\ \phi_{n1} \end{pmatrix}, \qquad \phi_2 = \begin{pmatrix} \phi_{12} \\ \phi_{22} \\ \cdots \\ \phi_{n2} \end{pmatrix}, \qquad \ldots, \qquad \phi_n = \begin{pmatrix} \phi_{1n} \\ \phi_{2n} \\ \cdots \\ \phi_{nn} \end{pmatrix} \qquad (2)$$

$$\Phi = \begin{pmatrix} \phi_{11} & \phi_{12} & \cdots & \phi_{1n} \\ \phi_{21} & \phi_{22} & \cdots & \phi_{2n} \\ & & \cdots & \\ \phi_{n1} & \phi_{n2} & \cdots & \phi_{nn} \end{pmatrix} \qquad (3)$$

If $\phi_1, \ldots, \phi_n$ are solutions of Eq. (1), their Wronski matrix is a solution matrix of Eq. (1).

THEOREM 12.6 Let $\phi_1, \phi_2, \ldots, \phi_n$ be continuous vector functions on J. If the Wronskian of these functions is nonzero for at least one point in J, then $\{\phi_1, \phi_2, \ldots, \phi_n\}$ is a linearly independent set.

PROOF: If the Wronskian is nonzero at x_0, the column vectors form a linearly independent set $\{\phi_1(x_0), \phi_2(x_0), \ldots, \phi_n(x_0)\}$. The desired result follows immediately from Lemma 12.1. ▌

THEOREM 12.7 Let $\phi_1, \phi_2, \ldots, \phi_n$ be solutions of the equation $y' = A(x)y$ on an interval J where $A(x)$ is continuous. Then the following statements are equivalent:

1. The Wronskian det $\Phi(x)$ of these solutions is nonzero for every x in J.
2. There exists a point x_0 in J such that det $\Phi(x_0) \neq 0$.
3. Φ is a fundamental matrix for the equation on J.

PROOF: By Theorem 12.6, (2) implies (3). Obviously (1) implies (2). We shall now prove that (3) implies (1). Suppose that Φ is a fundamental matrix and let $\phi_1, \phi_2, \ldots, \phi_n$ be its column vectors. Take any x_0 in J and consider the set $\{\phi_1(x_0), \phi_2(x_0), \ldots, \phi_n(x_0)\}$. Suppose that $c_1, c_2, \ldots, c_n$ are scalars for which

$$c_1\phi_1(x_0) + c_2\phi_2(x_0) + \cdots + c_n\phi_n(x_0) = 0 \tag{4}$$

Define

$$\phi(x) = \sum_{j=1}^{n} c_j\phi_j(x), \qquad x \in J$$

Then ϕ is a solution of Eq. (1) and $\phi(x_0) = 0$. By the Corollary of Theorem 12.3, $\phi(x)$ is identically zero; that is,

$$\sum_{j=1}^{n} c_j\phi_j(x) = 0, \qquad x \in J \tag{5}$$

Because Φ is fundamental, $\{\phi_1, \phi_2, \ldots, \phi_n\}$ is linearly independent on J, and therefore the last equation implies that $c_1, \ldots, c_n$ are all zero. We have shown that Eq. (4) holds only if all the scalars are zero, which proves that $\{\phi_1(x_0), \phi_2(x_0), \ldots, \phi_n(x_0)\}$ is a linearly independent set. Since the Wronski matrix $\Phi(x_0)$ has these linearly independent vectors as its columns, the matrix is nonsingular and its determinant is not zero. Since x_0 was an arbitrary point in J, it has been proved that (3) implies (1).

It follows from the above that each of (1), (2), (3) implies the other two, and so the theorem is proved.   ∎

COROLLARY A solution matrix $\Phi(x)$ of $y' = A(x)y$ is a fundamental matrix on J if and only if det $\Phi(x)$ is never zero on J.

Exercises

1. Prove Lemma 12.1. What is a correct statement of the lemma in the complex case? What changes in the proof are needed?

2. Using Theorem 12.6, prove linear independence of each set of vector functions

(a) $\phi_1 = \begin{pmatrix} 1 \\ x \end{pmatrix}$ $\phi_2 = \begin{pmatrix} x \\ 0 \end{pmatrix}$

(b) $\phi_1 = \begin{pmatrix} e^{2x} \\ e^{-x} \end{pmatrix}$ $\phi_2 = \begin{pmatrix} e^{4x} \\ 2e^x \end{pmatrix}$

3. Show that

$$\Phi(x) = \begin{pmatrix} e^x & e^{-x} \\ 2e^x & -e^{-x} \end{pmatrix}$$

is a solution matrix and a fundamental matrix for the system

$$3y_1' = -y_1 + 2y_2$$
$$3y_2' = 4y_1 + y_2$$

4. Use Theorem 12.7 to establish the conclusions in Examples 1 and 2 of Section 12.7.

5. Show that

$$\Phi(x) = \begin{pmatrix} 1 & x & x^2 \\ 0 & 1 & 2 \\ 0 & 0 & 0 \end{pmatrix}$$

has linearly independent columns, yet det $\Phi(x) \equiv 0$. Does this contradict Theorem 12.7?

6. Let Φ be a solution matrix of Eq. (1) and let C be a constant $n \times n$ matrix.

    (a) Let $\phi_1, \ldots, \phi_n$ be the column vectors of Φ. Show that the jth column of ΦC is the linear combination

$$c_{1j}\phi_1 + c_{2j}\phi_2 + \cdots + c_{nj}\phi_n$$

    where $c_{1j}, c_{2j}, \ldots, c_{nj}$ are the elements of the jth column of C.

    (b) Prove that ΦC is a solution matrix of Eq. (1).

    (c) Prove that if Φ is a fundamental matrix and if det $C \neq 0$, then ΦC is a fundamental matrix.

    (d) Give an example to show that $C\Phi$ need not be a solution matrix.

7. Let Φ be a fundamental matrix for Eq. (1). Prove that a matrix Ψ is:

    (a) a solution matrix if and only if there is a constant matrix C such that $\Psi = \Phi C$;

    (b) a fundamental matrix if and only if there is a nonsingular constant matrix C such that $\Psi = \Phi C$.

8. Let $\phi_1, \ldots, \phi_n$ be solutions of Eq. (1) on J and let $W(x) = \det \Phi(x)$ be their Wronskian. Prove that

$$\frac{dW}{dx} = \left[\sum_{i=1}^{n} \phi_{ii}(x) \right] W$$

where ϕ_{ii} is the ith component of the vector ϕ_i. [*Hint*: Use the rule for differentiating a determinant, Exercise 6 of Section 7.7, and simplify the determinants by use of the differential equation (1).]

9. Let $\phi_1, \ldots, \phi_n$ be functions in $\mathbf{C}_n{}^0(a, b)$ given by Eq. (2). Let

$$g_{ij} = \int_a^b \phi_i^T(x)\phi_j(x)\, dx \qquad (i, j = 1, \ldots, n)$$

and let G be the $n \times n$ matrix $G = (g_{ij})$. Is $\det G \neq 0$ a necessary and sufficient condition for linear independence of $\{\phi_1, \ldots, \phi_n\}$? (See Exercise 10 in Section 8.6.)

10. The equation $z' = -zA(x)$ is called the *adjoint* equation to $y' = A(x)y$. Here z is a row vector. Prove that if $\Phi(x)$ is a fundamental solution of $y' = A(x)y$, then $\Phi^{-1}(x)$ is a fundamental matrix solution of the adjoint equation. That is, the rows of $\Phi^{-1}(x)$ are linearly independent solutions of the adjoint equation.

12.9 General solutions and initial value problems

Given a system of n linear equations $y' = A(x)y$, we can first try to find a set of n linearly independent solutions $\phi_1, \phi_2, \ldots, \phi_n$, or equivalently, a fundamental matrix Φ. Then every solution must be of the form:

$$c_1\phi_1 + c_2\phi_2 + \cdots + c_n\phi_n \quad \text{or} \quad \Phi c \quad \text{where } c = \begin{pmatrix} c_1 \\ c_2 \\ \cdots \\ c_n \end{pmatrix} \qquad (1)$$

We can refer to the set of all solutions or to the expression in Eq. (1) as a *general solution*, as usual.

If a general solution can be found, the solution of a given initial value problem can be produced by algebraic means. Suppose the initial condition is

$$y(x_0) = \eta \qquad (2)$$

To satisfy this with the function in Eq. (1), we require

$$\Phi(x_0)c = \eta \qquad (3)$$

Since Φ is fundamental, $\Phi^{-1}(x_0)$ exists, and Eq. (3) determines the components $c_1, \ldots, c_n$ of c uniquely.

Example 1 As shown above

$$\phi_1(x) = \begin{pmatrix} e^{-x} \\ -e^{-x} \end{pmatrix} \qquad \phi_2(x) = \begin{pmatrix} 2e^{-2x} \\ -3e^{-2x} \end{pmatrix}$$

are independent solutions of $y' = Ay$ where

$$A = \begin{pmatrix} 1 & 2 \\ -3 & -4 \end{pmatrix}$$

To find a solution ϕ satisfying $\phi(0) = \begin{pmatrix} 0 \\ 3 \end{pmatrix}$, we must solve

$$c_1 \phi_1(0) + c_2 \phi_2(0) = \Phi(0) \begin{pmatrix} c_1 \\ c_2 \end{pmatrix} = \begin{pmatrix} 1 & 2 \\ -1 & -3 \end{pmatrix} \begin{pmatrix} c_1 \\ c_2 \end{pmatrix} = \begin{pmatrix} 0 \\ 3 \end{pmatrix}$$

Since

$$\{\Phi(0)\}^{-1} = \begin{pmatrix} 3 & 2 \\ -1 & -1 \end{pmatrix}$$

we find

$$\begin{pmatrix} c_1 \\ c_2 \end{pmatrix} = \{\Phi(0)\}^{-1} \begin{pmatrix} 0 \\ 3 \end{pmatrix} = \begin{pmatrix} 6 \\ -3 \end{pmatrix}$$

Hence, the required solution is

$$\phi(x) = 6 \begin{pmatrix} e^{-x} \\ -e^{-x} \end{pmatrix} - 3 \begin{pmatrix} 2e^{-2x} \\ -3e^{-2x} \end{pmatrix} = \begin{pmatrix} 6e^{-x} - 6e^{-2x} \\ -6e^{-x} + 9e^{-2x} \end{pmatrix}$$

Exercises

1. Using $\phi(x)$ as given in Exercise 3 of Section 12.8, find the solution of

$$\begin{aligned} 3y_1' &= -y_1 + 2y_2 \\ 3y_2' &= 4y_1 + y_2 \end{aligned} \qquad y_1(0) = 1, \quad y_2(0) = 8$$

2. Show that

$$\phi_1 = \begin{pmatrix} e^{-x} \\ -2e^{-x} \end{pmatrix} \qquad \phi_2 = \begin{pmatrix} e^{3x} \\ 2e^{3x} \end{pmatrix}$$

are solutions of the system $y' = Ay$ where $A = \begin{pmatrix} 1 & 1 \\ 4 & 1 \end{pmatrix}$.

Find the solution ϕ for which $\phi(0) = \begin{pmatrix} -1 \\ -10 \end{pmatrix}$.

3. Show that

$$\phi_1 = \begin{pmatrix} e^x \\ 0 \\ e^x \end{pmatrix}, \qquad \phi_2 = \begin{pmatrix} e^x \\ -e^x \\ 0 \end{pmatrix}, \qquad \phi_3 = \begin{pmatrix} e^{-x} \\ e^{-x} \\ e^{-x} \end{pmatrix}$$

are solutions of $y' = Ay$, where

$$A = \begin{pmatrix} -1 & -2 & 2 \\ -2 & -1 & 2 \\ -2 & -2 & 3 \end{pmatrix}$$

and find the solution ϕ for which $\phi(0) = \begin{pmatrix} 6 \\ 1 \\ 4 \end{pmatrix}$.

12.10   Linear nonhomogeneous systems

The general theory of the nonhomogeneous system

$$y' = A(x)y + b(x) \quad \text{or} \quad (\mathscr{D} - A(x))y = b(x) \tag{1}$$

is completely analogous to the theory of the nonhomogeneous scalar nth order equation as given in Section 8.8, and we merely state the basic theorem here.

THEOREM 12.8   Consider Eq. (1) wherein $A(x)$ and $b(x)$ are continuous on an interval. If y_p is a solution of Eq. (1) and y is a solution of the corresponding homogeneous system $y' = A(x)y$, then $y_p + y$ is a solution of Eq. (1). Conversely, if y_p is any fixed solution of Eq. (1), then every solution of Eq. (1) can be expressed as a sum $y_p + y$ where y is a solution of $y' = A(x)y$.

This theorem can again be expressed in geometrical language as follows. The set of all solutions of the nonhomogeneous equation (1) is an *affine sub-space* in $\mathbf{C}_n{}^1(J)$ (or $\mathbf{C}_n{}^1(J, \mathbf{C})$), the translation of the null space of $\mathscr{D} - A(x)$ by y_p, where y_p is any one solution of Eq. (1).
   If Φ is a fundamental matrix of $y' = A(x)y$, the expression Φc represents all solutions, where c is an arbitrary constant vector. (See Section 12.9.) Hence, the expression

$$\Phi c + y_p \tag{2}$$

represents all solutions of the nonhomogeneous equation.

Exercises

1.   (a) Consider the system $y' = Ay + b$, where A is the matrix of Exercise 2, Section 12.9, and

$$b(x) = \begin{pmatrix} 0 \\ -4e^x \end{pmatrix}$$

Show that every solution has the form

$$
\begin{pmatrix} e^{-x} & e^{3x} \\ -2e^{-x} & 2e^{3x} \end{pmatrix} \begin{pmatrix} c_1 \\ c_2 \end{pmatrix} + \begin{pmatrix} e^x \\ 0 \end{pmatrix}
$$

(b) Find the solution ϕ which satisfies

$$
\phi(0) = \begin{pmatrix} 3 \\ 0 \end{pmatrix}
$$

2. Consider the system $y' = Ay + b$, where A is the matrix of Exercise 3, Section 12.9, and

$$
b(x) = \begin{pmatrix} (2 - 2x)e^x \\ (2 - 2x)e^x \\ (3 - 2x)e^x \end{pmatrix}
$$

(a) Show that every solution has the form

$$
\begin{pmatrix} e^x & e^x & e^{-x} \\ 0 & -e^x & e^{-x} \\ e^x & 0 & e^{-x} \end{pmatrix} \begin{pmatrix} c_1 \\ c_2 \\ c_3 \end{pmatrix} + \begin{pmatrix} 0 \\ e^x \\ xe^x \end{pmatrix}
$$

(b) Find the solution ϕ which satisfies

$$
\phi(0) = \begin{pmatrix} 2 \\ 0 \\ 1 \end{pmatrix}
$$

12.11 The *n*th order equation as a special case

The *n*th order equation

$$
Lz \equiv z^{(n)} + a_{n-1}(x)z^{(n-1)} + \cdots + a_0(x)z = f(x) \tag{1}
$$

was treated in Chapter 8. We are now able to show that its theory is but a special case of the theory of systems $y' = A(x)y + b(x)$ developed in this chapter. To show this, we introduce new variables

$$
y_1 = z, y_2 = z', \ldots, y_n = z^{(n-1)} \tag{2}
$$

and

$$
y = \begin{pmatrix} y_1 \\ \cdots \\ y_n \end{pmatrix}
$$

Then Eq. (1) becomes

$$y' = A(x)y + b(x) \tag{3}$$

where

$$b(x) = \begin{pmatrix} 0 \\ 0 \\ \vdots \\ 0 \\ f(x) \end{pmatrix}$$

$$A(x) = \begin{pmatrix} 0 & 1 & 0 & \cdots & 0 \\ 0 & 0 & 1 & \cdots & 0 \\ & & \cdots & & \\ 0 & 0 & 0 & \cdots & 1 \\ -a_0(x) & -a_1(x) & -a_2(x) & \cdots & -a_{n-1}(x) \end{pmatrix} \tag{4}$$

Equation (3) with A, b defined by Eqs. (4) is sometimes called the *companion equation* of Eq. (1).

THEOREM 12.9 If $z(x)$ is a solution of Eq. (1) on an interval J, then the vector function

$$y(x) = \begin{pmatrix} z(x) \\ z'(x) \\ \vdots \\ z^{(n-1)}(x) \end{pmatrix}$$

is a solution of Eq. (3) on J. Conversely, if $y(x)$ is a vector solution of Eq. (3) on J and if we let $z(x) = y_1(x)$, the first component of y, then the derivatives of z satisfy Eq. (2) and z is a solution of Eq. (1).

In other words, Eq. (1) is entirely equivalent to Eq. (3) with $A(x)$ and $b(x)$ defined by Eqs. (4).

PROOF: This is left as an exercise.   ∎

Example 1 The equation $z'' + \omega^2 z = 0$ (ω being a constant) corresponds to a system of form (3) where

$$A = \begin{pmatrix} 0 & 1 \\ -\omega^2 & 0 \end{pmatrix}$$

The function $z = c \sin \omega x$ is a solution of the scalar equation and if we set

$$y = \begin{pmatrix} c \sin \omega x \\ c\omega \cos \omega x \end{pmatrix}$$

then

$$y' = \begin{pmatrix} c\omega \cos \omega x \\ -c\omega^2 \sin \omega x \end{pmatrix} = \begin{pmatrix} 0 & 1 \\ -\omega^2 & 0 \end{pmatrix} \begin{pmatrix} c \sin \omega x \\ c\omega \cos \omega x \end{pmatrix} = Ay$$

Let us illustrate how our theorems on Eq. (3) immediately yield corresponding theorems on Eq. (1). For example, suppose $a_0(x), \ldots, a_{n-1}(x), f(x)$ are continuous functions on an interval. Then clearly, $b(x)$ and $A(x)$ as defined by Eqs. (4) are continuous vector and matrix functions, respectively. It then follows from Theorem 12.3 that Eq. (3) has a unique solution satisfying an initial condition $y(x_0) = \eta$. Hence, there is a unique solution $z(x) = y_1(x)$ of Eq. (1) which satisfies the initial conditions

$$z(x_0) = \eta_1, \quad z'(x_0) = \eta_2, \quad \ldots, \quad z^{(n-1)}(x_0) = \eta_n$$

where η_i is the ith coordinate of η. In other words, Theorem 8.4 on existence and uniqueness of solutions of Eq. (1) is a corollary of Theorem 12.3.

From Theorem 12.4 we can deduce that the solution space of

$$z^{(n)} + a_{n-1}(x)z^{(n-1)} + \cdots + a_0(x)z = 0 \tag{5}$$

is an n-dimensional subspace of $C^n(J)$, which is Theorem 8.5. (See Exercise 4.)

Suppose $\phi_1, \ldots, \phi_n$ are solutions of the homogeneous system

$$y' = A(x)y \tag{6}$$

where A is given by Eqs. (4). Let $z_1, \ldots, z_n$ be the first components of $\phi_1, \ldots, \phi_n$, respectively. Then for each i, we know from Theorem 12.9 that

$$\phi_i(x) = \begin{pmatrix} z_i(x) \\ z_i'(x) \\ \vdots \\ z_i^{(n-1)}(x) \end{pmatrix}$$

and that z_i is a solution of Eq. (5). The Wronski matrix of the system (6) is therefore

$$\begin{pmatrix} z_1(x) & \cdots & z_n(x) \\ z_1'(x) & \cdots & z_n'(x) \\ & \cdots & \\ z_1^{(n-1)}(x) & \cdots & z_n^{(n-1)}(x) \end{pmatrix}$$

We see that this is also the Wronski matrix of the n solutions $z_1, \ldots, z_n$ of the scalar equation (5).

Example 2   Two independent solutions of $z'' + \omega^2 z = 0$ are $z_1 = \sin \omega x$ and $z_2 = \cos \omega x$. The Wronski matrix as defined in Chapter 8 is

$$\mathscr{W} = \begin{pmatrix} z_1 & z_2 \\ z_1' & z_2' \end{pmatrix} = \begin{pmatrix} \sin \omega x & \cos \omega x \\ \omega \cos \omega x & -\omega \sin \omega x \end{pmatrix}$$

Corresponding to z_1 and z_2 we have the two vector solutions

$$\phi_1 = \begin{pmatrix} \sin \omega x \\ \omega \cos \omega x \end{pmatrix} \qquad \phi_2 = \begin{pmatrix} \cos \omega x \\ -\omega \sin \omega x \end{pmatrix}$$

of the companion system. The Wronski matrix for ϕ_1 and ϕ_2 has the elements of ϕ_1 in the first column and the elements of ϕ_2 in the second column and is obviously the same as above.

This argument establishes the following theorem.

THEOREM 12.10   A Wronski matrix of solutions of Eq. (5) is a solution matrix of Eq. (6), and conversely. A Wronski matrix of n independent solutions of Eq. (5) is a fundamental matrix of Eq. (6), and conversely.

It is now easy to deduce Theorem 8.7 from Theorem 12.7. (See Exercise 5.)

Exercises

1.   Prove Theorem 12.9.

2.   Find the companion system of each of the following:

   (a) $z''' - 2z'' + 6z' = 0$

   (b) $z'' - xz = 0$

   (c) $z^{iv} - 2z'' + z = 0$

3.   Each of the following systems is the companion equation of a scalar equation. Find the scalar equation, find the general solution for it, and thereby construct a general solution for the system.

   (a) $y' = \begin{pmatrix} 0 & 1 \\ -10 & 7 \end{pmatrix} y$

   (b) $y' = \begin{pmatrix} 0 & 1 & 0 \\ 0 & 0 & 1 \\ -2 & -1 & -2 \end{pmatrix} y$

   (c) $y' = \begin{pmatrix} 0 & 1 \\ -\lambda_1\lambda_2 & \lambda_1 + \lambda_2 \end{pmatrix} y$

4.   Prove Theorem 8.5 from Theorem 12.4, and Theorem 9.2 from Theorem 12.5.

5.   Explain how Theorem 8.7 can be deduced from Theorem 12.7.

12.12   Infinite series of matrices

The previous discussion has established a satisfactory theory for first order normal linear systems, and we are now ready to tackle the practical problem of constructing a fundamental matrix. As might be expected from Chapter 9, this is easiest when all the coefficients in the system are constants. Accordingly, we consider the system

$$y' = Ay \tag{1}$$

where A is a given $n \times n$ matrix of scalars.

If Eq. (1) were a scalar equation, its general solution would be $e^{Ax}c$, where c is a constant. It turns out that this is also the solution when Eq. (1) is a system if we define e^{Ax} in an appropriate way. In the rest of this section we shall therefore examine the problem of defining e^A and e^{Ax} when A is a matrix. The reader may wish to read Appendix C for a review on infinite sequences and series before proceeding.

DEFINITION 12.15   A *sequence* of matrices is a matrix-valued function whose domain is the positive integers. Sequences can be denoted $\{A_1, A_2, \dots\}$ or briefly $\{A_m\}$. Usually all the matrices $A_1, A_2, \dots$ have the same dimension, that is, all are $n \times n$ for the same n.

DEFINITION 12.16   Let $\{A_m\}$ be a sequence of $n \times n$ matrices and let the ijth element of A_m be $a_{ij}^{(m)}$. The sequence $\{A_m\}$ is said to *converge* or to be *convergent* if the sequence of scalars $\{a_{ij}^{(m)}\}$ converges as $m \to \infty$ for each pair i, j. In this case we let $\lim_{m \to \infty} a_{ij}^{(m)} = a_{ij}$, define $A = (a_{ij})$, and write

$$\lim_{m \to \infty} A_m = A \tag{2}$$

The matrix A is called the *limit* of the sequence $\{A_m\}$. A sequence $\{A_m\}$ is called *divergent* if it is not convergent.

Example 1   Let

$$A = \begin{pmatrix} \dfrac{1}{m} & \dfrac{m}{m+1} \\ 2 & \sin\dfrac{1}{m} \end{pmatrix}$$

Then

$$A = \lim_{m \to \infty} A_m = \begin{pmatrix} 0 & 1 \\ 2 & 0 \end{pmatrix}$$

On the other hand, the sequence $\{B_m\}$ where $B_m = mA_m$ does not converge since $m^2/(m+1)$ does not have a finite limit.

DEFINITION 12.17   Consider a sequence $\{A_1, A_2, \dots\}$ and the associated *infinite series*

$$\sum_{k=1}^{\infty} A_k = A_1 + A_2 + \cdots \tag{3}$$

Let $S_m = A_1 + A_2 + \cdots + A_m$. Then $\{S_m\}$ is called the sequence of mth *partial sums*. If the sequence $\{S_m\}$ converges with limit S, we say that the infinite series (3) *converges*, call S the *sum* of the series, and write

$$S = \sum_{k=1}^{\infty} A_k \tag{4}$$

If the sequence $\{S_m\}$ diverges, we say that the infinite series *diverges*.

It is easy to show (Exercise 1) that the series (3) converges if and only if the series $\sum_{k=1}^{\infty} a_{ij}^{(k)}$ all converge $(i, j = 1, \ldots, n)$.

Example 2   Let

$$A_k = \begin{pmatrix} \dfrac{1}{k!} & \dfrac{1}{k+1} & -\dfrac{1}{k} \\ 0 & 2^{-k} \end{pmatrix} \qquad (k = 1, 2, \ldots)$$

Then

$$S_m = \begin{pmatrix} \dfrac{1}{1!} + \dfrac{1}{2!} + \cdots + \dfrac{1}{m!} & \dfrac{1}{m+1} - 1 \\ 0 & \dfrac{1}{2} + \dfrac{1}{4} + \cdots + \dfrac{1}{2^m} \end{pmatrix}$$

and clearly

$$\lim_{m \to \infty} S_m = \begin{pmatrix} e - 1 & -1 \\ 0 & 1 \end{pmatrix}$$

We are now able to define e^A, where A is a matrix, imitating the familiar series definition in the scalar case.

DEFINITION 12.18   Let A be any $n \times n$ matrix with elements in $\mathbf{C}^1$. Then we define e^A by the equation

$$e^A = I + \frac{A}{1!} + \frac{A^2}{2!} + \cdots \tag{5}$$

Since A is a square matrix, each of the powers A^k exists. Multiplication by the scalar $1/(k!)$ is then permissible. Therefore Eq. (5) makes sense for every A for which the infinite series converges. The next theorem shows that this is true for every A.

THEOREM 12.11   For every square matrix A, with real or complex elements, the exponential series $I + A/1! + A^2/2! + \cdots$ converges.

PROOF: Let $A = (a_{ij})$ and let

$$c = \max_{ij} |a_{ij}|$$

Thus $|a_{ij}| \le c$ for all i, j. Let $B = A^2 = (b_{ij})$. By the rule for matrix multiplication,

$$|b_{ij}| = \left| \sum_{k=1}^{n} a_{ik} a_{kj} \right| \le \sum_{k=1}^{n} |a_{ik}||a_{kj}| \le \sum_{k=1}^{n} c^2 = nc^2$$

This shows that every element in A^2 is bounded by nc^2. By induction it can be shown that every element in A^k is bounded by $n^{k-1}c^k$ ($k = 1, 2, 3, \ldots$). Let S_m be the mth partial sum

$$S_m = I + \frac{A}{1!} + \cdots + \frac{A^m}{m!}$$

The ijth element of S_m is a sum $\sum_{k=0}^{m} \alpha_{ij}^{(k)}$, where $\alpha_{ij}^{(k)}$ is the ijth element in $A^k/(k!)$. We see that

$$\sum_{k=0}^{m} |\alpha_{ij}^{(k)}| \le 1 + \frac{c}{1!} + \frac{nc^2}{2!} + \cdots + \frac{n^{m-1}c^m}{m!}$$

$$\le 1 + \frac{nc}{1!} + \frac{(nc)^2}{2!} + \cdots + \frac{(nc)^m}{m!}$$

Since the expression on the right is a partial sum of the scalar series for e^{nc}, which is convergent for any n and c, it follows from the comparison test that the infinite series $\sum_{k=0}^{\infty} \alpha_{ij}^{(k)}$ converges. Since this is true for every i and j, S_m converges to a limit, which is to say that the series for e^A converges.   ∎

Although the exponential e^A exists for every matrix A, not all the familiar properties of the exponential are preserved. The next theorem indicates some which are.

THEOREM 12.12   If O denotes the zero $n \times n$ matrix and I the identity $n \times n$ matrix,

$$e^O = I \tag{6}$$

Let A be any $n \times n$ matrix and let s and t be scalars. Then

$$e^{sA + tA} = e^{(s+t)A} = e^{sA} e^{tA} \tag{7}$$

The matrix e^A is always invertible and

$$(e^A)^{-1} = e^{-A} \tag{8}$$

For any integer k,

$$(e^A)^k = e^{kA} \tag{9}$$

PROOF: Equation (6) is obvious from the definition of e^A. Equation (7) is obtained as follows.

Since $sA + tA = (s + t)A$, we have

$$e^{sA+tA} = e^{(s+t)A} = \sum_{k=0}^{\infty} \frac{(s + t)^k A^k}{k!}$$

On the other hand,

$$e^{sA}e^{tA} = \left(\sum_{i=0}^{\infty} \frac{s^i A^i}{i!}\right)\left(\sum_{j=0}^{\infty} \frac{t^j A^j}{j!}\right) = \sum_{k=0}^{\infty} \sum_{i=0}^{k} \frac{s^i}{i!} \frac{t^{k-i}}{(k - i)!} A^k$$

$$= \sum_{k=0}^{\infty} \frac{A^k}{k!} \left(\sum_{i=0}^{k} \frac{k!}{i! (k - i)!} s^i t^{k-i}\right)$$

$$= \sum_{k=0}^{\infty} (s + t)^k \frac{A^k}{k!} = e^{sA+tA}$$

In the above we have used the fact that absolutely convergent series may be rearranged in arbitrary fashion.[3]

Equations (8) and (9) can be obtained directly from Eq. (7). (See Exercise 2.)   ∎

Example 3   Let

$$A = \begin{pmatrix} 1 & 0 \\ 0 & 3 \end{pmatrix}$$

Then

$$A^k = \begin{pmatrix} 1 & 0 \\ 0 & 3^k \end{pmatrix}, \qquad k = 1, 2, \ldots$$

Hence

$$e^A = \begin{pmatrix} 1 & 0 \\ 0 & 1 \end{pmatrix} + \frac{1}{1!}\begin{pmatrix} 1 & 0 \\ 0 & 3 \end{pmatrix} + \frac{1}{2!}\begin{pmatrix} 1 & 0 \\ 0 & 3^2 \end{pmatrix} + \cdots$$

$$= \begin{pmatrix} 1 + \dfrac{1}{1!} + \dfrac{1}{2!} + \cdots & 0 \\ 0 & 1 + \dfrac{3}{1!} + \dfrac{3^2}{2!} + \cdots \end{pmatrix}$$

$$= \begin{pmatrix} e & 0 \\ 0 & e^3 \end{pmatrix}$$

---

[3] This is proved in most advanced calculus textbooks. For example, see Angus E. Taylor and W. Robert Mann, *Advanced Calculus*, 2nd ed., Waltham, Mass., Xerox College Publishing, 1972.

Although $e^{sA+tA} = e^{sA}e^{tA}$, it is not generally true that $e^{A+B} = e^{A}e^{B}$ for arbitrary $n \times n$ matrices A and B. In fact, it can be shown that $e^{A+B} = e^{A}e^{B}$ if and only if $AB = BA$; that is, the matrices A and B commute.

The calculation of e^{A} is in general quite difficult. We will discuss this further in subsequent sections.

Exercises

1. Prove that the series (3) converges if and only if the series $\sum_{k=1}^{\infty} a_{ij}^{(k)}$ converges for $i, j = 1, \ldots, n$.

2. Derive Eqs. (8) and (9) from Eq. (7).

3. Prove by induction that if $|a_{ij}| \le c$ for $i, j = 1, \ldots, n$, then every element in A^{k} is bounded by $n^{k-1}c^{k}$.

4. Let $\{A_m\}$ and $\{B_m\}$ be convergent sequences of $n \times n$ matrices. Prove that if C is any $n \times n$ matrix and b is any scalar, the sequences $\{bA_m\}$, $\{CA_m\}$, $\{A_m C\}$, $\{A_m + B_m\}$ all converge. What are the limits?

5. Let

$$D = \begin{pmatrix} d_{11} & 0 & \cdots & 0 \\ 0 & d_{22} & \cdots & 0 \\ & & \cdots & \\ 0 & 0 & \cdots & d_{nn} \end{pmatrix}$$

be a diagonal matrix. Prove that e^{D} is a diagonal matrix with entries $e^{d_{11}}, e^{d_{22}}, \ldots, e^{d_{nn}}$ on the diagonal.

6. For what matrices A do the series

$$\sin A = A - \frac{A^3}{3!} + \frac{A^5}{5!} - \cdots$$

$$\cos A = I - \frac{A^2}{2!} + \frac{A^4}{4!} - \cdots$$

converge?

7. Prove $e^{A+B} = e^{A}e^{B}$ if and only if $AB = BA$.

12.13 Matrix functions of a real variable, e^{Ax}

The considerations of Section 12.2 can easily be generalized to matrix-valued functions of a real variable.

DEFINITION 12.19   A function f with domain in $\mathbf{R}^1$ and range in the space of all $n \times n$ matrices is called a *matrix function* of a scalar variable. For each x in the domain of f, $f(x)$ is a matrix and can be written $f(x) = (f_{ij}(x))$ or

$$f(x) = \begin{pmatrix} f_{11}(x) & f_{12}(x) & \cdots & f_{1n}(x) \\ f_{21}(x) & f_{22}(x) & \cdots & f_{2n}(x) \\ & & \cdots & \\ f_{n1}(x) & f_{n2}(x) & \cdots & f_{nn}(x) \end{pmatrix}$$

The range of each function f_{ij} is a subset of $\mathbf{C}^1$. A matrix function f is said to be *continuous* at x_0 if each function $f_{ij}(x)$ is continuous at x_0. It is said to be *differentiable* at x_0 if each $f_{ij}(x)$ is differentiable at x_0, and its derivative is $f'(x_0) = (f_{ij}'(x_0))$. It is integrable over $a < x < b$ if each $f_{ij}(x)$ is integrable, and

$$\int_a^b f(x)\,dx = \begin{pmatrix} \int_a^b f_{11}(x)\,dx & \cdots & \int_a^b f_{1n}(x)\,dx \\ & \cdots & \\ \int_a^b f_{n1}(x)\,dx & \cdots & \int_a^b f_{nn}(x)\,dx \end{pmatrix} \qquad (2)$$

Our primary interest here is in the particular matrix function e^{Ax}. Since Ax is an $n \times n$ matrix, for each fixed x, and $(Ax)^k = x^k A^k$, it follows from Definition 12.18 that

$$e^{Ax} = I + \frac{x}{1!} A + \frac{x^2}{2!} A^2 + \cdots \qquad (3)$$

If A is a fixed $n \times n$ matrix, this series converges for all x. Hence the domain of the function e^{Ax} is all of $\mathbf{R}^1$.

THEOREM 12.13   If A is a constant $n \times n$ matrix, the function $\Phi(x) = e^{Ax}$ is differentiable for all x and satisfies

$$\Phi'(x) = A\Phi(x), \qquad \Phi(0) = I$$

Thus $\Phi(x)$ is a fundamental matrix for the system $y' = Ay$.

PROOF: In the proof of Theorem 12.11 it was shown that the ijth component of the matrix $\Phi(x)$ is

$$\phi_{ij}(x) = \alpha_{ij}^{(0)} + \alpha_{ij}^{(1)} \frac{x}{1!} + \alpha_{ij}^{(2)} \frac{x^2}{2!} + \cdots \qquad (5)$$

where $\alpha_{ij}^{(k)}$ is the ijth element of A^k. Since the series for e^{Ax} converges for all x, the series in (5) converges for all x. This is a (scalar) power series and can be differentiated term-by-term to yield

$$\phi_{ij}'(x) = \alpha_{ij}^{(1)} + \alpha_{ij}^{(2)} \frac{x}{1!} + \alpha_{ij}^{(3)} \frac{x^2}{2!} + \cdots \qquad (6)$$

Hence $\Phi(x)$ is differentiable and $\Phi'(x)$ is the matrix with ijth element given by Eq. (6). This shows that

$$\Phi'(x) = A + A^2\frac{x}{1!} + A^3\frac{x^2}{2!} + \cdots = A\left(I + A\frac{x}{1!} + A^2\frac{x^2}{2!} + \cdots\right)$$

In other words, the matrix power series in (3) can be differentiated termwise, and $\Phi'(x) = A\Phi(x)$.

Finally, let $\phi_1, \phi_2, \ldots, \phi_n$ be the n columns of Φ. Since $\Phi' = A\Phi$ it is readily seen that $\phi_1' = A\phi_1, \ldots, \phi_n' = A\phi_n$. Thus Φ is a solution matrix for the system $y' = Ay$. Since $\det \Phi(0) = \det I = 1$, the columns of Φ are linearly independent and Φ is a fundamental matrix. ∎

According to Theorem 12.13, the matrix function e^{Ax} is a fundamental matrix. As shown in Section 12.9, every solution of $y' = Ay$ has the form

$$c_1\phi_1(x) + c_2\phi_2(x) + \cdots + c_n\phi_n(x) = \Phi(x)c = e^{Ax}c$$

where $\phi_1, \phi_2, \ldots, \phi_n$ are the columns of $\Phi = e^{Ax}$ and $c_1, c_2, \ldots, c_n$ are arbitrary scalars. In short, the matrix e^{Ax} enables us to write a general solution of the system.

Example 1   Consider the system

$$\begin{aligned} y_1' &= y_1 \\ y_2' &= 3y_2 \end{aligned} \qquad (7)$$

Here

$$A = \begin{pmatrix} 1 & 0 \\ 0 & 3 \end{pmatrix} \qquad A^k = \begin{pmatrix} 1 & 0 \\ 0 & 3^k \end{pmatrix}$$

Hence

$$e^{Ax} = I + \begin{pmatrix} 1 & 0 \\ 0 & 3 \end{pmatrix}x + \begin{pmatrix} 1 & 0 \\ 0 & 3^2 \end{pmatrix}\frac{x^2}{2!} + \cdots = \begin{pmatrix} e^x & 0 \\ 0 & e^{3x} \end{pmatrix}$$

A general solution is

$$y = e^{Ax}c = \begin{pmatrix} e^x & 0 \\ 0 & e^{3x} \end{pmatrix}\begin{pmatrix} c_1 \\ c_2 \end{pmatrix} = \begin{pmatrix} c_1 e^x \\ c_2 e^{3x} \end{pmatrix} = c_1\begin{pmatrix} e^x \\ 0 \end{pmatrix} + c_2\begin{pmatrix} 0 \\ e^{3x} \end{pmatrix}$$

Of course, for this simple example, the solution $y_1 = c_1 e^x$, $y_2 = c_2 e^{3x}$ can be obtained directly from Eq. (7) without using any matrix theory.

In the next section, we will take up the problem of constructing e^{Ax} for an arbitrary matrix A.

Exercises

1. A series of the form

$$S(x) = A_0 + A_1(x - x_0) + A_2(x - x_0)^2 + \cdots$$

where A_0, A_1, and so on, are $n \times n$ matrices, can be called a matrix power series. Using Theorem C.1 of Appendix C, prove that if such a series converges at x', then it converges absolutely for all x such that $|x - x_0| < |x' - x_0|$. Here absolute convergence of $S(x)$ means that each of the component scalar series converges absolutely.

2.   Prove that there is a real number ρ such that $S(x)$ converges for all x such that $|x - x_0| < \rho$, and diverges for all x such that $|x - x_0| > \rho$.

3.   Extend Theorem C.2(1) of Appendix C to matrix power series.

4.   Discuss multiplication of matrix power series. See Theorem C.2(2) of Appendix C.

5.   Extend Theorem C.2(4) to matrix power series.

6.   Extend Theorem C.3 and its corollary to matrix power series.

7.   Show that

$$\int_0^t e^{As} \, ds = A^{-1}(e^{At} - I) = (e^{At} - I)A^{-1}$$

if A^{-1} exists.

12.14   Calculation of e^{Ax} when A is diagonable

For a general matrix A, it may be difficult or of little use to calculate e^{Ax} from the defining formula

$$e^{Ax} = I + \frac{x}{1!} A + \frac{x^2}{2!} A^2 + \cdots \tag{1}$$

Not only will a large amount of arithmetic be required, but also it will usually be difficult to extract any information about the analytic nature of e^{Ax}.

   Example 1   Let

$$A = \begin{pmatrix} 1 & 1 \\ 0 & 2 \end{pmatrix}$$

The reader can show that

$$A^k = \begin{pmatrix} 1 & a_k \\ 0 & 2^k \end{pmatrix}$$

where the sequence $\{a_k\}$ is defined by the recurrence relation $a_k = a_{k-1} + 2^{k-1}$, $a_1 = 1$. The first few values in the sequence $\{a_k\}$ are 1, 3, 7, 15, 31, ... Hence

$$e^{Ax} = \begin{pmatrix} 1 & 0 \\ 0 & 1 \end{pmatrix} + x \begin{pmatrix} 1 & 1 \\ 0 & 2 \end{pmatrix} + \frac{x^2}{2!} \begin{pmatrix} 1 & 3 \\ 0 & 2^2 \end{pmatrix} + \cdots = \begin{pmatrix} e^x & g(x) \\ 0 & e^{2x} \end{pmatrix}$$

where

$$g(x) = x + \frac{3x^2}{2!} + \frac{7x^3}{3!} + \cdots + \frac{a_k x^k}{k!} + \cdots$$

Even in this relatively simple case, it is not easy to see what the nature of the function $g(x)$ is. With the method to be developed now, it can be shown that $g(x) = e^{2x} - e^x$. (See Exercise 2a.)

If the matrix is diagonal, the calculation of its exponential matrix is extremely simple. In fact, if

$$D = \begin{pmatrix} d_{11} & 0 & \cdots & 0 \\ 0 & d_{22} & \cdots & 0 \\ & & \cdots & \\ 0 & 0 & \cdots & d_{nn} \end{pmatrix}$$

then (see Exercise 5 in Section 12.12)

$$e^{Dx} = \begin{pmatrix} e^{d_{11}x} & 0 & \cdots & 0 \\ 0 & e^{d_{22}x} & \cdots & 0 \\ & & \cdots & \\ 0 & 0 & \cdots & e^{d_{nn}x} \end{pmatrix} \tag{2}$$

Moreover, if A is diagonalizable, e^{Ax} can be found by the method suggested in the following theorem.

THEOREM 12.14 Let A be diagonable over the field $\mathbf{K}$ and let $A = PDP^{-1}$ where D is a diagonal matrix. Then $e^{Ax} = Pe^{Dx}P^{-1}$.

PROOF: Since $A = PDP^{-1}$, we get $A^k = PD^kP^{-1}$, $k = 1, 2, \ldots$. Therefore

$$e^{Ax} = I + A\frac{x}{1!} + A^2\frac{x^2}{2!} + \cdots$$

$$= PIP^{-1} + P\frac{(Dx)}{1!}P^{-1} + P\left(\frac{D^2x^2}{2!}\right)P^{-1} + \cdots$$

$$= P\left(I + \frac{Dx}{1!} + \frac{D^2x^2}{2!} + \cdots\right)P^{-1} = Pe^{Dx}P^{-1} \qquad \blacksquare$$

Example 2   The matrix

$$A = \begin{pmatrix} 1 & 1 \\ 4 & 1 \end{pmatrix}$$

has eigenvalues $-1, 3$. A matrix P which has the corresponding eigenvectors as columns is

$$P = \begin{pmatrix} 1 & 1 \\ -2 & 2 \end{pmatrix}$$

It can be verified that

$$P^{-1}AP = \begin{pmatrix} \frac{1}{2} & -\frac{1}{4} \\ \frac{1}{2} & \frac{1}{4} \end{pmatrix}\begin{pmatrix} 1 & 1 \\ 4 & 1 \end{pmatrix}\begin{pmatrix} 1 & 1 \\ -2 & 2 \end{pmatrix} = \begin{pmatrix} -1 & 0 \\ 0 & 3 \end{pmatrix}$$

Therefore

$$e^{Ax} = Pe^{Dx}P^{-1} = \begin{pmatrix} 1 & 1 \\ -2 & 2 \end{pmatrix}\begin{pmatrix} e^{-x} & 0 \\ 0 & e^{3x} \end{pmatrix}\begin{pmatrix} \frac{1}{2} & -\frac{1}{4} \\ \frac{1}{2} & \frac{1}{4} \end{pmatrix}$$

$$= \begin{pmatrix} \frac{1}{2}e^{-x} + \frac{1}{2}e^{3x} & -\frac{1}{4}e^{-x} + \frac{1}{4}e^{3x} \\ -e^{-x} + e^{3x} & \frac{1}{2}e^{-x} + \frac{1}{2}e^{3x} \end{pmatrix}$$

This is a fundamental matrix for the system

$$\begin{aligned} y_1' &= y_1 + y_2 \\ y_2' &= 4y_1 + y_2 \end{aligned}$$

A general solution can be obtained as follows.

$$y = e^{Ax}\begin{pmatrix} c_1 \\ c_2 \end{pmatrix} = \begin{pmatrix} c_1(\frac{1}{2}e^{-x} + \frac{1}{2}e^{3x}) + c_2(-\frac{1}{4}e^{-x} + \frac{1}{4}e^{3x}) \\ c_1(-e^{-x} + e^{3x}) + c_2(\frac{1}{2}e^{-x} + \frac{1}{2}e^{3x}) \end{pmatrix}$$

This rather cumbersome form can be simplified by making the substitution

$$k_1 = \tfrac{1}{2}c_1 - \tfrac{1}{4}c_2, \quad k_2 = \tfrac{1}{2}c_1 + \tfrac{1}{4}c_2$$

Then

$$y = k_1 \begin{pmatrix} 1 \\ -2 \end{pmatrix} e^{-x} + k_2 \begin{pmatrix} 1 \\ 2 \end{pmatrix} e^{3x} \tag{3}$$

where k_1 and k_2 are arbitrary constants.

Observe that the vector multiplying e^{-x} in Eq. (3) is an eigenvector associated with the eigenvalue -1 and the vector multiplying e^{3x} is an eigenvector associated with the eigenvalue 3. The following theorem shows that this situation holds generally.

THEOREM 12.15   Let A be diagonable over the field $\mathbf{K}$ and let $\{v_1, v_2, \ldots, v_n\}$ be a linearly independent set of n eigenvectors. Let $\lambda_1, \lambda_2, \ldots, \lambda_n$ be the associated eigenvalues (not necessarily distinct). Then a fundamental matrix for the system $y' = Ay$ is the matrix with the vector $e^{\lambda_k x}v_k$ in its kth column $(k = 1, 2, \ldots, n)$. Equivalently, the set

$$\{v_1 e^{\lambda_1 x}, v_2 e^{\lambda_2 x}, \ldots, v_n e^{\lambda_n x}\} \tag{4}$$

is a basis for the solution space of the system.

PROOF: Let P be the matrix with v_k in the kth column ($k = 1, 2, \ldots, n$). As we know, $P^{-1}AP = D$, where D is a diagonal matrix with diagonal entries $\lambda_1, \lambda_2, \ldots, \lambda_n$. By Theorem 12.14, $e^{Ax} = Pe^{Dx}P^{-1}$ is a fundamental matrix. Moreover, $e^{Ax}P$ is a fundamental matrix since P is nonsingular (see Exercise 6, Section 12.8); that is, a fundamental matrix is

$$e^{Ax}P = Pe^{Dx} = (v_1 \; v_2 \ldots \; v_n) \begin{pmatrix} e^{\lambda_1 x} & & 0 \\ & \cdots & \\ 0 & & e^{\lambda_n x} \end{pmatrix}$$

An easy calculation shows that Pe^{Dx} has $v_k e^{\lambda_k x}$ as its kth column ($k = 1, 2, \ldots, n$). This completes the proof. ∎

Theorem 12.15 provides a simple procedure for constructing a basis for the solutions of $y' = Ay$ as follows: first, compute the eigenvalues and eigenvectors of A; then (4) provides a basis for the solution space. Thus it is not necessary to write P or to compute e^{Ax} explicitly. Indeed, the *solution of the differential system is accomplished by the wholly algebraic procedure of finding the eigenvalues and eigenvectors.* In Example 2, for instance, once the eigenvalues $\lambda = -1, 3$ and corresponding eigenvectors $v_1 = \begin{pmatrix} 1 \\ -2 \end{pmatrix}$, $v_2 = \begin{pmatrix} 1 \\ 2 \end{pmatrix}$ have been found, the general solution in Eq. (3) can be written at once.

There is a slightly different way of obtaining the result in Theorem 12.15. We know that a matrix P for which $P^{-1}AP = D$ represents a change of coordinates. This suggests that in the system $y' = Ay$ we make the change of variable $y = Pz$. Then $y' = Pz'$ and the system takes the form $Pz' = APz$ or

$$z' = P^{-1}APz = Dz \tag{5}$$

This system is

$$z_k' = \lambda_k z_k \qquad (k = 1, 2, \ldots, n) \tag{6}$$

Each of these equations involves only one z_k and can easily be solved to yield $z_k = c_k e^{\lambda_k x}$, where c_k is an arbitrary constant. Therefore $e^{\lambda_1 x}e_1$, $e^{\lambda_2 x}e_2, \ldots,$ $e^{\lambda_n x}e_n$ are linearly independent solutions of $z' = Dz$. The corresponding solutions of $y' = Ay$ are $(Pe_1)e^{\lambda_1 x}, \ldots, (Pe_n)e^{\lambda_n x}$. It is easy to see that Pe_k is the kth column of P; that is, the kth eigenvector v_k. Moreover, these n solutions are independent since their Wronskian is

$$\det (e^{\lambda_1 x}v_1 \;\; e^{\lambda_2 x}v_2 \;\; \cdots \;\; e^{\lambda_n x}v_n) = e^{(\lambda_1 + \lambda_2 + \cdots \lambda_n)x} \det P \neq 0$$

Thus the set in Eq. (4) is a basis for the solution space.

Exercises

1. Find a fundamental matrix for the diagonal system $y' = Dy$ if:

    (a)
    $$D = \begin{pmatrix} 1 & 0 & 0 \\ 0 & 2 & 0 \\ 0 & 0 & 3 \end{pmatrix}$$

    (b)
    $$D = \begin{pmatrix} 1 + \sqrt{5} & 0 \\ 0 & 1 - \sqrt{5} \end{pmatrix}$$

2. For each system $y' = Ay$, compute n independent eigenvectors, form the matrices D and P, and find the fundamental matrix e^{Ax} for the system $y' = Ay$.

    (a)
    $$A = \begin{pmatrix} 1 & 1 \\ 0 & 2 \end{pmatrix}$$

    (b)
    $$\begin{pmatrix} 0 & 1 & 0 \\ 0 & 0 & 1 \\ 0 & 1 & 0 \end{pmatrix}$$

    (c)
    $$\begin{aligned} y_1' &= y_3 \\ y_2' &= 4y_3 \\ y_3' &= y_2 \end{aligned}$$

    (d)
    $$A = \begin{pmatrix} -1 & -2 & 2 \\ -2 & -1 & 2 \\ -2 & -2 & 3 \end{pmatrix}$$

3. For each system in Exercise 2 write a general solution using Eq. (4) without explicitly calculating e^{Ax}.

4. Solve the initial value problems:

    (a) The system in Exercise 2b with
    $$y(0) = \begin{pmatrix} 1 \\ 0 \\ 2 \end{pmatrix}$$

    (b) The system in Exercise 2c with $y_1(0) = 3$, $y_2(0) = 4$, $y_3(0) = 1$.

    (c) The system in Exercise 2d with $y_1(0) = 4$, $y_2(0) = 7$, $y_3(0) = 6$.

5. Find a general solution of the system $y' = Ay$, where A is as follows, by solving the three scalar equations in turn, starting with the last.

    (a)
    $$A = \begin{pmatrix} 0 & 1 & 0 \\ 0 & 0 & 1 \\ 0 & 0 & 0 \end{pmatrix}$$

    (b)
    $$A = \begin{pmatrix} 2 & 1 & 0 \\ 0 & 2 & 1 \\ 0 & 0 & 2 \end{pmatrix}$$

6. (a) Convert the linear scalar equation
    $$Lz = z''' + a_2 z'' + a_1 z' + a_0 z = 0$$
    to a system of the form $y' = Ay$. Prove that the characteristic polynomial of A is $-(\lambda^3 + a_2\lambda^2 + a_1\lambda + a_0)$ and therefore that the eigenvalues of A are the characteristic roots of L.

(b) Prove that the dimension of each eigenspace for A is 1, and exhibit an eigenvector.

(c) Assume that there are 3 distinct characteristic roots, $\lambda_1, \lambda_2, \lambda_3$. Exhibit a general solution of the system $y' = Ay$. From this deduce a general solution of $Lz = 0$.

(d) Show that if any characteristic root is multiple, the companion matrix A is not similar to a diagonal matrix.

12.15   Complex eigenvalues

Example 1   Consider the system

$$y_1' = y_2$$
$$y_2' = -\omega^2 y_1$$

where ω is a real constant. This system has matrix

$$A = \begin{pmatrix} 0 & 1 \\ -\omega^2 & 0 \end{pmatrix}$$

The eigenvalues are $\pm i\omega$. To determine an eigenvector associated with $i\omega$ we must solve the system

$$\begin{pmatrix} -i\omega & 1 \\ -\omega^2 & -i\omega \end{pmatrix} \begin{pmatrix} a \\ b \end{pmatrix} = \begin{pmatrix} 0 \\ 0 \end{pmatrix}$$

A solution is $a = 1$, $b = i\omega$, and thus all eigenvectors are of the form $c \begin{pmatrix} 1 \\ i\omega \end{pmatrix}$.

Similarly we find that eigenvectors associated with the eigenvalue $-i\omega$ must be of the form $c \begin{pmatrix} 1 \\ -i\omega \end{pmatrix}$. By Theorem 12.15 a basis for the complex solution space is

$$\left\{ e^{i\omega x} \begin{pmatrix} 1 \\ i\omega \end{pmatrix}, \ e^{-i\omega x} \begin{pmatrix} 1 \\ -i\omega \end{pmatrix} \right\}$$

and a fundamental matrix is

$$P e^{Dx} = \begin{pmatrix} 1 & 1 \\ i\omega & -i\omega \end{pmatrix} \begin{pmatrix} e^{i\omega x} & 0 \\ 0 & e^{-i\omega x} \end{pmatrix}$$

If real-valued solutions are desired, we can deduce them as follows. We have two independent solutions

$$y = e^{i\omega x} \begin{pmatrix} 1 \\ i\omega \end{pmatrix} \qquad \bar{y} = e^{-i\omega x} \begin{pmatrix} 1 \\ -i\omega \end{pmatrix}$$

which are complex conjugates. Let $u = (y + \bar{y})/2$, $v = (y - \bar{y})/2i$. Each of these vectors is a solution, since it is a linear combination of solutions. Moreover we have

$$u = \begin{pmatrix} \cos \omega x \\ -\omega \sin \omega x \end{pmatrix} \qquad v = \begin{pmatrix} \sin \omega x \\ \omega \cos \omega x \end{pmatrix}$$

and these are real and linearly independent. Thus a basis for the space of real solutions is

$$\left\{ \begin{pmatrix} \cos \omega x \\ -\omega \sin \omega x \end{pmatrix}, \begin{pmatrix} \sin \omega x \\ \omega \cos \omega x \end{pmatrix} \right\}$$

This technique can be applied generally, as the following theorem shows.

THEOREM 12.16 Let A be a real $n \times n$ matrix. Let a basis for the complex solution space of $y' = Ay$ be

$$\{z_1, \ldots, z_p, w_1, \bar{w}_1, \ldots, w_q, \bar{w}_q\} \tag{1}$$

where $z_1, \ldots, z_p$ are real solutions, $w_1, \bar{w}_1, \ldots, w_q, \bar{w}_q$ are complex solutions, and $\bar{w}_j$ is the conjugate of w_j. Let $w_j(x) = u_j(x) + iv_j(x)$, $\bar{w}_j(x) = u_j(x) - iv_j(x)$, $j = 1, \ldots, q$. Then

$$\{z_1, \ldots, z_p, u_1, v_1, \ldots, u_q, v_q\} \tag{2}$$

is a basis for the real solution space of $y' = Ay$.

PROOF: Since $u_j = (w_j + \bar{w}_j)/2$, $v_j = (w_j - \bar{w}_j)/2i$, the functions $u_j(x)$ and $v_j(x)$ are real solutions. To show that Eq. (2) is a basis, linear independence must be proved. This can be done much as in the proof of Theorem 9.4, and is left as an exercise.   ∎

Exercises

1. First find a complex fundamental matrix and then a real fundamental matrix for each system $y' = Ay$.

    (a) $A = \begin{pmatrix} 1 & 1 \\ -5 & -1 \end{pmatrix}$

    (b) $A = \begin{pmatrix} 0 & 1 & 0 \\ 0 & 0 & 1 \\ 1 & -1 & 1 \end{pmatrix}$

2. Suppose Φ is a fundamental matrix of $y' = Ay$, where A is real. Let Ψ be the matrix of real parts of the elements of Φ. Show that Ψ is a solution matrix. Give an example to show that Ψ need not be a fundamental matrix.

3. Complete the proof of Theorem 12.16.

4. Let A be a real constant 2×2 matrix with complex conjugate eigenvalues $\alpha \pm i\beta$ for $\beta \neq 0$. Let $u + iv$ be an eigenvector corresponding to $\alpha + i\beta$.

(a) Prove that $u \neq 0$ and that $v \neq 0$; that is, the eigenvector cannot be purely real or purely imaginary.

(b) Let

$$u = \begin{pmatrix} u_1 \\ u_2 \end{pmatrix} \qquad v = \begin{pmatrix} v_1 \\ v_2 \end{pmatrix}$$

and define

$$P = \begin{pmatrix} u_1 & v_1 \\ u_2 & v_2 \end{pmatrix}$$

Show that P is nonsingular by showing that its columns are linearly independent.

(c) Show that

$$P^{-1}AP = \begin{pmatrix} \alpha & \beta \\ -\beta & \alpha \end{pmatrix}$$

Thus there is a real linear transformation under which A is similar to a real matrix of this special form.

(d) Show that

$$e^{\alpha x} \begin{pmatrix} \cos \beta x & \sin \beta x \\ -\sin \beta x & \cos \beta x \end{pmatrix}$$

is a fundamental matrix for $z' = Bz$, where

$$B = \begin{pmatrix} \alpha & \beta \\ -\beta & \alpha \end{pmatrix}$$

(e) Using the result in (d), write a real general solution of $y' = Ay$.

12.16   Calculation of e^{Ax} when A is not diagonable

As seen in the last two sections, the calculation of e^{Ax} or the solution of $y' = Ay$ is simple when A is diagonable and its eigenvalues and eigenvectors are known. In this section we shall explain a procedure for the calculation in the general, not necessarily diagonable, case. This procedure is based on the decomposition of n-dimensional space into the direct sum of generalized eigenspaces, as explained in Section 11.7.

We begin by recalling the notation of Section 11.7. Let $\mathbf{C}^n$ be complex n-dimensional space, and let A be an $n \times n$ matrix of real or complex scalars. Let $\lambda_1, \ldots, \lambda_s$ be the distinct eigenvalues of A. It then follows from Section 11.7 that for each λ_i the null spaces of the operators $T - \lambda_i I$, $(T - \lambda_i I)^2, \ldots$, form an increasing chain, and that there is a positive integer p_i such that equality occurs in the chain for all higher powers. More precisely,

$$\mathbf{N}_{T-\lambda_i I} \subset \mathbf{N}_{(T-\lambda_i I)^2} \subset \cdots \subset \mathbf{N}_{(T-\lambda_i I)^{p_i}} = \mathbf{N}_{(T-\lambda_i I)^{p_i+1}} = \cdots \qquad (1)$$

The null space of $(T - \lambda_i I)^{p_i}$ is called the generalized eigenspace of λ_i, and is denoted by $\mathbf{M}_{\lambda_i}$. By Theorem 11.20

$$\mathbf{C}^n = \mathbf{M}_{\lambda_1} \oplus \mathbf{M}_{\lambda_2} \oplus \cdots \oplus \mathbf{M}_{\lambda_s} \tag{2}$$

Let y_0 be any fixed vector in $\mathbf{C}^n$. By Eq. (2),

$$y_0 = \sum_{i=1}^{s} y_0^{(i)} \quad \text{where} \quad y_0^{(i)} \in \mathbf{M}_{\lambda_i} \tag{3}$$

and this representation is possible in exactly one way. Then

$$e^{Ax} y_0 = \sum_{i=1}^{s} e^{Ax} y_0^{(i)} \tag{4}$$

and so we only need to show how to compute $e^{Ax} y_0^{(i)}$.

Dropping the superscript for the moment, for convenience of notation, we suppose that y_0 lies in a particular generalized eigenspace $\mathbf{M}_\lambda$. Then

$$e^{Ax} y_0 = e^{\lambda x} e^{(A - \lambda I)x} y_0$$

$$= e^{\lambda x} \left[I + \frac{x}{1}(A - \lambda I) y_0 + \frac{x^2}{2!}(A - \lambda I)^2 y_0 + \cdots \right] \tag{5}$$

Since $y_0 \in \mathbf{M}_\lambda$, we have $(A - \lambda I)^i y_0 = 0$ for $i \geq p$, where p is the index of λ. Therefore the above series is actually terminating. Thus

$$e^{Ax} y_0 = e^{\lambda x} \sum_{j=0}^{p-1} \frac{x^j}{j!}(A - \lambda I)^j y_0$$

We note that from $y_0 \in \mathbf{M}_\lambda$ follows $e^{Ax} y_0 \in \mathbf{M}_\lambda$, since the null space of any power of $(A - \lambda I)$ is invariant under $(A - \lambda I)$, by Lemma 11.5. Thus, any solution curve of $y' = Ay$, which starts at a point in a generalized eigenspace $\mathbf{M}_\lambda$, must remain in $\mathbf{M}_\lambda$ for all x. This can also be seen in the following more geometrical way. If $y_0 \in \mathbf{M}_\lambda$, then $Ay_0 \in \mathbf{M}_\lambda$, since $\mathbf{M}_\lambda$ is invariant under any polynomial in A. Therefore the direction field at y_0 points in a direction which lies in the subspace $\mathbf{M}_\lambda$, and so the solution curve through y_0 must remain in $\mathbf{M}_\lambda$. The following theorem summarizes our results.

THEOREM 12.17   Let $\lambda_1, \ldots, \lambda_s$ be the distinct eigenvalues of the $n \times n$ matrix A, let $M_{\lambda_1}, \ldots, M_{\lambda_s}$ be their respective generalized eigenspaces, and let $p_1, \ldots, p_s$ be their respective indices. Then the solution of the initial value problem

$$y' = Ay \qquad y(0) = y_0 \tag{6}$$

is given by

$$y(x) = \sum_{i=1}^{s} e^{\lambda_i x} \sum_{j=0}^{p_i - 1} \frac{x^j}{j!}(A - \lambda_i I)^j y_0^{(i)} \tag{7}$$

where the $y_0^{(i)}$ are vectors in $\mathbf{M}_{\lambda_i}$, determined by the decomposition of y_0 given in Eq. (3).

The usefulness of the formula (7) arises from the fact that we need calculate only a finite number of powers of matrices, rather than the infinitely many required if we apply Eq. (5) directly.

A way in which one may apply Theorem 12.17 to a specific system is as follows. Let λ be one of the eigenvalues. Compute successive powers $(A - \lambda I)$, $(A - \lambda I)^2, \ldots$, and their null spaces, and stop when two successive spaces are the same. This determines the index p, and all the powers needed in Eq. (7). In the process a basis for $\mathbf{M}_\lambda$, the null space of $(A - \lambda I)^p$, has been found. Having done this for each eigenvalue, we have a basis for $\mathbf{C}^n$ and the initial vector y_0 can be decomposed as in Eq. (3). Finally, $(A - \lambda_i I) y_0^{(i)}$ is computed.

It is also known (see Theorem 11.21) that the dimension of $\mathbf{M}_\lambda$ is equal to the algebraic multiplicity m of λ. If this is known, we can, alternatively, say that the calculation of the successive powers $(A - \lambda I)$, $(A - \lambda I)^2, \ldots$, is to stop when the dimension of the null space equals m.

Example 1   Consider $y' = Ay$ where $A = \begin{pmatrix} 0 & 1 \\ 0 & 0 \end{pmatrix}$. The characteristic equation is $\lambda^2 = 0$, so $\lambda = 0$ is a root of algebraic multiplicity 2. We have $A - \lambda I = A$, and easily find that the null space of A (the eigenspace of $\lambda = 0$) has a basis consisting of the single vector $\begin{pmatrix} 1 \\ 0 \end{pmatrix}$. Since this has dimension 1, we go on to a higher power. We find that $(A - \lambda I)^2 = A^2$ is the zero matrix. Its null space is all of $\mathbf{C}^2$. Thus $\lambda = 0$ has index $p = 2$, and its generalized eigenspace is $\mathbf{C}^2$. Any initial vector y_0 lies in $\mathbf{M}_\lambda$, and Eq. (7) reduces in this case to

$$y(x) = \sum_{j=0}^{1} \frac{x^j}{j!} A^j y_0 = y_0 + xAy_0$$

If y_0 has components y_{01}, y_{02}, we get

$$y(x) = \begin{pmatrix} y_{01} + y_{02}x \\ y_{02} \end{pmatrix}$$

Example 2   Consider $y' = Ay$, where

$$A = \begin{pmatrix} 0 & -1 & 2 \\ 0 & 1 & 0 \\ 1 & 1 & -1 \end{pmatrix}$$

We find that the characteristic equation is $\lambda^3 - 3\lambda + 2 = (\lambda + 2)(\lambda - 1)^2 = 0$. Since $\lambda = -2$ is a simple eigenvalue, its generalized eigenspace is the same as its eigenspace, and we find that

$$\mathbf{M}_{-2} = \mathbf{N}_{A+2I} = [(1, 0, -1)]$$

For $\lambda = 1$, we find that $\mathbf{M}_1$ is two-dimensional, and a basis is given by $(0, 2, 1)$ and $(2, 0, 1)$, for example. Thus

$$\mathbf{M}_1 = \mathbf{N}_{A-I} = [(0, 2, 1), (2, 0, 1)]$$

Let y_0 be any vector in $\mathbf{C}^n$ and write

$$y_0 = \begin{pmatrix} a \\ b \\ c \end{pmatrix} = k_1 \begin{pmatrix} 1 \\ 0 \\ -1 \end{pmatrix} + k_2 \begin{pmatrix} 0 \\ 2 \\ 1 \end{pmatrix} + k_3 \begin{pmatrix} 2 \\ 0 \\ 1 \end{pmatrix}$$

Then

$$\begin{aligned} k_1 \quad\quad + 2k_3 &= a \\ 2k_2 \quad\quad &= b \\ -k_1 + k_2 + k_3 &= c \end{aligned}$$

from which we find

$$k_1 = \tfrac{1}{3}(a + b - 2c), \quad k_2 = \tfrac{1}{2}b, \quad k_3 = \tfrac{1}{6}(2a - b + 2c) \quad\quad (8)$$

The part of y_0 in $\mathbf{M}_{-2}$ is $k_1 \begin{pmatrix} 1 \\ 0 \\ -1 \end{pmatrix}$ and the corresponding part of the solution (7) is

$$e^{-2x} k_1 \begin{pmatrix} 1 \\ 0 \\ -1 \end{pmatrix} = \tfrac{1}{3}e^{-2x}(a + b - 2c) \begin{pmatrix} 1 \\ 0 \\ -1 \end{pmatrix}$$

The part of y_0 in $\mathbf{M}_{-1}$ is

$$k_2 \begin{pmatrix} 0 \\ 2 \\ 1 \end{pmatrix} + k_3 \begin{pmatrix} 2 \\ 0 \\ 1 \end{pmatrix}$$

and the corresponding part of the solution (7) is

$$e^{-x} \left[k_2 \begin{pmatrix} 0 \\ 2 \\ 1 \end{pmatrix} + k_3 \begin{pmatrix} 2 \\ 0 \\ 1 \end{pmatrix} \right] + e^{-x}x\,(A - I) \left[k_2 \begin{pmatrix} 0 \\ 2 \\ 1 \end{pmatrix} + k_3 \begin{pmatrix} 2 \\ 0 \\ 1 \end{pmatrix} \right]$$

$$= e^{-x} \left[k_2 \begin{pmatrix} 0 \\ 2 \\ 1 \end{pmatrix} + k_3 \begin{pmatrix} 2 \\ 0 \\ 1 \end{pmatrix} \right] + e^{-x}x \begin{pmatrix} 0 \\ 0 \\ 0 \end{pmatrix}$$

Consequently,

$$y(x) = k_1 e^{-2x} \begin{pmatrix} 1 \\ 0 \\ -1 \end{pmatrix} + k_2 e^{-x} \begin{pmatrix} 0 \\ 2 \\ 1 \end{pmatrix} + k_3 e^{-x} \begin{pmatrix} 2 \\ 0 \\ 1 \end{pmatrix} \quad\quad (9)$$

where k_1, k_2, k_3 are expressed in terms of the components of the initial vector y_0 by Eq. (8). Equation (9) represents a general solution, since k_1, k_2, k_3 can assume any values.

Theorem 12.17 shows us the general form of the solution of any linear system $y' = Ay$ with A constant. The solution is a sum of exponential terms $e^{\lambda_i x}$ multiplied by coefficients which are vector polynomials of x of degree at most $p_i - 1$. Another procedure for constructing the solution is based on this observation. The part of the solution containing a given exponential $e^{\lambda x}$ must have the form

$$e^{\lambda x}(v_0 + v_1 x + \cdots + v_{p-1}x^{p-1})$$

where $v_0, v_1, \ldots, v_{p-1}$ are vectors which are unknown. By substituting this expression into the system, one obtains a set of equations for the determination of these vectors.

The structure of the solutions of $y' = Ay$ can be made more explicit with the aid of the Jordan canonical form. According to Theorem 11.21, A is similar to a matrix J in Jordan normal form, say

$$J = P^{-1}AP = \begin{pmatrix} J_1 & & \\ & J_2 & \\ & & \ddots \end{pmatrix} \tag{10}$$

The form of the matrices $J_1, J_2, \ldots,$ on the diagonal is given in Theorem 11.21. We see that

$$e^{Ax} = Pe^{Jx}P^{-1} = P\begin{pmatrix} e^{J_1 x} & & \\ & e^{J_2 x} & \\ & & \ddots \end{pmatrix}P^{-1} \tag{11}$$

Therefore, to calculate the fundamental matrix e^{Ax}, it suffices to compute $e^{J_i x}$ for each Jordan block J_i. We can write $J_i = \lambda_i I + Q_i$, where

$$Q_i = \begin{pmatrix} 0 & 1 & 0 & \cdots & 0 & 0 \\ 0 & 0 & 1 & \cdots & 0 & 0 \\ & & & \cdots & & \\ 0 & 0 & 0 & \cdots & 0 & 1 \\ 0 & 0 & 0 & \cdots & 0 & 0 \end{pmatrix}$$

Q_i has 1's on the "superdiagonal," and 0's everywhere else. Now it is easy to see that

$$e^{J_i x} = e^{\lambda_i x I}e^{Q_i x} = e^{\lambda_i x}e^{Q_i x} = e^{\lambda_i x}\begin{pmatrix} 1 & x & \dfrac{x^2}{2!} & \cdots & \dfrac{x^{r-1}}{(r-1)!} \\ 0 & 1 & x & \cdots & \dfrac{x^{r-2}}{(r-2)!} \\ & & & \cdots & \\ 0 & 0 & 0 & \cdots & 1 \end{pmatrix} \tag{12}$$

where r is the order of J_i. Equations (11) and (12) give an explicit representation of the fundamental matrix e^{Ax}, assuming that the Jordan form has been computed. However, for the calculation of solutions of $y' = Ay$ it is simpler to use Eq. (7) directly.

In applications, it is sometimes the case that we do not need an explicit formula for the solutions, but only need certain qualitative properties of the solutions. As an example, suppose that the vector $y(x)$ represents the deviation of a certain physical system from an equilibrium state at time x. For example for a mass-spring system, y could measure the distance, positive or negative, from the rest position. Suppose that $y' = Ay$. Our principal interest may be in knowing whether all solutions y tend to zero as x increases. If so, the rest point is stable in the sense that all solutions tend to it, whereas in the contrary case there are some possible motions which do not approach the equilibrium. The following theorem is of use in discussing such problems.

THEOREM 12.18   (1) A necessary and sufficient condition in order that all solutions of $y' = Ay$ approach 0 as $x \to +\infty$ is that all eigenvalues of A have negative real parts. (2) A necessary and sufficient condition in order that every solution remain bounded as $x \to +\infty$ is that all eigenvalues of A have negative or zero real parts, and further that if λ is an eigenvalue with zero real part, the generalized eigenspace $\mathbf{M}_\lambda$ have dimension 1.

PROOF: If $\operatorname{Re} \lambda_i < 0$ for all i, then $e^{\lambda_i x} \to 0$ as $x \to \infty$ and it is clear from Eq. (7) that every solution tends to zero. On the other hand, one solution is always $e^{\lambda_i x} v$ where v is an eigenvector of λ_i (see Exercise 3). Hence, if $\operatorname{Re} \lambda_i \geq 0$, there is a solution which does not tend to zero as $x \to +\infty$. This proves (1).

To prove (2) we note that all terms in Eq. (7) with $\operatorname{Re} \lambda_i < 0$ tend to zero. Terms with $\operatorname{Re} \lambda_i = 0$ will certainly remain bounded if $p_i = 1$, since then no positive powers of x appear. On the other hand, suppose $p_i > 1$. It is possible to choose $y_0^{(i)}$ such that $(A - \lambda_i I)^{p_i-1} y_0^{(i)}$ is nonzero. Hence

$$x^{p_i-1}(A - \lambda_i I)^{p_i-1} y_0^{(i)}$$

is unbounded as $x \to \infty$, and therefore the solution $y(x)$ given by Equation (7) is unbounded as $x \to \infty$.   ∎

Exercises

1.   Find a general solution of each system $y' = Ay$, using Theorem 12.17.

(a) $A = \begin{pmatrix} 0 & 1 \\ -3 & 4 \end{pmatrix}$

(b) $A = \begin{pmatrix} -2 & 1 \\ -1 & -4 \end{pmatrix}$

(c) $A = \begin{pmatrix} 7 & 4 & -4 \\ 4 & -8 & -1 \\ -4 & -1 & -8 \end{pmatrix}$

(d) $A = \begin{pmatrix} 6 & 1 & 1 \\ 1 & 6 & 1 \\ 3 & 3 & 6 \end{pmatrix}$

(e)
$$A = \begin{pmatrix} 0 & 1 & 0 \\ 4 & 3 & -4 \\ 1 & 2 & -1 \end{pmatrix}$$

(f)
$$A = \begin{pmatrix} 0 & 1 & 0 \\ 0 & 0 & 1 \\ 2 & -5 & 4 \end{pmatrix}$$

(g)
$$A = \begin{pmatrix} 0 & 1 & 1 \\ 1 & 0 & 1 \\ 1 & 1 & 0 \end{pmatrix}$$

2. Find the solution of each initial value problem. In each part the matrix A is given in the corresponding part of Exercise 1.

(a)
$$y(0) = \begin{pmatrix} 1 \\ 1 \end{pmatrix}$$

(b)
$$y(0) = \begin{pmatrix} 1 \\ 2 \end{pmatrix}$$

(c)
$$y(0) = \begin{pmatrix} 1 \\ -1 \\ 0 \end{pmatrix}$$

(d)
$$y(0) = \begin{pmatrix} 0 \\ 1 \\ 1 \end{pmatrix}$$

(e)
$$y(0) = \begin{pmatrix} 1 \\ 1 \\ 1 \end{pmatrix}$$

(f)
$$y(0) = \begin{pmatrix} 2 \\ 1 \\ 0 \end{pmatrix}$$

(g)
$$y(0) = \begin{pmatrix} -1 \\ 0 \\ 1 \end{pmatrix}$$

3. Let λ be an eigenvalue of A and v an eigenvector corresponding to λ. Show that $e^{\lambda x}v$ is a solution of $y' = Ay$.

4. For each system in Exercise 1 find whether all solutions remain bounded as $x \to +\infty$, or all solutions tend to zero, using Theorem 12.18.

12.17 The variation of constants formula

In the previous sections we have for the most part considered homogeneous linear systems

$$y' = A(x)y \tag{1}$$

We now want to examine nonhomogeneous systems

$$y' = A(x)y + b(x) \tag{2}$$

From Section 12.10, we know that the totality of solutions of Eq. (2) is given by the expression $\Phi(x)c + y_p(x)$, where Φ is a fundamental matrix of Eq. (1), c is any constant vector, and y_p is a particular solution of Eq. (2). However, the problem of determining y_p still has to be considered.

We recall that for scalar linear equations, we have given two principal methods for computing a particular solution. The more general of these is the method of variation of parameters or variation of constants. We will now show that this method generalizes to the matrix case. Following the usual idea, we replace the general solution of the homogeneous equation, $\Phi(x)c$, by $\Phi(x)z(x)$, where z is a function to be determined. Let $y(x) = \Phi(x)z(x)$. Then, if z is differentiable, $y' = \Phi'z + \Phi z'$. Since Φ is a fundamental matrix, $\Phi' = A\Phi$. Therefore

$$y' = A\Phi z + \Phi z' = Ay + \Phi z'$$

It is evident that if y satisfies Eq. (2), then $\Phi z' = b$. Since Φ is a fundamental matrix, $\det \Phi(x)$ is nonzero for every x by Theorem 12.7. Therefore $\Phi^{-1}(x)$ exists and we have $z'(x) = \Phi^{-1}(x)b(x)$. Therefore

$$z(x) = c + \int_{\xi}^{x} \Phi^{-1}(s)b(s)\,ds \tag{3}$$

$$y(x) = \Phi(x)\left[c + \int_{\xi}^{x} \Phi^{-1}(s)b(s)\,ds\right] \tag{4}$$

for some constant c. Conversely, let c be any constant vector and let y be defined by Eq. (4). Then if we define z by Eq. (3) we obtain $z' = \Phi^{-1}b$, $y = \Phi z$, and it is easily verified that y satisfies Eq. (2). We have proved the following theorem when we observe that from Eq. (4), c must equal $\Phi^{-1}(\xi)y(\xi)$.

THEOREM 12.19 Let $A(x)$ and $b(x)$ be continuous on an interval J. Let $\Phi(x)$ be any fundamental matrix for Eq. (1) on J, and let ξ be any point in J. Then a general solution of Eq. (2) is given by

$$y(x) = \Phi(x)\left[\Phi^{-1}(\xi)y(\xi) + \int_{\xi}^{x} \Phi^{-1}(s)b(s)\,ds\right], \qquad x \in J \tag{5}$$

Equation (5) is of the expected form $\Phi(x)c + y_p(x)$, where $c = \Phi^{-1}(\xi)y(\xi)$ and $y_p(x)$ is $\Phi(x)$ times the integral. The reader may feel that Eq. (5) was obtained by a simpler calculation than the analogous result for scalar nth order equations. This illustrates the economy of thought and writing achieved with the vector-matrix notation. Equation (5) is referred to as the *variation of constants formula* or *variation of parameters formula* for Eq. (2).

Example 1 In applying Eq. (5) to a particular case, lengthy calculations may be required. It is necessary to compute a fundamental matrix, find its inverse, and then perform the indicated multiplications and integration. We illustrate for the system $y' = Ay + b$, where

$$A = \begin{pmatrix} 1 & 1 \\ 4 & 1 \end{pmatrix} \qquad b = \begin{pmatrix} e^{-x} \\ -2e^{-x} \end{pmatrix}$$

In Section 12.14, we found that a fundamental solution is

$$\Phi(x) = e^{Ax} = \begin{pmatrix} \tfrac{1}{2}e^{-x} + \tfrac{1}{2}e^{3x} & -\tfrac{1}{4}e^{-x} + \tfrac{1}{4}e^{3x} \\ -e^{-x} + e^{3x} & \tfrac{1}{2}e^{-x} + \tfrac{1}{2}e^{3x} \end{pmatrix}$$

If we take $\xi = 0$, then $\Phi^{-1}(0) = I$, but still it is necessary to compute the inverse of $\Phi(x)$, which looks a bit tedious. Instead, let us recall that we can form a fundamental matrix out of any linearly independent solutions. As shown in Section 12.14, $\begin{pmatrix} 1 \\ -2 \end{pmatrix} e^{-x}$ and $\begin{pmatrix} 1 \\ 2 \end{pmatrix} e^{3x}$ are independent solutions. This suggests taking

$$\Phi(x) = \begin{pmatrix} e^{-x} & e^{3x} \\ -2e^{-x} & 2e^{3x} \end{pmatrix}$$

Then we easily find

$$\Phi^{-1}(x) = \tfrac{1}{4}e^{-2x} \begin{pmatrix} 2e^{3x} & -e^{3x} \\ 2e^{-x} & e^{-x} \end{pmatrix}$$

Equation (5) then gives, for $\xi = 0$,

$$\Phi^{-1}(0) = \begin{pmatrix} \tfrac{1}{2} & -\tfrac{1}{4} \\ \tfrac{1}{2} & \tfrac{1}{4} \end{pmatrix}$$

$$\Phi^{-1}(s)b(s) = \tfrac{1}{4}e^{-2s} \begin{pmatrix} 4e^{2s} \\ 0 \end{pmatrix} = \begin{pmatrix} 1 \\ 0 \end{pmatrix}$$

$$y(x) = \Phi(x)\left[\Phi^{-1}(0)y(0) + \int_0^x \begin{pmatrix} 1 \\ 0 \end{pmatrix} ds \right]$$

If $y(0) = \begin{pmatrix} a_1 \\ a_2 \end{pmatrix}$, we get

$$y_1(x) = e^{-x}\left(x + \frac{a_1}{2} - \frac{a_2}{4} \right) + e^{3x}\left(\frac{a_1}{2} + \frac{a_2}{4} \right)$$

$$y_2(x) = e^{-x}\left(-2x - a_1 + \frac{a_2}{2} \right) + e^{3x}\left(a_1 + \frac{a_2}{2} \right)$$

We now wish to reformulate Theorem 12.19 in a slightly different way.

DEFINITION 12.20 For any ξ, the *principal matrix at* ξ for Eq. (1) is the unique fundamental solution $\Phi(x)$ which at $x = \xi$ is equal to the identity matrix. The principal matrix at ξ is a function of both x and ξ and will be denoted by $\Phi(x; \xi)$. It satisfies

$$\frac{d}{dx}\Phi(x; \xi) = A(x)\Phi(x; \xi), \qquad \Phi(\xi; \xi) = I \tag{6}$$

THEOREM 12.20   Let $A(x)$ and $b(x)$ be continuous on an interval J, let ξ be any point in J, and let $\Phi(x; \xi)$ be the principal matrix at ξ. Then a general solution of Eq. (2) is given by

$$y(x) = \Phi(x; \xi)y(\xi) + \int_{\xi}^{x} \Phi(x; s)b(s)\, ds, \qquad x \in J \tag{7}$$

PROOF: We use Eq. (5) with $\Phi(x; \xi)$ as the fundamental matrix. This yields

$$y(x) = \Phi(x; \xi)\Phi^{-1}(\xi; \xi)y(\xi) + \Phi(x; \xi)\int_{\xi}^{x} \Phi^{-1}(s; \xi)b(s)\, ds \tag{8}$$

We have $\Phi^{-1}(\xi; \xi) = I$. If we can show that for any x, s, and ξ in J,

$$\Phi(x; \xi) = \Phi(x; s)\Phi(s; \xi) \tag{9}$$

then Eq. (7) will follow at once from Eq. (9). To prove Eq. (9), we introduce the matrix function

$$\Psi(x; s; \xi) = \Phi(x; s)\Phi(s; \xi)$$

Then

$$\frac{d}{dx}\Psi(x; s; \xi) = \left[\frac{d}{dx}\Phi(x; s)\right]\Phi(s; \xi)$$

$$= A(x)\Phi(x; s)\Phi(s; \xi) = A(x)\Psi(x; s; \xi)$$

Therefore Ψ satisfies the differential equation. Also, since $\Phi(s; s) = I$, $\Psi(s; s; \xi) = \Phi(s; s)\Phi(s; \xi) = \Phi(s; \xi)$. Since $\Psi(x; s; \xi)$ and $\Phi(x; \xi)$ are both solutions (as functions of x), and are equal at $x = s$, it follows from the uniqueness of solution of initial value problems that they are equal. In view of the definition of Ψ, this proves Eq. (9).   ∎

Theorem 12.20 is of more theoretical than practical interest. It shows that y is the sum of

$$y_h(x) = \Phi(x; \xi)y(\xi)$$

and

$$y_p(x) = \int_{\xi}^{x} \Phi(x; s)b(s)\, ds$$

The function y_h is the solution of the homogeneous equation which equals $y(\xi)$ at $x = \xi$. The function y_p is the solution of the nonhomogeneous equation which equals 0 at $x = \xi$. Both y_h and y_p are expressed in terms of the principal matrix Φ. Note that Φ depends on $A(x)$ but not on $b(x)$. Moreover, in case $b(x) = 0$, we get $y(x) = \Phi(x; \xi)y(\xi)$. Therefore, $\Phi(x; \xi)$ is the operator which must act on an initial vector $y(\xi)$ in order to yield the solution $y(x)$ (of the homogeneous equation) emanating from $y(\xi)$; that is,

$$\Phi(x;\ \xi) \text{ maps } y(\xi) \text{ into } y(x) \tag{10}$$

In view of this meaning for $\Phi(x;\ \xi)$, Eq. (9) has a simple interpretation. Since

$$\Phi(s;\ \xi) \text{ maps } y(\xi) \text{ into } y(s)$$
$$\Phi(x;\ s) \text{ maps } y(s) \text{ into } y(x)$$
$$\Phi(x;\ \xi) \text{ maps } y(\xi) \text{ into } y(x)$$

it is clear that the composition of the first two maps is the third.

The principal matrix $\Phi(x;\ \xi)$ plays the role of the one-sided Green's function introduced in Section 9.5. In fact, it is shown in Exercise 3 that $\Phi(x;\ \xi)$ reduces to $G(x;\ \xi)$ for a system corresponding to an nth order scalar equation.

Exercises

1.   Find a general solution of each system $y' = Ay + b$

(a)
$$A = \begin{pmatrix} 0 & 1 & 0 \\ 0 & 0 & 1 \\ 0 & 1 & 0 \end{pmatrix} \qquad b = \begin{pmatrix} 1 \\ -1 \\ 1 \end{pmatrix}$$

(b)
$$A = \begin{pmatrix} -1 & -2 & 2 \\ -2 & -1 & 2 \\ -2 & -2 & 3 \end{pmatrix} \qquad b = \begin{pmatrix} 3e^x \\ e^x \\ 4e^x \end{pmatrix}$$

(c)
$$A = \begin{pmatrix} -2 & 1 \\ -1 & -4 \end{pmatrix} \qquad b = \begin{pmatrix} e^{-3x} \\ -e^{-3x} \end{pmatrix}$$

(d)
$$A = \begin{pmatrix} 6 & 1 & 1 \\ 1 & 6 & 1 \\ 3 & 3 & 6 \end{pmatrix} \qquad b = \begin{pmatrix} e^{9x} \\ e^{9x} \\ 2e^{9x} \end{pmatrix}$$

2.   For each x and ξ, the matrix $\Phi(x;\ \xi)$ defines a transformation T of $\mathbf{R}^n$ into $\mathbf{R}^n$ by $T: y_0 \to \Phi(x;\ \xi)y_0$; that is, the image of y_0 is the solution, at x, starting from the initial condition y_0 at ξ.

(a) Show that T is a linear transformation.

(b) Is T one-to-one and onto?

(c) Consider the solutions $y(x)$ of $y' = Ay$ emanating at $x = 0$ from initial points y_0 in a q-dimensional subspace of $\mathbf{R}^n$. Show that for any $x_1 > 0$, the set of vectors $y(x_1)$ forms a subspace of $\mathbf{R}^n$ of the same dimension q.

3.   Let $y' = A(x)y + b(x)$ be the companion equation of the linear equation

$$z^{(n)} + a_{n-1}(x)z^{(n-1)} + \cdots + a_0(x)z = f(x) \tag{11}$$

so that $A(x)$, $b(x)$ are given by Eqs. (4) of Section 12.11. Let

$$\mathscr{W}(x) = \begin{pmatrix} z_1 & z_2 & \cdots & z_n \\ z_1' & z_2' & \cdots & z_n' \\ & & \cdots & \\ z_1^{(n-1)} & z_2^{(n-1)} & \cdots & z_n^{(n-1)} \end{pmatrix}$$

be the Wronski matrix formed from any n independent solutions $z_1(x), \ldots,$ $z_n(x)$ of Eq. (11).

(a) Show that $\Phi(x; \xi) = \mathscr{W}(x)\mathscr{W}^{-1}(\xi)$ is the principal matrix at ξ for the system.

(b) Using $\mathscr{W}^{-1}(\xi) = \operatorname{adj}\mathscr{W}(\xi)/\det\mathscr{W}(\xi)$, show that the first element of the vector $\int_{\xi}^{x} \Phi(x; s)b(s)\, ds$ is

$$\int_{\xi}^{x} \frac{z_1(x)W^{n1}(s) + \cdots + z_n(x)W^{nn}(s)}{W(s)} f(s)\, ds = \int_{\xi}^{x} G(x, s)f(s)\, ds$$

where $G(x, s)$ is the one-sided Green's function of Eq. (11).

COMMENTARY

Thorough discussions of linear systems $y' = A(x)y + b(x)$, similar to the one given here, can be found in many texts on differential equations. We mention, for example:

BOYCE, W., and R. DIPRIMA, *Elementary Differential Equations and Boundary Value Problems*, 2nd ed., New York, Wiley, 1969.

BRAND, L., *Differential and Difference Equations*, New York, Wiley, 1966.

HALE, J. K., *Ordinary Differential Equations*, New York, Wiley-Interscience, 1969.

Section 12.1   For a more complete discussion of the drug transport model, see:
HANSCH, C., et al. "Passive permeation of organic compounds through biological tissue: a non-steady-state theory," *Molecular Pharmacology*, vol. 5, 333–341 (1969).

Sections 12.5 and 12.7   It is possible to give a proof of Theorem 12.2 which is in a form almost identical to the proof of Theorem 3.3, once the concept of matrix norms has been introduced. Readers should consult more advanced books on differential equations for this proof. Similarly, a proof of Theorem 12.3 can be based on the method of successive approximations, but is omitted here.

Section 12.12   More extensive information on infinite series of matrices can be found in more advanced books on matrix theory. For example, one may ask for what real functions $f(x)$ can one ascribe a meaning to $f(A)$ "in a reasonable way." See:

BELLMAN, R., *Introduction to Matrix Analysis*, 2nd ed., New York, McGraw-Hill, 1970.

GANTMACHER, F. R., *The Theory of Matrices*, vols. I and II, New York, Chelsea, 1959.

Section 12.16   There are several methods known for calculating the solution of $y' = Ay$, where A is not diagonable, without previously finding the Jordan form of A. See:

APOSTOL, T. M., "Some explicit formulas for the exponential matrix e^{tA}," *Amer. Math. Monthly*, vol. 76, 289–292 (1969).

KIRCHNER, R. B., "An explicit formula for e^{tA}," *Amer. Math. Monthly*, vol. 74, 1200–1204 (1967).

PUTZER, E. J., "Avoiding the Jordan canonical form in the discussion of linear systems with constant coefficients," *Amer. Math. Monthly*, vol. 73, 2–7 (1966).

RICE, N. M., "More explicit formulas for the exponential matrix e^{tA}," *Queen's Mathematical Preprints* No. 1970–21, Kingston, Ontario, Queen's University.

The problem of stability and instability for differential equations was identified by the Russian mathematician A. Liapunoff (Liapunov) and has been studied assiduously in recent years. A good place to begin to study this problem is in:

BELLMAN, R., *Stability Theory of Differential Equations*, New York, McGraw-Hill, 1953.

Section 12.17   The idea of considering $\Phi(x; \xi)$ as an operator which takes an initial vector at ξ into the solution at x is valuable in further studies of differential systems. It fits in naturally with the point of view that a differential equation defines a *flow* in n-dimensional space; that is, the solution $y(x)$ starts at y_0 when $x = \xi$ and the point $y(x)$ moves along as x increases. Equation (9) expresses the composition property of the flow, or, as is said, the *semigroup property*. In studies of other kinds of functional equations (such as difference equations or differential-difference equations) one can often establish the existence of such an operator and of a variation of constants formula analogous to Eq. (7). It is important to be able to give an effective method for computing the quantities in such a formula.

Chapter 13

Linear difference equations

13.1 Calculus of differences and antidifferences

Many problems in areas such as biology, physics, engineering, and economics can be formulated in terms of difference equations rather than differential equations. The theory of difference equations bears a striking analogy to that of differential equations and provides another illustration of the applications of linear algebra. Accordingly, we shall give an introduction to this theory.

In the last section of the chapter, we briefly introduce differential equations with delay, which comprise another class of functional equations which is important in application.

This section is an introduction to the calculus of differences.

DEFINITION 13.1   Let J be an interval on the real line, and let $f(x)$ be a real or complex function defined on J. Then the (first forward) *difference* of f is the function Δf defined by

$$(\Delta f)(x) = f(x + 1) - f(x) \qquad (1)$$

Note that Δf is defined only for values of x for which x and $x + 1$ are both in J. In order to avoid this minor inconvenience, we shall usually assume that J is all of R, or a half-infinite interval of the form $J = [a, \infty)$ for some a. Also, as a notational convenience we shall often write $\Delta f(x)$ rather than $(\Delta f)(x)$.

In Definition 13.1, we have taken the difference interval of the independent variable to be 1. Sometimes one wishes to allow this interval to be h for some positive h different from 1 (h is sometimes called the *span* or *mesh size*). Then the difference can be defined by the equation

$$\Delta_h f(x) = f(x + h) - f(x)$$

The corresponding theory is no different, but some of the formulas must be modified. We leave these modifications for the reader to work out.

DEFINITION 13.2   Higher order differences $\Delta^2 f$, $\Delta^3 f$, and so on, are defined by

$$\Delta^k f = \Delta(\Delta^{k-1} f), \qquad k = 2, 3, \dots \tag{2}$$

Also we define

$$\Delta^0 f = f \tag{3}$$

We can think of Δ as an operator or mapping in the sense that Δ transforms a function f into a function Δf (see Section 5.1 on mappings). The difference operator Δ is *linear*;[1] that is,

$$\Delta(f + g) = \Delta f + \Delta g \quad \text{and} \quad \Delta(cf) = c\Delta f \tag{4}$$

for any functions f, g for which Δf and Δg are defined and for any scalar c. To prove this, observe that

$$(\Delta(f + g))(x) = (f + g)(x + 1) - (f + g)(x)$$

Since $f + g$ is defined to be the function whose value at x is $f(x) + g(x)$, we have

$$(\Delta(f + g))(x) = f(x + 1) + g(x + 1) - f(x) - g(x) = (\Delta f)(x) + (\Delta g)(x)$$

The other part of Eq. (4) is proved similarly. It is also easy to show that the higher order operators Δ^k are linear.

Example 1   Let $f(x) = x^2$. Then

$$\Delta f(x) = (x + 1)^2 - x^2 = 2x + 1$$
$$\Delta^2 f(x) = \Delta(2x + 1) = 2(x + 1) + 1 - 2x - 1 = 2$$

If $f(x) = e^x$, then

$$\Delta f(x) = e^x(e - 1), \quad \Delta^2 f(x) = e^x(e - 1)^2, \quad \Delta^3 f(x) = e^x(e - 1)^3$$

and so on.

---

[1] The concept of linearity for a mapping, as defined in Section 5.2, involves the structure of the underlying vector space. To be precise, therefore, we ought to specify the space of functions under consideration in Eq. (4). However, for any of the function spaces we have considered the notion of addition and scalar multiplication is the same, so we need not specify a particular space here.

DEFINITION 13.3   Let F and f be functions such that $\Delta F(x) = f(x)$, for x in an interval J. Then F is called an *antidifference* of f and we write $\Delta^{-1}f = F$.

DEFINITION 13.4   A function p is said to be *periodic* with *period* α if

$$p(x + \alpha) = p(x)$$

for all x. (Here α is a fixed number, real or complex.) Sometimes p is more briefly called α-*periodic*.

The terminology in Definition 13.3 is patterned after that in calculus, where we call F an antiderivative of f if $F'(x) = f(x)$. We know that if any antiderivative of F exists, then there is a whole set of them, differing only by constants. The following is an analogous theorem for the calculus of differences.

THEOREM 13.1   If a function f has an antidifference F, then $F + p$ is also an antidifference of f for any function p which is periodic with period 1. There are no other antidifferences.

PROOF: By hypothesis, $\Delta^{-1}f = F$, or $\Delta F = f$. Hence

$$\Delta(F + p)(x) = F(x + 1) + p(x + 1) - F(x) - p(x) = \Delta F(x) = f(x)$$

Thus, $F + p$ is an antidifference of f, for any p. Now suppose G is any antidifference of f. Then $\Delta F = f$ and $\Delta G = f$. By the linearity of Δ, $\Delta(G - F) = 0$. Let $H = G - F$. We have

$$\Delta H(x) = H(x + 1) - H(x) = 0$$

for all x. By definition, H is periodic with period 1. Since $G = F + H$, this shows that every antidifference of f is the sum of F and a function of period 1. ∎

Example 2 $\Delta^{-1}1 = x + p(x)$. To be more precise, let $f(x) \equiv 1$, $F(x) \equiv x$. Then, clearly, $\Delta F = f$. Hence, the family of antidifferences of f is $F + p$, or $x + p$ as we may write in notation which is not strictly correct but is common and helpful.

Example 3   If a is a constant, $a \neq 1$, $\Delta^{-1}a^x = a^x/(a - 1) + p$. To show this, compute

$$\Delta\left(\frac{a^x}{a - 1}\right) = \frac{1}{a - 1}\,\Delta a^x = \frac{1}{a - 1}\,(a^{x+1} - a^x) = a^x$$

Note that for $a = 2$, these formulas take the simplest form,

$$\Delta^{-1}2^x = 2^x + p, \qquad \Delta 2^x = 2^x \tag{5}$$

Thus in the difference calculus (with span $h = 1$), the behavior of the exponential 2^x is analogous to that of e^x in the differential calculus.

Example 4   From Example 3 we have

$$\Delta\, \Delta^{-1} a^x = \Delta\left(\frac{a^x}{a-1} + p\right) = a^x$$

On the other hand,

$$\Delta^{-1}\, \Delta a^x = \Delta^{-1} a^x(a-1) = (a-1)\,\Delta^{-1} a^x = (a-1)\left(\frac{a^x}{a-1} + p(x)\right)$$

is not uniquely defined. One of the antidifferences is a^x, but in general,

$$\Delta^{-1}\Delta a^x = a^x + q(x),$$

where q has period 1. One can similarly see that always

$$\Delta\, \Delta^{-1} f = f \tag{6}$$

but

$$\Delta^{-1}\, \Delta F = F + p, \tag{7}$$

where p is 1-periodic.

   In the ordinary calculus, it is not true that every function f has an anti-derivative F, although most functions of practical interest do. There are various theorems which describe classes of functions which have antiderivatives. For example, in elementary calculus the theorem is usually stated (and sometimes proved) that every function continuous on an interval does have an antiderivative. As the following theorem shows, the situation is simpler in the difference calculus.

THEOREM 13.2   Let $f(x)$ be any function defined on an interval $J = [a, b)$. Then $f(x)$ has an antidifference, defined on $[a, b + 1)$.

PROOF: We shall construct an antidifference $F(x)$ by what we call the *method of steps*.
   First, define $F(x)$ in a completely arbitrary way for $a \le x < a + 1$. We now extend F to the domain $[a, b + 1)$ by defining, for $a + 1 \le x < b + 1$,

$$F(x) = F(x - 1) + f(x - 1) \tag{8}$$

It is clear that $F(x)$ is now defined for $a \le x < b + 1$, and

$$\Delta\, F(x) = F(x + 1) - F(x) = f(x), \qquad a \le x < b \qquad \blacksquare$$

   The reason for the name associated with the above technique is that actual computation of F will first be on the interval $a \le x < a + 1$, then this is extended to the interval $a + 1 \le x < a + 2$, and so on. (See Exercise 14.)

   The alert reader will no doubt ask whether in difference calculus there is an analogue of the Fundamental Theorem of Calculus; that is, is there an operation inverse to the difference operation in the same way that integration is inverse to differentiation? There is such an analogue, the operation of summing, as the following simple theorem will show.

THEOREM 13.3   Let F be an antidifference of f and let a, b be integers with $b > a$. Then

$$\sum_{j=a}^{b} f(j) = F(j) \Big|_{j=a}^{j=b+1} = F(b+1) - F(a) \tag{9}$$

PROOF: Since $f(x) = F(x+1) - F(x)$,

$$\sum_{j=a}^{b} f(j) = \sum_{j=a}^{b} F(j+1) - \sum_{j=a}^{b} F(j)$$

$$= \sum_{k=a+1}^{b+1} F(k) - \sum_{j=a}^{b} F(j)$$

$$= F(b+1) - F(a) \qquad \blacksquare$$

In differential calculus, the analogous result is of great importance in calculating definite integrals in instances where an antiderivative can be found. Equation (9) is less important, since the sum can frequently be computed by simple addition; even in cases where there are many terms in the sum, the calculation may be feasible with the aid of a high-speed computer. Still, Eq. (9) sometimes leads to great simplifications or elegant formulas. This is illustrated in the Exercises.

Exercises

1. Compute Δf and if possible simplify, where $f(x)$ is

(a) $3x^2 - 4x$ (b) x^3 (c) $\sin(ax+b)$

(d) 10^{2x-5} (e) $\cos(ax+b)$ (f) $xa^x \quad (a \neq 1)$

(g) $\sec x$ (h) x^{-1}

2. Using tables if necessary, compute

(a) $\Delta^3 \sin x$ at $x = \pi/6$

(b) $\Delta^2 2^x$ at $x = 2$

(c) $\Delta \log_{10} x$ at $x = 100$

3. The *factorial power function* $x^{(n)}$ is defined by

$$x^{(1)} = x, \quad x^{(2)} = x(x-1)$$

and generally,

$$x^{(n)} = x(x - 1) \cdots (x - n + 1),$$
$$(x - x_0)^{(n)} = (x - x_0)(x - x_0 - 1) \cdots (x - x_0 - n + 1)$$

for any positive integer n and any real x_0. Show that

$$x^{(m+n)} = x^{(m)}(x - m)^{(n)} = x^{(n)}(x - n)^{(m)} \tag{10}$$

for any positive integers m and n.

4. The definition of the factorial power is extended to zero and negative integers by the equations

$$x^{(0)} = 1, \qquad x \neq 0$$

$$x^{(-n)} = \frac{1}{(x + n)^{(n)}}, \qquad n = 1, 2, \ldots, \quad x \neq -1, -2, \ldots, -n$$

Show that

$$x^{(-n)} = \frac{1}{(x + 1)(x + 2) \cdots (x + n)}$$

and show that Eq. (10) is valid for all integers m, n for all x for which $x^{(m+n)}$ is defined.

5. (a) Show that

$$\Delta(x + c)^{(n)} = n(x + c)^{(n-1)}$$

for any integer n and any scalar c. Thus, the factorial power differences behave in a fashion analogous to differentiation of the ordinary power function.

(b) Show that if p is periodic

$$\Delta^{-1}(x + c)^{(n)} = \frac{(x + c)^{(n+1)}}{n + 1} + p(x), \qquad n \neq -1$$

6. Show that

$$\Delta \frac{\sin\left(ax + b - \dfrac{a}{2}\right)}{2 \sin \frac{1}{2}a} = \cos(ax + b)$$

$$\Delta \frac{\cos\left(ax + b - \dfrac{a}{2}\right)}{2 \sin \frac{1}{2}a} = -\sin(ax + b)$$

for real numbers a for which the denominator is not zero.

7. (a) Prove that for any function f and positive integer n

$$\Delta^n f(x) = f(x + n) - \binom{n}{1} f(x + n - 1)$$

$$+ \binom{n}{2} f(x + n - 2) + \cdots + (-1)^n f(x)$$

(b) Prove that for any function f and positive integer n

$$f(x + n) = f(x) + n \, \Delta f(x) + \binom{n}{2} \Delta^2 f(x) + \cdots + \Delta^n f(x)$$

$$\left[Hint: \text{Show } \binom{n}{k} + \binom{n}{k + 1} = \binom{n + 1}{k + 1} \right]$$

8. Prove the following.

(a) If f is a constant function, $\Delta f = 0$.

(b) If m, n are nonnegative integers, and f is any function,

$$\Delta^m \Delta^n f(x) = \Delta^{m+n} f(x)$$

(c) For any two functions f and g

$$\Delta(fg)(x) = f(x) \, \Delta g(x) + (\Delta f(x)) g(x + 1)$$

where $fg(x) = f(x)g(x)$

$$\Delta \left(\frac{f}{g} \right) (x) = \frac{g(x) \, \Delta f(x) - f(x) \, \Delta g(x)}{g(x)g(x + 1)}$$

9. The *backward difference operator* ∇ is defined by the equation

$$(\nabla f)(x) = f(x) - f(x - 1) \tag{11}$$

and on certain occasions is more convenient than the *forward* operator Δ. Compute ∇f for each of the functions in Exercise 1.

10. Prove that any polynomial $f(x)$ of degree n can be expanded in a finite series of factorial powers by the formula

$$f(x) = f(0) + \frac{\Delta f(0)}{1!} x^{(1)} + \frac{\Delta^2 f(0)}{2!} x^{(2)} + \cdots + \frac{\Delta^n f(0)}{n!} x^{(n)}$$

the analogue of the finite Maclaurin expansion. Compute this expansion for $f(x) = x^3 - 3x^2 + x - 5$.
[*Hint:* See Exercise 5.]

11. Expand each function f in a series of factorials and then compute $\Delta^{-1}f$.

(a) x^2 (b) x^3 (c) $(2x)^{(3)}$

12. Express each function f as a series of negative factorials and thus find $\Delta^{-1}f$.

(a) $\dfrac{1}{x(x + 1)}$ (b) $\dfrac{1}{x(x + 2)}$

13. It is natural to try to extend the formula in Exercise 10 to functions other than polynomials by the use of infinite expansions. Compute such an expansion for $f(x) = 2^x$. Try to deduce something about the convergence of the series.

14. Let $f(x) = \log_2 x$ on $1 \le x < 3$. Compute an antidifference of f by the method of steps, starting with $F(x) = 0$ for $1 \le x < 2$.

15. Let $f(x) = x^2 = x^{(1)} + x^{(2)}$. Thus, an antidifference of f is $F(x) = \frac{1}{2}x^{(2)} + \frac{1}{3}x^{(3)}$. Using this in Eq. (9), show that for any integer $n \ge 1$,

$$\sum_{x=1}^{n} x^2 = \frac{1}{2}x^{(2)} + \frac{1}{3}x^{(3)}\big|_{x=1}^{x=n+1} = \frac{1}{6}n(n + 1)(2n + 1)$$

16. Using the method of Exercise 15, evaluate each sum.

(a) $\displaystyle\sum_{x=1}^{n} x^3$ (b) $\displaystyle\sum_{x=1}^{n} \frac{1}{x(x + 1)}$

(c) $1 \cdot 3 \cdot 4 + 2 \cdot 4 \cdot 5 + 3 \cdot 5 \cdot 6 + \cdots + n(n + 2)(n + 3)$

17. Show that

$$\sum_{x=1}^{n} \cos x = \frac{(\sin \frac{1}{2}n)(\cos \frac{1}{2}(n + 1))}{\sin \frac{1}{2}}$$

18. In ordinary calculus, the inverse nature of differentiation and definite integration is expressed by the formulas

$$\int_{a}^{x} F'(t)\, dt = F(x) - F(a), \qquad \frac{d}{dx}\int_{a}^{x} f(t)\, dt = f(x)$$

Show that if a and x are integers with $x \ge a + 1$, and if F and f are arbitrary functions,

$$\sum_{j=a}^{x-1} \Delta F(j) = F(x) - F(a)$$

and

$$\Delta \sum_{j=a}^{x-1} f(j) = f(x)$$

19. For any fixed number λ in the range $0 \leq \lambda \leq 1$, define a *generalized difference operator*

$$\Delta_\lambda f(x) = f(x + \lambda) - f(x + \lambda - 1)$$

Thus Δ_1 is the forward difference operator, Δ_0 is the backward difference operator ∇, and $\Delta_{1/2}$ is the central or symmetric difference operator.

$$\Delta_{1/2} f(x) = f(x + \tfrac{1}{2}) - f(x - \tfrac{1}{2})$$

Compute Δ_λ for $f(x) = x, x^2, 2^x$, and $\sin x$. (This definition is due to Roger Folsom, "Linear Constant Coefficient Difference or Differential Equations for Economic Models," *Thesis*, Claremont Graduate School, Claremont, California, 1973.)

13.2   First order difference equations

Our objective is to develop a theory of difference operators and difference equations, aided by the concepts from linear algebra which are known to us. Before doing this, we present some basic terminology and show how to solve certain difference equations. This will provide a basis for more theoretical developments later.

DEFINITION 13.5   An equation involving a function $f(x)$ and differences $\Delta f(x)$, $\Delta^2 f(x)$, and so on, of that function is called a *difference equation*.

Example 1   The following equations are difference equations:

$$f(x) + 2 \Delta f(x) - 3 \Delta^2 f(x) = 2^x \tag{1}$$

$$[\Delta f(x)]^2 - [\Delta f(x)] f(x) = x \tag{2}$$

Also,

$$\Delta_x f(x, y) + \Delta_y f(x, y) = x + y \tag{3}$$

where Δ_x denotes differencing in x and Δ_y denotes differencing in y, is a difference equation; it is naturally termed a *partial difference equation*. We shall restrict attention here to ordinary difference equations; that is, equations involving differences only with respect to a single independent variable.

We note that Eqs. (1) and (2) can be rewritten without the use of the Δ symbol, since we can replace $\Delta f(x)$ by $f(x + 1) - f(x)$ and $\Delta^2 f(x)$ by $f(x + 2) - 2f(x + 1) + f(x)$. Thus Eq. (1) is the same as

$$3 f(x + 2) - 8 f(x + 1) + 4 f(x) = -2^x \tag{4}$$

and Eq. (2) is the same as

$$[f(x + 1)]^2 - 3f(x + 1)f(x) + 2[f(x)]^2 = x \tag{5}$$

Clearly, every difference equation can be written without the use of Δ symbols; the resulting equation contains the unknown function evaluated at arguments $x, x + 1, x + 2, \ldots$ and will be called the *dated form* of the difference equation.

Example 2   Consider the equation

$$f(x + 2) - f(x)f(x + 1) + 2f(x) = 0$$

By Exercise 7b of the preceding section, we can replace $f(x + 2)$ by $\Delta^2 f(x) + 2f(x + 1) - f(x)$ and so the equation can be written

$$\Delta^2 f(x) + 2f(x + 1) - f(x)f(x + 1) + f(x) = 0$$

Replacing $f(x + 1)$ by $\Delta f(x) + f(x)$, we get

$$\Delta^2 f(x) + 2\,\Delta f(x) - f(x)\Delta f(x) - [f(x)]^2 + 3f(x) = 0$$

By such a procedure, if we are given any equation in dated form, we can find an equivalent equation in differenced form.

DEFINITION 13.6   The *order* of a difference equation on $a \le x < b$ (in differenced form) is the order of the highest order difference appearing in the equation.

Usually it is required that when the equation is written in dated form the coefficient of $f(x)$ is nonzero. In our treatment, this will not be required.

Thus Eq. (1) is of second order and Eq. (2) is of first order. A first order equation is of the form $g(x, f(x), \Delta f(x)) = 0$ for some function g, or if we can solve this for $\Delta f(x)$, of the form

$$\Delta f(x) = G(x, f(x)) \tag{6}$$

When $f(x)$ is replaced by y, this can also be written in the form

$$\Delta y = G(x, y)$$

The basic problem for Eq. (6) is to find f, when G is given.

DEFINITION 13.7   A *solution* of a difference equation (6) is a function f which satisfies the equation over an interval $a \le x < b$.

Example 3   Consider the equation

$$\Delta f(x) = G(x)$$

All solutions are given by

$$f(x) = \Delta^{-1}G(x) + p(x)$$

where $\Delta^{-1}G$ denotes any antidifference of G, and p is an arbitrary function of period one.

We shall now consider the case in which G is a linear function. Then Eq. (6) takes the form

$$\Delta f(x) = a(x)f(x) + b(x) \tag{7}$$

where a and b are given functions. Alternatively, by defining $c(x) = a(x) + 1$, we can put this in the dated form

$$f(x + 1) = c(x)f(x) + b(x) \tag{8}$$

DEFINITION 13.8   An equation of the form (7) or (8) is called a *first order linear* difference equation. If $b(x)$ is identically zero, the equation is called *homogeneous*, otherwise it is called *nonhomogeneous*.

The most general solution of Eq. (8) is readily found by the method of steps. Assume that b and c are given for $x \geq \alpha$. Let

$$f(x) = \phi(x), \qquad \alpha \leq x < \alpha + 1, \tag{9}$$

where ϕ is an arbitrary function. Then from Eq. (8) we get

$$f(x) = c(x - 1)\phi(x - 1) + b(x - 1), \qquad \alpha + 1 \leq x < \alpha + 2 \tag{10}$$

Since f is now known on $[\alpha + 1, \alpha + 2)$, we can use Eq. (8) again to obtain f on the next interval, thus

$$
\begin{aligned}
f(x) &= c(x - 1)f(x - 1) + b(x - 1) \\
&= c(x - 1)[c(x - 2)\phi(x - 2) + b(x - 2)] + b(x - 1) \\
&\qquad\qquad\qquad\qquad\qquad \alpha + 2 \leq x < \alpha + 3
\end{aligned}
$$

In general, on the interval $\alpha + k \leq x < \alpha + k + 1$ we have

$$
\begin{aligned}
f(x) = &\left\{ \prod_{i=1}^{k} c(x - i) \right\} \phi(x - k) \\
&+ b(x - 1) + c(x - 1)b(x - 2) + \cdots \\
&+ \left\{ \prod_{i=1}^{k-1} c(x - i) \right\} b(x - k)
\end{aligned} \tag{11}
$$

The formula is cumbersome, but completely general so far as Eq. (8) is concerned. (For $k = 0$, we must interpret $\prod_{i=1}^{0} = 1$ and $\prod_{i=1}^{-1} = 0$.)

Example 4   Consider the equation with constant coefficients

$$f(x + 1) = cf(x), \qquad x \geq \alpha \tag{12}$$

Applying the method of steps we get

$$f(x) = c^k \phi(x - k), \qquad \alpha + k \leq x < \alpha + k + 1 \tag{13}$$

where ϕ is an arbitrary function given on $\alpha \le x < \alpha + 1$. For the nonhomogeneous equation with constant coefficients,

$$f(x + 1) = cf(x) + b, \qquad x \ge \alpha \tag{14}$$

we get

$$f(x) = c^k \phi(x - k) + b + cb + \cdots + c^{k-1}b,$$
$$\alpha + k \le x < \alpha + k + 1 \tag{15}$$

The results in Eqs. (11) and (15) can be rewritten in a helpful way. Let $p(x)$ be an arbitrary function of period one. Then we can identify $\phi(x)$ with $p(x)$ on $\alpha \le x < \alpha + 1$. Furthermore, for $\alpha + k \le x < \alpha + k + 1$ we have

$$\phi(x - k) = p(x - k) = p(x)$$

Therefore, we can replace $\phi(x - k)$ by $p(x)$ in Eq. (11). Also, an equation on $x \ge \alpha$ can be converted to one on $x \ge 0$ by a change of variable $g(x) = f(x + \alpha)$. Thus our result is expressed by the following theorem.

THEOREM 13.4 Let $b(x)$ and $c(x)$ be defined for all $x \ge 0$. Then the function f defined by

$$f(x) = p(x), \qquad\qquad\qquad\qquad 0 \le x < 1$$

$$f(x) = \left\{ \prod_{i=1}^{k} c(x - i) \right\} p(x) + b(x - 1) + c(x - 1)b(x - 2) + \cdots$$

$$+ \left\{ \prod_{i=1}^{k-1} c(x - i) \right\} b(x - k), \quad k \le x < k + 1, \quad k = 1, 2, \ldots$$

$$\tag{16}$$

where p is an arbitrary 1-periodic function is a solution of Eq. (8) on $x \ge 0$. Any solution on $x \ge 0$ is of the form in Eq. (16).

Considerable insight can be gained by careful inspection of Eq. (16). This formula expresses $f(x)$ as the sum of several terms. The first term contains an arbitrary 1-periodic function and is a general solution of the homogeneous difference equation. The sum of the others is a particular solution of the nonhomogeneous equation. In fact, we shall prove below that for any nth order linear difference equation, the solution is the sum of a general solution of the homogeneous equation and a particular solution of the nonhomogeneous equation. (This result should not surprise the reader.)

For a first order differential equation $y' = f(x, y)$, a unique solution from the infinite set of all solutions can be specified by giving an initial condition, the value of y at a particular point x_0. In the same way, for the first order equation (8) we can formulate an initial value problem which has a unique solution.

DEFINITION 13.9   Let $b(x)$, $c(x)$ be defined for $x \geq \alpha$, and let $\phi(x)$ be defined for $\alpha \leq x < \alpha + 1$. The *initial value problem* for the equation $f(x + 1) = c(x)f(x) + b(x)$, corresponding to ϕ, is to find a solution $f(x)$, defined for $x \geq \alpha$ and satisfying the *initial condition* $f(x) = \phi(x)$ for $\alpha \leq x < \alpha + 1$. The pair α, ϕ is called the *initial data*.

It is immediately clear from Eq. (11) that this initial value problem has a unique solution. We shall later extend the idea to nth order linear equations. For nonlinear equations such as (6), an initial value problem is obtained by giving $\phi(x)$ on $\alpha \leq x < \alpha + 1$, but existence of a solution is not guaranteed (see Exercise 8).

Example 5   We shall solve the initial value problem

$$f(x + 1) = cf(x), \qquad x \geq 0,$$
$$f(x) = 1, \qquad 0 \leq x < 1$$

The solution is

$$f(x) = \begin{cases} c, & 1 \leq x < 2 \\ c^2, & 2 \leq x < 3 \\ \cdots \end{cases}$$

Note in particular that if $c = 0$, then

$$f(x) = \begin{cases} 1, & 0 \leq x < 1 \\ 0, & 1 \leq x \end{cases}$$

Example 6   Consider the initial value problem

$$f(x + 1) = xf(x), \qquad x \geq 0,$$
$$f(x) = x, \qquad 0 \leq x < 1$$

By Eq. (16), the solution is of the form

$$f(x) = (x - 1) \cdots (x - k)p(x) = (x - 1)^{(k)}p(x), \qquad k \leq x < k + 1$$

where p is 1-periodic. Taking $k = 0$ and using the initial condition we obtain $p(x) = x$ for $0 \leq x < 1$. Thus, $f(x) = (x - 1)^{(k)}p(x)$, where p is the "sawtooth function"

$$p(x) = \begin{cases} x, & 0 \leq x < 1 \\ x - 1, & 1 \leq x < 2 \\ \cdots \end{cases}$$

Exercises

1.   Give the order of the following difference equations:

(a) $(\Delta y)(\Delta^2 y) + x^2 y = (\Delta y)^2$

(b) $x(\Delta^3 y)^2 + y\Delta y = 2^x$

2. Find all solutions of $\Delta f(x) = x^{(-2)}$ on $-1 < x < \infty$. Does this also define solutions for $x < -1$?

3. Prove, by mathematical induction, that Eq. (11) gives the most general solution of Eq. (8).

4. Prove that every equation $f(x + 2) = G(x, f(x), f(x + 1))$ can be written in the form $\Delta^2 f(x) = H(x, f(x), \Delta f(x))$, and vice versa. If G is given, is H unique? If H is given, is G unique?

5. Let

$$C(x) = c(x - 1)c(x - 2) \cdots c(x - k), \qquad \alpha + k \leq x < \alpha + k + 1$$

Since k is uniquely determined by x, we can regard C as a function of x alone; in fact, we can write

$$C(x) = \prod_{i=1}^{[x-\alpha]} c(x - i)$$

where $[z]$ is the integer part of z. Show that the equation $f(x + 1) = c(x)f(x)$ is equivalent to

$$\Delta \left(\frac{f(x)}{C(x)} \right) = \frac{f(x + 1)}{C(x + 1)} - \frac{f(x)}{C(x)} = 0$$

as long as the denominators are nonzero. Thus $1/C(x + 1)$ is a *multiplier* which converts the equation into one that is an exact difference. This leads to the solution $f(x)/C(x) = p(x)$ or $f(x) = C(x)p(x)$, in agreement with Eq. (16). Also, show how to solve the nonhomogeneous equation (8) with the aid of the multiplier $1/C(x + 1)$.

6. Find a general solution of each linear difference equation. Except where otherwise indicated take $\alpha = 0$, and simplify the sums and products where possible.

(a) $f(x + 1) = 3f(x)$ (b) $f(x + 1) = xf(x)$

(c) $f(x + 1) - 3f(x) = x$ (d) $f(x + 1) = xf(x) + x$

(e) $xf(x + 1) = f(x)$ (f) $f(x + 1) = \dfrac{x + 2}{x + 1} f(x)$

(g) $f(x + 1) = cf(x) + x^{(2)}$ (h) $f(x + 1) = f(x) + 2^x$

(i) $f(x + 1) = 2f(x) + x2^x$

(j) $f(x + 1) = f(x) + \dfrac{1}{x(x + 1)}$ (Take $\alpha = 1$.)

(k) $f(x + 1) = f(x) + \sin(ax + b)$

     (See Section 13.1, Exercise 6.)

7. Find the unique solution of each initial value problem, giving a general formula for f on $k \leq x < k + 1$, if possible. Graph f for $0 \leq x \leq 3$.

(a) $f(x + 1) = 3f(x)$, $x \geq 0$

$$f(x) = \sin 2\pi x, \qquad 0 \leq x < 1$$

(b) $f(x + 1) = xf(x)$, $x \geq 0$

$$f(x) = 1, \qquad 0 \leq x < 1$$

(c) $f(x + 1) = \dfrac{x + 2}{x + 1} f(x)$, $x \geq 0$

$$f(x) = x, \qquad 0 \leq x < 1$$

(d) $f(x + 1) = f(x) + 2^x$, $x \geq 0$

$$f(x) = 2^x, \qquad 0 \leq x < 1$$

8. (a) Consider the initial value problem consisting of the equation $f(x + 1) = [f(x)]^{1/2}$ and the initial condition $f(x) = -1$ for $0 \leq x < 1$. Show that there is no real solution existing on $0 \leq x \leq b$, where $b \geq 1$.

(b) Consider the initial value problem $f(x + 1) = G(f(x))$, $x \geq 0$, and the initial condition $f(x) = 1, 0 \leq x < 1$. Show that a necessary and sufficient condition for the existence of a unique solution $f(x)$ on $0 \leq x < 2$ is that $G(1)$ be defined.

13.3   Discrete difference equations

In Sections 13.1 and 13.2, we have dealt with the difference operator Δ defined by $\Delta f(x) = f(x + 1) - f(x)$, and we have assumed that x is a continuous real variable, $x \in J$ for some interval J. There is another context in which difference operators appear. Let us assume that the functions in question are defined only at the points $x = 1, 2, \ldots$ (or, if more convenient, $0, 1, 2, \ldots$). We shall now show that much of what we have done is still meaningful, and in some cases is simpler than before.

A function f defined only on the set J of positive integers[2] is in fact a sequence, and instead of writing $f(n)$ it is customary to write a symbol such as f_n for the value of the function at n. The sequence itself will be denoted by the symbols $\{f_n\}$ or $\{f_1, f_2, \ldots\}$ or f, interchangeably. Our definition of Δf is still valid if f has domain J, but in our new notation Δf is the sequence $\{\Delta f_n\}$ given by

$$\Delta f_n = f_{n+1} - f_n, \qquad n \in J \tag{1}$$

---

[2] We could more generally take $J = \{\alpha, \alpha + h, \alpha + 2h, \ldots\}$ for arbitrary real α and $h > 0$. Using differences with span h, we then obtain parallel results.

where Δf_n is an abbreviation for $(\Delta f)(n)$. From Definition 13.2 we get the higher differences

$$\Delta^k f_n = \Delta(\Delta^{k-1} f_n), \qquad k = 2, 3, \ldots$$
$$\Delta^0 f_n = f_n \tag{2}$$

To illustrate the meaning of Δ in this context, consider the sequence $\{f_n\} = \{1, 4, 9, \ldots, n^2, \ldots\}$, and place alongside it the sequences $\{\Delta f_n\}$ and $\{\Delta^2 f_n\}$, as in Table 13.1.

Table 13.1

n	f_n	Δf_n	$\Delta^2 f_n$
1	1	3	2
2	4	5	2
3	9	7	2
4	16	9	2
...	...	...	...

The entries in the Δf_n column can be obtained by subtracting adjacent entries in the f_n column and the entries in the $\Delta^2 f_n$ column can be obtained by subtracting adjacent entries in the Δf_n column (since $\Delta^2 f_n = \Delta f_{n+1} - \Delta f_n$). Tables of this kind are called *finite difference tables* and are helpful in the study of curve-fitting, numerical approximation, and so on. When dealing with difference operators applied to sequences, authors often refer to *finite difference operators*. These operators are useful in biological and economic modeling, in the study of numerical solution methods for differential equations, and elsewhere (see Section 13.8).

The concept of an antidifference as given in Section 13.1 needs no change when dealing with sequences. If $\{F_n\}$ and $\{f_n\}$ are sequences and $\Delta F_n = f_n$, then $\{F_n\}$ is called an antidifference of $\{f_n\}$ and we write $\Delta^{-1} f_n = F_n$. Theorems 13.1 and 13.2 now take a somewhat simpler form. The reason is that if $p(x)$ need be defined only over the domain of f, and if p has period one, then p is constant on this domain. Hence we can specialize to the following theorem.

THEOREM 13.5 Every sequence $\{f_n\}$ has an antidifference $\{F_n\}$. If $\{F_n\}$ is one antidifference, then $\{F_n + C\}$ is also an antidifference, for any constant C. There are no other antidifferences.

PROOF: The construction of F_n can be made by the method of steps, since we must have $F_{n+1} = F_n + f_n$. Taking $F_1 = C$ we obtain $F_2 = C + f_1$, $F_3 = C + f_1 + f_2$, and so on. Thus the only latitude in the choice of $\{F_n\}$ lies in the arbitrary constant C. ∎

We leave it to our readers to verify that the remaining equations and ideas in Section 13.1 carry over to the present situation. We turn now to a discussion of difference equations in which a set of integers J is the underlying domain of the functions. A general first order equation is now

$$\Delta y_n = G(n, y_n), \qquad n \in J \tag{3}$$

and a solution is a sequence $\{y_n\}$ satisfying Eq. (3) for $n \in J$. One can, of course, also consider higher order equations such as

$$n^2 \Delta^2 y_n + 3n \Delta y_n + y_n = 0 \tag{4}$$

DEFINITION 13.10   A difference equation with underlying domain J, a set of positive integers, will be called a *discrete difference equation* or a *finite difference equation*.

As before the equations in which we are interested can be written in dated form if desired. For example, Eq. (4) is equivalent to

$$n^2 y_{n+2} + (3n - 2n^2) y_{n+1} + (n^2 - 3n + 1) y_n = 0$$

The first order linear equation now has either of the following forms:

$$\Delta y_n = a_n y_n + b_n \tag{5}$$

$$y_{n+1} = c_n y_n + b_n \tag{6}$$

By following the derivation of Eqs. (11) and (16) in Section 13.2, or merely reinterpreting these, we obtain the following.

THEOREM 13.6   Let $\{b_n\}$, $\{c_n\}$ be given sequences. Then the sequence $\{y_n\}$ defined by $y_1 = C$, $y_2 = Cc_1 + b_1$,

$$y_n = C \prod_{i=1}^{n-1} c_i + b_{n-1} + c_{n-1} b_{n-2} + c_{n-1} c_{n-2} b_{n-3} + \cdots$$

$$+ \left\{ \prod_{i=2}^{n-1} c_i \right\} b_1 \qquad n = 3, 4, \ldots \tag{7}$$

is a solution of Eq. (6), for any constant C. Any solution of Eq. (6) is of this form for some constant C ($n = 1, 2, \ldots$).

Example 1   By Eq. (7) the equation $y_{n+1} = c_n y_n$ has as its most general solution

$$y_n = C \prod_{i=1}^{n-1} c_i$$

If the equation has constant coefficients, that is $y_{n+1} = cy_n$, then $y_n = Cc^{n-1}$. In this case, y_n increases exponentially if $c > 1$, decreases exponentially if $0 < c < 1$, or is constant if $c = 1$.

Example 2 Consider the initial value problem

$$y_{n+1} = (n + 1)y_n + n + 1, \qquad y_2 = 2$$

It was shown in Exercise 4 of Section 10.5 that y_n is the number of multiplications (of two numbers) which must be performed to evaluate an $n \times n$ determinant by complete expansion in cofactors. According to Eq. (7) above, since $c_n = n + 1 = b_n$, a general solution is

$$y_2 = 2C + 2$$
$$y_n = C(n!) + n + n(n - 1) + n(n - 1)(n - 2) + \cdots + n! \qquad (n \geq 3)$$

The initial condition $y_2 = 2$ implies $C = 0$, and so

$$\frac{y_n}{n!} = \frac{1}{(n - 1)!} + \frac{1}{(n - 2)!} + \cdots + \frac{1}{2!} + 1$$

Exercises

1. Compute Δy_n if $\{y_n\}$ is the sequence

   (a) $(-1)^n$ (b) e^n

   (c) $\sin \dfrac{n\pi}{3}$

2. Prove that each formula is valid.

   (a) $\Delta(ny_n) = (n + 1)\Delta y_n + y_n$
   (b) $\Delta(2^n y_n) = 2^{n+1}(\Delta y_n + \frac{1}{2} y_n)$

3. For each equation in differenced form, find the corresponding dated form, and vice versa.

   (a) $f_{n+2} - nf_{n+1} + f_n = n^2$
   (b) $\Delta^2 y_n + \tan(\Delta y_n) = 1$
   (c) $f_{n+2} = F(n, f_n, f_{n+1})$
   (d) $\Delta^2 y_n = F(n, y_n, \Delta y_n)$

4. Find a general solution of each finite difference equation. If possible, simplify the sums and products.

   (a) $y_{n+1} = 2y_n + 2^n$

   (b) $y_{n+1} = ny_n + \dfrac{1}{n}$

   (c) $y_{n+1} = (n!)y_n$

5. Find the unique solution of each initial value problem. Find the first few terms and if possible a general formula for y_n.

(a) $y_{n+1} - y_n = n$ $\qquad$ $y_1 = 1$

(b) $y_{n+1} = \left(1 + \dfrac{a}{n}\right) y_n$ $\qquad$ $y_1 = 1$

(c) $y_{n+1} - ay_n = b$ $\qquad$ $y_1 = c$

(d) $y_{n+2} - ny_n = 0$ $\qquad$ $y_1 = 0, \; y_2 = 1$

6. Let y_n denote the number of permutations of L different letters taken n at a time $(n \le L)$; that is, the number of distinct sequences of n letters selected in an arbitrary way from L letters. Show that

$$y_{n+1} = (L - n)y_n, \qquad y_1 = L$$

and deduce that

$$y_n = L(L - 1) \cdots (L - n + 1)$$

13.4 Systems of linear difference equations

We now wish to study more general classes of linear difference equations. One possibility is to study the scalar equation of kth order, attempting to find a theory analogous to that developed for differential equations in Chapters 8 and 9. Alternatively, we can study systems of first order equations and attempt to construct a theory analogous to that in Chapter 12.[3] For differential equations the latter approach is more elegant and more general, and in fact we think it is simpler when one is familiar with the notation of vectors and matrices. The same is true for difference equations of the kind we want to consider.

DEFINITION 13.11 A system of k difference equations of the form

$$\begin{aligned}
y_1(x + 1) &= f_1(x, y_1(x), \ldots, y_k(x)) \\
y_2(x + 1) &= f_2(x, y_1(x), \ldots, y_k(x)) \\
&\cdots \\
y_k(x + 1) &= f_k(x, y_1(x), \ldots, y_k(x))
\end{aligned} \qquad (1)$$

will be called a *first order system in normal form*.

---

[3] We assume the reader has read Chapters 11 and 12. However, those who have not done so can turn to Section 13.5 where we translate the results of this section to kth order scalar difference equations in such a way that the meaning and use of theorems should be clear without reading this section.

System (1) can be rewritten in the form

$$
\begin{aligned}
\Delta y_1(x) &= F_1(x, y_1(x), \ldots, y_k(x)) \\
\Delta y_2(x) &= F_2(x, y_1(x), \ldots, y_k(x)) \\
&\cdots \\
\Delta y_k(x) &= F_k(x, y_1(x), \ldots, y_k(x))
\end{aligned}
\tag{2}
$$

where $F_j = f_j - y_j(x)$, $j = 1, \ldots, k$. Conversely, a system of the form in Eq. (2) can be converted to one of the form in Eq. (1). We shall generally work with the dated form of System (1).

The use of vectors and matrices provides a simplification of notation and thought. We introduce the vectors

$$
y = \begin{pmatrix} y_1 \\ y_2 \\ \cdots \\ y_k \end{pmatrix} \qquad f = \begin{pmatrix} f_1 \\ f_2 \\ \cdots \\ f_k \end{pmatrix}
$$

and then Eq. (1) can be written

$$
y(x + 1) = f(x, y(x))
\tag{3}
$$

In this concise form the system has the appearance of the scalar first order equation studied in Section 13.2.

DEFINITION 13.12   A *solution* of system (3) on an interval $\alpha \le x < \beta$ is a vector function y defined on the interval $\alpha \le x < \beta + 1$ such that

1. The point $(x, y_1(x), \ldots, y_k(x))$ is in the domain of the function f for $\alpha \le x < \beta$, and
2. System (3) is satisfied for $\alpha \le x < \beta$.

We shall specialize at once to linear systems. If $f(x, y)$ is linear in y, then $f(x, y) = \mathscr{A}(x)y$ for some matrix $\mathscr{A}(x)$, as shown in Section 12.6. This suggests the following definitions.

DEFINITION 13.13   A first order normal system of difference equations is called a *linear homogeneous* system if it has the form

$$
y(x + 1) = \mathscr{A}(x)y(x)
\tag{4}
$$

and a *linear nonhomogeneous* system if it has the form

$$
y(x + 1) = \mathscr{A}(x)y(x) + b(x)
\tag{5}
$$

for $b(x) \not\equiv 0$.

In Eqs. (4) and (5), $\mathscr{A}(x)$ is a $k \times k$ matrix and $b(x)$ is a $k \times 1$ matrix or column vector. Equation (5) has the same form as Eq. (8) in Section 13.2, and some of the previous results have rather obvious generalizations. First, though, we want to be more precise about the space of functions to be considered in our discussion and the interval of x values to be allowed. Suppose Eq. (5) is to be considered for $x \geq \alpha$. The change of variable $z(x) = y(x + \alpha)$ then yields a new equation $z(x + 1) = \mathscr{\hat{A}}(x)z(x) + \hat{b}(x)$ to be considered for $x \geq 0$. Thus it is no real loss of generality to suppose from the beginning that Eq. (5) is to be satisfied for $x \geq 0$. Also, in order to avoid minor complications of notation, we suppose that $b(x)$ and $\mathscr{A}(x)$ are defined for all $x \geq 0$. If they are defined only for $0 \leq x \leq \beta < \infty$, minor changes in the developments below are required. Finally, there is no reason to restrict the allowable functions b, $\mathscr{A}$, y in any way. Therefore we consider b, y, and each column of $\mathscr{A}$ to lie in the vector space $\mathbf{V}_k$ of all k-vector[4] functions defined on $0 \leq x < \infty$, with range in $\mathbf{K}^n$ where $\mathbf{K}$ will be either the real or complex field.

With the above preparation we can now begin to develop a general theory for Eqs. (4) and (5).

THEOREM 13.7 The set of all solutions of Eq. (4) is a subspace of $\mathbf{V}_k$.

PROOF: The proof is left as an exercise for the reader. ∎

THEOREM 13.8 Let $\phi(x)$ be a given function defined for $0 \leq x < 1$. Then, there is one and only one function y in $\mathbf{V}_k$ satisfying the condition

$$y(x) = \phi(x), \qquad 0 \leq x < 1 \tag{6}$$

and satisfying Eq. (5) for $x \geq 0$. This solution is given by the formula

$$
\begin{aligned}
y(x) = {}& \mathscr{A}(x - 1)\phi(x - 1) + b(x - 1), \\
& \qquad\qquad\qquad 1 \leq x < 2, \\
y(x) = {}& \mathscr{A}(x - 1)\,\mathscr{A}(x - 2) \cdots \mathscr{A}(x - m)\phi(x - m) \\
& + b(x - 1) + \mathscr{A}(x - 1)b(x - 2) + \cdots \\
& + \mathscr{A}(x - 1)\,\mathscr{A}(x - 2) \cdots \mathscr{A}(x - m + 1)b(x - m), \\
& \qquad\qquad\qquad m \leq x < m + 1, \quad m = 2, 3, \ldots
\end{aligned}
\tag{7}
$$

PROOF: This theorem can be established by the method of steps in the same way that Eq. (11) of Section 13.2 was proved. In fact, the previous proof carries over without any change. The point is that if we are given $\phi(x)$ on $0 \leq x < 1$, we can generate $y(x)$ step-by-step (or as one says, *recursively*) on subsequent intervals. ∎

---

[4] k-dimensional vector functions.

In order to simplify Eq. (7) and others below, we can define the matrix functions $\mathcal{A}^{(j)}$ by

$$\mathcal{A}^{(j)}(x) = \prod_{i=0}^{j-1} \mathcal{A}(x-i), \quad x \geq j - 1, \quad j = 1, 2, \ldots$$

Note that j indicates the number of factors. Then Eq. (7) becomes

$$y(x) = \mathcal{A}(x-1)\phi(x-1) + b(x-1),$$
$$1 \leq x < 2$$
$$y(x) = \mathcal{A}^{(m)}(x-1)\phi(x-m) + b(x-1) \qquad (8)$$
$$+ \mathcal{A}(x-1)b(x-2) + \cdots + \mathcal{A}^{(m-1)}(x-1)b(x-m)$$
$$m \leq x < m + 1, \quad m = 2, 3, \ldots$$

The latter formula is valid for $m = 1$ if we define $\mathcal{A}^{(0)}(x) = I$.

DEFINITION 13.14 The *initial value problem* for Eq. (5) corresponding to a given vector function ϕ is the problem of finding the solution which satisfies condition (6). The function ϕ is called the *initial function* and Eq. (6) is called the *initial condition*.

COROLLARY 1 The only solution of the initial value problem

$$y(x+1) = \mathcal{A}(x)y(x), \qquad x \geq 0,$$
$$y(x) = 0, \qquad\qquad 0 \leq x < 1,$$

is $y(x) \equiv 0$ (for $x \geq 0$).

COROLLARY 2 If b and all columns of $\mathcal{A}$ are in V_k, the function y defined by

$$y(x) = p(x), \qquad\qquad\qquad\qquad 0 \leq x < 1,$$
$$y(x) = \mathcal{A}^{(m)}(x-1)p(x) + b(x-1)$$
$$+ \mathcal{A}(x-1)b(x-2) + \cdots + \mathcal{A}^{(m-1)}(x-1)b(x-m)$$
$$m \leq x < m + 1 \qquad (9)$$

where p is an arbitrary 1-periodic vector function, is a solution of Eq. (5) on $x \geq 0$. Any solution on $x \geq 0$ is of this form.

PROOF: Let p be any vector function of period 1. Regard $p(x)$ for $0 \leq x < 1$ as the initial function. Then the solution is given by Eq. (8). Since now $\phi(x-m) = p(x-m) = p(x)$, for $m \leq x < m + 1$, Eq. (8) reduces to Eq. (9). ∎

Of particular interest is the case in which $\mathscr{A}$ is a constant matrix. Then since $\mathscr{A}^{(j)} = \mathscr{A}^j$ Eq. (9) becomes

$$y(x) = p(x), \qquad 0 \le x < 1$$

$$y(x) = \mathscr{A}^m p(x) + \sum_{j=0}^{m-1} \mathscr{A}^j b(x - j - 1), \tag{10}$$

$$m \le x < m + 1, m = 1, 2, \ldots$$

As we see, the problem under consideration is simpler than that for differential equations in that we have a recursive formula for the solutions. Let us now see what theoretical consequences can be derived from this formula. When we were dealing with systems of differential equations $y' = A(x)y$, our basic theoretical result was that the solution space has finite dimension equal to the order of the system (Theorem 12.4). This is no longer true for the difference system $y(x + 1) = \mathscr{A}(x)y(x)$. Since the solution contains an arbitrary periodic function $p(x)$, the space of solutions has infinite dimension; that is, there is no finite set of 1-periodic functions with the property that their span is the space of all 1-periodic functions. On the other hand, Eq. (9) contains only one arbitrary $p(x)$, with k components, which suggests that there is a kind of k-dimensional property here. Can we define precisely what this means? The following shows one way to do this.

DEFINITION 13.15   Let S_k be the space of all sequences $\{a_i\} = (a_0, a_1, a_2, \ldots)$ where each a_i is in $\mathbf{K}^k$. Multiplication of a sequence by any c in $\mathbf{K}$ is defined by $c\{a_i\} = \{ca_i\}$ and addition of sequences by $\{a_i\} + \{b_i\} = \{a_i + b_i\}$. The zero sequence is the one in which every a_i is the zero vector.

THEOREM 13.9   Consider the equation $y(x + 1) = \mathscr{A}(x)y(x)$ on the sequence of points $x = x_0, x_0 + 1, x_0 + 2, \ldots$, for any fixed $x_0, 0 \le x_0 < 1$. There is a unique solution sequence satisfying the initial condition $y(x_0) = y_0$, where y_0 is a given k-vector. This solution has the form

$$y(x_0 + n) = \mathscr{A}(x_0 + n - 1) \cdots \mathscr{A}(x_0)y_0 = \mathscr{A}^{(n)}(x_0 + n - 1)y_0,$$
$$n = 1, 2, \ldots \tag{11}$$

The family of all solution sequences (with x_0 fixed but y_0 arbitrary) is a k-dimensional subspace of the space S_k.

PROOF: Applying Eq. (8) or (9) with $b(x) \equiv 0$ and $x = x_0 + n$, we obtain Eq. (11). This equation says, in effect, that if we restrict attention to the points $x = x_0, x_0 + 1, \ldots$, the solution depends on one arbitrary k-vector y_0.

Hence, the solution space must have dimension k. More precisely, from Eq. (11) we can write $y(x_0 + n)$ as a sum of vectors

$$y(x_0 + n) = \sum_{j=1}^{k} \mathscr{A}(x_0 + n - 1) \cdots \mathscr{A}(x_0) y_{0j} e_j$$

where e_j is the vector with all zero entries except for 1 in the jth component, and y_{0j} is the jth component of y_0. Therefore,

$$y(x_0 + n) = \sum_{j=1}^{k} y_{0j} y^{(j)}(x_0 + n) \qquad (12)$$

where $y^{(j)}$ is the solution starting from the special initial condition $y^{(j)}(x_0) = e_j$. By Eq. (12), the sequence $\{y(x_0 + n)\}$, $n = 0, 1, \ldots$, is a linear combination of the special solution sequences $\{y^{(j)}(x_0 + n)\}$ for $j = 1, \ldots, k$. It is left as an exercise to show that these k sequences are linearly independent in S_k. This will complete the proof. ∎

The meaning of Theorem 13.9 is as follows. The values of the solution at $x_0 + 1$, $x_0 + 2$, and so on, depend only on the value at x_0 (and on $\mathscr{A}$); the initial values in $0 \leq x < 1$ at points other than x_0 do not affect the solution at $x_0 + 1$, $x_0 + 2, \ldots$. The initial values do not "interact." Another way to say this is that $y(x_0 + n)$ depends only on $y(x_0)$ and on the mapping matrix $\mathscr{A}(x_0 + n - 1) \cdots \mathscr{A}(x_0)$. We can think of the initial value problem in the continuous case as generating a family of initial value problems, one for each initial point x_0, and each of these is for a finite difference equation.

Example 1 Consider the system with constant matrix

$$\mathscr{A} = \begin{pmatrix} 2 & 0 \\ 0 & 1 \end{pmatrix}$$

Suppose the initial condition is

$$y(x) = \phi(x) = \begin{pmatrix} \phi_1(x) \\ \phi_2(x) \end{pmatrix} \quad \text{for} \quad 0 \leq x < 1$$

Then by Eq. (11)

$$y(x_0 + n) = \mathscr{A}^n \begin{pmatrix} \phi_1(x_0) \\ \phi_2(x_0) \end{pmatrix} \qquad n = 0, 1, 2, \ldots, \quad 0 \leq x_0 < 1$$

Evidently, a knowledge of all powers of the matrix $\mathscr{A}$ will provide complete information about the solution for any initial ϕ. We shall have more to say below about equations with constant coefficients.

Our next observation in this section concerns the nonhomogeneous system. Just as for systems of differential equations, every solution of Eq. (5) is of the form $y_p + y$ where y_p is a particular solution of Eq. (5) and y is some solution of Eq. (4).

Example 2   Consider the system

$$y(n + 1) = \begin{pmatrix} 2 & 0 \\ 0 & 1 \end{pmatrix} y(n) - \begin{pmatrix} n - 1 \\ 0 \end{pmatrix}, \qquad n = 0, 1, \ldots$$

By Example 1, every solution of the homogeneous system is of the form $\mathscr{A}^n$ times a constant vector. It can be verified that $\begin{pmatrix} n \\ 3 \end{pmatrix}$ is a particular solution of the nonhomogeneous equation. Therefore the set of all solutions is given by the expression

$$y(n) = \begin{pmatrix} 2^n & 0 \\ 0 & 1 \end{pmatrix} \begin{pmatrix} a \\ b \end{pmatrix} + \begin{pmatrix} n \\ 3 \end{pmatrix} = \begin{pmatrix} 2^n a + n \\ b + 3 \end{pmatrix}$$

where a and b are arbitrary constants. Methods for finding particular solutions are discussed below.

For difference systems on a discrete set, it is easy to introduce the concept of a fundamental matrix and to develop a theory similar to that in Section 12.8.

DEFINITION 13.16   Consider the difference system $y(x + 1) = \mathscr{A}(x)y(x)$ on $x = x_0, x_0 + 1, x_0 + 2, \ldots, 0 \leq x_0 < 1$. A set of k solution sequences linearly independent in S_k is called a *fundamental set* of solutions. A $k \times k$ matrix sequence

$$\Phi(x_0 + n) = [\phi^{(1)}(x_0 + n) \quad \cdots \quad \phi^{(k)}(x_0 + n)], \qquad n = 0, 1, 2, \ldots$$

in which each of the k columns is a vector solution sequence is called a *solution matrix*. It is called a *fundamental matrix* if its k column sequence form a basis for the solution space.

COROLLARY 3   Let Φ be defined by $\Phi(x_0) = I$ and

$$\Phi(x_0 + n) = \mathscr{A}^{(n)}(x_0 + n - 1) \qquad n = 1, 2, \ldots$$

Then Φ is a fundamental matrix.

PROOF: By Eq. (11), every solution is a linear combination of the columns of Φ. Since the solution space has dimension k, all columns of Φ are necessary to obtain a basis.   ∎

It is possible to develop these concepts further, in analogy to those in Section 12.8, but we omit this here because of space restrictions.

Exercises

1. Consider again the problem of drug transport through tissue, discussed in Example 2 of Section 12.1. Now, however, let us imagine that we consider only a sequence of discrete times $t = 0, h, 2h, 3h, \ldots$ $(h > 0)$. Let $y_i(n)$ be the drug concentration in the ith compartment at time nh. Define α and β to be the rates of transfer as before. For small h the change of concentration in compartment 1 in the time from nh to $(n + 1)h$ will be approximately

$$y_1(n + 1) - y_1(n) = \beta y_2(n)h - \alpha y_1(n)h$$

Find a system of four difference equations which describes the four-compartment model. Do you think that the system of differential equations or the system of difference equations should be a better model? Which do you expect to be easier to use for numerical calculations?

2. Consider the finite difference system

$$\Delta y_1(n) = f_1(n, y_1(n), y_2(n))$$
$$\Delta y_2(n) = f_2(n, y_1(n), y_2(n))$$

At each point (n, y_1, y_2) in three-dimensional space at which n is an integer, the equation determines a unique direction, that of the vector

$$(1, f_1(n, y_1, y_2), f_2(n, y_1, y_2))$$

Show that this vector drawn from $(n, y_1(n), y_2(n))$ terminates at $(n + 1, y_1(n + 1), y_2(n + 1))$. Thus, the difference equations define a *vector field* at all points (n, y_1, y_2) such that solutions are obtained by following the field direction.

3. Verify that the sequences

$$y_n = \begin{pmatrix} 1 \\ -1 \end{pmatrix}(-1)^n \quad \text{and} \quad y_n = \begin{pmatrix} 2 \\ -3 \end{pmatrix}(-2)^n$$

are solutions of the system

$$y_{n+1} = \begin{pmatrix} 1 & 2 \\ -3 & -4 \end{pmatrix} y_n$$

4. Find the solution of the initial value problem

$$y_{n+1} = -y_n$$
$$z_{n+1} = y_n + z_n$$
$$y_1 = 2, \quad z_1 = 1$$

by first solving for y_n.

5. Solve this initial value problem for $1 \leq x < 3$:

$$y(x + 1) = \begin{pmatrix} 1 & -1 \\ 2 & 0 \end{pmatrix} y(x) + \begin{pmatrix} 0 \\ -2 \end{pmatrix}$$

$$y(x) = \begin{pmatrix} 1 \\ x \end{pmatrix} \quad \text{on} \quad 0 \leq x < 1$$

6. Show that $\begin{pmatrix} 1 \\ 1 \end{pmatrix}$ is a particular solution of the system

$$y(n + 1) = \begin{pmatrix} -1 & 0 \\ 0 & 2 \end{pmatrix} y(n) + \begin{pmatrix} 2 \\ -1 \end{pmatrix}$$

and find the most general solution.

7. Let δ be the operator on the space $\mathbf{V}_k$ such that

$$\delta y(x) = \begin{pmatrix} y_1(x + 1) \\ \cdots \\ y_k(x + 1) \end{pmatrix}, \quad 0 \leq x$$

Show that δ is a linear operator on $\mathbf{V}_k$. Is it one-to-one and onto?

8. Let $\mathscr{A}(x)$ be a $k \times k$ matrix and let T be the mapping on $\mathbf{V}_k$ such that

$$Ty(x) = \delta y(x) - \mathscr{A}(x)y(x), \quad 0 \leq x$$

Is this a linear transformation of $\mathbf{V}_k$ into $\mathbf{V}_k$ for every $\mathscr{A}(x)$?

9. Find a subspace of $\mathbf{V}_k$ such that the operator T of Exercise 8 is one-to-one from the subspace onto $\mathbf{V}_k$. [*Hint*: Consider Theorem 13.8.]

10. Complete the proof of Theorem 13.9 by showing that the solutions with initial conditions e_j $(j = 1, \ldots, k)$ are linearly independent in $\mathbf{S}_k$.

11. Prove Theorem 13.7.

13.5   The *k*th order scalar equation

This section is devoted to the scalar kth order equation

$$z(x + k) + a_{k-1}(x)z(x + k - 1) + \cdots$$
$$+ a_1(x)z(x + 1) + a_0(x)z(x) = f(x), \quad x \geq 0 \quad (1)$$

where $a_0, \ldots, a_{k-1}, f$ are given scalar functions. We treat this equation by converting it to a system of first order equations, but the theorems and examples are framed in such a way as to be understandable even if Section 13.4 has not been read.

To change Eq. (1) to a system, we introduce new variables $y_1, \ldots, y_k$ by the equations

$$y_1(x) = z(x), \quad y_2(x) = z(x + 1), \ldots, y_k(x) = z(x + k - 1) \qquad (2)$$

If $z(x)$ is defined for $x \geq 0$, then $y_j(x)$ is defined for $x \geq 1 - j$. Moreover, it is easy to see that if z satisfies Eq. (1) for $x \geq 0$, then the k-vector y satisfies

$$y(x + 1) = \mathcal{A}(x)y(x) + b(x), \qquad x \geq 0 \qquad (3)$$

where

$$\mathcal{A}(x) = \begin{pmatrix} 0 & 1 & 0 & \cdots & 0 \\ 0 & 0 & 1 & \cdots & 0 \\ & & \cdots & & \\ 0 & 0 & 0 & \cdots & 1 \\ -a_0 & -a_1 & -a_2 & \cdots & -a_{k-1} \end{pmatrix}, \quad b(x) = \begin{pmatrix} 0 \\ 0 \\ \cdots \\ 0 \\ f(x) \end{pmatrix} \qquad (4)$$

We shall refer to Eq. (3), with $\mathcal{A}$ and b as in Eq. (4), as the *companion equation* of Eq. (1). As noted above, if $z(x)$ satisfies Eq. (1) for $x \geq 0$, then $y(x)$ satisfies Eq. (3) for $x \geq 0$. Conversely, suppose $y(x)$ is defined and is a solution of Eq. (3) for $x \geq 0$. Define $z(x) = y_1(x)$ for $x \geq 0$. Then it is easy to see that $z(x)$ satisfies Eq. (1) for $x \geq 0$. In other words, Eqs. (1) and (3) through (4) are equivalent.

We know from Theorem 13.8 that in order to determine a unique solution of Eq. (3) we need only specify that $y(x) = \phi(x)$ on $0 \leq x < 1$, where ϕ is a given vector function. Looking at Eq. (2) we observe that this corresponds to giving $z(x)$ on $0 \leq x < 1$, $z(x + 1)$ on $0 \leq x < 1, \ldots, z(x + k - 1)$ on $0 \leq x < 1$, or in other words giving $z(x)$ on $0 \leq x < k$. This proves the following existence-uniqueness theorem.

THEOREM 13.10 Let ζ be a given function defined for $0 \leq x < k$. Then there is one and only one function z which is a solution of Eq. (1) for $x \geq 0$ and which satisfies the initial condition

$$z(x) = \zeta(x), \qquad 0 \leq x < k \qquad (5)$$

For systems of difference equations, Eqs. (7), (8), and (9) of Section 13.4 provide explicit representations of solutions, from which we may try to obtain a formula for the solutions of Eq. (1). For the homogeneous equation we have $f(x) = 0$, $b(x) = 0$, and all that is needed is the first row of $\mathcal{A}^{(m)}(x - 1)$. It would be fortunate if we could compute this directly in terms of the coefficients $a_0, a_1, \ldots, a_{k-1}$ in Eq. (1), but the formulas obtained are too complicated to be of much value (see Exercise 3). For constant $\mathcal{A}$ we will show below how to obtain explicit solutions, but for the moment we proceed as follows. Consider the homogeneous equation and let

$$\mathcal{A}^{(m)}(x) = \begin{pmatrix} a_{11}^{(m)}(x) & \cdots & a_{1k}^{(m)}(x) \\ \vdots & \cdots & \vdots \\ a_{k1}^{(m)}(x) & \cdots & a_{kk}^{(m)}(x) \end{pmatrix} \qquad x \geq m - 1 \qquad (6)$$

Then from Eq. (8) of Section 13.4 we obtain

$$z(x) = y_1(x) = \sum_{j=1}^{k} a_{1j}{}^{(m)}(x-1)\phi_j(x-m),$$

$$m \leq x < m+1, \, m = 1, 2, \ldots$$

By Eq. (5) for $m \leq x < m+1$ we have

$$\phi(x-m) = \begin{pmatrix} y_1(x-m) \\ y_2(x-m) \\ \cdots \\ y_k(x-m) \end{pmatrix} = \begin{pmatrix} \zeta(x-m) \\ \zeta(x-m+1) \\ \cdots \\ \zeta(x-m+k-1) \end{pmatrix}$$

and so we can write the solution as in the following theorem.

THEOREM 13.11   The unique solution of Eq. (1) with $f(x) \equiv 0$, satisfying the initial condition (5), is

$$z(x) = \sum_{j=1}^{k} a_{1j}{}^{(m)}(x-1)\zeta(x-m+j-1),$$

$$m \leq x < m+1, \, m = 1, 2, \ldots \qquad (7)$$

where $a_{1j}{}^{(m)}$ is the jth entry in the first row of $\mathscr{A}^{(m)}$.

If we start with Eq. (9) instead of Eq. (8) in Section 13.4, we obtain the following:

THEOREM 13.12   Let $p_1, p_2, \ldots, p_k$ be arbitrary 1-periodic scalar functions. Let

$$z(x) = p_1(x), \qquad 0 \leq x < 1$$

$$= \sum_{j=1}^{k} a_{1j}{}^{(m)}(x-1)p_j(x), \qquad m \leq x < m+1, \, m = 1, 2, \ldots \qquad (8)$$

Then z is a solution of Eq. (1) for $x \geq 0$. Every solution is of this form.

This last theorem is helpful in showing that the general solution of the homogeneous problem involves k arbitrary functions of period one.

Now let us examine a solution on a sequence of points $x_0, x_0 + 1, \ldots$. We expect that, as before, the theory will take a simpler form. From Theorem 13.11 we deduce the following.

THEOREM 13.13   Consider Eq. (1), with $f(x) \equiv 0$, on the sequence of points $x = x_0, x_0 + 1, x_0 + 2, \ldots$ only. There is a unique solution sequence satisfying the initial condition

$$z(x_0) = \zeta_1, \qquad z(x_0 + 1) = \zeta_2, \qquad \ldots, \qquad z(x_0 + k - 1) = \zeta_k \qquad (9)$$

where $\zeta_1, \zeta_2, \ldots, \zeta_k$ are given numbers. This solution has the form

$$z(x_0 + n) = \sum_{j=1}^{k} a_{1j}{}^{(n)}(x_0 + n - 1)\zeta_j, \qquad n = 1, 2, \ldots \qquad (10)$$

Thus, the family of all solution sequences is a k-dimensional subspace of the space S_k of all sequences of k-vectors.

From a practical point of view, the trouble with all these results lies in the difficulty of computing $a_{1j}{}^{(m)}(x - 1)$ or $a_{1j}{}^{(n)}(x_0 + n - 1)$. With the aid of modern computers, however, this can be done with sufficient speed and accuracy in many cases. In the next section, we shall show how to obtain solutions in "closed form" when the equation has constant coefficients. Also, we shall give below a few indications of how to find a particular solution of the nonhomogeneous equation.

Exercises

1. Find the companion system of each of the following equations.

    (a) $z(x + 2) - 2z(x + 1) + 5z(x) = x$
    (b) $z(x + 2) - x^2 z(x) = 0$
    (c) $z(x + 4) - 2z(x + 2) + z(x) = 0$

2. The system

    $$y(n + 1) = \begin{pmatrix} 0 & 1 \\ 2 & 3 \end{pmatrix} y(n)$$

    is the companion to a scalar second order difference equation. Find the equation. Find the first four elements $z(1), z(2), z(3), z(4)$ in its solution and verify that

    $$y_1(n) = z(n), \qquad y_2(n) = z(n + 1), \qquad n = 1, 2, 3$$

3. Let $\mathscr{A}(x)$ and $b(x)$ be defined by Eq. (4) and let

    $$\mathscr{A}^{(j)}(x) = \prod_{i=0}^{j-1} \mathscr{A}(x - i), \qquad x \geq j - 1$$

    (a) Compute $\mathscr{A}^{(2)}(x)$.
    (b) Show that the first row of $\mathscr{A}^{(j)}(x)$ is the row vector e_{j+1}^T for $j \leq k - 1$.
    (c) Show that Eq. (7) reduces to $z(x) = \zeta(x)$ for $0 \leq x < k$.
    (d) Show that Eq. (8) reduces to $z(x) = p_{m+1}(x)$ for $m \leq x < m + 1$, $m = 1, 2, \ldots, k - 1$.

4. (a) Deduce from Eq. (8) of Section 13.4 that the solution of Eq. (1) satisfying Eq. (5) is, for any $f(x)$, given by the formula

$$z(x) = \sum_{j=1}^{k} a_{1j}{}^{(m)}(x - 1)\zeta(x - m + j - 1)$$
$$+ a_{1k}(x - 1)f(x - 2) + a_{1k}{}^{(2)}(x - 1)f(x - 3)$$
$$+ \cdots + a_{1k}^{(m-1)}(x - 1)f(x - m), \qquad m \le x < m + 1$$

where $a_{1k}{}^{(j)}(x - 1)$ is the element in row 1, column k, of $\mathscr{A}^{(j)}(x - 1)$.

(b) Using the result in Exercise 3, show that

$$a_{1k}(x - 1)f(x - 2) + \cdots + a_{1k}{}^{(m-1)}(x - 1)f(x - m) = 0$$

for $m \le x < m + 1$ if $m \le k - 1$.

13.6   Systems with constant coefficients

It was shown in Chapter 12 that e^{Ax} is a fundamental solution for systems of differential equations $y' = Ay$ with constant coefficient matrix A. Moreover, it was shown that the function e^{Ax} can be calculated in a purely algebraic way by computing the eigenvalues and eigenvectors of A. It is not surprising that a similar procedure should work for systems of difference equations. As a matter of fact, Eq. (10) of Section 13.4 shows at once that all we have to do is to compute the exponential function $\mathscr{A}^m$, $m = 1, 2, \ldots$.

Suppose that $\mathscr{A}$ is a diagonable matrix. Then, there is a nonsingular matrix P, such that $\mathscr{A} = PDP^{-1}$ where

$$D = \begin{pmatrix} \lambda_1 & 0 & & 0 \\ 0 & \lambda_2 & & \\ & & \ddots & \\ 0 & & & \lambda_k \end{pmatrix}$$

that is, D is the diagonal matrix with eigenvalues on the main diagonal. It was shown in Chapter 11 that P can be taken to be a matrix with the eigenvectors as its columns. Now clearly $\mathscr{A}^2 = PDP^{-1}PDP^{-1} = PD^2P^{-1}$, and in general,

$$\mathscr{A}^m = PD^mP^{-1} = P\begin{pmatrix} \lambda_1{}^m & & 0 \\ & \ddots & \\ 0 & & \lambda_k{}^m \end{pmatrix}P^{-1} \qquad (1)$$

where if $\lambda_i = 0$ we take $\lambda_i{}^0 = 1$. From Eq. (10) of Section 13.4, it follows that the system $y(x + 1) = \mathscr{A}y(x)$ has a general solution given for $m \le x < m + 1$, $m = 0, 1, 2, \ldots$, by

$$y(x) = P\begin{pmatrix} \lambda_1{}^m & & 0 \\ & \ddots & \\ 0 & & \lambda_k{}^m \end{pmatrix}P^{-1}p(x) \qquad (2)$$

where $p(x)$ is an arbitrary function of period 1.

Example 1 Let

$$\mathcal{A} = \begin{pmatrix} 1 & 1 \\ 4 & 1 \end{pmatrix}$$

The eigenvalues of $\mathcal{A}$ are -1, 3, and a matrix P which has the corresponding eigenvectors as columns is

$$P = \begin{pmatrix} 1 & 1 \\ -2 & 2 \end{pmatrix}$$

Also

$$P\begin{pmatrix} (-1)^m & 0 \\ 0 & 3^m \end{pmatrix}P^{-1} = \begin{pmatrix} 1 & 1 \\ -2 & 2 \end{pmatrix}\begin{pmatrix} (-1)^m & 0 \\ 0 & 3^m \end{pmatrix}\begin{pmatrix} \frac{1}{2} & -\frac{1}{4} \\ \frac{1}{2} & \frac{1}{4} \end{pmatrix}$$

$$= \begin{pmatrix} \frac{1}{2}(-1)^m + \frac{1}{2}3^m & -\frac{1}{4}(-1)^m + \frac{1}{4}3^m \\ -(-1)^m + 3^m & \frac{1}{2}(-1)^m + \frac{1}{2}3^m \end{pmatrix}$$

From Eq. (2) we obtain, for $m \leq x < m + 1$,

$$y(x) = \begin{pmatrix} [\frac{1}{2}(-1)^m + \frac{1}{2}3^m]p_1(x) + [-\frac{1}{4}(-1)^m + \frac{1}{4}3^m]p_2(x) \\ [-(-1)^m + 3^m]p_1(x) + [\frac{1}{2}(-1)^m + \frac{1}{2}3^m]p_2(x) \end{pmatrix}$$

If we let

$$\tilde{p}_1(x) = \tfrac{1}{2}p_1(x) - \tfrac{1}{4}p_2(x)$$
$$\tilde{p}_2(x) = \tfrac{1}{2}p_1(x) + \tfrac{1}{4}p_2(x)$$

then we can write, for $m \leq x < m + 1$, $m = 0, 1, \ldots$,

$$y(x) = \tilde{p}_1(x)\begin{pmatrix} 1 \\ -2 \end{pmatrix}(-1)^m + \tilde{p}_2(x)\begin{pmatrix} 1 \\ 2 \end{pmatrix}3^m \tag{3}$$

The functions $\tilde{p}_1, \tilde{p}_2$ have period 1. In Eq. (3), $y(x)$ is expressed as a combination (with periodic multipliers) of the eigenvectors multiplied by the mth power of the eigenvalues. The following theorem shows that the procedure in this example is always applicable when $\mathcal{A}$ is diagonable.

THEOREM 13.14 Let $\mathcal{A}$ be diagonable over the field **K** and let $\{v_1, v_2, \ldots, v_k\}$ be a linearly independent set of k eigenvectors. Let $\lambda_1, \ldots, \lambda_k$ be the associated eigenvalues (not necessarily distinct). Then a general solution of the system $y(x + 1) = \mathcal{A}y(x)$ is

$$y(x) = \sum_{j=1}^{k} v_j \lambda_j^m \tilde{p}_j(x), \qquad m \leq x < m + 1, m = 0, 1, 2, \ldots \tag{4}$$

where $\tilde{p}_1(x), \ldots, \tilde{p}_k(x)$ are arbitrary functions with period 1.

PROOF: As noted, we can take $P = (v_1 v_2 \cdots v_k)$; that is, the columns of P are the k independent eigenvectors. Let $\tilde{p}(x) = P^{-1}p(x)$. Then $\tilde{p}(x)$ has period 1. From Eq. (2) we obtain, for $m \le x < m + 1$, $m \ge 0$,

$$y(x) = (v_1 \; v_2 \cdots v_k) \begin{pmatrix} \lambda_1{}^m & & 0 \\ & \ddots & \\ 0 & & \lambda_k{}^m \end{pmatrix} \tilde{p}(x)$$

$$= (v_1\lambda_1{}^m \quad v_2\lambda_2{}^m \quad \cdots \quad v_k\lambda_k{}^m)\,\tilde{p}(x)$$

$$= \sum_{j=1}^{k} v_j\lambda_j{}^m\tilde{p}_j(x) \tag{5}$$

where the $\tilde{p}_j(x)$ are the components of $\tilde{p}(x)$. Thus, every solution is of the form in Eq. (4). [Conversely, every function y of the form in Eq. (5) is also of the form in Eq. (2), where $p(x) = P\tilde{p}(x)$, and so is a solution.]   ∎

Theorem 13.14 enables us to write the solution immediately, when the eigenvalues and eigenvectors of $\mathcal{A}$ have been computed.

Now we ask what happens if $\mathcal{A}$ is real but some of its eigenvalues and eigenvectors are complex. As before, every solution has the form in Eq. (4) for some choice of periodic functions $\tilde{p}_j(x)$. In general, $\tilde{p}(x) = P^{-1}p(x)$ will be complex since P^{-1} will be. However, if the $\tilde{p}_j(x)$ are arbitrary complex periodic functions, Eq. (4) may produce complex functions. Suppose $y(x)$ is a real solution. Then

$$\bar{y}(x) = y(x) = \sum_{j=1}^{k} \bar{v}_j\bar{\lambda}_j{}^m\bar{\tilde{p}}_j(x) \tag{6}$$

where $\bar{y}$ denotes the complex conjugate of y, and so on. Adding Eqs. (4) and (6) we get

$$y(x) = \frac{1}{2} \sum_{j=1}^{k} [v_j\lambda_j{}^m\tilde{p}_j(x) + \bar{v}_j\bar{\lambda}_j{}^m\bar{\tilde{p}}_j(x)]$$

$$= \sum_{j=1}^{k} \mathrm{Re}\,[v_j\lambda_j{}^m\tilde{p}_j(x)] \tag{7}$$

Let $\tilde{p}_j(x) = \tilde{q}_j(x) + i\tilde{r}_j(x)$. Then, $\tilde{q}_j$ and $\tilde{r}_j$ are real and 1-periodic and

$$y(x) = \sum_{j=1}^{k} \mathrm{Re}\,[v_j\lambda_j{}^m]\tilde{q}_j(x) - \sum_{j=1}^{k} \mathrm{Im}\,[v_j\lambda_j{}^m]\tilde{r}_j(x) \tag{8}$$

Now if λ_j is real and v_j is a corresponding real eigenvector, $\mathrm{Im}\,[v_j\lambda_j{}^m] = 0$, and

$$y(x) = \sum_{j=1}^{k} v_j\lambda_j{}^m\tilde{q}_j(x)$$

But if λ_j and v_j are not real, then we must retain the separate terms. We have proved the first half of the following theorem.

THEOREM 13.15 Let $\mathscr{A}$ be a real $k \times k$ matrix. Let its real eigenvalues be $\lambda_1, \ldots, \lambda_s$ and its complex eigenvalues be $\mu_1 \pm iv_1, \ldots, \mu_t \pm iv_t$. Let real eigenvectors of $\lambda_1, \ldots, \lambda_s$ be $u_1, \ldots, u_s$ and let the complex eigenvalues have complex eigenvectors $v_1, \ldots, v_t, \bar{v}_1, \ldots, \bar{v}_t$. Then every real solution of $y(x + 1) = \mathscr{A}y(x)$ has the form (on $m \leq x < m + 1, m = 0, 1, 2, \ldots$)

$$y(x) = \sum_{j=1}^{s} u_j \lambda_j{}^m p_j(x) + \sum_{j=1}^{t} \text{Re}\, [v_j(\mu_j + iv_j)^m]q_j(x)$$

$$- \sum_{j=1}^{t} \text{Im}\, [v_j(\mu_j + iv_j)^m]r_j(x) \tag{9}$$

where p_j, q_j, and r_j are real and 1-periodic. Every function of the form in Eq. (9) is a real solution.

PROOF: To prove the last assertion, we note that any function of the form in Eq. (9) can be written in the form of Eq. (8), with q_j and r_j redefined in an appropriate way. From Eq. (8), we obtain Eq. (7). Each term in the sum in Eq. (7) is a solution, by the eigenvalue property; that is, for $m \leq x \leq m + 1$

$$\mathscr{A}y(x) = \tfrac{1}{2} \Sigma[\mathscr{A}v_j \lambda_j{}^m \tilde{p}_j(x) + \mathscr{A}\bar{v}_j \bar{\lambda}_j{}^m \bar{\tilde{p}}_j(x)]$$
$$= \tfrac{1}{2} \Sigma[\lambda_j^{m+1} v_j \tilde{p}_j(x) + \bar{\lambda}_j^{m+1} \bar{v}_j \bar{\tilde{p}}_j(x)] = y(x + 1) \qquad \blacksquare$$

Example 2 Let

$$\mathscr{A} = \begin{pmatrix} 0 & 1 \\ -w^2 & 0 \end{pmatrix}$$

The eigenvalues and eigenvectors, found in Section 12.15, are

$$\lambda = iw \qquad v = \begin{pmatrix} 1 \\ iw \end{pmatrix}$$

$$\bar{\lambda} = -iw \qquad \bar{v} = \begin{pmatrix} 1 \\ -iw \end{pmatrix}$$

From Eq. (2), a general solution is, for $m \leq x < m + 1$,

$$y(x) = P\begin{pmatrix} (iw)^m & 0 \\ 0 & (-iw)^m \end{pmatrix} P^{-1}p(x), \qquad P = \begin{pmatrix} 1 & 1 \\ iw & -iw \end{pmatrix}$$

From Eq. (4), a general solution is

$$y(x) = \begin{pmatrix} 1 \\ iw \end{pmatrix} (iw)^m \tilde{p}_1(x) + \begin{pmatrix} 1 \\ -iw \end{pmatrix} (-iw)^m \tilde{p}_2(x) \tag{10}$$

To exhibit a real general solution clearly, we use Eq. (9) of Theorem 13.15, and get

$$y(x) = \text{Re}\left[\begin{pmatrix} 1 \\ iw \end{pmatrix}(iw)^m\right]q(x) - \text{Im}\left[\begin{pmatrix} 1 \\ iw \end{pmatrix}(iw)^m\right]r(x)$$

If m is even, $m = 2n$, we have

$$\text{Re}\left[\begin{pmatrix} 1 \\ iw \end{pmatrix}(iw)^m\right] = \begin{pmatrix} (-1)^n w^{2n} \\ 0 \end{pmatrix}$$

$$\text{Im}\left[\begin{pmatrix} 1 \\ iw \end{pmatrix}(iw)^m\right] = \begin{pmatrix} 0 \\ (-1)^n w^{2n+1} \end{pmatrix}$$

Therefore for $2n \leq x < 2n + 1$,

$$y(x) = \begin{pmatrix} (-1)^n w^{2n} q(x) \\ (-1)^{n+1} w^{2n+1} r(x) \end{pmatrix} = (-1)^n w^{2n}\begin{pmatrix} q(x) \\ -wr(x) \end{pmatrix}$$

Similarly, for $2n - 1 \leq x < 2n$,

$$y(x) = \begin{pmatrix} (-1)^n w^{2n-1} r(x) \\ (-1)^n w^{2n} q(x) \end{pmatrix} = (-1)^n w^{2n-1}\begin{pmatrix} r(x) \\ wq(x) \end{pmatrix}$$

Here q and r are 1-periodic real functions.

Once again, when we pass to difference equations on a sequence x_0, $x_0 + 1, \ldots$, we can state the results more simply. For example, Theorem 13.14 is replaced by the following.

THEOREM 13.16 Let $\mathscr{A}$ be diagonable over the field $\mathbf{K}$ and let $\{v_1, v_2, \ldots, v_k\}$ be a linearly independent set of k eigenvectors. Let $\lambda_1, \lambda_2, \ldots, \lambda_k$ be the associated eigenvalues (not necessarily distinct). Then the solution space of $y(x + 1) = \mathscr{A} y(x)$, considered on $x = x_0, x_0 + 1, x_0 + 2, \ldots, 0 \leq x_0 < 1$, has a basis consisting of the k sequences

$$\{v_j \lambda_j^n\} \quad j = 1, \ldots, k; \qquad n = 0, 1, 2, 3, \ldots \tag{11}$$

Proof: By Theorem 13.14, a general solution is

$$y(x_0 + n) = \sum_{j=1}^{k} v_j \lambda_j^n \tilde{p}_j(x_0 + n), \qquad n = 0, 1, \ldots$$

Since $\tilde{p}_j(x_0 + n) = \tilde{p}_j(x_0)$ is a constant, this shows that the sequence $\{y(x_0 + n)\}$ is a linear combination of the k sequences in Eq. (11). ∎

We leave it to the reader to reformulate Theorem 13.15 for this situation.

Because of space limitations, we omit a general discussion of the case of nondiagonable matrices $\mathscr{A}$. Also, we will not consider the nonhomogeneous system further. See the references given at the end of the chapter for additional information.

Exercises

1. For each system $y(x + 1) = \mathscr{A} y(x)$, compute three independent eigenvectors and find a real general solution.

    (a)
    $$\mathscr{A} = \begin{pmatrix} 0 & 1 & 0 \\ 0 & 0 & 1 \\ 0 & 1 & 0 \end{pmatrix}$$

    (b)
    $$\mathscr{A} = \begin{pmatrix} 0 & 0 & 1 \\ 0 & 0 & 4 \\ 0 & 1 & 0 \end{pmatrix}$$

    (c)
    $$\mathscr{A} = \begin{pmatrix} -1 & -2 & 2 \\ -2 & -1 & 2 \\ -2 & -2 & 3 \end{pmatrix}$$

2. Solve the initial value problem

    $$\begin{aligned} y_1(n + 1) &= y_2(n) & y_1(0) &= 1 \\ y_2(n + 1) &= y_3(n) & y_2(0) &= 0 \\ y_3(n + 1) &= y_2(n) & y_3(0) &= 2 \end{aligned}$$

3. Assume that $k = 2$ and that the eigenvalues of $\mathscr{A}$ are complex conjugates. Take the eigenvectors to be complex conjugates and let P have these eigenvectors as columns. Prove that

    $$P\begin{pmatrix} \lambda^m & 0 \\ 0 & \bar{\lambda}^m \end{pmatrix} P^{-1}$$

    is real.

4. Find the eigenvalues and eigenvectors of $\mathscr{A}$, and write a general solution in both complex and real form for $y(x + 1) = \mathscr{A} y(x)$ if

    (a)
    $$\mathscr{A} = \begin{pmatrix} 1 & 1 \\ -5 & -1 \end{pmatrix}$$

    (b)
    $$\mathscr{A} = \begin{pmatrix} 0 & 1 & 0 \\ 0 & 0 & 1 \\ 1 & -1 & 1 \end{pmatrix}$$

5. Reformulate and prove Theorem 13.15 for the equation $y(n + 1) = \mathscr{A} y(n)$, $n = 1, 2, \ldots$.

13.7 The scalar equation with constant coefficients

In this section we shall give a brief discussion of the kth order scalar difference equation with constant coefficients,

$$z(x + k) + a_{k-1}z(x + k - 1) + \cdots + a_1 z(x + 1) + a_0 z(x) = 0 \quad (1)$$

From Section 13.5 we know that this equation is equivalent to its companion equation $y(x + 1) = \mathscr{A} y(x)$, where $\mathscr{A}$ is given by Eq. (4) of Section 13.5. It follows that the eigenvalues are the roots of the equation

$$\det(\lambda I - \mathscr{A}) = \lambda^k + a_{k-1}\lambda^{k-1} + \cdots + a_1\lambda + a_0 = 0 \qquad (2)$$

Let us suppose that $\mathscr{A}$ is diagonable. Then from Theorem 13.14 and the fact that $z(x) = y_1(x)$, the first component of $y(x)$, we deduce that

$$z(x) = \sum_{j=1}^{k} v_{j1}\lambda_j^m \tilde{p}_j(x), \qquad m \leq x < m + 1, \ m = 0, 1, 2, \ldots$$

Here v_{j1} is the first component of an eigenvector of λ_j. This constant can be absorbed into the arbitrary periodic function $\tilde{p}_j(x)$, and thus a general solution of Eq. (1) is

$$z(x) = \sum_{j=1}^{k} \lambda_j^m p_j(x), \qquad m \leq x < m + 1, \ m = 0, 1, \ldots \qquad (3)$$

where the p_j are arbitrary functions of period one. If the λ_j are real positive numbers, there is a simpler form of this result. We define functions $q_j(x)$ by

$$\lambda_j^m p_j(x) = \lambda_j^x q_j(x), \qquad m \leq x < m + 1$$

Then for any x_0 in $[m, m + 1)$, we have

$$q_j(x_0) = \lambda_j^{m-x_0} p_j(x_0)$$
$$q_j(x_0 + 1) = \lambda_j^{(m+1)-(x_0+1)} p_j(x_0 + 1) = q_j(x_0)$$

which shows that q_j has period 1. Therefore Eq. (3) can be replaced by

$$z(x) = \sum_{j=1}^{k} \lambda_j^x q_j(x), \qquad x \geq 0 \qquad (4)$$

where the q_j are arbitrary periodic functions. If λ_j is negative or complex, we have not shown how to define the power λ_j^x for real x, and there is in fact some inherent ambiguity. We will therefore use Eq. (4) only when all λ_j are positive, or when x is restricted to integer values.

From Theorem 13.15 we see that $z(x) = y_1(x)$ is given by Eq. (9) of Section 13.6 with u_j and v_j replaced by their first components. Since these can be taken to be real, we can write

$$z(x) = \sum_{j=1}^{s} \lambda_j^m p_j(x) + \sum_{j=1}^{t} \text{Re}\,(\mu_j + iv_j)^m q_j(x)$$

$$+ \sum_{j=1}^{t} \text{Im}\,(\mu_j + iv_j)^m r_j(x) \qquad (5)$$

where p_j, q_j, r_j are real 1-periodic functions. A modification like that in Eq. (4) is possible if all $\lambda_j > 0$ or if x is restricted to integer values. Finally, from

Theorem 13.16 we see that if there are k independent eigenvectors, the k sequences $\{\lambda_j^n\}$, $n = 0, 1, 2, \ldots$, form a basis for the solution space of Eq. (1) considered on $x = 0, 1, 2, \ldots$.

Example 1 $z(x + 2) - 2z(x + 1) - 3z(x) = 0$. The *characteristic equation* $\lambda^2 - 2\lambda - 3 = 0$ has roots $\lambda = 3, -1$. Therefore, a general solution is

$$z(x) = 3^m p_1(x) + (-1)^m p_2(x), \qquad m \leq x < m + 1$$

Suppose that we are given an initial condition, for example,

$$z(x) = x \quad \text{for} \quad 0 \leq x < 2$$

Then p_1 and p_2 must be chosen to satisfy

$$\begin{aligned} p_1(x) + p_2(x) &= x, & 0 \leq x < 1 \\ 3p_1(x) - p_2(x) &= x, & 1 \leq x < 2 \end{aligned}$$

According to the required periodicity, this pair of equations is equivalent to the pair

$$\begin{aligned} p_1(x) + p_2(x) &= x, & 0 \leq x < 1 \\ 3p_1(x) - p_2(x) &= x + 1, & 0 \leq x < 1 \end{aligned}$$

Therefore

$$p_1(x) = \tfrac{1}{4}(2x + 1), \quad p_2(x) = \tfrac{1}{4}(2x - 1), \qquad 0 \leq x < 1$$

For $x > 1$, p_1 and p_2 are determined by periodicity.

Example 2 For the equation

$$z(x + 2) - 2z(x + 1) + 2z(x) = 0$$

the eigenvalues are roots of the characteristic equation $\lambda^2 - 2\lambda + 2 = 0$. The roots are $\lambda = 1 \pm i$. Therefore a general solution is

$$z(x) = (1 + i)^m p_1(x) + (1 - i)^m p_2(x), \qquad m \leq x < m + 1$$

By Eq. (5), a real solution is

$$z(x) = \operatorname{Re} \left[(1 + i)^m\right] q(x) + \operatorname{Im} \left[(1 + i)^m\right] r(x)$$

Since

$$(1 + i)^m = (\sqrt{2}e^{i\pi/4})^m = 2^{m/2}\left(\cos\frac{m\pi}{4} + i \sin\frac{m\pi}{4}\right)$$

we get

$$z(x) = 2^{m/2}\left(\cos\frac{m\pi}{4}\right) q(x) + 2^{m/2}\left(\sin\frac{m\pi}{4}\right) r(x)$$

on $m \leq x < m + 1$, where q and r are arbitrary periodic functions of period one.

Example 3   Consider the discrete difference equation

$$z_{n+2} + 3z_{n+1} + 2z_n = 0, \qquad n = 1, 2, \ldots$$

The characteristic equation $\lambda^2 + 3\lambda + 2 = 0$ has roots $\lambda = -1, -2$. Hence by Eq. (3) or (4),

$$z_n = (-1)^n p_1(n) + (-2)^n p_2(n)$$

or since we are interested only in integer values of n,

$$z_n = c_1(-1)^n + c_2(-2)^n$$

where c_1 and c_2 are constants.

We now wish to discuss the solution of the kth order equation in the case in which the characteristic equation has multiple roots. Since we have omitted a discussion of systems with nondiagonable matrix $\mathscr{A}$, we must rely on some other approach. We shall use the method of factorization of operators in a manner similar to that in Section 9.3. For the sake of simplicity, we consider only the discrete equation

$$z(n + k) + a_{k-1}z(n + k - 1) + \cdots + a_1 z(n + 1) + a_0 z(n) = 0 \qquad (6)$$

Let E be the *shift operator* defined by

$$Ez(n) = z(n + 1) \qquad (7)$$

The operator E takes any sequence in $\mathbf{S}_1$ into another sequence in $\mathbf{S}_1$. ($\mathbf{S}_1$ denotes the space of all sequences.) In terms of it, Eq. (6) can be written

$$(E^k + a_{k-1}E^{k-1} + \cdots + a_1 E + a_0 I)z(n) = 0 \qquad (8)$$

where, of course,

$$E^j z(n) = EE^{j-1}z(n) = z(n + j) \qquad (9)$$

Suppose that the characteristic polynomial p of Eq. (6) factors

$$
\begin{aligned}
p(r) &= r^k + a_{k-1}r^{k-1} + \cdots + a_1 r + a_0 \\
&= (r - \lambda_1)^{m_1}(r - \lambda_2)^{m_2} \cdots (r - \lambda_s)^{m_s}
\end{aligned}
$$

where $\lambda_1, \ldots, \lambda_s$ are the distinct roots and $m_1, \ldots, m_s$ are their multiplicities. Then we can factor the operator and write Eq. (8) as

$$(E - \lambda_1 I)^{m_1}(E - \lambda_2 I)^{m_2} \cdots (E - \lambda_s I)^{m_s} z(n) = 0 \qquad (10)$$

Consider first the equation with one factor

$$(E - \lambda I)^m z(n) = 0 \qquad (11)$$

We claim that Eq. (11) has solutions

$$z(n) = n^{(j)}\lambda^n, \qquad j = 0, 1, \ldots, m - 1, \quad n = 0, 1, \ldots \qquad (12)$$

where $n^{(j)}$ is the factorial power (see the Exercises in Section 13.1) and where $n^{(0)} = 1$ for all n. In fact

$$(E - \lambda I)z(n) = z(n + 1) - \lambda z(n)$$
$$= (n + 1)^{(j)}\lambda^{n+1} - \lambda n^{(j)}\lambda^n$$
$$= \lambda^{n+1}\Delta n^{(j)} = jn^{(j-1)}\lambda^{n+1}$$

and in general,

$$(E - \lambda I)^m(n^{(j)}\lambda^n) = j(j - 1)\cdots(j - m + 1)n^{(j-m)}\lambda^{n+m}, \quad 1 \le m \le j$$
$$= 0, \qquad\qquad\qquad\qquad\qquad\qquad m \ge j + 1$$

It can also be shown that every solution of Eq. (11) is a linear combination of the solutions (12). Next we have the following lemma.

LEMMA 13.1   If for some integer l, $1 \le l \le s$, $z(n)$ is in the null space of $(E - \lambda_l I)^{m_l}$, then $z(n)$ is in the null space of

$$L = (E - \lambda_1 I)^{m_1} \cdots (E - \lambda_s I)^{m_s}$$

The proof is the same as that for Lemma 2 in Section 9.3. It follows that the null space of L contains the sequences (in n)

$$\{\lambda_1^n\}, \{n^{(1)}\lambda_1^n\}, \ldots, \{n^{(m_1-1)}\lambda_1^n\}, \ldots, \{\lambda_s^n\}, \{n^{(1)}\lambda_s^n\}, \ldots, \{n^{(m_s-1)}\lambda_s^n\} \tag{13}$$

There are $m_1 + m_2 + \cdots + m_s = k$ sequences in the set (13). It can be proved (but we omit the proof) that these are linearly independent. Therefore, we have the following result.

THEOREM 13.17   Let L be the difference operator

$$L = (E - \lambda_1)^{m_1}(E - \lambda_2)^{m_2} \cdots (E - \lambda_s)^{m_s},$$

where $\lambda_1, \lambda_2, \ldots, \lambda_s$ are the distinct roots of the characteristic polynomial of L and $m_1, m_2, \ldots, m_s$ are their respective multiplicities. Then, the sequences in (13) form a basis for the solution space of the discrete difference equation $Lz(n) = 0$.

Example 4   Consider the equation

$$z(n + 3) - 4z(n + 2) + 5z(n + 1) - 2z(n) = 0, \qquad n = 0, 1, \ldots$$

In factored form,

$$(E - I)^2(E - 2I)z(n) = 0$$

Therefore, $\{1^n\}$, $\{n \cdot 1^n\}$, $\{2^n\}$ forms a basis. Every solution has the form

$$z(n) = c_1 + c_2 n + c_3 2^n$$

We refer the reader to texts on difference equations for techniques for finding particular solutions of nonhomogeneous equations.

Exercises

1. Solve each initial value problem on $x \geq 0$. Graph the periodic functions in each case.

    (a) $z(x + 2) - 5z(x + 1) + 6z(x) = 0$
    $z(x) = 4$ for $0 \leq x < 2$

    (b) $z(x + 2) - 2z(x + 1) - 8z(x) = 0$
    $z(x) = 4^x = e^{x\ln 4}$ for $0 \leq x < 2$

2. Consider the problem

    $$z(x + 2) + a_1 z(x + 1) + a_2 z(x) = 0$$
    $$z(x) = \phi(x) \quad \text{for} \quad 0 \leq x < 2$$

    Let ϕ be a continuous function. What condition must ϕ satisfy in order that the solution z be continuous for all $x \geq 0$?

3. Find a real general solution for $x \geq 0$.

    (a) $z(x + 2) + z(x) = 0$
    (b) $z(x + 2) + 2z(x + 1) + 2z(x) = 0$
    (c) $z(x + 2) - 4z(x + 1) + 5z(x) = 0$

4. Prove that the sequences $\{\lambda^n\}$, $\{n^{(1)}\lambda^n\}$, $\{n^{(2)}\lambda^n\}$ are linearly independent, if $\lambda \neq 0$.

5. Prove that the sequences $\{\lambda_1^n\}$, $\{\lambda_2^n\}$, $\ldots$, $\{\lambda_s^n\}$ are linearly independent, if $\lambda_1, \lambda_2, \ldots, \lambda_s$ are distinct.

6. Prove that the sequences $\{\lambda_1^n\}$, $\{n^{(1)}\lambda_1^n\}$, $\{n^{(2)}\lambda_1^n\}$, $\{\lambda_2^n\}$ are linearly independent if $\lambda_1 \neq 0, \lambda_2 \neq 0, \lambda_1 \neq \lambda_2$.

7. What is the justification for the factoring in Eq. (10)? Is there a unique factorization?

8. Find a general solution of each equation.

    (a) $z(n + 3) - z(n + 2) - 5z(n + 1) - 3z(n) = 0$
    (b) $z(n + 4) + 2z(n + 2) + z(n) = 0$

9. The equations

    $$z(n + 2) - 2z(n + 1) + 0z(n) = 0, \qquad n = 0, 1, \ldots$$
    $$z(n + 1) - 2z(n) = 0, \qquad n = 1, 2, \ldots$$

are equivalent but have characteristic equations $\lambda^2 - 2\lambda = 0$ and $\lambda - 2 = 0$, respectively. Do their general solutions differ?

13.8 Application: The gambler's ruin

Difference equations have applications in many areas of the physical and social sciences. The interested reader should consult the Commentary for references. In this section we give a special application to a classical problem in probability theory. In considering this application the reader should keep in mind that although the description of the problem is contained in a setting (that of gambling), which may not be of much interest, the general structure can be interpreted in many areas of application.

The gambler's ruin problem is as follows. We enter into a sequence of games of chance with an initial holding of c monetary units (for example, dollars). Our adversary begins with d units. At each game in the sequence we will win a unit from our opponent with probability p, and lose a unit with probability $q = 1 - p$, where $0 < p < 1$. For example, we can flip a coin which has probability p of falling heads. If a head appears we take one unit from our foe. If it falls tails we pay him one unit. We are interested in' determining the probability of our ultimate ruin, that is, the probability we eventually end up holding zero units.

We can express the problem in a difference equation. Let P_n denote the probability of eventual ruin, given we now have n units. It is obvious that the path (sequence of wins and losses) taken to get us to a holding of n units has no influence on the probability of ruin. The only factors influencing this probability are the size of our present reserve (n units) and the probability of success on each trial (p).[5] Given that we now hold n units, after the next trial we will with probability p hold $n + 1$ units, and with probability $q = 1 - p$ hold $n - 1$ units. Our probability of ruin P_n is equal to the probability of winning the next game followed by eventual ruin plus the probability of losing the next game followed by eventual ruin, and is therefore given as the solution to the difference equation

$$P_n = pP_{n+1} + qP_{n-1} \qquad 0 < n < c + d \qquad (1)$$

subject to the conditions

$$\begin{aligned} P_0 &= 1 \\ P_{c+d} &= 0 \end{aligned} \qquad (2)$$

We are now faced with solving a discrete difference equation, and will use the

---

[5] This property of a random or stochastic process is known as the Markov property. Much interesting work has been done in studying such processes.

shift operator as described in Example 3 of Section 13.7. Equation (1) can be rewritten as

$$P_{n+2} - \frac{1}{p} P_{n+1} + \frac{q}{p} P_n = 0 \qquad 0 \le n < c + d - 1 \tag{3}$$

$$P_0 = 1, \qquad P_{c+d} = 0$$

or equivalently,

$$\left(E^2 - \frac{1}{p} E + \frac{q}{p} I \right) P_n = 0 \qquad 0 \le n < c + d - 1 \tag{4}$$

The roots of the characteristic equation are $\lambda_1 = 1$ and $\lambda_2 = 1/p - 1$, which are distinct, provided $p \ne \frac{1}{2}$. Applying Theorem 13.17, we find that if $p \ne \frac{1}{2}$ the sequences $\{1\}$, $\{(q/p)^n\}$ form a basis for the solution space. If $p = \frac{1}{2}$ the sequences $\{1\}$, $\{n\}$ form a basis for the solution space.

To find the particular solution subject to the conditions in Eqs. (2) we first treat the case when $p \ne \frac{1}{2}$. The solution is

$$P_n = k_1 + k_2 \left(\frac{q}{p} \right)^n$$

Since $P_0 = 1$ and $P_{c+d} = 0$, we have the simultaneous equations for k_1 and k_2,

$$k_1 + k_2 = 1$$

$$k_1 + \left(\frac{q}{p} \right)^{c+d} k_2 = 0$$

which yield

$$k_1 = - \frac{(q/p)^{c+d}}{1 - (q/p)^{c+d}} \quad \text{and} \quad k_2 = \frac{1}{1 - (q/p)^{c+d}}$$

The particular solution to our problem is

$$P_n = \frac{(q/p)^n - (q/p)^{c+d}}{1 - (q/p)^{c+d}} \qquad 0 \le n \le c + d$$

for $p \ne \frac{1}{2}$.

If $p = \frac{1}{2}$ the general solution is

$$P_n = k_1 + n k_2$$

To obtain the particular solution we again impose the conditions $P_0 = 1$ and $P_{c+d} = 0$, which yield $k_1 = 1$ and $k_2 = -1/(c + d)$. Therefore our probability of ruin in a fair game is given by

$$P_n = 1 - \frac{n}{c + d}, \qquad 0 \le n \le c + d$$

and in particular $P_c = d/(c + d)$.

Many questions can be asked in connection with the gamblers ruin problem. One of particular interest is the expected duration (number of games) of the sequence of games leading to ultimate ruin for one of the players. For a more complete discussion the reader should consult Feller's text which is listed in the commentary.

Exercises

In this exercise, we formulate a version of the Domar economic growth model (see Section 2.3) employing difference equations instead of differential equations. Let

Y_n = national income during the nth time period
C_n = consumption of goods
I_n = investment during the nth time period
F_n = productive potential

Assumptions like those in Section 2.3 lead to the equations

$$\Delta I_n = a\Delta Y_n \qquad (0 \le a \le 1)$$
$$\Delta F_n = \sigma I_n \Delta t$$

The first of these equations states that the increase in investment is proportional to the increase in national income during the same time period; the second states that the increase in production capacity in one time period is proportional to the length of time and to the investment. In order to keep $\Delta Y_n = \Delta F_n$, we must have

$$\Delta I_n = a\sigma I_n \Delta t$$

Solve this first order difference equation.

13.9 Differential-difference equations

The concepts of linear algebra can be applied not only to differential equations and to difference equations but to other *functional equations* (equations for unknown functions) as well. Space does not allow us to mention a large number of the possibilities, nor to examine any one of them in detail. In this section we have a brief glance at one other class of equations, which are called ordinary *differential-difference equations*. These equations have applications in biology, economics, probability theory, control theory, and so on. As the name implies, these are equations containing both derivatives and differences of an unknown function. We restrict attention here to equations containing at most first order derivatives and first order differences. In difference form, these will have the form

$$g(x, y(x), \Delta y(x), y'(x), \Delta y'(x)) = 0 \tag{1}$$

We shall assume that this can be solved for $\Delta y'(x)$,

$$\Delta y'(x) = G(x, y(x), \Delta y(x), y'(x)) \tag{2}$$

Writing this in dated form, we have

$$y'(x + 1) = F(x, y(x), y(x + 1), y'(x)) \tag{3}$$

We shall not attempt to discuss a general theory of these equations here, but rather will illustrate a few simple ideas by means of an example.

Example 1   We consider equations in which F in Eq. (3) is independent of $y'(x)$ and $y(x + 1)$ and is linear in $y(x)$. A simple example of such an equation is

$$y'(x + 1) = ay(x), \qquad x > 0 \tag{4}$$

We suppose that y also satisfies the *initial condition*

$$y(x) = \phi(x), \qquad 0 \le x < 1 \tag{5}$$

where ϕ is an arbitrary function. We can construct a solution of Eq. (4) by the method of steps, much as we did for difference equations above. From Eqs. (4) and (5) we get

$$y'(x) = a\phi(x - 1), \qquad 1 < x < 2$$

Integrating this equation, we obtain $y(x)$ on $1 \le x < 2$, up to an arbitrary additive constant of integration:

$$y(x) = a \int_1^x \phi(s - 1) \, ds + C, \qquad 1 \le x < 2$$

Here, of course, we must assume that ϕ is an integrable function. Then $y(x)$ is continuous on $1 < x < 2$. It is natural to impose the condition $y(1) = \phi(1)$, which determines the constant C. We then have

$$y(x) = \phi(1) + a \int_1^x \phi(s - 1) \, ds, \qquad 1 \le x < 2 \tag{6}$$

and $y(x)$ is continuous at $x = 1$. As a matter of fact, in most applications ϕ will be continuous and we will be seeking a continuous solution y. Having obtained $y(x)$ on $1 \le x \le 2$, another application of Eq. (4) gives

$$y(x) = y(2) + a \int_2^x y(s - 1) \, ds, \qquad 2 \le x \le 3 \tag{7}$$

That is, $y(x)$ has been determined on $2 \le x \le 3$. It is clear that this stepwise process can be continued and generates a continuous function $y(x)$ on $0 \le x$, which satisfies Eq. (4) for $x > 0$ and for which the *initial condition* (5) is fulfilled. Note that $y'(1^+) = ay(0) = a\phi(0)$, by Eq. (4). On the other hand, $\phi'(1^-)$

may not exist, and if it does exist it may not equal $a\phi(0)$. Thus $y'(1)$ will not in general exist in the true sense; that is, Eq. (4) is satisfied at $x = 0$ only in the sense that the right-hand derivative $y'(1^+)$ equals $ay(1)$. But if $\phi'(1^-) = a\phi(0)$, then $y'(1)$ exists in the two-sided sense.

Although the method of steps shows that Eqs. (4) and (5) possess a unique continuous solution, when ϕ is continuous, it does not lead to simple formulas as it did for the pure difference equations above. For example, inserting Eq. (6) into (7) we get

$$
y(x) = \phi(1) + a \int_1^2 \phi(s - 1) \, ds
$$

$$
+ a \int_2^x \left\{ \phi(1) + a \int_1^{s-1} \phi(s_1 - 1) \, ds_1 \right\} ds \qquad 2 \le x \le 3
$$

Although it is possible to continue in this way for a few steps (and in the present simple example even to obtain a general term for the expansion obtained), very little insight seems to result, and so we seek a different approach. First, we note that the solutions of Eq. (4) (for all possible continuous ϕ) form a subspace of the vector space of continuous functions on $0 \le x$. Thus we should attempt to find a basis for this solution space. For ordinary linear differential equations, with constant coefficients, we succeeded in finding a basis for the solution space made up of exponential functions, and a similar result was found for difference equations. Could the same be true for Eq. (4)? Let us try $y(x) = e^{\lambda x}$. Substituting into Eq. (4), we obtain

$$
\lambda e^\lambda = a \tag{8}
$$

which we call the *characteristic equation* of Eq. (4).

Equation (8) is a transcendental equation for λ, and there is no simple way to find its roots. It can be shown, however, that the following statements are true.

1. Equation (8) has a (countable) infinity of roots $\lambda_1, \lambda_2, \lambda_3, \ldots$. All roots are simple, with the one exception that if $a = -e^{-1}$, $\lambda = -1$ is a double root. All roots are complex, except perhaps for one real root. The complex roots occur in conjugate pairs.

2. As $n \to \infty$, $|\lambda_n| \to \infty$ and Re $(\lambda_n) \to -\infty$.

It is clear from this that Eq. (4) has infinitely many simple exponential solutions $e^{\lambda_i x}$. It can be proved under quite general conditions that every continuous solution of Eq. (4) can be expanded in the convergent infinite series

$$
y(x) = \sum_{i=1}^{\infty} c_i e^{\lambda_i x}, \qquad x > 1 \tag{9}
$$

for some choice of the constants c_i. In this sense, the exponentials $e^{\lambda_i x}$ form a basis for the solution space. This problem is more difficult than those we have considered before because the solution space has infinite dimension. Even for difference equations with continuous domain x, where the solution space was of infinite dimension, we were able to extract finite dimensional subproblems by considering sequences $x = x_0, x_0 + 1, x_0 + 2, \ldots$. No such decomposition is readily apparent here. Nevertheless, our problem is solvable, since there is a known general procedure for calculating the constants c_i in (9), based on Laplace transform methods. To pursue the matter further is beyond the scope of this text. Our hope is that in this text we have provided an adequate basis for our readers to pursue infinite dimensional problems later if they wish or need to do so.

Exercises

1. Compute the solution of Eq. (4) by the method of steps when $a = 1$, $\phi(x) = 1$ for $0 \le x < 1$. Show that

    $$y(x) = \sum_{j=0}^{N} \frac{(x - j)^j}{j!} \qquad N \le x \le N + 1, N = 0, 1, \ldots$$

2. Use the method of steps to calculate the solution of

    $$y'(x + 1) = 1 + y(x), \qquad x > 0$$
    $$y(x) = 1, \qquad 0 \le x \le 1$$

    Find a formula valid for $N \le x \le N + 1$, $N = 0, 1, 2, 3$.

3. Use the method of steps to calculate the solution of each equation for $0 \le x \le 4$.

    (a) $y'(x + 1) = y(x), \qquad x > 0$
    $\qquad y(x) = x, \qquad 0 \le x \le 1$,

    (b) $y'(x + 2) = 2y(x) - y(x + 1), \qquad x > 0$
    $\qquad y(x) = 1, \qquad 0 \le x \le 2$

4. Show that the problem

    $$y'(x + 1) - y(x + 1) = ay(x), \qquad x > 0$$
    $$y(x) = \phi(x), \qquad 0 \le x \le 1$$

    has a unique solution by making the substitution

    $$y(x)e^{-x} = z(x)$$

5. By graphing the functions λ and $ae^{-\lambda}$, show that if $a > 0$, Eq. (8) has a unique positive root. What is the situation if $a < 0$?

6. Find a characteristic equation for each functional equation.

(a) $y'(x + 1) = ay(x) + by(x + 1) + cy'(x)$

(b) $\displaystyle\sum_{i=0}^{k} [a_i y'(x + i) + b_i y(x + i)] = 0$

(c) $y(x) = \displaystyle\int_0^1 a(s) y(x - s) \, ds$

COMMENTARY

The following books are among those devoted to difference operators and difference equations:

BATCHELDER, P. M., *An Introduction to Linear Difference Equations*, Cambridge, Mass., Harvard University Press, 1927. (Reprint by Dover Publications, New York, 1967.)

BRAND, L., *Differential and Difference Equations*, New York, Wiley, 1966.

FORT, T., *Finite Differences and Difference Equations in the Real Domain*, Oxford, Clarendon Press, 1948.

GOLDBERG, S., *Introduction to Difference Equations*, New York, Wiley, 1958.

JORDAN, C., *Calculus of Finite Differences*, 3rd ed., New York, Chelsea, 1965 (First edition, Budapest, 1939).

LEVY, H. and F. LESSMAN, *Finite Difference Equations*, London, Pitman, 1959.

MILLER, K. S., *An Introduction to the Calculus of Finite Differences and Difference Equations*, New York, Holt Rinehart and Winston, 1960.

MILLER, K. S., *Linear Difference Equations*, New York, Benjamin, 1968.

MILNE-THOMSON, L. M., *The Calculus of Finite Differences*, London, Macmillan, 1933.

Section 13.1   Fairly extensive tables of differences and antidifferences can be found in some of the books listed above. These are analogous to tables of derivatives and integrals. Since we are here not primarily concerned with developing skill in the calculation of differences and antidifferences, we have omitted such tables. In the same spirit, many of the usual formulas are given only in the exercises or are omitted entirely.

The terminology "method of steps" has been adapted from the theory of differential-difference equations where it is in common use.

For a discussion of backward and other difference operators, see the above books and the thesis of R. Folsom (cited in Exercise 19).

The formula in Exercise 10 is widely used as an interpolation formula under the name *Newton's interpolation formula*. For a discussion, see textbooks on numerical analysis, such as:

HILDEBRAND, F. B., *Introduction to Numerical Analysis*, New York, McGraw-Hill, 1956.

This text also includes a discussion of the infinite Newton interpolation series.

Section 13.2 In the more advanced study of difference equations, equations of the more general form

$$\sum_{j=1}^{n} a_j(x)f(x + h_j) = 0$$

are sometimes considered, where $h_1, \ldots, h_n$ are arbitrary real (or complex) numbers. For example,

$$f(x + \sqrt{2}) - f(x + 1) + 2f(x) = 0$$

is of this type. Such equations cannot, in general, be written in differenced form using a single difference operator. The study of these equations, which involves difficult analysis, is beyond the scope of this text.

For a discussion of "exact" difference equations, see the book by Brand listed above.

Section 13.3 For the application of finite difference operators and tables in numerical analysis, see texts on the subject. The book by Goldberg cited above includes many examples of the applications of finite difference equations in psychology, economics, and so on.

Sections 13.4 and 13.5 The book by Brand contains an extensive discussion of second order linear difference equations on sets $x_0, x_0 + 1, x_0 + 2, \ldots$. The theory of linear systems is treated thoroughly in the books by Miller and Fort, cited above. These books introduce the Casorati matrix, (which is the analogue of the Wronski matrix) and cover many topics omitted here.

Section 13.6 The solution of the system $y(x + 1) = \mathcal{A}y(x)$ when $\mathcal{A}$ is not diagonable can be carried out by a method similar to that of Section 12.16. See:

LINFIELD, B. Z., On the explicit solution of simultaneous linear difference equations with constant coefficients, *Amer. Math. Monthly*, vol. 47(1940), 552–554.

For an alternative construction of real solutions when $\mathcal{A}$ is real, see:

TURRITTIN, H. L., Linear differential or difference equations with constant coefficients, *Amer. Math. Monthly*, vol. 66(1959), 869–875.

Section 13.7 For a discussion of how to define λ^x when λ is complex, refer to texts on functions of a complex variable.

Section 13.8 For further applications of difference equations, the reader should consult the following:

BUSH, R. R. and William K. ESTES., *Studies in Mathematical Learning Theory*, Stanford, Stanford University Press, 1956.

CHIANG, A. C., *Fundamental Methods of Mathematical Economics*, New York, McGraw-Hill, 1967.

FELLER, W. J., *An Introduction to Probability Theory and Its Applications*, vol. I, 3rd ed., New York, Wiley, 1968.

GOLDBERG, S., *Introduction to Difference Equations*, New York, Wiley, 1958.

Section 13.9   The theory of differential-difference equations is treated in:

BELLMAN, R., and K. L. COOKE, *Differential Difference-Equations*, New York, Academic
Press, 1963.

EL'SGOLT'S, L. E., *An Introduction to the Theory of Differential Equations with Deviating
Arguments*, San Francisco, Holden-Day, 1966 (translation from the Russian).

For a recent work on a more general type of equation, including differential-
difference equations as a special case, see:

HALE, J. K., *Functional Differential Equations,* Berlin, Springer-Verlag, 1971.

For a discussion of differential-difference equations in economics, see:

GRANDOLFO, G., *Mathematical Methods and Models in Economic Dynamics*, Amsterdam and London, North-Holland, and New York, American Elsevier, 1971.

One topic which we have reluctantly omitted from this chapter is the use of
difference equations as approximations to differential equations; for example, since

$$\lim_{h \to 0} \frac{y(h + h) - y(x)}{h} = y'(x)$$

the difference equation

$$y(x + h) - y(x) = h f(x, y(x))$$

in a certain sense "converges" to the differential equation $y'(x) = f(x, y(x))$ as $h \to 0$.
Therefore, if h is small, solutions of the difference equation should be good approximations to solutions of the differential equation. This is the basic idea behind many
methods for numerical solution of differential equations. For another point of view
on the convergence of difference equations to differential equations, see the thesis of
Folsom cited above.

Chapter 14

Inner products and quadratic forms

14.1 The inner product

In developing the concepts of vectors and vector spaces we were careful to avoid depending upon geometric considerations (see Section 4.1). One reason for this is that the notions of length and direction were assumed not to have been defined. In this chapter we will define an *inner product* which can be used in turn to define concepts of length and direction in arbitrary finite dimensional vector space, and which will be consistent with what is known for $\mathbf{G}^2$ and $\mathbf{G}^3$.

DEFINITION 14.1 Let $\mathbf{V}$ be a real vector space.[1] An (real) inner product on $\mathbf{V}$ is a mapping of ordered pairs of elements of $\mathbf{V}$ into $\mathbf{R}^1$, that is, a mapping of $\mathbf{V} \times \mathbf{V}$ into $\mathbf{R}^1$, which satisfies the following axioms.

Let $\langle \, , \, \rangle$ denote the value of the function. Then it is required that for any v_1, v_2 and v_3 in $\mathbf{V}$ and any scalars a and b, the following hold:

1. $\langle v_1, v_2 \rangle = \langle v_2, v_1 \rangle$
2. $\langle av_1 + bv_2, v_3 \rangle = a\langle v_1, v_3 \rangle + b\langle v_2, v_3 \rangle$
3. $\langle v_1, v_1 \rangle > 0$ if $v_1 \neq \mathbf{0}$; and $\langle \mathbf{0}, \mathbf{0} \rangle = 0$

Before illustrating this concept we make several observations. First, we are restricting our attention to *real* inner products defined on *real* vector spaces. An analogous theory can be developed for inner products defined on complex vector spaces which are then mappings into the complex field. Second, combining properties (1) and (2) yields

$$\langle v_1, av_2 + bv_3 \rangle = a\langle v_1, v_2 \rangle + b\langle v_1, v_3 \rangle$$

---

[1] A real vector space is a vector space defined over the field of real numbers.

Finally, if we consider a specific element v_0 of V, and the mapping $Tv = \langle v, v_0 \rangle$, that is, T is the mapping of V into R^1 which is the inner product of elements of V with the specific element v_0, then by (2) T is a linear transformation. This can be expressed in an abbreviated way by saying that the inner product is linear in its first variable. Property (2) is therefore referred to as the *linearity* property of $\langle \, , \, \rangle$. As observed, the inner product is similarly linear in its second variable, and so it is often called *bilinear*. Property (1) is called *symmetry* and property (3) *positive definiteness* of the inner product.

Example 1 Consider the vector space R^3. If $x = (x_1, x_2, x_3)$ and $y = (y_1, y_2, y_3)$, then $\langle x, y \rangle = x_1 y_1 + x_2 y_2 + x_3 y_3$ is an inner product on R^3. To verify this result we need to test the mapping $\langle \, , \, \rangle$ against the three conditions of Definition 14.1. Certainly, $\langle x, y \rangle = \langle y, x \rangle$ for any vectors x and y in R^3. Next, let x, y and z be three vectors in R^3. Then,

$$ax + by = (ax_1 + by_1, ax_2 + by_2, ax_3 + by_3)$$

and

$$\langle ax + by, z \rangle = (ax_1 + by_1)z_1 + (ax_2 + by_2)z_2 + (ax_3 + by_3)z_3$$
$$= a\langle x, z \rangle + b\langle y, z \rangle$$

Finally, $\langle x, x \rangle = x_1{}^2 + x_2{}^2 + x_3{}^2$ is positive unless $x_1 = x_2 = x_3 = 0$, that is, unless $x = \mathbf{0}$. The reader will recall that $x_1 y_1 + x_2 y_2 + x_3 y_3$ is the usual dot product of two vectors as used in elementary geometry and physics. Thus, the usual dot product is an instance of an inner product on R^3.

Example 2 Again consider R^3 and define

$$\langle x, y \rangle = \sqrt{2} x_1 y_1 - \frac{x_1 y_2 + x_2 y_1}{2} + x_2 y_2 + 2 x_3 y_3,$$

where $x = (x_1, x_2, x_3)$ and $y = (y_1, y_2, y_3)$. We now show that $\langle \, , \, \rangle$ is an inner product. The symmetry property (1) is immediate. For a fixed y it is evident that $\langle x, y \rangle$ is linear in x and hence (2) is satisfied. Finally, note that

$$\langle x, x \rangle = \sqrt{2} x_1{}^2 - x_1 x_2 + x_2{}^2 + 2 x_3{}^2$$

$$= \left(\frac{x_1}{2} - x_2\right)^2 + \left(\sqrt{2} - \frac{1}{4}\right) x_1{}^2 + 2 x_3{}^2$$

Since the second and third terms of the last sum are positive unless $x_1 = 0$ and $x_3 = 0$, respectively, and the first term is positive unless $x_2 = x_1/2$, we can see that $\langle x, x \rangle$ is positive unless $x_1 = x_2 = x_3 = 0$, that is, unless $x = \mathbf{0}$. Therefore $\langle \, , \, \rangle$ is an inner product in R^3. Examples 1 and 2 show that it is possible to have several different inner products on the same vector space.

Next we consider two examples of inner products on spaces other than R^n.

Example 3 Consider the space $C^0(J)$ of all continuous functions on a closed interval $J = [x_0, x_1]$, $x_0 \neq x_1$. Define

$$\langle f_1, f_2 \rangle = \int_{x_0}^{x_1} f_1(x) f_2(x)\, dx$$

for f_1 and f_2 elements of $C^0(J)$. Then $\langle \ , \ \rangle$ is an inner product. For,

$$\langle f_1, f_2 \rangle = \langle f_2, f_1 \rangle$$

$$\langle af_1 + bf_2, f_3 \rangle = \int_{x_0}^{x_1} [af_1(x) + bf_2(x)] f_3(x)\, dx$$

$$= a \int_{x_0}^{x_1} f_1(x) f_3(x)\, dx + b \int_{x_0}^{x_1} f_2(x) f_3(x)\, dx$$

$$= a\langle f_1, f_3 \rangle + b\langle f_2, f_3 \rangle$$

and finally,

$$\langle f_1, f_1 \rangle = \int_{x_0}^{x_1} f_1{}^2(x)\, dx > 0$$

unless $f_1(x) = 0$ for $x_0 \leq x \leq x_1$, that is, unless f_1 is the zero vector of $C^0(J)$.

DEFINITION 14.2   A vector space **V** together with a specified inner product is called an *inner product space*. When the vector space is real and the inner product is real, it is often called more specifically a *Euclidean space*.

In Examples 1 and 2, $\mathbf{R}^3$ was the vector space upon which the inner products were defined. However the inner product spaces in Examples 1 and 2 are distinct, because the inner products are different.

Exercises

1.   Prove the following. Let $\langle \ , \ \rangle$ be an inner product on **V**, and v_1, v_2 elements of **V** such that v_2 is not the zero vector. Then there is a unique v in **V** and scalar c such that $v_1 = cv_2 + v$, and $\langle v, v_2 \rangle = 0$. Determine the scalar c.

In Exercises 2 through 7 determine whether the given expression defines an inner product.

2. $\langle f, g \rangle = \int_0^t f(x) g(t - x)\, dx$ for f and g elements of $C^0[0, t]$.

3. $\langle M, N \rangle = 1^T M^T N 1$ where M and N are elements of $\mathbf{M}_{pp}$, the space of $p \times p$ matrices, and 1 is the $p \times 1$ vector with each element the scalar 1.

4. $\langle x, y \rangle = x_1 y_1 + x_2 y_2 + x_3 y_3 + (x_2 y_3 + x_3 y_2)/3$ for $x = (x_1\ x_2, x_3)$, $y = (y_1, y_2, y_3)$ elements of $\mathbf{R}^3$.

5. $\langle M, N \rangle = m_{11}n_{11} + m_{12}n_{12} + m_{21}n_{21} + m_{22}n_{22}$, where M and N are elements of $\mathbf{M}_{22}$.

6. $\langle P, Q \rangle = \int_2^3 P(x)Q(x)\,dx$, where P and Q are arbitrary elements of the space of all polynomials of degree at most 5.

7. $\langle x, y \rangle = |x_1y_1| + |x_2y_2|$, where $x = (x_1, x_2)$, $y = (y_1, y_2)$ are elements of $\mathbf{R}^2$.

14.2   Matrix representation of inner products

In this section we will illustrate how each inner product has matrix representations, and determine the properties required of a matrix so that it represents an inner product. An important tool will be the coordinate map discussed in Section 5.3.

First we will justify restricting our attention to matrix representations on $\mathbf{R}^n$.

LEMMA 14.1   Let $B = (v_1, v_2, \ldots, v_n)$ be a basis of an inner product space **V**. Then the inner product $\langle \ , \ \rangle$ is uniquely determined by the inner product of elements of B; that is, there is no other inner product $\langle \ , \ \rangle^*$ on **V** such that

$$\langle v_i, v_j \rangle = \langle v_i, v_j \rangle^* \qquad (i, j = 1, 2, \ldots, n)$$

PROOF: Let u and w be vectors in **V**. Then

$$u = \sum_{i=1}^n x_i v_i, \qquad w = \sum_{i=1}^n y_i v_i$$

where the x_i and y_i are the unique coordinates of u and w with respect to B. Hence

$$\langle u, w \rangle = \left\langle \sum_{i=1}^n x_i v_i, w \right\rangle = \sum_{i=1}^n x_i \langle v_i, w \rangle$$

$$= \sum_{i=1}^n x_i \left\langle v_i, \sum_{j=1}^n y_j v_j \right\rangle = \sum_{i=1}^n \sum_{j=1}^n x_i y_j \langle v_i, v_j \rangle \qquad (1)$$

From Eq. (1) we see that if $\langle v_i, v_j \rangle = \langle v_i, v_j \rangle^*$, then $\langle u, w \rangle = \langle u, w \rangle^*$.   ∎

THEOREM 14.1   Let **V** be an n-dimensional vector space. There is a one-to-one correspondence between the set of all inner products on **V** and the set of all inner products on $\mathbf{R}^n$.

PROOF: Consider any isomorphism[1] between **V** and $\mathbf{R}^n$. We shall indicate

---

[1] See Section 5.3 for the definition of isomorphism.

the correspondence between elements v of **V** and r of $\mathbf{R}^n$ by writing $v \leftrightarrow r$. Suppose that $\langle \, , \, \rangle$ is an inner product on **V**. We then define a function $(\, , \,)$ on $\mathbf{R}^n \times \mathbf{R}^n$ into $\mathbf{R}^1$ by

$$(r, s) = \langle u, v \rangle \tag{2}$$

where $r \leftrightarrow u$ and $s \leftrightarrow v$. That is, for each pair of elements r, s in $\mathbf{R}^n$ we find the corresponding elements u, v in **V** and form their inner product. We shall show that the function $(\, , \,)$ is an inner product. First, $(s, r) = \langle v, u \rangle$ by definition, and $\langle v, u \rangle = \langle u, v \rangle$ since $\langle \, , \, \rangle$ is an inner product. Thus $(s, r) = (r, s)$ and $(\, , \,)$ is symmetric. Next, let r, s, t be in $\mathbf{R}^n$ and let $r \leftrightarrow u$, $s \leftrightarrow v$, $t \leftrightarrow w$, where u, v, w are in **V**. Then, for any scalars a and b we have $ar + bs \leftrightarrow au + bv$ since an isomorphism preserves the algebraic relations in a vector space. Hence

$$(ar + bs, t) = \langle au + bv, w \rangle = a\langle u, w \rangle + b\langle v, w \rangle = a(r, t) + b(s, t)$$

Finally, $(r, r) = \langle u, u \rangle$ and since $\langle \, , \, \rangle$ is an inner product it follows that $(r, r) \geq 0$. Since the isomorphism associates the zero vector in $\mathbf{R}^n$ with the zero vector in **V**, and $\langle u, u \rangle > 0$ unless $u = 0$, we see that $(r, r) > 0$ unless $r = \mathbf{0}$; and clearly $(\mathbf{0}, \mathbf{0}) = 0$. We have in this way shown that by Eq. (2) we define an inner product $(\, , \,)$. Conversely, if we are given an inner product $(\, , \,)$, then Eq. (2) determines an inner product $\langle \, , \, \rangle$ on **V**. Thus Eq. (2) establishes a one-to-one correspondence between inner products. ∎

We now restrict our attention to inner products on $\mathbf{R}^n$ since we will be able to generalize our results on $\mathbf{R}^n$ to arbitrary n-dimensional inner product spaces.

THEOREM 14.2 Consider the inner product space consisting of $\mathbf{R}^n$ and $\langle \, , \, \rangle$, and let $B = \{r_1, r_2, \ldots, r_n\}$ be a basis of $\mathbf{R}^n$. Then there exists a unique symmetric matrix M such that $\langle x, y \rangle = X^T M Y$, where X is the $n \times 1$ matrix with elements that are the coordinates of x with respect to B, and Y is the $n \times 1$ matrix with elements that are coordinates of y with respect to B.

PROOF: We begin by constructing the matrix M. First, if the equation $\langle x, y \rangle = X^T M Y$ is to hold, then in particular

$$\langle r_i, r_j \rangle = (0 \cdots 1 \cdots 0) \, M \begin{pmatrix} 0 \\ \cdots \\ 1 \\ \cdots \\ 0 \end{pmatrix} = m_{ij}$$

where m_{ij} is the ijth element of M. Therefore, there is only one possible choice of M, $m_{ij} = \langle r_i, r_j \rangle$. Further, suppose $m_{ij} = \langle r_i, r_j \rangle$ and let x and y be arbitrary elements of $\mathbf{R}^n$. Then

$$x = \sum_{i=1}^{n} x_i r_i \quad \text{and} \quad y = \sum_{j=1}^{n} y_j r_j$$

By Eq. (1),

$$\langle x, y \rangle = \sum_{i=1}^{n} \sum_{j=1}^{n} x_i y_j \langle r_i, r_j \rangle = \sum_{i=1}^{n} \sum_{j=1}^{n} x_i y_j m_{ij} = X^T M Y$$

showing that the desired property $\langle x, y \rangle = X^T M Y$ is obtained. Since $\langle r_i, r_j \rangle = \langle r_j, r_i \rangle$, M must be a symmetric matrix. ∎

The matrix M in Theorem 14.2 is called the matrix of the inner product with respect to the basis B.

Example 1   Consider the space $\mathbf{P}_2$ of polynomials of degree no greater than two. Let $p(x) = a_0 + a_1 x + a_2 x^2$ and $q(x) = b_0 + b_1 x + b_2 x^2$. Then if x_0 is a specified real number,

$$\langle p(\cdot), q(\cdot) \rangle = 2a_0 b_0 + (a_1 b_0 + a_0 b_1)x_0 + 3a_1 b_1 x_0^2$$
$$+ (a_2 b_1 + a_1 b_2)x_0^3 + 2a_2 b_2 x_0^4$$

defines an inner product on $\mathbf{P}_2$. Note that we cannot define the inner product to be a function of x, since the inner product must map pairs of polynomials into $\mathbf{R}^1$. (We leave to the reader to verify that $\langle p(\cdot), q(\cdot) \rangle$ is an inner product.) Consider the basis $\{v_1, v_2, v_3\} = \{1, 1 + x, 1 + x^2\}$ of $\mathbf{P}_2$, and the standard basis $\{e_1, e_2, e_3\}$ of $\mathbf{R}^3$ where $e_1 = (1, 0, 0)$, and so on. The relation $v_1 \leftrightarrow e_1$, $v_2 \leftrightarrow e_2$, $v_3 \leftrightarrow e_3$ determines an isomorphism of $\mathbf{R}^3$ and $\mathbf{P}_2$. We now define the inner product $(\ ,\ )$ on $\mathbf{R}^3$ to be such that

$$(e_1, e_1) = \langle v_1, v_1 \rangle = 2,$$
$$(e_1, e_2) = \langle v_1, v_2 \rangle = 2 + x_0 = (e_2, e_1)$$
$$(e_1, e_3) = \langle v_1, v_3 \rangle = 2 = (e_3, e_1)$$
$$(e_2, e_2) = \langle v_2, v_2 \rangle = 2 + 2x_0 + 3x_0^2$$
$$(e_2, e_3) = \langle v_2, v_3 \rangle = 2 + x_0 + x_0^3 = (e_3, e_2)$$

and

$$(e_3, e_3) = \langle v_3, v_3 \rangle = 2 + 2x_0^4$$

The inner product $(\ ,\ )$ on $\mathbf{R}^3$ is that inner product corresponding to $\langle \ , \ \rangle$ on $\mathbf{V}$ with respect to the bases $\{v_1, v_2, v_3\}$ of $\mathbf{V}$ and $\{e_1, e_2, e_3\}$ of $\mathbf{R}^3$.

Example 2   Consider the inner product $(\ ,\ )$ defined on $\mathbf{R}^3$ in Example 1. The matrix of $(\ ,\ )$ with respect to $\{e_1, e_2, e_3\}$ is then

$$M = \begin{pmatrix} 2 & 2 + x_0 & 2 \\ 2 + x_0 & 2 + 2x_0 + 3x_0^2 & 2 + x_0 + x_0^3 \\ 2 & 2 + x_0 + x_0^3 & 2 + 2x_0^4 \end{pmatrix}$$

For a particular $x = (x_1, x_2, x_3)$ and $y = (y_1, y_2, y_3)$,

$$(x, y) = X^T M Y$$

where

$$X = \begin{pmatrix} x_1 \\ x_2 \\ x_3 \end{pmatrix} \quad \text{and} \quad Y = \begin{pmatrix} y_1 \\ y_2 \\ y_3 \end{pmatrix}$$

are column vectors containing the coordinates of x and y, respectively, with respect to the basis $\{e_1, e_2, e_3\}$.

The next question to be investigated is: What characteristics of a matrix will allow it to be the matrix of an inner product? Certainly not every $n \times n$ matrix represents an inner product on $\mathbf{R}^n$, since we have already seen that the matrix must be symmetric. The symmetry requirement follows from the symmetry condition on inner products.

A second condition which an inner product must satisfy is that $\langle x, x \rangle$ must be positive for all nonzero vectors x. This is equivalent to $X^T M X > 0$ for all nonzero X, where M is the matrix of the inner product relative to the basis B and X is the coordinate vector of x relative to the basis B. (Actually the coordinate vector of x relative to a basis B will be an element of $\mathbf{R}^n$. We will use capital letters to denote the $n \times 1$ matrix in which the elements are these coordinates.)

DEFINITION 14.3   A symmetric matrix M is called *positive definite* if $X^T M X > 0$ for every nonzero column vector X. M is *positive semidefinite* if $X^T M X \geq 0$ for all X.

The above discussion shows that if M is to be the matrix of some inner product relative to B, M must be positive definite. Theorem 14.3 indicates that these conditions are both necessary and sufficient.

THEOREM 14.3   Let $B = \{r_1, r_2, \ldots, r_n\}$ be a basis of $\mathbf{R}^n$. An $n \times n$ matrix M is a matrix of an inner product on $\mathbf{R}^n$ relative to B if and only if M is positive definite.

PROOF: We leave the proof as an exercise for the reader.   ∎

In concluding this section we again point out that the above theorems concerning matrix representations of inner products on $\mathbf{R}^n$ hold equally for inner products on arbitrary n-dimensional real vector spaces. Also, a point worth emphasizing is that given an n-dimensional inner product space $\mathbf{V}$ and $\langle\,,\,\rangle$, two bases B and B' of $\mathbf{V}$ will lead to two distinct matrix representations of $\langle\,,\,\rangle$. If we consider a fixed n-dimensional vector space $\mathbf{W}$ and a fixed basis

B of **W**, then two distinct positive definite matrices imply two distinct inner products on **W**.

Exercises

1. Prove Theorem 14.3.

   In Exercises 2 through 4, give the symmetric matrix which represents the specified inner product with respect to the specified basis.

2. Let $B = \{1, x, x^2\}$ be a basis for the vector space of polynomials of degree at most 2, and $\langle f, g \rangle = \int_0^1 f(t)g(t) \, dt$, where f and g are elements of this vector space.

3. Let $B = \{(1, 0, 1), (0, 1, 0), (1, -1, -1)\}$ be a basis for $\mathbf{R}^3$. Define $\langle (a_1, a_2, a_3), (b_1, b_2, b_3) \rangle = a_1 b_1 + a_2 b_2 + a_3 b_3 - 2(a_1 b_2 + a_2 b_1)/3$.

4. Let

$$B = \left\{ \begin{pmatrix} 1 & 0 \\ 0 & 0 \end{pmatrix}, \begin{pmatrix} 0 & 1 \\ 0 & 0 \end{pmatrix}, \begin{pmatrix} 0 & 0 \\ 1 & 0 \end{pmatrix}, \begin{pmatrix} 0 & 0 \\ 0 & 1 \end{pmatrix} \right\}$$

be a basis for $\mathbf{M}_{22}$, and

$$\langle M, N \rangle = (m_{11} n_{11} + m_{12} n_{12} + m_{21} n_{21} + m_{22} n_{22})$$

for $M = (m_{ij})$ and $N = (n_{ij})$.

14.3 Angles, lengths, and inequalities

In this section we will introduce concepts of angle and length in n-dimensional inner product spaces, being careful to keep our definitions consistent with the definitions already established for geometric vectors in $\mathbf{G}^2$ and $\mathbf{G}^3$. The introduction of geometric terminology into the study of arbitrary vector spaces in this way has proved to be of great value, even in the study of spaces of infinite dimension. We also present several important inequalities.

DEFINITION 14.4 The *standard inner product* on $\mathbf{R}^n$ is $\langle x, y \rangle = X^T Y$, where X and Y are the $n \times 1$ matrices of coordinates of x and y, respectively, relative to the standard basis; that is, if $x = (x_1, x_2, \ldots, x_n)$ and $y = (y_1, y_2, \ldots, y_n)$, then the standard inner product is $\langle x, y \rangle = x_1 y_1 + x_2 y_2 + \cdots + x_n y_n$.

To relate inner products to our concepts of length and direction we will consider the elements of $\mathbf{R}^3$ as the coordinates, with respect to the rectangular coordinate system, of geometric vectors in $\mathbf{G}^3$. We further consider the standard inner product $\langle \ , \ \rangle$ on $\mathbf{R}^3$ which for any vectors v, w in $\mathbf{G}^3$ yields the usual dot product $\langle v, w \rangle = |v|\,|w| \cos\theta$. Let $x = (x_1, x_2, x_3)$ be the coordinate vector of an arbitrary vector v_1 in $\mathbf{G}^3$. Then, by the Pythagorean theorem, the length of v_1 is $\sqrt{x_1{}^2 + x_2{}^2 + x_3{}^2}$. (See Figure 14.1.) But this quantity is just $\langle v_1, v_1 \rangle^{1/2}$. This suggests the following definition of length for vectors in arbitrary inner product spaces.

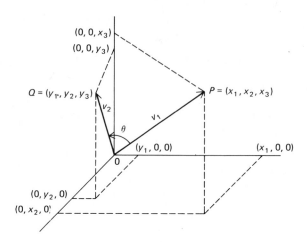

FIGURE 14.1

DEFINITION 14.5 Let $\mathbf{V}$ and $\langle\ ,\ \rangle$ be an inner product space and v a vector in $\mathbf{V}$. Then the *length* or *norm* of v (relative to the inner product $\langle\ ,\ \rangle$), denoted by $|v|$, is $|v| = \langle v, v \rangle^{1/2}$.

We leave to the reader the verification that this definition of length is consistent with our previously conceived notions for the standard inner products on $\mathbf{R}^1$ and $\mathbf{R}^2$.

To treat the direction of vectors we make use of the law of cosines. In Figure 14.1, θ is the angle between v_1, represented by $\overline{OP}$, and v_2, represented by $\overline{OQ}$. By the law of cosines

$$|\overline{PQ}|^2 = |\overline{OP}|^2 + |\overline{OQ}|^2 - 2|\overline{OP}|\,|\overline{OQ}| \cos\theta$$

First, note that $|\overline{OP}|^2 = \langle v_1, v_1 \rangle$ and $|\overline{OQ}|^2 = \langle v_2, v_2 \rangle$. Next, $|\overline{PQ}|^2 = |v_3|^2$ where v_3 is a geometrical vector of length and direction the same as $\overline{PQ}$. Thus

$$|\overline{PQ}|^2 = \langle v_3, v_3 \rangle = (x_1 - y_1)^2 + (x_2 - y_2)^2 + (x_3 - y_3)^2$$
$$= \langle v_1, v_1 \rangle + \langle v_2, v_2 \rangle - 2\langle v_1, v_2 \rangle$$

Applying the law of cosines we obtain

$$\langle v_1, v_2 \rangle = \langle v_1, v_1 \rangle^{1/2} \langle v_2, v_2 \rangle^{1/2} \cos\theta \quad \text{or} \quad \cos\theta = \frac{\langle v_1, v_2 \rangle}{|v_1|\,|v_2|}$$

This suggests the following definition of the angle between vectors in arbitrary vector spaces.

DEFINITION 14.6   Let V and $\langle\ ,\ \rangle$ be an inner product space. The *angle θ between two nonzero vectors* v_1 and v_2 of V, relative to $\langle\ ,\ \rangle$, is given by the relationship

$$\cos\theta = \frac{\langle v_1, v_2\rangle}{|v_1|\,|v_2|}$$

This definition determines a number $\cos\theta$, but it is not yet clear that there is a real angle θ determined, since we have not shown that the number $\cos\theta$ lies between -1 and $+1$. This will be proved below. Of course, θ is then determined up to multiples of 2π.

The reader should keep in mind that the length of a vector, or angle between two vectors, in an inner product space is dependent upon the particular inner product being used.

We now turn to the concept of a projection of one vector onto another. Consider Figure 14.2. In terms of our usual geometric notions of projections,

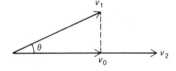

FIGURE 14.2

the projection of v_1 onto v_2 is the vector v_0, which has direction the same as that of v_2, and length obtained from the equation $|v_0| = |v_1|\cos\theta$. Since $v_2/|v_2|$ is the vector in the direction of v_2 but with unit length, we obtain

$$v_0 = \cos\theta\,\frac{|v_1|}{|v_2|}\,v_2$$

Replacing $\cos\theta$ by $\langle v_1, v_2\rangle/|v_1|\,|v_2|$ we are led to the following definition.

DEFINITION 14.7   Let V and $\langle\ ,\ \rangle$ be an inner product space, and let v_1 and v_2 be two vectors in V, $v_2 \neq 0$. The *projection of v_1 onto v_2* is the vector v_0 given by

$$v_0 = \frac{\langle v_1, v_2\rangle v_2}{|v_2|^2}$$

The idea of projecting vectors onto subspaces of vector spaces will be treated in our discussions dealing with orthogonality. We now turn to two basic inequalities.

THEOREM 14.4 (Cauchy-Schwarz Inequality) Let **V** be an inner product space. Then for any vectors u and v in **V**, $\langle u, v \rangle^2 \leq |u|^2 |v|^2$.

PROOF: If the vector v is the zero vector, the result is immediate. Suppose $v \neq 0$. Let w be the vector u minus the projection of u onto v; that is,

$$w = u - \frac{\langle u, v \rangle v}{|v|^2}$$

Then

$$0 \leq \langle w, w \rangle = \left\langle u - \frac{\langle u, v \rangle}{|v|^2} v, u - \frac{\langle u, v \rangle v}{|v|^2} \right\rangle$$

$$= |u|^2 - 2 \frac{\langle u, v \rangle^2}{|v|^2} + \frac{\langle u, v \rangle^2}{|v|^4} \langle v, v \rangle$$

$$= |u|^2 - \frac{\langle u, v \rangle^2}{|v|^2}$$

Hence

$$\langle u, v \rangle^2 \leq |u|^2 |v|^2 \qquad \blacksquare$$

It follows from this theorem that for any vectors v_1, v_2 of **V**,

$$-1 \leq \frac{\langle v_1, v_2 \rangle}{|v_1| |v_2|} \leq 1$$

This shows that the angle θ in Definition 14.6 exists.

Example 1 Consider the space $C^\infty(J)$, where J is an interval $[x_0, x_1]$, and define

$$\langle f, g \rangle = \int_{x_0}^{x_1} f(x)g(x)\, dx$$

Then, by the Cauchy–Schwarz inequality,

$$\left[\int_{x_0}^{x_1} f(x)g(x)\, dx \right]^2 \leq \left[\int_{x_0}^{x_1} f^2(x)\, dx \right]\left[\int_{x_0}^{x_1} g^2(x)\, dx \right]$$

(This particular example is a special case of another well-known result known as Hölder's inequality.)

Example 2 Consider $\mathbf{R}^3$ with the standard inner product and let $x = (x_1, x_2, x_3)$ and $y = (y_1, y_2, y_3)$ be two vectors in $\mathbf{R}^3$. Then

$$|\langle x, y \rangle| = |x_1 y_1 + x_2 y_2 + x_3 y_3|$$
$$\leq (x_1^2 + x_2^2 + x_3^2)^{1/2} (y_1^2 + y_2^2 + y_3^2)^{1/2} = |x| |y|$$

The second inequality which we consider is known as the triangle inequality, since in $\mathbf{R}^2$ and $\mathbf{R}^3$ it can be proved through geometric considerations on lengths of legs of triangles.

THEOREM 14.5 (Triangle Inequality) Let u and v be elements of an inner product space determined by $\mathbf{V}$ and $\langle \, , \, \rangle$. Then

$$|u + v| \leq |u| + |v|$$

PROOF: $|u + v|^2 = \langle u + v, u + v \rangle = \langle u, u \rangle + 2\langle u, v \rangle + \langle v, v \rangle$. By the Cauchy–Schwarz inequality, this leads to

$$|u + v|^2 \leq |u|^2 + |v|^2 + 2|u|\,|v| = (|u| + |v|)^2$$

Hence $|u + v| \leq |u| + |v|$ and the theorem is proved.   ∎

Example 3 Consider vectors in $\mathbf{G}^2$, the space of arrows in geometric 2-space. Then if u, v and hence $u + v$ are as in Figure 14.3, the triangle inequality implies that the sum of the lengths of two legs of a triangle is at least as great as the length of the third leg.

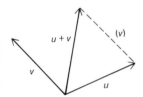

FIGURE 14.3

Example 4 Consider the space $\mathbf{C}^\infty(J)$, $J = [x_0, x_1]$ as used in Example 1, and the inner product

$$\langle f, g \rangle = \int_{x_0}^{x_1} f(x)g(x)\,dx$$

Then the triangle inequality verifies

$$\left[\int_{x_0}^{x_1} (f(x) + g(x))^2\,dx\right]^{1/2} \leq \left[\int_{x_0}^{x_1} f^2(x)\,dx\right]^{1/2} + \left[\int_{x_0}^{x_1} g^2(x)\,dx\right]^{1/2}$$

(This example is a special case of another famous inequality known as Minkowski's inequality.)

The reader can now verify that length for vectors in an inner product space has the properties

1. $|cv| = |c|\,|v|$, for $v \in \mathbf{V}$, c real
2. $|v| > 0$ if $v \neq \mathbf{0}$ and $|\mathbf{0}| = 0$
3. $|u + v| \leq |u| + |v|$ for all $u, v \in \mathbf{V}$

Thus the length which results from an inner product has three of the most basic properties of length as ordinarily understood in geometry.

Exercises

1. Determine the projection of v_1 onto v_2 for the following vectors in $\mathbf{R}^n$ using the standard inner product.

   (a) $v_1 = (1, 2, 3)$, $v_2 = (1, 1, 2)$
   (b) $v_1 = (1, 2, 1)$, $v_2 = (2, 1, -1)$
   (c) $v_1 = (1, 0, 1, 0)$, $v_2 = (0, 2, 2, 0)$

2. Compute the length of each vector and the cosine of the angle between each pair of vectors in Exercise 1.

3. Consider the space $C^\infty(J)$ where $J = [0, 1]$ and the inner product is

$$\langle f, g \rangle = \int_0^1 f(x)g(x)\, dx$$

   Compute the length of each vector and the cosine of the angle between each pair of vectors in each of the following.

   (a) $f(x) = 1$, $g(x) = x$
   (b) $f(x) = x^n$, $g(x) = x^m$ (*n, m* nonnegative integers)
   (c) $f(x) = e^{ax}$, $g(x) = e^{bx}$
   (d) $f(x) = \sin \pi n x$, $g(x) = \sin \pi m x$ (*n, m* integers)

4. Show that

$$\langle v_1, v_2 \rangle = x_1 y_1 + 2x_2 y_2 + x_1 y_2 + x_2 y_1$$

   where $v_1 = (x_1, x_2)$, $v_2 = (y_1, y_2)$, defines an inner product on $\mathbf{R}^2$. Compute the cosine of the angle between each pair of vectors, first using this inner product and then using the standard inner product.

   (a) $v_1 = (3, 4)$, $v_2 = (5, 12)$
   (b) $v_1 = (1, 1)$, $v_2 = (2, 1)$

5. Write the Cauchy–Schwarz and triangle inequalities for $\mathbf{R}^2$ with the inner product given in Exercise 4.

6. Show that

$$\langle f, g \rangle = \int_1^2 f(x)g(x)\, x^2 dx$$

   defines an inner product on $C[1, 2]$. Write the Cauchy–Schwarz and triangle inequalities in this case.

7. Prove that in an inner product space

$$|cv| = |c| |v| \quad \text{for} \quad v \in V, \quad c \text{ real}$$
$$|v| > 0 \quad \text{if} \quad v \neq 0 \quad \text{and} \quad |0| = 0$$

8. Let $u = (1, 1, \ldots, 1, 1)$ and $v = (n, 1, \ldots, 1, 1)$ be vectors in $\mathbf{R}^n$, and let θ be the angle between u and v relative to the standard inner product.

(a) Find the limit of θ as $n \to \infty$.

(b) Find the projection of v on u and its limiting form as $n \to \infty$.

14.4 Orthogonal vectors and Gram-Schmidt orthogonalization

The concepts of perpendicularity are assumed to be well understood in $\mathbf{G}^2$ and $\mathbf{G}^3$, and the usefulness of these concepts is reflected by the fact that the rectangular coordinate systems are most commonly employed in these spaces. It is the notion of perpendicular vectors that we will generalize in this section. In many cases it is easier to solve problems and obtain results in general vector spaces when the notion of perpendicularity can be exploited.

Example 1 Consider the space $\mathbf{G}^3$ with the usual Cartesian coordinate system; that is, each vector in $\mathbf{G}^3$ is represented by an element in $\mathbf{R}^3$, its coordinate vector with respect to some rectangular coordinate system. We now consider the inner product space of $\mathbf{R}^3$ with the standard inner product. If

$$x = (x_1, x_2, x_3) \quad \text{and} \quad y = (y_1, y_2, y_3)$$

then $\langle x, y \rangle = X^T Y$, where

$$X = \begin{pmatrix} x_1 \\ x_2 \\ x_3 \end{pmatrix} \quad Y = \begin{pmatrix} y_1 \\ y_2 \\ y_3 \end{pmatrix}$$

Therefore if g_1 and g_2 are two geometric vectors with coordinate vectors x and y, respectively, the inner product of g_1 with g_2 is given by $X^T Y$. We have seen that in this case the angle θ between two nonzero vectors g_1 and g_2 is given by

$$\cos \theta = \frac{\langle x, y \rangle}{|x| \, |y|}$$

Since g_1 and g_2 are perpendicular if $\theta = \pm \pi/2$, we can conclude that g_1 and g_2 are perpendicular (or orthogonal) if $\langle x, y \rangle = 0$. Hence the following definition is adopted.

DEFINITION 14.8 Consider the inner product space V and $\langle\,,\,\rangle$. Two vectors v_1 and v_2 in V are *orthogonal* if $\langle v_1, v_2 \rangle = 0$, and *orthonormal* if they are orthogonal and $|v_1| = |v_2| = 1$. (A vector of length one is called a *unit vector*.)

By the above definition we see that the zero vector is orthogonal to every vector in an inner product space.

Example 2   Consider the inner product space of $\mathbf{R}^n$ with the standard inner product. We can see that the standard basis vectors are orthonormal. Consider $\langle e_i, e_j \rangle$, where e_i is the *i*th basis vector $(0, 0, \ldots, 1, \ldots, 0)$. By E_i we denote the $n \times 1$ column vector with 1 in the *i*th row and zero elsewhere, $\langle e_i, e_j \rangle = E_i^T E_j = \delta_{ij}$, where δ_{ij} is the *Kronecker delta* defined by $\delta_{ij} = 0$ if $i \neq j$ and $\delta_{ii} = 1$. Therefore distinct basis vectors are orthogonal ($\langle e_i, e_j \rangle = 0$ for $i \neq j$) and they each have unit length ($\langle e_i, e_i \rangle = 1$). Hence they form an orthonormal set.

Example 3   Consider the space of continuous functions on $[-1, 1]$; that is, $\mathbf{C}^0(J)$ where $J = [-1, 1]$. Define an inner product

$$\langle f, g \rangle = \int_{-1}^{1} f(x)g(x)\, dx$$

Let f be an even function; that is, $f(x) = f(-x)$. Let g be an odd function; that is, $g(x) = -g(-x)$. Then

$$\langle f, g \rangle = \int_{-1}^{1} f(x)g(x)\, dx = -\int_{-1}^{1} f(y)g(y)\, dy$$

where $y = -x$. Hence $\langle f, g \rangle = 0$ and f is orthogonal to g. If we set $f_1(x) = f(x)/|f|$ and $g_1(x) = g(x)/|g|$, then f_1 and g_1 will be orthonormal vectors (provided $|f| \neq 0, |g| \neq 0$).

In Example 2 we showed that the standard basis of $\mathbf{R}^n$ forms an orthonormal set, or orthonormal basis. We turn now to the question of whether every finite dimensional real vector space has such a basis.

THEOREM 14.6   Let $\mathbf{V}$ be an inner product space, and let $\{v_1, v_2, \ldots, v_k\}$ be an orthogonal set of nonzero vectors; that is, $\langle v_i, v_j \rangle = 0$ for $i \neq j$. Then $\{v_1, v_2, \ldots, v_k\}$ is a linearly independent set of vectors.

PROOF: Suppose the set to be dependent. Then there exists a vector, say v_1, such that

$$v_1 = \sum_{i=2}^{k} a_i v_i$$

Therefore

$$\langle v_1, v_1 \rangle = \left\langle v_1, \sum_{i=2}^{k} a_i v_i \right\rangle = \sum_{i=2}^{k} a_i \langle v_1, v_i \rangle$$

But since $\langle v_1, v_i \rangle = 0$ due to the fact the v_i form an orthogonal set, we have $\langle v_1, v_1 \rangle = 0$, which contradicts the assumption $v_1 \neq 0$. The theorem is proved. ∎

Theorem 14.6 shows that orthogonal vectors are linearly independent. However, the converse is not true; that is, linearly independent vectors are not necessarily orthogonal. The question then arises as to whether every inner product space has an orthogonal basis. We show in the next theorem that this is indeed the case, and we prove the theorem by constructing such a basis for an arbitrary space. The method of construction is known as the *Gram–Schmidt process*.

THEOREM 14.7 Every finite dimensional inner product space has an orthonormal basis.

PROOF: Consider an arbitrary n-dimensional inner product space $\mathbf{V}$ and $\langle \ , \ \rangle$, and let $B = \{v_1, v_2, \ldots, v_n\}$ be a basis of $\mathbf{V}$. We now construct an orthonormal basis $B' = \{u_1, u_2, \ldots, u_n\}$ for $\mathbf{V}$ as follows. Set $u_1 = v_1/|v_1|$. Then u_1 has unit length. We now select u_2 of unit length and orthogonal to u_1. This is done by setting

$$w_2 = v_2 - \frac{\langle v_2, u_1 \rangle}{|u_1|^2} u_1$$

that is, w_2 is v_2 minus the projection of v_2 onto u_1. Then

$$\langle u_1, w_2 \rangle = \langle u_1, v_2 \rangle - \langle v_2, u_1 \rangle \frac{|u_1|^2}{|u_1|^2} = 0$$

and hence u_1 and w_2 are orthogonal. Set $u_2 = w_2/|w_2|$ and we have $\{u_1, u_2\}$ forming an orthonormal set. Next we set

$$w_3 = v_3 - \frac{\langle v_3, u_1 \rangle u_1}{|u_1|^2} - \frac{\langle v_3, u_2 \rangle u_2}{|u_2|^2}$$

(note $|u_1| = |u_2| = 1$, but we express w_3 in this form to point out we are taking projections) and $u_3 = w_3/|w_3|$. Again we will have an orthonormal set $\{u_1, u_2, u_3\}$. For any k, $2 \leq k \leq n$, we define

$$u_k = \frac{w_k}{|w_k|}$$

where

$$w_k = v_k - \sum_{j=1}^{k-1} \frac{\langle v_k, u_j \rangle u_j}{|u_j|^2}$$

and assume $\{u_1, u_2, \ldots, u_k\}$ forms an orthonormal set. (We have shown this result for $k = 2$. We now show it true for $k \leq n$ by induction.) Let $k < n$. Then

$$w_{k+1} = v_{k+1} - \sum_{j=1}^{k} \frac{\langle v_{k+1}, u_j \rangle u_j}{|u_j|^2}$$

$$\langle u_i, w_{k+1} \rangle = \langle u_i, v_{k+1} \rangle - \sum_{j=1}^{k} \frac{\langle v_{k+1}, u_j \rangle \langle u_i, u_j \rangle}{|u_j|^2}$$

$$= \langle u_i, v_{k+1} \rangle - \langle v_{k+1}, u_i \rangle = 0, \qquad 1 \leq i \leq k,$$

since $\langle u_i, u_j \rangle = 0$ for $i \neq j$, i and $j \leq k$. Hence u_i and w_{k+1} are orthogonal, for $i \leq k$, and since

$$u_{k+1} = \frac{w_{k+1}}{|w_{k+1}|}$$

u_i and u_{k+1} are orthonormal. Hence, $\{u_1, u_2, \ldots, u_n\}$ is an orthonormal set. Since **V** is of dimension n and an orthonormal set of vectors is independent (Theorem 14.6), we can conclude that $\{u_1, u_2, \ldots, u_n\}$ is an orthonormal basis of **V**.   ∎

The first step in the Gram–Schmidt process is illustrated geometrically in Figure 14.4. By subtracting from v_2 the projection of v_2 on u_1, a vector w_2 is obtained which is orthogonal to u_1. This result depends, of course, on the fact that projection as defined in Definition 14.7 corresponds to our usual geometric idea of perpendicular projection. Since the word "projection" is often used with other meanings, we may sometimes refer to the operation in Definition 14.7 as *orthogonal projection*.

Example 4   Let us again treat $\mathbf{R}^3$ with the standard inner product. Set $v_1 = (-1, 0, 1)$, $v_2 = (-1, 1, 3)$, and $v_3 = (-1, 4, -1)$. The reader can verify that $\{v_1, v_2, v_3\}$ is a basis of $\mathbf{R}^3$. We now proceed to form an orthogonal basis for $\mathbf{R}^3$ by the Gram–Schmidt process. First,

$$u_1 = \frac{v_1}{|v_1|} = \frac{1}{\sqrt{2}}(-1, 0, 1)$$

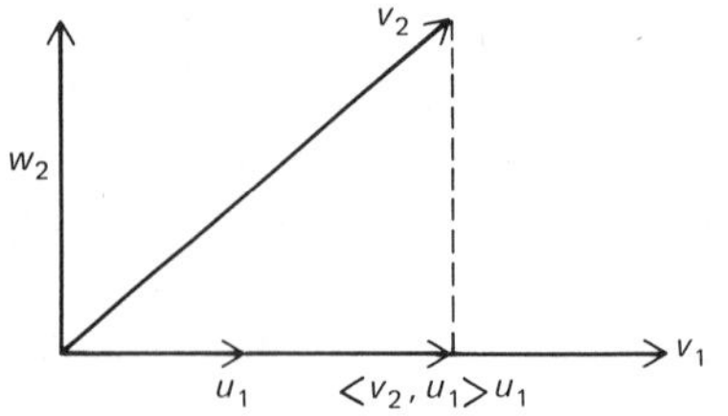

FIGURE 14.4

Next,

$$w_2 = v_2 - \langle v_2, u_1 \rangle u_1$$

$$= (-1, 1, 3) - 2\sqrt{2}\left(-\frac{1}{\sqrt{2}}, 0, \frac{1}{\sqrt{2}}\right) = (1, 1, 1).$$

Then

$$u_2 = \frac{w_2}{|w_2|} = \frac{1}{\sqrt{3}}(1, 1, 1)$$

Finally,

$$w_3 = v_3 - \langle v_3, u_1 \rangle u_1 - \langle v_3, u_2 \rangle u_2$$

$$= (-1, 4, -1) - 0 \cdot u_1 - \frac{2}{\sqrt{3}} u_2$$

$$= (-1, 4, -1) - \frac{2}{3}(1, 1, 1) = \left(\frac{-5}{3}, \frac{10}{3}, \frac{-5}{3}\right)$$

Thus

$$|w_3| = 5\sqrt{\frac{2}{3}} \quad \text{and} \quad u_3 = \left(-\frac{1}{\sqrt{6}}, \frac{2}{\sqrt{6}}, \frac{-1}{\sqrt{6}}\right)$$

The reader can verify that $\{u_1, u_2, u_3\}$ forms an orthonormal basis for $\mathbf{R}^3$.

As a consequence of Theorem 14.7 we get the following.

THEOREM 14.8 Consider the n-dimensional inner product space determined by $\mathbf{V}$ and $\langle \, , \, \rangle$. There exists a basis $B = \{v_1, \ldots, v_n\}$ of $\mathbf{V}$ such that the matrix of the inner product, with respect to B, is the identity matrix I_n.

PROOF: The proof is left as an exercise. ∎

The reader should consider the following implications of this theorem. Let v_1 and v_2 be vectors in $\mathbf{V}$ and let their coordinate vectors in terms of the orthonormal basis B be

$$X = \begin{pmatrix} x_1 \\ \cdots \\ x_n \end{pmatrix} \quad \text{and} \quad Y = \begin{pmatrix} y_1 \\ \cdots \\ y_n \end{pmatrix}$$

Then

$$\langle v_1, v_2 \rangle = X^T Y = \sum_{i=1}^{n} x_i y_i$$

That is, $\langle v_1, v_2 \rangle$ is equal to the dot product of X and Y. Moreover,

$$|v_1| = \left(\sum_{i=1}^{n} x_i^2\right)^{1/2}$$

and the angle between v_1 and v_2 is given by

$$\cos \theta = \frac{x_1 y_1 + \cdots + x_n y_n}{(x_1^2 + \cdots + x_n^2)^{1/2}(y_1^2 + \cdots + y_n^2)^{1/2}}$$

Thus, relative to an orthonormal basis, the formulas for length and angle have the form familiar from analytic geometry.

COROLLARY   Let M be a positive definite matrix. There exists a nonsingular matrix P such that $M = P^T P$. Conversely, if P is any nonsingular matrix and $M = P^T P$, M is positive definite.

PROOF: Consider a vector space V, and let $B = \{v_1, \ldots, v_n\}$ be a basis of V. Then M defines an inner product $\langle \, , \, \rangle$ on V by $m_{ij} = \langle v_i, v_j \rangle$. By Theorem 14.8 we know there is a basis B' such that the matrix of $\langle \, , \, \rangle$ with respect to B' is I_n. Let P be the transition matrix from B to B'. If u_1 and u_2 are arbitrary vectors in V, and X_1 and X_2 the coordinates with respect to B of u_1 and u_2, respectively, and Y_1 and Y_2 the coordinates with respect to B', then $\langle u_1, u_2 \rangle = X_1^T M X_2 = Y_1^T I_n Y_2$ for all vectors X_1 and X_2 (that is, all pairs of vectors in V). But $Y_1 = PX_1$ and $Y_2 = PX_2$. Since $Y_1^T Y_2 = X_1^T P^T P X_2 = X_1^T M X_2$ for all X_1, X_2 we can conclude that $M = P^T P$. (See Exercise 6.)

We now prove the converse. Let P be nonsingular and $M = P^T P$. Then for any X,

$$X^T M X = X^T P^T P X = (PX)^T (PX) = \sum_{i=1}^{n} y_i^2$$

where y_i is the ith component of PX. It follows that $X^T M X \geq 0$, with equality if and only if $PX = 0$. Since P is nonsingular, $PX = 0$ implies $X = 0$ and so M is positive definite.   ∎

Exercises

1.   Prove Theorem 14.8.

2.   Consider the basis $\{v_1, v_2, v_3\}$ of $\mathbf{G}^3$, where the coordinates with respect to the usual Cartesian coordinate system are $\sigma v_1 = (1, 1, 1)$, $\sigma v_2 = (1, -1, 1)$ and $\sigma v_3 = (1, 1, 0)$. Apply the Gram–Schmidt process to this set so as to form an orthonormal basis for $\mathbf{G}^3$.

3.   Let the inner product space be defined as

$$\langle f, g \rangle = \int_0^1 f(t)g(t) \, dt$$

for elements in the space generated by the basis $\{1, x, x^2\}$. Using the Gram–Schmidt process, find an orthonormal basis for this space.

4. Apply the Gram-Schmidt process on the columns of the following matrices, and thus find an orthonormal basis for the column spaces.

(a) $\begin{pmatrix} 2 & 4 & 3 \\ 1 & 1 & 1 \\ 2 & 0 & 1 \end{pmatrix}$

(b) $\begin{pmatrix} 1 & 4 & 0 \\ -2 & -3 & 1 \\ 0 & 0 & 2 \end{pmatrix}$

(c) $\begin{pmatrix} 1 & 1 & 0 & 0 \\ 1 & 0 & 1 & 0 \\ 1 & 0 & 0 & 1 \\ 1 & 0 & 0 & 0 \end{pmatrix}$

5. Repeat Exercise 5 operating on rows rather than columns.

6. Prove that $X_1^T P^T P X_2 = X_1^T M X_2$ for all X_1, X_2 implies $M = P^T P$.

14.5 Orthocomplements

In previous sections we have extended geometric concepts of angle and length in G^2 and G^3 to arbitrary inner product spaces. In the geometry of G^2 and G^3 one is also accustomed to talking about lines perpendicular to planes, planes perpendicular to planes, and the like. We now show how to develop analogous ideas in any inner product space.

DEFINITION 14.9 Let S be a nonempty set of vectors in the inner product space V, $\langle \ , \ \rangle$. The set of all vectors in V which are orthogonal to S (that is, to each vector in S) is the *orthogonal complement* or *orthocomplement* of the set S and is denoted $S^\perp$. (The symbol $S^\perp$ is usually read "S perp.").

Note in the above definition that S need not be a subspace of V. Also, if S is the zero vector, then $S^\perp = V$. S could be any other single vector if we desired. However, even if S is not a subspace, the set $S^\perp$ will be a subspace as is shown in the following theorem.

THEOREM 14.9 Let S be a nonempty set of vectors in V, $\langle \ , \ \rangle$. Then $S^\perp$ is a subspace of V.

PROOF: The zero vector is in $S^\perp$ as it is orthogonal to every vector in V. Let v_1 and v_2 be in $S^\perp$. Then, for every vector u in S, $\langle v_1, u \rangle = \langle v_2, u \rangle = 0$, and hence $\langle av_1 + bv_2, u \rangle = 0$ for all scalars a and b. Hence, $av_1 + bv_2$ is in $S^\perp$ and we have S as a subspace of V. ∎

We apply the above theorem in the proof of the following important result.

THEOREM 14.10 Let S be a nonempty set of vectors in V, $\langle \, , \, \rangle$ and let W be the subspace of V spanned by S. Then $S^\perp$ is the orthocomplement of W, that is, $S^\perp = W^\perp$, and moreover $W = (S^\perp)^\perp$.

PROOF: Let $B_1 = \{v_1, v_2, \ldots, v_k\}$ be a maximal linearly independent subset of S. Then, B_1 forms a basis of W. Let u be any vector in W and v any vector in $S^\perp$. Then

$$u = \sum_{i=1}^{k} a_i v_i \quad \text{and} \quad \langle u, v \rangle = \left\langle \sum_{i=1}^{k} a_i v_i, v \right\rangle = \sum_{i=1}^{k} a_i \langle v_i, v \rangle = 0$$

Hence $W^\perp \supseteq S^\perp$. But $S \subseteq W$ since W is the space generated by S. Hence each vector orthogonal to every vector of W (that is, each vector in $W^\perp$) must necessarily be orthogonal to every vector in S. Hence $W^\perp \subseteq S^\perp$ and so equality must obtain.

To show $(S^\perp)^\perp = W$, or equivalently $(W^\perp)^\perp = W$, proceed as follows. Let $B_1 = \{v_1, \ldots, v_k\}$ be a basis for W, and $B = \{v_1, \ldots, v_k, v_{k+1}, \ldots, v_n\}$ be a basis of V. By the Gram–Schmidt process we can construct an orthonormal basis

$$B' = \{w_1, w_2, \ldots, w_k, w_{k+1}, \ldots, w_n\}$$

such that for any r, $\{w_1, w_2, \ldots, w_r\}$ spans the same subspace of V as does $\{v_1, v_2, \ldots, v_r\}$. In particular $B_1' = \{w_1, \ldots, w_k\}$ is an orthonormal basis of W. Further, the space spanned by $B_2' = \{w_{k+1}, \ldots, w_n\}$ is $W^\perp$. Since every vector orthogonal to $(W^\perp)$ must be orthogonal to every element of B_2', such vectors are in the space spanned by B_1'. We can therefore conclude that $(W^\perp)^\perp \subseteq W$. And since every vector in W is orthogonal to $W^\perp$ we can conclude $(W^\perp)^\perp \supseteq W$. Hence $(W^\perp)^\perp = W$. ∎

We close this section with the following theorem.

THEOREM 14.11 Let W be a subspace of V, $\langle \, , \, \rangle$. Then $V = W \oplus W^\perp$.

PROOF: Recall from Section 4.12 that V is the direct sum of W and $W^\perp$ if, for each element $v \in V$, there are unique elements $w_1 \in W$ and $w_2 \in W^\perp$ such that $v = w_1 + w_2$. We leave the proof as an exercise. ∎

Example 1 Consider $\mathbf{R}^3$ with the standard inner product and let $S = \{v_1, v_2\}$, where $v_1 = (-1, 0, 1)$, $v_2 = (-1, 1, 3)$. Let W be the subspace spanned by S. We shall find $S^\perp$. We begin by extending S to a basis for $\mathbf{R}^3$. For example, choose $v_3 = (-1, 4, -1)$. Then $B = \{v_1, v_2, v_3\}$ is a basis. We now apply the Gram–Schmidt process to obtain an orthonormal basis. By Example 4 in Section 14.4 we obtain $\{u_1, u_2, u_3\}$, where $u_1 = (-1, 0, 1)/\sqrt{2}$, $u_2 = (1, 1, 1)/\sqrt{3}$, $u_3 = (-1, 2, -1)/\sqrt{6}$. Moreover, since u_1 is a multiple of

v_1 and u_2 is a linear combination of v_1 and v_2, the vectors u_1, u_2 span the same subspace $\mathbf{W}$ as $\{v_1, v_2\}$; that is, $S^{\perp} = \mathbf{W}^{\perp} = [u_1, u_2]^{\perp}$. Obviously, $[u_1, u_2]^{\perp}$ is the subspace spanned by u_3, so $S^{\perp} = [u_3] = [(-1, 2, -1)]$. In geometrical language, we can say that the line containing $(-1, 2, -1)$ is perpendicular to the plane determined by $(-1, 0, 1)$ and $(-1, 1, 3)$.

Exercises

1. Prove Theorem 14.11.

2. Let $S = \{v_i\}$, where $\{v_i\}$ is specified below. For each S, find $S^{\perp}$, relative to the standard inner product on $\mathbf{R}^n$.

   (a) $v_1 = (0, 1, 0)$, $v_2 = (0, 0, 1)$
   (b) $v_1 = (1, 1, 0)$, $v_2 = (1, 1, 1)$
   (c) $v_1 = (1, 0, 1, 2)$, $v_2 = (1, 1, 1, 1)$, $v_3 = (2, 2, 0, 1)$

3. Let $\mathbf{V} = \mathbf{W} + \mathbf{W}^{\perp}$ and $v = u_1 + u_2$, where $u_1 \in \mathbf{W}$, $u_2 \in \mathbf{W}^{\perp}$. Prove that $|v - u_1| \le |v - w|$ for every w in $\mathbf{W}$. [*Note:* A special case of this is the geometrical proof that the shortest line segment connecting a point and a hyperplane is the perpendicular to the hyperplane.]

4. Let $\mathbf{S}$ and $\mathbf{T}$ be two subspaces of vectors in $\mathbf{V}$.

   (a) Prove that $(\mathbf{S} + \mathbf{T})^{\perp} = \mathbf{S}^{\perp} \cap \mathbf{T}^{\perp}$.
   (b) If $\mathbf{V}$ is three-dimensional space, and $\mathbf{S}$ and $\mathbf{T}$ are distinct lines, what is the geometrical interpretation of (a)?

5. Let $\langle \, , \, \rangle$ be an inner product on $\mathbf{R}^n$. Consider a subspace $\mathbf{W}$ of $\mathbf{R}^n$ with orthonormal basis $\{v_1, v_2, \ldots, v_k\}$. The *orthogonal projection* of $v \in \mathbf{R}^n$ onto $\mathbf{W}$ is defined as

$$\text{proj } v = \langle v, v_1 \rangle v_1 + \langle v, v_2 \rangle v_2 + \cdots + \langle v, v_k \rangle v_k$$

   (a) Prove that the projection operator (that is, the operator which takes v into proj v) is a linear transformation and hence has an $n \times n$ matrix representation.
   (b) Prove that proj $v = v$ for every v in $\mathbf{W}$.

6. Consider $\mathbf{R}^n$ with an inner product $\langle \, , \, \rangle$ and let $\mathbf{W}$ be a subspace. By Theorem 14.11, $\mathbf{R}^n = \mathbf{W} \oplus \mathbf{W}^{\perp}$; thus every $v \in \mathbf{R}^n$ can be represented as $v = w + z$ where $w \in \mathbf{W}$ and $z \in \mathbf{W}^{\perp}$. Show that w is the projection of v on $\mathbf{W}$ and z is the projection of v on $\mathbf{W}^{\perp}$.
   [*Hint:* Let $\{v_1, v_2, \ldots, v_k\}$ be an orthonormal basis for $\mathbf{W}$ and let $w = \sum_{i=1}^{k} a_i v_i$. Find the a_i.]

7. Let P be the matrix of a transformation of $\mathbf{R}^n$ into $\mathbf{R}^n$ with respect to the standard basis. Prove that P is the matrix of a projection (onto some subspace) if and only if P is symmetric ($P = P^T$) and idempotent ($PP = P$).

8. Prove the Pythagorean Theorem for general inner product spaces; that is, if v_1 and v_2 are orthogonal, $|v_1 + v_2|^2 = |v_1|^2 + |v_2|^2$.

9. Let $\mathbf{V} = \mathbf{W} \oplus \mathbf{W}^{\perp}$ as in Theorem 14.11. Let $v \in \mathbf{V}$, $v = w + z$, where $w \in \mathbf{W}$, $z \in \mathbf{W}^{\perp}$. Prove that

$$|v|^2 = |w|^2 + |z|^2$$

10. If $\langle \ , \ \rangle$ denotes the standard inner product in $\mathbf{R}^n$, A is an $n \times n$ matrix, and X and Y are $n \times 1$ vectors, show that

$$\langle AX, Y \rangle = \langle X, A^T Y \rangle.$$

14.6   Orthogonal matrices and transformations

In this section we will introduce a very useful concept in matrix algebra—the concept of orthogonal matrix. We will then attempt to relate orthogonal matrices and orthogonal transformations and give geometric interpretations to each.

DEFINITION 14.10   Let M be an $r \times s$ matrix and let

$$M_i = \begin{pmatrix} m_{1i} \\ \cdots \\ m_{ri} \end{pmatrix}$$

denote the ith column vector of M. The *standard inner product on the columns of M* is given by

$$\langle M_i, M_j \rangle = \sum_{k=1}^{r} m_{ki} m_{kj}$$

We define the standard inner product on the rows of M in a like fashion.

DEFINITION 14.11   Let P be an $n \times n$ matrix. Then P is *orthogonal* if the columns of P form an orthonormal set (relative to the standard inner product on columns).

We now derive properties of orthogonal matrices. In particular we will show that if a matrix is orthogonal its rows will form an orthonormal set also.

THEOREM 14.12   Let P be an orthogonal matrix. Then $PP^T = P^T P = I$. Conversely, if $PP^T = P^T P = I$, then P is an orthogonal matrix.

PROOF: If P is orthogonal the columns of P are orthonormal. Hence, the ijth element of $P^T P$ is

$$\sum_{k=1}^{n} p_{ki} p_{kj} = \delta_{ij}$$

where δ_{ij} is the previously defined Kronecker symbol. Therefore $P^T P = I$. But this means P is invertible and $P^T = P^{-1}$. Since $P^{-1}P = PP^{-1} = I$, we can conclude $P^T P = PP^T = I$. The reader is asked to prove the converse. ∎

COROLLARY If P is an orthogonal matrix, the rows of P form an orthonormal set; that is,

$$\sum_{k=1}^{n} p_{ik} p_{jk} = \delta_{ij}$$

PROOF: This result is a direct consequence of the fact $PP^T = I$. ∎

We have seen in Theorem 14.12 that an orthogonal matrix is nonsingular. A further result is as follows.

THEOREM 14.13 If P is orthogonal, the determinant of P is plus or minus one; that is, $|\det P| = 1$.

PROOF: By Theorem 14.12 we know $P^T P = I$. Hence, $\det P^T P = 1$. But $\det P^T P = \det P^T \det P = [\det P]^2 = 1$ by Theorems 7.5 and 7.7. Hence $|\det P| = 1$. ∎

We now turn our attention to orthogonal transformations, which we will relate to orthogonal matrices.

DEFINITION 14.12 Let $\mathbf{V}, \langle\,,\,\rangle$ be an inner product space. A linear transformation $T: \mathbf{V} \to \mathbf{V}$ is an *orthogonal transformation* if $\langle Tv_1, Tv_2 \rangle = \langle v_1, v_2 \rangle$ for every pair of vectors v_1, v_2 in $\mathbf{V}$.

Let us consider the geometric interpretation of this definition. First, for any vector v, $\langle Tv, Tv \rangle = \langle v, v \rangle$ implies that lengths are unchanged by orthogonal transformations. Further, since $\langle Tv_i, Tv_j \rangle = \langle v_i, v_j \rangle$ for every pair v_i, v_j we can conclude

$$\frac{\langle Tv_i, Tv_j \rangle}{|Tv_i|\,|Tv_j|} = \frac{\langle v_i, v_j \rangle}{|v_i|\,|v_j|}$$

That is, the angle between the transformed vectors, Tv_i and Tv_j, is the same as the angle between v_i and v_j. Therefore, angles are unchanged (remain invariant) under the orthogonal transformation T.

We now turn to a theorem which relates orthogonal transformations and orthogonal matrices.

THEOREM 14.14   Let T be a linear transformation on an n-dimensional inner product space $\mathbf{V}$, $\langle \, , \, \rangle$ and let $B = \{v_1, v_2, \ldots, v_n\}$ be a basis of $\mathbf{V}$ such that the matrix of the inner product with respect to B is I_n. Then T is an orthogonal transformation if and only if the matrix of T with respect to the basis B is an orthogonal matrix.

PROOF: Let us begin by verifying that we can find a basis B satisfying the condition required in the theorem. We know by Theorem 14.7 that we can find an orthonormal basis for $\mathbf{V}$. Let such a basis be $B = \{v_1, v_2, \ldots, v_n\}$. The matrix of $\langle \, , \, \rangle$ with respect to B has as its i, jth element $\langle v_i, v_j \rangle = \delta_{ij}$. Hence the basis B is as required.

Let P be the matrix of T with respect to the basis B. This implies that if u and v are vectors in $\mathbf{V}$, and the coordinates of u and v with respect to B are given by

$$X = \begin{pmatrix} x_1 \\ x_2 \\ \vdots \\ x_n \end{pmatrix}, \qquad Y = \begin{pmatrix} y_1 \\ y_2 \\ \vdots \\ y_n \end{pmatrix}$$

respectively, then the coordinates, with respect to B, of Tu and Tv are given by PX and PY, respectively. Since the matrix, with respect to B, of the inner product is I_n, $\langle u, v \rangle = X^T I_n Y = X^T Y$. Also $\langle Tu, Tv \rangle = (PX)^T I_n (PY) = X^T P^T P Y$. But then $\langle Tu, Tv \rangle = \langle u, v \rangle$ if and only if $X^T P^T P Y = X^T Y$ for all X and Y, since u and v are arbitrary. Hence T is orthogonal if and only if $P^T P = I$; that is, if and only if P is orthogonal.   ∎

Orthogonal matrices and transformations have many applications throughout mathematics due to their property of preserving distance and angle. One of these applications is in mathematical statistics where they are used to show the independence of the sample mean and sample variance in a set of normally (Gaussian) distributed random variables. We will make use of them in our development of the important Principal Axis Theorem in the next section.

Exercises

1. Prove that an orthogonal matrix has full rank.

2. Verify whether each of the following is an orthogonal matrix.

(a) $\begin{pmatrix} 1/\sqrt{2} & 1/\sqrt{2} \\ -1/\sqrt{2} & 1/\sqrt{2} \end{pmatrix}$ (b) $\begin{pmatrix} 1/\sqrt{6} & 2/\sqrt{6} & 1/\sqrt{6} \\ 1/\sqrt{3} & -1/\sqrt{3} & 1/\sqrt{3} \\ 0 & 0 & 0 \end{pmatrix}$

(c) $\begin{pmatrix} -1/\sqrt{2} & 0 & 1/\sqrt{2} \\ 1/\sqrt{3} & 1/\sqrt{3} & 1/\sqrt{3} \\ 1/\sqrt{6} & -2/\sqrt{6} & 1/\sqrt{6} \end{pmatrix}$

(d) $\begin{pmatrix} 1/2 & 1/2 & 1/2 & 1/2 \\ -1/\sqrt{2} & 0 & 0 & 1/\sqrt{2} \\ 2/\sqrt{10} & 1/\sqrt{10} & -1/\sqrt{10} & -2/\sqrt{10} \\ -1/2 & 1/2 & 1/2 & -1/2 \end{pmatrix}$

(e) $\begin{pmatrix} -3/2\sqrt{5} & 1/2\sqrt{5} & 3/2\sqrt{5} & -1/2\sqrt{5} \\ 0 & 1/\sqrt{2} & 1/\sqrt{2} & 0 \\ 1/2 & 1/2 & 1/2 & 1/2 \\ 1/\sqrt{2} & 0 & 0 & -1/\sqrt{2} \end{pmatrix}$

(f) $\begin{pmatrix} 1/\sqrt{n} & 1/\sqrt{n} & 1/\sqrt{n} & \cdots & 1/\sqrt{n} \\ -1/\sqrt{2} & 1/\sqrt{2} & 0 & \cdots & 0 \\ -1/\sqrt{6} & -1/\sqrt{6} & 2/\sqrt{6} & \cdots & 0 \\ & & \cdots & & \\ -1/\sqrt{n(n-1)} & -1/\sqrt{n(n-1)} & -1/\sqrt{n(n-1)} & \cdots & (n-1)/\sqrt{n(n-1)} \end{pmatrix}$

3. Which of the following transformations are orthogonal with respect to the standard inner product?

(a) $T(x, y, z) = (z, x, y)$
(b) $T(x) = kx$, where k is a constant
(c) $T(x, y, z) = (x + z, y)$
(d) $T(x, y, z) = (-x/\sqrt{2} + y/\sqrt{3} + z/\sqrt{6}, \; y/\sqrt{3} - 2z/\sqrt{6},$

$$x/\sqrt{2} + y/\sqrt{3} + z/\sqrt{6})$$

4. Let $\{v_1, v_2, \ldots, v_n\}$ be an orthonormal basis of V and let T be an orthogonal transformation on V. Show that $\{Tv_1, Tv_2, \ldots, Tv_n\}$ is an orthonormal basis of V.

5. Show that if an orthogonal matrix is triangular, then it is diagonal.

6. Find a matrix representation, relative to the standard basis, for the linear transformation which rotates each vector of the real plane through a fixed angle θ. Prove that this matrix is orthogonal.

7. Let A be any real $n \times n$ matrix and let P be any orthogonal $n \times n$ matrix. Show that A and PAP^T have the same characteristic polynomial.

8. Let $T: V \to V$ be a linear transformation such that $\langle Tv, Tv \rangle = \langle v, v \rangle$ for every v in V. Prove that T is an orthogonal transformation. [*Hint:* Show that

$$2\langle v_1, v_2 \rangle = \langle v_1 + v_2, v_1 + v_2 \rangle - \langle v_1, v_1 \rangle - \langle v_2, v_2 \rangle]$$

9. Let B be an orthonormal basis for V and let P be a transition matrix to a new basis B'. Prove that B' is orthonormal if, and only if, P is orthogonal.

10. Prove that if P is an orthogonal matrix and λ is an eigenvalue of P, then $|\lambda| = 1$.

14.7   Quadratic forms

In this chapter we have introduced a "bilinear" transformation which we have called the inner product. We have used the inner product to define length (or norm) of vectors, and angle between vectors, and have therefore introduced a *metric* concept into our discussion of vector spaces.

We will now extend our discussion of inner products to a treatment of what are known as quadratic forms and, in so doing, arrive at the important Principal Axis Theorem.

DEFINITION 14.13   Consider an n-dimensional vector space V. Let B be a basis of V and let $X = (x_1, x_2, \ldots, x_n)$ be the coordinates of an arbitrary vector $v \in V$ relative to B. Then a function g on V such that

$$g(v) = \sum_{i=1}^{n} \sum_{j=1}^{n} m_{ij} x_i x_j, \quad m_{ij} \in \mathbf{R}^1$$

is called a *quadratic function* on V, and the expression

$$Q(X) = \sum_{i=1}^{n} \sum_{j=1}^{n} m_{ij} x_i x_j$$

is called a *quadratic form*.

Note that a quadratic function may have many associated quadratic forms depending upon choice of basis for V. Every quadratic form has a matrix representation, for if

$$X = \begin{pmatrix} x_1 \\ x_2 \\ \vdots \\ x_n \end{pmatrix}$$

where the x_i are the coordinates of v with respect to B, then $g(v) = X^T M X$, where the matrix $M = (m_{ij})$. Moreover, formal multiplication of $X^T M X$, with $X^T = (x_1 \ x_2 \ \dots \ x_n)$, yields the form $Q(X)$. Ordinarily it will not be necessary to make a sharp distinction between a quadratic form and its related quadratic function.

THEOREM 14.15 Every quadratic form can be represented by a unique symmetric matrix. Conversely every symmetric matrix represents a unique quadratic form.

PROOF: That every symmetric matrix represents a unique quadratic form is immediate. Consider the quadratic form

$$Q(X) \equiv \sum_{i=1}^{n} \sum_{j=1}^{n} m_{ij} x_i x_j$$

Then $Q(X) = X^T M X$ (where M is not necessarily symmetric). Construct the symmetric matrix N such that $N = \frac{1}{2}(M + M^T)$; that is, $n_{ij} = (m_{ij} + m_{ji})/2 = n_{ji}$. Now

$$X^T M^T X = \sum_{i=1}^{n} \sum_{j=1}^{n} x_i m_{ji} x_j = \sum_{j=1}^{n} \sum_{i=1}^{n} x_j m_{ij} x_i$$

by interchanging the symbols i and j. Hence $X^T M^T X = Q(X)$. Therefore

$$X^T N X = \frac{1}{2}[X^T M X + X^T M^T X] = Q(X)$$

and N is a symmetric matrix representation for Q. If $X^T N_1 X = X^T N_2 X$ for all X, where N_1 and N_2 are symmetric, then $X^T(N_1 - N_2)X = 0$ for all X. Hence $N_1 = N_2$ (see Exercise 10) so there is a unique symmetric matrix which represents $Q(X)$. ∎

Example 1 Consider the quadratic form in three variables

$$Q(X) = \sum_{i=1}^{3} \sum_{j=1}^{3} (4i - j)x_i x_j$$

$$= 3x_1{}^2 + 2x_1 x_2 + x_1 x_3 + 7x_2 x_1 + 6x_2{}^2$$

$$+ 5x_2 x_3 + 11x_3 x_1 + 10x_3 x_2 + 9x_3{}^2$$

Taking $m_{ij} = 4i - j$, $Q(X)$ has matrix

$$M = \begin{pmatrix} 3 & 2 & 1 \\ 7 & 6 & 5 \\ 11 & 10 & 9 \end{pmatrix}$$

Many other matrix representations are possible. In fact, all that is required is

that $m_{11} = 3$, $m_{22} = 6$, $m_{33} = 9$, $m_{12} + m_{21} = 9$, $m_{13} + m_{31} = 12$,
$m_{23} + m_{32} = 15$. The unique symmetric representative is

$$N = \begin{pmatrix} 3 & 4.5 & 6 \\ 4.5 & 6 & 7.5 \\ 6 & 7.5 & 9 \end{pmatrix}$$

Thanks to the preceding theorem, we will limit our attention to symmetric matrices. We leave it to the reader to verify that a quadratic function for which the matrix representation (relative to a given basis) is positive definite establishes a norm on a vector space. That is, the matrix representation of the corresponding quadratic form is that of an inner product. We now make use of properties of quadratic forms to investigate properties of symmetric matrices, and conversely.

THEOREM 14.16   The eigenvalues of any real symmetric matrix are real.

PROOF: Let M be a real symmetric $n \times n$ matrix and suppose $\lambda = \alpha + \beta i$ is an eigenvalue of M. Since λ is an eigenvalue, there must exist a vector $Z = (X + iY)$ such that $MZ = \lambda Z$. But $MZ = MX + iMY$, and $\lambda Z = \alpha X - \beta Y + i(\alpha Y + \beta X)$, where X and Y are real vectors. Equating real and imaginary parts gives

$$MX = \alpha X - \beta Y \tag{1}$$

and

$$MY = \alpha Y + \beta X \tag{2}$$

Multiplying Eq. (1) by Y^T and Eq. (2) by X^T, and noting $X^T M Y = (X^T M Y)^T = Y^T M X$, due to the symmetry of M, we obtain

$$\alpha Y^T X - \beta Y^T Y = \alpha X^T Y + \beta X^T X$$

Since $X^T Y = Y^T X$, we obtain $\beta(X^T X + Y^T Y) = 0$. But Z is an eigenvector, and therefore either X or Y must be nonzero. Hence $X^T X + Y^T Y$ is positive, implying $\beta = 0$, and the theorem is proved.   ∎

THEOREM 14.17   The eigenvalues of a positive definite (positive semidefinite) matrix are positive (nonnegative).

PROOF: Let M be positive definite. Then M is, by definition, symmetric and hence has real eigenvalues. Let λ be any eigenvalue, and let X be an eigenvector associated with λ. (For any eigenvalue λ there must be at least one eigenvector, although we cannot say anything specific at this point about the dimension of the eigenspace $\mathbf{W}_\lambda$.) Then $MX = \lambda X$. Hence $X^T M X = \lambda X^T X$. Since M is positive definite, $X^T M X$ is positive, as is $X^T X$. Therefore $\lambda > 0$. The proof for M positive semidefinite follows the same argument and is left to the reader.   ∎

THEOREM 14.18 Let M be a symmetric matrix. The eigenvectors of M associated with distinct eigenvalues are orthogonal (with respect to the standard inner product).

PROOF: Let λ_1 and λ_2 be distinct eigenvalues of M, and let X and Y be eigenvectors of M associated with λ_1 and λ_2, respectively. Then, $MX = \lambda_1 X$ and $MY = \lambda_2 Y$. Premultiplying by Y^T and X^T, respectively, we obtain $Y^T M X = \lambda_1 Y^T X$ and $X^T M Y = \lambda_2 X^T Y$. But $X^T Y = Y^T X$, and since M is symmetric, $X^T M Y = Y^T M X$. Hence, by subtraction, $0 = (\lambda_1 - \lambda_2) X^T Y$. Since λ_1 and λ_2 are distinct, we can conclude $X^T Y = 0$. ∎

We are now in a position to attack the main theorem of this section. We will delay presenting a geometrical interpretation until after the proof. (The reader may want to turn to this discussion prior to tackling the proof.)

THEOREM 14.19 (Principal Axis Theorem) Every symmetric matrix can be diagonalized by an orthogonal matrix; that is, if M is an $n \times n$ symmetric matrix there exists an orthogonal matrix P such that $P^T M P = D$, where D is a diagonal matrix. Furthermore, the diagonal elements of D are the eigenvalues of M.

PROOF: Note that $P^T = P^{-1}$ since P is orthogonal. If we could assume the eigenvalues of M to be distinct we could conclude that M is diagonable by Theorem 11.10, and that the diagonalizing matrix is orthogonal from Theorem 14.18. However, we do not generally have distinct eigenvalues. Therefore we must show that we can find n linearly independent eigenvectors.

Let $\lambda_1, \lambda_2, \ldots, \lambda_n$ denote the n (not necessarily distinct) eigenvalues of M, and let Y_1 be an eigenvector associated with λ_1; that is, $MY_1 = \lambda_1 Y_1$. Define $X_1 = Y_1 / |Y_1|$ so that X_1 has unit length. (We assume the standard inner product so that $|Y_1|^2 = Y_1^T Y_1$.) By the Gram–Schmidt process we can extend X_1 to an orthonormal set $(X_1, X_2, \ldots, X_n)$. Let P be the orthogonal matrix $P = (X_1 \ X_2 \cdots X_n)$.

Consider the matrix $N = P^T M P$. Since M is symmetric, $N^T = P^T M^T P = N$ and hence N is symmetric. Since $MX_1 = \lambda_1 X_1$, we can see that

$$P^T M X_1 = \lambda_1 P^T X_1 = \lambda_1 \begin{pmatrix} X_1^T X_1 \\ X_2^T X_1 \\ \vdots \\ X_n^T X_1 \end{pmatrix} = \lambda_1 \begin{pmatrix} 1 \\ 0 \\ \vdots \\ 0 \end{pmatrix}$$

since the X_i form an orthonormal set. Hence, by the symmetry of N,

$$N = (P^T M X_1 \cdots P^T M X_n) = \begin{pmatrix} \lambda_1 & 0 & \cdots & 0 \\ 0 & & & \\ \vdots & & N_1 & \\ 0 & & & \end{pmatrix}$$

where N_1 is an $(n - 1) \times (n - 1)$ symmetric matrix. Since M and N are similar they have the same eigenvalues (by Theorem 11.12). Hence λ_2 is an eigenvalue of N regardless of whether λ_1 is equal to λ_2. We now show λ_2 is an eigenvalue of N_1. For let

$$Y_2 = \begin{pmatrix} y_{12} \\ y_{22} \\ \vdots \\ y_{n2} \end{pmatrix}$$

be an eigenvector of N associated with λ_2. Then $NY_2 = \lambda_2 Y_2$. Therefore, by equating corresponding elements of NY_2 and $\lambda_2 Y_2$, we can conclude that

$$\lambda_1 y_{12} = \lambda_2 y_{12} \quad \text{and} \quad \lambda_2 y_{i2} = \sum_{k=2}^{n} n_{ik} y_{k2} \quad \text{for} \quad i = 2, \ldots, n$$

But the first equality holds only if $\lambda_1 = \lambda_2$ or $y_{12} = 0$. The second equality holds for $i = 2, \ldots, n$ only if λ_2 is an eigenvalue with eigenvector

$$\begin{pmatrix} y_{22} \\ \vdots \\ y_{n2} \end{pmatrix} \quad \text{of} \quad N_1$$

Finally, if $\lambda_1 = \lambda_2$ we can set $y_{12} = 0$ and still have Y_2 an eigenvector of N. Redefine the X_i as follows. Set

$$X_1 = \begin{pmatrix} 1 \\ 0 \\ \vdots \\ 0 \end{pmatrix}, \quad X_2 = \frac{Y_2}{|Y_2|}$$

where X_2, an eigenvector of N associated with λ_2, will now be of the form

$$\begin{pmatrix} 0 \\ x_{22} \\ \vdots \\ x_{n2} \end{pmatrix}$$

Therefore X_1 and X_2 are orthonormal. Extend X_1 and X_2 to form an orthonormal set of n vectors, and define $Q = (X_1 \; X_2 \cdots X_n)$. Then

$$Q^T N Q = Q^T P^T M P Q = \begin{bmatrix} \lambda_1 & 0 & \cdots & 0 \\ 0 & \lambda_2 & \cdots & 0 \\ \vdots & \vdots & & N_2 \\ 0 & 0 & & \end{bmatrix}$$

Now note that PQ is an orthogonal matrix, for $(PQ)(PQ)^T = PQQ^T P^T = PIP^T = I$ since both P and Q are orthogonal matrices.

We can repeat the above procedure n times and will have created an orthogonal matrix R and a diagonal matrix

$$D = \begin{bmatrix} \lambda_1 & 0 & & 0 \\ 0 & \lambda_2 & & \\ & & \ddots & \\ 0 & & & \lambda_n \end{bmatrix}$$

such that $R^T M R = D$. Both by the method of constructing R and D, and the fact that similar matrices have the same eigenvalues, we can conclude that the diagonal elements of D are the eigenvalues of M. Further, we can conclude that the columns of R form an orthonormal set of eigenvectors of M. Our theorem is proved. ∎

The point to remember concerning the above proof is that until we had actually constructed the matrix R, we could not be assured that we could find n linearly independent eigenvectors of M, and hence could not construct the orthonormal set of vectors. Now that we know this can be done we can turn to simpler methods of constructing such vector sets, as is illustrated in the following example.

Example 2 Consider the matrix

$$M = \begin{pmatrix} -\frac{1}{2} & 0 & \frac{3}{2} \\ 0 & 1 & 0 \\ \frac{3}{2} & 0 & -\frac{1}{2} \end{pmatrix}$$

The characteristic polynomial for M is

$$p(\lambda) = -(1 - \lambda)^2 (2 + \lambda)$$

and therefore M has eigenvalues $\lambda_1 = 1$, $\lambda_2 = 1$ and $\lambda_3 = -2$. We know by the Principal Axis Theorem that there exists an orthogonal matrix P such that $P^T M P = D$, where

$$D = \begin{pmatrix} 1 & 0 & 0 \\ 0 & 1 & 0 \\ 0 & 0 & -2 \end{pmatrix}$$

To find P we proceed as follows. We first find linearly independent eigenvectors X_1, X_2, and X_3 associated with λ_1, λ_2, and λ_3, respectively. We then apply the Gram–Schmidt process to obtain an orthonormal set of eigenvectors P_1, P_2, P_3 which form the columns of the diagonalizing matrix P. Note that although the eigenvalues are not distinct we can find a basis for each eigenspace and use the Gram–Schmidt procedure to construct an orthonormal set for this eigenspace. Solving the algebraic system $(M - \lambda_i I)X_i = 0$ we see that the coordinates of the eigenvectors associated with $\lambda = 1$ must satisfy $-\frac{3}{2}x_1 + \frac{3}{2}x_3 = 0$, which implies $x_1 = x_3$ for any choice of x_2. We take

$$X_1 = \begin{pmatrix} 1 \\ 1 \\ 1 \end{pmatrix}, \quad X_2 = \begin{pmatrix} 1 \\ 0 \\ 1 \end{pmatrix}$$

The coordinates of an eigenvector associated with $\lambda_3 = -2$ must satisfy $\frac{3}{2}x_1 + \frac{3}{2}x_3 = 0$ and $x_2 = 0$. We take

$$X_3 = \begin{pmatrix} 1 \\ 0 \\ -1 \end{pmatrix}$$

and now apply the Gram–Schmidt process to the set $\{X_1, X_2, X_3\}$ to form an orthonormal basis for $\mathbf{R}^3$. Set

$$P_1 = \frac{X_1}{|X_1|} = \begin{pmatrix} 1/\sqrt{3} \\ 1/\sqrt{3} \\ 1/\sqrt{3} \end{pmatrix}$$

$$R_2 = \begin{pmatrix} 1 \\ 0 \\ 1 \end{pmatrix} - \frac{2}{\sqrt{3}}\begin{pmatrix} 1/\sqrt{3} \\ 1/\sqrt{3} \\ 1/\sqrt{3} \end{pmatrix} = \begin{pmatrix} \dfrac{1}{3} \\ \dfrac{-2}{3} \\ \dfrac{1}{3} \end{pmatrix} \quad \text{and} \quad P_2 = \frac{R_2}{|R_2|} = \begin{pmatrix} 1/\sqrt{6} \\ -2/\sqrt{6} \\ 1/\sqrt{6} \end{pmatrix}$$

Since distinct eigenspaces are orthogonal, we set

$$P_3 = \frac{X_3}{|X_3|} = \begin{pmatrix} 1/\sqrt{2} \\ 0 \\ -1/\sqrt{2} \end{pmatrix}$$

Therefore the matrix

$$P = \begin{pmatrix} 1/\sqrt{3} & 1/\sqrt{6} & 1/\sqrt{2} \\ 1/\sqrt{3} & -2/\sqrt{6} & 0 \\ 1/\sqrt{3} & 1/\sqrt{6} & -1/\sqrt{2} \end{pmatrix}$$

is an orthogonal matrix such that

$$P^T M P = \begin{pmatrix} 1 & 0 & 0 \\ 0 & 1 & 0 \\ 0 & 0 & -2 \end{pmatrix}$$

In order to discover the geometric interpretation of the Principal Axis Theorem we will discuss the following example. We consider the space $\mathbf{G}^3$ with the usual Cartesian coordinate system and the basis B with coordinates $\{(1, 0, 0), (0, 1, 0), (0, 0, 1)\}$ with respect to this coordinate system. We now specify a quadratic function g on $\mathbf{G}^3$ with associated quadratic form $Q(X) = X^T M X$ with respect to B where M is given as

$$M = \begin{pmatrix} \frac{5}{9} & -\frac{8}{9} & -\frac{10}{9} \\ -\frac{8}{9} & \frac{11}{9} & -\frac{2}{9} \\ -\frac{10}{9} & -\frac{2}{9} & \frac{2}{9} \end{pmatrix}$$

A level surface of a quadratic form, that is, the set of points with coordinates satisfying $X^T M X = C$ for a given scalar C, is known as a *quadric curve* or *quadric surface*. The surfaces are ellipsoids, hyperboloids, and paraboloids, depending upon the matrix M. (Note the correspondence to conic sections in $\mathbf{G}^2$.) If we now apply the Principal Axis Theorem to the matrix M we observe that there exists an orthogonal matrix P such that $P^T M P = D$, where D is diagonal. The reader can verify that the characteristic polynomial for the above matrix M is $p(\lambda) = -(1 + \lambda)(1 - \lambda)(2 - \lambda)$. Hence the eigenvalues of M are $\lambda_1 = -1$, $\lambda_2 = 1$ and $\lambda_3 = 2$. Corresponding eigenvectors, obtained by solving the system

$$(M - \lambda_i I) X_i = 0$$

are

$$X_1 = \begin{pmatrix} 2 \\ 1 \\ 2 \end{pmatrix} \quad X_2 = \begin{pmatrix} 1 \\ 2 \\ -2 \end{pmatrix} \quad \text{and} \quad X_3 = \begin{pmatrix} 2 \\ -2 \\ -1 \end{pmatrix}$$

Setting $P_i = X_i / |X_i|$ gives us the matrix

$$P = (P_1\ P_2\ P_3) = \begin{pmatrix} \frac{2}{3} & \frac{1}{3} & \frac{2}{3} \\ \frac{1}{3} & \frac{2}{3} & -\frac{2}{3} \\ \frac{2}{3} & -\frac{2}{3} & -\frac{1}{3} \end{pmatrix} \quad \text{where} \quad P^T M P = D = \begin{pmatrix} -1 & 0 & 0 \\ 0 & 1 & 0 \\ 0 & 0 & 2 \end{pmatrix}$$

But by premultiplying by P and postmultiplying by P^T, we obtain $M = PDP^T$. Therefore $X^T M X = X^T P D P^T X$. Since P is orthogonal and hence nonsingular, P^T can be considered a transition matrix associated with a change from B to a new basis B'. And the quadratic function which has representation $X^T M X$, with X the coordinates of a vector v with respect to B, will have representation $Y^T D Y$, where $Y = P^T X$ are the coordinates of the same vector with respect to $B' = \{v_1', v_2', v_3'\}$. Since P^T is an orthogonal matrix, and hence represents an orthogonal transformation, we can conclude that our new coordinate system is just a rotation and/or reflection of the original (that is, angles and lengths are left invariant under the change of coordinates). Further, our quadratic function now has representation $Y^T D Y = \lambda_1 y_1^2 + \lambda_2 y_2^2 + \lambda_3 y_3^2$, where

$$Y = \begin{pmatrix} y_1 \\ y_2 \\ y_3 \end{pmatrix}$$

is the coordinate vector with respect to B' of v, and λ_1, λ_2, and λ_3 are the eigenvalues of the matrix M. However, the new representation $Y^T D Y$ of our quadratic form is such that the quadric surfaces have their principal axes in the directions of the new basis vectors v_1', v_2', and v_3'.

To complete our discussion we note that the ith column of P^T gives the coordinates of the ith vector in B with respect to B'. Similarly, the columns of $P = (P^T)^{-1}$, as the matrix of the inverse change of basis, give the coordinates of the elements of B' with respect to B. Hence, we can conclude that the principal axes of our quadric surfaces as represented by M will be in the direction of the elements of

$$B' = \left\{ \left(\frac{2}{3}, \frac{1}{3}, \frac{2}{3} \right), \left(\frac{1}{3}, \frac{2}{3}, \frac{-2}{3} \right), \left(\frac{2}{3}, \frac{-2}{3}, \frac{-1}{3} \right) \right\}$$

Exercises

1. Prove that the eigenvalues of a positive semidefinite matrix are nonnegative.

2. Verify the remark in the text which states that if a quadratic function on a vector space **V** is positive definite, it specifies an inner product on **V**.

For each of the matrices in Exercises 3 through 7, find matrices P and D such that P is orthogonal and D is diagonal and $P^T M P = D$.

3. $M = \begin{pmatrix} 3 & 1 \\ 1 & 3 \end{pmatrix}$

4. $M = \begin{pmatrix} 2 & 1 & 1 \\ 1 & 4 & 2 \\ 1 & 2 & 4 \end{pmatrix}$

5. $M = \begin{pmatrix} 1 & 1 & 0 \\ 1 & -1 & -1 \\ 0 & -1 & 1 \end{pmatrix}$

6. $M = \begin{pmatrix} 1 & 2 & -1 \\ 2 & 1 & 1 \\ -1 & 1 & 1 \end{pmatrix}$

7. $M = \begin{pmatrix} 3 & 0 & 1 \\ 0 & 2 & 0 \\ 1 & 0 & 1 \end{pmatrix}$

8. Let g be a quadratic function on $\mathbf{R}^3$, and let

$$M = \begin{pmatrix} 7 & 4 & -5 \\ 4 & -2 & 4 \\ -5 & 4 & 7 \end{pmatrix}$$

be the matrix of g with respect to the basis $B = \{(1, 0, 1), (1, 1, 0), (0, 0, 1)\}$.

(a) Find an orthogonal matrix P and diagonal matrix D such that $P^T M P = D$.

(b) Construct a basis B' for $\mathbf{R}^3$ such that the elements of B' are the principal axes of the quadratic surface.

9. Prove the following. For any real matrix M, the matrix $M^T M$ is symmetric, positive semidefinite, and has the same rank as the matrix M.

10. Prove that if N_1 and N_2 are symmetric matrices, and $X^T(N_1 - N_2)X = 0$ for all X, then $N_1 = N_2$.

11. Prove that if all the eigenvalues of a symmetric matrix are positive, the matrix is positive definite (the converse of Theorem 14.17).

COMMENTARY

Section 14.1 The theory of complex inner product spaces is treated in many books on algebra. The definition of inner product is altered by requiring

$$\langle v_1, v_2 \rangle = \overline{\langle v_2, v_1 \rangle}$$

where the bar denotes the complex conjugate. If an inner product has been defined on a complex vector space, the space is often called *unitary*. See:

HALMOS, P. R., *Finite-Dimensional Vector Spaces*, 2nd ed., Princeton, N.J., Van Nostrand Reinhold, 1958.

Section 14.2 The analogue of the concept of matrix transpose for a real matrix is the concept of *conjugate transpose* for a complex matrix. The conjugate transpose of $M = (m_{ij})$ is the matrix, denoted M^*, with elements $(\overline{m}_{ji})$. A matrix M for which $M = M^*$ is called *Hermitian*.

An interesting discussion of positive definite matrices is given in the following article, which was the winning essay in the Contest for undergraduates announced in the January, 1969, issue of the *American Mathematical Monthly*.

JOHNSON, C. R., "Positive Definite Matrices," *American Math. Monthly*, vol. 77 (1970), 259–264.

Section 14.3 For Hölder's and Minkowski's inequalities, see:

TOLSTED, E., "An Elementary Derivation of the Cauchy, Hölder, and Minkowski Inequalities from Young's Inequality," *Mathematics Magazine*, vol. 37 (1964), 2–12.

A classic work, which includes a panoply of inequalities and their applications, is:

HARDY, G. H., J. E. LITTLEWOOD, and G. POLYA, *Inequalities*, Cambridge, 1934.

Section 14.5 Theorem 14.11 is sometimes called the *Projection Theorem*. The reason for this terminology is that the theorem shows that each vector v in **V** can be represented in the form $v = w + z$ where $w \in$ **W** and $z \in$ **W**$^\perp$. Moreover, as shown in Exercise 6, w can be regarded as the orthogonal projection of v onto **W** and z as the orthogonal projection of v onto **W**$^\perp$.

For a more general discussion of geometric concepts such as projection, particularly in their relation to linear algebra, see:
YALE, P. B., *Geometry and Symmetry*, San Francisco, Holden-Day, 1968.

Section 14.6 In the complex case, the concept of orthogonal matrix is replaced by that of unitary matrix. A matrix M is called *unitary* if $MM^* = M^*M = I$. A linear transformation is called unitary if it preserves inner products. The term *isometry* denotes a transformation which is orthogonal or unitary, according as the inner product space is real or complex.

Section 14.7 Theorem 14.19 is a special case of what is called the *Spectral Theorem*. See the book of Halmos, *op. cit.*, for further information.

Chapter 15

The Laplace transform

15.1   Introduction

The concept of linear transformation was introduced in Chapter 5. As particular examples of linear transformations, we treated the differential operator D and the integral operator S. We then applied properties of linear transformations to certain linear differential equations, relating these equations to polynomial equations. This equivalent algebraic representation was a valuable tool in our development of methods for solving linear differential equations.

In this chapter we introduce a new operator which is quite helpful in obtaining solutions to many problems in applied mathematics and probability theory. This operator is the Laplace transform, which we shall see is a special case of what is known as a *linear integral transformation.*

DEFINITION 15.1   Let J be an interval in $\mathbf{R}^1$, and $H(t, s)$ a prescribed function of the variable t and a parameter s for $t \in J$ and s in some prescribed set. The *linear integral transform* of a function f, with respect to the *kernel H,* is the function of s given by

$$L_H(f; s) = \int_J H(t, s)f(t) \, dt \tag{1}$$

Where there is no chance for confusion, we will shorten the notation to $L_H(f)$, assuming the dependence on s to be understood.

One obvious point is that in speaking of transforms of functions with respect to a specific kernel H, we must restrict our attention to the class of functions for which the integral in Eq. (1) exists. And in so doing, we are restricting ourselves to a vector space. (See Exercise 1.)

A second point is contained in the following theorem.

THEOREM 15.1   Let H be a given kernel and let s be fixed. Consider the transformation L_H, which takes f into $L_H(f; s)$. Then L_H is a linear transformation over the space of functions for which the integral in Eq. (1) exists.

PROOF: The proof follows immediately from the properties of Riemann integration and is left as an exercise.   ∎

It should be observed that the term *transform* is used both for the linear transformation L_H and for the image function $L_H(f)$.

Exercises

1.   Consider a kernel $H(t, s)$ and a particular value of the parameter s. Prove that the set of all functions f for which the transform $\int_J H(t, s)f(t)\, dt$ exists forms a vector space.

2.   Prove Theorem 15.1.

15.2   The Laplace transform

DEFINITION 15.2   Let $f^*(s)$ denote the linear integral transform $L(f)$ with kernel e^{-st} over the interval $J = (0, \infty)$; that is,

$$f^*(s) = \int_0^\infty e^{-st} f(t)\, dt \tag{1}$$

where s is real.[1] Then $f^*(s)$ is the *Laplace transform* of f with (real) *argument s*.

In order for f^* to exist, f must be defined for all t in $0 < t < \infty$, and f must be such that the integral in (1) converges. This integral is improper because the interval of integration is infinite. Also, it may be improper because f has discontinuities at one or more finite values of t, including possibly $t = 0$. For much of our work, f will be a well-behaved function for all t except $t = 0$, and in this case the integral in (1) is defined by the double limit

$$\lim_{T \to \infty} \lim_{\varepsilon \to 0} \int_\varepsilon^T e^{-st} f(t)\, dt \tag{2}$$

If this double limit exists, its value is $f^*(s)$.

----

[1] The theory of Laplace transforms has been developed for complex arguments. For more complete treatment the reader should consult the references at the end of the chapter.

Example 1 We shall consider the function $f(t) = e^{t^2}$. For fixed ε and T, $0 < \varepsilon < T < \infty$,

$$\int_{\varepsilon}^{T} e^{-st} f(t)\, dt = \int_{\varepsilon}^{T} e^{t^2 - st}\, dt = e^{-s^2/4} \int_{\varepsilon}^{T} e^{(t-s/2)^2}\, dt$$

$$= e^{-s^2/4} \int_{\varepsilon - s/2}^{T - s/2} e^{u^2}\, du$$

But for any fixed ε and s, the last integral has no limit as $T \to \infty$. Hence there is no Laplace transform for the function $f(t) = e^{t^2}$ for any value of the argument s.

Since it is now clear that the Laplace transform does not exist for some functions, it would be helpful to have some test to determine when the transform does exist. Hence we give the following theorem.

THEOREM 15.2 Let $f(t)$ be a function defined on $(0, \infty)$ such that

1. For every a and b, $0 < a < b < \infty$, $\int_a^b f(t)\, dt$ exists.
2. For some $a > 0$, $\lim_{\varepsilon \to 0} \int_{\varepsilon}^{a} f(t)\, dt$ exists.
3. There exist constants M, s_0, and t_0 such that $|f(t)|e^{-s_0 t} \le M$ for every $t \ge t_0$.

Then the Laplace transform $f^*(s)$ exists for every $s > s_0$.

PROOF: We omit a formal proof, but present a heuristic discussion to explain the three conditions. In order for the Laplace transform to exist, the limit in Eq. (2) must exist. If there exists an interval $J_0 = [a_0, b_0]$, with $0 < a_0 < b_0 < \infty$, such that $\int_{J_0} e^{-st} f(t)\, dt$ does not exist for a particular s, then certainly the integral over any interval containing J_0 will not exist and hence $f^*(s)$ will not exist. Since e^{-st} is a well-behaved function for any value of s, condition (1) is sufficient to guarantee the existence of $\int_{a_0}^{b_0} e^{-st} f(t)\, dt$ for any interval J_0. The condition (2) makes use of the fact that, for small values of t, e^{-st} is very close to one, and so the existence or nonexistence of $\lim_{\varepsilon \to 0} \int_{\varepsilon}^{a} e^{-st} f(t)\, dt$ depends only upon the behavior of $f(t)$. Finally, suppose $f(t)$ grows at a rate no greater than $e^{s_0 t}$ for large t; that is, $|f(t)|e^{-s_0 t} \le M$ for $t \ge t_0$. Then $|f(t)|e^{-st} \le e^{-(s-s_0)t} M$. Since

$$\lim_{T \to \infty} \int_{t_0}^{T} e^{-(s-s_0)t} M\, dt$$

exists for $s > s_0$, we can conclude that

$$\lim_{T \to \infty} \int_{t_0}^{T} |f(t)|e^{-st}\, dt$$

exists, which implies that

$$\lim_{T \to \infty} \int_{t_0}^{T} f(t)e^{-st} \, dt$$

exists. Thus we have established the existence of

$$\lim_{\varepsilon \to 0} \int_{\varepsilon}^{a_0} e^{-st} f(t) \, dt + \int_{a_0}^{t_0} e^{-st} f(t) \, dt + \lim_{T \to \infty} \int_{t_0}^{T} e^{-st} f(t) \, dt$$

for every $s > s_0$, and hence the existence of the integral in Eq. (1). ∎

Condition (3) is sometimes expressed by saying that $f(t)$ must be of *exponential order* as $t \to \infty$. This means that $|f(t)|$ can grow no faster than $Me^{s_0 t}$ as $t \to \infty$ for some values of M and s_0.

Theorem 15.2 gives sufficient conditions for the existence of the Laplace transform. These conditions are not necessary.

By an argument similar to that given for condition (3) in the preceding theorem, we can obtain the following result.

THEOREM 15.3 If the Laplace transform $f^*(s_0)$ exists for argument s_0, then $f^*(s)$ will exist for every $s > s_0$.

PROOF: See Exercise 1. ∎

We close this section by calculating several elementary Laplace transforms. The results are given for quick reference in Table A in the appendix.

Example 2   Let $f(t) = e^t$ for $0 < t < \infty$. Then

$$f^*(s) = \int_{0}^{\infty} e^{-st} f(t) \, dt = \int_{0}^{\infty} e^{-(s-1)t} \, dt = \frac{1}{s-1} \quad \text{for} \quad s > 1$$

and $f^*(s)$ does not exist for $s \le 1$.

Example 3   Let $f(t) = 1$ for $0 < t < \infty$. Then,

$$f^*(s) = \int_{0}^{\infty} e^{-st} f(t) \, dt = \int_{0}^{\infty} e^{-st} \, dt = \frac{1}{s} \quad \text{for} \quad s > 0$$

Example 4   Let

$$f(t) = \begin{cases} 1 & 0 \le t \le 1 \\ 0 & t > 1 \end{cases}$$

Although this function is not continuous, its integral over $a < t < b$ exists for any a and b, $0 < a < b < \infty$. The function is bounded as $t \to 0^+$ and therefore condition (2) of Theorem 15.2 is satisfied. Also, (3) is satisfied with any s_0, any

$M > 0$, and $t_0 = 1$. It follows that the Laplace transform exists for all s. In fact, the transform is

$$f^*(s) = \int_0^\infty e^{-st} f(t)\, dt = \int_0^1 e^{-st}\, dt = \frac{(1 - e^{-s})}{s} \quad \text{for } s \neq 0$$

$$= 1 \quad \text{for } s = 0$$

Example 5 Let $f(t) = cg_1(t) + dg_2(t)$, where c and d are scalars and g_1 and g_2 are functions possessing Laplace transforms $g_1^*(s)$ for $s > s_1$ and $g_2^*(s)$ for $s > s_2$, respectively. Then $f^*(s) = cg_1^*(s) + dg_2^*(s)$ for $s > \max [s_1, s_2]$. We leave the proof of this result as an exercise.

Example 6 Let $f(t) = t^{\alpha-1}$ for $t > 0$, where $\alpha > 0$. Then

$$f^*(s) = \int_0^\infty t^{\alpha-1} e^{-st}\, dt = \frac{\Gamma(\alpha)}{s^\alpha}, \ s > 0$$

Here $\Gamma(\alpha)$ is the gamma function, defined as $\int_0^\infty t^{\alpha-1} e^{-t}\, dt$. This function has the property

$$\Gamma(\alpha) = (\alpha - 1)\Gamma(\alpha - 1)$$

for $\alpha > 1$. The reader unfamiliar with the properties of this function can find them treated in many calculus texts.

Exercises

1. Prove Theorem 15.3 under the assumption that $\int_0^\infty e^{-s_0 t} |f(t)|\, dt$ converges.

2. Determine which conditions of Theorem 15.2 are violated by the function $f(t) = e^{t^2}$. (See Example 1.)

3. Compute the Laplace transforms for the following functions, and specify the interval in which the parameter may lie for the transform to exist.

   (a) $f(t) = \sin at$ for $t > 0$.
   (b) $f(t) = t^a e^{-bt}$ for $t > 0$, $a > -1$

4. Verify the result given in Example 5, namely if

$$f(t) = cg_1(t) + dg_2(t), \quad \text{then} \quad f^*(s) = cg_1^*(s) + dg_2^*(s)$$

5. Consider the function

$$f(t) = \begin{cases} 1 & 0 \leq t \leq 1 \\ 0 & \text{otherwise} \end{cases}$$

   described in Example 4. Discuss the behavior of $f^*(s)$ in the neighborhood of $s = 0$.

15.3   Basic properties of Laplace transforms

This section contains a series of basic and useful properties of Laplace transforms. These properties are also listed in Table A.

Property 1 (*Linearity*)   We have already noted the linearity property of the Laplace transform; namely if $f(t) = cg_1(t) + dg_2(t)$, and if $g_1{}^*(s)$ for $s > s_1$ and $g_2{}^*(s)$ for $s > s_2$ are the Laplace transforms of g_1 and g_2, respectively, then $f^*(s) = cg_1{}^*(s) + dg_2{}^*(s)$ for $s > \max(s_1, s_2)$.

Property 2 (*Change of Scale*)   Let $g(t)$ be a function with Laplace transform $g^*(s)$, $s > s_0$, and suppose we desire the Laplace transform of a function $f(t) = g(t/a)$ for some positive scalar a. Then

$$f^*(s) = \int_0^\infty e^{-st} f(t)\, dt = \int_0^\infty e^{-st} g\left(\frac{t}{a}\right) dt = \int_0^\infty e^{-sa(t/a)} g\left(\frac{t}{a}\right) dt$$

$$= a \int_0^\infty e^{-sau} g(u)\, du = ag^*(sa)$$

which exists for $sa > s_0$.

Example 1   To illustrate, consider $f(t) = e^{ct}$ for some arbitrary scalar $c > 0$. By Example 2 of the previous section we know, if we set $g(t) = e^t$, that $g^*(s) = 1/(s - 1)$ for $s > 1$. Setting $a = 1/c$, we obtain

$$f^*(s) = \frac{1}{c} g^*\left(\frac{s}{c}\right) = \frac{1}{c}\frac{1}{(s/c - 1)} = \frac{1}{s - c} \quad \text{for} \quad \frac{s}{c} > 1 \quad \text{or} \quad s > c$$

Property 3 (*Shift of Origin*)   Let $g(t)$ be a function with known Laplace transform $g^*(s)$ for $s > s_0$ and suppose $f(t)$ is given by

$$f(t) = \begin{cases} g(t - a) & t > a > 0 \\ 0 & t \le a \end{cases}$$

Then

$$f^*(s) = \int_0^\infty e^{-st} f(t)\, dt = \int_a^\infty e^{-st} g(t - a)\, dt$$

$$= e^{-sa} \int_a^\infty e^{-s(t-a)} g(t - a)\, dt = e^{-sa} g^*(s) \quad \text{for} \quad s > s_0$$

Example 2   As a particular illustration of the above properties, suppose we desire the Laplace transform of the function:

$$f(t) = \begin{cases} c & t_0 \le t \le t_1, \text{ where } 0 \le t_0 < t_1 \\ 0 & \text{otherwise} \end{cases}$$

From Example 4 of the previous section, we know the Laplace transform of the function

$$g(t) = \begin{cases} 1 & 0 \le t \le 1 \\ 0 & \text{otherwise} \end{cases}$$

is

$$g^*(s) = \begin{cases} \dfrac{1 - e^{-s}}{s} & \text{for } s \ne 0 \\ 1 & \text{for } s = 0 \end{cases}$$

To obtain f from g we must perform not only a shift of origin, but also a scale change. In particular, define $a = t_1 - t_0$ and $g_1(t) = g(t/a)$. Then

$$g_1(t) = \begin{cases} 1 & 0 \le t \le a = t_1 - t_0 \\ 0 & \text{otherwise} \end{cases}$$

Next define $g_2(t) = g_1(t - t_0)$. Then

$$g_2(t) = \begin{cases} 1 & t_0 \le t \le t_1 \\ 0 & \text{otherwise} \end{cases}$$

Finally, note that $f(t) = cg_2(t)$. We now obtain the Laplace transform $f^*(s)$ of $f(t)$ as follows. By the properties derived in the above three examples,

$$f^*(s) = cg_2^*(s) = ce^{-st_0}g_1^*(s) = ace^{-st_0}g^*(as)$$

where $a = t_1 - t_0$. But

$$g^*(as) = \frac{(1 - e^{-as})}{as} \qquad s \ne 0$$

and hence

$$f^*(s) = \frac{ce^{-st_0}\left[1 - e^{-(t_1 - t_0)s}\right]}{s} = \frac{c}{s}\left[e^{-st_0} - e^{-st_1}\right]$$

We leave to the reader a verification of the equation $f^*(0) = c(t_1 - t_0)$.

The above properties are of value in algebraic manipulations with Laplace transforms. The following two properties are useful in treating integral and differential equations. We will omit a discussion of precise conditions necessary to ensure the validity of these properties.

Property 4   Let f be a function with Laplace transform f^* and $g(t) = df(t)/dt$. Then the Laplace transform of g is

$$g^*(s) = \int_0^\infty e^{-st}g(t)\, dt = \int_0^\infty e^{-st}\frac{df(t)}{dt}\, dt$$

Integrating by parts gives

$$g^*(s) = f(t)e^{-st}\Big|_0^\infty + s \int_0^\infty e^{-st} f(t)\, dt = -f(0) + sf^*(s)$$

The result can be extended to nth order derivatives. (See Exercise 2.)

Property 5   Let g be a function for which the Laplace transform exists (for $s >$ some s_0), and such that

$$g(t) = \int_0^t f(u)\, du$$

where f has known transform $f^*(s)$. Then

$$g^*(s) = \frac{1}{s} f^*(s)$$

We leave the derivation of this result as an exercise.

Example 3   Consider the function

$$g(t) = \begin{cases} t & 0 \le t \le 1 \\ 1 & t > 1 \end{cases}$$

Then

$$g(t) = \int_0^t f(x)\, dx \quad \text{where} \quad f(x) = \begin{cases} 1 & 0 \le x \le 1 \\ 0 & x > 1 \end{cases}$$

We have already determined the transform $f^*(s)$ of f to be $(1 - e^{-s})/s$ for $s \ne 0$ and $f^*(0) = 1$. By Property 5, we can conclude $g^*(s) = f^*(s)/s = (1 - e^{-s})/s^2$ over the range of s for which g^* exists. We leave to the reader the verification that $g^*(s)$ exists on the interval $0 < s < \infty$. (See Exercise 6.)

Example 4   Suppose we have verified that the Laplace transform of $f(t) = \sin t$ is $f^*(s) = 1/(s^2 + 1)$. (See Exercise 3 in Section 15.2.) To compute the transform of $g(t) = \cos t$ we need only recognize that $g(t) = f'(t)$. Applying Property 4, we obtain

$$g^*(s) = -f(0) + sf^*(s) = -\sin 0 + \frac{s}{s^2 + 1} = \frac{s}{s^2 + 1}$$

Exercises

1.   Using Property 4, derive Property 5.

2.   Suppose $g(t) = d^n f(t)/dt^n$. Prove that

$$g^*(s) = s^n f^*(s) - s^{n-1} f(0) - \cdots - f^{(n-1)}(0)$$

where

$$f^{(k)}(0) = \frac{df^k(t)}{dt^k}\Big|_{t=0}$$

3. Show that the Laplace transform of

$$g(t) = \int_0^t \int_0^u f(r) \, dr \, du$$

is $g^*(s) = s^{-2}f^*(s)$.

4. Prove that if $f^*(s)$ is the Laplace transform of a function $f(t)$, then the Laplace transform of $e^{ct}f(t)$ is $f^*(s - c)$.

5. Suppose f is a periodic function; that is, there exists a constant a such that $f(t + a) = f(t)$ for $t \geq 0$. Assume that the transform of f exists. Prove that

$$f^*(s) = \frac{\int_0^a e^{-st} f(t) \, dt}{1 - e^{-as}}$$

6. Verify that the interval of convergence of $g^*(s)$ in Example 3 is $(0, \infty)$ by computing g^* directly from g.

7. Define the *unit step function at a* to be the function $H_a(t)$ which is 0 for $t < a$ and 1 for $t \geq a$. Find the transform of $f(t) = cH_a(t) + dH_b(t)$ where $0 < a < b$ and c and d are constants. Sketch a graph of $f(t)$.

8. Let $f(t)$ be a function with derivative everywhere except t_0. Let $f(t_0^+) = f(t_0^-) + c$. Derive $g^*(s)$ where

$$g(t) = \begin{cases} df(t)/dt & \text{for } t \neq t_0 \\ 0 & \text{for } t = t_0 \end{cases}$$

15.4   The inverse of the Laplace transform

Let us attempt to solve some differential equations with the aid of the Laplace transform by making use of the properties discussed in the previous section. We begin with an example in which the solution function exists and has a continuous first derivative throughout $\mathbf{R}^1$.

Example 1   Consider the linear differential equation

$$y'(t) + 3y(t) = 2t \tag{1}$$

subject to the initial condition $y(0) = -1$. By Theorem 3.3 on existence and uniqueness for first order differential equations, we know a unique solution exists throughout $\mathbf{R}^1$. We now attack the problem with Laplace transforms. Assume that the Laplace transform of the solution exists. Taking transforms on both sides of Eq. (1) yields

$$sy^*(s) - y(0) + 3y^* = \frac{2}{s^2}$$

and solving for y^*, noting $y(0) = -1$, yields

$$y^*(s) = \frac{2/s^2 - 1}{s + 3} = \frac{2}{s^2(s + 3)} - \frac{1}{(s + 3)} \tag{2}$$

We have now transformed the differential equation (1) into an algebraic equation (2). If we can find a function y, which has the Laplace transform given in Eq. (2), it seems reasonable that such a function would lead to a solution of our differential equation.

We can determine that, if $y = y_1 + y_2$, where y_1 has transform $2/[s^2(s + 3)]$ and y_2 has transform $-1/(s + 3)$, then $y^*(s)$ will be as in Eq. (2). Noting $x(t) = 2e^{-3t}$ has transform $2/(s + 3)$, then (see Exercise 3 of Section 15.3)

$$\int_0^t \int_0^u 2e^{-3v}\, dv\, du = \tfrac{2}{3}[t - \tfrac{1}{3}(1 - e^{-3t})] = y_1(t)$$

has transform $2/[s^2(s + 3)]$. Also, $y_2(t) = -e^{-3t}$ has transform $-1/(s + 3)$. Therefore if we take

$$y(t) = \tfrac{2}{3}[t - \tfrac{1}{3}(1 - e^{-3t})] - e^{-3t} = \frac{6t - 2 - 7e^{-3t}}{9} \tag{3}$$

$y(t)$ will have the desired transform. If we knew, further, that $y(t)$ is the only function with the transform in Eq. (2), we would know we have the desired unique solution to Eq. (1). But if more than one function can have the transform in Eq. (2), then we cannot assume Eq. (3) to be the desired solution.

Example 1 illustrates the advantage which would accrue if the Laplace transform had a unique inverse; that is, if $f^*(s)$ is any specific Laplace transform, there is only one function $f(t)$ with

$$f^*(s) = \int_0^\infty e^{-st} f(t)\, dt$$

We can see immediately that this is not the case.

Example 2   Let

$$g(t) = \begin{cases} 1 & \text{for} \quad t \neq 1, 2, 3 \ldots \\ 0 & \text{for} \quad t = 1, 2, 3 \ldots \end{cases}$$

Then

$$\int_0^\infty e^{-st} g(t)\, dt = \frac{1}{s}$$

But this is the transform of $f(t) = 1$, $0 \le t < \infty$. Hence $g(t)$ and $f(t)$ have the same transform. It is illustrative to note that f and g differ only at points of discontinuity of g.

Although Example 2 illustrates that the inverse of a Laplace transform is not unique, we might be willing to consider the two functions f and g used in the example as equal for all practical purposes. In fact, if we wished to restrict our attention to continuous functions it could be shown that $f(t)$ would be the only continuous function with the specified transform. It is this property we now generalize.

DEFINITION 15.3   A function $N(t)$ is a *null function* if for every $t > 0$,

$$\int_0^t N(u)\,du = 0$$

With this definition, we state (without proof) a theorem which makes Laplace transforms a very powerful tool.

THEOREM 15.4   Two functions with the same Laplace transform can differ, at most, by a null function.[2]

Since two continuous functions which differ, at most, by a null function must be identical (see Exercise 1), we have the following.

COROLLARY   Let $f^*(s)$ be a Laplace transform. There exists at most one continuous function $f(t)$, $t \geq 0$, with transform f^*.

The value of the above corollary is that it specifies the uniqueness of inverse transforms with respect to continuous functions. We can now use this corollary in Example 1 to determine that Eq. (3) is in fact the solution to the differential equation (1).

The problem of inverting Laplace transforms is, in general, not easy.[3] We have already seen how certain properties of transforms can be used to find the inverse transform. In Example 1 we made use of the linearity property, and also the fact that $(1/s)f^*(s)$ is the transform of $\int_0^t f(u)\,du$. By recognizing the function with transform $f^*(s)$, we can determine the function with transform $(1/s)f^*(s)$. The following examples illustrate an additional trick which can be employed in trying to invert Laplace transforms. This trick is the method of partial fractions.

Example 3   Suppose

$$f^*(s) = \frac{2}{s^2 - 3s - 4} = \frac{2}{(s - 4)(s + 1)}$$

---

[2] For a proof of this theorem, which is attributed to Lerch, consult M. G. Smith, *Laplace Transforms*, New York, Van Nostrand-Reinhold, 1966.

[3] An integral formula for determining f from f^*, known as the inversion integral of the Laplace transform, can be found in most texts on Laplace transforms. However, application of this integral is, in general, extremely difficult and requires a knowledge of complex analysis beyond that assumed for this text. In particular, see Smith, *op. cit.*, Chapter 4.

We seek to find A and B such that

$$\frac{2}{(s-4)(s+1)} = \frac{A}{s-4} + \frac{B}{s+1} = \frac{A(s+1) + B(s-4)}{(s-4)(s+1)}$$

Equating the numerators of the outer terms gives us $(A+B)s + (A-4B) = 2$. We now have two equations in A and B by equating the coefficients of like powers of s.

$$A + B = 0 \quad \text{and} \quad A - 4B = 2$$

Hence, $A = \frac{2}{5}$, and $B = -\frac{2}{5}$. Our transform can now be written as

$$f^*(s) = \frac{\frac{2}{5}}{s-4} - \frac{\frac{2}{5}}{s+1}$$

Recalling that the transform of $f(t) = e^{ct}$ is $1/(s-c)$, we see immediately

$$f(t) = \tfrac{2}{5}(e^{4t} - e^{-t})$$

Example 4   Suppose we are trying to determine the function with Laplace transform $f^*(s) = (s^2 + 3s + 2)/[(s^2 + 4)(s^2 + 3)]$. We proceed as follows:

$$\frac{s^2 + 3s + 2}{(s^2 + 4)(s^2 + 3)} = \frac{s^2 + 3}{(s^2 + 4)(s^2 + 3)} + \frac{3s - 1}{(s^2 + 4)(s^2 + 3)}$$

$$= \frac{1}{s^2 + 4} + \frac{3s - 1}{(s^2 + 4)(s^2 + 3)}$$

We now treat the second term by partial fractions:

$$\frac{3s - 1}{(s^2 + 4)(s^2 + 3)} = \frac{As + B}{s^2 + 4} + \frac{Cs + D}{s^2 + 3}$$

Clearing fractions yields

$$3s - 1 = As^3 + 3As + Bs^2 + 3B + Cs^3 + 4Cs + Ds^2 + 4D$$

Equating like coefficients of s,

$$A + C = 0$$
$$B + D = 0$$
$$3A + 4C = 3$$
$$3B + 4D = -1$$

Solving for A, B, C, and D we obtain $A = -3$, $B = 1$, $C = 3$, $D = -1$.

Hence

$$f^*(s) = \frac{1}{s^2 + 4} - \frac{3s}{s^2 + 4} + \frac{1}{s^2 + 4} + \frac{3s}{s^2 + 3} - \frac{1}{s^2 + 3}$$

Recalling if $g_1(t) = \sin at$, then $g_1^*(s) = a/(s^2 + a^2)$, and if

$$g_2(t) = \cos at = \frac{1}{a} \frac{dg_1(t)}{dt}$$

then $g_2^*(s) = s/(s^2 + a^2)$, we obtain

$$f(t) = \sin 2t - 3 \cos 2t + 3 \cos \sqrt{3}t - \frac{1}{\sqrt{3}} \sin \sqrt{3}t$$

Although the above examples do not present a unified approach to inverting Laplace transformations, they do illustrate the techniques which are most practical—namely a good table of Laplace transforms[4], and much ingenuity.

Exercises

1. Prove that a continuous null function cannot be nonzero at any point. Hence, two continuous functions that differ by a null function are identical for every $t > 0$.

2. Suppose $f(t)$ has transform $f^*(s)$. Show that $df^*(s)/ds$ is the Laplace transform of $-tf(t)$, provided f^* can be differentiated under the integral sign. Obtain a similar relationship for $d^n f^*(s)/ds^n$ and $\int_s^\infty f^*(u) \, du$.

15.5 Convolutions

Consider two function f and g defined on $J = [0, \infty)$. Define $F(x, y) = f(x)g(y)$, the product of f and g, for $x \in J$, $y \in J$. A problem not uncommon in applications is to find the area of the region in the plane $x + y = t$, which is bounded by $x = 0$, $y = 0$, $z = 0$, and the graph of F. Denote this area by $h(t)$. (See Figure 15.1.) We then have $h(t)$ as the shaded area in the plane $x + y = t$ bounded by $z = 0$, $x = 0$, $y = 0$, and $z = F(x, y)$. Note the height $F(x, y_0)$ is proportional to $f(x)$; namely $g(y_0)f(x)$. Similarly, $F(x_0, y)$ is proportional to $g(y)$.

DEFINITION 15.4 Consider two functions f and g defined on $J = [0, \infty)$. The *convolution integral* $h(t)$ of f and g is

$$h(t) = \int_0^t f(t - u)g(u) \, du, \qquad t \geq 0$$

The function $h(t)$ is called the *convolution* of f and g.

---

[4] A. Erdelyi, W. Magnus, F. Oberhettinger, and F. Tricomi, *Tables of Integral Transforms*, vol. 1, New York, McGraw-Hill, 1954.

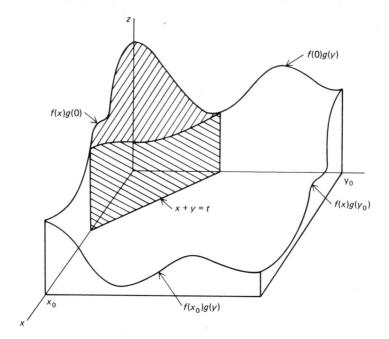

FIGURE 15.1

We leave for the reader to verify that the area described in Figure 15.1 is equal to $\sqrt{2}\, h(t)$. (See Exercise 1.)

Example 1   Let

$$f(x) = \frac{x^{a-1}e^{-x}}{\Gamma(a)}, \; 0 < x < \infty, a > 0$$

$$g(y) = \frac{x^{b-1}e^{-x}}{\Gamma(b)}, \; 0 < y < \infty, b > 0$$

where

$$\Gamma(\alpha) = \int_0^\infty x^{\alpha-1}e^{-x}\,dx, \quad \alpha > 0$$

Then for $t > 0$

$$h(t) = \int_0^t \frac{(t-u)^{a-1}e^{-(t-u)}u^{b-1}e^{-u}}{\Gamma(a)\Gamma(b)}\,du$$

$$= \frac{e^{-t}}{\Gamma(a)\Gamma(b)}\int_0^t (t-u)^{a-1}u^{b-1}\,du$$

$$= \frac{t^{a+b-2}e^{-t}}{\Gamma(a)\Gamma(b)}\int_0^t \left(1-\frac{u}{t}\right)^{a-1}\left(\frac{u}{t}\right)^{b-1}\,du$$

Setting $z = u/t$, we obtain

$$h(t) = \frac{t^{a+b-1}e^{-t}}{\Gamma(a)\Gamma(b)} \int_0^1 (1 - z)^{a-1}z^{b-1} \, dz$$

The function $\int_0^1 (1 - z)^{a-1}z^{b-1} \, dz$ is the *beta function*, which is equal to $\Gamma(a)\Gamma(b)/\Gamma(a + b)$.[5] Hence

$$h(t) = \frac{t^{a+b-1}e^{-t}}{\Gamma(a + b)}$$

(We have just displayed what probabilists call the reproductive property of the gamma distribution.)

Example 2 Consider

$$f(x) = g(x) = \begin{cases} 1 & 0 \le x \le 1 \\ 0 & x > 1 \end{cases}$$

Then, $F(x, y) = f(x)g(y)$ gives the unit cube shown in Figure 15.2.

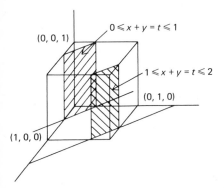

FIGURE 15.2

Care must be used in evaluating the convolution integral, as follows:

For $0 \le t \le 1$

$$h(t) = \int_0^t f(t - u)g(u) \, du = \int_0^t du = t$$

For $1 \le t \le 2$

$$h(t) = \int_0^t f(t - u)g(u) \, du = \int_0^{t-1} f(t - u)g(u) \, du$$

$$+ \int_{t-1}^1 f(t - u)g(u) \, du + \int_1^t f(t - u)g(u) \, du$$

---

[5] See Watson Fulks, *Advanced Calculus*, 2nd ed., New York, Wiley, 1969.

On $0 \le u \le t - 1$, the argument of f, namely $t - u$, varies between 1 and t, over which region f is zero. Hence

$$\int_0^{t-1} f(t - u)g(u) \, du = 0 \quad \text{for} \quad t > 1$$

Similarly, for $1 < u \le t$, $g(u) = 0$ and hence

$$\int_1^t f(t - u)g(u) \, du = 0$$

Therefore

$$h(t) = \int_{t-1}^1 f(t - u)g(u) \, du = 2 - t \quad \text{for} \quad 1 \le t \le 2$$

For $t > 2$, $h(t) = 0$. Therefore $h(t)$ is the triangular function

$$h(t) = \begin{cases} t & 0 \le t \le 1 \\ 2 - t & 1 \le t \le 2 \\ 0 & t > 2 \end{cases}$$

We now compute the Laplace transform of the convolution integral. The result is given in the following theorem.

THEOREM 15.5   Let h denote the convolution of two functions f and g; that is,

$$h(t) = \int_0^t f(t - u)g(u) \, du$$

Then if f^* and g^* are the respective Laplace transforms of f and g, the Laplace transform of h is given by $h^*(s) = f^*(s)g^*(s)$.

PROOF:

$$h^*(s) = \int_0^\infty e^{-st}h(t) \, dt$$

$$= \int_0^\infty e^{-st} \int_0^t f(t - u)g(u) \, du \, dt$$

$$= \int_0^\infty \int_0^t e^{-s(t-u)} f(t - u)e^{-su}g(u) \, du \, dt$$

Changing the order of integration, a process which we assume is valid, we get

$$h^*(s) = \int_0^\infty e^{-su}g(u) \int_u^\infty e^{-s(t-u)} f(t - u) \, dt \, du$$

If we set $z = t - u$, we obtain

$$h^*(s) = \int_0^\infty e^{-su} g(u) \int_0^\infty e^{-sz} f(z)\, dz\, du$$

$$= \int_0^\infty e^{-su} g(u)\, du \int_0^\infty e^{-sz} f(z)\, dz$$

$$= f^*(s) g^*(s) \qquad\blacksquare$$

More rigorous analyses of this proof can be found in the references given at the end of the chapter.

To close this section let us again consider the convolutions treated in Examples 1 and 2, but with Laplace transforms.

Example 3 Let

$$f(x) = \frac{x^{a-1} e^{-x}}{\Gamma(a)} \qquad 0 < x < \infty$$

$$g(x) = \frac{x^{b-1} e^{-x}}{\Gamma(b)} \qquad 0 < x < \infty$$

By the table of transforms given in Table A, we can determine that $f^*(s) = (s + 1)^{-a}$ and $g^*(s) = (s + 1)^{-b}$. By Theorem 15.5, $h^*(s) = (s + 1)^{-(a+b)}$ which, again making use of the table of transforms, implies

$$h(x) = \frac{x^{a+b-1} e^{-x}}{\Gamma(a + b)} \qquad 0 < x < \infty$$

Example 4 Let

$$f(x) = g(x) = \begin{cases} 1 & 0 \le x \le 1 \\ 0 & \text{otherwise} \end{cases}$$

Then

$$f^*(s) = g^*(s) = \int_0^1 e^{-sx}\, dx = \frac{1 - e^{-s}}{s}$$

Therefore

$$h^*(s) = \left(\frac{1 - e^{-s}}{s}\right)^2 = \frac{1 - 2e^{-s} + e^{-2s}}{s^2}$$

$$= \frac{1 - e^{-s}}{s^2} - \frac{(e^{-s} - e^{-2s})}{s^2}$$

$$= \frac{1 - e^{-s}}{s^2} - \frac{e^{-s}(1 - e^{-s})}{s^2}$$

Since $(1 - e^{-s})/s$ is the transform of

$$f(x) = \begin{cases} 1 & 0 \le x \le 1 \\ 0 & \text{otherwise} \end{cases}$$

we can conclude that $(1 - e^{-s})/s^2$ is the transform of

$$h_1(t) = \int_0^t f(x)\, dx = \begin{cases} t & 0 \le t \le 1 \\ 1 & t \ge 1 \end{cases}$$

Further, $e^{-s}(1 - e^{-s})s^{-2}$ is the transform of the function $h_2(t)$ obtained by shifting the origin of $h_1(t)$ by one unit; that is,

$$h_2(t) = h_1(t - 1) = \begin{cases} 0 & 0 \le t \le 1 \\ t - 1 & 1 \le t \le 2 \\ 1 & 2 \le t \end{cases}$$

Hence

$$h(t) = h_1(t) - h_2(t) = \begin{cases} t & 0 \le t \le 1 \\ 2 - t & 1 \le t \le 2 \\ 0 & 2 \le t \end{cases}$$

as was obtained in Example 2.

Exercises

1. Show that the area in Figure 15.1 is $\sqrt{2}h(t)$ where $h(t)$ is the convolution integral.

2. Compute the inverse transform by using Theorem 15.5.

    (a) $(s^2 + a^2)^{-2}$.
    (b) $(s - 1)^{-1}(s^2 + 4)^{-1}e^{-s}$.

3. The convolution h of f and g is often denoted $f * g$. Verify

    (a) $f_1 * f_2 = f_2 * f_1$.
    (b) $(f_1 * f_2) * f_3 = f_1 * (f_2 * f_3)$.
    (c) $f_1 * (f_2 + f_3) = f_1 * f_2 + f_1 * f_3$.

15.6   Applications of the Laplace transform

This section is devoted to illustrating the use of Laplace transforms in obtaining solutions of certain differential and integral equations. We have encountered an example of a first order nonhomogeneous linear differential equation in Example 1 of Section 15.4. We now treat higher order linear equations.

Example 1   Consider the second order nonhomogeneous equation

$$y'' + 3y' + 2y = t \tag{1}$$

with the initial condition $y(0) = 1$, $y'(0) = -2$. The methods of Chapter 9 could be used to solve this equation. However, we choose to attack it with Laplace transforms. Referring to Table A, and taking Laplace transforms term-by-term in Eq. (1), we obtain

$$s^2 y^*(s) - sy(0) - y'(0) + 3sy^*(s) - 3y(0) + 2y^*(s) = \frac{1}{s^2}$$

which, making use of the initial conditions, can be written as

$$(s^2 + 3s + 2)y^*(s) = \frac{1}{s^2} + sy(0) + (y'(0) + 3y(0)) = \frac{1}{s^2} + (s + 1)$$

Therefore

$$y^*(s) = \frac{1}{(s + 2)} + \frac{1}{s^2(s + 1)(s + 2)}$$

To obtain y we must invert y^*. We first set

$$y_1^*(s) = \frac{1}{(s + 2)} \quad \text{and} \quad y_2^*(s) = \frac{1}{s^2(s + 1)(s + 2)}$$

Our solution is then the sum of the inverses of y_1^* and y_2^*. It is easy to recognize the inverse of y_1^* as $y_1(t) = e^{-2t}$. To obtain y_2 we can use the method of partial fractions to express

$$y_2^* = \frac{1}{2s^2} - \frac{3}{4s} + \frac{1}{(s + 1)} - \frac{1}{4(s + 2)}$$

Hence

$$y_2 = \frac{t}{2} - \frac{3}{4} + e^{-t} - \frac{1}{4} e^{-2t}$$

Combining our results gives the solution

$$y(t) = \frac{t}{2} - \frac{3}{4} + e^{-t} + \frac{3}{4} e^{-2t}$$

The reader can verify this is a solution of Eq. (1).

Example 2 We next consider the equation

$$y^{(4)} - 3y^{(3)} + 3y^{(2)} - y' = 0 \tag{3}$$

subject to the initial condition $y(0) = 1$, $y'(0) = 3$, $y^{(2)}(0) = 3$, $y^{(3)}(0) = 3$. Taking Laplace transforms term-by-term, we obtain the equation

$$(s^4 - 3s^3 + 3s^2 - s)y^* - s^3 y(0) - s^2(y'(0) - 3y(0)) - s(y''(0) - 3y'(0) + 3y(0)) + (-y^{(3)}(0) + 3y^{(2)}(0) - 3y'(0) + y(0)) = 0$$

and, substituting the initial conditions,

$$s(s - 1)^3 y^* = s^3 - 3s + 2$$

Solving for y^* yields

$$y^* = \frac{s + 2}{s(s - 1)} = \frac{1}{(s - 1)} + \frac{2}{s(s - 1)}$$

Making use of the fact that the transform of $\int_0^t f(x)\, dx$ is $f^*(s)/s$, we obtain

$$y(t) = e^t + 2 \int_0^t e^x\, dx$$

Hence the solution to Eq. (3) is

$$y(t) = 3e^t - 2$$

Example 3   Consider the system of linear differential equations

$$3y_1' + y_2' + y_1 + 3y_2 = t$$
$$y_1' + y_2' + 3y_1 + 3y_2 = 0$$

subject to the initial condition $y_1(0) = a_1$, $y_2(0) = a_2$. Taking transforms in both equations yields

$$3sy_1^* - 3a_1 + sy_2^* - a_2 + y_1^* + 3y_2^* = \frac{1}{s^2}$$

$$sy_1^* - a_1 + sy_2^* - a_2 + 3y_1^* + 3y_2^* = 0$$

or

$$(3s + 1)y_1^* + (s + 3)y_2^* = \frac{1}{s^2} + (3a_1 + a_2)$$

$$(s + 3)y_1^* + (s + 3)y_2^* = a_1 + a_2 \tag{4}$$

Eliminating y_2^* by subtraction gives

$$(2s - 2)y_1^* = \frac{1}{s^2} + 2a_1$$

or

$$y_1^* = \frac{1}{2s^2(s - 1)} + \frac{a_1}{s - 1}$$

Hence

$$y_1(t) = \tfrac{1}{2}(e^t - 1 - t) + a_1 e^t$$

Referring to Eq. (4) we obtain

$$y_2^* = \frac{a_1 + a_2}{s + 3} - y_1^*$$

To obtain y_2 we need only find the inverse of $(a_1 + a_2)/(s + 3)$ and subtract y_1. Hence

$$y_2 = (a_1 + a_2)e^{-3t} - \left(\frac{1}{2} + a_1\right)e^t + \frac{t + 1}{2}$$

Example 4 Our last example is an application of Laplace transforms as an aid in solving integral equations. Suppose we are interested in determining $y(x)$ satisfying the equation

$$\int_0^t y(t - x)e^{-ax} \, dx - \int_0^t y(x) \, dx = t^k e^{-at} \qquad (5)$$

Since $\int_0^t y(t - x)e^{-ax} \, dx$ is the convolution of $y(t)$ with e^{-at}, Eq. (5) leads us to

$$\frac{y^*(s)}{s + a} - \frac{y^*(s)}{s} = \frac{\Gamma(k + 1)}{(s + a)^{k+1}}$$

which can be expressed as

$$
\begin{aligned}
y^*(s) &= -\frac{s}{a} \frac{\Gamma(k + 1)}{(s + a)^k} = -\frac{(s + a - a)}{a} \frac{\Gamma(k + 1)}{(s + a)^k} \\
&= \frac{\Gamma(k + 1)}{(s + a)^k} - \frac{\Gamma(k + 1)}{a(s + a)^{k-1}} \\
&= \frac{k\Gamma(k)}{(s + a)^k} - \frac{k(k - 1)\Gamma(k - 1)}{a(s + a)^{k-1}} \qquad (6)
\end{aligned}
$$

Inverting Eq. (6) yields

$$y(t) = kt^{k-1}e^{-at} - \frac{k(k - 1)}{a} t^{k-2}e^{-at}$$

as the solution to Eq. (5).

Exercises

Solve the following equations or system of equations.

1. $y'' - 4y' - 5y = e^{3t}$, $y(0) = 1$, $y'(0) = 0$

2. $y''' + 3y'' - 4y = 3t^2$, $y(0) = 2$, $y'(0) = -1$, $y''(0) = 5$

3. $y'' - 3y' + 2y = e^t + \cos 2t$, $y(0) = 2$, $y'(0) = 1$

4. $y_1' + 2y_2' + 3y_1 - 2y_2 = e^{2t}$
    $2y_1' + y_2' - y_1 - y_2 = 0$, $y_1(0) = 2$, $y_2(0) = 2$

5. $2y_1' + y_2' + 2y_1 + y_2 = t^2$
    $y_1' + y_2' + 2y_1 + y_2 = e^{2t}$, $y_1(0) = -1$, $y_2(0) = 1$

6. $3y_1' + 2y_2' + 2y_1 + y_3 = 0$
    $y_2' - y_3' + y_1 + y_3 = 0$
    $y_1' + 2y_3' + y_2 = 0$, $y_1(0) = 2$, $y_2(0) = 1$, $y_3(0) = -1$

7. $4\int_0^t (t - x)y(x) \, dx - 3y(t) + y'(t) = e^{2t}$, $y(0) = 0$

8. $\int_0^t 3^3(t - x)^2 e^{-3(t-x)}y(x) \, dx = 3^5 t^4 e^{-3t}$

COMMENTARY

General references on the theory and applications of the Laplace transform include the following:

CHURCHILL, R. V., *Operational Mathematics*, New York, McGraw-Hill, 1963.

DOETSCH, G., *Guide to the Application of Laplace Transforms*, New York, Van Nostrand Reinhold, 1961.

SMITH, M. G., *Laplace Transform Theory*, New York, Van Nostrand Reinhold, 1966.

For an interesting description of the historical ambiguity in the development of the field of "operational calculus" the reader should consult:

COOPER, J. L. B., "Heaviside and the Operational Calculus," *Mathematics Gazette*, vol. 36 (1952), 5–19.

While the Laplace transform is used where interest is restricted to the interval $J = [0, \infty)$, other transforms, such as the exponential Fourier transform, can be applied when interest is over the real line. For a brief description of this transform, the Mellin transform, and complete tables of a variety of transforms and their inverses, consult:

BATEMAN MANUSCRIPT PROJECT, *Table of Integral Transforms*, A. Erdélyi (ed.), New York, McGraw-Hill, 1954.

In recent years there has been considerable interest in inverting Laplace transforms by numerical techniques. See:

BELLMAN, R., R. E. KALABA, and J. A. LOCKETT, *Numerical Inversion of the Laplace Transform: Applications to Biology, Economics, Engineering, and Physics*, New York, American Elsevier, 1966.

Chapter 16

Series solutions of ordinary differential equations

16.1 Introduction

We have previously developed a theory concerning solutions of linear differential equations. Included was a theorem (Theorem 8.4) which, under rather general conditions, guaranteed the existence and uniqueness of a solution to an nth order linear differential equation satisfying a given set of initial conditions. We also discussed a method for obtaining solutions of homogeneous linear differential equations provided the coefficients of the $y^{(k)}$, $k = 0, 1, \ldots, n$ are constants.[1] For these equations, the solutions can be expressed as combinations of some of the "elementary" functions of analysis: the polynomials, exponential functions, sines, and cosines.

For equations in which the coefficients are not constant, including simple ones such as $y'' - xy = 0$, the solutions cannot be expressed in a simple way in terms of the elementary functions. The purpose of this chapter is to develop one of the most powerful methods for solving such equations—the method of *solution by infinite series*. Of course, it is fair to say that this is not so different in principle from what we did before, since the well-known exponential, sine, and cosine functions are themselves usually defined by their power series. The difference now is that we will be dealing with new and unfamiliar functions.

At this point the reader is advised to read Appendix C on power series. We shall assume familiarity with this material, and proceed to its application to differential equations.

---

[1] Or in a form which could be transformed to create an equation with constant coefficients.

DEFINITION 16.1   A function f is said to be *analytic at* x_0 if $f(x)$ has a Taylor series expansion about x_0, with positive radius of convergence ρ, which converges to $f(x)$; that is, f is analytic at x_0 if it can be represented in the form

$$f(x) = \sum_{n=0}^{\infty} c_n(x - x_0)^n \quad \text{for} \quad |x - x_0| < \rho$$

We shall encounter some cases in which $f(x)$ is undefined at $x = x_0$ but it is possible to redefine f at x_0 in such a way that $f(x)$ is then analytic at x_0. (See Example 3.) In these cases, $f(x)$ is said to have a *removable singularity* at x_0 and we shall call $f(x)$ analytic at x_0.

DEFINITION 16.2   Consider the linear differential equation

$$y^{(n)} + a_{n-1}(x)y^{(n-1)} + \cdots + a_0(x)y = 0 \tag{1}$$

A point x_0 at which the $a_k(x_0)$ are analytic, $k = 0, 1, \ldots, n-1$, is called an *ordinary point* of the differential equation. Those points which are not ordinary points are called *singular points* of the equation.

Our concern with ordinary and singular points stems partly from the condition in Theorem 8.4 that $a_n(x)$ be nonzero in the interval J in order to guarantee a solution to the equation in J. In Eq. (1) we have divided through by this leading coefficient.

Example 1   Consider the differential equation

$$y''' + (3 \sin x)y'' - \frac{2x}{1-x} y' + y = 0 \tag{2}$$

Since $a_3(x) \equiv 1$, the equation is in the form of Eq. (1). Further, $a_2(x) = 3 \sin x$, $a_1(x) = -2x/(1-x)$, and $a_0(x) = 1$. First consider $a_2(x)$. Since $\sin x$ can be expanded around any x_0,

$$\sin x = \sin x_0 + (x - x_0) \cos x_0$$
$$- \frac{(x - x_0)^2}{2!} \sin x_0 - \frac{(x - x_0)^3}{3!} \cos x_0 + \cdots$$

the function $3 \sin x = a_2(x)$ is analytic for any x_0, $-\infty < x_0 < \infty$. Further, the radius of convergence for the series is ∞. Next consider $a_1(x) = -2x/(1-x) = 2 - 2/(1-x)$. This function, $2 - 2/(1-x)$, can be expanded about x_0 as (see Appendix C, Exercise 2(e)).

$$a_1(x) = 2 - 2\left[\frac{1}{1-x_0} + \frac{x-x_0}{(1-x_0)^2} \right.$$
$$\left. + \frac{(x-x_0)^2}{(1-x_0)^3} + \cdots + \frac{(x-x_0)^n}{(1-x_0)^{n+1}} + \cdots \right]$$

for $|x - x_0| < |1 - x_0|$ and $x_0 \neq 1$. Hence for $x_0 \neq 1$, $a_1(x)$ is analytic with radius of convergence $\rho = |1 - x_0|$. However, $a_1(x)$ is not analytic at $x_0 = 1$, and hence $x_0 = 1$ is a singular point of the differential equation given in Eq. (2). Finally, since $a_0(x) \equiv 1$ is analytic everywhere, $x_0 = 1$ is the only singular point of Eq. (2).

It can be proved that a rational function $a(x) = p(x)/q(x)$, where p and q are polynomials with no common factor, is analytic at every point x_0, where $q(x_0) \neq 0$.

Example 2 Consider the function $a(x) = [(x^2 + 1)(x - 2)]^{-1}$. This function is analytic except at $x = 2$, i, and $-i$. As we shall see later, it is sometimes necessary to take account of the nonanalyticity of functions at complex values of x.

Example 3 Consider the function $a(x) = x^{-1} \sin x$. This function is not defined at $x = 0$. However, if we extend the domain of the function by defining $a(0) = 1$, then it is analytic since

$$a(x) = 1 - \frac{x^2}{3!} + \frac{x^4}{5!} - \cdots$$

for all x.

Exercises

Determine whether each of the following is analytic at the specified point, and if so give the corresponding radius of convergence.

1. $\ln x$ at $x = 1$, $x = e$

2. $e^{|x|}$ at $x = 0$, $x = 1$, $x = -1$

3. $x/(e^x - 1)$ at $x = 1$

4. $[\sin (2x)]/x$ at $x = \pi/4$, $x = \pi/2$

   Find all singular points for each of the following equations.

5. $(e^x - e)y'' + 2xe^x y' + y \cos x = 0$

6. $(x^2 - 1)y'' + \sin (\pi x)y = 0$

7. $x^2(3x - 1)y'' + 2xy' + x^2 y = 0$

8. $(x^2 - 2x)y'' + y' - 3y = 0$

16.2   Series solutions near ordinary points

We begin this section with a theorem concerning existence of series solutions for nth order linear differential equations. The reader is advised to consider the relationships between this theorem and Theorem 8.4. After a statement and a brief discussion of the theorem (and a corollary) we will present the method of series solutions by example. We omit the proof because of its complexity.

THEOREM 16.1   Consider the nth order linear differential equation

$$Ly = y^{(n)} + a_{n-1}(x)y^{(n-1)} + \cdots + a_0(x)y = 0 \qquad (1)$$

where $a_0, a_1, \ldots, a_{n-1}$ are analytic at x_0 with radii of convergence $\rho_0, \rho_1, \ldots, \rho_{n-1}$, respectively. Every solution (in a neighborhood of x_0) of $Ly = 0$ is analytic at x_0 with radius of convergence at least as large as $\min_{0 \le k \le n-1} \rho_k$. There is a general solution $y = b_0 y_1 + b_1 y_2 + \cdots + b_{n-1} y_n$ which has the form

$$y = \sum_{k=0}^{\infty} b_k(x - x_0)^k \qquad (2)$$

COROLLARY   Consider the differential equation (1). If $a_0, a_1, \ldots, a_{n-1}$ are polynomials, every solution can be represented as a power series which converges everywhere.

The full significance of this theorem will become more evident as we work through several examples. However, there are several points worth elaborating. First, note that the radius of convergence is at least as large as the smallest radius of convergence of the a_k. A second and somewhat surprising result is that Eq. (2) gives a general solution of Eq. (1), with the first n coefficients in the expansion of y, namely, $b_0, b_1, \ldots, b_{n-1}$ arbitrary. Once these coefficients have been selected, a particular solution is determined (see Section 7).

The basic idea in the series method of solution is that if y has a series expansion, we should be able to replace y in the equation by a series, and from the resulting equation (which will involve the yet unknown coefficients b_k) obtain a recursive relationship for the b_k, $k = 0, 1, \ldots$. The method of series solution is illustrated in the following example (see Example 1 of Section 1.4).

Example 1   Consider the differential equation $y' + 3x^2 y = 0$. Since this equation is of the form of Eq. (1) and $3x^2 = a_0(x)$ is analytic for $-\infty < x < \infty$, we know there is a series solution for y of the form given in Eq. (2). In particular, let $x_0 = 0$. Then there exist $b_0, b_1, \ldots$ such that $y = \sum_{k=0}^{\infty} b_k x^k$ is a solution to the given equation for $-\infty < x < \infty$. We now substitute for y its series. To do this notice that

$$y' = \sum_{k=1}^{\infty} k b_k x^{k-1} = \sum_{k=0}^{\infty} (k+1)b_{k+1}x^k$$

and

$$3x^2 y = \sum_{k=0}^{\infty} 3b_k x^{k+2} = \sum_{k=2}^{\infty} 3b_{k-2}x^k$$

Note that we have changed the index of summation in each case so as to have the index of summation correspond to the power of x. Our given differential equation becomes

$$y' + 3x^2y = \sum_{k=0}^{\infty} (k + 1)b_{k+1}x^k + \sum_{k=2}^{\infty} 3b_{k-2}x^k = 0 \qquad (3)$$

Combining like powers of x in Eq. (3) gives

$$b_1 + 2b_2x + \sum_{k=2}^{\infty} [(k + 1)b_{k+1} + 3b_{k-2}]x^k = 0$$

We now have a power series in x which is equal to zero for all x. It follows (see Corollary to Theorem C.3) that all coefficients equal zero. Therefore, $b_1 = b_2 = 0$. Further,

$$(k + 1)b_{k+1} + 3b_{k-2} = 0 \quad \text{for} \quad k = 2, 3, \ldots \qquad (4)$$

Rewriting Eq. (4) gives

$$b_{k+3} = \frac{-3}{k + 3} b_k \qquad k = 0, 1, 2, \ldots \qquad (5)$$

We first notice that, since $b_1 = b_2 = 0$, the only coefficients which need not be zero are b_{3j} for $j = 0, 1, 2, \ldots$. In particular

$$b_3 = -b_0$$

$$b_6 = -\frac{3b_3}{6} = \frac{b_0}{2}$$

$$b_9 = -\frac{3b_6}{9} = -\frac{b_0}{3!}$$

and in general

$$b_{3j} = \frac{(-1)^j b_0}{j!} \qquad j = 0, 1, 2, \ldots$$

Hence

$$y = \sum_{k=0}^{\infty} b_k x^k = \sum_{j=0}^{\infty} b_{3j}x^{3j} = b_0 \sum_{j=0}^{\infty} \frac{(-1)^j x^{3j}}{j!} = b_0 e^{-x^3}$$

as was obtained in Example 1 of Section 1.4.

The purpose of the next example is to illustrate the series method when expansion is around a point other than zero. We call upon Example 2 of Section 1.4.

Example 2 Consider the first order linear equation $xy' + y = 0$. Rewriting this equation in the form of Eq. (1) gives

$$y' + \frac{1}{x} y = 0$$

where $1/x$ is not analytic at $x = 0$. However, by taking $x_0 = 1$ we can expand

$$\frac{1}{x} = \frac{1}{1 - (1 - x)} = \sum_{k=0}^{\infty} (1 - x)^k \quad \text{for} \quad |1 - x| < 1$$

Hence $1/x$ is analytic at $x_0 = 1$ and a series solution of the differential equation will exist and be of the form

$$y = \sum_{k=0}^{\infty} b_k(x - 1)^k$$

Further, the radius of convergence of this series will be at least one; that is, the series will converge for at least those x such that $|x - 1| < 1$. Assuming

$$y = \sum_{k=0}^{\infty} b_k(x - 1)^k$$

we obtain

$$y' = \sum_{k=1}^{\infty} k b_k(x - 1)^{k-1}$$

Since the original equation can be written

$$xy' + y = (x - 1)y' + y' + y = 0$$

substitution of the series for y and y' gives

$$\sum_{k=1}^{\infty} k b_k(x - 1)^k + \sum_{k=1}^{\infty} k b_k(x - 1)^{k-1} + \sum_{k=0}^{\infty} b_k(x - 1)^k = 0$$

Changing the indices of summation so that the power of $(x - 1)$ is the index of summation, and noting that $\sum_{k=1}^{\infty} k b_k(x - 1)^k = \sum_{k=0}^{\infty} k b_k(x - 1)^k$, gives

$$\sum_{k=0}^{\infty} k b_k(x - 1)^k + \sum_{k=0}^{\infty} (k + 1) b_{k+1}(x - 1)^k + \sum_{k=0}^{\infty} b_k(x - 1)^k = 0$$

or

$$\sum_{k=0}^{\infty} [(k + 1) b_k + (k + 1) b_{k+1}](x - 1)^k = 0$$

Since this series must equal zero for all x for which $|x - 1| < 1$, we obtain $(k + 1) b_k + (k + 1) b_{k+1} = 0$ for all k and hence $b_{k+1} = -b_k$. Hence

$$y = b_0 \sum_{k=0}^{\infty} (-1)^k(x - 1)^k = b_0 \left[\frac{1}{1 + (x - 1)} \right] = \frac{b_0}{x}$$

Note that this series converges for $|x - 1| < 1$ but does not converge for $|x - 1| = 1$. This answer is consistent with that obtained in Section 1.4.

We now turn our attention to several examples involving second order differential equations which play important roles in applications.

Example 3   An equation that arises frequently in mathematical physics, such as in the investigation of the Schrödinger equation for a harmonic oscillator, is the *Hermite equation*

$$y'' - 2xy' + \lambda y = 0 \tag{6}$$

where λ is a specified constant. First note that $a_1(x) = -2x$ and $a_0(x) = \lambda$ are analytic for $-\infty < x < \infty$.

By Theorem 16.1 we know a solution exists of the form

$$\sum_{k=0}^{\infty} b_k(x - x_0)^k$$

for any x_0, in particular for $x_0 = 0$. Let us now determine $b_0, b_1, \ldots, b_k, \ldots$ such that

$$y = \sum_{k=0}^{\infty} b_k x^k$$

is a solution of Eq. (6). We obtain

$$y' = \sum_{k=1}^{\infty} k b_k x^{k-1}, \qquad xy' = \sum_{k=1}^{\infty} k b_k x^k = \sum_{k=0}^{\infty} k b_k x^k$$

$$y'' = \sum_{k=2}^{\infty} k(k-1) b_k x^{k-2} = \sum_{k=0}^{\infty} (k+2)(k+1) b_{k+2} x^k$$

Substituting into Eq. (6) yields

$$\sum_{k=0}^{\infty} (k+2)(k+1) b_{k+2} x^k - 2 \sum_{k=0}^{\infty} k b_k x^k + \lambda \sum_{k=0}^{\infty} b_k x^k$$

$$= \sum_{k=0}^{\infty} [(k+2)(k+1) b_{k+2} - 2k b_k + \lambda b_k] x^k = 0$$

Since this series is equal to zero for all x, we can conclude each coefficient of x^k must be zero and hence

$$(k+2)(k+1) b_{k+2} + (\lambda - 2k) b_k = 0 \quad \text{for} \quad k = 0, 1, 2, \ldots \tag{7}$$

so that $b_2 = (-\lambda/2)b_0$, and, in general,

$$b_{k+2} = \frac{(2k - \lambda) b_k}{(k+2)(k+1)} \qquad k = 0, 1, \ldots$$

An immediate observation is that once b_0 is specified, so is b_2, which in turn determines b_4, and so on. Similarly, b_1 determines b_3 which determines b_5 *ad infinitum*. However, b_0 and b_1 can be chosen independently. Setting $b_0 = 1$ and $b_1 = 0$ yields a solution y_1, and setting $b_0 = 0$ and $b_1 = 1$ yields a second solution, y_2. These are linearly independent solutions and hence form a basis for the solution space. The desired general solution is

$$y = b_0 y_1 + b_1 y_2$$

where b_0, b_1 are arbitrary scalars.

In this case, we can work out an explicit form for y_1 and y_2. We have

$$b_4 = \frac{4 - \lambda}{4 \cdot 3} b_2 = \frac{(4 - \lambda)(0 - \lambda)}{4!} b_0$$

$$b_6 = \frac{8 - \lambda}{6 \cdot 5} b_4 = \frac{(8 - \lambda)(4 - \lambda)(0 - \lambda)}{6!} b_0$$

In general,

$$b_{2m} = \frac{(4(m - 1) - \lambda) \cdots (4 - \lambda)(0 - \lambda)}{(2m)!} b_0 = \frac{b_0}{(2m)!} \prod_{j=0}^{m-1} (4j - \lambda)$$

Similarly

$$b_{2m+1} = \frac{(4(m - 1) + 2 - \lambda) \cdots (6 - \lambda)(2 - \lambda)}{(2m + 1)!} b_1$$

$$= \frac{b_1}{(2m + 1)!} \prod_{j=0}^{m-1} (4j + 2 - \lambda)$$

Therefore

$$y_1 = 1 + \frac{-\lambda}{2} x^2 + \frac{(4 - \lambda)(0 - \lambda)}{4!} x^4 + \cdots$$

$$= 1 + \sum_{m=1}^{\infty} \left\{ \prod_{j=0}^{m-1} (4j - \lambda) \right\} \frac{x^{2m}}{(2m)!}$$

$$y_2 = x + \frac{2 - \lambda}{3!} x^3 + \frac{(6 - \lambda)(2 - \lambda)}{5!} x^5 + \cdots$$

$$= x + \sum_{m=1}^{\infty} \left\{ \prod_{j=0}^{m-1} (4j + 2 - \lambda) \right\} \frac{x^{2m+1}}{(2m + 1)!}$$

These formulas may seem formidable to the reader. However, it is a routine task to calculate numerical values for series such as these for a range of values of x and λ with the aid of a digital computer. Techniques for doing this efficiently and with desired accuracy are discussed in courses in "numerical methods" or "numerical analysis."

Example 4   A second important differential equation, known as the *Legendre equation*, is

$$(1 - x^2)y'' - 2xy' + \alpha(\alpha + 1)y = 0 \qquad (8)$$

where α is a constant. We see immediately that $x = \pm 1$ are singular points of the equation. Hence if we rewrite Eq. (8) in the form

$$y'' - \frac{2x}{1 - x^2} y' + \frac{\alpha(\alpha + 1)}{1 - x^2} y = 0$$

we see that the radii of convergence about $x_0 = 0$ for $a_0(x)$ and $a_1(x)$ are both 1. Since a_0 and a_1 are both analytic at $x = 0$, we know by Theorem 16.1 that a series solution exists with radius of convergence equal to at least one. As in the previous examples, let us perform the substitution

$$y = \sum_{k=0}^{\infty} b_k x^k$$

We obtain

$$y' = \sum_{k=0}^{\infty} b_{k+1}(k + 1)x^k, \qquad y'' = \sum_{k=0}^{\infty} b_{k+2}(k + 2)(k + 1)x^k$$

Multiplying these results by the appropriate power of x yields

$$xy' = \sum_{k=0}^{\infty} b_k k x^k, \qquad x^2 y'' = \sum_{k=2}^{\infty} b_k k(k - 1)x^k = \sum_{k=0}^{\infty} b_k k(k - 1)x^k$$

Substituting into Eq. (8) gives

$$(1 - x^2)y'' - 2xy' + \alpha(\alpha + 1)y = \sum_{k=0}^{\infty} [b_{k+2}(k + 2)(k + 1) - b_k k(k - 1)$$

$$- 2b_k k + \alpha(\alpha + 1)b_k]x^k = 0$$

Therefore

$$b_{k+2}(k + 2)(k + 1) = b_k[k(k - 1) + 2k - \alpha(\alpha + 1)]$$

or

$$b_{k+2} = \frac{k(k + 1) - \alpha(\alpha + 1)}{(k + 2)(k + 1)} b_k = \frac{(k - \alpha)(k + \alpha + 1)}{(k + 2)(k + 1)} b_k \qquad (9)$$

Note that as in Example 3 (and as determined by Theorem 16.1) b_0 and b_1 can be chosen independently but together determine the total set of coefficients.

Before leaving this example let us consider various values for α. In particular, if α is a nonnegative integer there will be some k (namely, $k = \alpha$) such that the numerator on the right-hand side of Eq. (9) is zero. Hence $b_{\alpha+2} = 0$. But then $b_{\alpha+2} = b_{\alpha+4} = \cdots = 0$. Hence if α is even we find one solution to Eq. (8) is

$$y_1 = \sum_{k=0}^{\alpha/2} b_{2k} x^{2k}$$

where the b_{2k} are determined by choice of b_0. The important thing to notice about this result is that even though the original Eq. (8) is not analytic at $x = \pm 1$, we have found solutions which are analytic everywhere. Similarly, if α is odd we obtain a solution

$$y_2 = \sum_{k=0}^{(\alpha-1)/2} b_{2k+1} x^{2k+1}$$

which is a finite series and again is analytic at ± 1 and indeed everywhere.

Exercises

Solve Exercises 1 through 4 by the series method with expansions around the specified point. Give a minimum for the radius of convergence of the series in each case.

1. Give a basis for the solution space of

$$(x - 2)y' + 3x^2 y = 0 \quad \text{(around } x_0 = 0)$$

2. Give a basis for the solution space of

$$(x - 1)^2 y'' + 2xy = 0 \quad \text{(around } x_0 = 0)$$

3. Give a general solution to

$$xy' - x(x + 1)y = 2(x + 1)^2 \quad \text{(around } x_0 = 1)$$

4. Give a general solution to

$$(3x^2 + 2x)y'' - 2xy = x^2 + 1 \quad \text{(around } x_0 = 1)$$

5. In Exercises 3 and 4, why could we not ask for a basis for the solution space?

6. Give the first four terms in the series solution for

$$e^x y'' - 2xy' + y = 0 \quad \text{(around } x_0 = 0)$$

7. Solve by the series method and derive the solutions $\sin x$ and $\cos x$ from the series.

$$y'' + y = 0$$

8. Using Eq. (9), establish the following formulas for the solution of Legendre's equation.

$$y = b_0 y_1 + b_1 y_2$$

where

$$y_1 = 1 + \sum_{m=1}^{\infty} (-1)^m \left\{ \frac{\prod_{j=1}^{m} (\alpha + 2j - 1)(\alpha - 2j + 2)}{(2m)!} \right\} x^{2m}$$

$$y_2 = x + \sum_{m=1}^{\infty} (-1)^m \left\{ \frac{\prod_{j=1}^{m} (\alpha + 2j)(\alpha - 2j + 1)}{(2m + 1)!} \right\} x^{2m+1}$$

9. Show that if $\alpha = n$ is a nonnegative integer (even or odd), Legendre's equation has a polynomial solution which can be written in the form

$$y = c_n \left[x^n - \frac{n(n-1)}{2(2n-1)} x^{n-2} + \frac{n(n-1)(n-2)(n-3)}{2 \cdot 4(2n-1)(2n-3)} x^{n-4} - \cdots \right]$$

where c_n is an arbitrary constant.

10. Let

$$c_n = \begin{cases} 1 & n = 0 \\ \dfrac{(2n)!}{2^n(n!)^2} & n = 1, 2, \ldots \end{cases}$$

The polynomial solution of Exercise 9 is then called the *Legendre polynomial* of degree n,

$$P_0(x) = 1$$

$$P_n(x) = \frac{(2n)!}{2^n(n!)^2} \left[x^n - \frac{n(n-1)}{2(2n-1)} x^{n-2} + \cdots \right]$$

Explicitly calculate P_2, P_3, P_4.

11. By the Binomial Theorem,

$$(x^2 - 1)^n = x^{2n} - \binom{n}{1} x^{2n-2} + \binom{n}{2} x^{2n-4} - \cdots$$

Show by induction that

$$\frac{d^n}{dx^n} (x^2 - 1)^n = 2n(2n-1) \cdots (n+1)x^n$$

$$- n(2n-2)(2n-3) \cdots n(n-1)x^{n-2}$$

$$+ \frac{1}{2!} n(n-1)(2n-4)(2n-5) \cdots (n-3)x^{n-4} - \cdots$$

Hence show that

$$\frac{d^n}{dx^n} (x^2 - 1)^n = 2^n(n!)P_n(x)$$

This is called *Rodrigues's formula*.

12. Using Rodrigues's formula, show that $P_n(1) = 1$. Thus, $P_n(x)$ is the unique polynomial solution of Legendre's equation which has value 1 at $x = 1$.

13. Prove that $P_n(-x) = (-1)^n P_n(x)$.

14. (a) Prove that $\{P_0(x), P_1(x), P_2(x)\}$ is a basis for the vector space of all polynomials of degree at most 2.
    (b) Generalize to the space of all polynomials of degree at most n.

15. Show that the series method for the equation $y'' + y' + xy = 0$ leads to the *three-term recurrence relation* (or difference equation)

$$(k + 2)(k + 1)b_{k+2} + (k + 1)b_{k+1} + b_{k-1} = 0$$

Find the terms up to x^4 in the general solution of the differential equation.

16. Show that

$$\int_{-1}^{1} P_n(x)P_m(x) \, dx = 0, \quad n \neq m$$

[*Hint:* Show that

$$\frac{d}{dx}[(1 - x^2)P_n'(x)] = -n(n + 1)P_n(x)$$

$$\frac{d}{dx}[(1 - x^2)P_m'(x)] = -m(m + 1)P_m(x)$$

Multiply the first equation by P_m, the second by P_n, subtract, and integrate.]

17. Let y be a solution of Eq. (1) having the expansion in Eq. (2). Show that

$$y = b_0 y_1 + b_1 y_2 + \cdots + b_{n-1} y_n$$

where $y_1, y_2, \ldots, y_n$ are the solutions of Eq. (1) determined by the initial conditions

$$y_{j+1}^{(j)}(x_0) = j!$$
$$y_i^{(j)}(x_0) = 0 \quad i \neq j + 1$$

for $i = 1, 2, \ldots, n$ and $j = 0, 1, \ldots, n - 1$. Consequently, Eq. (2) is a general solution of Eq. (1), with $b_0, b_1, \ldots, b_{n-1}$ being arbitrary constants; that is, $y_1, y_2, \ldots, y_n$ form a basis for the solution space of Eq. (1).

16.3   Taylor series method

There is a variation of the power series method which can sometimes be more convenient, particularly if initial conditions are given. We call this the *Taylor series method*. Consider an initial value problem

$$Ly = y^{(n)} + a_{n-1}(x)y^{(n-1)} + \cdots + a_0(x)y = 0 \qquad (1)$$

with initial conditions $y(x_0) = c_0$, $y'(x_0) = c_1, \ldots, y^{(n-1)}(x_0) = c_{n-1}$, where x_0 is an ordinary point of Eq. (1). We know, by Theorem 16.1, there is a solution of Eq. (1) of the form

$$y(x) = \sum_{k=0}^{\infty} b_k(x - x_0)^k$$

where the coefficients b_k are given by

$$b_k = \frac{y^{(k)}(x_0)}{k!}$$

If we rewrite Eq. (1) as

$$y^{(n)} = -a_{n-1}(x)y^{(n-1)} - \cdots - a_0(x)y \tag{2}$$

we can compute the b_k for our initial value problem as follows. We have been given

$$y^{(k)}(x_0) = c_k, \quad k = 0, 1, \ldots, n-1$$

and hence

$$b_k = \frac{c_k}{k!} \quad \text{for} \quad k = 0, 1, \ldots, n-1$$

From Eq. (2) we get

$$y^{(n)}(x_0) = -a_{n-1}(x_0)y^{(n-1)}(x_0) - \cdots - a_0(x_0)y(x_0)$$

$$= -\sum_{k=0}^{n-1} c_k a_k(x_0)$$

This gives the value of $y^{(n)}(x_0)$ and so of b_n. Differentiating Eq. (2) yields

$$y^{(n+1)}(x) = -\sum_{k=0}^{n-1} a_k(x)y^{(k+1)}(x) - \sum_{k=0}^{n-1} a_k'(x)y^{(k)}(x)$$

from which we can obtain $y^{(n+1)}(x_0)$ and hence b_{n+1}. Proceeding in this fashion we obtain as many b_k as desired, and if we are lucky a general expression for b_k.

Example 1 Consider again the Hermite equation

$$y'' - 2xy' + \lambda y = 0 \tag{3}$$

with initial conditions $y(0) = 1$, $y'(0) = 0$. We obtain immediately $b_0 = 1$, $b_1 = 0$. Rewriting Eq. (3) as in Eq. (2) we obtain

$$y'' = 2xy' - \lambda y \tag{4}$$

Substituting $x = 0$ gives

$$y''(0) = -\lambda y(0) = -\lambda$$

and hence $b_2 = -\lambda/2!$. Differentiating Eq. (4) yields

$$y''' = 2xy'' + (2 - \lambda)y'$$

and hence

$$y'''(0) = (2 - \lambda)y'(0) = 0, \quad \text{implying} \quad b_3 = 0$$

Similarly,

$$y^{(iv)} = 2xy''' + (4 - \lambda)y'', \qquad y^{(iv)}(0) = -(4 - \lambda)\lambda$$

and hence $b_4 = -(4 - \lambda)\lambda/4!$. In general,

$$y^{(n)} = 2xy^{(n-1)} + [2(n - 2) - \lambda]y^{(n-2)}$$

For derivatives of even order, we have

$$y^{(2n)}(0) = (4n - 4 - \lambda)y^{(2n-2)}(0) = \prod_{k=1}^{n} [4(k - 1) - \lambda]$$

$$b_{2n} = \frac{\prod_{k=1}^{n} [4(k - 1) - \lambda]}{(2n)!}$$

Also $y^{(2n+1)}(0) = 0$, which implies that $b_{2n+1} = 0$. Therefore

$$y(x) = 1 - \frac{\lambda}{2!} x^2 - \frac{(4 - \lambda)\lambda}{4!} x^4 + \cdots$$

$$= 1 + \sum_{n=1}^{\infty} \frac{(\prod_{k=1}^{n} [4(k - 1) - \lambda])x^{2n}}{(2n)!}$$

which is the series found for $y_1(x)$ in Example 3 of Section 16.2.

Exercises

In each of the following, give the first four nonzero terms of the solution by using the Taylor series method.

1. $y'' + [2 \ln (1 - x)]y' + x^2 y = 0$, $y(0) = 0$, $y'(0) = 1$.

2. $y''' + 3(\sin x)y' + (\cos x)y = 0$, $y(0) = 1$, $y'(0) = 1$, $y''(0) = 0$.

3. $e^x y'' + 2xy' + 3x^2 y = 0$, $y(0) = 1$, $y'(0) = 0$.

4. $e^x y'' + [\ln (1 + x)]y' + (\sin x)y = 0$, $y(0) = 0$, $y'(0) = 0$.

5. $dy/dx = y^2$, $y(0) = 1$. [*Note:* This equation is not linear.]

16.4   Equations with singular points

In the previous sections we have considered the solution of linear differential equations on an interval on which the coefficient functions are analytic. For the equation

$$Ly = y^{(n)} + a_{n-1}(x)y^{(n-1)} + \cdots + a_0(x)y = 0 \qquad (1)$$

in which $a_{n-1}, \ldots, a_0$ are analytic at an ordinary point x_0, we found that the solutions must also be analytic at x_0 and that the series which represent them can be calculated by straightforward techniques.

It might be supposed that there would be no interest in seeking solutions of equations with nonanalytic coefficient functions, but the contrary is true. Consider the following example.

Example 1 Suppose that a thin circular disc of heat-conducting material is insulated perfectly on its flat faces and that heating (or cooling) elements have been attached to the circumference for an infinite length of time. One may then ask what the temperature is at an arbitrary point within the disc. Placing the origin of a polar coordinate system at the center of the disc, we can represent a position within the disc by a pair (r, θ) and the steady-state temperature by a function $u(r, \theta)$ (see Figure 16.1). The assumption of perfect insulation on the faces makes it possible to idealize so as to make the problem two-dimensional rather than three-dimensional. It can be shown from physical principles that the function u must satisfy the partial differential equation

$$r^2 \frac{\partial^2 u}{\partial r^2} + r \frac{\partial u}{\partial r} + \frac{\partial^2 u}{\partial \theta^2} = 0 \qquad (0 \le r \le a, \ 0 \le \theta \le 2\pi)$$

and the *boundary condition* $u(a, \theta) = f(\theta)$, where $f(\theta)$ describes the temperature, as a function of θ, which is maintained on the circumference by the heating elements.

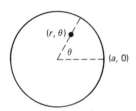

(r, θ)

θ

(a, 0)

FIGURE 16.1

It can be shown that this problem can be solved in the form of an infinite series, provided one can solve the ordinary differential equation

$$r^2 \frac{d^2 y}{dr^2} + r \frac{dy}{dr} - n^2 y = 0 \qquad (0 \le r \le a) \tag{2}$$

for each integer $n = 0, 1, 2, \ldots$. Writing Eq. (2) in the form

$$\frac{d^2 y}{dr^2} + \frac{1}{r} \frac{dy}{dr} - \frac{n^2}{r^2} y = 0 \tag{3}$$

we see that the coefficient functions $a_1(r) = r^{-1}$ and $a_0(r) = -n^2 r^{-2}$ are not analytic at $r = 0$ and not continuous on $0 \le r \le a$. Thus, Theorem 16.1 is not applicable so far as series expansion around the singular point $r = 0$ is concerned. Moreover, the basic existence theorem, Theorem 8.4, is also not applicable. Consequently, we are not even sure whether Eq. (3) has a solution $y(r)$ that exists on the interval $0 \le r \le a$.

On the other hand, if we consider Eq. (3) on any interval $\varepsilon \le r \le a$, where ε is small but positive, the coefficients are analytic and from Theorem 8.4 there must be two linearly independent solutions—for example, $y_1(r)$ and $y_2(r)$, satisfying $y_1(a) = 1$, $y_1'(a) = 0$, $y_2(a) = 0$, $y_2'(a) = 1$. Since ε is arbitrary, the solutions $y_1(r)$ and $y_2(r)$ must, in fact, exist on the whole interval $0 < r \le a$, open on the left. The questions to answer are: How can these solutions be constructed? What is the behavior of these solutions as $r \to 0^+$? Do they approach well-defined limits, tend to infinity, or oscillate?

In the present example, it is reasonable on physical grounds to suppose that the temperature is defined at $r = 0$ just as much as at any other point. We anticipate that at least one of the functions $y_1(r)$, $y_2(r)$ will be defined even at $r = 0$. Let us therefore try to find a power series solution of the form

$$y = \sum_{k=0}^{\infty} b_k r^k$$

even though Theorem 16.1 does not guarantee success in this case. By the usual procedure we obtain

$$\sum_{k=0}^{\infty} [k(k - 1) + k - n^2] b_k r^k = 0$$

Hence

$$(k^2 - n^2) b_k = 0, \qquad k = 0, 1, 2, \ldots$$

This equation implies that $b_k = 0$ for every $k \ne n$ since then $k^2 - n^2 \ne 0$. For $k = n$, no restriction is placed on b_k. Thus the series solution reduces to the simple polynomial $y = b_n r^n$, where b_n is arbitrary. It is easy to check this solution in Eq. (2).

In this example, we have found an analytic function (a polynomial) which is a solution on the whole interval $0 \le r \le a$ even though $r = 0$ is a singular point of the equation. The nature of other independent solutions as r approaches zero has not yet been ascertained. The reader can verify that $y = r^{-n}$ is an independent solution of Eq. (2) on $0 < r \le a$; this solution is unbounded as $r \to 0^+$ if $n > 0$.

Example 2   It is not always true that an equation with a singular point has even one nontrivial analytic solution. For example, if we substitute a power series into the equation

$$x^2 y'' + \frac{3}{2} xy' - \frac{y}{2} = 0$$

we obtain

$$\sum_{k=0}^{\infty} k(k - 1) b_k x^k + \frac{3}{2} \sum_{k=0}^{\infty} k b_k x^k - \frac{1}{2} \sum_{k=0}^{\infty} b_k x^k = 0$$

Therefore $b_0 = b_1 = 0$ and

$$[k(k - 1) + \tfrac{3}{2}k - \tfrac{1}{2}]b_k = (k - \tfrac{1}{2})(k + 1)b_k = 0, \qquad k = 2, 3, \ldots$$

It follows that every $b_k = 0$. Thus the only analytic solution is $y(x) \equiv 0$.

Example 2 shows that we cannot generally find families of solutions which are analytic at a singular point x_0. On the other hand, Example 1 suggests that there are solutions analytic for x near x_0 but not necessarily for x equal to x_0, and so our objective will be to find series representations for such solutions. Occasionally, these representations will turn out to be valid for $x = x_0$.

It may occur to the reader that problems of this type can be handled simply by constructing power series expansions about a point different from the singular point but near it. Further discussion of Example 2 shows that there are serious drawbacks to this approach. The two series

$$y_1(x) = \sum_{k=0}^{\infty} (-1)^k (x - 1)^k$$

$$y_2(x) = 1 + \tfrac{1}{2}(x - 1) + \sum_{k=2}^{\infty} (-1)^{k-1} \frac{1 \cdot 3 \cdots (2k - 3)}{2^k k!} (x - 1)^k \qquad (4)$$

converge for $0 < x < 2$. It can be verified that the sum functions are linearly independent solutions by substitution directly into the equation (see Exercise 1), or more easily, by observing that

$$y_1(x) = \frac{1}{1 + (x - 1)} = \frac{1}{x}, \qquad y_2(x) = [1 + (x - 1)]^{1/2} = \sqrt{x} \qquad (5)$$

and substituting these expressions into the equation. Let us suppose that the series solutions are known, but that the simple analytic forms in Eq. (5) are not known. In most problems, this would be the usual situation. Since the series for $y_1(x)$ does not converge at $x = 0$, it is not at all obvious from the series that $y_1(x) \to \infty$ but $xy_1(x) \to$ constant as $x \to 0^+$. Nor is it clear from the series that $y_2(x) \to 0$ as $x \to 0^+$. Furthermore, the use of these series to obtain accurate numerical values for x near zero may be quite difficult. For example, for $x = 0.1$, we have $y_1(0.1) = \Sigma(-1)^k(-0.9)^k$. In order to ensure an error of at most 10^{-5}, we could retain K terms, where $10(0.9)^{K+1}$, the sum of the omitted terms, is less than 10^{-5}. This yields K on the order of 130, indicating slow convergence.

In summary, we see that for purposes of certain applications, as well as of general curiosity, we may want to treat equations with singular points. To do so, we need a new and more effective method for constructing solutions. Also, we would like to obtain general theorems concerning the analytic nature of solutions near singular points.

Exercises

1. Show that the function $y_1(x)$ defined in Eq. (4) is a solution of the equation in Example 2 by differentiating the power series and substituting in the equation.

2. Verify that $y = b_n r^n$, where b_n is arbitrary, is a solution to Eq. (2).

3. If the series $\sum_{k=0}^{\infty} (-1)^k (0.5)^k$ is used to compute $y_1(1.5)$, how many terms are required to ensure an error of at most 10^{-27}?

4. Verify that $u(r, \theta) = r^n (A \cos n\theta + B \sin n\theta)$ is a solution of the partial differential equation in Example 1 for any constants A, B and any integer $n \geq 0$. Assuming that the solutions form a vector space, what can be said about the dimension of the space?

16.5   Regular singular points

Recall that a point x_0 is an ordinary point of

$$Ly = y^{(n)} + a_{n-1}(x)y^{(n-1)} + \cdots + a_0(x)y \tag{1}$$

if all the coefficients $a_0(x), \ldots, a_{n-1}(x)$ are analytic at x_0. Otherwise, x_0 is a singular point of the equation. It happens that a very satisfactory theory of solutions can be obtained if the singular point is of the following type.

DEFINITION 16.3 x_0 is called a *regular singular point* for Eq. (1) if x_0 is a singular point and $(x - x_0)^k a_{n-k}(x)$ is analytic at x_0 for $k = 1, \ldots, n$. It is called an *irregular singular point* if it is a singular point but not a regular singular point.[1]

For the second order equation $y'' + a_1(x)y' + a_0(x)y = 0$, to have a regular singular point at x_0, $(x - x_0)a_1(x)$ and $(x - x_0)^2 a_0(x)$ must be analytic functions, although $a_1(x)$ and $a_0(x)$ are not themselves both analytic at x_0. Equivalently, the equation must have the form

$$y'' + \frac{b_1(x)}{x - x_0} y' + \frac{b_2(x)}{(x - x_0)^2} y = 0$$

where $b_1(x)$ and $b_2(x)$ are analytic at x_0.

Example 1   The Legendre equation

$$y'' - \frac{2x}{1 - x^2} y' + \frac{\alpha(1 + \alpha)}{1 - x^2} y = 0$$

---

[1] Some authors use these terms with somewhat different meanings. See the references at the end of the chapter.

has a singular point at $x_0 = 1$. The function $a_1(x) = -2x/(1 - x^2)$ is not analytic but

$$(x - 1)a_1(x) = \frac{2x}{x + 1}$$

is analytic at $x_0 = 1$ as is

$$(x - 1)^2 a_0(x) = -\alpha(1 + \alpha)\frac{x - 1}{x + 1}$$

Example 2 For the equation introduced in Example 1 of Section 16.4

$$\frac{d^2 y}{dr^2} + \frac{1}{r}\frac{dy}{dr} - \frac{n^2}{r^2} y = 0$$

$a_1(r) = r^{-1}$, $a_0(r) = -n^2 r^{-2}$. Since $ra_1(r) = 1$, and $r^2 a_0(r) = -n^2$ are analytic, $r = 0$ is a regular singular point.

We may say in somewhat imprecise language that $x = x_0$ is a regular singular point if $a_{n-1}(x)$ has a singular point that is "cancelled out" by $(x - x_0)$, or $a_{n-2}(x)$ a singular point that is "canceled out" by $(x - x_0)^2$, and so on. If a singularity is "worse" than this, it is called an irregular singular point. The following two equations have irregular singular points at $x = 0$.

$$x^3 y'' + xy' + y = 0 \quad \text{and} \quad \sqrt{x}\, y'' + y' + y = 0$$

Exercises

In each of the following equations locate all singular points and classify as regular or irregular.

1. $(x - 2)(x + 3)^2 y'' + 3x^2 y' - 2(x + 3)y = 0$

2. $(x^3 - x^2)y'' + 3(x - 1)y' + 4y = 0$

3. $(x^2 - 9)^3(x + 2)y^{(iv)} + 3(x - 3)y''' + x^2(x + 2)y'' + 4x^2 y' = 0$

4. $x^2 y'' + (\sin x)y' + (\cos x)y = 0$

5. $3x^2 y'' + 2xy' + (x + 1)y = 0$

6. $(x - 3)^2 y'' + 2xy' + (x - 3)y = 0$

7. $(e^x - 1)^2 y'' + 2(\sin x)y' + 3y = 0$

8. $y'' + 3y' + \sqrt{x}\, y = 0, x \geq 0$

16.6 Method of Frobenius

In order to find a series solution around x_0 of

$$Ly = y^{(n)} + a_{n-1}(x)y^{(n-1)} + \cdots + a_0(x)y = 0 \tag{1}$$

when x_0 is a regular singular point, we can argue as follows. The coefficient functions are not all analytic, but when multiplied by suitable powers of $(x - x_0)$ the resulting functions are analytic. Thus, for each k, $(x - x_0)^k a_{n-k}(x)$ is analytic, and therefore $a_{n-k}(x)$ is of the form

$$a_{n-k}(x) = (x - x_0)^{-k} \sum_{j=0}^{\infty} c_j(x - x_0)^j \qquad (x \neq x_0)$$

where the series converges near $x = x_0$. Thus, if the coefficients are analytic functions multiplied by negative powers of $(x - x_0)$, it is reasonable to guess that the solutions of Eq. (1) have the same form. Therefore, we may try to find solutions in the form

$$y = (x - x_0)^r \sum_{j=0}^{\infty} b_j(x - x_0)^j \tag{2}$$

In order to allow the maximum possible flexibility, we shall leave the number r as well as the coefficients b_j to be determined. We do not exclude the possibility that r may turn out to be irrational, or even complex. Our procedure will be to assume a solution of the form in Eq. (2), substitute into the differential equation, and equate coefficients of like powers of x to zero. We shall call this procedure the method of Frobenius.[2] Consider the following example.

Example 1

$$Ly = x^2 y'' + x y' + (x^2 - \tfrac{1}{9})y = 0 \tag{3}$$

There is a regular singular point at $x = 0$. Let us assume that there is a solution of the form

$$y = x^r \sum_{j=0}^{\infty} b_j x^j = \sum_{j=0}^{\infty} b_j x^{r+j} \qquad (x \neq 0) \tag{4}$$

Without loss of generality, we can assume that $b_0 \neq 0$ since otherwise we could factor out a power of x, increasing r and beginning the series with a nonzero constant term. Differentiation of Eq. (4) yields (see Exercise 4)

$$y' = \sum_{j=0}^{\infty} (r + j)b_j x^{r+j-1}, \quad y'' = \sum_{j=0}^{\infty} (r + j)(r + j - 1)b_j x^{r+j-2}$$

Substituting into Eq. (3), we get

$$\sum_{j=0}^{\infty} (r + j)(r + j - 1)b_j x^{r+j} + \sum_{j=0}^{\infty} (r + j)b_j x^{r+j}$$

$$+ \sum_{j=0}^{\infty} (b_j x^{r+j+2} - \tfrac{1}{9}b_j x^{r+j}) = 0$$

or

$$\sum_{j=0}^{\infty} [(r + j)(r + j - 1) + (r + j) - \tfrac{1}{9}]b_j x^{r+j} + \sum_{j=2}^{\infty} b_{j-2} x^{r+j} = 0$$

---

[2] Some authors use the phrase *method of Frobenius* with a different meaning.

Removing the common factor x^r, and then using the principle that a power series is zero only if every coefficient is zero, we obtain

$$[r(r - 1) + r - \tfrac{1}{9}]b_0 = 0$$
$$[(r + 1)r + (r + 1) - \tfrac{1}{9}]b_1 = 0$$
$$[(r + j)(r + j - 1) + (r + j) - \tfrac{1}{9}]b_j + b_{j-2} = 0, \qquad j = 2, 3, \ldots$$

These relations can be put in the form

$$(r^2 - \tfrac{1}{9})b_0 = 0$$
$$[(r + 1)^2 - \tfrac{1}{9}]b_1 = 0 \tag{5}$$
$$[(r + j)^2 - \tfrac{1}{9}]b_j = -b_{j-2}, \qquad j = 2, 3, \ldots$$

Since we assumed that $b_0 \neq 0$, the first of these equations implies

$$r^2 - \tfrac{1}{9} = 0 \tag{6}$$

Equation (6) is called the *indicial equation*, since it determines the indices $r = \pm\tfrac{1}{3}$. Note that the *indicial polynomial* $r^2 - \tfrac{1}{9}$ is the coefficient of x^r in Ly when y is the series in Eq. (4). Let us now show how the other coefficients can be determined. First suppose r is chosen as $\tfrac{1}{3}$. Then from the second relation in Eq. (5), we have

$$[(\tfrac{4}{3})^2 - \tfrac{1}{9}]b_1 = 0$$

Hence, $b_1 = 0$. The last relation in Eq. (5) gives us the general recurrence relation

$$b_j = \frac{-b_{j-2}}{(\tfrac{1}{3} + j)^2 - \tfrac{1}{9}} = \frac{-b_{j-2}}{j^2 + \tfrac{2}{3}j}, \qquad j = 2, 3, \ldots$$

since the denominator can never be zero. Clearly, $b_3, b_5, \ldots$ are zero, whereas

$$b_{2k} = (-1)^k b_0 \prod_{j=1}^{k} \frac{1}{4(j^2 + \tfrac{1}{3}j)}$$

Thus, we appear to have obtained a solution (putting $b_0 = 1$)

$$y_1(x) = x^{1/3}\left[1 - \frac{1}{4(1 + \tfrac{1}{3})}x^2 + \frac{1}{4^2(1 + \tfrac{1}{3})(4 + \tfrac{2}{3})}x^4 - \cdots\right]$$

The same kind of calculation starting from $r = -\tfrac{1}{3}$ yields

$$y_2(x) = x^{-1/3}\left[1 - \frac{1}{4(1 - \tfrac{1}{3})}x^2 + \frac{1}{4^2(1 - \tfrac{1}{3})(4 - \tfrac{2}{3})}x^4 - \cdots\right]$$

If $y_1(x)$ and $y_2(x)$ are truly solutions on an interval around $x = 0$ and are linearly independent, then $y(x) = c_1 y_1(x) + c_2 y_2(x)$ will be a general solution on this interval. But how can we be sure that y_1 and y_2 are solutions? First, since

$$\lim_{j \to \infty} \left| \frac{b_j x^j}{b_{j-2}x^{j-2}} \right| = \lim_{j \to \infty} \frac{x^2}{j^2 + \tfrac{2}{3}j} = 0$$

the series in the expression for $y_1(x)$ converges for all x by the ratio test. Likewise, the series in the expression for $y_2(x)$ converges for all x. Thus, $y_1(x)$ and $y_2(x)$ are well-defined for all $x \neq 0$ (and $y_1(x)$ even at $x = 0$). Moreover, since a power series $\Sigma b_j x^j$ can be differentiated term-by-term, a series $x^r \Sigma b_j x^j = \Sigma b_j x^{r+j}$ can also (see Exercise 4). Therefore, we can find the derivatives of y_1 or y_2, substitute back into Eq. (3), and verify that the equation is satisfied. In fact, this verification consists exactly of the steps we took in relation to the series in Eq. (4). We can therefore be certain that $y_1(x)$ and $y_2(x)$ are solutions. A proof of linear independence is outlined in the exercises in the next section.

Exercises

For each equation in Exercises 1 through 3, write the indicial polynomial and find its roots, and find a solution corresponding to each root. Verify each is a solution by substitution into the equation.

1. $x^2 y'' + x(x + \frac{1}{2})y' + x^3 y = 0.$

2. $2x^2 y'' + 3xy' + (x - 1)y = 0.$

3. $y'' + \dfrac{2}{x} y' + \dfrac{2}{9x^2} y = 0.$

4. A power series can be differentiated term-by-term; that is,

$$\frac{d}{dx}\left(\sum_{j=0}^{\infty} b_j x^j\right) = \sum_{j=0}^{\infty} jb_j x^{j-1}$$

in the interval of convergence, $|x| < \rho$. Deduce that

$$\frac{d}{dx}\left(x^r \sum_{j=0}^{\infty} b_j x^j\right) = \sum_{j=0}^{\infty} (j + r)b_j x^{r+j-1}$$

for $0 < x < \rho$ (and for $-\rho < x < 0$ if x^r is real).

16.7   General theorems

The example in the preceding section suggests that the method of Frobenius may be successfully applied in many cases. However, it is not immediately clear under what conditions, if any, it may fail; that is, will the coefficients b_j and the indices r always be determinable? If so, will the resulting series always converge and represent linearly independent solutions? The following theorems provide a satisfactory answer to these questions for second order linear equations. The proofs and extensions to higher order equations are beyond the scope of this text.

Consider the equation

$$Ly = y'' + a_1(x)y' + a_0(x)y = 0 \qquad (1)$$

and assume that $x = 0$ is a regular singular point (this is no loss of generality since a regular singular point at $x_0 \neq 0$ can be shifted to 0 by a translation).

DEFINITION 16.4   The *indicial polynomial* for Eq. (1) is

$$r(r - 1) + \beta_0 r + \alpha_0 \qquad (2)$$

where α_0 and β_0 are the constant terms in the expansions

$$xa_1(x) = \sum_{k=0}^{\infty} \beta_k x^k, \qquad x^2 a_0(x) = \sum_{k=0}^{\infty} \alpha_k x^k \qquad (3)$$

THEOREM 16.2   Suppose that $x = 0$ is a regular singular point for Eq. (1) and let the series expansions in Eq. (3) converge for $|x| < \rho_0$ ($\rho_0 > 0$). Let r_1 and r_2 be the roots of the indicial polynomial and let r_1 be the root with the larger real part, Re $(r_1) \geq$ Re (r_2). Then there is a solution y_1 of the form

$$y_1(x) = |x|^{r_1} \sum_{j=0}^{\infty} b_j x^j, \qquad b_0 = 1 \qquad (4)$$

convergent at least for $0 < |x| < \rho_0$. If $r_1 \neq r_2$ and if $r_1 - r_2$ is not a positive integer, there is a second linearly independent solution of the form

$$y_2(x) = |x|^{r_2} \sum_{j=0}^{\infty} c_j x^j, \qquad c_0 = 1 \qquad (5)$$

for $0 < |x| < \rho_0$. The coefficients b_j and c_j can be calculated by substitution into the differential equation.

In other words, this theorem assures us that the method of Frobenius will lead to a complete solution, provided $r_1 - r_2$ is not zero or a positive integer. Theorem 16.3 below describes the nature of solutions in the *exceptional case* when $r_1 - r_2$ is a nonnegative integer.

The series in Eqs. (4) and (5) are well-suited to the numerical calculation of the solutions near $x = 0$. In fact, the nearer x is to 0, the easier it is to compute accurate values for the power series. Also, the analytic behavior of the solution near $x = 0$ is apparent. For example, for the series solutions of the equation treated in the preceding section we have $y_1(x)$ approximately equal to $x^{1/3}$ and $y_2(x)$ approximately equal to $x^{-1/3}$ near $x = 0$.

THEOREM 16.3   Suppose that the hypotheses of Theorem 16.2 are satisfied except that $r_1 = r_2$ or $r_1 - r_2$ is a positive integer. Then there is a solution y_1 of the form in Eq. (4). If $r_1 = r_2$, there is a second linearly independent solution

y_2 of the form

$$y_2(x) = y_1(x) \ln |x| + |x|^{r_1+1} \sum_{j=0}^{\infty} c_j x^j \qquad (6)$$

where the series converges at least for $0 \le |x| < \rho_0$. If $r_1 - r_2$ is a positive integer, there is a second linearly independent solution y_2 of the form

$$y_2(x) = |x|^{r_2} \sum_{j=0}^{\infty} c_j x^j + dy_1(x) \ln |x|, \qquad c_0 = 1, \quad \text{for} \quad 0 < |x| < \rho_0 \qquad (7)$$

Here, d is a constant which may be zero.

In Theorem 16.3, the power series converge for $0 \le |x| < \rho_0$, but the expressions in Eqs. (6) and (7) are generally meaningless at $x = 0$.

Although these exceptional cases arise in certain applied problems, we shall confine our treatment of them to a single example, intended to show why the basic Frobenius method fails. The interested reader can consult the references at the end of the chapter for more details.

Example 1   Consider the equation

$$y'' - \frac{2}{x} y' + \frac{2+x}{x^2} y = 0$$

In this example, $xa_1(x) = -2$ and $x^2 a_0(x) = 2 + x$, so $\beta_0 = -2$, $\alpha_0 = 2$. The indicial polynomial $r(r-1) - 2r + 2$ has roots $r_1 = 2$, $r_2 = 1$. If we let $y = \sum_{j=0}^{\infty} b_j x^{r+j}$ we obtain

$$\sum_{j=0}^{\infty} b_j (r+j)(r+j-1)x^{r+j} - 2 \sum_{j=0}^{\infty} b_j (r+j)x^{r+j} + 2 \sum_{j=0}^{\infty} b_j x^{r+j}$$

$$+ \sum_{j=1}^{\infty} b_{j-1} x^{r+j} = 0$$

Hence,

$$b_0 [r(r-1) - 2r + 2] = 0$$
$$b_j [(r+j)(r+j-1) - 2(r+j) + 2] + b_{j-1} = 0, \qquad j = 1, 2, \ldots \qquad (8)$$

Taking $b_0 = 1$, the first equation yields the indicial equation. For $r = r_1 = 2$, the second equation becomes

$$(j^2 + j)b_j = -b_{j-1}, \qquad j = 1, 2, \ldots ,$$

from which we obtain the solution

$$y_1(x) = x^2 \left[1 - \frac{1}{1^2 + 1} x + \frac{1}{(1^2 + 1)(2^2 + 2)} x^2 - \cdots \right]$$

For $r = r_1 = 1$, Eq. (8) becomes

$$(j^2 - j)b_j = -b_{j-1}, \qquad j = 1, 2, \ldots$$

In particular, for $j = 1$ we get $0b_1 = -b_0 = -1$. It is impossible to satisfy this relation. This indicates that we cannot generate a second independent solution of the form $x^r \sum_{j=0}^{\infty} b_j x^j$ corresponding to the smaller indicial root $r = 1$. According to Theorem 16.3, there must be a second solution of the form in Eq. (7).

One point which should be made concerning the series solutions obtained in Theorems 16.2 and 16.3 is that the indicial roots r_1 and r_2 may be complex. In that case, the series will define complex-valued functions. In accordance with Theorem 9.4, linearly independent real solutions can be obtained by taking real and imaginary parts. Actually, the reader may question the meaning of $|x|^r$ when r is a complex number, $r = a + bi$. It can be shown that if this is defined, for real nonzero x, as

$$|x|^r = e^{r \ln |x|} = e^{a \ln |x|} \left[\cos (b \ln |x|) + i \sin (b \ln |x|) \right]$$

then our theorems remain valid.

Example 2

$$y'' + \frac{1}{x} y' + \frac{1 + x}{x^2} y = 0$$

In this example, $xa_1(x) = 1$ and $x^2 a_0(x) = 1 + x$. The indicial polynomial $r(r - 1) + r + 1$ has roots $r_1 = i, r_2 = -i$. Indeed, letting $y = \sum_{j=0}^{\infty} b_j x^{r+j}$, we obtain

$$(r^2 + 1)b_0 = 0$$

$$[(r + j)^2 + 1]b_j + b_{j-1} = 0, \qquad j = 1, 2, \ldots$$

If $r = i$, we get

$$b_j = \frac{-b_{j-1}}{1 + (i + j)^2} = \frac{-b_{j-1}}{2ij + j^2}, \qquad j = 1, 2, \ldots$$

Thus, one solution is

$$y_1(x) = |x|^i \left[1 - \frac{1}{1 + 2i} x + \frac{1}{(1 + 2i)(4 + 4i)} x^2 - \cdots \right]$$

$$= |x|^i \left[1 - \frac{1 - 2i}{5} x + \frac{(1 - 2i)(1 - i)}{40} x^2 - \cdots \right]$$

$$= [\cos (\ln |x|) + i \sin (\ln |x|)] \left[1 - \frac{1 - 2i}{5} x + \frac{-1 - 3i}{40} x^2 - \cdots \right]$$

By Theorem 16.2, the series converges for $|x| > 0$. The real and imaginary parts of $y_1(x)$ are real solutions, as follows:

$$\text{Re }(y_1(x)) = \cos(\ln|x|) - [\tfrac{1}{5}\cos(\ln|x|) + \tfrac{2}{5}\sin(\ln|x|)]x$$
$$+ \tfrac{1}{40}[-\cos(\ln|x|) + 3\sin(\ln|x|)]x^2 - \cdots$$

$$\text{Im }(y_1(x)) = \sin(\ln|x|) + [\tfrac{2}{5}\cos(\ln|x|) - \tfrac{1}{5}\sin(\ln|x|)]x$$
$$+ \tfrac{1}{40}[-3\cos(\ln|x|) - \sin(\ln|x|)]x^2 - \cdots$$

In this example, use of $r = -i$ will lead to a solution $y_2(x)$, which is the conjugate to $y_1(x)$; that is,

$$y_2(x) = [\cos(\ln|x|) - i\sin(\ln|x|)]\left[1 - \frac{1 + 2i}{5}x + \frac{-1 + 3i}{40}x^2 - \cdots\right]$$

Taking real and imaginary parts does not lead to any new independent real solutions.

Exercises

1. An equation of the form $x^2 y'' + \alpha x y' + \beta y = 0$, where α and β are real numbers, is often called an Euler equation. Find two linearly independent solutions of the equation in each of the following cases by applying the method of Frobenius.

   (a) $\alpha = \tfrac{1}{2}$, $\beta = -\tfrac{1}{2}$
   (b) $\alpha = -5$, $\beta = 9$

2. State and prove a general result on the solution of Euler's equation, including all possible cases.

3. Solve $x^2 y'' + x y' + (x - 2)y = 0$ in the neighborhood of zero.

4. Discuss the solution of Legendre's equation $(1 - x^2)y'' - 2xy' + \alpha(\alpha + 1)y = 0$ in the neighborhoods of $x = 1$ and $x = -1$.

5. Let y_1 and y_2 be the solutions constructed in the example in Section 16.6 and write

$$y_1(x) = x^{1/3}\sum_{k=0}^{\infty} b_{2k}x^{2k}, \qquad y_2(x) = x^{-1/3}\sum_{k=0}^{\infty} c_{2k}x^{2k}$$

Show that y_1 and y_2 are linearly independent. [*Hint:* If these are linearly dependent there are constants d_1 and d_2 such that $d_1 y_1(x) + d_2 y_2(x) = 0$ for all $x \neq 0$. Show that neither d_1 nor d_2 can be zero. Then $d_1 x^{1/3}y_1(x) + d_2 x^{1/3}y_2(x) = 0$ for $x \neq 0$. Let $x \to 0$ and deduce a contradiction.]

6. State and prove a general theorem concerning linear independence of functions

$$y_1(x) = x^{r_1} \sum_{j=0}^{\infty} b_j x^j, \qquad y_2(x) = x^{r_2} \sum_{j=0}^{\infty} c_j x^j \quad (b_0 = c_0 = 1)$$

on $x > 0$, if r_1 and r_2 are real and $r_1 \neq r_2$. Is the same argument valid for complex r_1 and r_2?
[*Hint:* See Exercise 5.]

For each equation in Exercises 7 through 9, write the indicial polynomial and find its roots. Write the form of two linearly independent solutions without computing the coefficients, and describe the limiting behavior as $x \to 0$.

7. $y'' + \left(\dfrac{5}{2x}\right) y' + \dfrac{1}{2x} y = 0.$

8. $x^2 y'' + 4xy' + (2 - x)y = 0.$

9. $x^2 y'' + (1 - 6x)y = 0.$

10. Find the first three nonzero terms in two independent solutions (real-valued) of the equation $x^2 y'' + (3x - x^2)y' + 2y = 0.$

11. Show that the following equation has an irregular singular point at $x = 0$. Try to apply the method of Frobenius and show in what respect it fails.

$$x^3 y'' + (x + 2x^2)y' - (1 + x)y = 0$$

12. The equation

$$Ly = x^2 y'' + xy' + x^2 y = 0$$

is called Bessel's equation of order zero. Find the indicial roots and show that there is a solution of the form

$$y = 1 + \sum_{j=1}^{\infty} \frac{(-1)^j x^{2j}}{2^{2j}(j!)^2}$$

This function is known as Bessel's function of the first kind of order zero and is customarily denoted by $J_0(x)$.

13. By Theorem 16.3, another solution of Bessel's equation of order zero will have the form, for $x > 0$,

$$y = J_0(x) \ln x + x \sum_{j=0}^{\infty} c_j x^j = J_0(x) \ln x + \sum_{j=1}^{\infty} b_j x^j$$

(a) Show that

$$Ly = 2x J_0'(x) + \sum_{j=1}^{\infty} j^2 b_j x^j + \sum_{j=3}^{\infty} b_{j-2} x^j$$

(b) Deduce that

$$y = J_0(x) \ln x + \sum_{j=1}^{\infty} \frac{(-1)^{j+1}h_j}{2^{2j}(j!)^2} x^{2j}, \qquad x > 0$$

where

$$h_j = 1 + \frac{1}{2} + \cdots + \frac{1}{j}$$

14. The equation

$$Ly = x^2 y'' + xy' + (x^2 - \tfrac{1}{4})y = 0$$

is called Bessel's equation of order one half. Show that

$$y_1(x) = \left(\frac{2}{\pi x}\right)^{1/2} \sin x$$

$$y_2(x) = \left(\frac{2}{\pi x}\right)^{1/2} \cos x$$

are linearly independent solutions for $x > 0$.

15. Show that for Bessel's equation of order one half the indicial roots are $r_1 = \frac{1}{2}$, $r_2 = -\frac{1}{2}$, which differ by a positive integer. Show, nevertheless, that both solutions in Exercise 14 can be derived by the method of Frobenius without the necessity of introducing a logarithm term.

16. The equation

$$x^2 y'' + xy' + (x^2 - \alpha^2)y = 0 \qquad (\alpha > 0)$$

is called Bessel's equation of order α. Show that the indicial roots are $r_1 = \alpha$, $r_2 = -\alpha$. By computing the series solution corresponding to the larger root, show that one solution on $x > 0$ is

$$x^\alpha \left[1 + \sum_{j=1}^{\infty} \frac{(-1)^j}{j!\,(j + \alpha)(j + \alpha - 1) \cdots (2 + \alpha)(1 + \alpha)} \left(\frac{x}{2}\right)^{2j} \right]$$

If 2α is not an integer, show that a second solution is

$$x^{-\alpha} \left[1 + \sum_{j=1}^{\infty} \frac{(-1)^j}{j!\,(j - \alpha)(j - \alpha - 1) \cdots (2 - \alpha)(1 - \alpha)} \left(\frac{x}{2}\right)^{2j} \right]$$

16.8 Series solutions for large *x*

In many applications, we are given a differential equation and are asked to furnish a description of the solutions as $x \to \pm\infty$; that is, we may want to know whether the solutions tend to a constant, or oscillate periodically, or tend to infinity, and so on. This kind of information is very hard to deduce from a power series expansion around, say, $x = 0$ or $x = 1$. The following device is often helpful in such a case. We shall illustrate for the second order equation

$$Ly = y'' + a_1(x)y' + a_0(x)y = 0 \tag{1}$$

Let us make the change of independent variable $s = 1/x$. There is clearly a one-to-one correspondence between x and s for $x > 0$, $s > 0$ (or for $x < 0$, $s < 0$). Under this correspondence any solution $y(x)$ of Eq. (1) is transformed into a function $z(s)$. Moreover,

$$y' = \frac{dz}{ds}\frac{ds}{dx} = -\frac{1}{x^2}\frac{dz}{ds} = -s^2\frac{dz}{ds}$$

$$y'' = -\frac{ds}{dx}\frac{d}{ds}\left(s^2\frac{dz}{ds}\right) = s^2\frac{d}{ds}\left(s^2\frac{dz}{ds}\right)$$

$$= s^4\frac{d^2z}{ds^2} + 2s^3\frac{dz}{ds}$$

From Eq. (1) we find that $z(s)$ satisfies

$$Mz = s^4\frac{d^2z}{ds^2} + (2s^3 - s^2 a_1(1/s))\frac{dz}{ds} + a_0(1/s)z = 0 \tag{2}$$

Here, of course, M is a linear differential operator.

DEFINITION 16.5 Equation (1) is said to have an *ordinary point at* $x = \infty$ if Eq. (2) has an ordinary point at $s = 0$, and a *singular point at* $x = \infty$ if Eq. (2) has a singular point at $s = 0$. The singular point at ∞ is called *regular* if Eq. (2) has a regular singular point at $s = 0$.

Example 1 The equation with constant coefficients $y'' + by' + cy = 0$ transforms into $s^4 z'' + (2s^3 - bs^2)z' + cz = 0$ (where now the prime means d/ds), for which $s = 0$ is an irregular singular point unless $b = c = 0$. On the other hand, Legendre's equation

$$y'' - \frac{2x}{1 - x^2}y' + \frac{\alpha(1 + \alpha)}{1 - x^2}y = 0$$

transforms into

$$\frac{d^2z}{ds^2} + \left[\frac{2}{s} + \frac{2}{s(s^2 - 1)}\right]\frac{dz}{ds} + \frac{\alpha(\alpha + 1)}{s^2(s^2 - 1)}z = 0$$

Since this has a regular singular point at $s = 0$, Legendre's equation has a regular singular point at $x = \infty$.

If the transformed Eq. (2) has an ordinary point at $s = 0$, its general solution is of the form

$$z = \sum_{k=0}^{\infty} b_k s^k = b_0 z_1(s) + b_1 z_2(s) \tag{3}$$

and the series converges in some interval, say $|s| < s_0$. It then follows that

$$y = \sum_{k=0}^{\infty} \frac{b_k}{x^k} = b_0 y_1(x) + b_1 y_2(x) \tag{4}$$

is convergent for $|x| > s_0^{-1}$ and defines a solution of Eq. (1). Since z_1, z_2 are linearly independent on $|s| < s_0$, y_1 and y_2 are linearly independent on $|x| > s_0^{-1}$. Therefore Eq. (4) defines a general solution of Eq. (1). Likewise, if Eq. (2) has a regular singular point at $s = 0$, we can use the method of Frobenius to obtain series solutions valid for $0 < s < s_0$. Replacing s by x^{-1}, we then get series solutions of Eq. (1) valid for $s_0^{-1} < x < \infty$.

Example 2   Consider the equation

$$y'' + \left(\frac{2}{x^3} + \frac{2}{x} \right) y' + \frac{\lambda}{x^4} y = 0$$

Here $a_1(1/s) = 2s^3 + 2s$, $a_0(1/s) = \lambda s^4$, and the transformed equation is

$$s^4 \frac{d^2 z}{ds^2} - 2s^5 \frac{dz}{ds} + \lambda s^4 z = 0$$

or

$$\frac{d^2 z}{ds^2} - 2s \frac{dz}{ds} + \lambda z = 0$$

This equation has an ordinary point at $s = 0$. In fact, it is the Hermite differential equation and, as shown in Example 3 of Section 16.2, it has two solutions

$$z_1(s) = 1 - \frac{\lambda}{2} s^2 + \frac{(4 - \lambda)(0 - \lambda)}{4!} s^4 + \cdots$$

and

$$z_2(s) = s + \frac{2 - \lambda}{3!} s^3 + \frac{(6 - \lambda)(2 - \lambda)}{5!} s^5 + \cdots$$

which converge for all s. Consequently, the functions

$$y_1(x) = 1 - \frac{\lambda}{2} x^{-2} + \frac{(4 - \lambda)(0 - \lambda)}{4!} x^{-4} + \cdots$$

$$y_2(x) = x^{-1} + \frac{2 - \lambda}{3!} x^{-3} + \frac{(6 - \lambda)(2 - \lambda)}{5!} x^{-5} + \cdots$$

are independent solutions of the original equation. These series converge for $x \neq 0$. These expansions make evident such relations as

$$\lim_{x \to \infty} y_1(x) = 1, \qquad \lim_{x \to \infty} xy_2(x) = 1$$

Exercises

In Exercises 1 through 3, determine whether $x = \infty$ is an ordinary, regular singular, or irregular singular point for each equation.

1. $xy'' + 3y' + 2xy = 0.$

2. $x^5 y'' + 2x^4 y' + y = 0.$

3. $x^2 y'' + xy' + (x^2 - \alpha^2)y = 0$ (Bessel's equation).

4. What conditions on $a_0(x)$ and $a_1(x)$ are necessary and sufficient in order that the equation $Ly = y'' + a_1(x)y' + a_0(x)y = 0$ have an ordinary point at $x = \infty$?

5. For the transformed form of Legendre's equation (see Example 1) compute the indicial roots and describe the behavior of solutions near $s = 0$. What can you say about the solutions of Legendre's equation as $x \to \infty$?

6. Show that the *hypergeometric equation* of Gauss

$$(x - x^2)y'' + [\gamma - (\alpha + \beta + 1)x]y' - \alpha\beta y = 0$$

where α, β, γ are real constants has regular singular points at $x = 0$, 1, ∞, and no other singular points.

7. Show that the equation $x^3 y'' + \tfrac{1}{2}x^2 y' + y = 0$ has a regular singular point at $x = \infty$. Compute two linearly independent solutions in the form of series in negative powers of x. For what values of x do these series converge? What is the behavior of the solutions as $x \to \infty$?

16.9   Nonlinear equations

Although Theorems 16.1, 16.2, and 16.3 are stated for linear differential equations, the principle of solution by infinite series is of great value for certain nonlinear equations.

Example 1 Consider the initial value problem, $y' = 1 + y^2$, $y(0) = 1$. If we let

$$y = \sum_{j=0}^{\infty} b_j x^j$$

and assume that this power series converges, we deduce

$$y^2 = \left(\sum_{j=0}^{\infty} b_j x^j\right)\left(\sum_{k=0}^{\infty} b_k x^k\right) = \sum_{j=0}^{\infty}\sum_{k=0}^{\infty} b_j b_k x^{j+k}$$

$$= b_0^2 + 2b_{1\ 0}x + (b_1^2 + 2b_0 b_2)x^2 + \cdots$$

$$y' = \sum_{j=0}^{\infty} jb_j x^{j-1}$$

Substituting in the equation, we obtain

$$b_1 + 2b_2 x + 3b_3 x^2 + \cdots = 1 + b_0^2 + 2b_1 b_0 x + (b_1^2 + 2b_0 b_2)x^2 + \cdots$$

Using $y(0) = b_0 = 1$ and equating coefficients, we have

$$b_1 = 1 + b_0^2 = 2$$
$$b_2 = b_0 b_1 = 2$$
$$b_3 = \tfrac{1}{3}(b_1^2 + 2b_0 b_2) = \tfrac{8}{3}$$

It is clear that each b_j depends on its known predecessors and therefore that all the b_j are uniquely determined. In fact, since we can write

$$y^2 = \sum_{n=0}^{\infty}\left(\sum_{j=0}^{n} b_j b_{n-j}\right) x^n$$

we find that the general recurrence relation is

$$(n + 1)b_{n+1} = \sum_{j=0}^{n} b_j b_{n-j}, \qquad n = 1, 2, \ldots$$

Unfortunately, this is a *nonlinear* relation connecting b_{n+1} to all the preceding b_j's. Although the b's can be computed one-by-one, it is not easy to find a general expression for b_n or to verify that the resulting series converges and defines a solution of the equation. In this particular problem, the differential equation is separable. We find $dy/(1 + y^2) = dx$, and obtain the solution, $y = \tan(x + \pi/4)$. It is left as an exercise to show that the series expansion of this function around $x = 0$ agrees (to the number of terms shown) with the series computed above,

$$y = \sum_{j=0}^{\infty} b_j x^j = 1 + 2x + 2x^2 + \tfrac{8}{3}x^3 + \cdots$$

In more advanced courses on differential equations, the following general result is proved.

THEOREM 16.4   Consider the problem

$$y' = f(x, y) \qquad y(0) = c \tag{1}$$

and assume that f is a polynomial in x and y. Then the initial value problem has a unique solution which exists and is analytic in some interval $|x| < \rho$. The coefficients in the series expansion

$$y = c + \sum_{j=1}^{\infty} b_j x^j$$

can be computed by substitution into the differential equation.

This theorem appears to guarantee the success of the power series method for a wide class of first order nonlinear equations, namely, those where f is a polynomial. There are analogous theorems for a more general class of functions f and for higher order equations. Unfortunately, there is no known method of determining, or even obtaining a very good estimate of, the radius of convergence ρ in advance. Thus although it is not too difficult to compute a few terms in the series, it is dangerous to use these to compute numerical values for the solution. In the realm of nonlinear equations, careful checks on accuracy are essential.

If one goes on to nonlinear equations with singular points, the theory of series solutions becomes quite complicated. This is a vast area of mathematics, and we can only refer the reader to treatises for more thorough discussions.

Exercises

1. Find the first few terms in the series representation of the solution:

   (a) $y' = 1 + xy$, $y(0) = 1$

   (b) $y' = e^y$, $y(0) = 0$ (compare with the exact solution)

   (c) $y' = \dfrac{1}{(1 - x)(1 - y)}$, $y(0) = 0$

   (d) $y'' = x + y^2$, $y(0) = 0$, $y'(0) = 0$

2. *Numerical study.* Find the series solution of $y' = y^2$, $y(0) = c$ for each of the values $c = 0.1, 1, 2, 10$. Write a computer program to compute the value of the series for values of x on, say $0 < x < 1$, and study the results obtained. In each case, compare with the exact solution.

3. Perform the verification suggested in Example 1, namely, that the first several terms in the series expansion of $\tan (x + \pi/4)$ are 1, $2x$, $2x^2$, and $8x^3/3$.

16.10 Perturbation series

The preceding sections have dealt with one basic idea, that of constructing a solution $y(x)$ of an equation in the form of an infinite series. The series were of several types—power series in $(x - x_0)$, power series multiplied by $(x - x_0)^r$, or power series in $(x - x_0)^{-1}$—depending on the particular problem under consideration. There is still another kind of series which is of great importance in many applications. This is illustrated in the following simple example.

Example 1 Consider the initial value problem

$$\frac{dy}{dx} + (1 + \varepsilon)y = 0, \qquad y(0) = 1 \tag{1}$$

where ε is a real number, a parameter. The exact solution of this equation is, of course, $y = e^{-(1+\varepsilon)x}$. However, for illustrative purposes we shall examine the series solution, which is

$$y(x) = 1 - \frac{(1 + \varepsilon)x}{1!} + \frac{(1 + \varepsilon)^2 x^2}{\cdot 2!} - \cdots \tag{2}$$

Let us suppose that $\varepsilon = 0.1$ and the solution is desired for $0 \le x \le 10$. A large number of terms will be required to obtain even modest accuracy. For example, for $n = 20$, $x = 10$, the error after n terms is

$$\text{error} = -\frac{11^{21}}{21!} + \frac{11^{22}}{22!} - \frac{11^{23}}{23!} + \cdots$$

Since this is an alternating series and each term is smaller than its predecessor, we have

$$|\text{error}| = \frac{11^{21}}{21!} - \frac{11^{22}}{22!} + \frac{11^{23}}{23!} - \cdots$$

$$> \frac{11^{21}}{21!} - \frac{11^{22}}{22!} = \frac{1}{2}\frac{11^{21}}{21!} \ge \frac{10^{21}}{2(21!)} = 9.79$$

Clearly, a better approach is desirable. One possibility is to use the series method to obtain several series for y, for example, one expansion around $x = 0$, another expansion around $x = 1$, another around $x = 2$, and so on. Each series might then converge rapidly enough to give accurate values near its expansion point.

However, we wish to introduce a different approach, based on expanding y in a power series in ε; that is, instead of regarding y as a function of x alone, regard y as a function of both x and ε. Write

$$y = \sum_{k=0}^{\infty} y_k(x)\varepsilon^k \tag{3}$$

where $y_0(x)$, $y_1(x)$, ... are functions of x which are to be determined. Differentiation of Eq. (3) yields

$$y' = \frac{dy}{dx} = \sum_{k=0}^{\infty} y_k'(x)\varepsilon^k$$

assuming that it is valid to differentiate the infinite series term-by-term. Now substitution into the differential equation (1) gives

$$\sum_{k=0}^{\infty} y_k{}'(x)\varepsilon^k + \sum_{k=0}^{\infty} y_k(x)\varepsilon^k + \sum_{k=0}^{\infty} y_k(x)\varepsilon^{k+1} = 0,$$

or

$$(y_0{}' + y_0) + (y_1{}' + y_1 + y_0)\varepsilon + (y_2{}' + y_2 + y_1)\varepsilon^2 + \cdots = 0$$

Equating the coefficients of powers of ε to zero gives

$$y_0{}' + y_0 = 0$$
$$y_1{}' + y_1 = -y_0$$

and, in general,

$$y_k{}' + y_k = -y_{k-1} \qquad k = 1, 2, \ldots \qquad (4)$$

Similarly, from Eq. (3) for $x = 0$ and the initial condition $y(0) = 1$, we get

$$1 = y(0) = y_0(0) + y_1(0)\varepsilon + y_2(0)\varepsilon^2 + \cdots$$

Again equating coefficients of like powers of ε, we obtain

$$y_0(0) = 1, \quad y_k(0) = 0, \qquad k = 1, 2, \ldots \qquad (5)$$

The procedure has led to an infinite sequence of differential equations, which at first appears to be a long leap backward. However, solving this sequence of initial value problems gives

$$y_0(x) = e^{-x}$$
$$y_k(x) = (-1)^k \frac{x^k}{k!} e^{-x}, \qquad k = 1, 2, \ldots$$

and hence Eq. (3) yields

$$y = e^{-x} - \frac{\varepsilon x}{1!} e^{-x} + \frac{\varepsilon^2 x^2}{2!} e^{-x} - \cdots + (-1)^k \frac{\varepsilon^k x^k}{k!} e^{-x} + \cdots \qquad (6)$$

If $0 \le x \le 10$, $\varepsilon = 0.1$, this is an alternating series with each term less than the term before. Hence the error after n terms is bounded by

$$|\text{error}| \le \frac{\varepsilon^n x^n e^{-x}}{n!} \le \frac{1}{n!}$$

For example, for $n = 5$, the error is less than $1/120$ over the whole range. A few terms in the series in Eq. (6) provide accurate results for all x in the interval $0 \le x \le 10$.

The series (6) is e^{-x} times the series for $e^{-\varepsilon x}$, and thus checks the known exact solution. The method works so well because it takes advantage of the fact that the solution for $\varepsilon = 0$ is known.

Example 2 Consider the problem

$$y'' + (1 + \varepsilon x)y = 0, \quad y(0) = 1, \quad y'(0) = \varepsilon \tag{7}$$

The method of power series expansion yields

$$y = 1 + \varepsilon x - \frac{1}{2} x^2 - \frac{1}{3} \varepsilon x^3 + \frac{1 - 2\varepsilon^2}{4!} x^4 + \cdots$$

On the other hand, if $\varepsilon = 0$ the problem is simply $y'' + y = 0$, $y(0) = 1$, $y'(0) = 0$, and the solution is $y = \cos x$. It is therefore reasonable to suppose that for small values of ε the solution will be a slight deviation or perturbation of the function $\cos x$. Put

$$y = y_0(x) + \varepsilon y_1(x) + \varepsilon^2 y_2(x) + \cdots$$
$$y'' = y_0''(x) + \varepsilon y_1''(x) + \varepsilon^2 y_2''(x) + \cdots$$

Then substitution gives

$$(y_0'' + \varepsilon y_1'' + \varepsilon^2 y_2'' + \cdots) + (y_0 + \varepsilon y_1 + \varepsilon^2 y_2 + \cdots)$$
$$+ \varepsilon x(y_0 + \varepsilon y_1 + \varepsilon^2 y_2 + \cdots) = 0$$
$$y_0(0) + \varepsilon y_1(0) + \varepsilon^2 y_2(0) + \cdots = 1$$
$$y_0'(0) + \varepsilon y_1'(0) + \varepsilon^2 y_2'(0) + \cdots = \varepsilon$$

Equating coefficients of like powers of ε yields

$$y_0'' + y_0 = 0$$
$$y_k'' + y_k = -xy_{k-1}, \quad k = 1, 2, \ldots$$
$$y_0(0) = 1, \qquad y_0'(0) = 0$$
$$y_1(0) = 0, \qquad y_1'(0) = 1$$
$$y_k(0) = y_k'(0) = 0, \qquad k = 2, 3, \cdots$$

Solving these equations, we get $y_0 = \cos x$ and then

$$y_1'' + y_1 = -x \cos x$$

By the methods of Chapter 9 we find a general solution of this equation is

$$y_1 = A \cos x + B \sin x - \tfrac{1}{4} x \cos x - \tfrac{1}{4} x^2 \sin x$$

The solution satisfying the given initial conditions is

$$y_1 = \tfrac{5}{4} \sin x - \tfrac{1}{4} x \cos x - \tfrac{1}{4} x^2 \sin x$$

Thus

$$y = \cos x + \frac{\varepsilon}{4} (5 \sin x - x \cos x - x^2 \sin x) + \cdots \tag{8}$$

Since

$$(5 \sin x - x \cos x - x^2 \sin x) \le 5 + |x| + x^2$$

and this is at most 35 for $0 \leq x \leq 5$, the term in ε has magnitude at most 0.1 if $\varepsilon \leq 0.01$. Since this is small relative to cos x for most x in the interval, it is a reasonable guess that the two terms written in Eq. (8) provide an approximation to the exact solution for $0 \leq x \leq 5$, $\varepsilon \leq 0.01$. If greater accuracy is desired or a larger x-interval is to be studied, it may be necessary to calculate the term in ε^2.

We shall now describe the new method in more general terms. Suppose that we are given a linear equation of the form

$$Ly = y'' + a_1(x, \varepsilon)y' + a_0(x, \varepsilon)y = f(x, \varepsilon) \tag{9}$$

where a_1, a_0, f are given functions of x and ε, along with initial conditions of the form

$$y(x_0, \varepsilon) = z_0(\varepsilon), \qquad y'(x_0, \varepsilon) = z_1(\varepsilon) \tag{10}$$

where z_0 and z_1 are given functions of ε (often these are constants, independent of ε). We then seek a solution in the form

$$y(x, \varepsilon) = \sum_{k=0}^{\infty} y_k(x)\varepsilon^k \tag{11}$$

Such a series is called a *perturbation series* because the method is usually applied where the problem in (9) and (10) reduces, for $\varepsilon = 0$, to one for which the solution is known exactly. The problem then can be regarded as a perturbation or deviation from a known situation.

The solution technique is to differentiate Eq. (11) term-by-term and substitute into Eq. (9). On expanding the $a_j(x, \varepsilon)$ and $f(x, \varepsilon)$ in series of powers of ε, we can then equate coefficients of like powers of ε. The result will be a sequence of differential equations for the unknowns $y_k(\varepsilon)$. Initial conditions for these equations are found by substituting Eq. (11) into (10) and equating coefficients of like powers of ε.

Any thorough study of the perturbation method would involve questions such as these: Under what conditions does the series (11) converge? For what range of values of x and ε does it converge? If it converges, is its sum really a solution of the original problem? If only a few terms in the series are retained, what error estimates are known? The following theorem answers some of these questions.

THEOREM 16.5 Consider the linear equation (9) with initial conditions (10). Assume that $a_0(x, \varepsilon)$, $a_1(x, \varepsilon)$ and $f(x, \varepsilon)$ are polynomials in x and ε and let $z_0(\varepsilon)$ and $z_1(\varepsilon)$ be polynomials in ε. Then there is a unique solution $y(x, \varepsilon)$ which possesses an expansion of the form in Eq. (11), convergent to $y(x, \varepsilon)$ for all x and ε. Moreover, $y(x, 0)$ or $y_0(x)$ is the solution of the problem (9), (10) for $\varepsilon = 0$; that is,

$$y_0'' + a_1(x, 0)y_0' + a_0(x, 0)y_0 = f(x, 0)$$
$$y_0(x_0) = z_0(0), \qquad y_0'(x_0) = z_1(0)$$

For a proof of this theorem, see the references given at the end of the chapter.

Theorem 16.5 provides assurance that the method of perturbation series leads to a solution for the linear equation (9), but tells us nothing about the rapidity of convergence of the series or error estimates. There is an extension of the theorem valid for nonlinear differential equations, but then the region of convergence for the series is not precisely known. Often, it is necessary to use the method in applied problems without complete a priori knowledge about the validity of the procedure. In such cases, confidence in calculations may come from comparisons with approximations obtained by other methods, or with experimental results, or from a knowledge that the method has worked well in similar cases.

Exercises

1. Construct a solution by the perturbation method and compare with the exact solution.

    (a) $y' + 2\varepsilon xy = 0, \quad y(0) = 1$
    (b) $y'' + \varepsilon y = 0, \quad y(0) = 1, y'(0) = 0$

2. Construct a solution by the perturbation method of the problem

    $$y'' + y + \varepsilon y^3 = 0, \qquad y(0) = \varepsilon, \quad y'(0) = 0$$

    Verify that for $\varepsilon = 0$ this solution reduces to the solution of $y'' + y = 0$, $y(0) = 0, y'(0) = 0$.

3. Calculate y_0 and y_1 in the perturbation series for Van der Pol's equation

    $$y'' + \varepsilon(y^2 - 1)y' + y = 0, \quad y(0) = 1, \quad y'(0) = 0$$

4. The basic idea of expanding in powers of a parameter can be applied to other kinds of equations. Obtain the first four terms in an expansion

    $$x = \sum_{k=0}^{\infty} c_k \varepsilon^k$$

    for a solution of the equation $x = 1 + \varepsilon x^2$ by substituting the series in the equation. Compare the result with series expansions of the exact solutions

    $$x_1 = \frac{1}{2\varepsilon}(1 + \sqrt{1 - 4\varepsilon}), \qquad x_2 = \frac{1}{2\varepsilon}(1 - \sqrt{1 - 4\varepsilon})$$

COMMENTARY

For proofs of the theorems, discussions of the "exceptional cases," and extensions to equations of higher order, see:

CODDINGTON, E. A., *An Introduction to Differential Equations*, Englewood Cliffs, N.J., Prentice-Hall, 1961.

INCE, E. L., *Ordinary Differential Equations*, London, Longmans, Green, 1927 (Reprint, New York, Dover, 1944 and 1953).

BIRKHOFF, G. D., and G.-C. ROTA, *Ordinary Differential Equations*, Boston, Ginn, 1962.

Section 16.8   Ince's text contains more thorough discussions of these topics.

Section 16.9   A somewhat more general form of Theorem 16.4 is proved in:
KAPLAN, W., *Ordinary Differential Equations*, Reading, Mass., Addison-Wesley, 1958.

Section 16.10   Theorem 16.5 provides a special case of Poincare's Expansion Theorem. See:
LEFSCHETZ, S., *Differential Equations: Geometric Theory*, New York, Interscience, 1957, Chapter II.

CODDINGTON, E. A., and N. LEVINSON, *Theory of Ordinary Differential Equations*, New York, McGraw-Hill, 1955, Chapter I, Section 8.

Extensive applications of the perturbation method are described in:
BELLMAN, R., *Perturbation Techniques in Mathematical Physics and Engineering*, New York, Holt, Rinehart and Winston, 1964.

COLE, J. D., *Perturbation Methods in Applied Mathematics*, Waltham, Mass., Blaisdell, 1968.

BOGOLIUBOV, N. N., and Y. A. MITROPOLSKY, *Asymptotic Methods in the Theory of Non-Linear Oscillations*, Delhi, Hindustan, 1961.

Algebra of complex numbers

A *complex number* is of the form $a + bi$, where a and b are real and $i^2 = -1$. Two complex numbers $a_1 + b_1i$ and $a_2 + b_2i$ are called *equal* if $a_1 = a_2$ and $b_1 = b_2$. There is an operation called *addition of complex numbers* defined as follows. Let $z_1 = a_1 + b_1i$ and $z_2 = a_2 + b_2i$. Then $z_1 + z_2 = (a_1 + a_2) + (b_1 + b_2)i$, which is again a complex number. The operation of *multiplication of two complex numbers* $z_1 = a_1 + b_1i$ and $z_2 = a_2 + b_2i$ is performed by the usual rules for algebraic multiplication with $i^2 = -1$. Hence

$$z_1 z_2 = (a_1 + b_1i)(a_2 + b_2i) = a_1a_2 - b_1b_2 + (a_1b_2 + a_2b_1)i$$

It can be shown that the set of all complex numbers, together with the above addition and multiplication, is a field (see Section 4.3 for definition). This field is denoted by $\mathbf{C}$ in this text.

Example 1 Let $z_1 = 3 + 2i$ and $z_2 = 2 - i$. Then

$$z_1 + z_2 = (3 + 2) + (2 - 1)i = 5 + i$$
$$z_1 z_2 = (3 + 2i)(2 - i) = (6 + 2) + (4 - 3)i = 8 + i$$

The *complex conjugate* of a number $z = a + bi$, denoted $\bar{z}$, is defined to be $\bar{z} = a - bi$. Note that $z\bar{z} = (a + bi)(a - bi) = a^2 + b^2$.

There is a geometrical representation of complex numbers. To a given complex number $z = a + bi$ we associate the point in a plane with abscissa a and ordinate b, relative to a rectangular coordinate system in the plane. In this way there is a one-to-one correspondence between the set of all complex numbers and the set of all points in the plane.

The *absolute value* or *modulus* of z, denoted $|z|$, is defined as $|z| = \sqrt{z\bar{z}} = (a^2 + b^2)^{1/2}$. Geometrically this is the polar distance r of (a, b) from the origin $(0, 0)$; that is, $|z| = r$. We also define the *argument* of z, denoted $\arg z$, to be

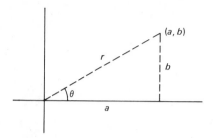

FIGURE 1

the polar angle θ in Figure 1. Of course, this definition is ambiguous since the polar pairs $(r, \theta + 2k\pi)$ represent the same point for any integer k. Whenever it is necessary to pick a particular angle, we shall so indicate. Now note that if $z = a + ib$ is any complex number, we have

$$a = r \cos \theta, \qquad b = r \sin \theta$$
$$z = r (\cos \theta + i \sin \theta) \tag{1}$$

Equation (1) is called the polar representation of z. Also

$$|z| = \sqrt{a^2 + b^2}$$
$$\arg z = \tan^{-1} \frac{b}{a} \tag{2}$$

Exercises

1. Perform the indicated operations on complex numbers.
    (a) $(2 - 3i) + (6 + 3i)$
    (b) $(3 + 0i)(10 - 2i)$
    (c) $(1 + i)(2 - 3i)$
    (d) $2(4 + i) - 3(-1 + 5i)$

2. Solve for x in the following equations.
    (a) $(x - i)(x + 2i) = 3 + i$
    (b) $x + i + i(x - 2i) = 2x + 3$

3. Let z_1 and z_2 be complex numbers such that $z_1 z_2 = 0$. (The zero complex number is $0 + 0i$, but there is no need to distinguish between this and the zero real number 0.) Show that either $z_1 = 0$ or $z_2 = 0$ or both.

4. The inverse of a real number a, $a \neq 0$, is that number b such that $ba = 1$. An inverse of a complex number z is a number w such that $zw = 1$. Prove that if $z \neq 0$, then $w = \bar{z}/|z|^2$ is an inverse of z. Also show that this is the only inverse of z. The inverse is denoted by $1/z$ or z^{-1}, as usual.

5. If $z = a + bi$, write z^{-1} in the form $c + di$, where c and d are expressed in terms of a and b. (See Exercise 4.)

6. A correspondence can be established between complex numbers and ordered pairs of real numbers by making $a + bi$ correspond to (a, b).

    (a) Show that addition of complex numbers corresponds to addition of pairs in the sense that if $z = a + bi$ corresponds to (a, b) and $w = c + di$ corresponds to (c, d), then $z + w$ corresponds to $(a, b) + (c, d)$.

    (b) If multiplication of ordered pairs is to correspond to multiplication of complex numbers, how must multiplication of ordered pairs be defined?

B

Polynomials

A polynomial can be defined as a function p from the complex numbers $\mathbf{C}$ into the complex numbers such that the value of p at z can be computed by the formula

$$p(z) = a_n z^n + a_{n-1} z^{n-1} + \cdots + a_1 z + a_0 \tag{1}$$

where n is a nonnegative integer and $a_n, \ldots, a_0$ are real or complex scalars.

DEFINITION B.1   A *root* or *zero* of a polynomial p is a number λ such that $p(\lambda) = 0$. The largest power of z with nonzero coefficient that appears in the expression defining a polynomial is called the *degree* of the polynomial.

Some basic facts about polynomials are stated without proof in the following theorem and corollaries.

THEOREM B.1   (Fundamental Theorem of Algebra)   Every polynomial of degree one or more has at least one root in $\mathbf{C}$.

COROLLARY 1   If p is a polynomial of degree $n(n \geq 1)$ with leading coefficient $a_n = 1$, and if λ is a root of p, then

$$p(z) = (z - \lambda)q(z) \tag{2}$$

where q is a polynomial of degree $n - 1$ with leading coefficient 1. Conversely, if Eq. (2) holds, then λ is a root of p.

COROLLARY 2   If p is a polynomial of degree n ($n \geq 1$), then there are exactly n numbers $\lambda_1, \lambda_2, \ldots, \lambda_n$ (not necessarily all different) which are roots of p and are such that

$$p(z) = a_n(z - \lambda_1)(z - \lambda_2) \cdots (z - \lambda_n) \tag{3}$$

This factorization is unique (except for the order of the factors).

DEFINITION B.2 If p has the factorization

$$p(z) = a_n(z - \lambda_1)^{m_1}(z - \lambda_2)^{m_2} \cdots (z - \lambda_k)^{m_k} \tag{4}$$

where $\lambda_1, \lambda_2, \ldots, \lambda_k$ are all different and $m_1, m_2, \ldots, m_k$ are positive integers, the root λ_j is said to have *multiplicity* m_j ($j = 1, 2, \ldots, k$). (The sum of the multiplicities is the degree of p.) A root with multiplicity $m_j = 1$ is called a *simple* root and a root with multiplicity m_j larger than one is called a *multiple root*.

THEOREM B.2 If p is a polynomial with real coefficients and $\lambda = c + id$ is a root, then the complex conjugate $\bar{\lambda} = c - id$ is also a root.

PROOF: We have

$$p(\lambda) = a_n(c + id)^n + \cdots + a_1(c + id) + a_0 = 0$$

Also since the product of two complex conjugates is the conjugate of the product,

$$\overline{(c + id)^2} = (c - id)^2$$

and, in general, by induction the conjugate of $(c + id)^k$ is $(c - id)^k$ for $k = 1, \ldots, n$. Hence

$$\overline{p(\lambda)} = a_n(c - id)^n + \cdots + a_1(c - id) + a_0 = \bar{0} = 0$$

Clearly, $c - id$ is a root of p. ∎

THEOREM B.3 If

$$p(z) = a_n z^n + \cdots + a_1 z + a_0, \qquad q(z) = b_n z^n + \cdots + b_1 z + b_0$$

and if $p(z) = q(z)$ at $n + 1$ distinct values of z, then p and q are the same polynomial; that is, $a_j = b_j$ ($j = 0, 1, \ldots, n$).

PROOF: Suppose $p(\lambda_j) = q(\lambda_j)$ for $j = 1, \ldots, n + 1$, where the λ_j are distinct. Then

$$p(z) - q(z) = (a_n - b_n)z^n + \cdots + (a_1 - b_1)z + (a_0 - b_0)$$

is a polynomial of degree at most n which has $n + 1$ distinct roots λ_j. This is impossible if p and q are different polynomials, since from Corollary 2, a polynomial of degree n can have only n roots. Hence, p and q are the same polynomial. ∎

If p is a real polynomial of degree n, we may want a factorization which involves only real numbers. In this case, we say that p is to be factored *over* **R**. It may not be possible to achieve a factorization involving only real linear factors. For example, $z^2 + 4$ cannot be factored into linear factors with real coefficients, and hence is called *irreducible over* **R**. The general situation is as follows.

THEOREM B.4 If p is a polynomial with real coefficients and degree n, then $p(z)$ has a unique factorization (except for order of factors) into linear and quadratic factors irreducible over **R**:

$$p(z) = a_n(z^2 + c_1 z + d_1) \cdots (z^2 + c_k z + d_k)(z - \lambda_{2k+1}) \cdots (z - \lambda_n)$$

where $c_1, \ldots, c_k, d_1, \ldots, d_k, \lambda_{2k+1}, \ldots, \lambda_n$ are real numbers, and $z^2 + c_j z + d_j$ has no real root ($j = 1, \ldots, k$).

THEOREM B.5 **(Division Algorithm)** Let $s(z)$ and $p(z)$ be given polynomials with $p(z) \neq 0$. Then there exist polynomials $q(z)$ (quotient) and $r(z)$ (remainder) such that

$$s(z) = p(z)q(z) + r(z)$$

and such that $r(z)$ is either zero or has degree less than the degree of $p(z)$.

For more information on polynomials the reader may consult textbooks on algebra, such as Neal McCoy, *Introduction to Modern Algebra*, Boston, Allyn and Bacon, 1968.

The practical determination of the roots of a polynomial is discussed in textbooks on numerical analysis, such as P. Henrici, *Elements of Numerical Analysis*, New York, Wiley, 1964.

Exercises

1. Using mathematical induction and Corollary 1, show that the polynomial p must have a factorization

    $$p(z) = a_n(z - \lambda_1)(z - \lambda_2) \cdots (z - \lambda_n)$$

2. Prove the uniqueness part of Corollary 2.

3. If $3 - 2i, 3 + 2i, 5, 2, \sqrt{2} - i, \sqrt{2} + i$ are the zeros of a polynomial p of degree 6, what are the irreducible factors of $p(z)$ over **R**? over **C**?

4. Let p be the polynomial $p(z) = a_2 z^2 + a_1 z + a_0$. In each part, show that the indicated function q is a polynomial and find its coefficients and its degree.

(a) $q(z) = [p(z)]^2$

(b) $q(z) = p(z + 1)$

(c) $q(z) = \displaystyle\int_0^z p(u)\, du$

(d) $q(z) = \dfrac{dp(z)}{dz}$

(e) $q(z) = p(z^2)$

5. Let p have a zero, λ, of multiplicity m. Show that the derivative p' has λ as a zero of multiplicity $m - 1$ if $m \geq 2$, but $p'(\lambda)$ is nonzero if $m = 1$.

Appendix

Infinite series

This appendix is devoted to a sketchy discussion of those results on infinite series necessary for an understanding of Chapter 16. For a more thorough discussion the reader is referred to any calculus text.[1]

DEFINITION C.1   A *sequence* is a function (real or complex-valued) whose domain is the positive integers. We denote the image of n by a_n, and call a_n the nth *term* in the sequence. The sequence will be denoted $\{a_1, a_2, \dots\}$ or briefly $\{a_n\}$.

DEFINITION C.2   A sequence is said to be *convergent* if $\lim_{n \to \infty} a_n$ exists; that is, $\{a_1, a_2, \dots\}$ converges if there is a number a such that for every $\varepsilon > 0$ there exists an N_ε such that $|a_n - a| < \varepsilon$ for all $n \geq N_\varepsilon$. The number a is called the *limit* of the sequence. A sequence which is not convergent is said to be *divergent*.

DEFINITION C.3   Consider a sequence $\{a_1, a_2, \dots\}$. Denote by s_n the nth *partial sum*, given by $a_1 + a_2 + \cdots + a_n$. Then $\{s_1, s_2, \dots\}$ forms a sequence. The limit of this sequence is a sum of the form $a_1 + a_2 + \cdots + a_n + \cdots$, which is called an *infinite series*.[2] If the sequence $\{s_1, s_2, \dots\}$ converges to a finite limit S we say the infinite series $\sum_{n=1}^{\infty} a_n$ is a convergent series and S is called the *sum* of the series and we write $S = \sum_{n=1}^{\infty} a_n$. Otherwise the series is said to be *divergent*.

---

[1] Lipman Bers, *Calculus*, New York, Holt, Rinehart and Winston, 1969; Robert James, *Advanced Calculus*. Belmont, Calif., Wadsworth, 1966.
[2] Note the ambiguity. If the sequence of partial sums is convergent, the infinite series exists; otherwise, the infinite series $a_1 + a_2 + \cdots + a_n + \cdots$ is a "formal series" without sum.

Example 1 Consider the sequence $\left\{ 1, \frac{1}{2}, \frac{1}{3}, \frac{1}{4}, \ldots, \frac{1}{n}, \ldots \right\}$, where

$a_n = 1/n$. It is easily seen that this sequence converges to zero, for take any $\varepsilon > 0$. Then for any $n \geq N_\varepsilon$, where N_ε is any integer greater than $1/\varepsilon$, $|a_n - 0| = 1/n \leq 1/N_\varepsilon < \varepsilon$. Next consider $S = \sum_{i=1}^{\infty} a_i$. It can be shown that this series diverges.

In contrast, the series

$$ 1 - \frac{1}{2} + \frac{1}{3} - \cdots + \frac{(-1)^{n-1}}{n} + \cdots $$

can be shown to converge. Hence, the following definition is appropriate.

DEFINITION C.4 A series $a_1 + a_2 + \cdots + a_n + \cdots$ is said to be *absolutely convergent* if $\sum_{i=1}^{\infty} |a_i|$ converges.

The series $1 - \frac{1}{2} + \frac{1}{3} - \cdots$ is convergent, but not absolutely convergent. We now turn to power series.

DEFINITION C.5 A series of the form

$$ S(x) = a_0 + a_1(x - x_0) + a_2(x - x_0)^2 + \cdots + a_n(x - x_0)^n + \cdots $$

is called a *power series* in $(x - x_0)$, or a power series in x *centered at* x_0, or a power series *about* x_0. The numbers a_n are called the coefficients of the power series. The power series $S(x)$ is said to converge at the point x_1 if the series $S(x_1)$ is a convergent series. Otherwise, the series is said to be *divergent*. Similarly, the power series is said to be *absolutely convergent* at x_1 if $S(x_1)$ is absolutely convergent.

Methods for determining whether series, and in particular power series, converge or diverge can be found in any elementary calculus text. We omit them here and turn instead to theorems which are of use in Chapter 16.

THEOREM C.1 If the power series $\sum_{n=0}^{\infty} a_n(x - x_0)^n$ converges at x_1, then it converges absolutely for all x such that $|x - x_0| < |x_1 - x_0|$. If the series diverges at x_1, then it diverges for all x such that $|x - x_0| > |x_1 - x_0|$.

A proof of this theorem depends upon the comparison test for convergence of infinite series.

Example 2 Consider the power series $1 + x^2/2 + x^3/3 + \cdots$. We have seen in Example 1 that when $x = 1$, the series diverges. Hence we can conclude that this series will diverge for all x for which $|x| > 1$. Further, the series converges for $x = -1$ and hence by Theorem C.1, the series converges absolutely for all x for which $|x| < 1$. However, at $x = +1$ the series diverges and at $x = -1$ the series converges.

We are led to the following definition.

DEFINITION C.6 The number ρ such that the series $\sum_{n=0}^{\infty} a_n(x - x_0)^n$ converges for all x such that $|x - x_0| < \rho$, and diverges for all x for which $|x - x_0| > \rho$, is called the *radius of convergence* of the series. If the series converges for all x, we define $\rho = \infty$.

Theorem C.1 can be used to show that the radius of convergence exists for each power series.

The following properties of power series are often useful.

THEOREM C.2 If $\sum_{n=0}^{\infty} a_n(x - x_0)^n$ converges to $f(x)$ for $|x - x_0| < \alpha$, and $\sum_{n=0}^{\infty} b_n(x - x_0)^n$ converges to $g(x)$ for $|x - x_0| < \beta$, then

1. $\sum_{n=0}^{\infty} (a_n + b_n)(x - x_0)^n$ converges to $f(x) + g(x)$ for $|x - x_0| < \min (\alpha, \beta)$.

2. $f(x)g(x) = [\sum_{n=0}^{\infty} a_n(x - x_0)^n][\sum_{n=0}^{\infty} b_n(x - x_0)^n] = \sum_{n=0}^{\infty} c_n(x - x_0)^n$, where

$$c_n = a_0 b_n + a_1 b_{n-1} + \cdots + a_{n-1} b_1 + a_n b_0 \quad \text{for} \quad |x - x_0| < \min (\alpha, \beta)$$

3. If $g(x_0)$ is not zero, $f(x)/g(x)$ can be expanded in a power series in $(x - x_0)$, convergent for $|x - x_0|$ sufficiently small.

4. $\sum_{n=1}^{\infty} n a_n(x - x_0)^{n-1}$ converges to $f'(x)$ for $|x - x_0| < \alpha$.

$\sum_{n=2}^{\infty} n(n - 1)a_n(x - x_0)^{n-2}$ converges to $f''(x)$ for $|x - x_0| < \alpha$, and so on.

We can make use of property (4) to derive the formula for what is known as the Taylor's expansion of a function. Suppose f is a function with a series expansion centered at x_0; that is,

$$f(x) = \sum_{n=0}^{\infty} a_n(x - x_0)^n \qquad (1)$$

By property (4) of Theorem C.2, the kth derivative of Eq. (1) is

$$f^{(k)}(x) = \sum_{n=k}^{\infty} \frac{n!}{(n - k)!} a_n(x - x_0)^{n-k} \qquad (2)$$

If we now evaluate $f^{(k)}(x_0)$ from Eq. (2), we obtain

$$f^{(k)}(x_0) = \frac{k!}{0!} a_k \quad \text{or} \quad a_k = \frac{f^{(k)}(x_0)}{k!}$$

Example 3   Every reader should be familiar with the series for $f(x) = e^x$. From elementary calculus, we know

$$f'(x) = f''(x) = \cdots = f^{(k)}(x) = e^x$$

Hence $f^{(k)}(x_0) = e^{x_0}$. Therefore our series expansion is $\sum_{k=0}^{\infty} a_k(x - x_0)^k$, where $a_k = e^{x_0}/k!$. Hence

$$e^x = \sum_{k=0}^{\infty} \frac{e^{x_0}}{k!} (x - x_0)^k$$

If we consider $x_0 = 0$, we obtain the usual form

$$e^x = \sum_{k=0}^{\infty} \frac{x^k}{k!}$$

THEOREM C.3   Suppose that two power series in $(x - x_0)$ are convergent and have the same sum for all x in some interval $|x - x_0| < r$; that is,

$$\sum_{n=0}^{\infty} a_n(x - x_0)^n = \sum_{n=0}^{\infty} b_n(x - x_0)^n, \quad |x - x_0| < r$$

Then $a_n = b_n$ for all n.

The following corollary plays a central role in series solutions of differential equations.

COROLLARY   Suppose

$$\sum_{n=0}^{\infty} a_n(x - x_0)^n = 0, \quad 0 \le |x - x_0| < r$$

Then $a_n = 0$ for all n.

It is strongly recommended that the reader work the exercises in this appendix before attempting Chapter 16.

Exercises

1.   Find the radius of convergence and interval of convergence for each of the following power series.

(a) $\displaystyle\sum_{n=1}^{\infty} \sqrt{n}\, x^n$

(b) $\displaystyle\sum_{n=1}^{\infty} \frac{3^n x^n}{n^2}$

(c) $\displaystyle\sum_{n=1}^{\infty} \frac{(-1)^n x^n}{n(2^n)}$

(d) $\displaystyle\sum_{n=1}^{\infty} \frac{(x + 4)^n}{2^n}$

(e) $\displaystyle\sum_{n=2}^{\infty} \frac{(\ln n)x^n}{n^2}$

(f) $\displaystyle\sum_{n=1}^{\infty} (n!)2^{-n}x^n$

(g) $\displaystyle\sum_{n=1}^{\infty} \frac{(2x - 3)^n}{n^p}$ $(p > 0)$

2. Find the Taylor series expansion about the given x_0 for each of the following functions.

(a) $\ln (1 + x)$, $x_0 = 0$

(b) $\ln \dfrac{1}{1 - x}$, $x_0 = 0$

(c) $\sin (2x)$, $x_0 = \dfrac{\pi}{4}$

(d) $e^{(e^x - 1)}$, $x_0 = 0$

(e) $1/(1 - x)$, $x_0 \ne 1$

3. (a) Using the addition of power series and the results of Exercise 2(a) and (b), find the Taylor series in powers of x for $\ln [(1 + x)/(1 - x)]$.

(b) Use this series to write down a few terms in a series for $\ln 10$.

4. Find the Taylor series for $f(x) = 5x^2 + 13x - 12$ about the point $x_0 = -3$.

5. Show that the power series for $\cosh x$ is

$$\cosh x = 1 + \frac{x^2}{2!} + \frac{x^4}{4!} + \cdots = \sum_{n=0}^{\infty} \frac{x^{2n}}{(2n)!}$$

and that the radius of convergence is infinite.

TABLE A

Table of Laplace Transforms

$f(t)$	$f^*(s) = \int_0^\infty e^{-st} f(t)\, dt$
1. $ag_1(t) + bg_2(t)$	$ag_1^*(s) + bg_2^*(s)$
2. $(1/a)g(t/a)$	$g^*(as)$
3. $g(t - a)\quad t \ge a$ $\quad\quad\quad 0 \quad\quad t < a$	$e^{-as}g^*(s)$
4. $dg(t)/dt = g'(t)$	$-g(0) + sg^*(s)$
5. $g^{(n)}(t)$	$-s^{n-1}g(0) - s^{n-2}g'(0) - \cdots - g^{(n-1)}(0)$ $\quad\quad + s^n g^*(s)$
6. $\int_0^t g(x)\, dx$	$(1/s)g^*(s)$
7. $e^{at}g(t)$	$g^*(s - a)$
8. $\int_0^t g_1(t - x)g_2(x)\, dx$	$g_1^*(s)g_2^*(s)$
9. f periodic; that is, $f(t) = f(t + a)$ for some scalar a	$\int_0^a e^{-st}f(t)\, dt/(1 - e^{-sa})$
10. $f(t) = a \quad t \ge 0$ $\quad\quad\quad 0 \quad t < 0$	$a/s \quad s > 0$
11. $f(t) = 1 \quad t_0 \le t \le t_1$ $\quad\quad\quad 0 \quad$ otherwise	$(e^{t_0 s} - e^{t_1 s})/s$
12. $f(t) = t^v \quad t \ge 0 \quad v > -1$ $\quad\quad\quad 0 \quad t < 0$	$\Gamma(v + 1)/s^{v+1}$
13. $t^v e^{-at}$	$\Gamma(v + 1)/(s + a)^{v+1}$
14. $\ln t$	$-[\ln s + c]/s$, where c is the Euler–Mascheroni constant, $\lim_{m\to\infty}(\sum_{n=1}^m 1/n - \ln m)$
15. $\sin at$	$a/(s^2 + a^2)$
16. $\cos at$	$s/(s^2 + a^2)$
17. $\sinh at$	$a/(s^2 - a^2)$
18. $\cosh at$	$s/(s^2 - a^2)$

Symbols

Symbols are listed in order of first appearance

P^2, P^3 85

$\mathbf{G}^2, \mathbf{G}^3$ 87, 88

R^m 88

$\mathbf{R}^2$ 88

$\mathbf{R}^m$ 89

$\in, \notin$ 90

$\cup$ 90

$\cap$ 90

$\varnothing$ 90

$\subset$ 90

$\times$ 90

$a \sim b$ 90

$a \approx b$ 91

$\mathbf{0}$ 92

$\mathbf{C}$ 95

$\mathbf{K}$ 95

$\mathbf{R}$ 95

$\mathbf{V}$ 96

$\mathbf{F}[a, b], \mathbf{F}[-\infty, \infty)$ 100

$\mathbf{z}$ 101

$\mathbf{C}^k[a, b], \mathbf{C}^k(a, b), \mathbf{C}^k(-\infty, \infty)$

$\mathbf{C}^\infty(a, b)$ 102

$[S]$ 107

$[v_1, v_2, \ldots, v_n]$ 107

e_i 111

$\mathbf{K}^n$ 112

$\mathbf{U} + \mathbf{W}$ 123

$\sum_{i=1}^n \mathbf{U}_i$ 124

$\mathbf{U} \oplus \mathbf{W}$ 124

$\sum_{i=1}^n \mathbf{U}_i$ 126

$F: A \to B$ 128

$a \to F(a)$ 128

D 131

S 131

$T: \mathbf{V} \to \mathbf{V}'$ 134

$\mathbf{Z}$ 137

σ 141

$I_\mathbf{V}$ 142

$\mathbf{N}_T$ 151

$\ker T$ 151

$v(\mathbf{T})$ 151

$\mathbf{R}_T$ 151

$\rho(\mathbf{T})$ 151

$\dim \mathbf{V}$ 154

T^{-1} 157

(m_{ij}) 164

O 165

$\mathscr{L}(\mathbf{V}, \mathbf{V}')$ 173

M_{qp} 173

I 181

M^T 185

$\rho(M)$ 188

M^{-1} 191

T^- 202

M^- 202

I_p 208

$\det M$ 208

$|M|$ 208

Answers to selected exercises

Chapter I

Section 1.2

1. (a) 2   (b) 3   (c) 3

5. $y = \ln |x| + c, \; x \neq 0$

   (a) $y = \ln x + 3, \; x > 0$

   (b) $y = \ln (-x) + 3, \; x < 0$

6. (a) $y = \pm \sqrt{c - \sin^2 x}, \; |\sin^2 x| \leq c$

   (b) $y = \pm \sqrt{2b(x - a)}, \; \cdots$

Section 1.4

1. (a) $y = ke^{-3x}$

   (c) $x = ke^{(t^2/2 - t)}$

   (e) $v = k \left(\dfrac{1 - u}{1 + u}\right)^{1/2}$

Section 1.5

1. (b) $y = (k + x)e^{-x^2/2}$

   (d) $v = k \csc u + (\sin^2 u)/3$

   (f) $y = k \csc x - x^2 \cot x$

          $+ 2x + 2 \cot x$

   (h) $y = kx^2 + x^4/2$

Section 1.6

1. (a) $y = -2x^2 + x^4$

2. $a = 5; \; k = \ln 6$

4. (a) $y = 2x^{-3} - x^{-1}$

Section 1.7

1. (b) $y = ke^{-3x^2/2 + 4x}$

   (d) $x = 3 \sec (c - 3 \sqrt{9 - y^2})$

   (f) $y = [-2 \ln(\ln x) + c]^{1/2}$

   (h) $y = \tan[x - \ln|x + 1| + c]$

   (j) $3y + y^3/3 = \dfrac{1}{3} \ln (\cos 3x) + c$

   (i) $x = ke^{(\ln y)^2/2}$

3. $y = x^{-1} \ln (c - x^2/2)$

5. $p(x) \equiv 0$ or $q(x) \equiv 0$ or

   $q(x) \equiv cp(x)$

6. $\tan y = 1 - 2x + \sin 2x$

Section 1.8

1. (a) $y = -x \ln (c - \dfrac{1}{2} \ln|x|)$

2. (b) $y + \sin x = cx^2$

603

(d) $\sqrt{x^2 + y^2} \ (y + 2) = cy$

9. (b) $(3y^2 - x^2)^2 = cx$

   (d) $c^2 x^2 = \dfrac{y^2}{x^2 + y^2}$

11. (a) $y = ke^{3x}$
    $- \dfrac{1}{9}(x^3 + x^2 + \dfrac{2}{3} x + c)$

   (b) $\dfrac{1}{3} x + C_2 = \int \dfrac{dy}{y^3 + C_1}$

Section 1.10

1. (b) $ax^2/2 + bxy + ay^2/2 = c$

   (c) $e^y x + \sin x = c$

   (d) $xy^3 + y^4/4 = c$

   (e) $x^3 \ln y = c$

   (g) $x^2 y^2 (x^2 - 4y^2) = c$

3. $(y + 1 - e^{-1})e^{-y} + e^{-(x + y)} = c$

5. (a) Let $v = y^{-2}$;
    $y + x/y = c$. Also $y \equiv 0$

   (c) $v = x$;
    $x^3 + 2x^2 y + y^2 = c$

8. $u(x) = k \ \exp[\int \dfrac{1}{N} (\dfrac{\partial M}{\partial y} - \dfrac{\partial N}{\partial x}) dx]$

9. $2x - \dfrac{y}{x} + \dfrac{1}{2} y^2 = c$

Miscellaneous Exercises

1. $y = ce^{1/x} - \dfrac{e^{1/x}}{x}$

3. $y = (c + x)e^x$

5. $x^3 \tan y + y \cos y - \sin y = c$

7. $y = c - x - 2 \ln x$

9. $e^u + x = \dfrac{1}{x + c}$

11. $x + y = c$

13. $\ln(x + y) - x/(x + y) = c$

15. $-x^2/(x^2 + y^2) + \ln|x| = c$

17. $y = k \tan^{-1} x/2 + 2$

19. $\sin(e^v) = cx - x^2$

21. $y = \ln(c - \sin x)/x^2$

23. $(\tan^{-1} 3y\sqrt{21})/\sqrt{21} + (3 - 2x^2)^{3/2}/24 = c$

25. $y = cx - x^3/2$

27. $y = c \sin x - \cos x$

29. $x/y + x^2/2 = c$

31. $x = ke^{y/x}$

33. $y = c$

35. $y = c/(e^{-x} + 1)$

37. $y = x \sin(c + \ln|x|)$

39. (a) $y = 3x/(k - x^3)$

   (b) $y^2 = ce^{-x^2} + x^2 - 1$

40. (a) $y = \sqrt{\dfrac{b}{-a}} \tan[\sqrt{-ab}\,(x + c)]$
    if $ab < 0$

    $y = \sqrt{\dfrac{b}{a}} \dfrac{ce^{2\sqrt{ab}\,x} - 1}{ce^{2\sqrt{ab}\,x} + 1}$

    if $ab > 0$

   (b) $x^3 (2 - xy) = c(1 + xy)$

43. $y = x + 1/(ke^x + 1)$

45. $y = c_1 \sin x + c_2 \cos x + c_3$

Chapter 2

Section 2.2

3. $\dfrac{\ln 3}{64} = 0.01717$

5. $t = (m/k)\ln(1 + v_0 k/mg)$
   $x = mv_0 k^{-1} - m^2 g k^{-2} \ln(1 + kv_0/mg)$

7. $\omega = \pm \ [2(g/L)\cos \theta + c]^{1/2}$

9. $\cosh t/10 = e^{1.5}$

   t is approximately 22 sec

Section 2.3

1. $I = (-1/c)\ln(e^{-cI_0} - cat)$

Section 2.4

1. $x = (x_0 - kP/c)e^{-ct} + kP/c$
    for $0 \le t \le t_0$

   $x = (x_0 - kP/c + kPe^{ct_0}/c)e^{-ct}$
    for $t > t_0$

3. $x = \left(\dfrac{kA\omega}{c^2 + \omega^2} - \dfrac{kB}{c}\right)e^{-ct}$

$\qquad + \dfrac{kB}{c} + \dfrac{kA}{c^2 + \omega^2}(c \sin \omega t - \omega \cos \omega t),\ \dfrac{ds}{dt} = cx$

4. (a) $y_0 = 3$

$\qquad y_1 = 3 - 3x$

5. $\dfrac{dx(t)}{dt} + cx(t - T) = kP(t)$

$\qquad y_2 = 3 - 3x + 3x^2/2!$

Section 2.5

$\qquad y_3 = 3 - 3x + 3x^2/2! - 3x^3/3!$

1. $i = I_0 e^{-(R/L)t}$

(c) $y_0 = 1$

4. $A = E_0(R^2 + \omega^2 L^2)^{-1/2},$

$\qquad y_1 = 2 - \cos x$

$\quad \sin \phi = L(R^2 + \omega^2 L^2)^{-1/2},$

$\qquad y_2 = 5/2 - 2 \cos x + (\cos^2 x)/2$

$\quad \cos \phi = R(R^2 + \omega^2 L^2)^{-1/2}$

(e) $y_0 = 1$

5. (b) $i = (E/R)(1 - e^{-(R/L)t})$

$\qquad y_1 = 1 + x^2/2$

$\qquad y_2 = 1 + x^2/2 + x^4/2\cdot 4$

$\qquad\qquad$ for $0 \le t < t_1$

$\qquad y_n = 1 + x^2/2 + \cdots + x^{2n}/2^n n!$

$\qquad = (E/R)(1 - e^{-(R/L)t_1})e^{-(R/L)(t-t_1)}$

$\qquad\qquad$ for $t > t_1$

6. (a) $q = CE_0 + (q_0 - CE_0)e^{-t/RC}$

(b) $q = \left(q_0 - \dfrac{CE_0}{1 + \omega^2 R^2 C^2}\right)e^{-t/RC} + \dfrac{CE_0}{1 + \omega^2 R^2 C^2}(\cos \omega t + \omega RC \sin \omega t)$

Chapter 3

Section 3.3

2. (a) $y_0 \ne 0$

(c) $x_0 \ne \pm 1,\ y_0 \ne x_0^2$

(e) $(x_0, y_0) \ne (0, 0)$

Section 3.4

1. $y = 0$ and $y = (3x/5)^{5/3}$

Section 3.5

3. (a) $2x\ dx + 8y\ dy = 0$

(c) $(\ln 4/y)dx/x^2 + dy/xy = 0$

4. (a) $-\infty < x < \infty$

(c) $x \ge 0$

Section 3.7

3. (a) $L = 6$

Section 3.8

1. (a) $u(x) \le C_1 \exp(C_2 x)$

2. $u(x) \ge C \exp\left[\int_a^x v(s)ds\right]$

6. (a) $-\infty < x < 1/3$

Chapter 4

Section 4.1

3. (a) $(-1, 17)$

(c) $(-1, 4, -10, 1)$

4. (c) $(8/\sqrt{113}, 7/\sqrt{113})$

Section 4.3

1. (a) group; zero element $(0, 0)$;

$\qquad$ additive inverse $(-x, -y)$

(d) not a group

2. All are groups with the zero matrix as the additive identity. The additive inverses are, respectively,

(a) $\begin{pmatrix} -a_{11} & -a_{12} \\ -a_{21} & -a_{22} \end{pmatrix}$

(d) $\begin{pmatrix} -a_{11} & -a_{11} \\ -a_{22} & -a_{22} \end{pmatrix}$

Section 4.4

5. The following are vector spaces: a, b, c, d.

10. No, because there is no additive identity.

Section 4.5

The following are vector spaces over **R**: 1,2,7,9,10,11,12,13,14.

Section 4.6

6. Condition (1) follows from (3). Conditions (2) and (3) are independent as shown by Exercises 1e and 2f.

Section 4.8

3. The following sets are linearly independent: a, b.

5. All are independent sets.

Section 4.9

5. There are 11 linearly independent subsets including $\emptyset$.

Section 4.10

3. (a) $\{v_1, v_2, v_4\}$ or

$\{v_1, v_3, v_4\}$ or

$\{v_2, v_3, v_4\}$

Section 4.11

2. (-2, 5, -9, 7)

4. (3/4, 0, -1/4)

6. Coordinates of $f'(x)$ are

(-2, 2, 0, 0).

Coordinates of $\int_1^x f(t)dt$ are

(- 7/3, 3, -1, 1/3).

Section 4.12

5. $W \subseteq U$

7. a, b are direct sums.

Chapter 5

Section 5.1

1. (a) $F(Z) = Z$

(c) $F(Z) = Z$

(e) $F(Z)$ is the projection of Z onto the x-axis.

2. (b) $Z = \{(x, y): -1 \leq 2x - y \leq 3\}$

(d) $Z = \{(x_1, x_2, x_3):$

$|x_1| + |x_2| \leq 1\}$

3. (a), (b), (d) are one-to-one onto.

4. (c) $n = 2; m = 2$; one-to-one;

onto.

(e) $n = 3; m = 3$; not one-to-one;

not onto.

5. (a) $Df = ae^{ax}$

(c) $2x/(1 + x^2)$

6. (a) $x^3/3$

7. (a) $Tf_1 = (ax + 1)e^{ax}$

$Tf_2 = (n + 1)x^n$

(d) $Tf_1 = xe^{ax}$ $Tf_2 = x^{n+1}$

Section 5.2

7. (a) $T:(x,y,z) = (x, x - z, x - y)$

10. linear $\mathbf{C}^\circ (-\infty, \infty)$

Section 5.3

3. T is not onto $\mathbf{C}^\circ [0, 1]$ and not onto $\mathbf{W}$. The function $x^{1/2} \in \mathbf{W}$ but there is no continuous f such that $xf(x) = x^{1/2}$.

Section 5.4

4. (b) $(S + T)(x,y) = (2x - y,y)$

       $(ST)(x,y) = (x,0)$

       $(TS)(x,y) = (x - y, 0)$

7. $Rp = 5$ $Sp = 5x - 3x^2 + x^4$

  $Tp = 5 + x^3$ $RSTp = 0$ $RTSp = 0$

  $STRp = 5x$ $SRTp = 5x$

Section 5.5

1. (a) $(D^2 - I)(2e^x) = 0$

    $(D^2 - I)e^x + e^{-x} = 0$

3. (a) $2(3D - 5I)(D + 2I)$

  (b) $(D - I)(D^2 + D + I)$

    $= (D - I)(D - \dfrac{-1 + i\sqrt{3}}{2})(D - \dfrac{-1 - i\sqrt{3}}{2})$

  (c) $(D^2 + \sqrt{2}D + I)(D^2 - \sqrt{2}D + I)$

    $= (D - \dfrac{1 + i}{\sqrt{2}})(D - \dfrac{1 - i}{\sqrt{2}})(D - \dfrac{-1 + i}{\sqrt{2}})(D - \dfrac{-1 - i}{\sqrt{2}})$

  (d) $(D - (1 + i)I)(D - (1 - i)I)$

4. (b) $T^n = (x, 0)$

6. (b) (i) $4t^2$

      (ii) $\dfrac{4t^3 - 2}{t}$

      (iii) $t^2 e^t + 2te^t - 2e^t$

Section 5.6

1. (d) $\nu(T) = 1$ $\rho(T) = 2$

    $N_T = [(1, -2, -2)]$

4. $N_T = [e^{3x}]$

7. (f) $N_T = [(1, 1, 0), (0, 2, 1)]$

10. (a) not linear

11. (b) (i) $\left\{\begin{pmatrix} 1 & 0 \\ 0 & 0 \end{pmatrix}, \begin{pmatrix} -1 & 1 \\ 0 & 0 \end{pmatrix}, \begin{pmatrix} 1 & 2 \\ 1 & 0 \end{pmatrix}, \begin{pmatrix} 0 & 0 \\ 0 & 1 \end{pmatrix}\right\}$

     (ii) $\begin{pmatrix} 1 & 0 \\ 0 & 0 \end{pmatrix} + \begin{pmatrix} 0 & 0 \\ 0 & 1 \end{pmatrix}$

     (iii) $(a - 2b)\begin{pmatrix} 1 & 0 \\ 0 & 0 \end{pmatrix} - b\begin{pmatrix} -1 & 1 \\ 0 & 0 \end{pmatrix} + b\begin{pmatrix} 1 & 2 \\ 1 & 0 \end{pmatrix} + a\begin{pmatrix} 0 & 0 \\ 0 & 1 \end{pmatrix}$

Section 5.7

1. (a) one-to-one and onto $T^{-1}(u, v) = (u + v, -v)$

Chapter 6

Section 6.2

1. (b) $\begin{pmatrix} 1 & 0 \\ 0 & -1 \end{pmatrix}$

  (d) $\begin{pmatrix} 0 & 1 & -1 \\ 2 & 1 & 0 \\ 0 & 0 & 0 \end{pmatrix}$

  (f) $(1 \quad -1 \quad 2)$

2. (a) $p = 2$, $T(x_1, x_2) = (x_2, x_1)$

  (c) $p = 1$, $Tx = (3x, 2x)$

  (e) $p = 4$, $4T(x_1, x_2, x_3, x_4) = (x_1, x_2)$

3. (b) $\begin{pmatrix} 0 & 1/2 & 1/2 \\ -1 & -1/2 & 1/2 \end{pmatrix}$

5. (a) $(1 \quad 0 \quad 0 \quad 0)$

  (c) $\begin{pmatrix} 2 & 2 & 8/3 & 4 \\ 1 & 0 & 0 & 0 \\ 0 & 1 & 0 & 0 \end{pmatrix}$

6. (b) $\begin{pmatrix} 0 & 1 & 0 \\ 0 & 0 & 2 \\ 0 & 0 & 0 \end{pmatrix}$ (d) $\begin{pmatrix} 0 & 1 & 0 \\ 0 & 0 & 1 \\ 0 & 0 & 0 \end{pmatrix}$

7. (a) $\begin{pmatrix} 1 & 1 & 1 \\ 0 & 1 & 2 \\ 0 & 0 & 1 \end{pmatrix}$ (c) $\begin{pmatrix} 1 & -1 & 1 \\ 1 & -1 & 1 \\ 0 & 0 & 0 \end{pmatrix}$

9. (b) $\begin{pmatrix} 1 & 0 \\ 0 & 4 \end{pmatrix}$ (d) $\begin{pmatrix} 0 & 0 \\ 0 & 0 \end{pmatrix}$

10. $\begin{pmatrix} 2/3 & 0 \\ -1/3 & -1 \end{pmatrix}$

Section 6.3

1. (b) $\begin{pmatrix} 1 & 6 & \sqrt{2} \\ 0 & -\sqrt{2} & -3 \end{pmatrix}$

2. (a) $\begin{pmatrix} 9 & 0 & 6 \\ -3 & 12 & 3 \\ 6 & 3 & -9 \end{pmatrix}$

3. (b) $\begin{pmatrix} 17 & 7 \\ -1 & -1 \\ 12 & 2 \end{pmatrix}$

4. $Sp(x) = (a_0 + a_1, a_0 - 2a_1)$
$Tp(x) = (a_1 + 3a_2, 2a_1 - 2a_0 + a_2)$

Section 6.4

1. (b)
$$M = \begin{pmatrix} 1 & 0 \\ 0 & 2 \\ 0 & -1 \end{pmatrix}$$

$$N = \begin{pmatrix} 1/2 & 1/2 & 1/2 \\ 1 & 0 & -1 \\ 1/2 & -1/2 & 1/2 \end{pmatrix}$$

$$P = \begin{pmatrix} 1/2 & 1/2 \\ 1 & 1 \\ 1/2 & -3/2 \end{pmatrix}$$

2.
$$M = \begin{pmatrix} 1 & 1 & 1 \\ 0 & 1 & 2 \\ 0 & 0 & 1 \end{pmatrix} \quad N = \begin{pmatrix} 1 & 0 & 0 \\ 0 & 1 & 0 \\ 0 & 0 & 2 \end{pmatrix}$$

$$P = \begin{pmatrix} 1 & 1 & 1 \\ 0 & 1 & 2 \\ 0 & 0 & 2 \end{pmatrix}$$

3. (a) $\begin{pmatrix} 2 & 9 \\ 0 & 18 \\ 5 & 26 \end{pmatrix}$ (c) (27)

4. A must be square.

5. (a) $2^{n-1} \begin{pmatrix} 1 & 1 \\ 1 & 1 \end{pmatrix}$

8. kI

12. (a) $Tv = 5u_1 - 3u_2 + 7u_3$

13. (b) $2x^2 + 4x + 3$
    (c) $-4x^2 + 8x + 1$

15. $A = \begin{pmatrix} 1 & 1 & 0 \\ 0 & 0 & 1 \end{pmatrix}$

$T(a,b,c) = (a + b, c)$

Section 6.5

1. (a) $A^T = \begin{pmatrix} 2 & 1 & 5 \\ 3 & 0 & -2 \end{pmatrix}$

(c) $C^T = \begin{pmatrix} 1 & 1/2 \\ 4 & 0 \\ \sqrt{2} & -5/2 \end{pmatrix}$

3. $\begin{pmatrix} 1 & 0 & 0 & 0 \\ 0 & 0 & 1 & 0 \\ 0 & 1 & 0 & 0 \\ 0 & 0 & 0 & 1 \end{pmatrix}$

6. (a) $\begin{pmatrix} 10 & 8 & -1 \\ 5 & 0 & 1 \end{pmatrix}$

(c) $\begin{pmatrix} 7/2 & 1 & 4 \\ 8 & 4 & 20 \\ 2\sqrt{2} & -15/2 & \sqrt{2} & 5(1 + \sqrt{2}) \end{pmatrix}$

8. No.

Section 6.6

2. row rank of A = column rank of A = 2
   row rank of B = column rank of B = 2

5. (a) rank = 1   (b) rank = 2
6. (a) rank = 3   (c) rank = 1
   (d) rank = 2

Section 6.7

1. (b) nonsingular

2. $A^{-1} = \begin{pmatrix} 3/2 & -5/2 \\ -1 & 2 \end{pmatrix}$

3. $M = \begin{pmatrix} 1 & 0 & 0 \\ 0 & 0 & 1 \\ 0 & 1 & 0 \end{pmatrix}$

$T^{-1}(a_0', a_1'', a_2')$

$= a_0' + a_2' x + a_1' x^2)$

$M^{-1} = \begin{pmatrix} 1 & 0 & 0 \\ 0 & 0 & 1 \\ 0 & 1 & 0 \end{pmatrix}$

4. $M^{-1} = \begin{pmatrix} 1/a_1 & 0 & \cdots & 0 \\ 0 & 1/a_2 & \cdots & 0 \\ 0 & & \cdots & 1/a_n \end{pmatrix}$

5. $M^{-1} = \begin{pmatrix} 1/a & -b/ad & (be-dc)/adf \\ 0 & 1/d & -e/df \\ 0 & 0 & 1/f \end{pmatrix}$

6. $v^2 \neq c^2$

14. (a) $A^{-1} = \begin{pmatrix} 1 & -2/5 & 0 & 0 \\ 0 & 1/5 & 0 & 0 \\ 0 & 0 & 3/2 & -5/2 \\ 0 & 0 & -1 & 2 \end{pmatrix}$

Section 6.8

3. $\begin{pmatrix} 1 & 1 & 1 \\ 0 & 1 & 1 \\ 0 & 0 & 1 \end{pmatrix}$

4. $\begin{pmatrix} -1 & -1/3 & 1/3 \\ 1 & 2/3 & 1/3 \\ 1 & 1 & 0 \end{pmatrix}$

5. $\begin{pmatrix} a_1 & b_1 & c_1 \\ a_2 & b_2 & c_2 \\ 0 & 0 & c_3 \end{pmatrix}$

6. $\begin{pmatrix} 2 & -1 & 0 \\ 1 & -1 & 1 \\ 1 & 0 & -1 \end{pmatrix}$

7. $\begin{pmatrix} 1 & -1/3 & -2/3 \\ -1 & 4/3 & 5/3 \\ 8 & 19/3 & 8/3 \end{pmatrix}$

Section 6.9

4. (a) $M^- = \begin{pmatrix} 1 & 0 & a \\ 0 & 1 & b \end{pmatrix}$,

   a and b arbitrary

Chapter 7

Section 7.1

1. (a) -1   (c) 9
5. (b) 10 by property (2)

Section 7.3

2. (a) minor $M = \begin{pmatrix} -2 & 13 & 8 \\ -7 & 13 & 28 \\ 13 & 13 & 13 \end{pmatrix}$

   cofactor $M = \begin{pmatrix} -2 & -13 & 8 \\ 7 & 13 & -28 \\ 13 & -13 & 13 \end{pmatrix}$

   det $M = -65$

Section 7.5

5. (a) $2^{(p^2 - p)/2}$

Section 7.6

4. $x = 0, 1, -2, 4$
6. (a) -1/3   (c) singular

Section 7.7

2. 116
4. (a) 44   (b) -220

Section 7.8

1. $M^{-1} = \dfrac{1}{m_{11}m_{22} - m_{12}m_{21}}\begin{pmatrix} m_{22} & -m_{12} \\ -m_{21} & m_{11} \end{pmatrix}$

3.
$$A^{-1} = \begin{pmatrix} -18 & 5 & 0 & -3 \\ 7/3 & 1 & -1/3 & 4/3 \\ 1/3 & -3 & 2/3 & -5/3 \\ -9 & -2 & 1 & -4 \end{pmatrix}$$

5.
$$M^{-1} = \begin{pmatrix} 1/m_{11} & 0 & \cdots & 0 \\ 0 & 1/m_{22} & \cdots & 0 \\ & & \cdots & \\ 0 & 0 & \cdots & 1/m_{nn} \end{pmatrix}$$

Section 7.9

2. $x = -3 \quad \rho = 2$

   $x \neq -3 \quad \rho = 3$

4. $p \begin{pmatrix} q \\ p - 1 \end{pmatrix}$

   $= p(q!)/[(p - 1)!(q - p + 1)!]$

Section 7.10

1. (a) $x = -4 \quad y = -5$

   (c) no solution

   (e) $x_1 = -1 \quad x_2 = 2$

   $x_3 = -1 \quad x_4 = 4$

4. (a) $\begin{pmatrix} 2 & -1 & 0 \\ -1/3 & 1/3 & -1/3 \\ -10/3 & 7/3 & -1/3 \end{pmatrix}$

6. (b) $Y = 2/3 \, M_1 + M_2 - 1/3 \, M_3$

Chapter 8

Section 8.2

1. (a) $\cos x$

   (c) $-3 + \ln x$

6. (b) $(x^2 + x)D + I$

8. (a) ce^x

   (c) polynomials of degree

   $n - 1$ or less

9. (c) $e^x D^2 + (e^x + 2)D + \ln x$

10. onto, not one-to-one

13. (b) $(D - I)(D + I)(D - iI)(D + iI)$

Section 8.3

1. (a) nonlinear

   (c) nonlinear

3. (b) $y = c_1 e^{x/2} + c_2 e^x$

Section 8.4

2. (b) unique solution on some

   interval $|x - 1| < \alpha$

   (d) unique solution on $(-1, 1)$

   (f) unique solution on $(0, e^\pi)$

Section 8.6

3. (a) linearly independent if and

   only if $a \neq b$

   (c) linearly independent

   (d) linearly independent

   (f) linearly independent if and

   only if $a \neq b$ or $j \neq k$

Section 8.7

2. $y = c_1 e^x + c_2 x e^x$;

   $c_1 = 0 \quad c_2 = 8$

4. $y = c_1 x^{-1} \sin x + c_2 x^{-1} \cos x$;

   $c_1 = -1 \quad c_2 = -\pi$

Section 8.8

1. (b) $y = c_1 e^x + c_2 x e^x + x^3$;

   $y = e^x + 6x e^x + x^3$

   (d) $y = c_1 x^{-1} \sin x + c_2 x^{-1} \cos x$

   $+ \sin x$;

   $y = -2\pi x^{-1} \sin x + \sin x$

2. (a) $y = c_1 e^x + c_2 e^{-x} - x$

Chapter 9

Section 9.2

7. $\cos^3 x = (\cos 3x + 3 \cos x)/4$

   $\sin^3 x = (-\sin 3x + 3 \sin x)/4$

9. (a) $4ix; \; 2ix^3/3 + c$

13. No, because fields are different.

Section 9.3

2. (b) e^{-x}, e^{-ix}, xe^{-ix}

3. (a) $y = k_1 e^{4x} + k_2 e^x$

   (c) $z = k_1 e^x + k_2 e^{3x/4}$

   (f) $y = (c_1 + c_2 x)e^x + c_3 e^{ix}$

      $+ c_4 e^{-ix}$, or

      $y = k_1 e^x + k_2 x e^x + k_3 \sin x$

      $+ k_4 \cos x$

   (j) $y = k_1 e^x + k_2 e^{2x} + k_3 e^{-3x/2}$

4. (b) $y = e^{-x/2}[\cos\sqrt{3}x/2$

      $+ \sqrt{3}/3 \sin \sqrt{3}x/2]$

   (c) $y = -26e^{x-8} + 36e^{-6+3x/4}$

Section 9.4

2. (a) $y_1 = x^2$, $y_2 = x$

   (b) $y = c_1 x + c_2 x^2 - x \ln x$

4. (a) $y = \cos x + 3 \sin x$

      $- [\ln(\sec x + \tan x)]\cos x$

   (c) $y = (x^2 e^{-x/2})/8$

   (e) $y = 7x - 3x^2 - x \ln x$

5. $y_p = e^x$

Section 9.5

1. (b) $\sinh (x - z)$

   (d) $\cosh (x - z) - 1$

   (f) $(e^{3(x-z)} - 3(x - z) - 1)/9$

6. $\int_1^x \dfrac{x^2 - xz}{z} f(z) dz$

Section 9.6

2. (a) $y_p = x - 1$

   (c) $y_p = \dfrac{x^{n+1}}{(n + 1)!} + \dfrac{5x^n}{n!}$

   (e) $y_p = x^2 e^{-x/2}/8$

   (g) $y_p = x^2 e^x/2$

   (j) $y_p = \dfrac{-x \cos 2x}{4}$

7. $y = -c_1 + x + 1 + e^x \int x(\ln x)e^{-x} dx$

8. (b) $y = c_1 e^{-x/2} + c_2 x e^{-x/2}$

      $+ x^2 e^{-x/2}/8$

   (e) $y = c_1 + c_2 e^{-x} + e^x \int^x e^{-t} \ln t \, dt$

Section 9.7

1. (a) $y = c_1 x^2 + c_2 (x^3 \ln x - x^3)$,

     $x > 1$

   (c) $y = (c_1 e^{-x^2} + c_2)/x$,

     $x > 0$

5. (b) $(c_1 + c_2 e^{-x^2} + x^2/4)/x = y$

   (d) $e^x \int^x \cot t (\int^t e^{-s} \tan s \, ds) dt$

     $+ c_1 e^x + c_2 e^x \ln(\sin x)$

7. (a) $y = (c_1 + c_2 e^x)e^{-e^x}$

12. $y = \dfrac{1}{2} x^{3/2} + c_1 x^{1/2} + c_2 x^{-1/2}$

Miscellaneous Exercises

1. $q = E_0 C (1 - \cos t/\sqrt{LC})$,

   $i = E_0 \sqrt{\dfrac{C}{L}} \sin t/\sqrt{LC}$

3. $q = E_0 C (1 - LC\omega^2)^{-1}(\cos \omega t - \cos t/\sqrt{LC})$

   $+ q_0 \cos t/\sqrt{LC} + i_0 \sqrt{LC} \sin t/\sqrt{LC}$

5. $q = c_1 \cos \omega t + c_2 \sin \omega t$

   $+ \dfrac{E_0}{2} \sqrt{\dfrac{C}{L}} t \sin \omega t$

Chapter 10

Section 10.2

4. (a) Add first row to third row

      (third column to first column)

   (c) Multiply third row (or column)

      by -2.

6. (b) $\begin{pmatrix} 1 & 0 & 0 \\ 0 & 1 & 0 \\ 0 & 0 & -1 \end{pmatrix}$

Section 10.3

3. (a) $\rho = 2$ (c) $\rho = 3$

4. (b) $\begin{pmatrix} 0 & 1 & 0 & -6 \\ 0 & 0 & 1 & 2 \\ 0 & 0 & 0 & 0 \end{pmatrix}$

Section 10.5

1. (a) dimension is 3.

   (c) dimension is 0.

2. (b) rank of A is 2;

      rank of $(A:Y)$ is 3;

      no solution.

   (d) rank of A is 3;

      rank of $(A:Y)$ is 4;

      no solution.

5. for 10 unknowns, $\sim$ 68 seconds;

   for 20 unknowns, $\sim$ 2.8 x 10^6 years.

7. for 10 unknowns, $\sim$ 3 x 10^{-4}

   seconds; for 20 unknowns,

   $\sim$ 2.7 x 10^{-3} seconds.

Section 10.6

1. (b) $\rho = 2$ (d) $\begin{pmatrix} 2 & -1 & 1 & 0 \\ -1 & 0 & -2 & 0 \\ -2 & 2 & 1 & 0 \\ -8 & 3 & -3 & 1 \end{pmatrix}$

Section 10.7

1. no.

4.

$M = \begin{pmatrix} 2 & 1 \\ -2 & 0 \\ -1 & -1 \end{pmatrix}$ $N = \begin{pmatrix} 3 & 1 \\ 1 & 0 \\ 0 & 1 \end{pmatrix}$

$P = \begin{pmatrix} 1 & 1 \\ 1 & 0 \end{pmatrix}$ $Q = \begin{pmatrix} 1 & 1 & -1 \\ 1 & 1 & 0 \\ 0 & -1 & 1 \end{pmatrix}$

Chapter 11

Section 11.2

3. (b) A represents a projection onto a subspace, namely, the space of all eigenvectors with eigenvalue 1.

Section 11.3

4. $\lambda_1 = 1$ with basis $\begin{pmatrix} 3 \\ 2 \end{pmatrix}$

   $\lambda_2 = 2$ with basis $\begin{pmatrix} 2 \\ 1 \end{pmatrix}$

6. $\lambda_1 = 2$ with basis $\begin{pmatrix} 1 \\ 0 \\ 0 \end{pmatrix}$

   $\lambda_2 = 1$ with basis $\left\{ \begin{pmatrix} 0 \\ 1 \\ 0 \end{pmatrix}, \begin{pmatrix} 0 \\ 0 \\ 1 \end{pmatrix} \right\}$

8. (b)

   $\lambda_1 = 2, \begin{pmatrix} 5 \\ -1 \\ -2 \end{pmatrix}$

   $\lambda_2 = 1, \begin{pmatrix} -1 \\ 1 \\ 1 \end{pmatrix}$

   $\lambda_3 = -1, \begin{pmatrix} 1 \\ 1 \\ -1 \end{pmatrix}$

   (d) $P = \begin{pmatrix} 0 & 0 & 1 \\ 0 & 1/2 & -1/2 \\ 1/2 & -1/4 & 1/4 \end{pmatrix}$

Section 11.4

3. $p(\lambda) = \lambda^2 - 4\lambda - 11$

   $\lambda = 2 \pm \sqrt{15}$

5. $p(\lambda) = -\lambda^3 + 4\lambda^2 + 24\lambda$

   $\lambda = 0, 2 \pm 2\sqrt{7}$

9. $p(\lambda) = -\lambda^3 + 3\lambda^2 + 3\lambda - 9$

   $\lambda = 3, \sqrt{3}, -\sqrt{3}$

Section 11.5

1. $\begin{pmatrix} 20 & 17 \\ 17 & 37 \end{pmatrix}$

3. $\begin{pmatrix} 6 & -7 & -7 \\ 3 & 12 & -3 \\ -6 & 2 & 5 \end{pmatrix}$

5. $p(\lambda) = \lambda^2 - 5\lambda + 5 = 0$

7. $p(\lambda) = -\lambda^3 + 2\lambda^2 - 3 + 6 = 0$

Section 11.6

2. no.

4. (a) no   (b) no

6. (a)
$$M = \begin{pmatrix} 1 & -1 & 0 \\ -1 & 1 & 0 \\ 1 & 1 & 2 \end{pmatrix}$$

Section 11.7

1. (b) $\lambda_1 = 1$ with an index of 1;
   $\lambda_2 = -1$ with an index of 2.

Chapter 12

Section 12.1

1. (b) $y_1' = y_2$, $y_2' = y_2^2 + x$

3. $y_1' = y_2$,
   $y_2' = y_3$,
   $y_3' = y_4$, $\cdots$ $y_{n-1}' = y_n$
   $y_n' = -a_1(x)y_n - a_2(x)y_{n-1}$
   $\quad - \cdots - a_n(x)y_1 + g(x)$

5. $\dfrac{dC_3}{dt} = \beta C_2 - 2\alpha C_3$

   $\dfrac{dC_4}{dt} = \alpha C_3$

Section 12.2

2. (a) $\begin{pmatrix} 2i \\ -2i \end{pmatrix}$

   (b)
   $$\begin{pmatrix} 2 - \dfrac{\pi}{2} + i(\dfrac{1}{2} - \ln 2) \\ \dfrac{1}{2}\ln 2 - i\,\dfrac{\pi}{4} \end{pmatrix}^{1/2}$$

Section 12.5

4. $\begin{pmatrix} 2e^{-x} \\ -e^{-x} + 2e^{x} \end{pmatrix}$

Section 12.6

1. (a) $\begin{pmatrix} 0 \\ 0 \end{pmatrix}$, $\begin{pmatrix} e^{x}(1 - x) \\ e^{x}(1 + x - x^2) \end{pmatrix}$

3. No.

Section 12.7

1. (a) unique solution exists on
   interval $0 < x < \pi$.

Section 12.9

2. $2\phi_1 - 3\phi_2$

3. $\phi_1 + 2\phi_2 + 3\phi_3$

Section 12.10

1. (b) $\begin{pmatrix} e^{-x} + e^{3x} + e^{x} \\ -2e^{-x} + 2e^{3x} \end{pmatrix}$

2. (b) $\begin{pmatrix} 2e^{x} \\ 0 \\ (1 + x)e^{x} \end{pmatrix}$

Section 12.11

2. (a)
   $$A(x) = \begin{pmatrix} 0 & 1 & 0 \\ 0 & 0 & 1 \\ 0 & -6 & 2 \end{pmatrix}$$

3. (b) $z''' + 2z'' + z' + 2z + 0$;
   $z = c_1 e^{-2x} + c_2 e^{ix} + c_3 e^{-ix}$
   $$y = \begin{pmatrix} e^{-2x} & e^{ix} & e^{-ix} \\ -2e^{-2x} & ie^{ix} & -ie^{-ix} \\ 4e^{-2x} & -e^{ix} & -e^{-ix} \end{pmatrix} \begin{pmatrix} c_1 \\ c_2 \\ c_3 \end{pmatrix}$$

Section 12.14

1. (a) $\begin{pmatrix} e^{x} & 0 & 0 \\ 0 & e^{2x} & 0 \\ 0 & 0 & e^{3x} \end{pmatrix}$

2. (a) $P = \begin{pmatrix} 1 & 1 \\ 0 & 1 \end{pmatrix}$ $e^{Dx} = \begin{pmatrix} e^{x} & 0 \\ 0 & e^{2x} \end{pmatrix}$

   $e^{Ax} = \begin{pmatrix} e^{x} & e^{2x} - e^{x} \\ 0 & e^{2x} \end{pmatrix}$

3. (a) $c_1 \begin{pmatrix} 1 \\ 0 \end{pmatrix} e^{x} + c_2 \begin{pmatrix} 1 \\ 1 \end{pmatrix} e^{2x}$

   (d) $c_1 \begin{pmatrix} 1 \\ 1 \\ 1 \end{pmatrix} e^{-x} + c_2 \begin{pmatrix} 1 \\ 0 \\ 1 \end{pmatrix} e^{x} + c_3 \begin{pmatrix} 0 \\ 1 \\ 1 \end{pmatrix} e^{x}$

4. (b) $\begin{pmatrix} 3e^{2x}/4 + e^{-2x}/4 + 2 \\ 3e^{2x} + e^{-2x} \\ 3e^{2x}/2 - e^{-2x}/2 \end{pmatrix}$

5. (a) $y_1 = c_1 + c_2\,x + \dfrac{1}{2}c_3 x^2$,
   $y_2 = c_2 + c_3 x$, $y_3 = c_3$

Section 12.15

1. (b)
$$Pe^{Dx} = \begin{pmatrix} e^{x} & e^{ix} & e^{-ix} \\ e^{x} & ie^{ix} & -ie^{-ix} \\ e^{x} & -e^{ix} & -e^{-ix} \end{pmatrix},$$

$$\begin{pmatrix} e^{x} & \cos x & \sin x \\ e^{x} & -\sin x & \cos x \\ e^{x} & -\cos x & -\sin x \end{pmatrix}$$

Section 12.16

2. (d)
$$\begin{pmatrix} (-e^{5x} + e^{9x})/2 \\ (e^{5x} + e^{9x})/2 \\ e^{9x} \end{pmatrix}$$

(f) $\begin{pmatrix} 2e^{x} & -xe^{x} \\ e^{x} & -xe^{x} \\ & -xe^{x} \end{pmatrix}$ (g) $\begin{pmatrix} -e^{x} \\ 0 \\ e^{x} \end{pmatrix}$

4. (a) not bounded

   (c) not bounded

   (e) not bounded

Section 12.17

1. (c) Let $y_1(0) = a_1$, $y_2(0) = a_2$.

   Then

   $y_1(x) = (a_1 + x + a_1 x + a_2 x)e^{-3x}$

   $y_2(x) = (a_2 - x - a_1 x - a_2 x)e^{-3x}$

   (d) Let $y_1(0) = a_1$, $y_2(0) = a_2$, $y_3(0) = a_3$.

   Then

   $y_1(x) = xe^{9x} + e^{9x}(3a_1 + 3a_2 + 2a_3)/10$
   $\qquad + e^{4x}(a_1 + a_2 - a_3)/5$
   $\qquad + e^{5x}(a_1 - a_2)/2$

   $y_2(x) = xe^{9x} + e^{9x}(3a_1 + 3a_2 + 2a_3)/10$
   $\qquad + e^{4x}(a_1 + a_2 - a_3)/5$
   $\qquad - e^{5x}(a_1 - a_2)/2$

   $y_3(x) = 2xe^{9x} + e^{9x}(3a_1 + 3a_2 + 2a_3)/5$
   $\qquad - 3e^{4x}(a_1 + a_2 - a_3)/5$

Chapter 13

Section 13.1

1. (a) $6x - 1$

   (c) $\sin[a(x + 1) + b] - \sin[ax + b]$

   (e) $\cos[a(x + 1) + b] - \cos[ax + b]$

   (g) $\sec(x + 1) - \sec x$

2. (b) 4

9. (b) $3x^2 - 3x + 1$

   (d) $10^{2x-5}(1 - 10^{-2})$

   (f) $a^{x-1}(ax - x + 1)$

   (h) $-1/x(x - 1)$

11. (a) $x^{(1)} + x^{(2)}$,

   $\frac{1}{3}x^{(3)} + \frac{1}{2}x^{(2)} + p(x)$

   (c) $8x^{(3)} + 12x^{(2)}$,
   $2x^{(4)} + 4x^{(3)}$

12. (a) $(x - 1)^{(-2)}$; $-1/x$

14. $F(x) = 0$, $1 \le x < 2$;
   $F(x) = \log_2(x - 1)$, $2 \le x < 3$;
   $F(x) = \log_2(x - 1)(x - 2)$,
   $\qquad 3 \le x < 4$

16. (b) $n/(n + 1)$

19. (c) $2^{x+\lambda-1}$

Section 13.2

1. (a) 2

4. Yes

6. (b) $(x - 1)^{(k)} p(x)$

    (d) $(x - 1)^{(k)} p(x) + (x - 1)$

       $+ (x - 1)(x - 2)$

       $+ \cdots + (x - 1)^{(k)}$

    (f) $(x + 1)p(x)/(x - k + 1)$

    (h) $p(x) + 2^x(1 - 2^{-k})$

7. (a) $f(x) = 3^k \sin 2\pi x$

    (c) $f(x) = \dfrac{(x + 1)(x - k)}{(x - k + 1)}$

Section 13.3

1. (b) $e^n(e - 1)$

3. (b) $y_{n+2} - 2y_{n+1} + y_n$
    $+ \tan(y_{n+1} - y_n) = 1$

4. (a) $y_n = (y_1 + n-1)2^{n-1}$

5. (b) $y_n = \prod\limits_{i=1}^{n-1} (1 + a/i)$

    (d) $y_{2n+1} = 0$, $y_{2n} = 2^{n-1}(n - 1)!$

Section 13.4

4. $y_n = 2(-1)^{n-1}$,
    $z_{2n} = 3$, $z_{2n+1} = 1$

6. $y_{n+1} = \begin{pmatrix} a(-1)^n + 1 \\ b2^n + 1 \end{pmatrix}$

Section 13.5

1. (a) $y(x + 1) = \begin{pmatrix} 0 & 1 \\ -5 & 2 \end{pmatrix} y(x) + \begin{pmatrix} 0 \\ x \end{pmatrix}$

(c)
$$y(x + 1) = \begin{pmatrix} 0 & 1 & 0 & 0 \\ 0 & 0 & 1 & 0 \\ 0 & 0 & 0 & 1 \\ -1 & 0 & 2 & 0 \end{pmatrix} y(x)$$

Section 13.6

1. (c) $v_1 p_1(x) + v_2 p_2(x) + (-1)^m v_3 p_3(x)$

    where $v_1 = \begin{pmatrix} 1 \\ 0 \\ 1 \end{pmatrix}$,

    $v_2 = \begin{pmatrix} 0 \\ 1 \\ 1 \end{pmatrix}$,

    $v_3 = \begin{pmatrix} 1 \\ 1 \\ 1 \end{pmatrix}$

2. $y_1(n) = y_3(n) = 1 + (-1)^n$,
    $y_2(n) = 1 + (-1)^{n+1}$,
    $n = 1, 2, 3, \cdots$

4. (a)
$$y(x) = \begin{pmatrix} -1 \\ 1 - 2i \end{pmatrix}(2i)^m p_1(x)$$
$$+ \begin{pmatrix} -1 \\ 1 - 2i \end{pmatrix}(-2i)^m p_2(x)$$

    For $2n \leq x < 2n + 1$
$$y(x) = (-1)^n 2^{2n} \begin{pmatrix} -q(x) \\ q(x) + 2r(x) \end{pmatrix}$$
    For $2n - 1 \leq x < 2n$
$$y(x) = (-1)^{n-1} 2^{2n-1} \begin{pmatrix} r(x) \\ 2q(x) - r(x) \end{pmatrix}$$

Section 13.7

2. $\emptyset(2-) = -a_1 \emptyset(1) - a_2 \emptyset(0)$

3. (b) $z(x) = 2^{m/2}[q(x)\cos(3\pi m/4)$
       $+ r(x)\sin(3\pi m/4)]$,
    $m \leq x < m + 1$

8. (a) $z(n) = (c_1 + c_2 n)(-1)^n + c_3(3^n)$

9. $z(n) = 2^n c$ $(n = 1, 2, \cdots)$
    for both

Section 13.9

2. $y(x) = 1 + 2(x - 1)$, $1 \leq x < 2$

   $\qquad = 3 + 2(x - 2) + (x - 2)^2$,

   $\qquad 2 \leq x < 3$

6. (a) $\lambda e^{\lambda} = a + b e^{\lambda} + c\lambda$

   (c) $1 = \int_0^1 a(s) e^{-\lambda s} ds$

Chapter 14

Section 14.1

1. $c = \dfrac{\langle v_1, v_2 \rangle}{\langle v_2, v_2 \rangle}$

3. Defines an inner product

5. Defines an inner product.

7. Does not define an inner product.

Section 14.2

2. $\begin{pmatrix} 1 & 1/2 & 1/3 \\ 1/2 & 1/3 & 1/4 \\ 1/3 & 1/4 & 1/5 \end{pmatrix}$

4. $\begin{pmatrix} 1 & 0 & 0 & 0 \\ 0 & 1 & 0 & 0 \\ 0 & 0 & 1 & 0 \\ 0 & 0 & 0 & 1 \end{pmatrix}$

Section 14.3

1. (a) $\dfrac{3}{2} v_2$ (c) $\dfrac{1}{4} v_2$

3. (a) 1, $\dfrac{\sqrt{3}}{3}$, $\dfrac{\sqrt{3}}{2}$

4. (b) $\dfrac{7}{5\sqrt{2}}$, $\dfrac{3}{\sqrt{10}}$

Section 14.4

3. $\{1, \ 2\sqrt{3}x - \sqrt{3}, \ 6\sqrt{5}x^2 - 6\sqrt{5}x + \sqrt{5}\}$

5. (a) $\left\{ \left(\dfrac{2}{\sqrt{29}} \quad \dfrac{4}{\sqrt{29}} \quad \dfrac{3}{\sqrt{29}} \right),$

   $\left(\dfrac{11}{\sqrt{174}} \quad \dfrac{-7}{\sqrt{174}} \quad \dfrac{2}{\sqrt{174}} \right),$

   $(0 \quad 0 \quad 0) \right\}$

Section 14.5

2. (a) $[(1, 0, 0)]$

   (c) $[(-3, 2, -1, 2)]$

Section 14.6

2. a, c, f are orthogonal matrices.

3. a, d are orthogonal matrices.

   b is orthogonal for $|k| = 1$.

Section 14.7

3. $P = \begin{pmatrix} 1/\sqrt{2} & 1/\sqrt{2} \\ -1/\sqrt{2} & 1/\sqrt{2} \end{pmatrix}$, $D = \begin{pmatrix} 2 & 0 \\ 0 & 4 \end{pmatrix}$

5. $P = \begin{pmatrix} \dfrac{1}{\sqrt{2}} & \dfrac{-1}{\sqrt{6-2\sqrt{3}}} & \dfrac{-1}{\sqrt{6+2\sqrt{3}}} \\ 0 & \dfrac{1-\sqrt{3}}{\sqrt{6-2\sqrt{3}}} & \dfrac{1+\sqrt{3}}{\sqrt{6+2\sqrt{3}}} \\ \dfrac{1}{\sqrt{2}} & \dfrac{1}{\sqrt{6-2\sqrt{3}}} & \dfrac{1}{\sqrt{6+2\sqrt{3}}} \end{pmatrix}$

   $D = \begin{pmatrix} 1 & 0 & 0 \\ 0 & \sqrt{3} & 0 \\ 0 & 0 & -\sqrt{3} \end{pmatrix}$

8. (b) $\{(2, 1, 2), (-1, -2, 2)$

   $(1, 0, 2)\}$

Chapter 15

Section 15.2

3. (a) $a/(s^2 + a^2)$, $s > 0$

   (b) $\Gamma(a + 1)/(s + b)^{a+1}$,

   $s > -b$

Section 15.4

2. $(-t)^n f(t)$, $-f(t)/t$

Section 15.6

1. $y = e^{5t}/4 + 7e^{-t}/8 - e^{3t}/8$

3. $y = \dfrac{9}{5} e^t + \dfrac{1}{4} e^{2t} - t e^t - \dfrac{1}{20} \cos 2t$

   $\qquad - \dfrac{3}{20} \sin 2t$

5. $y_1 = \frac{1}{3} t^3 - \frac{1}{2} - \frac{1}{2} e^{2t}$

$y_2 = 3 - 2t + t^2 - \frac{2}{3} t^3 - 3e^{-t} + e^{2t}$

7. $y = -\frac{1}{27} e^{-t} + \frac{1}{27} e^{2t} (1 + 24t + 18t^2)$

Chapter 16

Section 16.1

2. not analytic at zero

analytic at $x = 1$, $\rho = 1$

analytic at $x = -1$, $\rho = 1$

4. analytic at $x = \pi/4$, $\pi/2$, $\rho = \infty$

6. no singular points

8. $x = 2, 0$

Section 16.2

1. $y_1 = \sum_0^\infty b_k x^k$ where b_0 is arbitrary

$b_1 = b_2 \neq 0$, $b_k = \frac{(k-1)b_{k-1} + 3b_{k-3}}{2k}$ $(k = 3, 4, 5, \cdots)$, $\rho_{\min} = 2$

6. b_0 and b_1 are arbitrary, $b_2 = -b_0/2$, $b_3 = (b_0 + b_1)/6$

$b_4 = -(2b_0 + b_1)/12$

10. $P_2 = 3(x^2 - 1/3)/2$

$P_3 = 5(x^3 - 3x/5)/2$

$P_4 = 35(x^4 - \frac{6}{7} x^2 + \frac{3}{35})/8$

Section 16.3

1. $b_0 = 0$, $b_1 = 1$, $b_2 = 0$,

$b_3 = 1/3$, $b_4 = 1/12$, $b_5 = 1/12$

3. $b_0 = 1$, $b_1 = b_2 = b_3 = 0$,

$b_4 = -1/4$, $b_5 = 3/20$, $b_6 = 1/60$

Section 16.4

3. 89 terms

Section 16.5

2. 1 is regular singular,

0 is irregular singular.

4. 0 is regular singular.

6. 3 is irregular singular.

Section 16.6

3. $r^2 + r + 2/9 = 0$,

$r_1 = -1/3$, $r_2 = -2/3$

$y_1 = b_0 |x|^{-1/3}$

$y_2 = b_0 |x|^{-2/3}$

Section 16.7

1. (b) $y_1 (x) = |x|^3$

$y_2 (x) = |x|^3 \ln|x|$

7. $r^2 + 3r/2 = 0$,

$r_1 = 0$, $r_2 = -3/2$

$y_1 = \sum_{j=0}^\infty b_j x^j$

$y_2 = |x|^{-3/2} \sum_{j=0}^\infty c_j x^j$

as $x \to 0$, $y_1 \to b_0$ and $|x|^{3/2} y_2 \to c_0$

Section 16.8

1. irregular singular

3. irregular singular

Section 16.10

1. (a) $y = e^{-\epsilon x^2}$

3. $y_0 = \cos x$,

$y_1 = -3/2 \, x \cos x + 1/8 \sin 3x$

Appendix A

1. (a) 8   (c) $5 - i$ (d) $11 - 13i$

2. (b) $x = -1$

6. (b) $(a,b) \cdot (c,d)$

$= (ac - bd, \, ad + bc)$

Appendix B

4. (b) $a_2 z^2 + (2a_2 + a_1)z + (a_2 + a_0)$

(d) $2a_2 z + a_1$

Appendix C

1. (a) $\rho = 1$, $|x| < 1$

   (c) $\rho = 2$, $-2 < x \leq 2$

   (e) $\rho = 1$, $|x| \leq 1$

   (g) $\rho = 1/2$; if $0 < p \leq 1$
       then $1 \leq x < 2$; if
       $1 < p$; then $1 \leq x \leq 2$

2. (b) $\displaystyle\sum_{n=1}^{\infty} \frac{x^n}{n}$

   (d) $1 + x + \dfrac{2x^2}{2!} + \dfrac{5x^3}{3!} + \dfrac{15x^4}{4!} + \dfrac{52x^5}{5!} + \cdots$

3. (a) $2 \displaystyle\sum_{n=0}^{\infty} \frac{x^{2n+1}}{2n + 1}$

Index